OLFACTION AND TASTE

OLFACTION

PROCEEDINGS OF THE THIRD

AND TASTE

INTERNATIONAL SYMPOSIUM

(1968)

Edited by CARL PFAFFMANN

Published by The Rockefeller University Press

New York City 1969

This book has been brought to publication with the generous assistance of The Nutrition Foundation, Inc., and the National Science Foundation.

Standard Book Number 87470-0132

Library of Congress Catalog Card Number 67.2852

Printed in The United States of America

CONTENTS

II · PSYCHOPHYSICS AND SENSORY CODING

III · ROLE OF OLFACTION IN BEHAVIOR

TASTE

IV · RECEPTOR MECHANISMS

INTRODUCTION

This is the third symposium of a series that was launched after informal discussions among Yngve Zotterman, Lloyd Beidler, and me in the summer of 1959. The first, at which there were 67 participants, was held at the Wenner Gren Center in Stockholm in 1962, under the gracious hospitality of Yngve Zotterman. This number had grown to 107 by the time of the second symposium, which was held in Tokyo in 1965 under the cordial chairmanship of Dr. T. Hayashi. The current meeting was held at The Rockefeller University from August 19 to 21, 1968. Seventy-two persons presented papers, and many of the 142 attendees were students and observers who took part in the discussions.

All three meetings have been so-called satellite symposia of the International Congress of Physiology, and have usually been held just before the main Congress. Official recognition by the Congress Program Committee is appreciated, because the number of workers in these fields is relatively small and a triennial international gathering is most valuable. Our organizing committee likes to think that both the increase in the number of workers and the advance of knowledge in recent years can be attributed in part to these symposia. Already, plans are well along for the fourth meeting which will be held in conjunction with the next International Congress of Physiology in Munich. Dr. Dietrich Schneider has been elected Chairman.

At the Rockefeller meeting, effort was made to organize the program around several major themes. Some of the papers, usually 30 minutes long, were invited, and each author was asked to include not only his own data but those of other investigators, in order to give a broader review of the particular topic.

In addition, there was the usual call for papers. From these, certain ones were selected because of their pertinence to the several themes. For each receptor system, olfaction and taste, there was, on the first day, a consideration of the basic receptor mechanisms, their morphology, physiology, and biochemistry. On the second day, psychophysics and sensory coding were discussed. The third day was concerned with the role of chemoreceptors in behavior, including taste preferences, appetite, specific hungers, pheromones, and chemical communication in the control of behavior.

Another innovation in the present program was a scheduling of two roundtables on the second day, which focused particularly on some of the outstanding issues and debates on receptor specificity and sensory coding in both olfaction and taste. One central issue in gustation is the apparent contradiction between the implications from psychophysical and electrophysiological experiments. On the one hand, the psychophysical studies tend to reinforce the classical notions of the four primary tastes, each with its underlying basic receptor type. On the other hand, electrophysiological evidence shows that the afferent neurons are not highly specific, since individual afferent fibers can be found which may respond to one, two, three, or all four of the basic tastes. A series of formal background papers were presented just before the roundtable discussions. Such issues had, in a sense, been "under the carpet" at previous symposia, and time had not been allotted for their explicit discussion. In olfaction, Dr. Schneider's work on the so-called "odor specialists" and "odor generalists" in insects has brought some clarification, namely that some odor receptors, particularly the pheromone receptors are highly specific and others are non-specific. Most unit studies of mammals indicate that specific "specialists" have not, as yet, been seen, and something like an odor-generalist system seems to be the prototype. The mechanism by which the great variety of odors is coded still remains a puzzle, although it is possible to contrive a computer that could read and process the neural response matrix that is produced by a set of odor generalists. The discussions at the roundtables were transcribed, but are not reproduced verbatim. Instead, the editor prepared a summary of each, based on the tapes.

One point worth mentioning is the extraordinary range of professional fields represented at the symposium: anatomy, physiology, biochemistry, food chemistry, nutrition, medical science, neurology, neurobiology, endocrinology, animal behavior, behavioral biology, and psychology. This suggests one reason why the symposia play such an important role in the development of this segment of science. The opportunity for interdisciplinary exchange does not easily occur at the more usual scientific meetings. Workers in these related, yet diverse, fields often attend meetings with which they are professionally identified and where they can present data only to a more limited circle of their fellow professionals. Our international triennial symposia know no disciplinary bounds or limits. New techniques, data, and concepts from whatever source can be brought to the subject matter as needed.

Those of us at The Rockefeller University were pleased to serve as hosts. The University administration was happy to make a conference auditorium, lounges, smaller meeting rooms, dining and club facilities available. That they all were in one air-conditioned building, the Abby Aldrich Hall and associated Caspary Auditorium, did much to make the meeting in New York City in August pleasant and comfortable.

The Symposium Committee wishes especially to thank its Executive Secretary, Dr. Rudy Bernard, for his devoted and imaginative role in the preparations for and the actual conduct of the Symposium. The Committee is particularly grateful for the thorough and thoughtful attention to the many details of correspondence, to the planning before the meeting, and for the attention to details during the conference by Miss Carol Gallea, Miss Judith Lavicka, and Mrs. Mary Regan of the Rockefeller staff under Mrs. Nadja Schocken's direction. It is deeply appreciative of the care with which Mrs. Jane Cooper and her dining staff looked after our daily sustenance and the closing dinner. In addition, the committee wishes to thank Mr. Lewis Koster and his staff for arranging the many necessary visual aids. This editor, in particular, wishes to express his appreciation to The Rockefeller University Press, Mr. William Bayless, Mr. Reynard Biemiller, and especially Mrs. Helene Jordan, who saw all the manuscripts through the final editing process, and who played a major role in shaping the final book. The Committee also wants to express its gratitude to the Nutrition Foundation and to the National Science Foundation for their financial support of the symposium and their contributions to the cost of this publication. Their support was essential for the success of the symposium.

Carl Pfaffmann
Editor

ORGANIZING COMMITTEE FOR THE
THIRD INTERNATIONAL SYMPOSIUM ON OLFACTION AND TASTE

Seated, left to right: CARL PFAFFMANN, Chairman; YNGVE ZOTTERMAN
LORD ADRIAN, Honorary Chairman; T. HAYASHI.

Standing: RUDY A. BERNARD; DIETRICH SCHNEIDER; LLOYD BEIDLER

FOREWORD

When I was a small boy, my father would sometimes tell me about plays he had seen in London theaters in the 1860s. One that he liked very much was played, I think, by an American company and was based on Washington Irving's story about Rip van Winkle—the man who fell asleep in the Catskill mountains in New York and woke up 20 years later to go back to a quite unfamiliar world. When I woke up this morning in New York, I began to understand just what Rip van Winkle must have been feeling when he woke up, found his way back to his village, and discovered everyone there 20 years older than when he had known them before. It is now six years since the Stockholm meetings in the Wenner-Gren Center—my last real contact with taste and smell. (I could not get to the symposium that Professor Hayashi arranged in Tokyo three years later.) But today an absence of six years makes one far more behind the times than 20-years' absence would have done in the time of Rip van Winkle.

However, today I have only the pleasant and noncontroversial job of introducing the subject to an audience that is far more familiar with it than I am—that has, in fact, been up to the neck in it, at least for several years. So all I shall do is say that I think we have chosen a subject of real importance to biology, one that may well give some clues for solving the problems of intelligent, conscious behavior.

It is clearly right to begin at the periphery, as we shall today, with the receptors for taste and smell. Those of us who set out yesterday from Europe and are still resetting our circadian clocks may have hoped that the meetings today would be restful, would allow us to sit back with half-closed eyes, contemplating electron microscope pictures of cilia or semipermeable membranes or X-ray spectra of the molecules responsible for different classes of smell. But I fear we shall not be let off so lightly: we have a great deal to learn at the receptor level about chemical as opposed to mechanical or electrical stimulation. It will not only be the biophysicists among us who will not be allowed to sleep before the evening meeting is over.

It is unusual, I think, to dream about smells, and it will be interesting to find whether anyone does so tonight. Tomorrow we shall leave the receptors and get up to the olfactory bulb and then to the cerebral cortex, to the level of conscious sensation—taste in the morning and smell in the afternoon.

I hope someone will explain why taste has developed in the way it has done, with endings for acid and sweet and salt and bitter and apparently nothing more; and how are the different endings arranged in animals with different ways of feeding? Carnivorous animals that bolt their food have little time to taste it, and one would think that sheep and cattle could do with different sensory equipment in their mouths. Pigs and men and monkeys surely have varying needs, according to what they eat and the way they eat it. The structural species differences in smell are, of course, well recognized—the macrosmatic animals with large snouts and large olfactory connections to the brain, and the microsmatic, including man and apes.

I have often wondered whether some of the features of human olfactory sensation (like our persistent olfactory adaptation) are absent or quite different in macrosmatic animals. And what on earth is the organ of Jacobson for? But, of course, I am merely listing some of the puzzles that used to worry me, and I am quite prepared to be told that they have now been solved.

Tomorrow, in fact, we shall have reached the level at which the advances will be most rewarding. One of the most important modern additions to the technique of neurophysiology is the development of methods for making electrical records from the unanesthetized brain, particularly by the use of implanted electrodes in unrestrained animals. It may not be easy to fit electrodes in suitable parts of the olfactory pathway, but records made without anesthetics would be bound to give valuable information. We know that there are characteristic, rapid, electrical rhythms in the olfactory regions of the unanesthetized brain, and that these are modified by olfactory stimulation, but we cannot be sure how much they depend on the effect of the anesthetic.

So I think it is only a matter of time and trouble before we have as much information about the olfactory system of the brain as we have about the visual. At present, however, the visual system is certainly leading, so it is time that we started our meetings on taste and smell.

LORD ADRIAN

OLFACTION

I RECEPTOR MECHANISMS

I · RECEPTOR MECHANISMS

COMPARATIVE MORPHOLOGY OF OLFACTORY RECEPTORS

RUDOLF ALEXANDER STEINBRECHT
Max-Planck-Institut für Verhaltensphysiologie, Seewiesen, Germany

INTRODUCTION

Olfactory receptors are less conspicuous in their morphology than are light- and mechanoreceptive sense organs, and the proof of olfactory function of a given sensory element is especially difficult. These reasons—among others—may account, in the invertebrates, for the relatively small number of sense organs with proved olfactory function and well-studied morphology. Only the fine structure of odor receptors in insects is being dealt with in an increasing number of papers. The so-called aesthetasc hairs of crustaceans have been studied with the electron microscope,[20,32] yet their probable olfactory function still demands direct electrophysiological proof. The same holds for the primary sensory cells in the sucker of *Octopus*[22] and in the mucosa of the tentacle caps of a snail, *Vaginulus borellianus*.[42] (In a recent paper, Suzuki[62] gave proof of the olfactory function of the tentacle caps in the snail *Ezohelix flexibilis*; he used electrophysiological methods.) A wealth of comparative light-microscopic observations of vertebrates has accumulated during the last hundred years (reviewed by Allison[1]), but only a few investigators have used modern morphological techniques to study olfactory epithelia of species belonging to classes other than mammals. Much work has yet to be done before comparative morphology may find phylogenetic relationships and evolutionary pathways in olfactory receptors.

In this paper an attempt will be made to point out main features of the fine structure of vertebrate and insect olfactory receptors to provide a morphological basis for discussions on receptor function and specificity. For a more detailed survey of the literature, the reader may refer to the review of the vertebrate peripheral olfactory system by Moulton and Beidler[36] and to the checklist of insect olfactory sensilla by Schneider and Steinbrecht.[45]

VERTEBRATE OLFACTORY RECEPTORS

In vertebrates, the olfactory receptors are arranged in olfactory epithelia, the composition of which is relatively constant. Three constituent cell types were distinguished as early as 1856 by Schultze,[49] namely: receptor cells, supporting cells, and basal cells (Figure 1A). In fish, goblet cells intermingled among the supporting cells, produce the mucous layer which covers the free surface of the epithelium. In higher vertebrates, the mucus is secreted by the multicellular Bowman's glands, which underlie the epithelium.

The Receptor Cells

The receptor cells are bipolar, primary sense cells directly connected with the olfactory bulb by their axons, which are extremely thin (modal diameter 0.2 μ) and packed in bundles of naked fibers by glial cell processes. The fine structure, dimensions, and arrangement of olfactory nerve fibers are similar in all vertebrate species studied so far.[4,14,19,24,51] The perikarya contain little cytoplasm around the nucleus, a Golgi apparatus in supranuclear position, some loosely organized endoplasmic reticulum profiles, and other normal cell organelles in moderate concentration. Opposite the axon hillock, a single dendrite, which is always thicker than the axon but rarely exceeds 1 μ diameter, extends to the surface of the epithelium. Usually the dendrites are separated from each other by intercalated supporting cells, but occasionally dendrites in direct contiguity have been reported (H. Altner, unpublished, and references[18,66,71]). Contact between the perikarya of adjacent sense cells, however, seems to be common in the olfactory mucosa of the mouse.[18] Longitudinally arranged microtubules and an aggregation of mitochondria near the termination are found in the dendrite cytoplasm, which in most electron micrographs appears less dense than that of the surrounding supporting cells.

The termination of the dendrite is directly contiguous with the overlying mucus, and probably plays a key role in the stimulus-transducing process. This is also the part of the sense cell with the greatest structural variability (Figure 1 B-E). Usually the dendrite rises somewhat above the surrounding supporting cells to form a "terminal swelling" or "olfactory knob." In the dog, this olfactory knob is up to 4 μ long and 1.3–1.8 μ thick[4,40]; in the duck, it has the shape of a hemisphere up to 3.5 μ diameter[24]; in frogs and toads, it is a sphere of about 2 μ diameter.[9] In some species of fish, on the other hand, it is only a slightly convex membrane protrusion.[7,64,66] In the olfactory receptors of the carp *Carassius carassius*[71] and of Jacobson's organ in *Lacerta*,[2] the dendrite does not rise above the epithelium surface.

Numerous smooth-surfaced vesicles are consistently observed in the cytoplasm of the olfactory knob. It has been suggested that they build a continuous "labyrinth,"[4] which possibly connects with the overlying mucus via small openings. Microtubules may touch these vesicles and/or the basal feet of cilia.[4,7]

In the majority of cases, the olfactory knob bears cilia as the only surface differentiation. Exceptions are observed in:

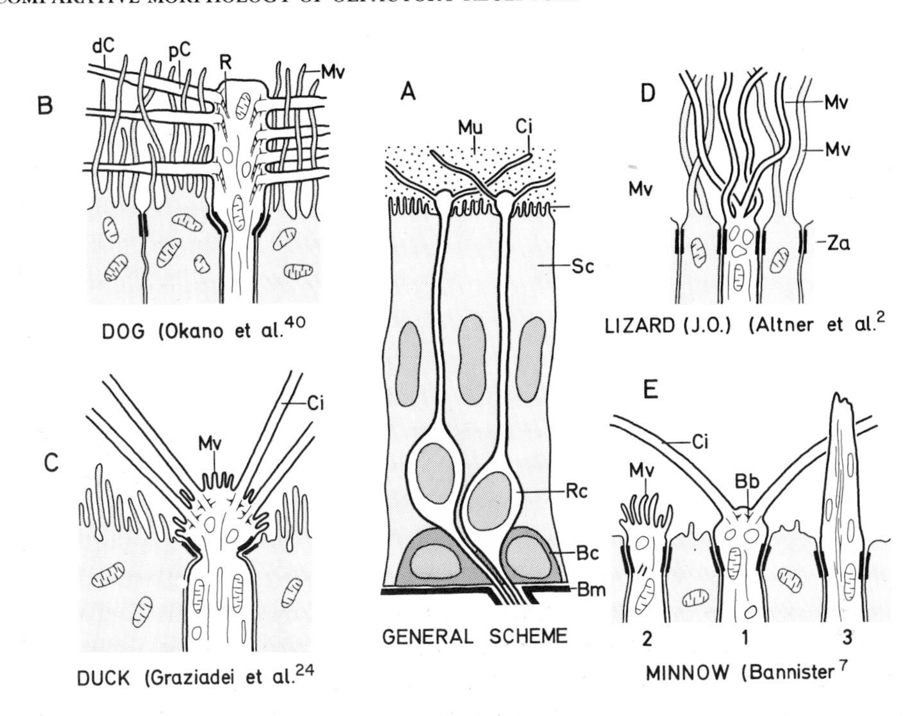

Figure 1 Schematic representation of vertebrate olfactory epithelia. A. The general composition of the olfactory mucosa. B–E. Various modifications of the mucosal surface and receptor endings (modified after original drawings by the authors indicated). The receptor cells and their processes are white, the supporting cells are light gray, the basal cells are darker gray. Basal bodies, ciliary rootlets, smooth surfaced vesicles and microtubules of the receptor endings, as well as the mitochondria, are schematically indicated (Bb, basal body; Bc, basal cell; Bm, basal membrane; Ci, cilium; dC, pC, distal and proximal part of cilium, respectively; J. O., Jacobson's organ; Mu, mucus; Mv, microvillus; R, ciliary rootlet; Rc, receptor cell; Sc, supporting cell; Za, zonula adhaerens; 1, 2, 3, the three types of receptor endings in the minnow).

1) The few species of birds examined so far and in the fish *Carassius*,[5,13,24,71] in which there are numerous microvilli in addition to cilia; 2) in Jacobson's organ of reptiles, studied in *Anguis fragilis*,[8] *Lacerta agilis*,[2] and *Natrix natrix* (H. Altner, unpublished), in which the dendrites bear no cilia but numerous microvilli; 3) in the minnow *Phoxinus*, in which either ciliated dendrite endings (Figure 1E, type 1) or those bearing microvilli (type 2) may occur, or in which dendrites sometimes bear neither cilia nor microvilli, but rise as a simple rod about 4 μ into the mucus (type 3).[7]

Estimates of cilium length are in the range of 3–10 μ for many species of mammals and fish. About 12 μ is the figure given for the duck. However, in some cases cilium length might have been underestimated, if thin sections were studied. A shortening of the cilia by breakage during preparation must also be taken into account. Under the light microscope, olfactory cilia as long as 200 μ have been observed in frogs[29,41,50] and in a bird, *Larus argentatus* (D. Drenkhahn, cited by Andres[6]), while in the cat cilia 80 μ long have been measured.[5,6] Six to 12 cilia per receptor ending are the values given most often, but in some mammals the number may occasionally be as much as 20[11] and even 40.[5] The number of 100 to 150 cilia per sense cell given by Okano and coworkers[40] is by no means confirmed by the published electron micrographs. Particularly few cilia (one to five) have been observed in olfactory cells of the mole.[23]

Motility of olfactory cilia has been observed in a fish, *Stenesthes chrysops*, in some species of frog, turtle, and tortoise,[29,52] and in tissue cultures of mammalian olfactory epithelia (Pomerat, cited by Le Gros Clark[34]). The movement is slow and disorganized in contrast to the directed synchronous beat of cilia in the nasal respiratory region. Reese[41] suggests that, at least in the frog, where long immotile cilia as well as shorter motile ones have been observed, prolonged motility might be induced if the fragile cilia are broken during preparation.

In electron micrographs of cross sections, olfactory cilia exhibit the typical pattern of motile cilia: nine pairs of peripheral tubules are arranged in a circle around two central tubules. In good preparations further substructures appear; these are known as "arms" from motile cilia (Figure 2).[41,64] Often the ciliary arrangement is lost in the distal part, and in successive sections a diminishing number of single tubules can be seen. The transition from the proximal to the distal cilium may be sharp and marked by a sudden decrease in diameter from 0.25 to 0.15 μ.[41,51,64] Toward the end, the distal segments may taper down to 600 Å and then contain only two tubules.[4,38,40,51] The relative length of the distal segment is about 98 per cent in the cat olfactory cilia, 80 per cent in those of the frog and of the gull, 50 per cent in those of the lamprey, and much less in those of the duck.[6,24,41,64] The peripheral ciliary tubules are continuous with a basal body consisting of nine tubule-triplets inside the olfactory knob. One, two, or nine basal feet are connected with the basal body,[7,40,64] but basal bodies, usually lacking basal feet, have also been reported.[18] In only two cases have cross-striated ciliary rootlets been described in olfactory cells,[40,41] but they are very faint, whereas such rootlets are well developed in ciliated cells of the nasal respiratory region.[39] Such structural diversity may account for the different mode of movement (see above), but there is still no way to prove this speculation. It is amazing that no observations have been made on human olfactory receptor endings since the pioneer work of Engström and Bloom,[16] for which they used teased preparations.

The morphogenic differentiation of ciliated receptor endings has been observed in newborn dogs and mice by Andres[3,4] and by Seifert and Ule.[51] Centrioles reduplicate in the outgrowing dendrite even before the epithelial surface has been reached. Later, they come to lie beneath the membrane of the olfactory knob as basal bodies, from which cilia grow out in the manner described by Sotelo and Trujillo-Cenóz[60] for developing cilia in neural-crest cells of chicks. From his study of the different appearance of olfactory knobs in adult mammals, Andres[6] suggests that the perikarya of the receptor cells are able to replace their peripheral processes in "a regular moulting." Degeneration of sense cells, as well as the differentiation of new ciliated endings, has also been reported in *Lampetra*.[64]

The ultrastructure of "olfactory microvilli" is much simpler. No marked differences from that of the supporting cell microvilli have been observed. In the duck, the

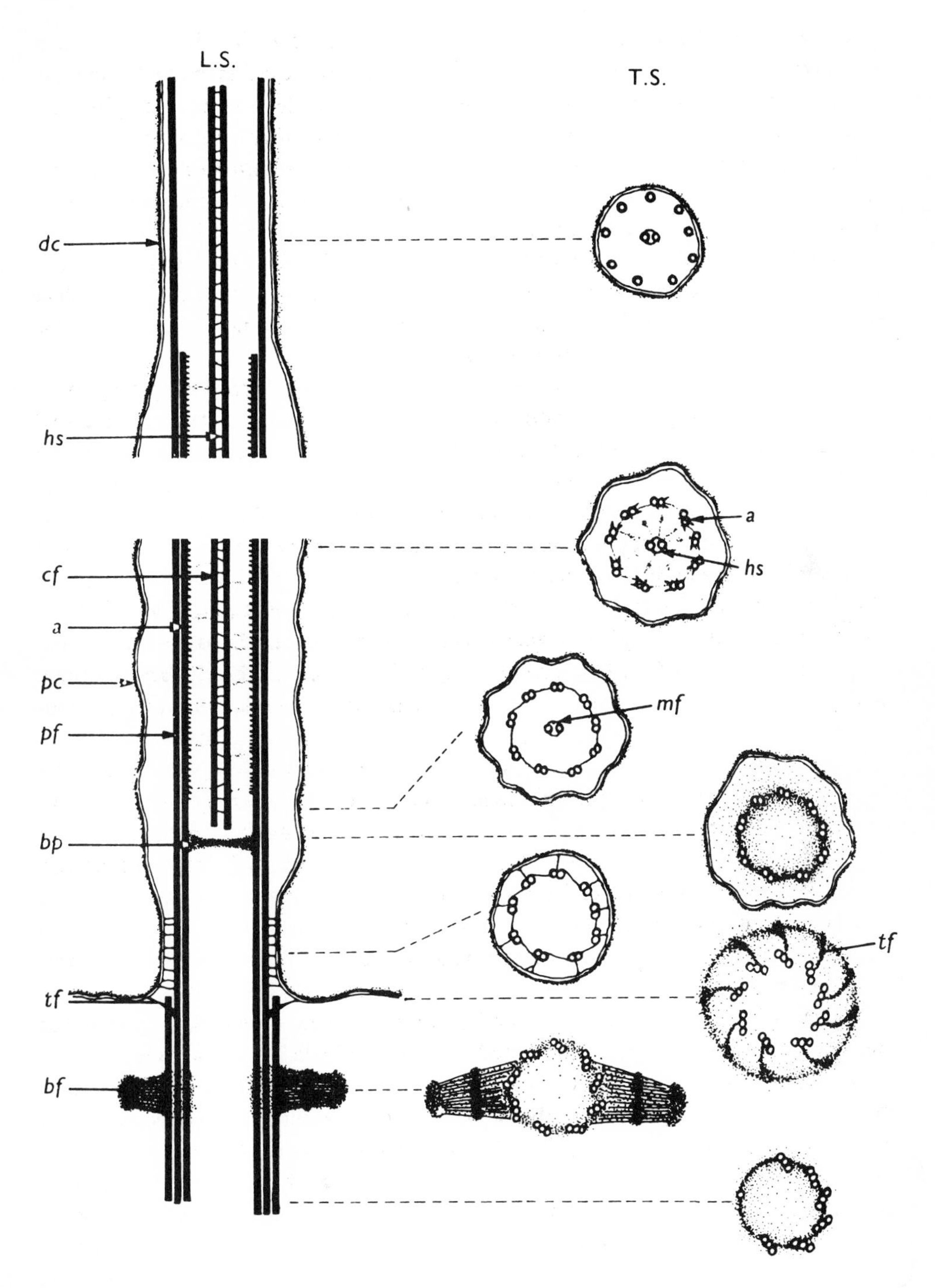

Figure 2 Diagram illustrating the structure of the sensory cilia (a, arms on peripheral fibers; bf, basal foot; bp, basal plate; cf, central fibers; dc, distal segment of sensory cilium; hs, helical sheath; mf, midfiber; pc, proximal segment of sensory cilium; pf, peripheral fibers; tf, transitional fibers).[64]

microvilli are 0.8 μ long and 0.13 μ in diameter. In Jacobson's organ of reptiles they are thinner (about 0.1 μ) but much longer—at least 4 μ in *Lacerta* (H. Altner, personal communication)—and contain delicate axial filaments. Such axial filaments, 50 Å in diameter, have also been described in the type 3 receptor ending of *Phoxinus*, where many of them are densely packed as fiber bundles.[7]

The Supporting Cells and the Basal Cells

The nuclei of the supporting cells generally lie nearer to the epithelial surface than those of the receptor cells. The apical cell processes are much thicker than the dendrites and bear microvilli that sometimes are short and rather scarce,[7] but sometimes are long and numerous, forming a dense feltwork around the olfactory knobs and their cilia.[40] Dense granules occur in the cytoplasm of supporting cells in frogs, representing material that is possibly secreted into the mucus.[9,41] Clear vesicles are observed in mammalian supporting cells, which may undergo micropinocytosis, either absorbing or extruding material,[4,14,40] and sometimes secretory granules are found.[18] Histochemical studies also suggest secretory activity.[26] Many authors report high concentrations of mitochondria in the apical part of the supporting cells; large mitochondria sometimes line the border lateral to the dendrites, in which small vesicles appear.[51]

There is general agreement that the function of the supporting cells is not only to support and ensheath the distal sensory processes. It has been suggested[4] that, besides secretory and nutritional activity, a regenerative interaction takes place between the microvilli of supporting cells and the olfactory cilia. Furthermore, the possibility of electrical activity has been discussed.[63,68] The latter speculation is based on the assumption that the desmosome-like membrane junctions between supporting and receptor cells in the region of the terminal bars are bridges with low electrical resistance, analogous to the suggestion of Loewenstein and coworkers.[35]

The basal cells provide the sheath for the olfactory axons before they penetrate the basement membrane and leave the mucosa. Slender processes of basal cells can be seen to extend up to the level of the sense cell perikarya. Andres[3,4] observed in young dogs a special type of basal cell undergoing mitosis and transforming into receptor cells, thus providing a pool for receptor regeneration.

The Mucus

The thickness of the mucous layer may vary much more from region to region in the same individual than between different species. Yet, it should be noted that there is no safe method for such measurements. Estimates range from 10 to 60 μ.[29] In electron microscope preparations, probably most of the mucus has been washed away by the fixing and dehydrating agents; the rest is condensed to a layer of fluffy material some microns thick. The best preservation of the mucus has been obtained by perfusion fixation. In this case, the outer mucous layers and a dense terminal film could also be demonstrated.[6] There is disagreement in the literature as to whether the mucus moves over the olfactory epithelium.[12,26–28] Also, the data concerning the chemical nature of the mucus diverge considerably. Herberhold[26] assumes a locally changing composition of the mucus, caused by a changing enzyme pattern among different Bowman's glands. According to that author, the mucus in species of Selachii, as well as of mammals, is free of enzymes, its main components being mucopolysaccharides; in mammals it contains lipids and phosphatids additionally.

Final Remarks

There have been many speculations on what structure of the olfactory epithelium may be responsible for olfaction. Even the pigments, which account for the yellow to brown color of many vertebrate olfactory mucosae, have been assumed to be essential for the perception of stimuli, and there is vast and controversial literature on this subject (see Moulton and Beidler[36]). However, it should be borne in mind that the pigment is generally situated in the supporting cells and in Bowman's glands, not in the receptor cells, and, further, that there are olfactory mucosae that lack any dispersed pigment.

There is not much evidence for the direct involvement of the ciliary apparatus in the stimulus-transducing process, since there are olfactory receptors without cilia or basal bodies. Furthermore, ablation of cilia in the receptor cells that bear them does not inhibit excitability.[67] This suggests that the membrane of the olfactory knob may be the most likely receptor site. The cilia may then serve as "stirrers" to accelerate the exchange of mucus that carries odor particles to the olfactory knob. This does not exclude the possibility that the membrane of the cilia is also excitable by odor particles. In this case, cilia could enlarge the receptive surface, an assumption which makes it readily understandable that in some cases microvilli complement or substitute for them.

Graziadei and Bannister[24] estimated the surface of the bare olfactory knob of the duck as 14 μ^2. The microvilli (about 160) enlarge the surface about four times and the eight cilia alone enlarge it seven times, together making a more than tenfold increase of receptor surface. In *Phoxinus*, the ciliated receptor (type 1) with no prominent olfactory knob and four to six short cilia (5 μ length) has about the same free surface as the huge rod-shaped receptor ending (type 3).[7] These considerations also show the effectiveness of surface enlargement by cilia. Furthermore, long cilia would bring a part of the receptive membrane nearer to the mucous surface, thus shortening the time it takes odor molecules to diffuse through the mucus. Nevertheless, all these speculations deserve further confirmation.

For a long time morphologists have sought for structural differences between the receptors of a given olfactory epithelium that might account for their different sensitivities to odors.[4,33] Yet, as long as the molecular architecture of the receptor membrane remains unknown, such distinctions—even in terms of modern electron microscope methods—appear too superficial and may be misleading.

INSECT OLFACTORY RECEPTORS

The olfactory receptors of insects are not arranged as epithelia, but in distinct units called the sensilla, which are mainly, but not completely, located on the antennal flagellum. However, sensilla may be densely packed, thus forming a "sensillum epithelium," as on parts of insect antennae or in the compound eye. A sensillum is characterized by a special differentiation of the cuticular exoskeleton, mostly a hair- or peg-shaped protrusion in combination with a small group of cells: the receptor cell(s), the trichogen cell, and the tormogen cell. The latter two are formative cells, responsible for the morphogenesis of the cuticular part of the sensillum. All these cells are derived from one epidermal mother cell in a strictly determined series of mitoses.[25]

Sensilla have been classified by earlier light microscopists according to the shape of the cuticular apparatus (see Snodgrass[59]; Weber[69]), but even with refined modern morphological methods, the structural arrangement often cannot predict their function with certainty. Electrophysiological methods have demonstrated that olfactory sense cells exist in the sensilla trichodea, basiconica, placodea, and coeloconica. But receptors responding to other modalities of stimuli may also occur within these categories. For instance, there are hygroreceptive coeloconic sensilla in addition to the olfactory ones in the antenna of the honeybee[31] and the locust (U. Waldow, unpublished).

The Receptor Cells

In insect olfactory sensilla, from two to six sense cells are most common. There is rarely a single receptor cell, as in one type of basiconic sensilla in the beetle *Necrophorus.*[10,17] The pore plates of the honeybee and the thin-walled pegs on the locust antennae are examples of sensilla with high numbers of sense cells (15 to 30 and 20 to 50, respectively)[54,58] (also R. A. Steinbrecht, unpublished).

In many respects, insect olfactory receptor cells resemble those of vertebrates. They are bipolar primary sense cells, with their axons directly connected to the central nervous system. Neither fusion of axons—as formerly proposed—nor synaptic contacts occur in the periphery.[61] Insect olfactory nerve fibers normally range from 0.1 to 0.3 μ in diameter and are assembled in bundles of naked axons. The dimensions and fine structure of the axons, the sense-cell perikarya, and the proximal part of the dendrites are not distinct from vertebrate olfactory cells. The receptor cells belonging to one sensillum are concentrically sheathed by the trichogen and the tormogen cells, up to the termination of the proximal segment of the dendrites. This termination is marked by an abrupt decrease of the diameter from about 1 μ to 0.35 μ and by the emergence of a modified cilium.

Nine peripheral pairs of tubules appear in cross sections of this ciliary segment. Slifer[53] negates the existence of a central pair for several thin-walled insect olfactory sensilla, although in some cases she describes two or more single tubules in the center of the cilium.[54,56,57] Ernst[17] also observes two central tubules, but admits that they are much shorter than the peripheral tubules and, therefore, can only be seen in favorable sections of a whole cross-section series (personal communication). No structures like the arms on peripheral tubules of motile cilia have been observed so far, but a good preservation of ciliary structure in insect sensilla is generally much more difficult to obtain than in ciliary epithelia. Two basal bodies lie one behind the other at the base of the cilium. The proximal one is enclosed by nine ciliary rootlets that emerge from the center of the distal one.[17,56] The ciliary segment is usually very short, about 3 μ (4 per cent of the total length of the dendrite) in *Melanoplus differentialis*,[57] but is said to be much longer in the honeybee.[53]

The dendrite gradually increases again in diameter distally from the ciliary segment, forming the distal, or outer, segment. This often contains numerous single tubules but no other cytoplasmic organelles except some smooth-surfaced vesicles. This is in sharp contrast with the inner segment, where a conspicuous accumulation of mitochondria can be observed. A division of the dendrite into an inner and an outer segment by a short and narrow ciliary portion is known to occur in most, if not all, bipolar primary receptor cells of arthropods: e.g., in the locust tympanal organ,[21] in mechanoreceptive bristles of the honeybee,[65] in crayfish chordotonal organs,[70] in statocyst sensory elements of *Astacus*,[48] and in the aesthetasc hairs of the hermit crab.[20]

While the outer segments of insect mechanoreceptive sensilla terminate at the hair base,[65] olfactory dendrites innervate the sense hairs up to the tip. They may remain unbranched, as in the sensilla trichodea of male Saturniid moths that contain the receptors for the female sex attractant odor,[46] (also R. A. Steinbrecht and K.-D. Ernst, unpublished), and in the sensilla coeloconica of the grasshopper.[58] In most sensilla basiconica, on the other hand, the outer segments divide into numerous dendritic branches.

The mechanism of this branching has been described for the basiconic sensilla of the carrion beetle, *Necrophorus*, by Ernst.[17] In the developing sensillum, the dendrite at first grows unbranched up to the tip of the hair. It contains 28 to 40 microtubules oriented parallel to the long axis. Later, chains of vesicles appear between the microtubules. Branching begins simultaneously at several points when the vesicles open and fuse in such a way that the membranes of the vesicles become the plasma membrane of the dendritic branches. Each of these branches gets one microtubule as an axial core. Thus, there are eventually as many branches as microtubules in the undivided part, and these emerge from approximately the same level at the hair base.

The ciliary and the outer segment are not sheathed by any cell processes, but are surrounded by an extracellular fluid of unknown composition, the so-called *Sensillenliquor*.[17] This liquor fills the lumen of the cuticular hair and the channel through which the outer segments penetrate the thick antennal cuticle. In adult insects, no cell processes except the dendrites are observed in the hair. Their possible relation to special structures in the cuticle, which are named "pore filaments" by Slifer and Sekhon[56] and are suggested to be the utmost dendritic endings, will be discussed later.

The Trichogen and the Tormogen Cells

The trichogen cell determines the shape of the cuticular process by secreting its cuticle in the early stages of development, whereas the tormogen cell is supposed to form the cuticle of the hair base, where it is inserted on the unmodified antennal cuticle.[69] There are but few electronmicroscopic data about these developmental stages: Noble-Nesbitt[37] describes the molting of *Podura aquatia* and shows electron micrographs of newly forming sensilla with unknown function. Ernst[17] reports the later stages of the development of an olfactory sensillum in *Necrophorus*: during the secretion of the sensillar cuticle the trichogen cell fills the whole lumen of the sense hair. The formation of the cuticle is finished four to five days before emergence of the adult beetle. At this stage, the cuticle is fully formed with all its special structures, which will be described in the next section, and no differences from later stages can be seen. Three to four days before emergence, the trichogen cell retracts from the hair, leaving the liquor-filled hair lumen, while the distally growing dendrite has not yet fully perforated the trichogen cell. One day before emergence, the dendrite has already reached the hair tip, whereas the trichogen cell is fully retracted from the hair. Some hours after emergence, the above-described branching of the dendrite begins.

The trichogen cell also secretes a sheath named cuticular sheath[53] or, better, dendrite sheath, because its composition is unknown. This sheath surrounds the outer segments of the dendrites as long as they travel through the antennal cuticle in the channel below the sense hair, and has been observed in most insect sensilla with the exception of the pore plate of the honeybee.[54] In the basiconic sensilla of grasshoppers, the sheath terminates at a certain point at the base of the peg, while the dendrites leave it via lateral openings and invade the peg lumen.[57] In *Necrophorus* there is no such insertion point, and the texture of the sheath is more fibrillar than in the grasshopper.[17] There are light-microscopic observations indicating that this sheath is molted together with the sensillum cuticle in some insects.[58,72]

In the adult insect, the trichogen and the tormogen cells provide the cellular sheath around the group of receptor cells belonging to one sensillum (Figure 3A). The innermost envelope is formed by the trichogen cell in a manner distinct from that of glial and Schwann cells. The trichogen cell does not fold around the receptor cell, but is perforated by the outgrowing dendrite (see above). Because of this perforation, no mesaxon-like membrane pair can be observed; otherwise, it is always present in glial sheaths. From the arrangement of cell membranes, it can be suggested that in sensilla with several sense cells, all dendrites perforate the trichogen cell as one group and later become separated from each other by trichogen-cell processes interdigitating between the dendrites (R. A. Steinbrecht, unpublished). This inner sheath is surrounded by an outer one formed by the tormogen cell in the usual way and exhibiting a mesaxon-like structure. Sometimes the picture as seen in cross section is more complex, and it is unknown whether other cells also contribute to the sheath system.

Where the trichogen and the tormogen cells border the liquor space, numerous slender microvilli protrude from their surface. Below this surface are many vesicles, while large mitochondria and profiles of a well-developed, rough-surfaced endo-

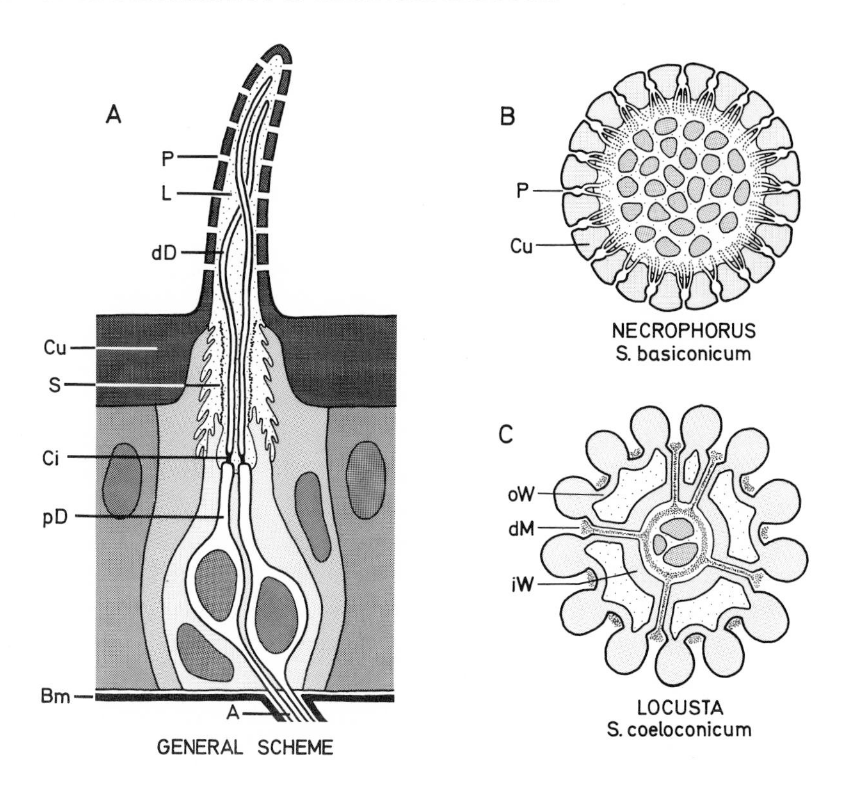

Figure 3 Schematic diagram of insect olfactory sensilla. A. General scheme of the insect olfactory sensillum in a longitudinal section: the receptor cells (white) are concentrically sheathed by the trichogen cell (light gray) and the tormogen cell (medium gray); unmodified epidermal cells are dark gray. B and C. Semischematic drawing of sense hairs in cross section. In B an example of a smooth-surfaced sense hair, the cuticle of which is penetrated by numerous pores with pore tubules. In the hair lumen the dendritic branches are seen cross-sectionally, surrounded by the Sensillenliquor. In C an example of the longitudinally-grooved pegs is shown, where no pore tubules have been observed. Three unbranched dendrites are located in the center, surrounded by a dense material, apparently continuous to the outside through hollow cuticular spokes. A, axon; Bm, basal membrane; Ci, ciliary segment of the dendrite; Cu, cuticle; dD, pD, distal and proximal segment of the dendrite; dM, dense material; L, Sensillenliquor; P, pore system consisting of pore channel, pore kettle, and several pore tubules (simplified in A); S, dendrite sheath; iW, oW, inner and outer cuticle wall of the grooved peg.

plasmic reticulum are distributed in all cell regions. This makes probable the conclusion that the two formative cells also have important functions in the fully formed sensillum. One of these functions might be the secretion of the liquor.[17]

A further feature of possible physiological importance is the occurrence of septate desmosomes nearly everywhere between the membranes of the trichogen and tormogen cells and between those of the trichogen cell and the dendrites, and of "zonulae adhaerentes" between these cells near the places at which they border the liquor space.[17] The septate desmosomes are known especially as connections with low electrical resistance.[35]

The Cuticular Apparatus

The cuticular apparatus is that part of insect olfactory sensilla that exhibits the greatest structural variety. There are the thick-walled hairs of the sensilla trichodea, which may be as long as 370 μ in male Saturniid moths, the thin-walled hairs and pegs of the sensilla basiconica, which usually range in length from 10 to 40 μ, and the pit pegs of the sensilla coeloconica, which rise about 8 μ from the floor of a cavity in the antennal wall of grasshoppers. While all these have a trichoid shape, there is no hair-like protrusion in the pore plates of the sensilla placodea.

The number of insect olfactory sensilla studied both electrophysiologically and electronmicroscopically is still very small.[45] The majority are the sensilla basiconica.[15,17,44,47,55–58] Sensilla trichodea have only been studied in the moths *Antheraea pernyi* and *Bombyx mori*[46] (also K.-D. Ernst and R. A. Steinbrecht, unpublished) and sensilla placodea only in the honeybee[30,43,54] (also R. A. Steinbrecht, unpublished).

The thickness of the cuticular hair wall of the basiconic sensilla is relatively constant over the whole length of the peg, and measures between 0.05 and 0.3 μ, depending on the species. This wall is penetrated by complex pore systems, consisting of an outer pore that often widens to a more or less spherical chamber, called "pore kettle" by Ernst.[17] Several "pore tubules" insert at the bottom of the pore kettle, penetrating the rest of the cuticle separately, and extending about 1000 Å into the liquor of the hair lumen (Figures 3B, 4). In different species the outer pores may be of different

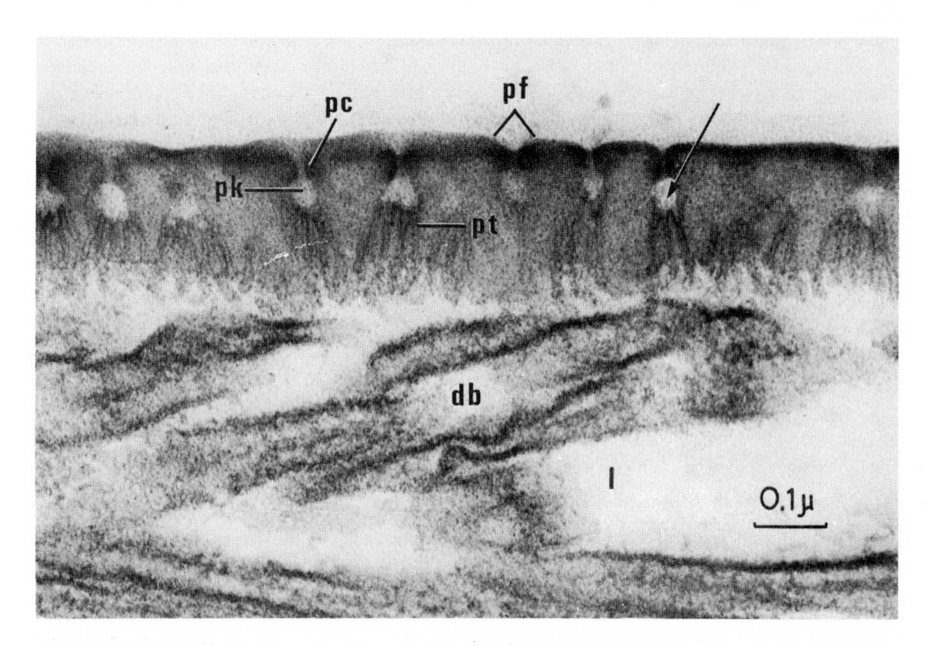

Figure 4 Longitudinal section through the wall of a basiconic sense hair of *Necrophorus vespilloides* at high magnification. Outside of the hair is up, the lumen down. Some pore tubules are seen opening to the pore kettle (arrow) (db, dendritic branches; 1, Sensillenliquor; pc, pore channel; pf, pore funnel; pk, pore kettle; pt, pore tubule). (Courtesy Dr. K.-D. Ernst)

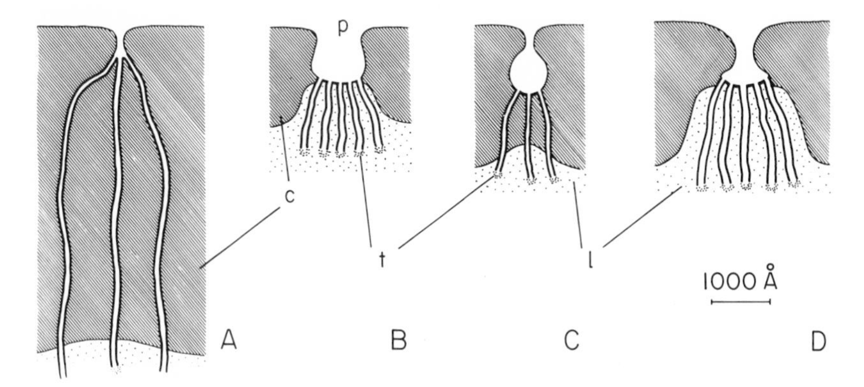

Figure 5 Various forms of pore-tubule systems of olfactory sensilla in longitudinal section: A. *Antheraea pernyi* ♂, sensillum trichodeum; B. *Locusta migratoria*, s. basiconicum; C. *Necrophorus vespillio*, s. basiconicum; D. *Bombyx mori*, s. basiconicum. The figures are drawn to scale after original electron micrographs, but slightly schematized. The outside of the hair is up, the lumen down. (c, cuticle; l, Sensillenliquor; p, pore; t, pore tubule). (Slightly modified after Schneider and Steinbrecht[45])

size and shape (Figure 5 B-D); the narrowest portion is only 100 Å wide in *Necrophorus vespilloides*,[17] but is up to 750 Å in *Locusta migratoria* (R. A. Steinbrecht, unpublished).

Also, the number of tubules per pore varies among different species, and among different sensillum types in the same species. For instance, the figure given for the thin-walled pegs on the antenna of *Sarcophaga argyrostoma* is 20 to 30 tubules per pore,[56] but in the basiconic sensillum of *Necrophorus* it is five to six (type 1 sensillum) or eight tubules per pore (type 2 sensillum).[17] The diameter of the pore tubules, however, is 100 to 200 Å in all cases studied so far. In cross sections they show a light center and a dense wall, which stands out against the less dense surrounding cuticle. In favorable longitudinal sections one can observe that the center of the pore tubules is continuous with the pore kettle, but the same is not yet demonstrated for the inner end, which always appears diffuse and somehow clogged (Figure 4). Direct contact of the pore tubules with the dendritic branches could not be observed in critical analysis of serial sections, but this does not exclude a contiguity in vivo.[17] Differential shrinking of the cuticle with its associated structures on one side, and of the outer segments of the dendrites on the other, might be expected during fixation, dehydration, and embedding, and could explain the separation and the fluffy appearance of the pore tubule endings as well. On the other hand, a solid connection is improbable for the following reasons: (1) no membrane destruction could be observed on the dendrite surface, and (2) in the type 1 sensillum of *Necrophorus* about 30 per cent of all tubules are necessarily at a considerable distance from the dendrites (because of the special architecture of the hair wall).[17]

The pore systems in sensilla trichodea and placodea are not different in principle from that described above. The main difference is in the thickness of the cuticle that is to be penetrated. This may be as much as 1 μ in the long sense hairs of *Antheraea*,

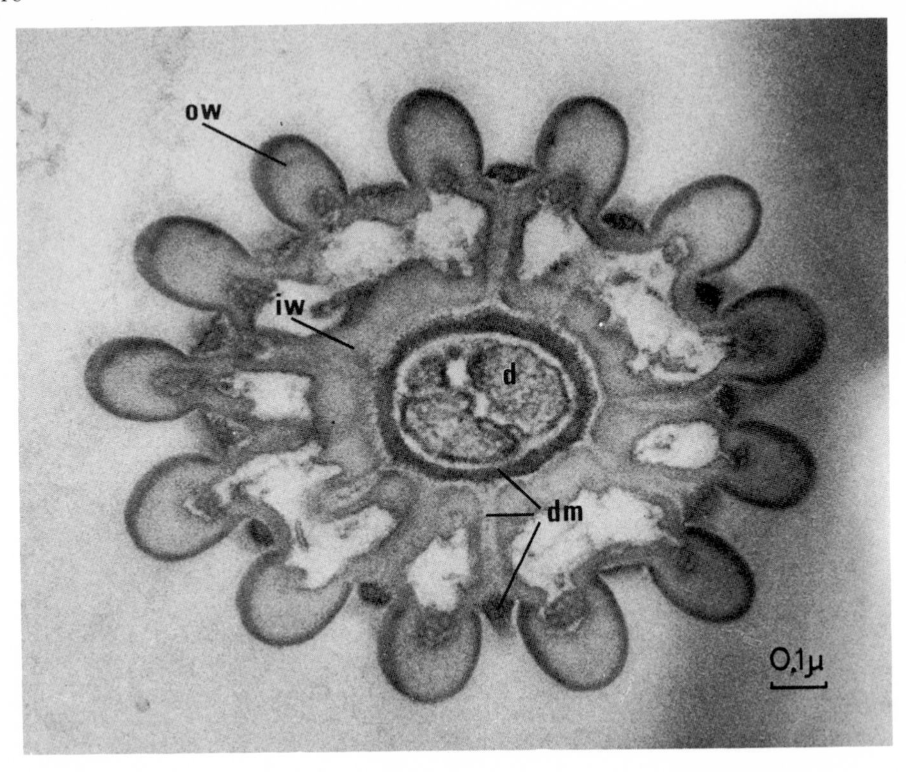

Figure 6 Cross section through longitudinally grooved olfactory pit peg of *Locusta migratoria*. (d, dendrite; dm, dense material around the dendrites, in the outside grooves, and in the core of the cuticular spokes; iw, ow, inner and outer cuticular wall of peg).

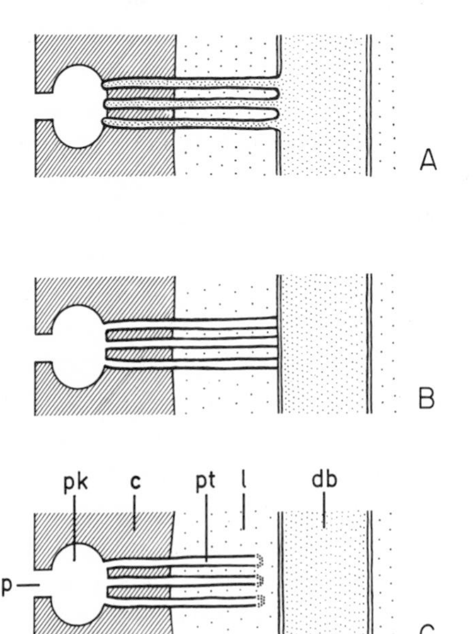

Figure 7 Hypotheses about the nature of the pore tubules and their connection with the dendrites. For further discussion, see text. (c, cuticle; db, dendritic branch; l, Sensillenliquor; p, pore; pk, pore kettle; pt, pore tubule).

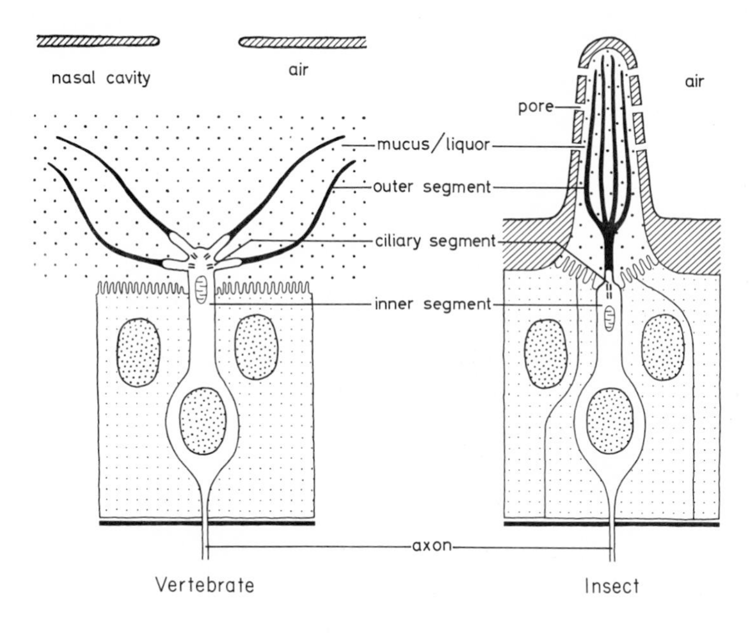

Figure 8 Schematic comparison between vertebrate and insect olfactory receptors. Possibly analogous structures are drawn similarly.

but less than 300 Å in the pore plate of *Apis mellifica*. While in the first case the major part of the cuticle is penetrated by the pore tubules (Figure 5A), the cuticular membrane in *Apis* seems to be perforated wholly by the pore, which then divides into about four pore tubules (R. A. Steinbrecht, unpublished). The delicacy of these structures in the honeybee may be the reason they were not detected in ultrathin sections of earlier studies,[30,54] although, as early as 1952, pores approximately 100 Å wide had been seen on free-hand tangential sections viewed normal to the plane of the sensillum surface.[43] These pores are located only on the thin, cuticular membrane at the circumference of the central plate, arranged in radial rows between radial thickenings of this cuticle, the "Balken."[43] The outer segments of the dendrites, seen immediately below the porous cuticle, follow the ellipsoidal circumference of the pore plate.

Rough estimates of pore density range from only 5 to 15 pores per μ^2 in sensilla trichodea to 40 to 100 pores per μ^2 in different sensilla basiconica, resulting in numbers of 1,000 to 50,000 pores per sensillum.[45] In spite of such high numbers, the relative cross-sectional area of these narrow pore channels is small—about 0.8 per cent of the total sensillar surface in *Necrophorus*. About the same value is obtained if the relative cross-sectional area of all pore tubules is estimated.[17]

The fine structure of a coeloconic sensillum has been briefly described by Slifer, Prestage, and Beams[58] in the grasshoppers *Romalea microptera* and *Melanoplus differentialis*, and in the honeybee.[54] These pit pegs have a longitudinally grooved surface which is similar to that of the surface pegs of some Dipteran species.[15,55,56]

From the penetration of dyes into the tip of the peg, Slifer and coworkers[58] concluded that the dendrites might be exposed to the air at their tips, and estimated the exposed surface per peg at 0.13 μ^2. However, no electron micrographs have shown such a relatively wide opening. Using refined methods, a search has been made in *Locusta migratoria* for such a tip opening, but it has never been observed in ultrathin serial sections of several pegs. In fact, the cuticle tapered at the tip to a thin, dense membrane (only 200 Å thick) that covered the dendrite endings (R. A. Steinbrecht and G. Thies, unpublished).

The complex structure of the fluted cuticular wall has been studied in detail (Figures 3C, 6). There are two cuticular walls, the inner one around the unbranched dendrites and the outer one longitudinally grooved. The connection is made by hollow cuticular spokes, the lumen of which is filled with a dense substance continuous with the material at the bottom of the grooves as well as with that around the dendrites. There it forms a dendrite sheath that reaches up to the peg tip. One could tentatively speculate that the hollow spokes serve the same function as the pore-tubule systems of other olfactory sensilla. Although on the locust antenna there are humidity receptors in sensilla that have a similar structure to those containing olfactory sense cells, it has been ascertained that there is a complex cuticular apparatus without a pore-tubule system in true olfactory sensilla. This was done by sectioning marked sensilla, the olfactory function of which had been proved electrophysiologically (R. A. Steinbrecht, G. Thies, and U. Waldow, unpublished).

Final Remarks

Slifer and her coworkers[53,54,56–58] and Richter[44] assume the pore tubules (pore filaments) to be extremely fine microvillus-like protrusions of the dendritic cell membrane (Figure 7A). As a consequence, the receptor cell would be in direct contact with the environment, its dendritic plasma membrane apposed to it at the bottom of the pore kettle.

Contrary to this hypothesis there is now evidence from Ernst's studies that the pore tubules are extracellular material, formed by the trichogen cell long before the dendrites invade the sense hair (p. 12). The proof that the pores and pore tubules are really hollow structures, open to the outside, has been obtained by applying hemolymph or protargol solutions (containing silver particles as small as 50 Å) to the outside sensillum surface. These substances were shown to penetrate into the lumen of the pores, pore kettles, and pore tubules, but not into the liquor.[17]

If we assume that odor molecules must reach the membrane of the receptor cell to elicit their stimulatory effects—so far the most probable hypothesis—then two possibilities must be discussed: either the tubules are contiguous with the dendritic membrane and the interaction between odor molecules and receptor surface occurs at the very end of the pore tubules (Figure 7B but cf. p. 15), or the pore tubules end in the liquor but are permeable for odor molecules, which then must pass through the liquor to reach the receptor membrane (Figure 7C).

CONCLUSION

If vertebrate and insect olfactory receptors are compared in terms of functional morphology, many analogous features appear (Figure 8). Bipolar primary receptor cells, surrounded by supporting or sheath cells, extend their dendrites into a liquid, the mucus or liquor. A tendency for surface enlargement is consistently observed in those parts of the receptor cell that are in contact with this liquid. It seems reasonable to assume that, in vertebrates as well as in insects, odor molecules must pass this liquid phase before they reach the receptor membrane.

The complex cuticular apparatus of insect sensilla need not imply a fundamental diversity of the olfactory process, but rather a device to protect the receptor endings from drying in these animals that generally lack mucosae because of their cuticular exoskeleton. Although still speculative, these considerations challenge further studies toward their proof.

REFERENCES

1. Allison, A. C. 1953. The morphology of the olfactory system in vertebrates. *Biol. Rev.* (*Cambridge*) **28**, pp. 195–244.
2. Altner, H., and W. Müller. 1968. Elektrophysiologische und elektronenmikroskopische Untersuchungen an der Riechschleimhaut des Jacobsonschen Organs von Eidechsen (*Lacerta*). *Z. Vergleich. Physiol.* **60**, pp. 151–155.
3. Andres, K. H. 1965. Differenzierung und Regeneration von Sinneszellen in der Regio olfactoria. *Naturwissenschaften* **52**, p. 500.
4. Andres, K. H. 1966. Der Feinbau der Regio olfactoria von Makrosmatikern. *Z. Zellforsch. Mikroskop. Anat.* **69**, pp. 140–154.
5. Andres, K. H. 1968. Neue Befunde zur Feinstruktur des olfactorischen Saumes. *J. Ultrastruct. Res.* **25**, p. 163.
6. Andres, K. H. 1969. Der olfactorische Saum der Katze. *Z. Zellforsch. Mikroskop. Anat.* **96**, pp. 250–274.
7. Bannister, L. H. 1965. The fine structure of the olfactory surface of teleostean fishes. *Quart. J. Microscop. Sci.* **106**, pp. 333–342.
8. Bannister, L. H. 1968. Fine structure of the sensory endings in the vomero-nasal organ of the slowworm *Anguis fragilis*. *Nature* **217**, pp. 275–276.
9. Bloom, G. 1954. Studies on the olfactory epithelium of the frog and the toad with the aid of light and electron microscopy. *Z. Zellforsch. Mikroskop. Anat.* **41**, pp. 89–100.
10. Boeckh, J. 1962. Elektrophysiologische Untersuchungen an einzelnen Geruchsrezeptoren auf den Antennen des Totengräbers (*Necrophorus*, Coleoptera). *Z. Vergleich. Physiol.* **46**, pp. 212–248.
11. Brettschneider, H. 1958. Elektronenmikroskopische Untersuchungen an der Nasenschleimhaut. *Anat. Anz.* **105**, pp. 194–204.
12. Bronshtein, A. A. 1964. Observations in vivo of the motion of cilia of olfactory cells. *Dokl. Akad. Nauk SSSR* **156**, pp. 715–718. (In Russian)
13. Brown, H. E., and L. M. Beidler. 1966. The fine structure of the olfactory tissue in the black vulture. *Fed. Proc.* **25**, Abstract 786, p. 329.
14. De Lorenzo, A. J. 1957. Electron microscopic observations of the olfactory mucosa and olfactory nerve. *J. Biophys. Biochem. Cytol.* **3**, pp. 839–850.
15. Dethier, V. G., J. R. Larsen, and J. R. Adams. 1963. The fine structure of the olfactory receptors of the blowfly. *In* Olfaction and Taste I (Y. Zotterman, editor). Pergamon Press, Oxford, etc., pp. 105–110.
16. Engström, H., and G. Bloom. 1953. The structure of the olfactory region in man. *Acta Oto-Laryngol.* **43**, pp. 11–21.
17. Ernst, K. D. 1969. Die Feinstruktur von Riechsensillen auf der Antenne des Aaskäfers, *Necrophorus*. *Z. Zellforsch. Mikroskop. Anat.* **94**, pp. 72–102.

18. Frisch,.D. 1967. Ultrastructure of mouse olfactory mucosa. *Am. J. Anat.* **121**, pp. 87–119.
19. Gasser, H. S. 1956. Olfactory nerve fibers. *J. Gen. Physiol.* **39**, pp. 473–496.
20. Ghiradella, H., J. Case, and J. Cronshaw. 1968. Fine structure of the aesthetasc hairs of *Coenobita compressus* Edwards. *J. Morphol.* **124**, pp. 361–385.
21. Gray, E. G. 1960. The fine structure of the insect ear. *Phil. Trans. Roy. Soc.* (*London*) *Ser. B* (*Biol. Sci.*) **243**, pp. 75–94.
22. Graziadei, P. 1964. Electron microscopy of some primary receptors in the sucker of *Octopus vulgaris. Z. Zellforsch. Mikroskop. Anat.* **64**, pp. 510–522.
23. Graziadei, P. 1966. Electron microscopic observations of the olfactory mucosa of the mole. *Proc. Zool. Soc. London* **149**, pp. 89–94.
24. Graziadei, P., and L. H. Bannister. 1967. Some observations on the fine structure of the olfactory epithelium in the domestic duck. *Z. Zellforsch. Mikroskop. Anat.* **80**, pp. 220–228.
25. Henke, K., and G. Rönsch. 1951. Über die Bildungsgleichheiten in der Entwicklung epidermaler Organe und die Entstehung des Nervensystems im Flügel der Insekten. *Naturwissenschaften* **38**, pp. 335–336.
26. Herberhold, C. 1968. Vergleichende histochemische Untersuchungen am peripheren Riechorgan von Säugetieren und Fischen. *Arch. Klin. Exp. Ohrenheilk. Nas.-u. Kehlkopfheilk.* **190**, pp. 166–182.
27. Hilding, A. 1932. The physiology of drainage of nasal mucus. I. The flow of the mucus currents through the drainage system of the nasal mucosa and its relation to ciliary activity. *Arch. Otolaryngol.* **15**, pp. 92–100.
28. Hilding, A. 1932. The physiology of drainage of nasal mucus. III. Experimental work on the accessory sinuses. *Am. J. Physiol.* **100**, pp. 664–670.
29. Hopkins, A. E. 1926. The olfactory receptors in vertebrates. *J. Comp. Neurol.* **41**, pp. 253–289.
30. Krause, B. 1960. Elektronenmikroskopische Untersuchungen an den Plattensensillen des Insektenfühlers. *Zool. Beitr. NF.* **6**, pp. 161–205.
31. Lacher, V. 1964. Elektrophysiologische Untersuchungen an einzelnen Rezeptoren für Geruch, Kohlendioxyd, Luftfeuchtigkeit und Temperatur auf den Antennen der Arbeitsbiene und der Drohne (*Apis mellifica* L.). *Z. Vergleich. Physiol.* **48**, pp. 587–623.
32. Laverack, M. S., and D. J. Ardill. 1965. The innervation of the aesthetasc hairs of *Panulirus argus. Quart. J. Microscop. Sci.* **106**, pp. 45–60.
33. Le Gros Clark, W. E. 1956. Observations on the structure and organization of olfactory receptors in the rabbit. *Yale J. Biol. Med.* **29**, pp. 83–95.
34. Le Gros Clark, W. E. 1957. The Ferrier Lecture: Inquiries into the anatomical basis of olfactory discrimination. *Proc. Roy. Soc.* (*London*), *Ser. B.* (*Biol. Sci.*) **146**, pp. 299–319.
35. Loewenstein, W. R., S. J. Socolar, S. Higashino, Y. Kanno, and N. Davidson. 1965. Intercellular communication: renal, urinary bladder, sensory, and salivary gland cells. *Science* **149**, pp. 295–298.
36. Moulton, D. G., and L. M. Beidler. 1967. Structure and function in the peripheral olfactory system. *Physiol. Rev.* **47**, pp. 1–52.
37. Noble-Nesbitt, J. 1963. The cuticle and associated structures of *Podura aquatica* at the moult. *Quart. J. Microscop. Sci.* **104**, pp. 369–391.
38. Okano, M. 1965. Fine structure of the canine olfactory hairlets. *Arch. Histol. Jap.* **26**, pp. 169–185.
39. Okano, M., and Y. Sugawa. 1965. Ultrastructure of the respiratory mucous epithelium of the canine nasal cavity. *Arch. Histol. Jap.* **26**, pp. 1–21.
40. Okano, M., A. F. Weber, and S. P. Frommes. 1967. Electron microscopic studies of the distal border of the canine olfactory epithelium. *J. Ultrastruct. Res.* **17**, pp. 487–502.
41. Reese, T. S. 1965. Olfactory cilia in the frog. *J. Cell Biol.* **25**, pp. 209–230.
42. Renzoni, A. 1968. Osservazioni istologiche, istochimiche ed ultrastrutturali sui tentacoli di *Vaginulus borellianus* (Colosi), Gastropoda soleolifera. *Z. Zellforsch. Mikroskop. Anat.* **87**, pp. 350–376.
43. Richards, A. G. 1952. Studies on arthropod cuticle. VIII. The antennal cuticle of honeybees, with particular reference to the sense plates. *Biol. Bull.* **103**, pp. 201–225.
44. Richter, S. 1962. Unmittelbarer Kontakt der Sinneszellen cuticularer Sinnesorgane mit der Aussenwelt. Eine licht- und elektronenmikroskopische Untersuchung der chemorezeptorischen Antennensinnesorgane der Calliphora-Larven. *Z. Morphol. Oekol. Tiere* **52**, pp. 171–196.
45. Schneider, D., and R. A. Steinbrecht. 1968. Checklist of insect olfactory sensilla. *Symp. Zool. Soc. London* **23**, pp. 279–297.
46. Schneider, D., V. Lacher, and K.-E. Kaissling. 1964. Die Reaktionsweise und das Reaktionsspektrum

von Riechzellen bei *Antheraea pernyi* (Lepidoptera, Saturniidae). *Z. Vergleich. Physiol.* **48**, pp. 632–662.
47. Schneider, D., R. A. Steinbrecht, and K.-D. Ernst. 1966. Antennales Riechhaar des Aaskäfers. *Naturw. Rundschau* **19**, H3, pp. 1 and 3.
48. Schöne, H., and R. A. Steinbrecht. 1968. Fine structure of statocyst receptor of *Astacus fluviatilis*. *Nature* **220**, pp. 184–186.
49. Schultze, M. 1856. Über die Endigungsweise des Geruchsnerven und der Epithelialgebilde der Nasenschleimhaut. *Monatsber. Deut. Akad. Wiss. Berlin* **21**, pp. 504–515.
50. Schultze, M. 1862. Untersuchungen über den Bau der Nasenschleimhaut, namentlich die Structur und Endigungsweise der Geruchsnerven bei dem Menschen und den Wirbelthieren. *Abh. Naturforsch. Ges. Halle* **7**, pp. 1–100.
51. Seifert, K., and G. Ule. 1967. Die Ultrastruktur der Riechschleimhaut der neugeborenen und jugendlichen weissen Maus. *Z. Zellforsch. Mikroskop. Anat.* **76**, pp. 147–169.
52. Shibuya, T. 1964. Dissociation of olfactory neural response and mucosal potential. *Science* **143**, pp. 1338–1340.
53. Slifer, E. H. 1967. Thin-walled olfactory sense organs on insect antennae. *In* Insects and Physiology (J. W. L. Beament and J. E. Treherne, editors). Oliver and Boyd Ltd., Edinburgh and London, pp. 233–245.
54. Slifer, E. H., and S. S. Sekhon. 1961. Fine structure of the sense organs on the antennal flagellum of the honeybee, *Apis mellifera* Linnaeus. *J. Morphol.* **109**, pp. 351–381.
55. Slifer, E. H., and S. S. Sekhon. 1962. The fine structure of the sense organs on the antennal flagellum of the yellow fever mosquito *Aedes aegypti* (Linnaeus). *J. Morphol.* **111**, pp. 49–67.
56. Slifer, E. H., and S. S. Sekhon. 1964. Fine structure of the sense organs on the antennal flagellum of a flesh fly, *Sarcophaga argyrostoma* R.-D. (Diptera, Sarcophagidae). *J. Morphol.* **114**, pp. 185–207.
57. Slifer, E. H., and S. S. Sekhon. 1964. The dendrites of the thin-walled olfactory pegs of the grasshopper (Orthoptera, Acrididae). *J. Morphol.* **114**, pp. 393–409.
58. Slifer, E. H., J. J. Prestage, and H. W. Beams. 1959. The chemoreceptors and other sense organs on the antennal flagellum of the grasshopper (Orthoptera, Acrididae). *J. Morphol.* **105**, pp. 145–191.
59. Snodgrass, R. E. 1935. Principles of Insect Morphology. McGraw-Hill Book Co., New York and London.
60. Sotelo, J. R., and O. Trujillo-Cenóz. 1958. Electron microscopic study on the development of ciliary components of the neural epithelium of the chick embryo. *Z. Zellforsch. Mikroskop. Anat.* **49**, pp. 1–12.
61. Steinbrecht, R. A. 1969. On the question of nervous syncytia: Lack of axon fusion in two insect sensory nerves. *J. Cell Sci.* **4**, pp. 39–53.
62. Suzuki, N. 1967. Behavioral and electrical responses of the land snail, *Ezohelix flexibilis* (Fulton), to odors. *J. Fac. Sci. Hokkaido Univ. Ser. VI* (*Zool.*) **16**, pp. 174–185.
63. Takagi, S. F., and T. Shibuya. 1961. Studies on the potential oscillation appearing in the olfactory epithelium of the toad. *Jap. J. Physiol.* **11**, pp. 23–37.
64. Thornhill, R. A. 1967. The ultrastructure of the olfactory epithelium of the lamprey *Lampetra fluviatilis*. *J. Cell Sci.* **2**, pp. 591–602.
65. Thurm, U. 1964. Mechanoreceptors in the cuticle of the honeybee: Fine structure and stimulus mechanism. *Science* **145**, pp. 1063–1065.
66. Trujillo-Cenóz, O. 1961. Electron microscope observations on chemo- and mechano-receptor cells of fishes. *Z. Zellforsch. Mikroskop. Anat.* **54**, pp. 654–676.
67. Tucker, D. 1967. Olfactory cilia are not required for receptor function. *Fed. Proc.* **26**, p. 544 (Abstract 1609).
68. Tucker, D., and T. Shibuya. 1965. A physiologic and pharmacologic study of olfactory receptors. *Cold Spring Harbor Symp. Quant. Biol.* **30**, pp. 207–215.
69. Weber, H. 1954. Grundriss der Insektenkunde. Fischer Verlag K. G., Stuttgart, Germany.
70. Whitear, M. 1962. The fine structure of crustacean proprioceptors. I. The chordotonal organs in the legs of the shore crab, *Carcinus maenas*. *Phil. Trans. Roy. Soc.* (*London*) *Ser. B* (*Biol. Sci.*) **245**, pp. 291–324.
71. Wilson, J. A. F., and R. A. Westerman. 1967. The fine structure of the olfactory mucosa and nerve in the teleost *Carassius carassius* L. *Z. Zellforsch. Mikroskop. Anat.* **83**, pp. 196–206.
72. Zacharuk, R. Y. 1962. Exuvial sheaths of sensory neurones in the larva of *Ctenicera destructor* (Brown) (Coleoptera, Elateridae). *J. Morphol.* **111**, pp. 35–47.

NOTES ON THE FINE STRUCTURE OF OLFACTORY EPITHELIUM

DONALD FRISCH
Department of Biostructure, Northwestern University School of Medicine, Chicago, Illinois

INTRODUCTION

The fine structure of vertebrate olfactory receptor organs has been presented in a number of recent reports.[1,3,4,8,13,14,16,18] While a comprehensive review of these papers is not appropriate here, it is useful to indicate that the observations of a number of investigators on a variety of species are all in agreement as to the principal structural features. This agreement is so general that, when an exception is presented—specifically the teleost fish *Carassius carassius*[18]—it raises the question whether chemoreception in this fish follows the predominant vertebrate mode. The present report is concerned with the fine structure of vertebrate olfactory epithelium in the mouse. Part of the material represents my thoughts on some preliminary observations of fetal mouse tissue obtained in collaboration with my colleague, Dr. Albert I. Farbman. These data are included in the hope of generating discussion concerning the possible advantage of fetal tissues for future physiological experimentation.

MATERIALS AND METHODS

We performed the electron microscopy reported here using standard procedures that have been fully described elsewhere.[8,9] Fixation of adult tissue was accomplished by flooding the nasal cavities with various concentrations of glutaraldehyde buffered to pH 7.2–7.4 with 0.1 M phosphate buffer. To obtain the fetal tissue, we dissected whole heads while they were immersed in the same fixatives. Quite arbitrarily, we chose 15-day-old fetuses as the starting point of this study, and all results reported here are of this age.

The light microscopy utilizes Nomarski interference optics.[2] This technique presents a number of operational advantages, including the ability to "section optically." That is, a sharp image with a depth of focus thinner than a single cell may be obtained at any level of focus from specimens as thick as 40 μ.[6]

OBSERVATIONS AND DISCUSSION

General Architecture

The olfactory epithelium of the mouse is a pseudostratified columnar epithelium (Figures 1, 2) that is considerably taller than the adjacent respiratory epithelium. The supporting cells (and presumably the olfactory receptor cells) are canted with respect to the free surface. This is visible in cross sections of the ovoid nuclei of supporting cells as seen in the light microscope. The profiles are round where the epithelium is thin and they are elongated only where it is thicker. The electron microscope makes this point clearer (Figure 3); cell boundaries may be determined with certainty, and thus the shape of the cytoplasmic profiles can confirm the interpretation.

Much of the significance of this cell packing awaits the future development of precise techniques for intraepithelial recording from single units. In using such techniques, the intraepithelial location of axons, perikarya, and dendrites becomes significant, as does the depth of penetration of a microelectrode. Regions with

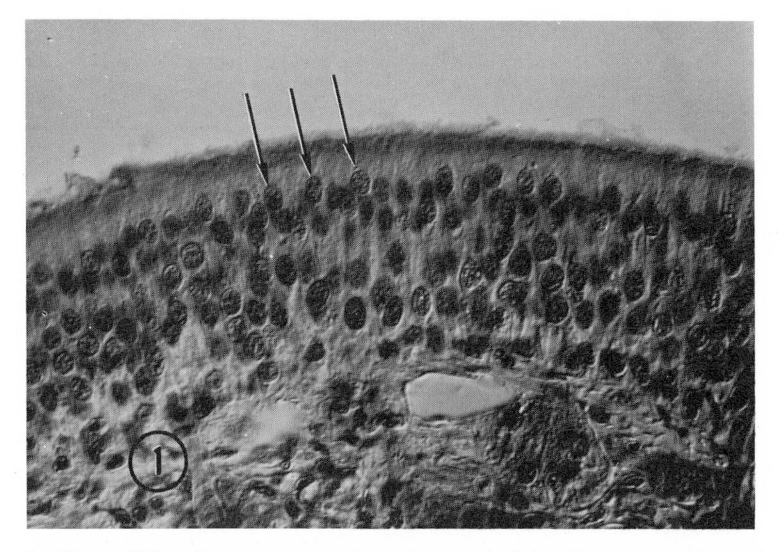

Figure 1 Nomarski interference contrast photomicrograph of the olfactory epithelium of the dog. The arrows indicate the superficial nuclei of the supporting cells and their roughly round shape is interpreted to mean the cells are nearly cross-sectioned. Magnification: 1,200×.

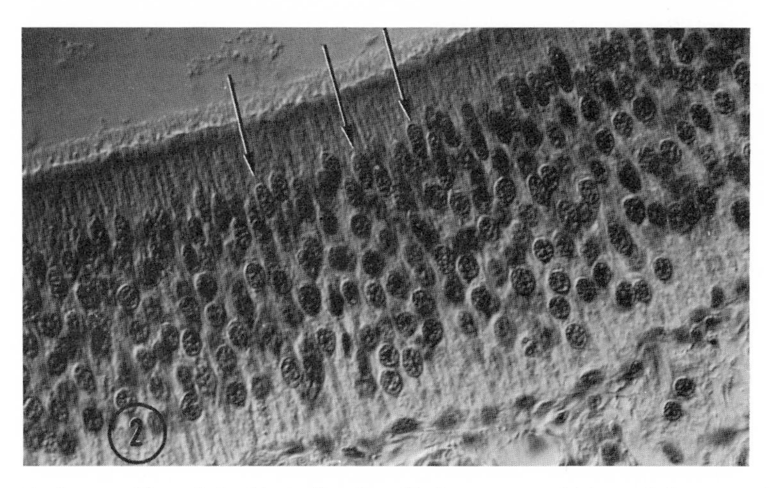

Figure 2 Same as Figure 1. In this section the epithelium appears thicker and the supporting cell nuclei show the typical ovoid shape. Magnification: 1,200×.

Figure 3 Low power electron micrograph of mouse olfactory epithelium. This section shows one of the shortest regions of the olfactory epithelium which, from the basement membrane to the free surface, measures approximately 40 μ. The supporting cell nuclei (SCN) are superficial and the olfactory cell nuclei (OCN) are more basal. At this magnification, olfactory vesicles (OV) and dendritic processes (DP) can be easily seen. Basal cell (BC) is at lower left. Almost all of the dendritic processes show as round profiles; indicating they have been cross-sectioned. Magnification: 4,000 $\times$.

inclined cells will not permit depth of penetration to be the only parameter of electrode position. It is feasible, however, that Nomarski interference optics can be manipulated to aid the investigator in selecting regions appropriate for electrode approach. Cells parallel to the optical axis and to the axis of electrode penetration will show round nuclear profiles. The Nomarski optical system allows this nuclear shape to be visualized in the thick, living mucosa. Moreover, the shallow depth of focus permits precise monitoring of the penetration of an electrode.

Throughout the epithelium, olfactory receptor cells are in juxtaposition.[7,8] That is, two adjacent receptor cells may be separated by as little as 200 Å with no intervening supporting cell cytoplasm to function in ionic insulation (Figure 4). This led to the suggestion that such morphology could be significant in recorded potentials,[8] as was suggested earlier for the special case of the olfactory nerves.[10] Ottoson and

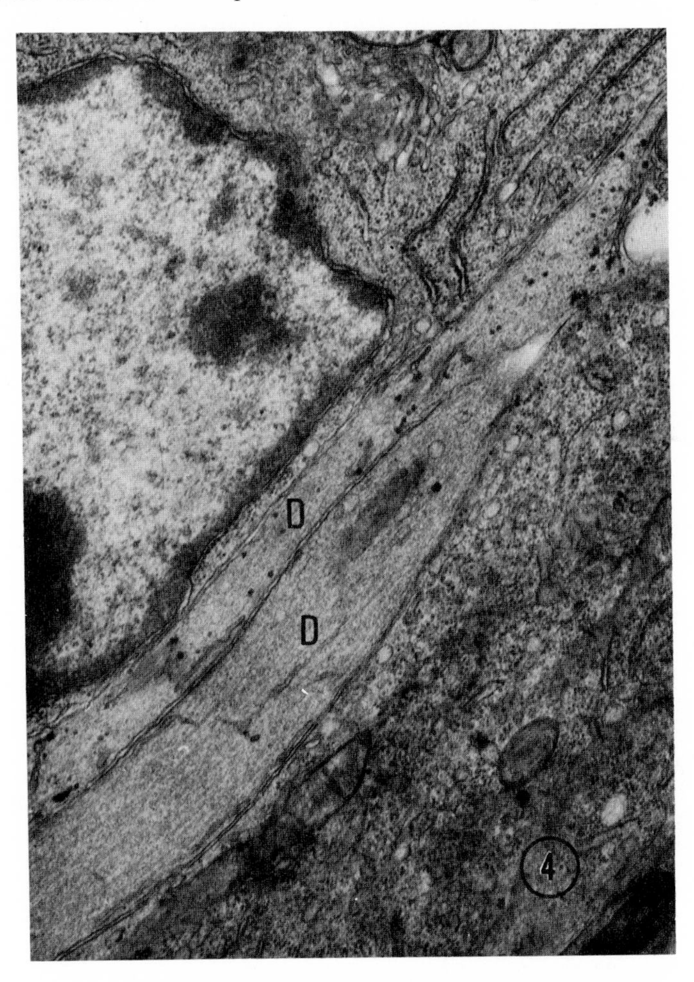

Figure 4 Two dendritic processes (D) are juxtaposed and have been sectioned at a level where they pass between the perikarya of two other receptor cells. In this instance, portions of four neurons are in apposition for at least a limited distance. Magnification: 25,500×.

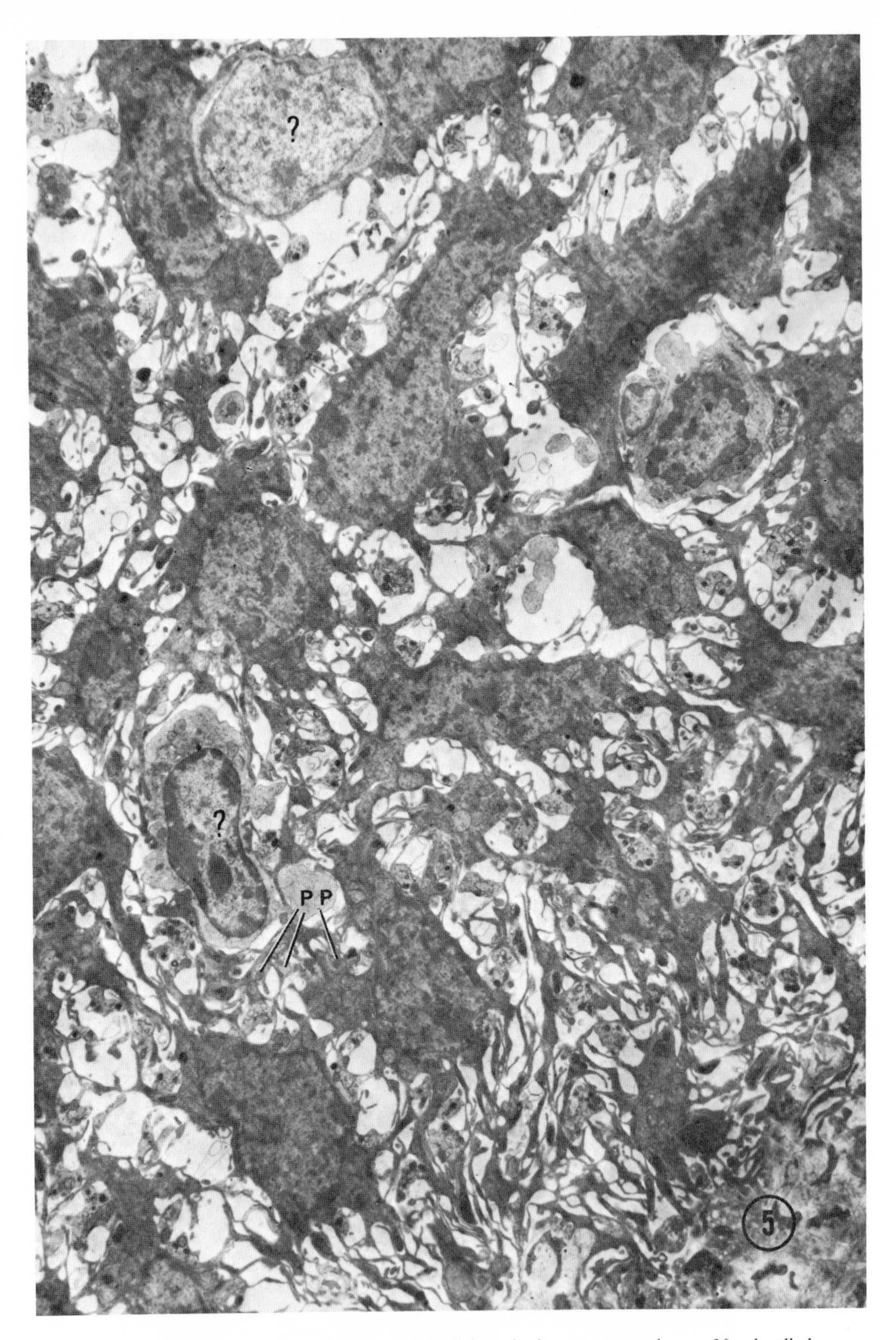

Figure 5 This plane of section is almost parallel to the basement membrane. Nearly all the cells included in the section are basal cells although two or three (?) may be wandering cells which occasionally seem to penetrate the epithelium. A relatively large percentage of this figure is extracellular space (white areas) which is interrupted by the interlaced basal cell processes. Some of the electron lucent cytoplasmic profiles are sectioned proximal processes of olfactory cells (PP) but the larger profiles are sections of the foot processes of the supporting cells. Magnification: 10,000 ×.

Shepherd[15] have indicated that electrical interaction between cells separated by a small gap might be limited, and they note a lack of evidence that such interaction does occur. Gesteland et al.[12] found no evidence of coupled behavior in their recordings from the olfactory system, nor have they observed any in their subsequent work.[11] Yet, the extracellular spaces in the basal region of the olfactory epithelium are far more extensive than in the apical regions or in the olfactory nerves. Basal cells do envelop bundles of axons, but the wrapping processes are finger-like, rather than plate-like. This conclusion derives from examination of sections cut both parallel and perpendicular to the basement membrane (Figures 5, 6). Sections in both planes show the same kind of thin, basal-cell, cytoplasmic processes, indicating a cylindrical shape. These finger-like processes can form only an interrupted sheath and so cannot completely shield axons from ionic currents. In functional terms, axons are separated only by extracellular space, and this space may be several thousand Å wide. Thus the possibility of electrical interaction in the basal region of the epithelium must still be considered, even if such interaction is not demonstrable in other areas.

It might be feasible to generate differences in the extracellular space and/or the amount of intraepithelial receptor-cell contacts via some experimental manipulation. Information on the significance of these morphologic features relative to function of the tissue might then be obtained. We can propose a natural "experiment" that may yield the same kind of information. In our laboratory, we have examined epithelium of 15-day-old mouse fetuses, and we find that the intraepithelial contacts are considerably more extensive than in adults (Figure 7). In fact, even olfactory vesicles may be separated by a gap of only 200 Å, and this is never observed in adult tissue. One possible interpretation of these data is that 15-day-old fetuses have either fewer supporting cells or a greater ratio of receptor to supporting cells.

At the other end of the scale, it is reasonable to expect that receptor cells will be lost as animals age, and there is no firm evidence that lost cells can be replaced. Hence a series of experiments including fetuses, young adults or juveniles, and aged animals could yield information about the relative contributions which receptor and supporting cells may make to potentials recorded from the epithelium. At the very least, such a series might yield data bearing upon the significance of intraepithelial cytoplasmic insulation and/or the lack of it.

Surface Architecture

Since all olfactory receptor cells are ciliated, and since these cilia lie above the surface of the epithelium where initial interaction with odorant molecules is most likely to occur, much attention has been directed to the elucidation of ciliary structure. However, one preliminary report[17] contends that cilia may not function in reception at all, but until confirmation of this conclusion is available, the study of cilia still seems worthy.

A recent study presented results of the examination of both rabbit and goldfish epithelia with scanning electron microscopy, yielding a direct three-dimensional image of the surface.[5] This investigation shows in a most striking fashion that we

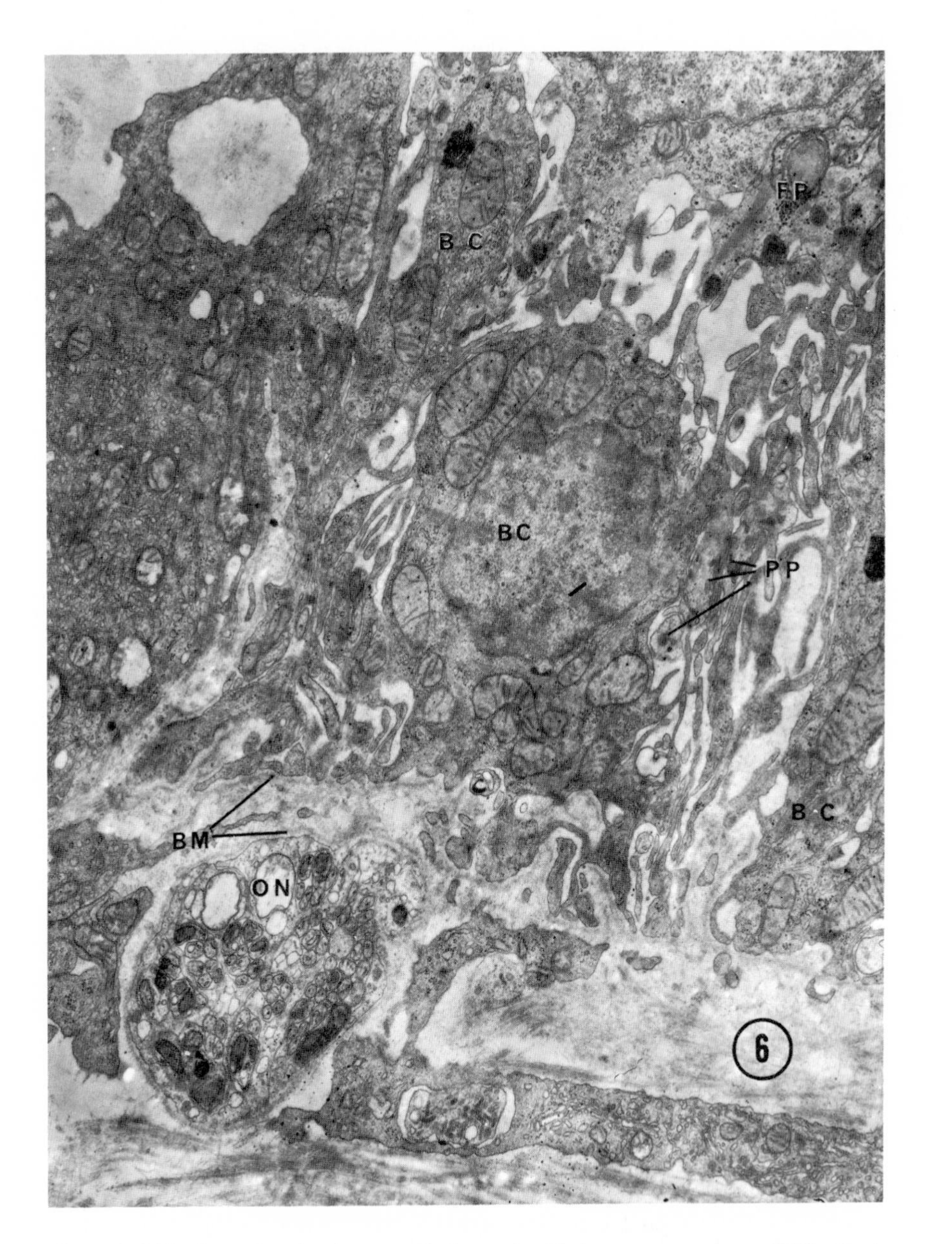

Figure 6 This plane of section is perpendicular to the thin basement membrane (BM). Basal cells (BC) and their thin processes may be seen, sometimes in association with olfactory cell proximal processes (PP). An olfactory nerve bundle (ON) is present in the subjacent connective tissue. Magnification: 12,000 ×.

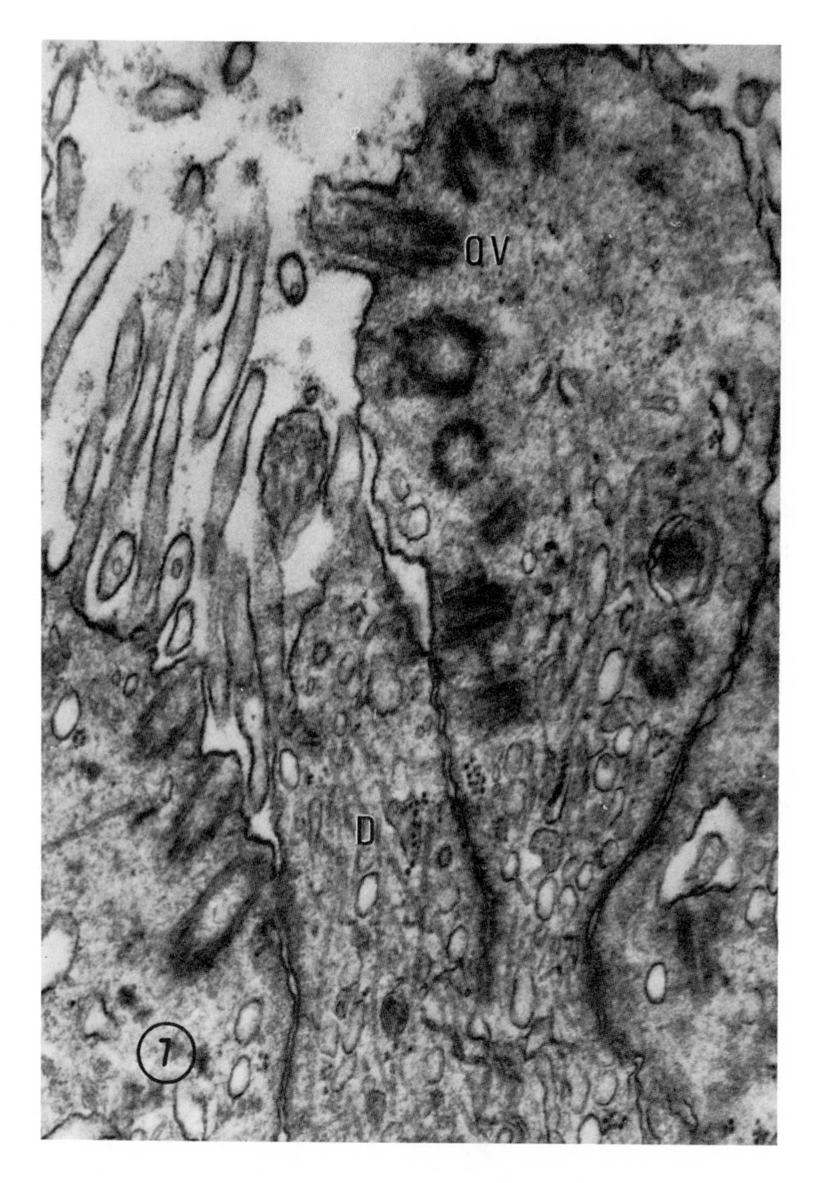

Figure 7 This electron micrograph shows a respiratory-olfactory junction in a 15-day-old mouse fetus nasal mucosa. A dendritic process (D) is adjacent to the ciliated respiratory epithelium cell on the left; an apposition not yet observed in adult tissue. This dendrite is also adjacent to an olfactory vesicle (OV) on the right of the figure. Magnification: 35,000 ×.

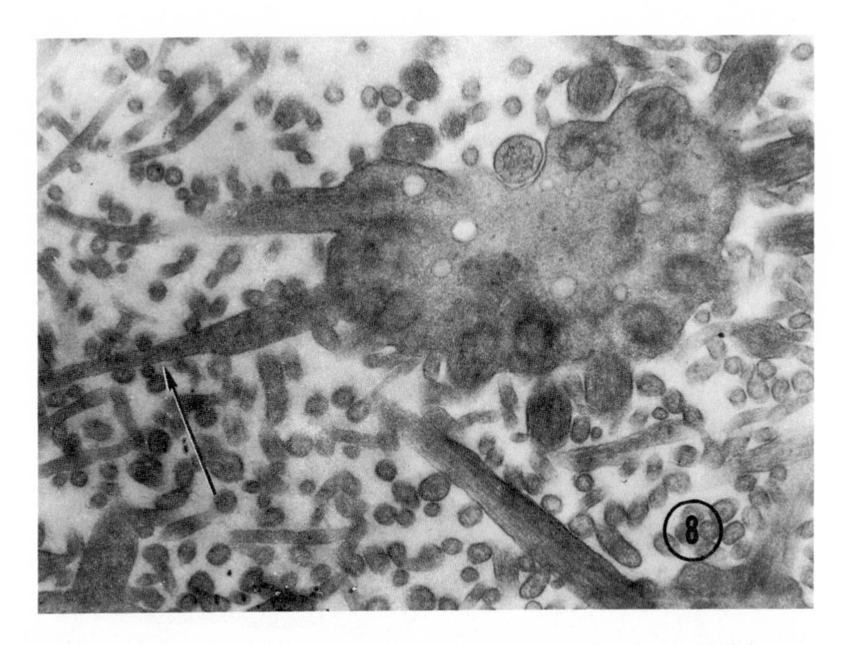

Figure 8 A total of 13 basal bodies and cilia are clearly present in this thin (750 Å) section through an olfactory vesicle. Three more cilia in the section probably have originated from the same vesicle. The arrow indicates the thin distal portion of one of the cilia. Magnification: 25,000×.

have grossly underestimated the number of cilia per cell. In the rabbit, each cell carries so many cilia the authors simply made no attempt to count them. This is supported by Figure 8, which shows 13 basal bodies or cilia in one thin section through an olfactory vesicle. Obviously these data are significant for future estimations of the amount of surface area available for olfactory stimulation, particularly if a role in reception can be established clearly for the cilia.

Olfactory cilia in the mouse, as in other animals,[1,4,14,16] are modified.[8] The proximal portion of each cilium shows the normal kinetocilium pattern of nine peripheral pairs of fused microtubules and a central nonfused pair. Then the cilium narrows to a long cylindrical process (Figure 8) containing as few as two separate microtubules. It is in this modified distal portion of the cilium that dilations have been observed.[8] There is no evidence as yet concerning the functional significance of these dilations, but it seems clear that they reflect a real difference between the membranes of the sensory cilia and the membranes of the adjacent respiratory cilia, which never show dilations under the same conditions.

In the 15-day-old mouse fetus, many of the olfactory cilia have approached adult morphology, including dilations along the length of the ciliary extension (Figures 9, 10). If anything, the dilations are more pronounced, which indicates a greater sensitivity to whatever conditions have caused the swelling.

This morphologic similarity suggests that fetal cilia may share other properties with adult cilia, perhaps including receptor membrane parameters. The mature aspects of the fetal olfactory receptor cell morphology contrast with the immature morphology of supporting cells. In adult tissue, the supporting cell is characterized by microvillous projections, secretory granules, numerous mitochondria, and extensive smooth-surfaced endoplasmic reticulum, both in intimate association with mitochondria and filling the bulk of the cytoplasm. The fetal cells show only microvilli, while the remainder of their morphology resembles that of undifferentiated epithelial cells.

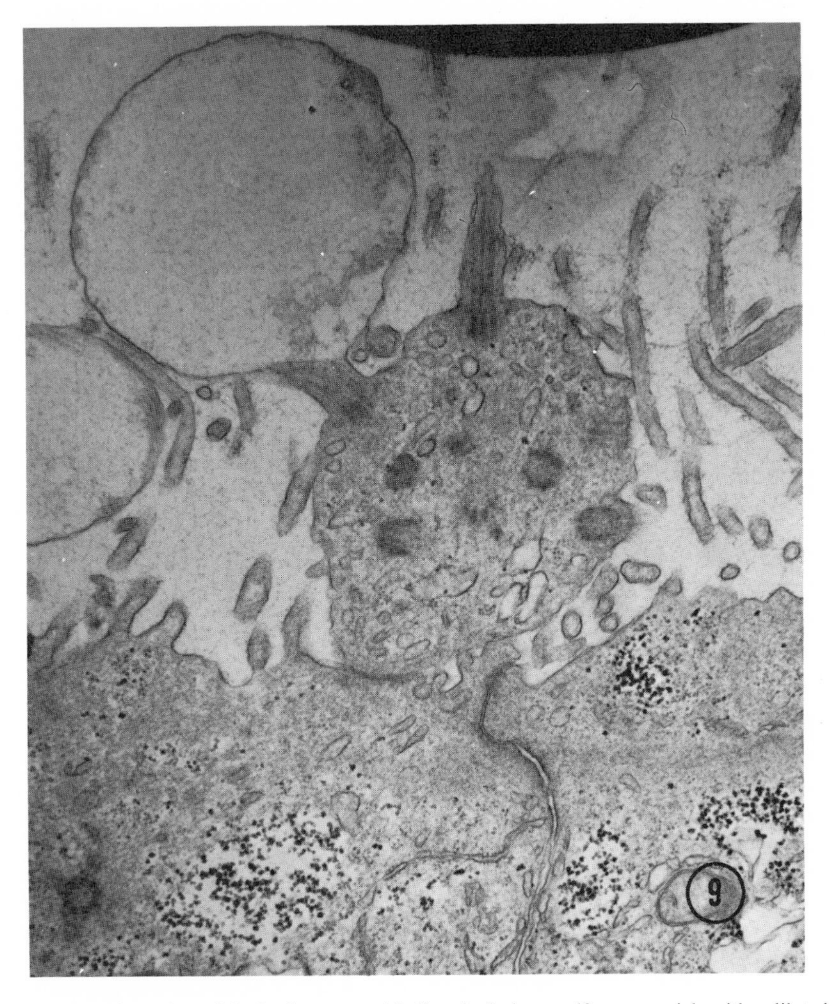

Figure 9 This section of the fetal mouse epithelium includes an olfactory vesicle with a dilated cilium. Another dilation is partly included in the figure. The supporting cell cytoplasm which occupies the lower third of the field seems to be relatively undifferentiated. The most striking feature of this cytoplasm is the dense glycogen particulates. Magnification: 25,000×.

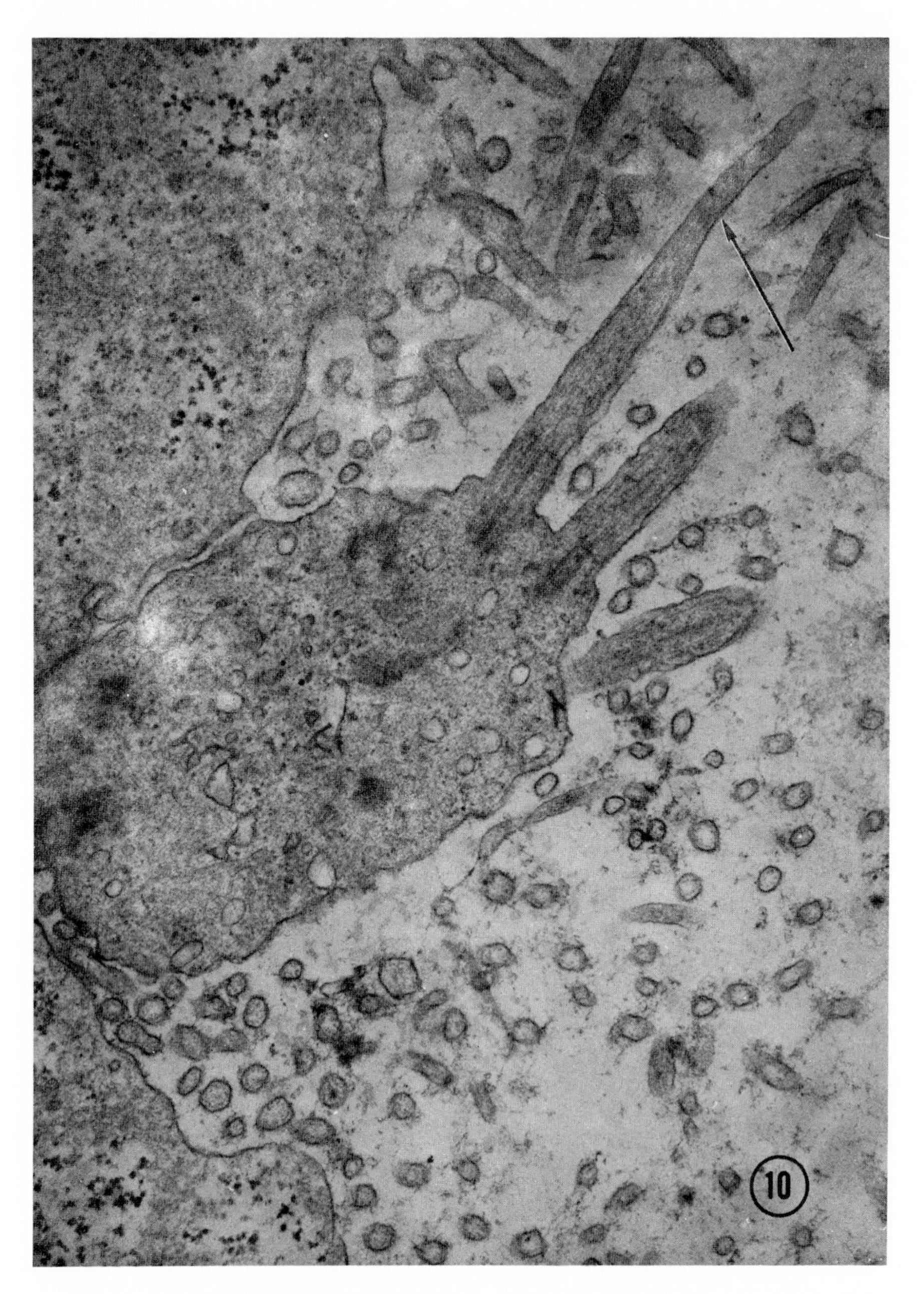

Figure 10 Another fetal olfactory vesicle is illustrated. The arrow indicates the thin distal portion of a cilium which may be compared to the adult structure (Figure 8). Magnification: 25,000×.

At least in the 15-day-old fetus, receptor cells seem to show adult characteristics to a significant degree, whereas supporting cells do not. This again enhances the possibility that the role of the receptor cell in the initial response to olfactant molecules might be clarified by physiological studies of the appropriate stages of development.

ACKNOWLEDGMENT

The author would like to express his gratitude to Dr. John W. Everingham for his generous assistance in the preparation of the light micrographs and for his critical review of the manuscript.

REFERENCES

1. Andres, K. H. 1966. Der Feinbau der Regio olfactoria von Makrosmatikern. *Z. Zellforsch. Mikroskop. Anat.* **69**, pp. 140–154.
2. Bajer, A., and R. D. Allen. 1966. Structure and organization of the living mitotic spindle of Haemanthus endosperm. *Science* **151**, pp. 572–574.
3. Balboni, G. C. 1967. L'ultrastruttura dell'epitelio olfattivo nel ratto e sue modificazioni in seguito a castrazione ed alla somministrazione, a ratti castrati, di testosterone. *Arch. Ital. Anat. Embriol.* **72**, pp. 203–223.
4. Bannister, L. H. 1965. The fine structure of the olfactory surface of teleostean fishes. *Quart. J. Microscop. Sci.* **106**, pp. 333–342.
5. Barber, V. C., and A. Boyde. 1968. Scanning electron microscopic studies of cilia. *Z. Zellforsch. Mikroskop. Anat.* **84**, pp. 269–284.
6. Everingham, J. W. 1967. Use of Nomarski-system optics in the study of thick embryonic tissues. *Anat. Rec.* **157**, p. 242.
7. Farbman, A. I. 1966. Structure of chemoreceptors. *In* Symposium on Foods-Chemistry and Physiology of Flavors (H. W. Schultz, E. A. Day, and L. M. Libbey, editors). Avi Publishing Co., Westport, Conn.
8. Frisch, D. 1967. Ultrastructure of mouse olfactory mucosa. *Amer. J. Anat.* **121**, pp. 87–119.
9. Frisch, D., and A. I. Farbman. 1968. Development of order during ciliogenesis. *Anat. Rec.* **162**, pp. 221–232.
10. Gasser, H. S. 1958. Comparison of the structure, as revealed with the electron microscope, and the physiology of the unmedullated fibers in the skin nerves and in the olfactory nerves. *Exp. Cell Res. Suppl.* **5**, pp. 3–17.
11. Gesteland, R. C. 1968. (Personal communication)
12. Gesteland, R. C., J. Y. Lettvin, and W. H. Pitts. 1965. Chemical transmission in the nose of the frog. *J. Physiol.* (*London*) **181**, pp. 525–559.
13. Graziadei, P. 1966. Electron microscopic observations of the olfactory mucosa of the mole. *J. Zool.* **149**, pp. 89–94.
14. Okano, M., A. F. Weber, and S. P. Frommes. 1967. Electron microscopic studies of the distal border of the canine olfactory epithelium. *J. Ultrastruct. Res.* **17**, pp. 487–502.
15. Ottoson, D., and G. M. Shepherd. 1967. Experiments and concepts in olfactory physiology. *In* Sensory Mechanisms: Progress in Brain Research, Vol. 23 (Y. Zotterman, editor). Elsevier Publishing Co., Amsterdam, The Netherlands, pp. 86–103.
16. Reese, T. S. 1965. Olfactory cilia in the frog. *J. Cell Biol.* **25**, pp. 209–230.
17. Tucker, D. 1967. Olfactory cilia are not required for receptor function. *Fed. Proc.* **26**, p. 544.
18. Wilson, J. A. F., and R. A. Westerman. 1967. The fine structure of the olfactory mucosa and nerve in the teleost *Carassius carassius* L. *Z. Zellforsch. Mikroskop. Anat.* **83**, pp. 196–206.

ELECTRICAL ACTIVITY IN OLFACTORY RECEPTOR CELLS

JÜRGEN BOECKH
Department of Zoology, University of Frankfurt, Germany

INTRODUCTION

Today we can look back on more than 60 years of history of electrophysiological research on olfactory systems, starting in 1900 with Garten's work on the olfactory nerve in the pike.[12] However, Garten was more interested in general questions about nerve function, and other nerves and sensory systems seemed to offer much better opportunities for electrophysiological studies. The electrical activity in olfactory receptor neurons did not become a center of interest until 1937, when Hosoya and Yoshida published their results on potentials from the olfactory mucosa of frogs.[18] In the fifties, a new impetus was provided when Ottoson[29] did his first experiments on the electroolfactogram (EOG) in frogs, Beidler and Tucker[3] investigated fiber bundles in olfactory nerves, and Gasser[13] studied the electrical properties of the fila olfactoria. At the same time, insects entered the story with the work of Schneider[31] on the silkmoth electroantennogram (EAG), and have continued to offer favorable conditions for recording from single cells. Since then there has been a great increase of activity. In the late nineteen fifties and early sixties, successful recordings were made from single receptor cells in several laboratories in Japan,[24] the United States,[14,38] and Europe.[35]

The number of publications and the diversity of interests in the field have increased so much that the organizing committee of this conference thought it appropriate to sum up existing knowledge of the electrical activity of olfactory receptors and to outline experimental results. Rather than give a summary of all the work done on olfactory receptors, I would like to provide a base for discussions and to introduce colleagues from different disciplines to the problems. Therefore, this paper is by no means a complete review. Excellent and competent articles have been published in recent years by Ottoson and Shepherd,[30] Moulton and Tucker,[26] Moulton and Beidler,[25] and Schneider.[32] A detailed discussion of electrical activity in insect olfactory receptors was given by Boeckh, Kaissling, and Schneider[8] (see also Schneider and Steinbrecht[37]).

Today we should integrate our results with those from other fields of sensory physiology; we should consider what we must do next and what we can expect from future work on olfactory receptors.

Why do we need receptor physiology for an understanding of olfaction? This question is not as trivial as it sounds, as was apparent in the discussions during the meetings in Istanbul on Theories of Odor and Odor Measurement.[43] All our theories on primary processes and peripheral events must be proved by evidence of receptor function, as was done in vision. We must know how molecules of a certain compound are translated into a message in the nervous system, with all the detailed consequences of nonlinear transfer functions. We must know the code and the coding in order to understand the events in higher centers through consideration of the peripheral events. It should also be emphasized here that electrophysiology of single receptors is merely one of many scientific methods in sensory physiology, but it is one that often helps in obtaining essential information.

However, olfactory receptors are also receptors in a general sense, and investigating them might contribute to our understanding and the development of general theories of receptor function and primary processes.

METHODS OF RECORDING

Summated potential responses of many cells that react simultaneously to a stimulus can be recorded as the EOG in vertebrates[29] or as the EAG in insect antennae.[31] In the latter, one often finds a "pure" sum of probable receptor potentials of many cells, all of which respond to a given odor in the same manner. These potentials appear as slow negative waves with an amplitude—proportional to the stimulus strength—measuring between several hundred μV and several mV. The EAG is of considerable value for determining thresholds and intensity response curves of a certain group of receptors[34] and for testing the efficiency of different stimuli, such as the purified extracts of natural odor sources (food, odor glands, etc.) or the odor source itself (cf. Priesner in this volume). The experimental method is simple and has often been described in detail.[31] The insect can be prepared inside of 15 minutes and can be used for days if it is in good condition. The reference electrode is placed in the base of the antenna, and the recording electrode—a saline-filled pipette—is placed into the tip. The resistance of the preparation is low, so there is no need for highly sophisticated electronic devices.

The investigation of the vertebrate EOG is more difficult. It shows complicated time courses and changing polarities,[16] and often it is not easy to discriminate decisively among receptor response, control, or even artifacts (cf. Takagi[42] and in this volume).

Another type of olfactory receptor response is the summated action potential recorded in fine bundles, which are hooked onto a platinum electrode.[3,47] Electrical integration by a special device can reveal a pattern of excitation in that bundle. With such preparations Tucker was able to correlate the activity of nasal receptors with different stimulus parameters as well as with influences from efferent control systems[47,48] (cf. also Figure 9).

Throughout the animal kingdom the olfactory receptor is a primary sense cell. Each neuron not only receives stimuli but transduces them into excitation and transforms that excitation into trains of action potentials, which are conducted directly to the brain along its axon. Figure 1 shows diagrammatically the structure of such cells in vertebrates and in insects, together with the recordings of their activity.

To record from single olfactory cells in insects is relatively easy (Figure 1A). A fine micropipette or tungsten needle with a tip diameter between 1 or a few μ, and a resistance of several kΩ to MΩ is introduced into the socket of a sensory hair with the aid of a micromanipulator. Recordings show slow "receptor" potentials and nerve impulses from the underlying receptor cell (cf. also Thurm[45]). In some hair types there are only one, two, or three sensory cells per sensillum; this permits fairly good observation of single-unit activity (cf. Figures 1, 12). Preparations can be kept a long time at average temperatures between 17° and 19°C; several times we had a single receptor cell under continuous observation for more than a day. Perhaps this is because there is no necessity for a dissection before placing the electrode near the receptor cells. In addition, the recording electrode is held in its position by the strong cuticle; this prevents damage to the sensory structures.

In vertebrates, the delicate structure of the receptor cells, which are embedded in a soft epithelium covered by mucus, require either microelectrodes with very fine tips or metal-filled pipettes especially designed for recording from bundles of extremely fine fibers.[15,16] The electrode is placed carefully in the vicinity of the cell bodies or into the fila near the basement membrane (Figure 1B). The neurons are not smaller than in insects, but it is more difficult to record from single units because of the unfavorable electrical conditions in a uniform epithelium that consists of a large number of closely packed neurons. (In insects, by comparison, we find sensory hairs of recognizable morphological types.) In addition, vertebrate experiments require a good preparation, and special care must be taken to prevent the olfactory epithelium from drying out or being flooded by mucus secretions or blood.

In spite of these difficulties, those of us conducting olfaction studies are better off than people in hearing or vision research. It is easier to get single-receptor activity from the nose than from the eyes or ears of vertebrates, from which no good recordings have yet been obtained. And it is certainly more difficult to record from retinal cells of insects than from antennal hairs.

REACTIONS OF RECEPTOR CELLS

Slow potentials. In insect recordings, the first detectable reaction is a slow negative-potential wave of rather high amplitude (up to 25 mV, cf. Figure 1A), considering the extracellular position of the recording electrode[4,24,35] (cf. also Thurm[44]). This is probably due to a favorable placement of the recording electrode in the small channel of conducting fluid near the receptor-cell dendrites, which is surrounded by material of higher resistance. The nerve impulses superimposed on that potential show a positive phase—a polarity which usually is found only in intracellular recordings. This has been discussed in detail by Morita,[21] Thurm,[44] and Wolbarsht and Hanson,[49] and is merely a problem of interpretation, not relevant in this context.

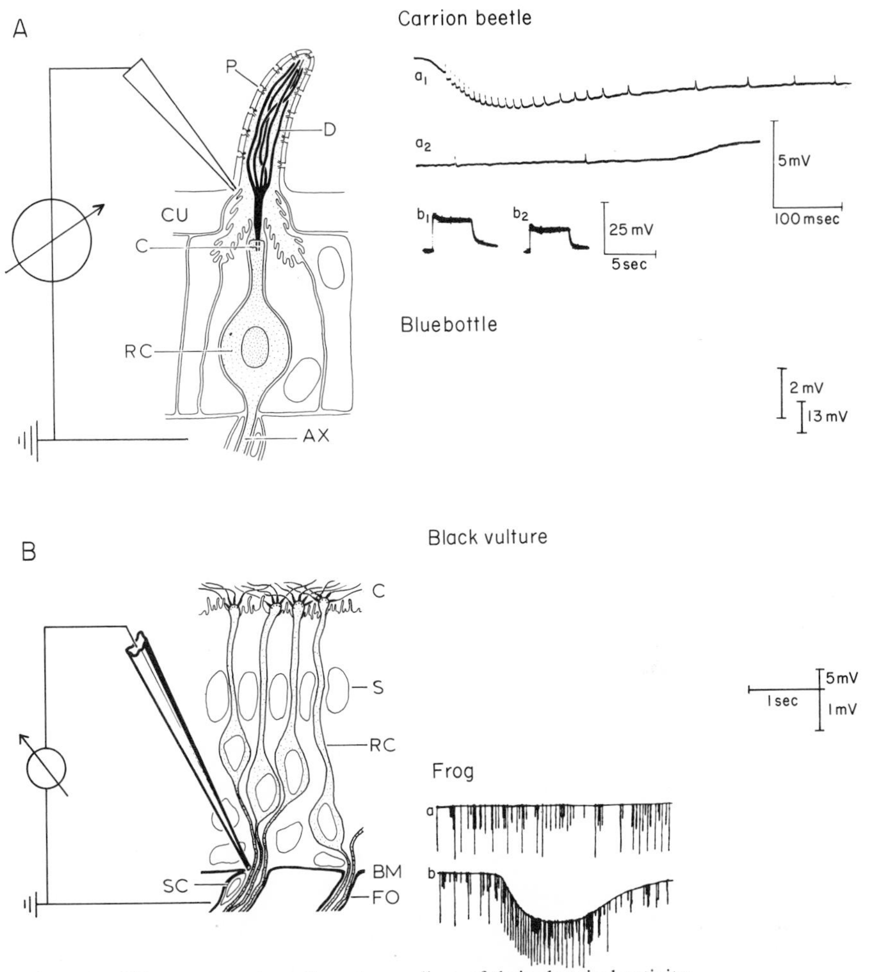

Figure 1 Olfactory receptor cells and recordings of their electrical activity.
A. Insect. At left, an antennal olfactory hair (sensillum basiconicum) of a carrion beetle (*Necrophorus*) with the recording electrode. AX = axon; RC = receptor cell; C = ciliar structure; CU = cuticle; D = dendrite; P = pore in the hair wall (after Ernst[11]). At right, recordings of electrical activity: a_1, start of excitation, stimulus carrion odor; negative downward. The slow potential goes in negative direction; the nerve impulses with their first phase go in positive direction. After several hundred msec, the slow potential and the impulse frequency have reached their plateau. End of the reaction, a_2; after offset of the stimulation the slow potential immediately drops, and the impulses disappear. Slow potentials are shown in b_1 and b_2. Stimulus is carrion odor, four times stronger at b_1 than at b_2 (after Boeckh[4]). Bluebottle (*Calliphora vicina*) sensillum basiconicum in an antennal pit. Reaction of a single unit to carrion odor. Upper trace, a-c recording; lower trace (slow potential), d-c recording (note the difference in amplification). Distance between time marks: 20 msec.
B. Vertebrate. At left is dog olfactory epithelium with the recording electrode. BM = basement membrane; FO = filum olfactorium; SC = Schwann cell; RC = receptor cell (dotted); S = supporting cell with microvilli; C = cilium. (Redrawn after Andres[1]) Right side, electrical activity. Black vulture (*Coragyps astratus*) recording from a single unit in the olfactory epithelium. Slow potentials (upper trace) indicate inspiration (first deflection) and expiration (second deflection). Stimulus in the upper record is 1%, in the lower record 10% saturated atmosphere (20°C) of hexylacetate. (From Shibuya and Tucker[39]). Frog (*Rana pipiens*) action potentials recorded with metal-filled micropipettes (amplitude several hundred μV). Several units are active. Upper record, "resting" activity; lower record, reaction to tetraethyl tin. The superimposed EOG is recorded with an extra electrode. (From Gesteland et al.[16])

We are much more concerned with the question of whether the slow potential represents a real receptor or is a generator potential, such as is generally found in most receptors. Because no intracellular recordings have yet been reported, our evidence rests upon a similarity to other sensory systems and correlations of the time course and intensity of the stimuli on the one hand and to the nerve impulses on the other hand. In many cases, the latter cannot be found in the vertebrate EOG.

It is still difficult to decide whether the potential is a summated response of many cells or the response of only one. If one records from a hair with several cells of different response specificities, one certainly gets a compound potential. But if only one cell is found under the sensillum the response is uniform. We do not know whether cells from neighboring hairs also contribute to that potential but, if they were all to be cells of equal sensitivity, the outcome would not be modified (cf. Boeckh[4]).

There is still no agreement on the nature of slow potentials in vertebrates.[16,30,42]

Nerve impulses.[4,8,15,19,24,35,36,38] During excitation, the olfactory receptor cells generate trains of impulses. The number of impulses per unit time is correlated with odor concentration. In several cases, one can detect a resting or background activity. This could be caused either by an odor contamination in the laboratory or by a true spontaneous activity. After an initial phasic peak, the frequency declines to a tonic plateau that frequently is kept more or less constant for the duration of the stimulus (Figures 3 and 4). Some receptors respond only with an initial burst of spikes and remain silent for the rest of the time. This is especially true for sex-attractant receptors of moths[36] (Priesner, this volume) and some sort of "switch-on" cells in the nostrils of the turkey vulture.[39] Sometimes the time course of excitation changes qualitatively with a change in stimulus intensity, e.g., from tonic into phasic (Figure 2, from Shibuya and Tucker[39]).

The peak frequency of impulses in insect antennae can occur at more than 500 per sec for a short time; during the plateaus, the level can drop to 50 or more impulses per sec. Vertebrate responses occur at much lower levels, reaching peak frequencies of about 30 to 50 per sec. The latencies also can differ between the two groups of animals, as can be observed in the times between the arrival of the stimulus at the olfactory organ and the first spikes. In insects, latencies are between 100 msec for weak stimuli and 5 msec for strong stimuli; in vertebrates it may take several hundred msec before the first spikes appear (Figure 2). This might be explained by the long distance the molecules must travel to reach the receptors, but often the slow potential has started some time before impulses appear.[15,39]

The strong peak excitation could be of some importance for an insect flying upwind over great distances, where the distribution of the odor is discontinuous and the animal must react to frequent changes in stimulus concentration rather than to a continuous odor flow of given strength.[4] Thus, we find only a high phasic reaction in the sex-odor receptors of the male moth (*Antheraea*).

Often the decline in frequency from peak to plateau is called adaptation. In sensory physiology this term also implies a shift in sensitivity under the influence of the

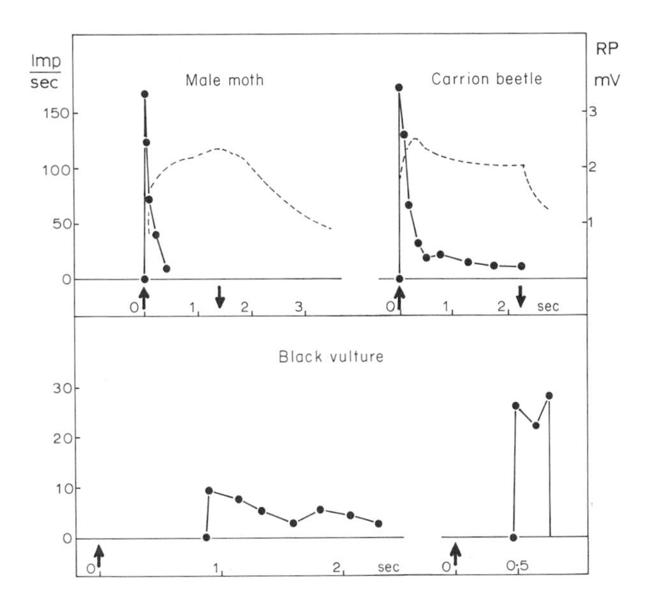

Figure 2 Time course of single olfactory unit activity. Reaction of male moth (*Antheraea pernyi*), top left, to a female lure gland. Top right, reaction of carrion beetle (*Thanatophilus rugosus*), to carrion odor. Solid lines with filled circles indicate the impulse frequency (ordinate at left); dotted lines represent receptor potentials (ordinate at right). Arrows mark onset and end of the stimulus. At bottom, reaction of black vulture (*Coragyps astratus*) to stimulus of amylacetate. At left, air contains $10^{-1.75}$ parts of air saturated with butylacetate (tonic reaction); at the right, the concentration is $10^{-1.25}$ (phasic reaction). The onset of the stimulus was estimated by the start of the slow potential. (Data from Shibuya and Tucker[39])

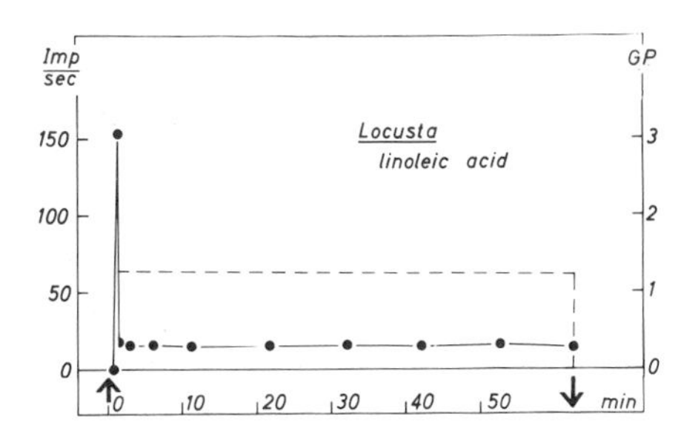

Figure 3 Extremely stable activity of an odor receptor under continuous stimulation, here the antennal sensillum coeloconicum (pit peg) of a locust (*Locusta migratoria*). Single-cell recording shows response to linoleic acid (contaminated substance). Note the time scale. Solid line indicates impulse frequency (left ordinate); the dotted line represents the slow potential (GP, right ordinate, calibration in mV). (From Boeckh, Kaissling and Schneider[9])

stimulus. But the transient from peak to plateau may be considered part of a nonlinear transformation that signals such parameters as rate of stimulus change. Whether a change in excitability occurs during that time must be tested by special procedures (cf. Thurm[46]).

The response of phasic receptors fades away during the stimulus, but it can be elicited repeatedly at short intervals. Some phasitonic odor receptors remain unchanged for an hour or more under continuous stimulation, and seem to lack adaptation completely during that time. Nevertheless, there is an initial decline of frequency

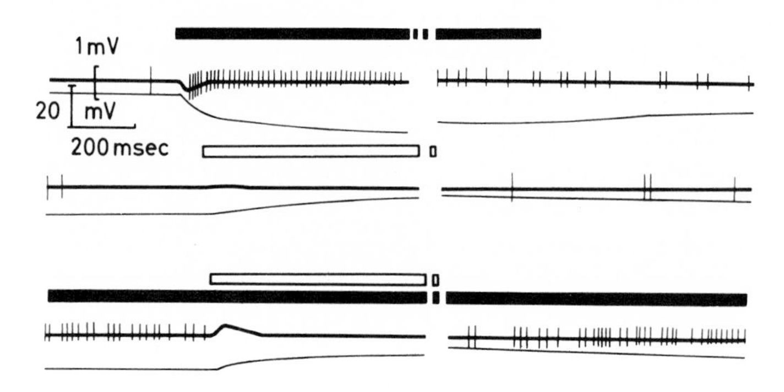

Figure 4 Excitation and inhibition in a single olfactory unit of the carrion beetle *Thanatophilus rugosus* (antennal sensillum basiconicum). Upper traces, a-c amplification; lower traces, d-c amplification (note the different amplification factors). The bars above each recording mark the stimulus. Downward deflection of the trace indicates a negative potential at the active electrode. The recordings are interrupted for several hundreds of msec. Top record: stimulus, carrion odor (filled bar); excitation. Middle record: stimulus, propionic acid (empty bar); inhibition. Bottom record: continuous stimulation with carrion odor (filled bar, only plateau is shown); then additional stimulation with propionic acid (empty bar); and the impulses disappear. After additional stimulus ends, the impulses return. (Redrawn from Boeckh[6])

after the peak excitation (Figure 4). Therefore, one should apply the term "adaptation" only to real changes in excitability (cf. Thurm[46]) and assign the other events to properties of receptor transfer functions (cf. Burkhardt[9]).

Excitation and Inhibition. (Figures 4–6) Chemoreceptors are the only receptors we know that can respond to stimuli of different quality with different polarities. (The change in polarity in the response of mechanoreceptors to a change in the direction of the acting stimulus is not considered to be a change in stimulus quality.) In both olfactory and taste receptors, some chemicals inhibit while others accelerate the firing of impulses.[4,5,8,15,17,19,36,39] Inhibition is accompanied—or, more exactly, preceded—by a positive, slow, inhibitory potential that rises about 10 times more slowly than that of the negative, "excitatory" potential (Figure 4).

Interactions. Inhibitory and excitatory potentials seem to sum up algebraically, if a mixture of excitatory and inhibitory odor stimuli are presented to a given cell. The resulting potential and impulse frequency is an average of the values obtained when the two stimuli are given separately (Figure 6). Presumably there is a summation of inhibitory and excitatory events, at least at the generator potential level.

Because of the slow rise time of the inhibition response, a compound potential of inhibitory and excitatory events will rise steeply in the beginning. The peak amplitude then will be somewhat diminished and will rapidly reach the average of the respective plateaus. The response to a mixture of a strong inhibitory and a strong excitatory stimulus will result in a quick rise in the impulse frequency, since the initial slope of the receptor potential is still steep enough to elicit the full peak frequency. When the slow potential decreases, the impulse frequency will rapidly reach its level, lowered by the inhibition (Figure 6). Even with a combination of a weak excitatory and a strong inhibitory stimulus we still find a peak frequency in the beginning (cf. Boeckh[6]). Subsequently, the slow potential decreases with rising inhibition and becomes positive, and the frequency drops to zero. If a strong inhibitory stimulus is given after the onset of an excitatory one, the potential is immediately reversed and the frequency returns to zero (Figure 4).

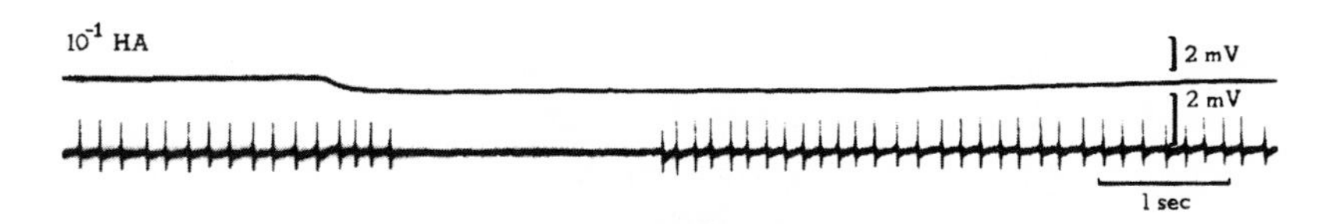

Figure 5 Inhibition of the "background" activity of an olfactory unit in the nose of the turkey vulture (*Cathartes aurea*) during hexylacetate stimulation. (From Shibuya and Tucker[39])

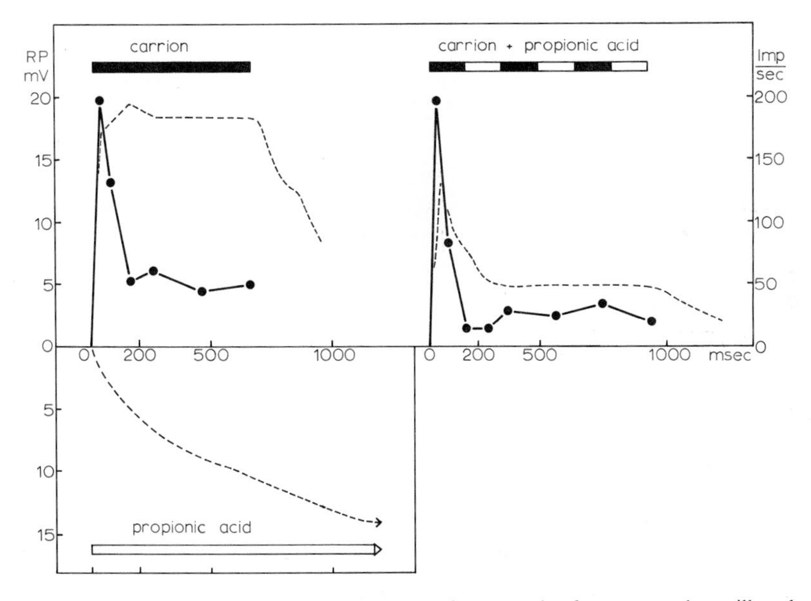

Figure 6 Excitation and inhibition in a single olfactory unit of an antennal sensillum basiconicum of the carrion beetle *Thanatophilus rugosus*. Upper left diagram shows the excitation under stimulation with carrion odor (black bar). Dotted line indicates the slow potential (left ordinate, RP); solid line represents the impulse frequency (right ordinate). Lower diagram, inhibitory slow potential going in positive direction under stimulation with propionic acid (white bar). Diagram at top right shows the excitation under a combined stimulation with carrion odor and propionic acid (black and white bar). Note the undiminished peak frequency and the lowered plateau.

Many vertebrate receptors show a wide variety of responses to odors, ranging from a purely phasic or phasic-tonic excitation (or inhibition) through on-off responses, to even more complicated changes of excitation and inhibition, often in the same time course of a reaction.[15,39] "Off" responses can be the results of rebound activity after inhibition. The other patterns are difficult to interpret; since intracellular recording has not been successful, we can only speculate.

While depolarization and excitation can be fairly well correlated in terms of impulse activity, it is still difficult to find a good hypothesis for the inhibitory events. We do not know whether they represent true hyperpolarization of the cell membrane or a repolarization of a depolarized cell (cf. Boeckh[4]; Gesteland et al.[16]). When an inhibitory stimulus was presented to an excited cell (cf. Figure 4), nerve impulses could be observed after the end of the inhibitory stimulus, even though the receptor potential was still below the extracellular zero level, i.e., there was no depolarization. Therefore, this was either a most unusual case of firing threshold level in the sensory neuron, or the zero level in the extracellular recording did not reflect the intracellular resting potential.

Inhibitory stimuli can elicit positive potentials of 25 mV and more (measured extracellularly, cf. Boeckh[4]). This seems to indicate a hyperpolarizing potential of approximately the same absolute amplitude as that which the depolarization can reach under strongly excitatory stimulation. This means either a very low resting potential or a great difference between the resting potential and a strongly negative membrane potential induced by certain ions permeating the membrane during inhibition.

Several speculative attempts have been made to interpret excitation and inhibition of olfactory receptors in terms of events at the membrane level. Gesteland et al.[16] assume that specific and independent sites at the receptor membrane are responsible only for the perception of excitatory stimuli, whereas other sites are affected only by inhibitory stimuli. An activation of these sites during odor stimulation induces permeation of certain ions; that, in turn, leads to de- (or re-) or hyperpolarization of the cell membrane, according to which sites are active. In the first case, Na^+ ions would permeate the membrane; in the second case K^+ or Cl^- ions. The olfactory receptor cell would then integrate these effects of different polarities in the same manner as a ganglion cell integrates inhibitory and excitatory synaptic inputs.[30]

Takagi and his collaborators investigated the influence of different extracellular ionic media on the olfactory epithelium of frogs. They found that EOG depolarization (excitation) depended on the presence of Na^+ ions, and that a similar correlation existed between Cl^- ions and hyperpolarization (inhibition).[41,42] (See also Takagi, in this volume.)

The role of inhibition for a coding of sensory quality is still theoretical because there have been no crucial experiments. One possibility is some sort of enhancement of "odor contours" when a certain (important) odor both stimulates cells which are especially excited by it and also inhibits all the background activity of other cells. Nevertheless, inhibition has been found in many receptor cells so far investigated (cf.

citations above), especially in those types presumably involved in odor discrimination (cf. p. 38, et seq.).

One special problem seems to be very important. From studies of vertebrate EOGs, it has been reported that sometimes the polarity of the response is reversed for different quantities of the same sort of odor (Takagi, in this volume). This has not been detected either in the EAG or the single receptors of insects, and further information on the subject would be interesting. For the coding of olfactory quality (see pp. 47–48), such an effect would introduce a certain difficulty, because, if a stable polarity of response for all the possible stimulus intensities did not exist, the nervous pattern of the whole sense organ would change in quality.

Range of reaction

Thresholds. It is difficult to determine reaction thresholds of single receptor cells in electrophysiological experiments. From behavioral experiments we know thresholds for a few important substances such as sexual attractants in insects and some other compounds in vertebrates and man.[27,28,34,40] The results show clearly that a few molecules can induce the full behavior pattern of the experimental animal. Therefore we must conclude that the activity of a few receptors can release behavioral responses and that only a few molecules can trigger the activity of the receptor cells.[8] We cannot expect that, under threshold conditions, our recording electrode happens to be situated near one of the responding cells (when perhaps only one out of 1000 or of 10,000 will be active). Furthermore, it is difficult to correlate single impulse events with a stimulus as badly defined as an odor. Here only statistical methods can help. The situation is still more complicated when there is an irregular resting activity or a shift in sensitivity during repetitive stimulation. That is one reason why the electrophysiological "threshold" has been found to be more than 10 times higher than is suggested by the behavior.[8]

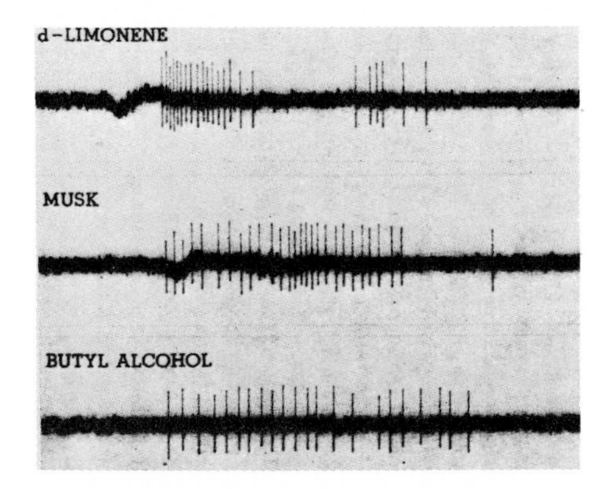

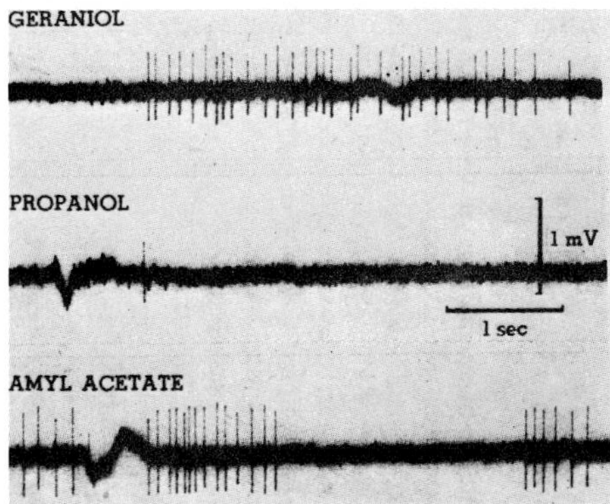

Figure 7 Different responses of a single olfactory unit in a turkey vulture (*Cathartes aurea*) to various odor substances. Odor stimuli given as puffs from squeeze bottles. (From Shibuya and Tucker[39])

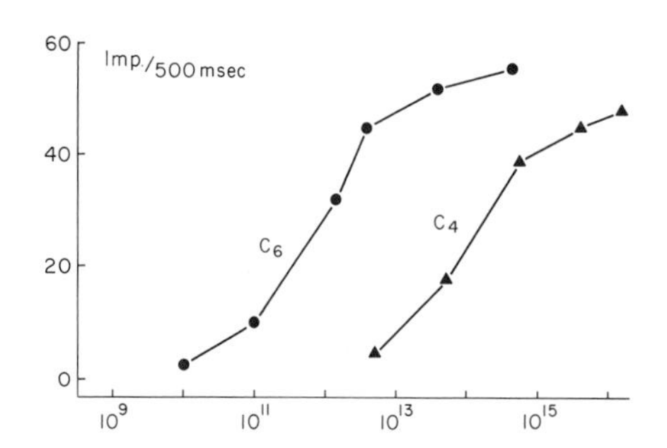

Figure 8 Responses of a single olfactory unit in the antennal coeloconic sensillum of a locust (*Locusta migratoria*) to different concentrations of two different compounds. C_6 = caproic acid; C_4 = butyric acid. Abscissa shows the number of molecules in the stimulating air (stimulus duration, 1 sec). The numbers of molecules are calculated from measurements with radioactive-labeled caproic acid; for butyric acid, from weight loss of the source. (From Boeckh.[5] For detailed information see also Boeckh[6])

Thresholds for biologically important substances, for which—to speak teleologically—the receptors are constructed, have, up to the present time, been tested only for the sexual attractants of the honeybee and the silkmoth.[34] (See also Kaissling, in this volume.) These substances have been extracted and purified, as well as synthesized. Electrophysiological evidence shows that values for single-cell thresholds are of the order of several hundred molecules per ml of air.[8,34] (See also Kaissling, this volume.) In other insects and in vertebrates we can apply only unpurified extracts of natural odor sources or surrogates, because either we do not know enough about the odor sources or we lack information about their chemical constituents (Figure 7).

Reaction range. (Figure 8) Above threshold, the reaction of the receptor increases with increasing odor concentration in the air up to a 500- to 1000-fold threshold intensity, which is the saturation point. Further increase in stimulus strength causes no increase in the reaction.[9,34] (Also Kaissling, this volume.)

The nose of the turkey vulture seems to contain units with an even narrower range between threshold and saturation.[39] But in fiber-strand preparations of the olfactory nerve of the tortoise (Figure 9) as well as in the insect EAG (Kaissling, in this volume) a somewhat wider intensity range can be found. Perhaps this is because, at higher concentrations, more and more of the less sensitive receptors are recruited. This would fit into the picture of a graded sensitivity of different receptors, where an overlap in individual sensitivities for a certain compound would be found in a cell population. How this matches the observations obtained in experiments on the quantitative range of the whole sense of olfaction must be learned through behavioral and psychophysical experiments.[10,20]

Data about the quantitative reaction range of single units can be of relevance in calculating the dynamics of primary processes, as was done by Beidler[2] and Kaissling (in this volume). Measures of such data under different physical conditions (e.g., temperature) or chemical conditions (e.g., pharmacological effects) can produce much information.

Graded chemical specificity. It is difficult to outline limits of the chemical specificity of olfactory cells. This is partly because of the almost infinite number of compounds that could act as stimulants. Man can smell thousands of compounds, and we cannot even guess at the number that can be detected by the whole animal

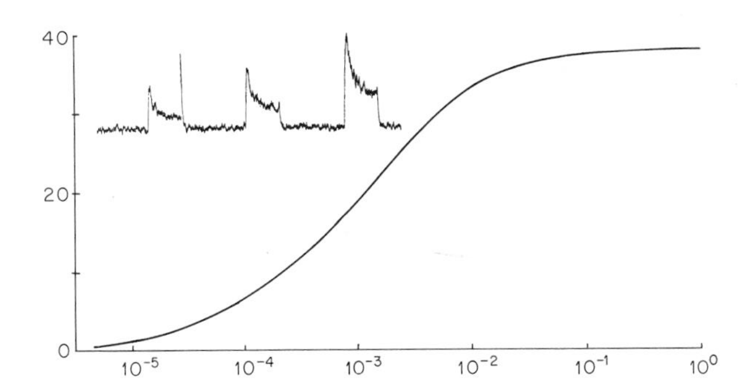

Figure 9 Activity in a twig of the olfactory nerve of a terrestrial tortoise (*Gopherus polyphemus*). Stimulus is amylacetate; concentration unity is saturated air at 20°C. Inset shows time course of integrated activity; stimuli (from left to right) 10^{-3}, $10^{-8/3}$, and $10^{-7/3}$ of the stimulus unity. Diagram: integrated activity in the twig in arbitrary units (ordinate) in relation to stimulus concentration at the receptors (abscissa calculated from air-current velocity and saturation). (From Tucker[47])

kingdom. When quantitatively comparing the stimulatory effects of different odorous compounds, one must be careful to deal with really comparable amounts of stimuli, i.e., one needs a scale of either partial pressures of the odorant or the numbers of molecules applied.

Olfactory stimuli cannot be produced and controlled as easily as light or sound (although if one goes into detail, these are not easy to control, either). Only the most recent techniques of gas chromatography and the application of radioactive-labeled compounds can provide good control of odor quality and quantity. In electrophysiological experiments, odors are usually presented as saturated and diluted odorous atmospheres, or as air blown over an odor source that is covered or soaked with different amounts or concentrations of the odor substance. An ideal controlling device for the stimulus would be some sort of gas-chromatographic detector placed close to the sense organ. This could control both the stimulus strength and the time course of the odor current. But even the most sensitive detector is defeated by the extremely low thresholds of olfactory receptors that respond to a molecule that makes a single impact. It would be difficult to control a stimulus at or near a threshold value. Another difficulty arises from the possible temporary absorption of odors at the hair cuticle or in the mucus that covers the sense organ. Here the stimulus control gets really difficult and we need much more experience.

In general, a receptor cell has been found to have a restricted qualitative reaction range, but it is not equally sensitive to all compounds that affect it. Thresholds can vary over the range of 10^5.[6] Such findings can be of utmost importance for the question of primary processes. If we could succeed in establishing a clear correlation between molecular properties of various compounds and their efficacy as odors on a certain cell or cell type, we would have a better base for speculations about primary processes.

Such an analysis requires the comparison of series of compounds that differ from one another only in a small number of molecular properties (however complexly we define this term). Figure 10 shows a series of fatty acids of varying chain lengths.

After comparing other parameters, as well, one might be able to predict what compound would be most efficient for the receptor under observation. This is not as simple as it sounds, and it might be a sophisticated task for an experienced molecular physicist and a computer.

Another similar program would be to alter, step by step, the structure of one compound known for optimal efficiency, e.g., a sex attractant (cf. Priesner, this volume). Several attempts have already been made, but the problem is still far from solved. Such work on single receptors has only begun.

Insects may be the most appropriate animals for this kind of investigation, because on one antenna there may be tens or hundreds of thousands of individual receptor cells of the same chemical specificity.[8] All one needs to do is to test a great number of receptors of the same type in many animals. In vertebrates, the situation is more complicated, because the odor spectrum of a single unit cannot be predicted. By the time one has tested a series of odors, the activity of the unit being investigated has diminished and the response of a neighboring cell with a different spectrum intrudes upon the record.

It is interesting, especially in view of the primary processes, that the intensity reaction relations show the same shape for odors of different efficiency. At least this is true in some insects (Figure 8, and Kaissling, this volume). The curves have nearly the same slope and cover the same relative range; they are only shifted along the intensity axis by the difference factor of their respective thresholds. Thus, it appears that 100 molecules of butyric acid elicit the same effect at the receptor membrane as does one molecule of caproic acid. Even more remarkable is that one molecule of caproic acid has the same effect as 10,000 molecules of caprylic acid (cf. Figure 10), and that this relation holds for the entire quantitative range between threshold and saturation. This shows again how carefully we must handle the control of our stimuli, because there could easily be a 10^{-4} contamination by some other substance mixed with the caprylic acid.

To classify a given olfactory organ by the particular substance or group of substances that offer the adequate stimulus is a delicate problem. At present it might be safer to distinguish only between more- and less-effective stimuli for a certain receptor or receptor type.

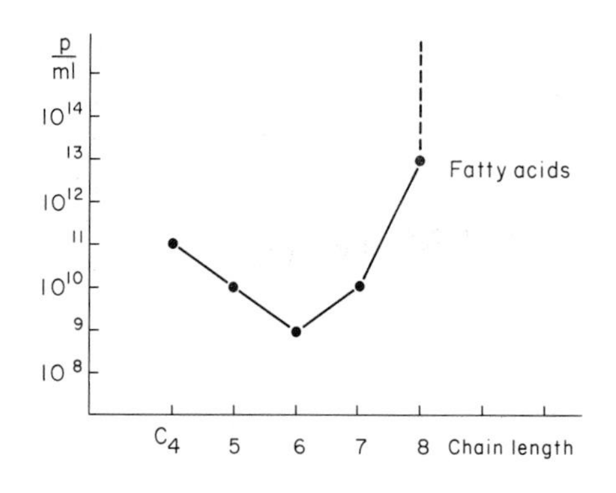

Figure 10 Relative chemical specificity of a single olfactory unit for members of a "homologous" series of aliphatic fatty acids. Abscissa: chain length of the molecule (number of C atoms). C_4 = butyric acid, C_8 = caprylic acid. Ordinate: threshold stimulus concentration (number of molecules per ml air calculated from weight-loss experiments, with higher concentrations by extrapolation). (From Boeckh[6])

The coding of olfactory quality

As I have already mentioned, insects have certain types of chemoreceptors that are most sensitive to such biologically significant stimuli as sexual or pheromone odors, which often release stereotyped, behavioral patterns.[33] Usually an olfactory organ contains thousands of these receptors, all of which respond alike to the same odors. Some male moths have more than 100,000 such receptor cells, whose only function is to detect the attractants of their females.[7,8] The large number of receptors gives the animal a better chance to capture the molecules from the surrounding air and provides for an even lower threshold. At higher stimulus concentration, a massive nervous input discharges into the CNS; this should be of some significance to the central processes.

Amazingly, a comparable receptor type has not yet been detected in vertebrates, probably because no investigations have yet been carried out with food odors or pheromones. However, a single population of these uniformly responding cells is of no use in discriminating among many odors. Such cells can signal response only in terms of excitation or inhibition of different amplitude or no response. There is no way to distinguish between a large amount of less effective and a small amount of highly effective odor, because the reaction is the same for both stimuli.

In some species there are also a few or several receptor types, each one represented by a great number of cells, but it is possible to observe some sort of cooperation between them. For instance, two such types have been studied in carrion beetles and in flies: one responds to carrion odor, the other to some odors that to us smell like resin (Figure 11). Not only is there an overlap in their spectra; there is also some sort of reciprocity in the responses.

This means that there are compounds that excite one receptor and simultaneously inhibit another. If we note the responses of the two cell types, we find odors that 1) excite both, 2) excite only one, 3) excite one and inhibit the other, 4) inhibit both. We can calculate the possibilities of chance combinations for the three types of response in two receptors, and we find 3^2; but because a "no" response of both cells cannot be regarded as a code, we are left with eight qualitatively different patterns of response.

This particular case is of great interest because it could represent a precursor to the response types found in some insects and in vertebrates, where every cell has an individual reaction spectrum with considerable overlap and where every odor quality elicits a distinct pattern of activity in the nerve (Figure 12).[15,19,36,39] To show this once more, the situation is illustrated in Figure 13 (cf. also Schneider[33]).

Imagine the antennal nerve of a honeybee as being a tube, whose cross-section shows many discrete light spots representing the single axons of the olfactory receptors. Although there are actually several tens or hundreds of thousands of them, only 15 are shown in the figure. Dim light in the spots indicates a resting activity. When an odor hits the receptors a certain number of fibers will respond with excitation, shown by a brightening in the corresponding spots. Others will be inhibited,

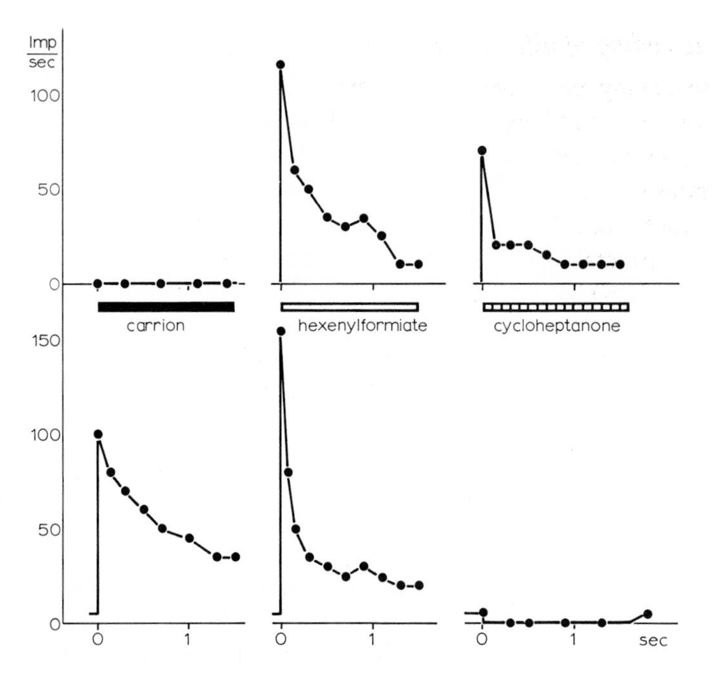

Figure 11 Overlapping reaction ranges of two types of olfactory receptors in one olfactory organ. Two different units in antennal basiconic sensilla of carrion beetle (*Thanatophilus rugosus*), simultaneous recording. Upper part of illustration shows responses of cell 1; lower part, responses of cell 2. Stimuli marked by bars. Solid lines, impulse frequency. There is no reaction in cell type 1 to carrion odor; type 2 is inhibited by cycloheptanone ("resting" activity disappears).

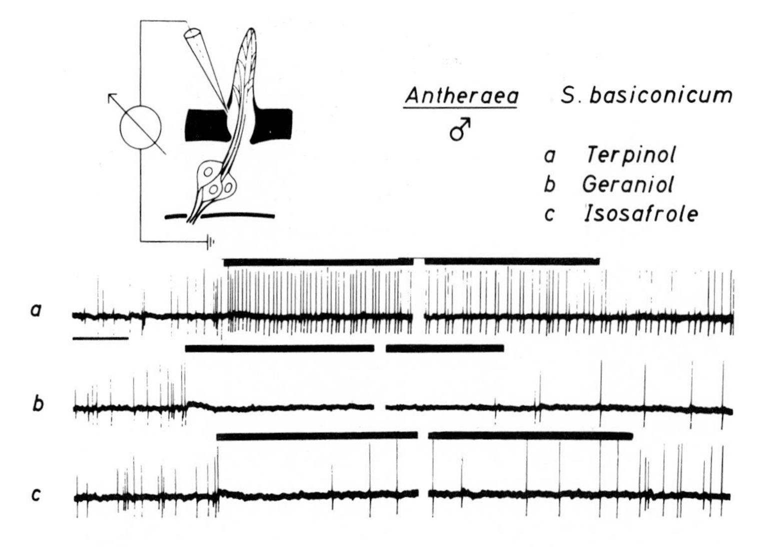

Figure 12 Individual responses of three receptor cells in an antennal basiconic sensillum of the moth *Antheraea pernyi*. (From Boeckh, Kaissling, and Schneider,[8] after data from Schneider, Lacher, and Kaissling[36])

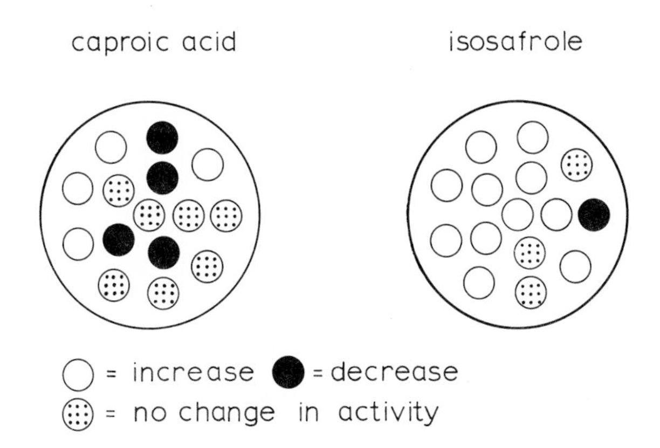

Figure 13 Nervous activity in single axons in the olfactory (antennal) nerve of a honeybee. Fifteen receptor neurons respond to two different odors (the total number of axons is in the order of 10^5). (After data from Lacher[19])

their light spots becoming darker, and other spots remaining unchanged. Another odor will elicit a different pattern of dim, light, and dark spots. If the CNS is able to read these patterns, the information channels could code an almost infinite number of olfactory qualities.

Finer resolution will be added by a graded sensitivity of the cells, which includes more than three steps of brightness in the spots.[19,36] With increasing odor intensity, the bright spots become brighter, the dark ones darker, and more of the dim ones either darker or brighter. If all the complicated time courses of the frog or vulture receptors were used for this kind of coding, the complexity of the resulting picture would go beyond our powers of imagination.[15,39]

In some insects we find a combination of the two types of receptors discussed so far. One (or possibly several) type of cell occurs in large numbers and is extremely sensitive to "alarming" odors. The other is an individualistic type of receptor (Schneider et al.[32]). Both types can be found in one and the same sensillum in the honeybee (Kaissling, this volume). The advantage is obvious: the cells which are most numerous provide the possibility of capturing many molecules and the others a well-graded scale for discriminating a large variety of odor qualities.

REFERENCES

1. Andres, K. H. 1966. Der Feinbau der Regio olfactoria von Makrosmatikern. *Histochemie.* **69**, 140–154.
2. Beidler, L. M. 1965. Taste receptor stimulation. *Progr. Biophys. Biophys. Chem.* **12**, pp. 107–151.
3. Beidler, L. M., and D. Tucker. 1955. Response of nasal epithelium to odor stimulation. *Science* **122**, p. 76.
4. Boeckh, J. 1962. Elektrophysiologische Untersuchungen an einzelnen Geruchsrezeptoren auf den Antennen des Totengräbers (*Necrophorus*, Coleoptera). *Z. Vergleich. Physiol.* **46**, pp. 212–248.
5. Boeckh, J. 1967. Inhibition and excitation of single insect olfactory receptor cells and their role as a primary sensory code. *In* Olfaction and Taste II (T. Hayashi, editor). Pergamon Press, Oxford, etc., pp. 721–735.

6. Boeckh, J. 1967. Reaktionsschwelle, Arbeitsbereich und Spezifität eines Geruchsrezeptors auf der Heuschreckenantenne. *Z. Vergleich. Physiol.* **55**, pp. 378–406.
7. Boeckh, J., K. E. Kaissling, and D. Schneider. 1960. Sensillen und Bau der Antennengeissel von *Telea polyphemus. Zool. Jahrb. Anat.* **78**, pp. 559–584.
8. Boeckh, J., K. E. Kaissling, and D. Schneider. 1965. Insect olfactory receptors. *Cold Spring Harbor Symp. Quant. Biol.* **30**, pp. 263–280.
9. Burkhardt, D. 1960. Die Eigenschaften und Funktionstypen der Sinnesorgane. *Ergebn. Biol.* **22**, pp. 226–267.
10. Engen, T., W. S. Cain, and C. K. Rovee. 1968. Direct scaling of olfaction in the newborn infant and the adult human observer. *In* Theories of Odors and Odor Measurement (N. Tanyolac, editor). Robert College Research Center, Istanbul, Turkey, pp. 271–294.
11. Ernst, K. Feinstruktur eines Geruchssensillums des Totengräbers *Necrophorus. Histochemie* (In press)
12. Garten, S. 1900. Physiologie der marklosen Nerven. G. Fischer, Jena.
13. Gasser, H. S. 1956. Olfactory nerve fibers. *J. Gen. Physiol.* **39**, pp. 473–496.
14. Gesteland, R. C. 1961. Action potentials recorded from olfactory receptor neurons. Ph.D. thesis, M.I.T., Cambridge, Mass.
15. Gesteland, R. C., J. Y. Lettvin, W. H. Pitts, and A. Rojas. 1963. Odor specificities of the frog's olfactory receptors. *In* Olfaction and Taste I (Y. Zotterman, editor). Pergamon Press, Oxford, etc., pp. 19–34.
16. Gesteland, R. C., J. Y. Lettvin, and W. H. Pitts. 1965. Chemical transmission in the nose of the frog. *J. Physiol.* (*London*) **181**, pp. 525–559.
17. Hodgson, E. S. 1957. Electrophysiological studies of arthropod chemoreception. II. Response of labellar chemoreceptors of the blowfly to stimulation with carbohydrates. *J. Insect Physiol.* **1**, pp. 240–247.
18. Hosoya, Y., and H. Yoshida. 1937. Über die bioelektrische Erscheinungen an der Riechschleimhaut. *Japan. J. Med. Sci. III. Biophys.* **5**, pp. 22–23.
19. Lacher, V. 1964. Elektrophysiologische Untersuchungen an einzelnen Rezeptoren für Geruch, Kohlendioxyd, Lufteuchtigkeit und Temperatur auf den Antennen der Arbeitsbiene und der Drohne (*Apis mellifica*). *Z. Vergleich. Physiol.* **48**, pp. 587–623.
20. Laffort, P. 1968. Some new data on the physico-chemical determinants of the relative effectiveness of odorants. *In* Theories of Odors and Odor Measurement (N. Tanyolac, editor). Robert College Research Center, Istanbul, Turkey, pp. 247–268.
21. Morita, H. 1959. Initiation of spike potentials in contact chemosensory hairs of insects. III. DC stimulation and generator potential of labellar chemoreceptor of Calliphora. *J. Cell. Comp. Physiol.* **54**, pp. 189–204.
22. Morita, H., and Takeda, K. 1959. Initiation of spike potentials in contact chemosensory hairs of Vanessa. *J. Cell. Comp. Physiol.* **54**, pp. 177–187.
23. Morita, H., and Yamashita, S. 1959. The back-firing of impulses in a labellar chemoreceptory hair of the fly. *Mem. Fac. Sci. Kyushu Univ., Ser. E.* **3**, pp. 81–87.
24. Morita, H. and Yamashita, S. 1961. Receptor potentials recorded from sensilla basiconica on the antenna of the silkworm larvae, *Bombyx mori. J. Exp. Biol.* **38**, pp. 851–861.
25. Moulton, D. G., and L. M. Beidler. 1967. Structure and function in the peripheral olfactory system. *Physiol. Rev.* **47**, pp. 1–52.
26. Moulton, D. G., and D. Tucker. 1964. Electrophysiology of the olfactory system. *Ann N.Y. Acad. Sci.* **116**, pp. 380–428.
27. Neuhaus, W. 1956. Die Riechschwelle von Duftgemischen beim Hund und ihr Verhältnis zu den Schwellen unvermischter Duftstoffe. *Z. Vergleich. Physiol.* **38**, pp. 238–258.
28. Neuhaus, W. 1956. Die Unterscheidungsfähigkeit des Hundes für Duftgemische. *Z. Vergleich. Physiol.* **39**, pp. 25–43.
29. Ottoson, D. 1956. Analysis of the electrical activity of the olfactory epithelium. *Acta Physiol. Scand.* **35**, Suppl. 122, pp. 1–83.
30. Ottoson, D., and G. M. Shepherd, 1967. Experiments and concepts in olfactory physiology. *In* Sensory Mechanisms: Progress in Brain Research, Vol 23 (Y. Zotterman, editor). Elsevier, Amsterdam, pp. 83–138.
31. Schneider, D. 1957. Elektrophysiologische Untersuchungen von Chemo- und Mechanorezeptoren der Antenne des Seidenspinners *Bombyx mori* L. *Z. Vergleich. Physiol.* **40**, pp. 8–41.

32. Schneider, D. 1964. Insect antennae. *Ann. Rev. Entomol.* **9**, pp. 103–122.
33. Schneider, D. 1966. Chemical sense communication in insects. *Soc. Exp. Biol. Symp.* 20 (Nervous and hormonal mechanisms of integration), pp. 273–297.
34. Schneider, D., B. C. Block, J. Boeckh, and E. Priesner. 1967. Die Reaktion der männlichen Seidenspinner auf Bombykol und seine Isomeren: Elektroantennogramm und Verhalten. *Z. Vergleich. Physiol.* **54**, pp. 192–209.
35. Schneider, D., and J. Boeckh. 1962. Rezeptorpotential und Nervenimpulse einzelner olfactorischen Sensillen der Insektenantenne. *Z. Vergleich. Physiol.* **45**, pp. 405–412.
36. Schneider, D., V. Lacher, and K. E. Kaissling. 1964. Die Reaktionsweise und das Reaktionsspektrum von Riechzellen bei *Antheraea pernyi* (Lepidoptera, Saturniidae). *Z. Vergleich. Physiol.* **48**, 632–662.
37. Schneider, D., and R. A. Steinbrecht. 1968. Checklist of insect olfactory sensilla. *Symp. Zool. Soc. Lond.*, **23**, pp. 279–297.
38. Shibuya, T., and Skibuya, S. 1963. Olfactory epithelium: Unitary responses in the tortoise. *Science* **140**, pp. 495–496.
39. Shibuya, T., and Tucker, D. 1967. Single unit response of olfactory receptors in vultures. *In* Olfaction and Taste II (T. Hayashi, editor). Pergamon Press, Oxford, etc., pp. 219–233.
40. Stuiver, M. 1958. Biophysics of the sense of smell. Diss. Natwiss. Fak. Univ. Groningen, Holland.
41. Takagi, S., and G. A. Wyse. 1965. Ionic mechanisms of olfactory receptor potential. *Proc. XXIII Int. Congr. Physiol. Sci.*, Tokyo. (Abstract)
42. Takagi, S. 1967. Are EOG's generator potentials? *In* Olfaction and Taste II (T. Hayashi, editor). Pergamon Press, Oxford, etc., pp. 167–179.
43. Tanyolac, N. (editor). 1968. Theories of Odors and Odor Measurement. Robert College Research Center, Istanbul, Turkey.
44. Thurm, U. 1962. Ableitung der Rezeptorpotentiale und Nervenimpulse einzelner Cuticualsensillen bei Insekten. *Z. Naturforsch*, **17 B**, pp. 258–286.
45. Thurm, U. 1964. Das Rezeptorpotential einzelner mechanorezeptorischer Zellen von Bienen. *Z. Vergleich. Physiol.* **48**, pp. 131–156.
46. Thurm, U. 1968. An insect mechanoreceptor. II. Receptor potentials. *Cold Spring Harbor Symp. Quant. Biol.* **30**, pp. 83–103.
47. Tucker, D. 1963. Physical variables in the olfactory stimulation process. *J. Gen. Physiol.*. **46**, pp. 453–489.
48. Tucker, D. Olfactory, vomeronasal and trigeminal receptor responses to odorants. *In* Olfaction and Taste I (Y. Zotterman, editor). Pergamon Press, Oxford, etc., pp. 45–69.
49. Wolbarsht, M. L., and F. E. Hanson. 1966. Electrical activity in the chemoreceptors of the blowfly. III. Dendritic action potentials. *J. Gen. Physiol.* **48**, pp. 673–683.

KINETICS OF OLFACTORY RECEPTOR POTENTIALS

KARL ERNST KAISSLING

Max-Planck-Institut fuer Verhaltensphysiologie, Seewiesen, Germany

INTRODUCTION

The mass-action law has often been used in chemoreception to prove the connection between the odor- or taste-substance concentration and the reaction of the chemoreceptors. The idea behind this formalism is that the receptor has a finite number of sites that reversibly bind the stimulating molecules or ions. Even if the mass-action law is fulfilled, one cannot yet answer the question of what type of binding occurs between the stimulating molecule and the receptor site. Also, it is unknown if the molecule after binding will be free, as such, or if it will be changed in its structure. Therefore, we must use other methods to find out if we are dealing with a pure adsorption, with an enzymatic process, or with something more complicated.

If we consider the mass-action law to be fulfilled for the primary process at the receptor site, we would not expect this for the relation between stimulus concentration and the electrical reactions of the receptor cell, except that the latter relation would be linear. One knows that the nerve membranes have nonlinear properties. Nevertheless, they may work linearly within certain ranges in the first approximation. Beidler[4–6] introduced this point of view into chemoreception work and found that the spike frequency of the afferent taste fibers can be directly related to the salt concentration by the mass-action law.

The purpose of this paper is, first, to extend the formalistic treatment from the equilibrium state to the transients, which occur at the beginning and at the end of a nervous reaction, elicited by a time controlled stimulus. A second use of the formalism will be the calculation of the number of the receptor sites, which is possible by measuring one additional parameter. The studied bioelectrical reactions are the receptor potentials of the male silkworm moth *Bombyx mori* and of the drone honeybee *Apis mellifera.* The olfactory stimuli are the pheromones bombykol (hexadeca-10 *trans*, 12 *cis*-dienol-1) and queen substance (9-oxodec-2 *trans*-enoic acid) (for references see Boeckh, Kaissling, and Schneider[8]).

THE ACCEPTOR MODEL

We suggest the name *acceptors* for the binding sites. The acceptor and the odor- or taste-substance are formally the same as the enzyme and the substrate in the Michaelis-Menten equations. The word acceptor may be more precise than binding site. Acceptor expresses the feeling of many investigators in chemoreception that we are indeed dealing with a number of special structures or even molecules, each of which reacts specifically to a single stimulus molecule. This assumption is based on two fundamental properties of chemoreceptors: sensitivity and specificity.

Sensitivity

Psychological and behavioral measurements in man and animals, as well as electrophysiological recordings from single olfactory cells in insects, show a very high *sensitivity* of the chemoreceptors. In several cases it seems that very few or even single molecules trigger an entire olfactory cell.[21,28,32,33] With few molecules necessary for a single cell, we are dealing with strong local effects of single molecules. We may note that, among chemicals, some odorants constitute the most powerful effects in all biology.

Specificity

From electrophysiological measurements in single insect olfactory cells we also know that high sensitivity can be paired with high *specificity*. In the pheromone receptors, especially, we find cells that seem to be developed for receiving a certain substance (e.g., bombykol or queen substance) and that also react to a series of other compounds, but mostly at much higher concentrations.[8,23,25] Contrary to the olfactory cells found in the frog and insects, these pheromone receptor cells seem to have a constant side spectrum.

An analogy to this sensitivity for single molecules, and high but not extreme specificity, is found in enzymes. The combination of these fundamental properties of chemoreceptors is difficult to imagine unless special acceptors are present on the receptor cell.

A probable proof for the existence of acceptors recently became known.[10,16,22] In these cases the acceptor seems to be a protein.

Alternatives of the acceptor model would be mass effects like pH changes, solution processes,[19] surface tension effects, or the membrane-puncturing theory of Davies.[11] For a comprehensive discussion of odor theories see Dravnieks.[13,14]

THE EQUILIBRIUM OR THE STEADY STATE

Unlike Beidler, we prefer to use a terminology similar to that of enzyme kinetics, which was introduced in 1913 by Michaelis and Menten (cited in Dixon and Webb,[12] Netter[20]). Instead of the enzyme symbol E we use A for the acceptor, because it is an open question as to whether chemoreceptors work like enzymes. The stimulus molecule or ion may be S, according to the substrate S in enzyme kinetics. Table I gives an explanation of and a comparison between Beidler's symbols and ours.

Table II contains some possible formulations to express the mass-action law in its equilibrium state. In addition, Figure 1 shows some of the graphical plots belonging to the formulae of Table I. These types of curves are called adsorption isotherms, because the same formalism is also valid for adsorption processes (Langmuir or

Freundlich, cited in Netter[20]). For proving whether data relations follow the adsorption isotherm, one should make several plots, because they are differently sensitive to the variance in different ranges of concentrations. For instance, plot *a* of Figure 1 is useful if one is interested in the linearity between the complex [AS] and [S] at very low concentrations of [S]. The semilog plot *b* demonstrates whether the curve has the correct steepness. The stimulus intensity range covers about 2.5 $\log_{10}$ units between 5 and 95 per cent saturation of the acceptors; *c* and *d* are plots in which the adsorption isotherm results in a straight line. In enzyme kinetics, *c* is known as the Lineweaver-Burk plot (formula 4 in Table II); it is insensitive to the low concentration range.

TABLE I
EXPLANATION OF SYMBOLS

Symbols used here	Beidler's symbols	
[A]	$S - n$	number of free acceptors (equal to free binding sites) at the receptor cell
[S]	c	concentration of the applied chemical stimulus
[AS]	$n = R/a$	number of occupied acceptors = number of molecules that simultaneously occupy the acceptors
$[A_{tot}] = [A] + [AS]$	$S = R_m/a$	total number of acceptors at the receptor cell
K_{eq}	$1/K$	equilibrium constant (=[S] at half saturation of the acceptors)
K_{st}	—	steady state constant (=[S] at half saturation of the acceptors)

All terms, including K_{eq} and K_{st}, have the dimension of concentration (e.g., molecules/cc) that is indicated by the brackets. For [A], [AS], and $[A_{tot}]$ one may also use the dimension number per receptor cell. Please, note that K_{eq} is the reciprocal of Beidler's K.

TABLE II
EXPRESSIONS OF THE MASS-ACTION LAW IN THE EQUILIBRIUM STATE

Formulations used here	Beidler's formulations
$[A] + [S] \underset{k_2}{\overset{k_1}{\rightleftharpoons}} [AS]$	$(S - n) + c \rightleftharpoons n$
(1) $\dfrac{[A] \times [S]}{[AS]} = K_{eq}$	$\dfrac{n}{(S - n) \times c} = K$
(2) $[A] + [AS] = [A_{tot}]$	
(3) $[AS] = [A_{tot}] \dfrac{[S]}{K_{eq} + [S]}$	$n = S \dfrac{c \times K}{1 + c + K}$
	$n = R/a,\ S = R_m/a$
(4) $\dfrac{1}{[AS]} = \dfrac{K_{eq}}{[S] \times [A_{tot}]} + \dfrac{1}{[A_{tot}]}$	$\dfrac{1}{R} = \dfrac{1}{c \times K \times R_m} + \dfrac{1}{R_m}$
(5) $\dfrac{[S]}{[AS]} = \dfrac{[S]}{[A_{tot}]} + \dfrac{K_{eq}}{[A_{tot}]}$	$\dfrac{c}{R} = \dfrac{c}{R_m} + \dfrac{1}{K \times R_m}$

Formula 5 in Table II (plot *d*) is the so-called fundamental taste equation of Beidler.[4,5] It is insensitive to high concentrations of [S] in comparison to the concentration K_{eq}. The last two plots are primarily used to determine the equilibrium constant K_{eq} (at $y = 0$) and the maximum or the saturation limit of the acceptors (at $x = 0$).

The constant $K_{eq} = k_2/k_1$ is the dissociation constant of a simple reversible process. In all cases, therefore, in which the unchanged stimulus molecule is desorbed, we may speak of *equilibrium*-type reactions. On the other hand, one could expect reactions in chemoreceptors that alter the molecular structure or irreversibly transfer the unchanged molecule from the acceptor to any unspecific site. This type of reaction can be called *steady state.*

$$[A] + [S] \underset{k_2}{\overset{k_1}{\rightleftharpoons}} [AS] \overset{k_3}{\rightarrow} [A] + [P].$$

Here $K_{st} = (k_2 + k_3)/k_1$ is a steady-state constant, according to the Michaelis constant K_m in enzyme kinetics. The reactions of the steady-state type as well as those of the equilibrium type may yield adsorption isotherms, even if more complicated reaction steps are involved.

In all subsequent equations in this paper we will use K_{st}, which includes decay constants k_2 and k_3. This may be the best compromise, because the relation of k_2 and k_3 is unknown. In the extreme case, one of the two constants would equal zero.

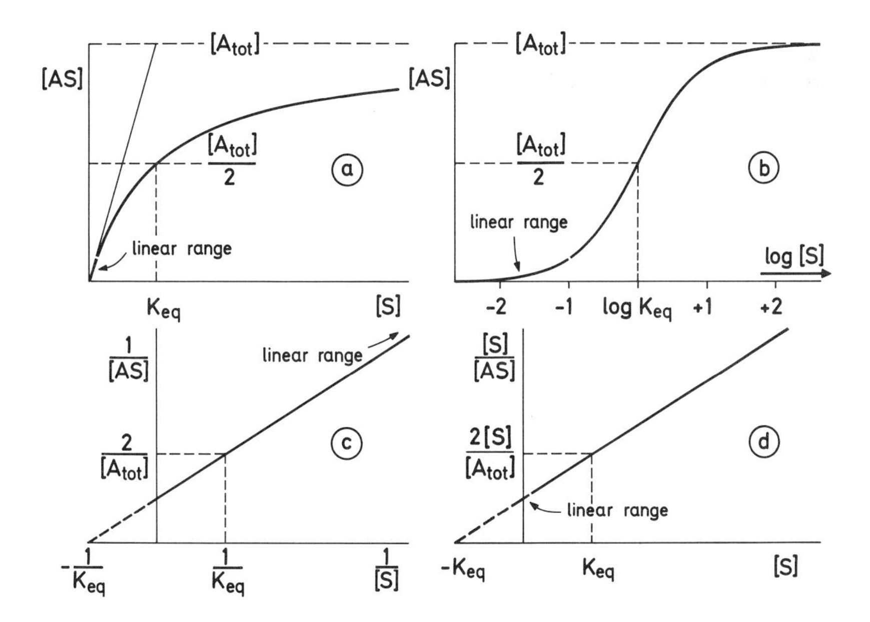

Figure 1 Four plots (a, b, c, d) of the equilibrium state of the mass-action law, which correspond to the adsorption isotherm of Langmuir and Freundlich. For further explanation see pages 53, 54. (d) Plot of Beidler's fundamental taste equation.

THE FORMATION PHASE AND THE DECAY PHASE

Further information on the kinetics of the two reacting materials, the acceptor and the stimulus molecule, is obtained by looking at the transients or the dynamic phases. If we start with a stimulus of concentration [S], which is delivered in a rectangular pulse, it will take some time until the acceptors are filled up to the steady state. During this initial phase the formation ($k_1 \cdot [A] \cdot [S]_{const.}$) of the complex [AS] prevails over the decay ($(k_2 + k_3) \cdot [AS]$). After the end of the stimulus only the decay of the complex takes place.

The differential equations of these phases are found in Figure 2. They are derived for a clamped rectangular stimulus [S]. This means that the stimulus concentration is kept constant during the stimulus and is zero after the end of the stimulus. Two parameters can be used easily to check if the dynamic phases also fit into the kinetic formalism: (1) both transients are exponential changes of [AS] against time, as we see from the integrated equations (Figure 2); (2) the concentration dependency is characteristic for the half-times of both transients.

The half-times can be determined from the integrated equations by setting [AS] equal to half of the steady-state concentration $[AS]_{st}$ (Figures 2, 3). While the formation half-time (τ_f) will be shorter with higher concentrations of [S], the decay half-time (τ_d) is independent of the stimulus concentration. Figure 3 shows that the concentration dependency of τ_f is similar to the adsorption isotherm, except for its

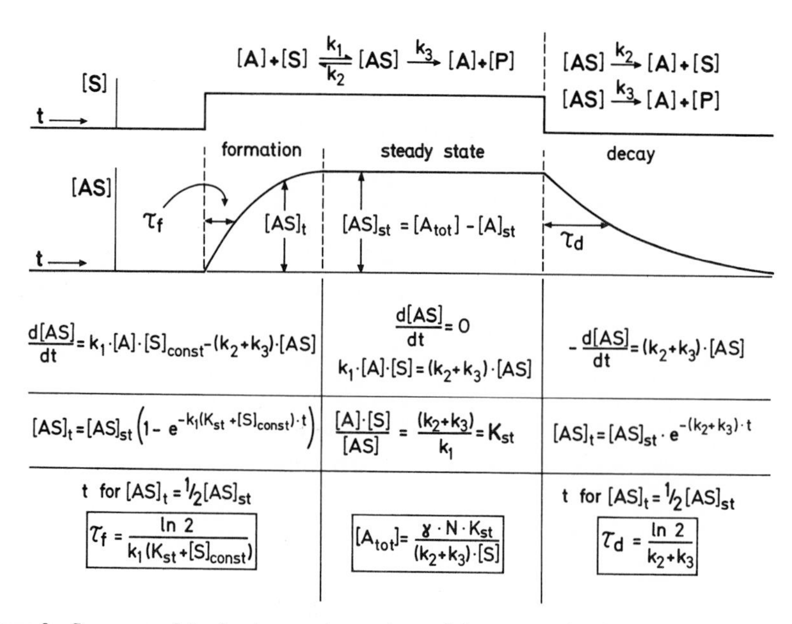

Figure 2 Summary of the fundamental equations of the mass-action kinetics in the dynamic phases and in the equilibrium or steady state, respectively. During the initial phase the formation of the complex [AS] prevails. After the end of the stimulus [S] only the decay of the complex takes place. The formalism is valid for a clamped stimulus concentration.

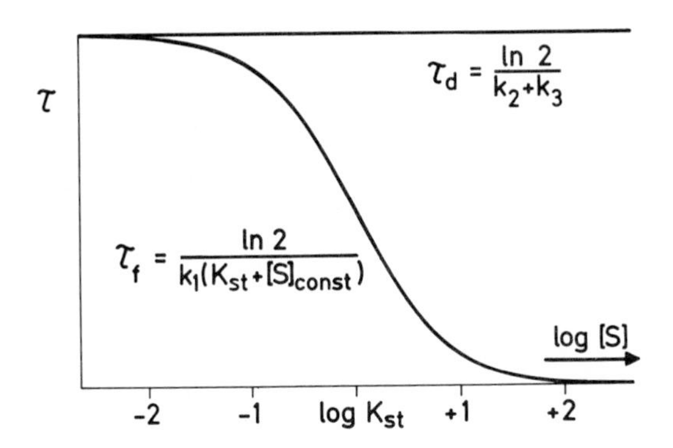

Figure 3 The half-time (τ) dependency on the stimulus concentration [S]. τ_f = half-time of the formation; τ_d = half-time of the decay of the complex [AS]. τ_f follows an adsorption isotherm, except for the sign.

sign. This will be clear if we combine τ_f and τ_d with formula 3 in Table II. The result will be:

$$\frac{\tau_f}{\tau_d} = 1 - \frac{[AS]}{[A_{tot}]} \quad \text{or} \quad \frac{\tau_f}{\tau_d} = \frac{[A]}{[A_{tot}]}. \tag{6}$$

One of the advantages of investigating the transients is that the formation half-time as well as [AS] can be used to determine K_{st} and the saturation limit of the acceptors. Because τ_f equals τ_d at low stimulus concentrations we can formulate:

$$K_{st} = [S], \text{ for } \tau_f = \frac{\tau_f \text{ (low conc.)}}{2}, \tag{7}$$

or also

$$K_{st} = [S], \text{ for } \tau_f = \frac{\tau_d}{2}. \tag{8}$$

Another advantage is that we can determine the turnover number of the stimulus molecules on the acceptor. The turnover number may be defined as k_2 in the equilibrium reaction and $(k_2 + k_3)$ in the steady-state reaction. This value says how many times per minute a single acceptor can form a complex. From the decay half-time (τ_d, Figure 2), it is possible to calculate k_2 or $(k_2 + k_3)$.

The third advantage is that we can calculate the number of acceptors $[A_{tot}]$ if we measure one additional parameter, the absolute rate of molecules N that hit the receptor organ. At *low concentrations* of [S] and at steady state, only a small fraction of $[A_{tot}]$ is occupied by [S]. Therefore, [A] nearly equals $[A_{tot}]$. The steady state can now be expressed as

$$k_1 \cdot [A_{tot}] \cdot [S] = (k_2 + k_3)\,[AS] \tag{9}$$

(compare with Figure 2).

Within the linear range of the adsorption isotherm, the number of formed and decaying complexes is a constant fraction γ of the rate of molecules N hitting the receptor per unit time:

$$k_1 \cdot [A_{tot}] \cdot [S] = \gamma \cdot N. \tag{10}$$

From this, one obtains

$$[A_{tot}] = \frac{\gamma \cdot N}{k_1 \cdot [S]} \tag{11}$$

or, if we measure $(k_2 + k_3)$ instead of k_1,

$$[A_{tot}] = \frac{\gamma \cdot N \cdot K_{st}}{(k_2 + k_3)\,[S]}. \tag{12}$$

γ is the quantum efficiency of chemoreception. If γ is unknown we can only determine the maximum possible number of acceptors, because γ in the maximum equals 1.

The rate N must always be determined for a certain value of [S]. If N is measured at the concentration $[S] = K_{st}$ we get

$$[A_{tot}] = \frac{\gamma \cdot N_{st}}{(k_2 + k_3)}. \tag{13}$$

RECEPTOR POTENTIALS

The slow receptor potentials are the first electrical reactions of the receptor cell we can measure. There are several ways in which they could be controlled by the primary process, which is the reaction between the stimulus molecule and the acceptors. Figure 4 shows the time course of four different parameters of the described reaction, which was elicited by a rectangular stimulus [S]. The first two belong to an equilibrium-type reaction; all four parameters could occur in a steady-state type reaction. The slow potentials recorded from the honeybee (see Figure 5) and the *Bombyx* antenna look like the first or the fourth curve of Figure 4. But in other insects slow potentials can be observed that look more like the third type of curve (Boeckh, Kaissling, and Schneider,[8] Figure 17). We may exclude d[AS]/dt, but we cannot definitely say from the formalism whether the concentration [AS], the uptake rate of

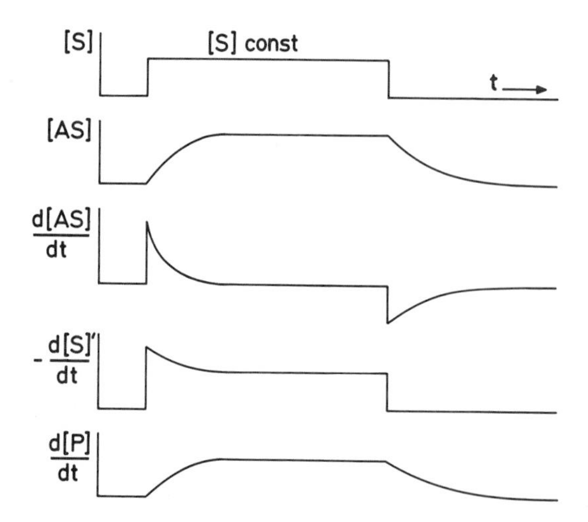

Figure 4 The time course of some parameters of mass-action kinetics induced by a rectangular stimulus.

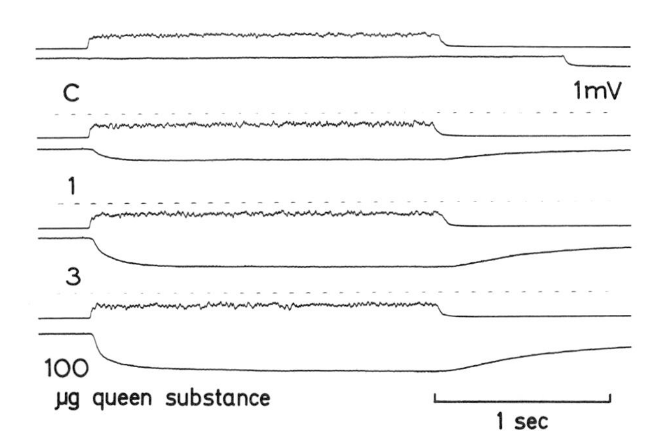

Figure 5 Olfactory receptor potentials (smooth lines) from the antenna of the drone honeybee. Extracellular microcapillary recordings with d-c amplification. Rough lines: registration of the airstream velocity (1.6 m/sec) with a fast anemometer developed by Dr. Ernst Kramer. C = control stimulus with fresh air. Stimuli: 1–100 g queen substance on glass plates. The *Bombyx* EAG (electroantennogram) has a similar time course.

the stimulus molecules $-\mathrm{d}[S]'/\mathrm{d}t$, or the rate of producing alterations $\mathrm{d}[P]/\mathrm{d}t$ triggers the slow potential. Nevertheless, we may plot the reaction amplitude R instead of [AS] on the ordinate to see if we find a typical adsorption isotherm such as Beidler found in 1954.[4]

In Figure 6 we plotted on the ordinate the amplitude of the slow potential, which is reached 1 sec after the onset of the stimulus. At this time the potential has approximated a stationary plateau that would remain constant for a much longer stimulus duration, except that the stimulus concentration is too high and causes adaptation (Boeckh, Kaissling, and Schneider,[8] Figure 16 b). The plot of Figure 6 corresponds to the semilog plot of Figure 1 b. We see that the drone pore-plate cell reacts to queen substance in a manner that roughly follows an adsorption isotherm (left side, solid line). The same cell reacts in a similar way to caproic acid, but at about 10^4 times higher concentrations (solid line, right side). Similar results were obtained from spike recordings of single cells. The proof that both substances work on the same receptor cell was made by cross-adaptation experiments (Boeckh, Kaissling, and Schneider,[8] Figure 19).

An exact comparison of these curves with the adsorption isotherm cannot yet be made, because the relative calibration of the odor concentration has not yet been obtained exactly enough. We put from 0.03 to 100 μg of the queen substance on small glass plates (7 mm × 18 mm). Each plate was placed in a 7-mm-wide glass tube. We then blew a calibrated airstream through the glass onto the animal's antennae. After that we divided the glass plate in two and tested the halves separately. If each half gave the same reaction, we assumed that each had produced half of the previous odor concentration. By this method we corrected the relative intensity scale for queen substance. For the absolute calibration we blew air over glass plates containing queen

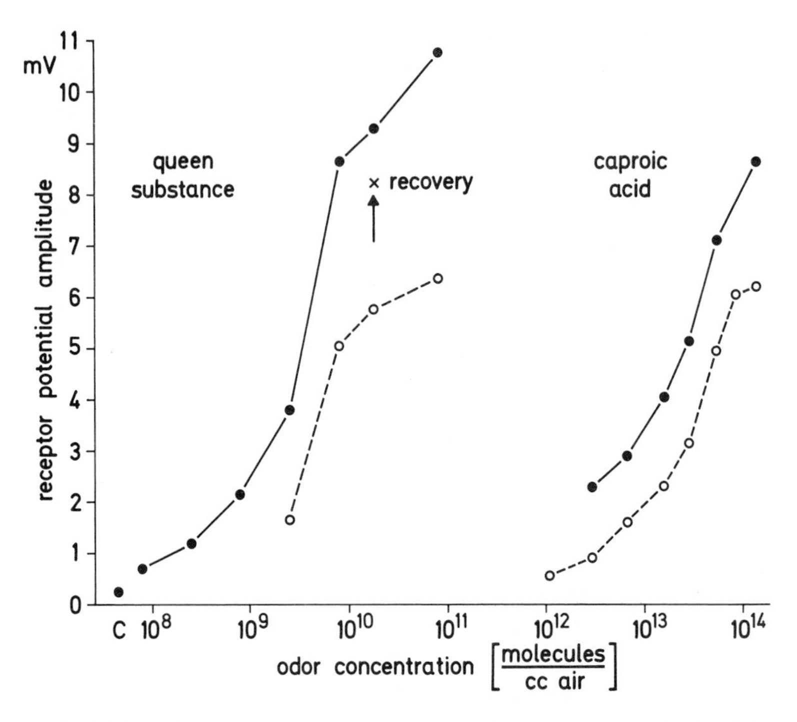

Figure 6 The intensity dependency of the same queen substance receptor, stimulated by queen substance and caproic acid. Solid lines: unadapted state. Broken lines: after adaptation with strong queen substance stimuli. Cross: amplitude after recovery. The receptor potential amplitude is measured 1 sec after the onset of the stimulus (compare Figure 5). For the calibration of the abscissa, see pages 59 to 60. Airstream velocity 1.6 m/sec.

substance and measured the loss of activity by means of the electrical reaction. For tests of caproic acid we used filter papers of the same size as the glass plates, and loaded them with caproic acid diluted in paraffin oil. For the absolute and relative intensity scale we used the calibration of Boeckh[7] with ^{14}C-labeled caproic acid.

Figure 7 demonstrates the concentration dependency of the formation and the decay half-times of the queen substance potentials from Figure 6. Both half-times fit into the formalism as predicted by Figure 3, except in the low concentration range. But, in fact, this exception does not exist. The relatively high values of both half-times at low concentrations are caused by the so-called control reaction (*C*), which is elicited by pure air. This control reaction is mainly a temperature effect that is additive to the olfactory reaction (Kaissling, unpublished). In the honeybee the effect has longer half-times than does the olfactory reaction. Because of this temperature effect, weak odor reactions have intermediate half-times.

While τ_d* is constant at 0.23 sec, τ_f* decreases to about 40 msec. That τ_f* does not

* All symbols applying to the kinetics of the receptor potentials are supplied with asterisks (*) because it has not been fully proved that they correspond to the respective symbols of the proposed acceptor kinetics.

seem to become shorter than 40 msec may be explained by the finite onset time of the applied stimulus. The fast feedback anemometer constructed by Dr. E. Kramer at Seewiesen enables us to measure changes of the airstream velocity up to 1000 Hz. The half-time of the onset of the applied airstream lies in the range of 10 msec. Nevertheless, the odor stream may rise more slowly. Another explanation is that the receptor potential in the honeybee olfactory cells is not able to rise faster than 40 msec (see page 67, et seq.).

The semilog plot of the stationary EAG amplitude in *Bombyx* shows a curve shape very different from the adsorption isotherm (Figure 8). The concentration range of the EAG covers almost six $\log_{10}$ units and shows no saturation at the highest concentrations used. The abscissa is calibrated in relative and absolute units by the use of ^{3}H-bombykol (Schneider, Kasang, and Kaissling,[28] Kaissling, unpublished).

The EAG is an extracellular summated response of the receptor potentials of many receptor cells. It may very well be that the EAG is the combination of several types of cells, each of which covers a smaller concentration range, but has a different sensitivity (cf. Tucker[37]). An almost identical curve shape was found in the electroretinogram of the fly *Calliphora* (Autrum, Autrum, and Hoffmann,[2] Figure 3).

The half-times of the *Bombyx* EAG show, in principle, the theoretically expected characteristics (Figure 9). The decay half-time, however, rises here at higher concentrations from 400 to about 600 msec. The formation half-time decreases to a minimum value of 20 msec. It is interesting that, even at low concentrations of bombykol, τ_f* is much smaller than τ_d*, which is in contradiction to the theory (Figure 3), but may be also explained by the control reaction. We know from radioactive measurements that the bombykol stimulus has, on the average, a five to 10 times higher concentration during the first 100 msec than afterwards. Generally, this may cause too short formation half-times. For exact measurements of the half-time, one must guarantee

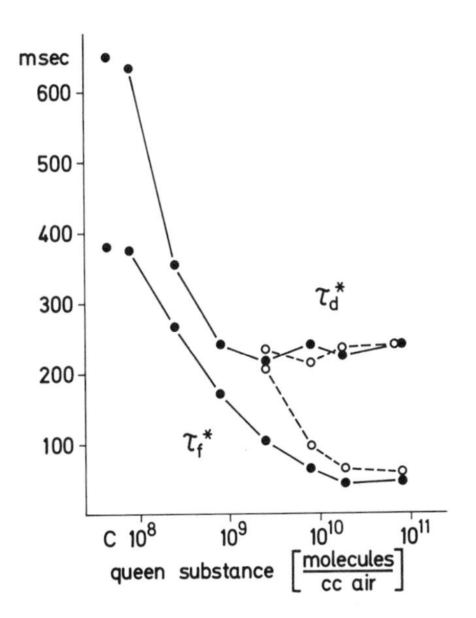

Figure 7 The intensity dependency of half-times (τ_f* = formation, τ_d* = decay). The values belong to the queen substance reactions of Figure 6. Solid lines: unadapted; broken lines: adapted state.

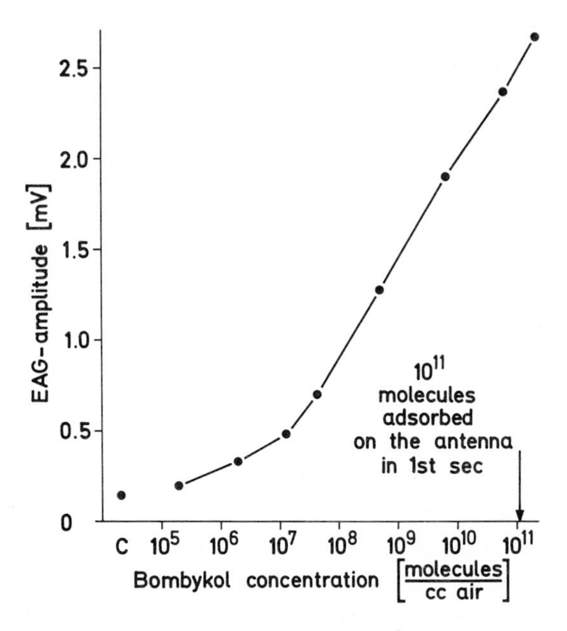

Figure 8 The intensity dependency of the *Bombyx* EAG. Average of four males. The stationary potential amplitude was measured 1 sec after the onset of the stimulus. Low concentration test stimuli in between the plotted reactions showed that no adaptation was involved. The abscissa is calibrated with ^{3}H-bombykol. C = control with fresh air. Airstream velocity 0.57 m/sec.

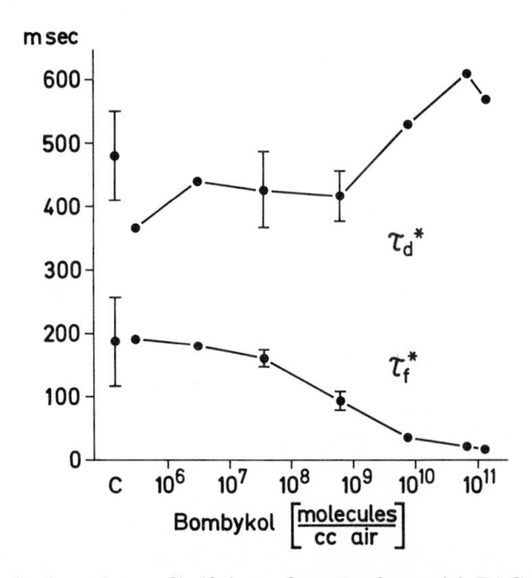

Figure 9 The intensity dependency of half-times of one *Bombyx* male's EAG (τ_f* = formation, τ_d* = decay). At six points the three standard deviations of 16 measurements each are indicated. The other points are averages of only two to five reactions each.

the rectangularity of the stimuli. At the end of the stimulus one must also make sure that the potential's decline does not come from a remaining and diminishing odor concentration in the air space around the antenna. This can be done by blowing fresh air immediately after the odor stream (Figure 10).

EQUILIBRIUM CONSTANTS AND THE TURNOVER NUMBERS

The steady-state constant K_{st}^* for queen substance in the honeybee can be provisionally fixed at $3 \cdot 10^9$ molecules/cc air, until the absolute calibration of the concentration has finally been determined by radioactively labeled material. The potential's stationary amplitude and the half-times give about the same result. For the EAG amplitude (Figure 8) we cannot determine the K_{st}. As the best approximation to a value of K_{st}^* we may take the concentration at

$$\tau_f^* = \frac{\tau_f^* \text{ (low conc.)}}{2}$$

(formula 7, Figure 9). The steady-state concentration then would be $6 \cdot 10^8$ molecules/cc air, but there may be cells with higher or lower values of K_{st}^*.

The turnover number (N) is very easy to calculate from τ_d^* (Figures 2, 3). For the queen substance receptor we get 180/min and for the bombykol receptor about 100/min. These numbers are small in comparison to enzymes, whose turnover rates are between 10^3 and 10^6 per min.

HIT RATE OF MOLECULES AND THE TOTAL NUMBER OF ACCEPTORS

To determine the total number of acceptors, one first needs the hit rate N of molecules against a receptor organ (formulae 11 to 13). The value N can be estimated from the molecule concentration [S], from the airstream velocity v_L, and from the receptive surface F_s that belongs to one receptor cell. The receptive surface may be defined as that part of the outer surface of a receptor organ from which a hit molecule has a chance to reach the acceptors. This is a functional definition. The question as to what part

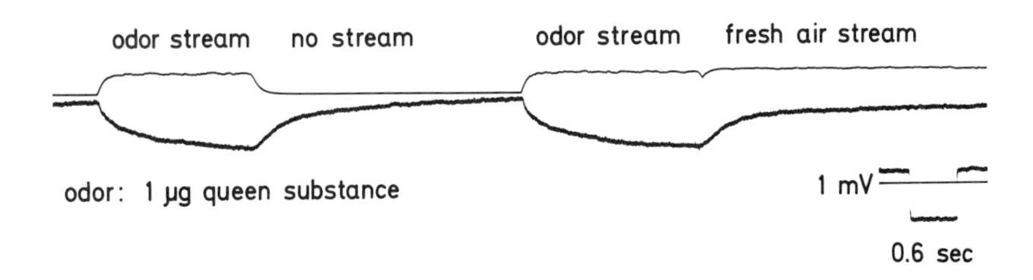

Figure 10 Experiment for proving that the slow decline of the receptor potential after the end of the stimulus is not produced by remaining odor in the surroundings. Recording is from the drone honeybee. Lower line: receptor potential; upper line: airstream velocity, recorded by a slow anemometer.

of the sense organ corresponds morphologically to the receptive surface will be discussed later. The hit rate of molecules for a single receptor cell is

$$N = \frac{[S] \cdot v_L \cdot F_s}{Q} \left(\frac{\text{number}}{\text{cell} \cdot \text{sec}}\right). \tag{14}$$

Q is a quotient that shows by how much the theoretical maximum of N is diminished by the geometry of the sense organ. This formula is valid if all molecules that touch the receptive surface are each adsorbed only once.

The maximum receptive surface of a queen substance receptor cell was calculated as 1.5 μ^2 (see page 66), and the maximum of Q as 3 (Kaissling, unpublished). From that and from the airstream velocity of 1.6 m/sec we get the rate

$$N_{st}^* = 2.4 \cdot 10^3 \frac{\text{molecules}}{\text{sec} \cdot \text{cell}}.$$

We measured the number of tritiated bombykol molecules adsorbed on the antenna of *Bombyx* in the first second. By dividing this number by the number of sense hairs, we derived the maximum rate of molecules per sense hair. One bombykol-sensitive hair has one or two sense cells.[26] For simplification, all further calculations are made for hairs with one sense cell. The maximum hit rate at K_{st}^* for the bombykol-sensitive cell is with $Q = 1$

$$N_{st}^* = 5.6 \cdot 10^4 \frac{\text{molecules}}{\text{sec} \cdot \text{cell}}.$$

Adam and Delbrück[1] calculated that the predominant portion of this number may be the right one, because only small parts of the molecules may be fixed at parts of the antenna other than the sensory hairs. Therefore, Q may not be more than 2.

All the important values necessary for the calculation of $[A_{tot}]^*$ are known (Table III), except for the quantum efficiency γ. In *Bombyx* the quantum efficiency may be unity, because the minimum number of molecules eliciting the behavior threshold of the animal is very small. On the average, less than one adsorbed molecule per receptor cell is sufficient (Schneider, Kasang, and Kaissling,[28] Kaissling, unpublished). Nevertheless, γ may be smaller than one in *Bombyx*, and the same may be true for the queen

TABLE III

OLFACTORY RECEPTOR CONSTANTS

	Drone honeybee	*Bombyx* male	Dimensions
K_{st}^*	3×10^9	6×10^8	molecules/cc air
N_{st}^*	2.4×10^3	5.6×10^{4}[a]	molecules/cc air
γ	1	1	
τ_d^*	0.23	0.42	sec
$(k_2 + k_3)^*$	181	104	1/min
$[A_{tot}]^*$	780	32.300[a]	number/cell
pore tubules	850	15.000	number/cell

[a] Valid if 100% of the molecules caught by the antenna are adsorbed on the sensory hairs (i.e., $Q = 1$). The right values may be higher than 50% of the maximum values (i.e., $1 < Q < 2$).

substance-sensitive cell, which seems not to be as sensitive as the bombykol cell. With $\gamma = 1$ we determine the maximum of the total acceptor number per receptor cell, which for *Bombyx* is 32,300 and for the drone 780. In the honeybee we took $Q = 3$ for this calculation. If we estimate $Q = 2$ for *Bombyx*, the acceptor number is about 16,000.

We calculated the acceptor number by combining formulae 12 and 14.

$$[A_{tot}]^* = \frac{\gamma \cdot F_s \cdot v_L \cdot K_{st}^*}{Q \cdot (k_2 + k_3)}. \tag{15}$$

This formula shows the relation among several constants that fit a chemoreceptor. Because the airstream velocity v_L is variable, at least one other parameter must be dependent on it. It can be shown experimentally that within certain limits the product $v_L \cdot K_{st}^*$ is a constant (Kaissling, unpublished). Keeping $v_L \cdot [S] =$ const., one finds constant reactions. Therefore, according to formula 14, the hit rate is constant. This means that the reaction amplitude is controlled by the hit rate and not by the concentration. Therefore, when concentration is specified in chemoreception papers, it is also necessary to give the airstream velocity.

THE NUMBER OF PORE TUBULES

If we look into the fine structure of the olfactory organs of *Bombyx* and of the honeybee we find in both cases a roughly 1:1 relation between the number of acceptors and the number of fine tubules located in the cuticular wall (Table III).

It was recently proved that these so-called pore tubules were not nerve processes, but belonged to the cuticle, and were open to the air space (Ernst,[15] carrion-beetle *Necrophorus*). About four to six pore tubules flow together into a chamber called the pore kettle, which has an opening (pore) to the outside of the exoskeletal wall of the sense organ (Figure 11). Presumably, this system of wall channels leads the odor

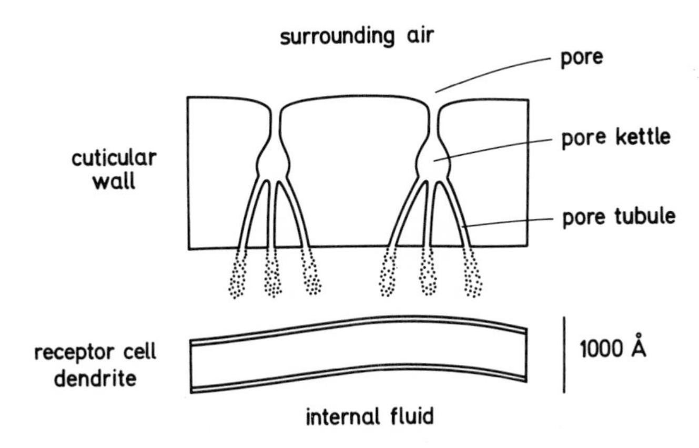

Figure 11 Fine structure of the olfactory hair wall of the beetle *Necrophorus*, investigated by Ernst in 1968.[15] Schematic drawing, validated by Ernst. The olfactory organs of *Bombyx* and of the drone honeybee are similarly constructed, but the geometrical proportions differ. For instance, the distance between two pores on the *Bombyx* hair is about 5000 Å (Steinbrecht, unpublished); on the honeybee pore plate it is only 400 Å (after Richards[24]).

molecules from the outside to the inside of the sensory hair. The inner end of these tubules is closed by a diffuse mass of material and is not connected with the nerve dendrites inside the hair.

The number of pores on the bombykol sense hair has been calculated from the density of pores on small pieces of hair cuticle prepared by Steinbrecht (unpublished) multiplied by the surface of the sense hair.[26] With a pore density of $5/\mu^2$ and a hair surface of 600 μ^2 one finds 3000 pores. Each pore in *Bombyx* has about five pore tubules.[29] One hair or one sense cell contains 15,000 tubules.

The corresponding structure of the sensory hair of the honeybee is the so-called pore plate. The pore plate also has numerous fine pores with pore kettles, and about four pore tubules.[29] Electron micrographs by Richards[24] show 3800 pores per pore plate. The total number of tubules per pore plate is 15,200, and should be divided by 18—the average number of receptor cells innervating each pore plate.[29] So, presumably, each queen substance receptor cell is supplied by a minimum of 850 pore tubules.

THE RECEPTIVE SURFACE

Considering the fine structure of the olfactory organs as described here, we may explain what we took to be the morphological correspondence of the receptive surface. One may think that the sum of the cross sections of all the pores together would reveal the receptive surface. We have evidence, however, that not only those molecules that directly hit a pore opening reach the acceptors; the ones that strike against the wall surface between the pores do so, as well. This assumption is based on the fact that the entire *Bombyx* antenna adsorbs about 14,000 molecules at the behavior threshold[28] (also Kaissling, unpublished). Because the pore openings (0.53 μ^2) comprise about one-thousandth of the hair surface (600 μ^2), only 14 molecules would reach the acceptors. Even this number is too high, because all 14,000 molecules may not be adsorbed on the sense hairs but on the nonreceptive surface parts of the antenna, as well. The number of molecules found on the antenna divided by 1000 seems to be too small to explain the behavioral and electrophysiological reactions (Kaissling, unpublished; Priesner, unpublished). Therefore, we are forced to assume that the receptive surface is much greater than the pore openings. In *Bombyx*, we used as a maximum the entire surface of the bombykol-sensitive hair (600 μ^2), the so-called *sensillum trichodeum.*[26] We took the whole region in which honeybee wall pores are located on the pore plate (26.5 μ^2, after Richards[24]). Each of the 18 receptor cells has a receptor surface of 1.5 μ^2.

The assumption of a receptive surface larger than the pore openings implies a transport mechanism that leads the molecules to the places where the acceptors are located. Adam and Delbrück[1] introduced the idea of surface diffusion, which would bring the molecules into the pores very quickly after adsorption on the outer wall surface.

LOCATION OF THE ACCEPTORS

Even if we take into account the uncertainties in all these preliminary calculations, it is surprising that the calculated number of acceptors and pore tubules lies within the same order of magnitude. If this agreement is not accidental, one could raise the question of whether each acceptor is located at the base of a tubule.

With this hypothesis, however, it is difficult to explain the observation that in some cases several receptor cells with different odor spectra supply one and the same sensory hair or pore plate.[17,27] This is easier to interpret if a specific acceptor type is located on each receptor-cell dendrite. One could divide the acceptor process into two steps. First, the tubules regulate the passage of the odor molecules; second, the nerve dendrites bear on their surface the structures that are responsible for the specificity of the receptor cell. The correspondence of the number of acceptors and pore tubules may now be interpreted as follows: the measured kinetics of the receptor potentials are dominated by the kinetics of the passage of the odor molecules through the pore tubules.

CRITICISMS

The most critical question we have to raise is: do the kinetics of the electrical responses come from the primary processes at the receptor cell or from other steps of the whole transducer process? Besides the one or two aforementioned acceptor steps we must number other steps, like diffusion processes, among the earlier, nonelectrical transducer steps. Subsequently, we can follow the two main electrical transducer steps—the receptor membrane, which produces the slow receptor potential, and that part of the membrane which converts the receptor potential into the spike frequency (Figure 12).

Furthermore, fine structure studies of the distal dendrite that innervates the sensory hairs generally show a subdivision into an outer and inner segment.[35,36] The parts are separated by a ciliary structure, which is indicated in Figure 12 by a bar within the distal dendrite. These morphological differentiations of the receptor membrane, which, in principle, are found in most chemoreceptors[15,30,31] as well as in many other sensory organs, may also have functional implications. As yet we do not know if the receptor potential originates in the outer or only in the inner segment of the receptor membrane.[35,36]

If we set aside the unknown, nonelectrical steps of the transducer process and look into the two electric ones, we would expect, from neurophysiological work, that they would have nonlinear properties. For instance, if one plots the membrane voltages under quasi-stationary conditions against the ratio G of membrane conductances, a curve like the adsorption isotherm is revealed.[35,36] On the other hand, Thurm[34] investigated a mechanoreceptor bristle of the honeybee and showed that the connection between the receptor potential and the spike frequency is nonlinear in the dynamic phases but rather linear in the stationary phase (cf. also Morita[18]).

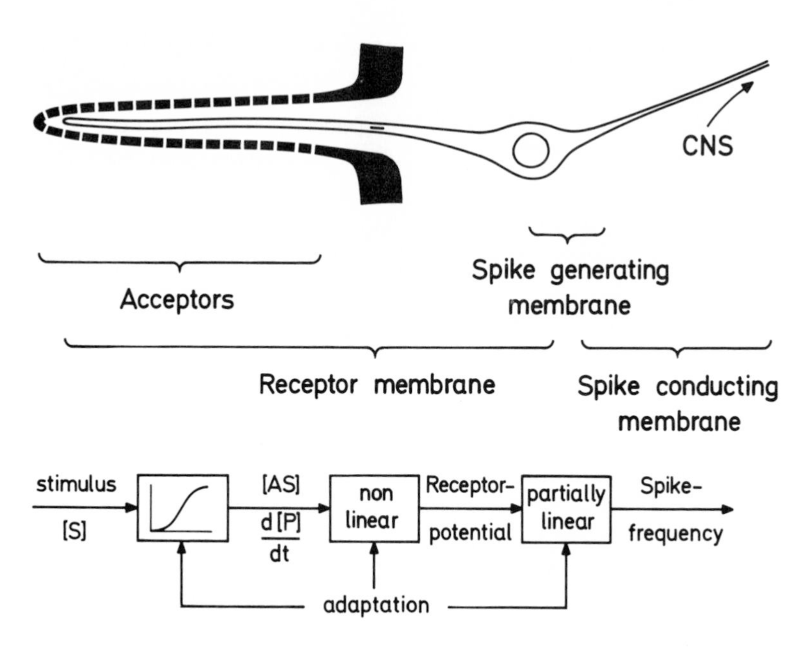

Figure 12 Morphological and functional steps in the transducer process of chemoreceptors. Each step has nonlinear properties, but may also work in linear parts of the whole working range (pp. 67–69). Adaptation may change the transducer properties of each step in a different manner. The bar inside the receptor cell indicates the ciliary region, which separates the outer and the inner segment of the receptor membrane.

Theoretically, it may very well be that the kinetics we investigated by measuring the receptor potential are the genuine kinetics of the receptor membrane. If this is so, one could assume that the cell has a large number of acceptors, of which only a small fraction will be occupied by the stimulus molecules, even with the highest stimulus concentrations. If the number of acceptor complexes remains under the limit of 10 per cent of the total acceptor number, the acceptor system would work only in the linear range (Figure 1). One knows that the bleaching of rhodopsin in the vertebrate visual receptors normally does not exceed this limit.[9] Generally, the step with the lowest saturation limit will dominate the stationary kinetics of the whole transducer process.

The dynamic phases of the response are controlled by the slowest link of the entire transducer chain. One argument, therefore, which may help find the dominating step in the receptors for queen substance and bombykol is the rather slow transients of the receptor potentials. The long half-times of the formation and the decay phase are unusual with respect to other receptors. The so-called "slow" receptor potential of a mechanosensitive bristle in the honeybee rises with a half-time of 2 msec and declines after the end of the stimulus with a half-time of about 5 msec.[35,36] The corresponding times in the visual receptors of the honeybee are both about 10 msec.[3] These examples may indicate that the 10 to 100 times slower transients of the olfactory receptors in the same animal are caused by the earlier, nonelectrical transducer steps. An additional proof could be made by verifying the number of acceptors by using another method.

Considering the whole transducer process, one has to be aware that each step may change its functional properties. Reversible changes induced by the stimulus are called adaptation, which may occur at each step in a different manner (Figure 12).

Figures 6 and 7 show adaptation at the queen substance receptor cell. In Figure 6, the amplitude of all reactions is reduced by 40 per cent, while the equilibrium concentration $K_{st}*$ seems not to change. On the basis of the acceptor-controlled kinetics, it would mean that 40 per cent of the acceptors are blocked. If $K_{st}*$ remains constant, the half-times also should not change, but τ_f* grows with adaptation (Figure 7). This cannot be explained by the acceptor model alone.

SUMMARY

The mass-action law is used to prove the connection between the odor concentration and the receptor potential's amplitude of the bombykol receptor and the queen substance receptor of two species of insects, *Bombyx mori* and *Apis mellifera.* Not only the stationary phase but also the transient dynamic phases after the onset and end of the stimulus are studied. From this, the steady-state constants of the stimulus and the turnover numbers of the receptor cells are determined. By additional determination of the hit rate of odor molecules against the receptive surface, it becomes possible to calculate the number of hypothetical odor acceptors per receptor cell. This number is in the same order of magnitude as the number of fine pore tubules in the wall of the sensory hairs. The criticisms which are discussed demonstrate that possibly the kinetics of the electrical response of the receptor cell are dominated by the nonlinear properties of the receptor membrane and not by the earlier, nonelectrical steps of the transducer process.

ACKNOWLEDGMENTS

The author wishes to thank Prof. Dr. Dietrich Schneider for his continuous interest and many discussions. Many thanks also go to all the colleagues of Mr. Schneider's group in Seewiesen and to Dr. K. Hansen and Dr. U. Thurm; to Miss I. Küderling for technical assistance; and to Stan Skordilis for correcting the manuscript.

REFERENCES

1. Adam, G., and M. Delbrück. 1968. Reduction of dimensionality in biological diffusion processes. *In* Structural Chemistry and Molecular Biology (A. Rich and N. Davidson, editors). W. H. Freeman and Co. Publishers, San Francisco, Calif. and London, England, pp. 198–215.
2. Autrum, H., I. Autrum, and C. Hoffmann. 1961. Komponenten im Retinogramm von Calliphora und ihre Abhängigkeit von der Spektralfarbe. *Biol. Zentralbl.* **80**, pp. 513–547.
3. Autrum, H., and V. v. Zwehl. 1964. Die spektrale Empfindlichkeit einzelner Sehzellen des Bienenauges. *Z. Vergleich. Physiol.* **48**, pp. 357–384.
4. Beidler, L. M. 1954. A theory of taste stimulation. *J. Gen. Physiol.* **38**, pp. 133–139.
5. Beidler, L. M. 1961. Taste receptor stimulation. *Progr. Biophys. Biophys. Chem.* **12**, pp. 107–151.
6. Beidler, L. M. 1965. Anion influences on taste receptor response. *In* Olfaction and Taste II (T. Hayashi, editor). Pergamon Press, Oxford, etc., pp. 509–534.
7. Boeckh, J. 1967. Reaktionsschwelle, Arbeitsbereich und Spezifität eines Geruchsrezeptors auf der Heuschreckenantenne. *Z. Vergleich. Physiol.* **55**, pp. 378–406.
8. Boeckh, J., K. E. Kaissling, and D. Schneider. 1965. Insect olfactory receptors. *Cold Spring Harbor Symp. Quant. Biol.* **30**, pp. 263–280.

9. Cone, R. A. 1965. The early receptor potential of the vertebrate eye. *Cold Spring Harbor Symp. Quant. Biol.* **30**, pp. 483–491.
10. Dastoli, F., D. V. Lopiekes, and A. R. Doig. 1968. Bitter-sensitive protein from porcine taste buds. *Nature* **218**, pp. 884–885.
11. Davies, J. T. 1965. A theory of the quality of odours. *J. Theor. Biol.* **8**, pp. 1–7.
12. Dixon, M., and E. C. Webb. 1958. Enzymes. Academic Press, Inc., New York.
13. Dravnieks, A. 1966. Current Status of Odor Theories. *Advan. Chem. Ser.* **56**, pp. 29–52.
14. Dravnieks, A. 1967. Theories of Olfaction. *In* Symp. Foods: The Chemistry and Physiology of Flavors (H. W. Schultz, editor). Avi Publ. Co., Westport, Conn., pp. 85–118.
15. Ernst, K. D. 1969. Die Feinstruktur von Riechsensillen auf der Antenne des Aaskäfers Necrophorus (Coleoptera). *Z. Zellforsch. mikroskop. Anat.* **94**, pp. 72–102.
16. Hansen, K. 1969. The mechanism of insect sugar perception: a biochemical investigation. *In* Olfaction and Taste III (C. Pfaffmann, editor). Rockefeller Univ. Press, New York, pp. 382–391.
17. Kaissling, K. E., and M. Renner. 1968. Antennale Rezeptoren für Queen Substance und Sterzelduft bei der Honigbiene. *Z. Vergleich. Physiol.* **59**, pp. 357–361.
18. Morita, H. 1969. Electrical signs of taste receptor activity. *In* Olfaction and Taste III (C. Pfaffmann, editor). Rockefeller Univ. Press, New York, pp. 370–381.
19. Mullins, L. J. 1955. Olfaction. *Ann. N.Y. Acad. Sci.* **62**, pp. 247–276.
20. Netter, H. 1959. Theoretische Biochemie. Springer-Verlag, Berlin, Göttingen, Heidelberg, Germany.
21. Neuhaus, W. 1956. Die Riechschwelle von Duftgemischen beim Hund und ihr Verhältnis zu den Schwellen unvermischter Duftstoffe. *Z. Vergleich. Physiol.* **38**, pp. 238–258.
22. Price, S., and R. M. Hogan. 1969. Glucose dehydrogenase activity of a "sweet-sensitive protein" from bovine tongues. *In* Olfaction and Taste III (C. Pfaffmann, editor). Rockefeller Univ. Press, New York, pp. 397–403.
23. Priesner, E. 1969. A new approach to insect pheromone specificity. *In* Olfaction and Taste III (C. Pfaffmann, editor). Rockefeller Univ. Press, New York, pp. 235–240.
24. Richards, A. G. 1952. Studies on arthropod cuticle, VIII. The antennal cuticle of honey bees, with particular reference to the sense plates. *Biol. Bull.* **103**, pp. 201–225.
25. Schneider, D. 1969. Insect olfaction: deciphering system for chemical messages. *Science* **163**, pp. 1031–1037.
26. Schneider, D., and K. E. Kaissling. 1957. Der Bau der Antenne des Seidenspinners *Bombyx mori* L. II. Sensillen, cuticulare Bildungen und innerer Bau. *Zool. Jahrb. Abt. Anat. Ontog. Tiere* **76**, pp. 223–250.
27. Schneider, D., V. Lacher, and K. E. Kaissling. 1964. Die Reaktionsweise und das Reaktionsspektrum von Riechzellen bei *Antheraea pernyi* (Lepidoptera, Saturniidae). *Z. Vergleich. Physiol.* **48**, pp. 632–662.
28. Schneider, D., G. Kasang, and K. E. Kaissling. 1968. Bestimmung der Riechschwelle von Bombyx mori mit Tritium-markiertem Bombykol. *Naturwissenschaften* **55**, p. 395.
29. Schneider, D., and R. A. Steinbrecht. 1968. Checklist of insect olfactory sensilla. *Symp. Zool. Soc. London* **23**, pp. 279–297.
30. Slifer, E. H. 1967. The thin-walled olfactory sense organs on insect antennae. *In* Insects and Physiology (J. W. L. Beament and J. E. Treherne, editors). Oliver and Boyd Ltd., Edinburgh and London, pp. 233–245.
31. Slifer, E. H., and S. S. Sekhon. 1960. The fine structure of the plate organs on the antenna of the honeybee, *Apis mellifera* Linnaeus. *Exp. Cell Res.* **19**, pp. 410–414.
32. Stuiver, M. 1958. Biophysics of the Sense of Smell. Doctoral Thesis. Diss. Natwiss. Fak. Univ. Groningen, Holland.
33. Teichmann, H. 1959. Über die Leistung des Geruchssinnes beim Aal [*Anguilla anguilla* L.]. *Z. Vergleich. Physiol.* **42**, pp. 206–254.
34. Thurm, U. 1963. Die Beziehungen zwischen mechanischen Reizgrößen und stationären Erregungszuständen bei Borstenfeld-sensillen von Bienen. *Z. Vergleich. Physiol.* **46**, pp. 351–382.
35. Thurm, U. 1965. An insect mechanoreceptor, Part I: fine structure and adequate stimulus. *Cold Spring Harbor Symp. Quant. Biol.* **30**, pp. 75–82.
36. Thurm, U. 1965. An insect mechanoreceptor, Part II: receptor potentials. *Cold Spring Harbor Symp. Quant. Biol.* **30**, pp. 83–94.
37. Tucker, D. 1963. Physical variables in the olfactory stimulation process. *J. Gen. Physiol.* **46**, pp. 453–489.

EOG PROBLEMS

SADAYUKI F. TAKAGI

Department of Physiology, School of Medicine, Gunma University, Maebashi, Japan

INTRODUCTION

Since Hosoya and Yoshida[21] began electrophysiological studies of the olfactory epithelium, five types of slow potential changes have been found in various animals. Ottoson[34] named these slow potentials electro-olfactograms, or EOGs, and that term will be used in this paper.

The first of the changes is the electronegative "on" type EOG. Hosoya and Yoshida[21] recorded them in the dog for the first time and Ottoson studied them extensively in the rabbit[33,36] and in the frog.[34] Since then, it has been studied most actively from various angles by Ottoson,[35,37] Takagi and his collaborators, MacLeod,[28] Kimura,[24] Mozell,[30] Gesteland and his coworkers, and Suzuki (references are given below).

The second is the negative EOG of the "off" type. It was first found by Takagi and Shibuya[50] in the frog, and then studied by Takagi and his coworkers.[51-54] The negative-on and -off EOGs were further studied by Shibuya,[40] Higashino, Takagi, and Yajima,[20] Ai and Takagi,[1] Shibuya and Takagi,[43,44] Higashino and Takagi,[19] Gesteland,[12] Byzov and Flerova,[7] Takagi and Yajima,[58,59] Gesteland et al.,[14] and Suzuki.[45] The ionic mechanisms of these EOGs were examined by Takagi and collaborators.[49,55,57]

The third type is the electropositive EOG of the "on" type. This EOG was found by Takagi et al.,[54] and studied later by Higashino and Takagi,[19] Takagi et al.,[56,57] and Takagi.[47] Recently, Suzuki[45] studied the electrical activity of the olfactory organ of the snail, and found negative and positive EOGs.

The fourth change is the positive EOG of the "off" type. It was found by Shibuya in the fish[40] and later by Gesteland et al. in the frog.[14]

The fifth change is the positive EOG of the after-potential type. It was found by MacLeod[28] in the olfactory epithelium of the rabbit. The ionic mechanism of this positive EOG was studied in the frog by Takagi and his colleagues.[47,48,55,56]

In the present review, the pioneer work on EOGs is reviewed and several problems are discussed.

EARLY STUDIES ON EOGS

Hosoya and Yoshida[21] found that the outer surface of the excised olfactory epithelium of dog is electronegative to the inner surface by about 6 to 1 mV, that this potential difference is larger the more the epithelium is pigmented, that it gradually decreases in magnitude and disappears with time, and that the potential difference shows various slow changes in response to the odors applied. These studies were performed in the physiological laboratory of the University of Taipei in Formosa. Unfortunately, their work was discontinued as a result of World War II, and most of the data obtained were destroyed by fire during the hostilities. Only the surviving data were published by Yoshida in 1950.[62] He found that pure odorants like lemon oil, naphthalene, guajacol, xylol, and menthol elicited monophasic action potentials (Figure 1A); weak irritating odorants like camphor and carbon bisulphide generated biphasic potentials (Figure 1B); and strong irritating odorants like dog urine and ammonia elicited triphasic or oscillatory potentials (Figure 1C). Latencies of these action potentials were 0.18 to 0.5 sec for pure odorants and 0.002 to 0.39 sec for irritating odorants. The amplitudes of the action potentials increased in relation to the increase in pigmentation of the olfactory epithelium. He further found that, among 122 dogs, the color of the olfactory epithelium darkened gradually in parallel to the range of body colors, from brown to light brown to yellow to blackish brown to black. The color of the olfactory epithelium darkened further when there were white areas in the body color of the dog.

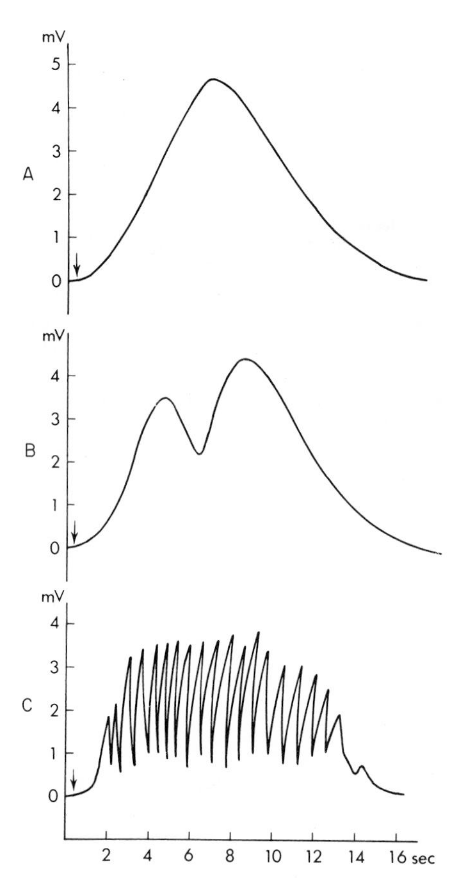

Figure 1 EOGs in the excised olfactory epithelia of dog.

Odors applied were lemon in A; camphor in B; acetic acid in C. Tracings of the original photographs, which were obtained by a mirror galvanometer (Shimazu Co. Kyoto). (Courtesy of Dr. H. Yoshida)

After the Hosoya-Yoshida work, this subject was left unstudied for about seventeen years until Ottoson took it up for electrophysiological research in 1954.[33] He studied the activity of the olfactory epithelium through the ethmoid bone in the rabbit; in 1956 he scrutinized it extensively in the frog.[34] In his studies, when a short puff of odorized air is applied onto the olfactory epithelium, a negative monophasic potential with a quick rise followed by slow decline is evoked (Figure 2A). It appears only in the yellow (pigmented) area of the olfactory eminentia, but not in the other nonpigmented (pink) region in the olfactory cavity (Figure 3). It is temporarily abolished by a puff of ethyl ether. The amplitude of the EOG is, within certain limits, proportional to the logarithm of the stimulus intensity (Figure 2B) and to the volume of the stimulating air at a given stimulus strength. Equal amounts of odorous materials distributed in different volumes of air evoke EOGs of equal amplitude. The EOG is different in

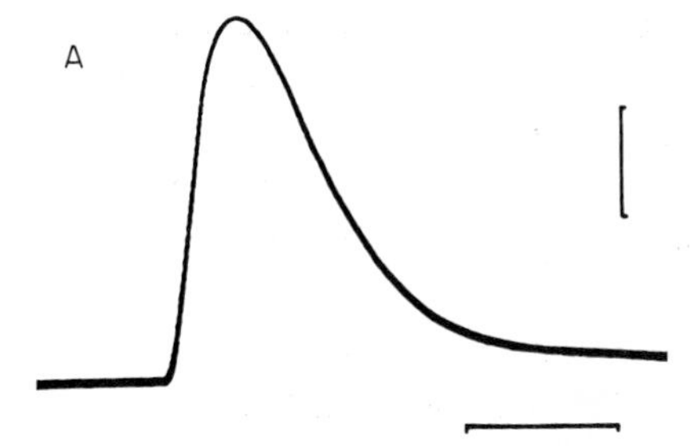

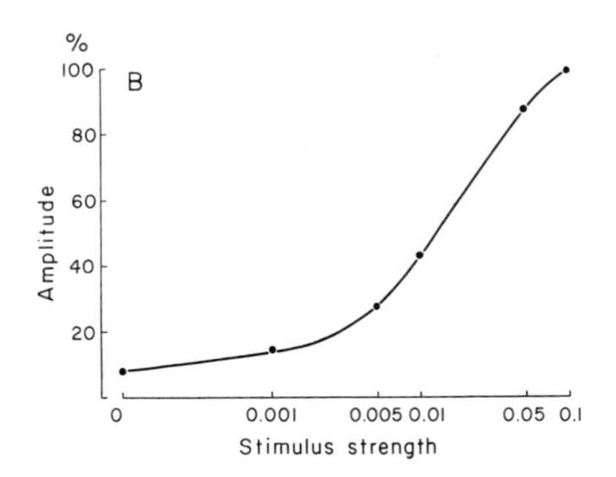

Figure 2 A: EOG to stimulation with butanol. Volume of stimulating air 0.5 cc. Vertical line 1 mV. Time bar 2 sec. Upward deflection in this and the following records indicates negativity of the exploring electrode.

B: Amplitude is proportional to logarithm of stimulus intensity. (Ottoson[34])

C: Relation between amplitude of potential and stimulus strength. Ordinate: amplitude in per cent of response to 0.1 M butanol. Abscissa: stimulus strength; zero, purified air. Stimulus: butanol. Volume of stimulating air 0.5 cc. The stimulus strengths in this curve are given in units corresponding to the concentrations of the solutions used for odorizing the stimulating air. (Ottoson[34])

Figure 3 Schematic diagram of the preparation showing responses to stimulation from restricted areas of the mucosa. Dotted regions indicate the sensory epithelium. (Ottoson[34])

shape and time course according to the strength of the stimulus, to the difference of the waveform of the stimulating air current, and to the difference of physicochemical properties of the odorants. With an increase of the odor intensity, the potential rises at a faster rate, the crest of the response broadens, and the decay time lengthens. When an olfactory stimulus of long duration is applied, the EOG shows a shape with an initial peak followed by a plateau that lasts throughout the stimulation at a reduced level, and finally by an exponential decline to the base line. When an odor is applied repetitively at short time intervals, the height of the response becomes successively smaller to a certain level, the reduction being greater the greater the stimulus strength. The sensitivity of the olfactory epithelium to different odorants can be selectively reduced, and phenomena of selective fatigue are easily observed. This indicates that there are receptors with specific sensitivity to particular substances or molecular configurations. Finally, by covering the olfactory epithelium with a thin plastic membrane which transmits only infrared radiation, Ottoson[34] proved that the EOG cannot be elicited unless the odorous material is brought into contact with the epithelium. Thus, the radiation hypothesis of olfaction was negated.

When these properties of the EOGs were compared with those of the other receptors, remarkable similarities were found. Consequently, Ottoson proposed a hypothesis that the EOG is a generator potential.[34,38] This will be discussed later.

Takagi et al.[54] studied the generative mechanism of the off EOG. They found indications that this is an independent potential, but not a final part of the sustained negative EOG, which remains because the initial part of the EOG is abolished by the anesthetic action of ethyl ether (Figure 4). The off EOG can be elicited not only in the frog, but also in the newt, fish, and other animals by the application of many organic solvents, and the corresponding off-induced potentials appear in the olfactory nerve and bulb.

With the progress of these studies, several questions have been raised as to the origins, natures and roles of the EOGs.

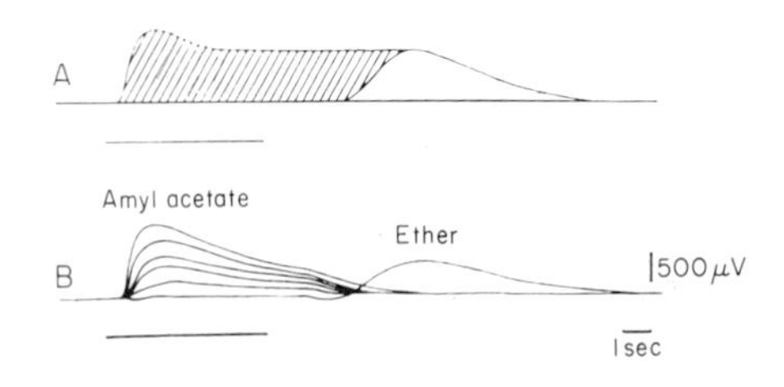

Figure 4 Off-type EOG.

A: A schematic drawing to show a hypothesis on the generative mechanism of the off EOG: A long sustained EOG is elicited by the stimulative action of ether, but the initial and main part of the potential (shadowed) is inhibited by the anesthetic action of ether. If the anesthetic action vanishes earlier than the stimulative action at the cessation of stimulation, a potential of the off-type remains.

B: Superimposed tracings of the on and the off EOGs obtained in the same preparation. When amyl acetate is applied at the concentrations of 1/16, 1/8, 1/4, 1/2 and 1/1 (dilution by purified air), the EOGs gradually increase in amplitude but not in duration. When ethyl ether is applied, the off EOG appears, but it appears only after the on EOG begins to disappear. The difference in the time of appearance shows that the two EOGs are of different origin. (Takagi et al.[54])

PROBLEM 1. ARE EOGs BIOLOGICAL PHENOMENA?

Hartman[18] reported that sustained potentials may be recorded from nonbiological systems in response to certain odorants. This suggested a possibility that the EOGs recorded by the above workers may be nonbiological phenomena. Mozell[30] showed that nonbiological potentials respond to increases in stimulus intensity, decrease with the introduction of cocaine, and even show adaptation. Thus, it was found impossible to differentiate between the real EOG and the nonbiological potential. However, Kimura[24] reported that if electrodes of the Ringer-agar types are used, such nonbiological potentials are not recorded. Kimura[24] and Mozell[30] warned that platinum-, tungsten-, nichrome-electrodes, and 3M KCl micropipettes are all able to record nonbiological potentials from Ringer-soaked cotton when an odorant is puffed toward the electrode-cotton junction, and recommended the use of Ringer-agar (silver-silver chloride) electrodes. Since Ottoson[34] and most other investigators used the electrodes of the recommended types, it is not necessary to consider the intervention of such nonbiological potentials.

Gesteland[12] studied the EOGs elicited by 62 odors in frogs immobilized with d-tubocurarine, and found that 35 odors elicit negative EOGs, while the remaining 27 elicit the strong initial positive potentials followed by the negative potentials (Table I). Recently, 122 odorants were applied to the excised olfactory epithelium of the bullfrog and were classified according to the shapes of the EOGs elicited (see Table I, next chapter). Among them, 106 odorants (87 per cent) elicit negative EOGs, and the remaining 16 elicit positive EOGs. These results are not coincident with Gesteland's results. Such a difference may, in part, be caused by differences in the qualities of the odorants, in the manufacturing companies, and in the impurities of the odorants, but mainly is the result of the condition of the experimental animals—whether they are immobilized with d-tubocurarine or decapitated.[48]

PROBLEM 2. ARE EOGs SIMPLE POTENTIALS?

Ottoson[34] presumed that the EOGs are composite potentials. At that time, however, only the EOG of the on type had been found. Since then, many data have accumulated.

Takagi et al.[54] found that ethyl ether and chloroform elicit the negative EOGs, and hence these substances have stimulating action at low concentration, but anesthetic effects at high concentrations. The anesthetic actions were found by applying these vapors onto the olfactory epithelium after using amyl acetate vapor. The negative EOG elicited by amyl acetate vapor or the negative-off EOG by ethyl ether vapor were abolished by subsequent application of ethyl ether or chloroform. Even in these cases, the on-induced waves appeared in the olfactory bulb (cf. Figure 1B in Takagi[46]). The waves could also be observed even when the positive-on EOGs were elicited by ethyl ether or chloroform (cf. Figure 1A, B in Takagi, loc. cit.). Since ether and chloroform are odorous, these discrepancies between the EOGs and the induced bulb waves were supposed to originate from the competition between the stimulative and anesthetic actions which ether and other organic solvents possess; in other words, from a competition between the negative and positive EOGs.[54]

TABLE I
NEGATIVE EOGs ELICITED BY 62 ODORANTS (GESTELAND[12])

A. Stimulants Which Cause a Little Initial Positive Potential

Benzothiazole	Isoquinoline
1,4-Butanediamine	Limonene
Butyrolactone	Linalyl acetate
Camphor	Methyl acetate
Citronellol	Menthol
Coumarin	Oil balsam peru
Cyclohexane	Oil peppermint
Cyclooctane	Oil spearmint
Diallyl disulfide	Pentadecanolide
Diamyl sulfide	Phenylacetic acid
2-Diethylaminoethanol	Phenylethyl alcohol
Ethylaminoethanol	Pond water
Ethyl butyrate	tert. Butyl alcohol
Eugenol	Tetraethyl tin
Geraniol	Undecalactone
Heptane	Vanillan
Hexachloroethane	Ylang ylang
Hexane	

B. Stimulants Which Cause a Strong Initial Positive Potential

Acetone	Human breath
Anisole	Indole
Benzonitrile	Isoamyl alcohol
n-Butanol	Linalool
Butyl acetate	Methanol
Butyric acid	Monochlorbenzene
n-Capric acid	Petroleum ether
Cyclopentadecanone	Piperonal
Cyclopentanone	Pyridine
Dichlorobenzaldehyde	Pyrrole
Ethanol	Rum ether
Ethyl acetate	Salicylaldehyde
Ethyl ether	2,4-Trinitro-3,5-dimethyl-tert. butylbeneze
Formic acid	

So far, five types of negative and positive EOGs have been found, and their generative mechanisms have been studied from various points of view. One such study examined the effect of polarizing currents upon EOGs.[19] When an anodal current was applied, the negative-on EOGs elicited by a general odor, such as amyl acetate, increased in magnitudes in accordance with increase of the current. When a cathodal current was applied, the magnitudes decreased with increase of the current (Figure 5A). Similar relations were observed in negative EOGs elicited by low concentrations of vapors of some organic solvents, such as ethyl ether or chloroform, and also in negative-off EOGs elicited by highly concentrated vapors of these organic

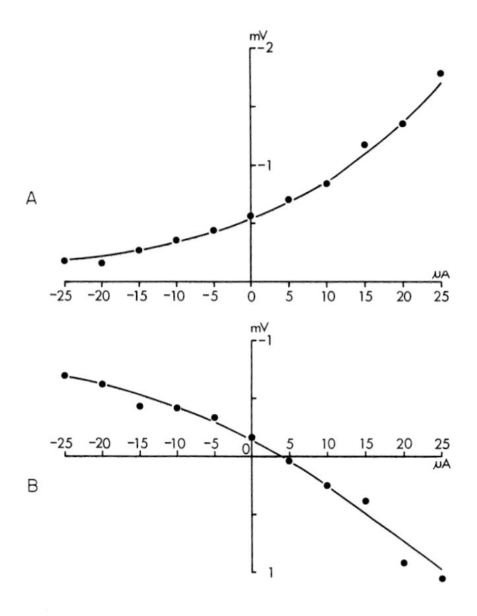

Figure 5 Analysis of EOG by polarizing currents. A: Relation between the EOGs elicited by amyl acetate (general odor) and the applied currents. Anodal currents on the right and cathodal currents on the left of the ordinate in A and B.
B: Relation between the EOGs elicited by ethyl ether (organic solvents) and the applied currents. An entirely opposite relation is found in this case. Further explanation in the text. (Higashino and Takagi[19])

solvents. This relation is commonly found in other nervous tissues. On the other hand, the negative- or positive-on EOGs elicited by these vapors showed an entirely opposite relation. They decreased in magnitude in accordance with the increase of anodal current, and increased with the increase of cathodal current (Figure 5B). From the difference in the behaviors of the two EOGs, there are probably two receptive processes in the olfactory epithelial cells. One is an ordinary excitatory process that produces the electronegative EOGs in response to general odors; the other is a process of a different kind, which is activated only by highly concentrated vapor of ether or chloroform, and which produces the electropositive EOGs.

The same conclusion was reached by an entirely different approach. Lettvin and Gesteland,[27] Gesteland, Lettvin, and Pitts,[14] and Gesteland[13] measured impedance change of the olfactory epithelium during EOGs. They found that the positive-drive process could be identified at a low frequency with a change in phase and the negative-drive process with a change in magnitude (Figure 6). They also found linear and non-linear interactions between the EOGs elicited by the simultaneous or successive applications of two different odors. From these findings, they distinguished the two processes that oppose each other.

In order to clarify the mechanisms of the EOGs further, we studied the ionic mechanisms of the negative and positive EOGs.[49,56,58] When the olfactory epithelium was immersed in Ringer's solutions in which Na^+ was replaced by sucrose, Li^+, choline$^+$, tetraethyl ammonium$^+$ (TEA), hydrazine or other monovalent or divalent cations, the negative EOGs decreased in amplitude, and in some cases nearly or completely disappeared (Figure 7). When the olfactory epithelium was returned into normal Ringer's solution, the EOGs recovered. When Na^+ was replaced by K^+ and

K^+ by Na^+ in the Ringer's solution, the negative EOGs reversed their polarity. Next, when the K^+ in the solution was replaced by equimolar Na^+, sucrose, Li^+, choline$^+$, TEA^+, hydrazine or other monovalent or divalent cations, the reversed potentials recovered completely only in Na^+-Ringer's solution, but never in the others (Figure 8). This demonstrated the essential role of Na^+ in negative EOGs.

When K^+ was removed from the bathing Ringer's solution, the negative EOGs

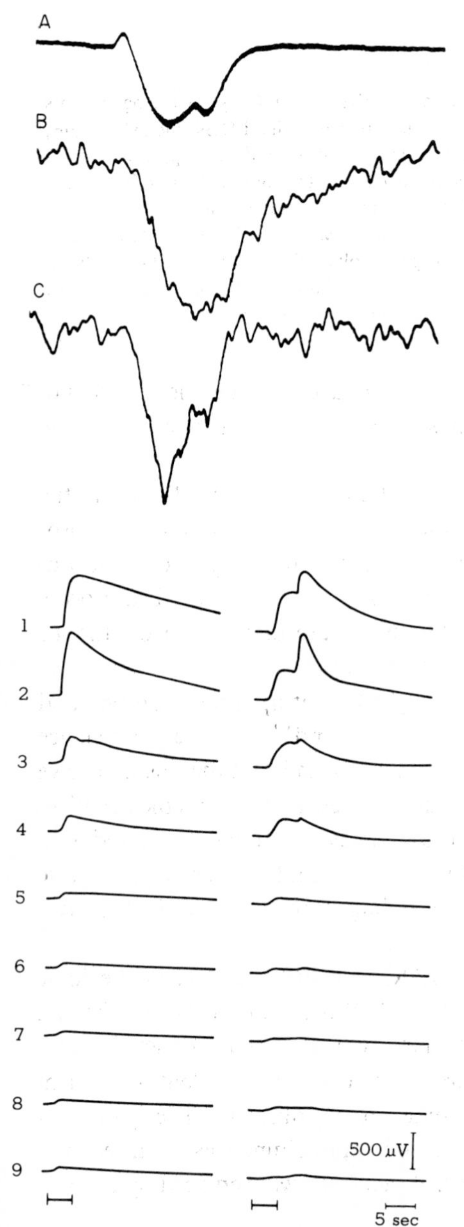

Figure 6 Impedance changes. A: The EOG for a puff of pyrrole. B: The magnitude change. C: Phase angle change. These records were obtained at 40 c/sec during the EOG. With pyrrole, the magnitude change is long lasting, while the phase-angle change is abruptly truncated at the cessation of the stimulus. Sweep time is 10 sec. (Gesteland et al.[13])

Figure 7 EOGs in sodium-free solution. The negative-on EOGs elicited by menthone vapor are shown in the left column, and the on-off EOGs by ethyl ether in the right one. 1: EOGs in normal Ringer's solution. 2 to 9: changes in magnitude of the negative-on and -off EOGs in sodium-free Ringer's solution. The records were taken at intervals of 20 min. Note the temporary increase in magnitude of the on and off EOGs in 2, and the decreases below 2. Short horizontal lines below columns give the time and duration (4 sec) of stimulation. (Takagi et al.[57])

first increased and then decreased in amplitude (Figure 9A). When K^+ was increased in Ringer's solution within a certain limit, the negative EOGs increased in amplitude (Figure 9B). The increase in the former case indicates a contribution of the increased membrane potential to the generation of the negative EOGs and the increase in the latter case is explained by a decrease of K^+ exit due to a decrease in the ratio of the

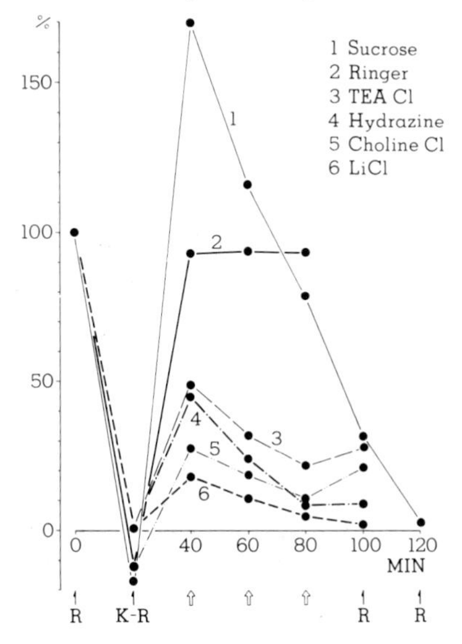

Figure 8 Comparison of recoveries of the reversed EOGs in various sodium-free solutions. After the negative EOGs were reversed in K^+ Ringer's solution, recoveries were compared in various sodium-free solutions in which Na^+ was replaced by Li^+, choline$^+$, TEA^+, sucrose, or hydrazine. Three hollow arrows indicate three immersions (during 60 min) in all these Na^+-free solutions except sucrose. Half arrows indicate, from left to right, one immersion in Ringer's solution (control), one immersion in K^+-Ringer's solution (K-R), and two immersions in Ringer's solution. In the case of sucrose-Ringer's solution (1), immersions were repeated 5 times. It is worthy of note that the EOG recovered completely or nearly completely and was maintained only in Na^+-Ringer's solution. (Takagi et al.[57])

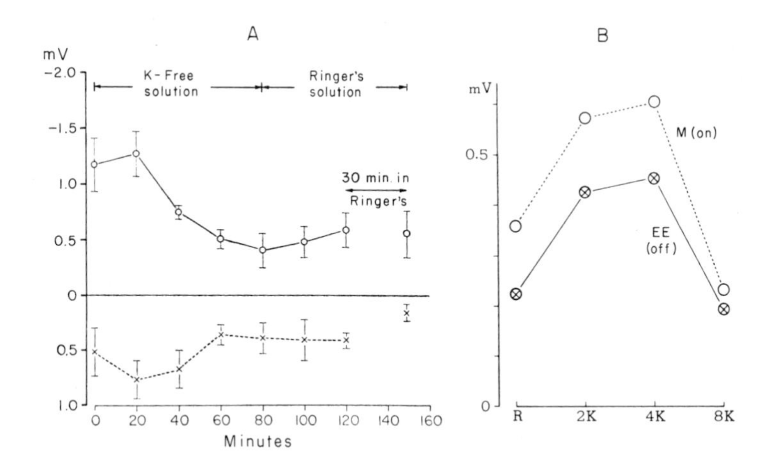

Figure 9 A: Increases of EOGs in K^+-free Ringer's solution. The negative and positive EOGs increased in amplitude just after Ringer's solution was replaced by K^+-free solution. (Takagi et al.[56])

B: Effects of high K^+. The negative EOGs of the on type (indicated by M, menthone) and the off type (indicated by EE, ethyl ether) increased in amplitude when the concentration of K^+ in Ringer's solution was increased two or four times normal (2K or 4K). At higher concentrations than four times normal (4K, 10 mM), the amplitude of the EOG decreased. Interval between trials, 20 min. (Takagi et al.[57])

external and internal K^+ concentrations. These two effects indicate the important role of K^+ in the generation of the negative EOGs.

When 1 mM Ba^{++} was added to Ringer's solution, the positive EOGs were depressed selectively, while the negative EOGs were barely depressed and never increased in amplitude (Figure 10). As will be shown, it was proved that the positive EOGs are elicited mainly by the entry of Cl^- and the exit of K^+.[56] If the exit of K^+ were affected by Ba^{++}, the negative EOGs should increase in amplitude. Since that is not the case, it follows that Cl^- entry is selectively inhibited by Ba^{++}. Then, if Cl^- entry contributes to the generation of the negative EOGs, and if it were inhibited by Ba^{++}, the negative EOGs should increase in amplitude. However, the negative EOGs never increased in the Ba^{++} solution. Thus, Cl^- probably does not contribute to the negative EOGs. It may well be concluded that the negative EOGs are elicited mainly by an increase in permeability of the olfactory receptive membrane to Na^+ and K^+. It is clear, then, that the negative EOGs resemble the receptor potentials in many respects, particularly the muscle endplate potential. Consequently, it is highly probable that most, if not all, of the negative EOGs are composed of the generator potentials of many receptor cells.

Similarly, by removing ions from, or adding them to, to the bathing Ringer's solution, the ionic mechanism of the positive EOGs was studied.[56] When Cl^- in the Ringer's solution was replaced by $SO_4^=$, the positive EOGs decreased remarkably in amplitude (Figure 11). Only when the $SO_4^=$ was replaced by Br^-, F^-, HCO_2^-, or Cl^-, did the positive EOGs recover. Next, when K^+ was removed from the Ringer's solution, the positive EOGs increased in amplitude, although only temporarily (Figure 9A). It was concluded, therefore, that the positive EOGs depend upon the entry of Cl^- and the exit of K^+, and that only Br^-, Cl^-, F^- and HCO_2^- can penetrate the membrane

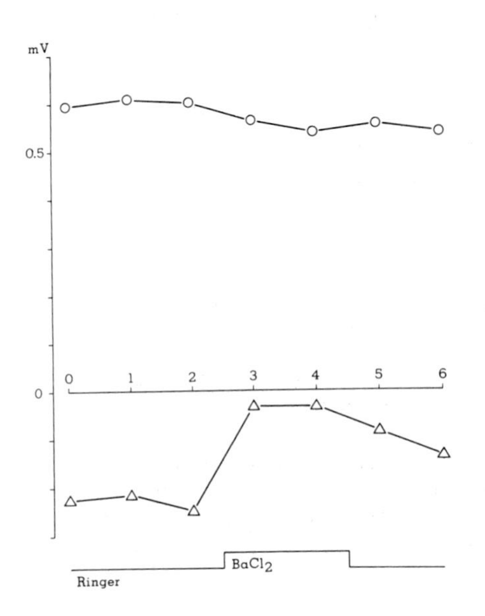

Figure 10 Effect of Ba^{++} on positive and negative EOGs. The positive EOGs (under the base line) decreased in amplitude strikingly when 1 mM Ba^{++} was added to Ringer's solution (3 and 4), while the negative EOGs (above the base line) were hardly affected. Thus, Ba^{++} specifically affects the positive EOGs. The positive EOGs recovered to some extent with repeated immersion in normal Ringer's solution (5 and 6). (Takagi et al.[57])

that elicits the positive EOGs. This demonstrated that the sieve hypothesis concerning the inhibitory postsynaptic membrane[2,8,22] is not applicable to the olfactory receptive membrane on the basis of the sizes of hydrated ions, but may be applicable on the basis of the sizes of naked ions (Table II). The ionic mechanism of the positive EOGs

TABLE II
PERMEABLE AND NONPERMEABLE ANIONS AND THEIR SIZES IN NAKED, SINGLY HYDRATED AND HYDRATED STATES
Hydration energies are added. (Takagi et al.[56])

	Ion	Naked ion size radii (Å)	Singly hydrated ion radii (Å) (Mullins)	Relative hydrated size	Hydration energy K cal/M
	F	1.36	3.59	1.33	94
Permeable	Cl	1.81	4.04	0.96	67
	Br	1.95	4.18	0.94	63
	HCO_2	2.03	4.26	1.35	
	NO_3	1.98	4.21	1.03	55
	I	2.16	4.39	0.96	49
Nonpermeable	ClO_3	2.16	4.42	1.14	61
	BrO_3	2.38	4.61	1.32	68
	BF_4	2.88	5.11	1.12	
	ClO_4	2.92	5.15	1.09	54
	NO_2	1.32	3.55	1.02	64

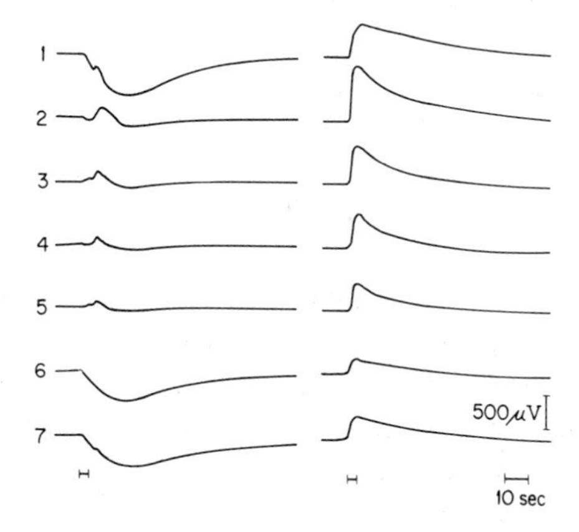

Figure 11 Changes of EOGs in chloride-free solution. The positive EOGs elicited by chloroform vapor are shown at left and the negative-on EOGs elicited by menthone vapor at right. Control potentials in Ringer's solution are shown in 1; 2 to 5, changes of the two potentials in chloride-free solution; 6 and 7, change and recovery of these two potentials in Ringer's solution. Although recovery in the positive EOGs was striking, recovery in the negative-on EOGs was slight. The short horizontal lines at the bottom show the duration of stimulation (4 sec). (Takagi et al.[56])

appearing in the degenerated olfactory epithelium is shown later.[48] Apparently, the ionic mechanisms of positive EOGs are entirely different from those of negative EOGs.

PROBLEM 3. ARE POSITIVE EOGs NONBIOLOGICAL POTENTIALS?

In the olfactory epithelia of several animals, olfactory stimulation has produced electropositive potentials. Ottoson, studying the olfactory epithelium of the frog,[34] found that a brief, small, positive deflection often precedes a slow negative deflection when the negative slow potentials (EOG) are elicited by odorous vapors (Figure 12–1B to D). He attributed this positive potential to the effect of humidity in the stimulating air. Then Takagi et al.[54] found in the same animal a slow positive potential that continued throughout stimulation and was accompanied by a negative potential at the cessation of stimulation (Figure 13B). This was called a positive-on EOG. Similar positive potentials (Figure 12–4) were later found in the frog,[12,14,20] in the newt,[43,44] and in the tortoise (Shibuya, personal communication). MacLeod[28] found a negative potential followed by a positive potential in the olfactory epithelium of the rabbit (Figure 12–2). MacLeod's discovery, i.e., the positive potential, is called a positive after-potential. Moreover, the positive-off EOG was found in the fish by Shibuya[40] (Figure 12–3), and later in the frog by Gesteland.[12] The effects of polarizing currents on the negative and positive potentials were studied in the olfactory epithelium of the frog,[19] and two different underlying processes were discovered which produce both potentials (Figure 5). Moreover, the ionic mechanism of the positive potential was clarified by Takagi.[56] Thus, data have been accumulated which indicate that true biological positive potentials appear in the olfactory epithelium in response to olfactory stimulation, although recently Ottoson and Shepherd[39] have again doubted the existence of the true positive EOG.

Since data on the positive EOGs were still felt to be insufficient, the properties of the positive potentials were further studied in the normal and degenerated olfactory epithelium. The results are discussed in the next chapter, where more findings are provided to indicate that the positive EOGs are true biological potentials in the olfactory epithelium. The experiments further support our view that the positive EOGs may be composites of at least two positive potentials: one elicited in the olfactory cell as an inhibitory potential, the other elicited in the degenerated olfactory epithelium and therefore believed to originate in the supporting cells. Because highly active secretory activity in the degenerated olfactory epithelia was proved in parallel with a slow positive potential, the positive potential may be attributed to the secretory activity of the supporting cell.[32]

PROBLEM 4. ARE NEGATIVE EOGs GENERATOR POTENTIALS?

Studying the properties of the negative EOGs, Ottoson[34] found striking similarities between the olfactory epithelium which elicits the EOGs and the retina, especially that of certain vertebrates, which elicits the electroretinogram (ERG). For instance,

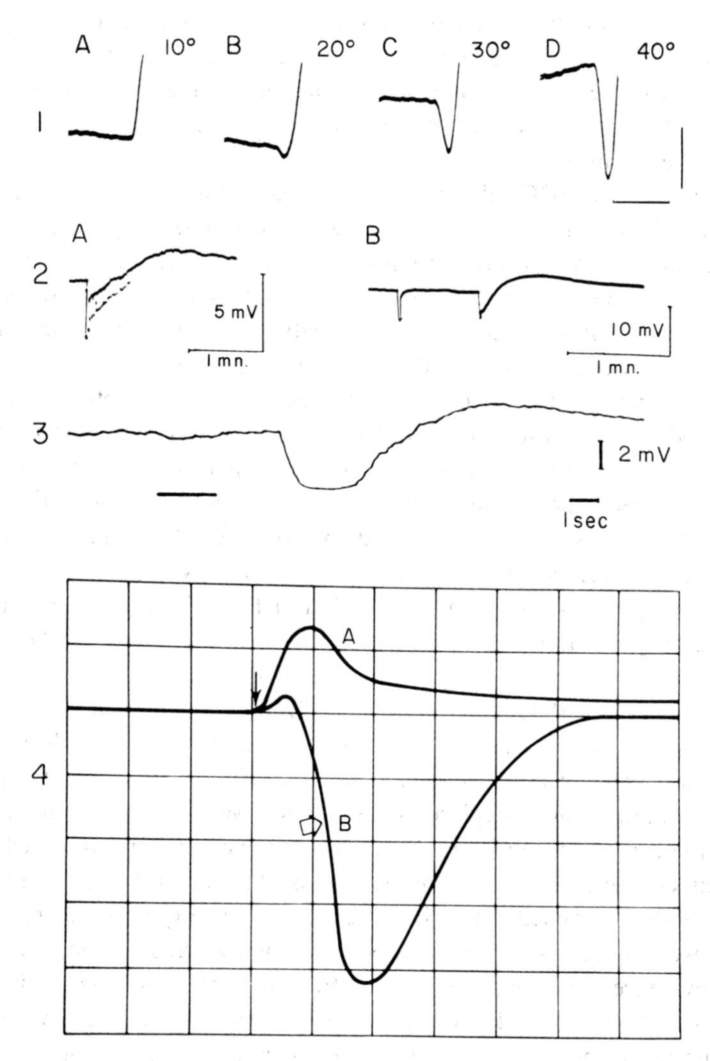

Figure 12 1. The effect of increasing humidity of the stimulating air. Records of initial phase of response to stimulation with butanol. Increase of humidity (A-D) obtained by increase of temperature of aqueous solution of the stimulating substance. Vertical line is 200 μV. Time bar, 2 sec. (Ottoson[34])

2. EOGs in the rabbit. A: Negative EOGs accompanied by positive after-potentials are elicited by repeated application of ethyl acetoacetate. B: Negative EOG by ethyl acetoacetate (left), and negative EOG followed by a positive after-potential by use of benzene (right). (MacLeod[28])

3. EOGs of the positive-off type. At the cessation of odorous stimulation, a positive potential appeared in Channa argus. (Shibuya[40])

4. A: Positive EOG to methanol. B: Negative EOG to n-butanol. The solid arrow indicates the times of odor application. The open arrow indicates an inflection point in the negative rise. (Gesteland et al.[13])

the retina of *Eledone moschata* consists of a single layer of visual cells, from which fine nerve fibers pass to the optic ganglion that is separated from the retina.[16] When the eye of *Eledone* is illuminated, the response is a slow negative monophasic potential almost identical in shape and time course with negative EOGs. Fröhlich[11] assumed that the response in the eye of *Eledone* comes from the visual cells. A similar potential is also obtained from the eye of the water beetle, and Bernhard[4] showed that it originates in the receptors.

In many respects, the EOGs also resemble the electroantennogram (EAG), named by Schneider and studied by his group.[5,6] However, the underlying structure is much simpler in the olfactory organ of the antenna than in the olfactory organ of vertebrates.

By using cocaine, Ottoson established that negative EOGs are not abolished in a concentration sufficiently high to block the activity in the olfactory nerve fibers. The same findings had been obtained in the compound eye of the water beetle,[4] in the isolated muscle spindle,[23] the Pacinian corpuscle,[15] and the stretch organ of the lobster,[9,25] and had been used to prove that the origins of these receptor potentials are located in the sensory terminals. Besides, antidromic stimulation of the nerves showed that the receptor in these tissues cannot be activated by antidromically arriving impulses. Thus, it was established that there are fundamental differences between the electrical response of receptors and that of nerve fibers. Since the EOGs are also unaffected by the electrical stimulation of the olfactory nerve, it is highly probable that negative EOGs are elicited in the sensory terminals of the olfactory receptor cells.

Ottoson[34] found that the amplitudes of the EOGs elicited by butanol are, within a certain limit, proportional to the logarithms of the intensities of odors. This resembles the properties of other generator potentials.[3,4,17] When Ottoson placed a microelectrode in the olfactory epithelium, the EOGs were found to be largest on the surface and to decrease gradually in amplitude as the electrode tip advanced into the deeper region. On the basis of these findings, he concluded that the negative EOG is a generator potential that elicits nerve impulses in the olfactory nerve fibers.[34,38]

However, we feel that the above findings are all indirect, and inconclusive. In fact, some later evidence was obtained that contradicts Ottoson's hypothesis. Takagi et al.[54] simultaneously recorded the EOGs with either the induced waves in the olfactory bulb or the induced discharges in the olfactory nerve, and found that often there is no parallel relationship between the amplitudes of the EOGs and those of the induced potentials (Figure 13). In extreme cases, the positive EOGs appeared together with the corresponding induced waves and, moreover, the off-induced waves in the bulb preceded the corresponding off EOGs (Figure 13). Mozell[30] recorded the EOGs simultaneously with the time-averaged neural discharges from the primary olfactory nerves, and indicated five circumstances under which the behavior of the EOGs does not parallel that of the neural summated discharge (Figure 14). He also suggested a need for caution in accepting the EOGs as the sole criterion of peripheral olfactory activity.

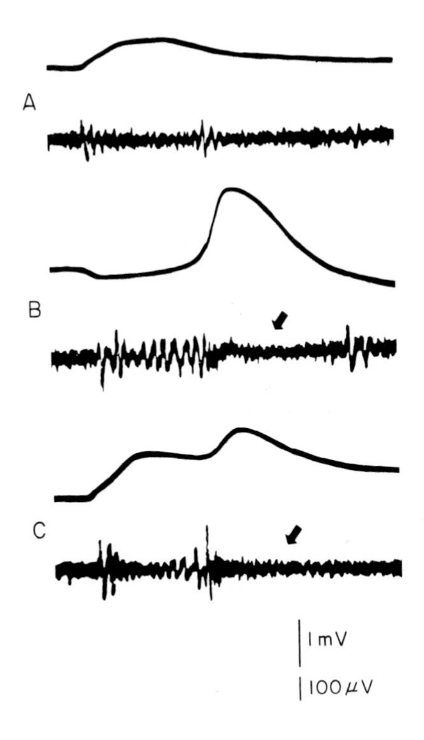

Figure 13 Relation between EOGs and olfactory bulb potentials.

A: An off response appears in the bulb (bottom) but not in the EOG (top).

B: Induced waves (bottom) appear corresponding to the positive-on EOG (top), while they subside (indicated by an arrow) corresponding to the negative-off EOG.

C: Induced waves of the on-off type (bottom) appear corresponding to the EOG of the same type (top). It is worthy of note that the off-induced wave precedes the negative off-EOG. An arrow indicates a decrease in amplitude of the bulb wave. (Takagi et al.[54])

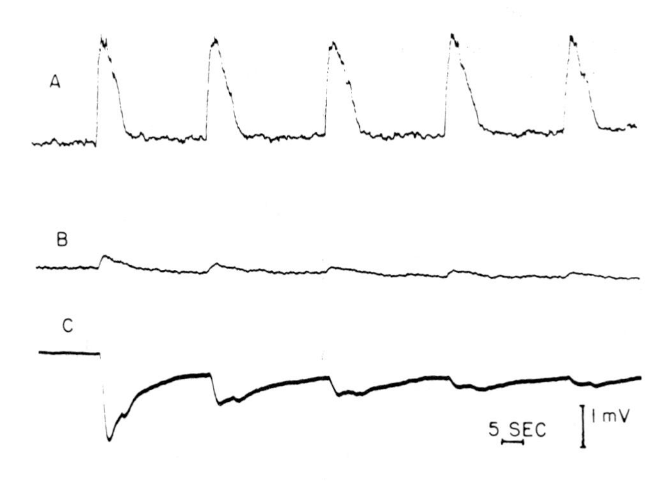

Figure 14 Differential effect of repetitive stimulation on the summated neural discharge and the sustained mucosal potential. Air was passed over concentrated geraniol and 2 cc was puffed onto the apical mucosa at 18.75 cc/min. The stimulus was given every 30 sec. Summated neural discharges (A) appear with nearly the same amplitudes, while the sustained mucosal potentials recorded from the lamina propria of the mucosa (B) and sustained mucosal responses recorded from the epithelium (C) gradually decrease in amplitude. (Mozell[30])

Furthermore, Shibuya[41] and, later, Tucker[60] succeeded in dissociating the negative EOGs from the olfactory nerve discharges. Shibuya isolated a very small olfactory nerve twig and applied an antidromic stimulus to it. By microelectrode, he recorded spike potentials elicited in the olfactory epithelium by antidromic stimulation. In this way, he could locate a small area that was innervated by the small nerve twig (cf. Figure 4 in Takagi[46]). When the olfactory mucus was removed with soft absorbent paper, the negative EOGs decreased in amplitude and disappeared, but the olfactory nerve discharges appeared without significant changes (cf. Figure 5 in Takagi[46]). From these findings he decided that the negative EOGs are not the generator potentials, and by using microelectrodes he demonstrated a potential that is hidden in the olfactory epithelium and that he believes to be a real generator potential.[42]

Ottoson and Shepherd[39] criticized Shibuya's 1964 experiment and explained that the nerve fibers contained in the olfactory nerve twig preparation originate not only in the restricted small region but also in other olfactory areas, and that the recorded olfactory nerve discharges originate from these other areas rather than from the restricted region from which the mucus is removed. Unfortunately, this explanation is not probable, because the olfactory nerve fibers from the neighboring cells in each of the small areas are grouped into small bundles surrounded by Schwann cells, and the fibers from all the olfactory areas intermingle in complicated ways only after they enter the olfactory bulb.[26,31] Tucker, in a personal communication, said that he repeated Shibuya's experiment. He located a small area in the olfactory epithelium that was innervated by an isolated olfactory nerve twig. When he covered the area with a thin plastic film, he found that the nerve-twig discharges were not elicited by the same odorous stimulation. Thus, it is difficult to support the Ottoson–Shepherd view that a small nerve twig contains nerve fibers from a broad olfactory area, and their criticism of the Shibuya experiment is unacceptable. I believe that the EOGs disappeared because some physical conditions around the receptive membrane were altered by removal of the olfactory mucus, causing the EOG recording to be unsuccessful.

On the other hand, several important results which support, or do not oppose, the generator potential hypothesis, were obtained in the experiment of polarizing currents and in the studies of ionic mechanisms, described in Problem 2. They were also obtained in a degeneration experiment, and in a Suzuki experiment, described below.

When the olfactory nerve is sectioned, retrograde degeneration occurs in the olfactory cells. Electrical activity (EOGs) and histological changes were studied in the degenerating olfactory epithelium of the bullfrog. After we sectioned the nerve, it was found that the negative EOGs disappeared in eight days in the summer, in 11 days in the early autumn, and in 16 days in the early winter. A striking decrease in the number of the olfactory cells was also found, but not in the number of the supporting cells. The ratio of the number of the olfactory cells to that of the supporting cells was found to decrease from five or six to below two after the nerve section. At a ratio below two, the EOGs disappeared. From these results we concluded that the negative EOGs originate from olfactory cell activity.[58,59]

However, there is another possibility: if odors depolarize the receptive membrane of the olfactory cell first, and this depolarization, in turn, induces the supporting cell to secrete olfactory mucus, the olfactory cell that has degenerated because of nerve section could no longer induce that secretion.[29,46,61] In the next chapter, we will show that positive, but not negative, EOGs can be elicited in the degenerated epithelium in which supporting cells remain. Since the positive EOGs can be observed in parallel with the secretory activity of the supporting cells, we believe that the possibility above is negated.

Recently, an entirely different approach to this problem was undertaken by Suzuki[45] in the olfactory organ of the snail. Under a dissecting microscope, he removed the receptor-cell layer of the tentacular epithelium with a pair of sharpened forceps. Before and after the removal, the EOGs were recorded from the region. After the EOG recording, the olfactory organs were examined histologically to see if the receptor-cell layer was removed (Figure 15). In the olfactory epithelium in which the olfactory cell layer was removed but the supporting cell layer left intact, the negative EOGs elicited by amyl acetate vapor were reduced to 18.5 ± 6.3 per cent of the original amplitude; the positive EOGs elicited by ethanol were reduced only to 64.5 ± 7.7 per cent of the original (Figure 16). Considering the difficulties in mechani-

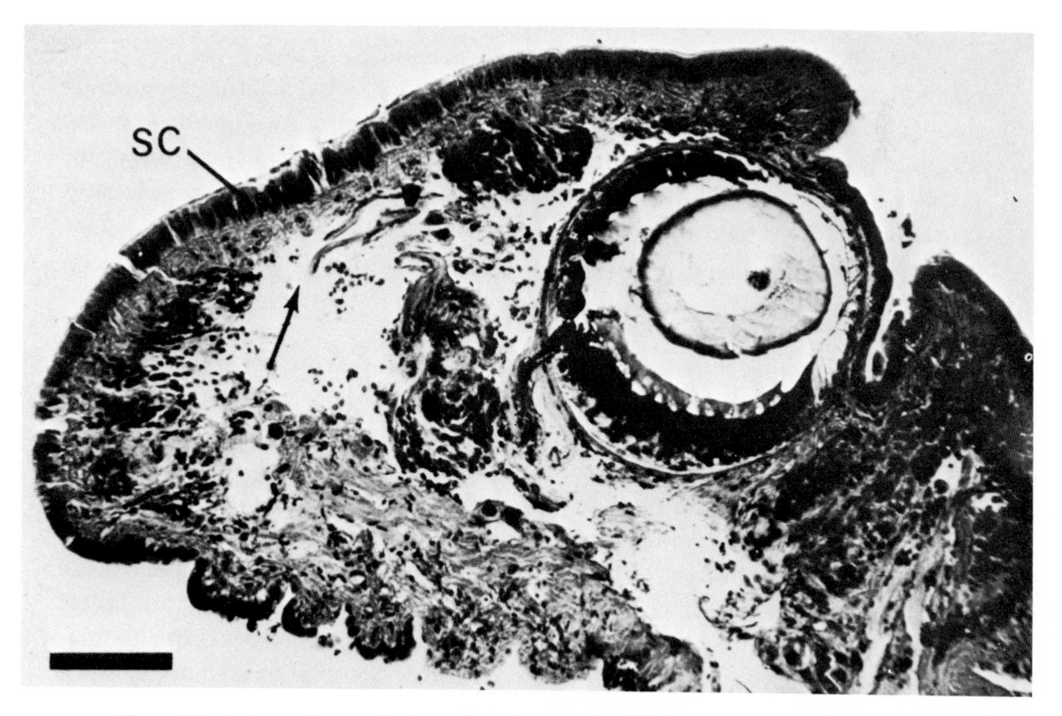

Figure 15 A photomicrograph of a section of tentacular epithelium after the receptor layer was removed. SC: supporting cell layer. Arrow indicates area from which olfactory cell layer was removed. EOGs were recorded on the surface corresponding to this area. Calibration mark indicates 100 μ. (Suzuki[45])

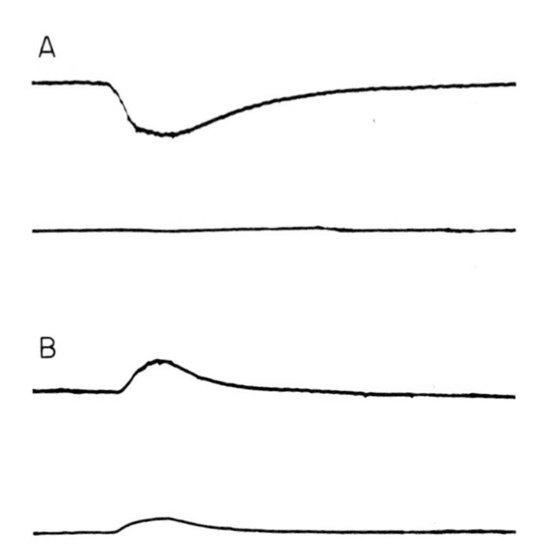

Figure 16 Decrease in amplitude of EOGs. A: 1/1 amyl acetate. B: 1/1 ethanol. The EOGs elicited in the normal olfactory epithelium by these odors are shown at the top of each paired record. The EOGs elicited after the receptor layer was removed are shown at the bottom. The negative EOG strikingly decreased in amplitude in A, while the positive EOG decreased much less in B. (Suzuki[45])

cal manipulation, the results strongly support the above conclusion by Takagi and others that negative EOGs originate in the olfactory cell. That true biologically positive EOGs, as well as negative EOGs are elicited in the normal and degenerated olfactory epithelia by many odorants will be shown in the next chapter. As for the origin of the positive EOGs, Suzuki's data support the supposition that they probably originate in both the olfactory and supporting cells, and this also will be treated in the subsequent paper.

CONCLUSION

After examining these opposing results, I have reached the conclusion that the negative EOGs are, in most cases, composites of the potentials of the opposite polarities: the negative potentials, which are true generator potentials, and the two positive potentials, which are inhibitory and very probably secretory. By assuming competition of the negative and the positive potentials, most of the discrepancies found between the EOGs and the olfactory nerve or bulb potentials can be explained. These two opposing potentials were also found in the EAG, and it was shown that one is a generator and the other an inhibitory potential.[5,6] Hosoya and Yoshida[21] indicated the presence of a standing potential of 1 to 6 mV across the olfactory epithelium. Its origin has not been made clear. Although the author has tried to explain the EOGs as composites of the negative and positive potentials originating in the olfactory and supporting cells, he has to add that this conclusion does not necessarily negate the possibility of other unknown potentials contributing to the EOGs.

REFERENCES

1. Ai, N., and S. F. Takagi. 1963. The effect of ether and chloroform on the olfactory epithelium. *Japan. J. Physiol.* **13**, pp. 454–465.
2. Araki, T., M. Ito, and O. Oscarsson. 1961. Anion permeability of the synaptic and non-synaptic motoneurone membrane. *J. Physiol.* (*London*) **159**, pp. 410–435.
3. Autrum, H. 1950. Die Belichtungspotentiale und das Sehen der Insekten (Untersuchungen an Calliphora und Dixippus). *Z. Vergleich. Physiol.* **32**, pp. 176–227.
4. Bernhard, C. G. 1942. Isolation of retinal and optic ganglion response in the eye of *Dytiscus*. *J. Neurophysiol.* **5**, pp. 32–48.
5. Boeckh, J. 1967. Inhibition and excitation of single insect olfactory receptors and their role as a primary sensory code. *In* Olfaction and Taste II (T. Hayashi, editor). Pergamon Press, Oxford, etc., pp. 721–735.
6. Boeckh, J., K. E. Kaissling, and D. Schneider. 1965. Insect olfactory receptors. *Cold Spring Harbor Symp. Quant. Biol.* **30**, pp. 263–280.
7. Byzov, A. L., and G. I. Flerova. 1964. Electrophysiological research on the olfactory epithelium of the frog. *Biofizika* **9**, pp. 217–225.
8. Coombs, J. S., J. C. Eccles, and P. Fatt. 1955. The specific ionic conductances and the ionic movements across the motoneuronal membrane that produce the inhibitory post-synaptic potential. *J. Physiol.* (*London*) **130**, pp. 326–373.
9. Eyzaguirre, C., and S. W. Kuffler. 1955. Process of excitation in the dendrites and in the soma of single isolated sensory nerve cells of the lobster and crayfish. *J. Gen. Physiol.* **39**, pp. 87–119.
10. Eyzaguirre, C., and S. W. Kuffler. 1955. Further study of soma, dendrite, and axon excitation in single neurons. *J. Gen. Physiol.* **39**, pp. 121–153.
11. Fröhlich, F. W. 1914. Beiträge zur allgemeinen Physiologie der Sinnesorgane. *Z. Sinnesphysiol.* **48**, pp. 28–164.
12. Gesteland, R. C. 1964. Initial events of the electro-olfactogram. *Ann. N.Y. Acad. Sci.* **116**, pp. 440–447.
13. Gesteland, R. C. 1967. Differential impedance changes of the olfactory mucosa with odorous stimulation. *In* Olfaction and Taste II (T. Hayashi, editor). Pergamon Press, Oxford, etc., pp. 821–831.
14. Gesteland, R. C., J. Y. Lettvin, and W. H. Pitts. 1965. Chemical transmission in the nose of the frog. *J. Physiol* (*London*) **181**, pp. 525–559.
15. Gray, J. A. B., and M. Sato. 1953. Properties of the receptor potential in Pacinian corpuscle. *J. Physiol.* (*London*) **122**, pp. 610–636.
16. Grenacher, H. 1886. *Abh. naturforsch. Ges. Halle* **16**, p. 207.
17. Hartline, H. K. 1928. A quantitative and descriptive study of the electrical response to illumination of the arthropod eye. *Amer. J. Physiol.* **83**, pp. 466–483.
18. Hartman, J. 1954. A possible objective method for the rapid estimation of flavors in vegetables. *Proc. Amer. Soc. Hort. Sci.* **64**, pp. 335–342.
19. Higashino, S., and S. F. Takagi. 1964. The effect of electrotonus on the olfactory epithelium. *J. Gen. Physiol.* **48**, pp. 323–335.
20. Higashino, S., S. F. Takagi, and M. Yajima. 1961. The olfactory stimulating effectiveness of homologous series of substances studied in the frog. *Japan. J. Physiol.* **11**, pp. 530–543.
21. Hosoya, Y., and H. Yashida. 1937. Ueber die bioelektrische Erscheinungen an der Riechschleimhaut. *Japan. J. Med. Sci.* III *Biophys.* **5**, pp. 22–23.
22. Ito, M., P. G. Kostyuk, and T. Oshima. 1962. Further study on anion permeability of inhibitory post-synaptic membrane of cat motoneurones. *J. Physiol.* (*London*) **164**, pp. 150–156.
23. Katz, B. 1950. Depolarization of sensory terminals and the initiation of impulses in the muscle spindle. *J. Physiol.* (*London*) **111**, pp. 261–282.
24. Kimura, K. 1961. Olfactory nerve response of frog. *Kumamoto J.* **14**, pp. 37–46.
25. Kuffler, S. W., and C. Eyzaguirre. 1955. Synaptic inhibition in an isolated nerve cell. *J. Gen Physiol.* **39**, pp. 155–184.
26. Le Gros Clark, W. E. 1957. Inquiries into the anatomical basis of olfactory discrimination. *Proc. Roy. Soc. B.* **146**, pp. 299–319.
27. Lettvin, J. Y., and R. C. Gesteland. 1965. Speculations on smell. *Cold Spring Harbor Symp. Quant. Biol.* **30**, pp. 217–225.

28. MacLeod, P. 1959. Première données sur l'électro-olfactogramme du lapin. *J. Physiol. (Paris)* **51**, pp. 85–92.
29. Moulton, D. G., and L. M. Beidler. 1967. Structure and function in the peripheral olfactory system. *Physiol. Rev.* **47**, pp. 1–52.
30. Mozell, M. M. 1962. Olfactory mucosal and neural responses in the frog. *Amer. J. Physiol.* **203**, pp. 353–358.
31. Mozell, M. M. 1967. The effect of concentration upon the spatiotemporal coding of odorants. *In* Olfaction and Taste II (T. Hayashi, editor). Pergamon Press, Oxford, etc., pp. 117–124.
32. Okano, M., and S. F. Takagi. 1969. (In preparation)
33. Ottoson, D. 1954. Sustained potentials evoked by olfactory stimulation. *Acta Physiol. Scand.* **32**, pp. 384–386.
34. Ottoson, D. 1956. Analysis of the electrical activity of the olfactory epithelium. *Acta Physiol. Scand.* **35**, Suppl. 122, pp. 1–83.
35. Ottoson, D. 1958. Studies on the relationship between olfactory stimulating effectiveness and physico-chemical properties of odorous compounds. *Acta Physiol. Scand.* **43**, pp. 167–181.
36. Ottoson, D. 1959. Studies on slow potentials in the rabbit's olfactory bulb and nasal mucosa. *Acta Physiol. Scand.* **47**, pp. 136–148.
37. Ottoson, D. 1959. Comparison of slow potentials evoked in the frog's nasal mucosa and olfactory bulb by natural stimulation. *Acta Physiol. Scand.* **47**, pp. 149–159.
38. Ottoson, D. 1963. Generation and transmission of signals in the olfactory system. *In* Olfaction and Taste I (Y. Zotterman, editor). Pergamon Press, Oxford, etc., pp. 35–44.
39. Ottoson, D., and C. M. Shepherd. 1967. Experiments and concepts in olfactory physiology. *In* Progress in Brain Research, Vol. 23 (Y. Zotterman, editor). Amsterdam, Elsevier Publishing Co., pp. 83–183.
40. Shibuya, T. 1960. The electrical responses of the olfactory epithelium of some fishes. *Japan. J. Physiol.* **10**, pp. 317–326.
41. Shibuya, T. 1964. Dissociation of olfactory neural response and mucosal potential. *Science* **143**, pp. 1338–1340.
42. Shibuya, T. 1969. Single activities of olfactory receptor cells. *In* Olfaction and Taste III (C. Pfaffmann, editor). Rockefeller Univ. Press, New York, pp. 109–116.
43. Shibuya, T., and S. F. Takagi. 1963. A note on the on- and off-responses observed in the olfactory epithelium of a newt. *Gunma J. Med. Sci.* **11**, pp. 63–68.
44. Shibuya, T., and S. F. Takagi. 1963. Electrical response and growth of olfactory cilia of the olfactory epithelium of the newt in water and on land. *J. Gen. Physiol.* **47**, pp. 71–82.
45. Suzuki, N. 1967. Behavioral and electrical responses of the land snail, *Ezohelix flexibilis* (Fulton), to odours. *J. Fac. Sci. Hokkaido Univ. Ser. VI. Zoology* **16**, pp. 174–185.
46. Takagi, S. F. 1967. Are EOG's generator potentials? *In* Olfaction and Taste II (T. Hayashi, editor). Pergamon Press, Oxford, etc., pp. 167–179.
47. Takagi, S. F. 1968. Ionic basis of olfactory receptor activity. *In* Theories of Odors and Odor Measurement (N. Tanyolac, editor). Robert College Research Center, Istanbul, Turkey, pp. 509–521.
48. Takagi, S. F., K. Aoki, M. Iino, and T. Yajima. 1969. The electropositive potential in the normal and degenerating olfactory epithelium *In* Olfaction and Taste III (C. Pfaffmann, editor). Rockefeller Univ. Press, New York, pp. 92–108.
49. Takagi, S. F., H. Kitamura, K. Imai, and H. Takeuchi. 1969. Further studies on the roles of sodium and potassium in the generation of the electro-olfactogram: effects of mono-, di-, and trivalent cations. *J. Gen. Physiol.* **53**, pp. 115–130.
50. Takagi, S. F., and T. Shibuya. 1959. "On"- and "off"-responses of the olfactory epithelium. *Nature* **184**, p. 60.
51. Takagi, S. F., and T. Shibuya. 1960. The "on" and "off" responses observed in the lower olfactory pathway. *Japan. J. Physiol.* **10**, pp. 99–105.
52. Takagi, S. F., and T. Shibuya. 1960. The electrical activity of the olfactory epithelium studied with micro- and macro-electrodes. *Japan. J. Physiol.* **10**, pp. 385–395.
53. Takagi, S. F., and T. Shibuya. 1960. Electrical activity of lower olfactory nervous system of toad. *In* Electrical Activity of Single Cells (Y. Katsuki, editor). Igaku-Shoin, Ltd., Tokyo, pp. 1–10.
54. Takagi, S. F., T. Shibuya, S. Higashino, and T. Arai. 1960. The stimulative and anaesthetic actions of ether on the olfactory epithelium of the frog and the toad. *Japan. J. Physiol.* **10**, pp. 571–584.

55. Takagi, S. F., and G. A. Wyse. 1965. The ionic mechanism of olfactory receptor potentials. Abstract of paper presented at XXIII International Congress of Physiological Sciences, Tokyo, p. 379.
56. Takagi, S. F., G. A. Wyse, and T. Yajima. 1966. Anion permeability of the olfactory receptive membrane. *J. Gen. Physiol.* **50**, pp. 473–489.
57. Takagi, S. F., G. A. Wyse, H. Kitamura, and K. Ito. 1968. The roles of sodium and potassium ions in the generation of the electro-olfactogram. *J. Gen. Physiol.* **51**, pp. 552–578.
58. Takagi, S. F., and T. Yajima. 1964. Electrical responses to odours of degenerating olfactory epithelium. *Nature* **202**, p. 1220.
59. Takagi, S. F., and T. Yajima. 1965. Electrical activity and histological change of the degenerating olfactory epithelium. *J. Gen Physiol.* **48**, pp. 559–569.
60. Tucker, D. 1967. Olfactory cilia are not required for receptor function. *Fed. Proc.* **26**, p. 544.
61. Tucker, D., and T. Shibuya. 1965. A physiologic and pharmacologic study of olfactory receptors. *Cold Spring Harbor Symp. Quant. Biol.* **30**, pp. 207–215.
62. Yoshida, H. 1950. Bioelectrical phenomena in the olfactory epithelium. *Hokkaido J. Med.* **25**, pp. 454–458. (In Japanese)

THE ELECTROPOSITIVE POTENTIAL IN THE NORMAL AND DEGENERATING OLFACTORY EPITHELIUM

SADAYUKI F. TAKAGI, KIYOSHI AOKI, MASAE IINO, *and* TOSHI YAJIMA

Department of Physiology, School of Medicine, Gunma University, Maebashi, Japan

INTRODUCTION

Since Ottoson[16] found a positive component in the electro-olfactograms elicited by an odorous vapor, positive potentials have been discovered in the olfactory epithelium by several investigators.[3,4,6,7,11,20–24,29,30,32] The types of the positive potentials and the animals in which they were found are stated in the preceding paper.[25]

Since data on the positive EOGs are still insufficient, the present experiment was undertaken. Odorants which elicit positive potentials are sought among 122 chemicals. The various properties of the positive EOGs elicited by these odorants are examined in the normal and degenerated olfactory epithelia in decapitated states and in vivo. The ionic mechanisms of these positive EOGs are also studied. In the end the origins of these positive EOGs are considered.

METHODS

Preparation

Bullfrogs (*Rana catesbiana*) were used. They were pithed and beheaded. After the skin and bones were removed, the olfactory epithelium was carefully excised at the ceiling of the olfactory cavity, exposing olfactory eminentia. The potentials elicited by odorous vapors were recorded both in the excised ceiling epithelium and in the eminential epithelium.

When the EOGs were recorded in vivo, the bullfrogs were immobilized with d-tubocurarine (0.25 gm per 100 g, or anesthetized with urethane (3 mg per 100 g). In this case the olfactory eminentia was exposed with minimal bleeding, and recording was performed here during good blood circulation.

Stimulants

Odorous chemicals were selected from Amoore's tables[1]: 13 chemicals from the camphoraceous odor group, eight pungent, 29 ethereal, 24 floral, seven pepperminty, four almond, four putrid, four rancid, three aromatic, three musky, two aniseed, and one each of amber, fruity, lemon, citrus, cedar, and garlic odors. In addition, 15 other miscellaneous odors were used (Table I). Among these 122 odorants, 8 chemicals were donated by the Givaudan Company. Saturated vapors of these chemicals were diluted to 1/6 and 1/36 by mixing with nonodorous air, which was obtained through silica gels and active charcoals.

Solutions

When the ionic mechanism of the positive EOG was studied, chloride-free Ringer's solutions were prepared as follows: 115 mM sodium disulfate, 2.5 mM potassium sulfate, 2.0 mM calcium sulfate, 1.1 mM sodium phosphate, and 0.4 mM sodium monophosphate. The same problem was also studied in solutions in which Cl^- was replaced by acetate or propionate.

Recording Apparatus

EOGs were recorded mostly by a pair of nonpolarizable (Zn-$ZnSO_4$-gelatin-Ringer) electrodes. For comparison, a pair of electrodes of Ag-AgCl-agar-agar type, a pair of silver wires inserted into glass pipettes filled with Ringer's solution, and a pair of calomel half-cell electrodes were used. For the sake of convenience, we will call these electrode types zinc, Ag-AgCl, silver wire, and calomel, respectively. They were amplified with a d-c amplifier and recorded with an inkwriting recorder (Nihon Kohden Co., Tokyo).

Experimental Procedure

Each odorant was stored in a 30-cc syringe. They were applied onto the olfactory epithelium from distance of 3 to 5 cm through Teflon tubes with inside diameters of 1 mm. All the Teflon tubes and syringes had been cleaned several times with soap and detergents, boiled in distilled water with active charcoal, dried in a clean room, and stored in desiccators. Odors were applied in order from low concentration (1/36) to higher concentrations (1/6 and 1/1). Each Teflon tube was used for only one kind of odor.

Procedures for the study of the ionic mechanism of the EOGs were stated in detail in a previous paper.[32] After the control EOGs were recorded several times in Ringer's solution and the stability of the EOGs was ascertained, Ringer's solution was replaced by a Cl^--free (SO_4) Ringer's solution. Changes in amplitude and shape of the positive EOGs were examined in the normal and degenerating epithelia. Then, the Cl^--free Ringer's solution was replaced by a modified Ringer's solution in which Cl^- was replaced by one of F^-, Br^-, HCO_2^-, NO_3^-, I^-, ClO_3^-, BF_4^- and NO_2^-, and the permeability of the positive-EOG-producing membrane to these anions was examined. At first, the positive EOG was depressed in the Cl^--free (SO_4) Ringer's solution. Next, if the positive EOG recovered in amplitude in one of these solutions, the anion in the solution was judged to be permeable; if not, impermeable.

Degeneration of olfactory cell

Several days after the olfactory nerve is sectioned, the olfactory cells in the olfactory epithelium degenerate, but the other elements survive. The negative EOGs then

TABLE I

122 ODORANTS USED IN THE PRESENT EXPERIMENTS

Classes of odorants are divided according to Amoore.[1] Odorants followed by * were donated by the Givaudan Co. Odorants followed by # are added from Döving's classification.[2] + indicates that ethylen dichloride (Kokusan Chemicals Co.) elicits EOGs different from those elicited by 1,2-dichloroethane (Matheson Co.), although the two chemicals have the same structure (cf. Table II). Numbers in the right two columns indicate the amplitudes of the EOGs elicited by the odorants. Amyl acetate and chloroform vapors were used as standards (100%) for calculation of the numbers. One through 5 indicate respectively 10–29% and 30–49%, 50–69%, 70–85%, and 86–100% of the amplitudes of the negative EOGs elicited by amyl acetate or the positive EOGs by chloroform. EOGs greater than 100% were included in 5. These results were obtained in the excised olfactory epithelia of the bullfrog.

Class	Chemicals	Negative EOG			Positive EOG		
		1/36	1/6	1/1	1/36	1/6	1/1
Camphoraceous	n-Amyl alcohol	1	2	2	0	0	0
	tert. Amyl alcohol	1	1	1	0	0	0
	Benzonitrile#	0	0	1	0	0	0
	tert. Butyl alcohol	1	1	1	1	2	5
	iso-Borneol*	0	1	2	0	0	0
	Cineole	3	4	4	0	0	0
	Cyclohexanol	1	1	1	0	0	0
	Cyclo-octane (strong)	0	0	1	0	0	0
	2-Methylcyclohexanol	0	1	2	0	0	0
	3-Methylcyclohexanol	0	0	1	0	0	0
	4-Methylcyclohexanol	0	0	1	0	0	0
	Methyl isobutyl ketone	2	3	3	0	0	0
	1,1,2,2-Tetrabromochloroethane	1	1	1	0	0	0
Pungent	Acrolein	1	1	1	0	0	0
	Acrylic acid	0	0	1	0	0	0
	Allyl alcohol	0	0	1	0	0	0
	Allylisothiocyanate (strong)	0	0	2	0	0	0
	Formic acid	2	0	0	0	2	3
	Phenylisocyanate	1	0	0	0	0	0
	Phenylisothiocyanate	1	1	1	0	0	0
	Propionic acid	0	1	1	0	0	0
Ethereal	Acetone	1	1	3	0	0	0
	Acetonitrile	0	0	0	1	2	3
	Acetylenetetrachloride	2	2	2	0	0	0
	Bromoethane	1	1	1	0	0	0
	Bromoform	0	1	2	0	0	0
	Carbon disulphide	0	0	0	1	1	1
	Carbon tetrachloride#	0	0	1	0	0	0
	Chloroform	0	0	0	2	3	5
	Dichloromethane	0	0	0	1	2	4
	1,2-Dichloroethane (Matheson)	0	0	0	1	2	3
	Dimethylsulphide	0	1	2	0	0	0
	Dioxan	0	0	0	0	0	0
	Ethyl acetate#	2	4	4	0	0	1
	Ethyl ether	3	4	5	0	0	0
	Ethylene dichloride+	3	0	0	0	1	2

TABLE I (*continued*)

Class	Chemicals	Negative EOG 1/36	1/6	1/1	Positive EOG 1/36	1/6	1/1
	Ethyl iodide	1	1	2	0	0	0
	Furan	1	1	2	0	0	0
	n-Hexane#	0	0	1	0	0	0
	Methyl acetate	1	2	2	0	0	0
	Methylchloride	1	2	2	0	0	0
	Methyl formate	0	1	2	0	0	0
	Methyl iodide	0	1	1	0	0	0
	Methyl malonate	0	1	1	0	0	0
	Pentachloroethane	1	2	2	0	0	0
	Pyrrole#	0	0	0	0	0	0
	1,1,2,2-Tetrabromoethane	1	1	1	0	0	0
	Tetrahydrofuran	1	3	4	1	2	4
	Trichloroethylene	0	0	0	1	1	1
	Triethylphosphite	1	3	3	0	0	0
Floral	Acetophenone	1	1	2	0	0	0
	Alphaterpineol*	0	0	1	0	0	0
	Anisole	2	3	4	0	0	0
	Coumarin*	0	0	1	0	0	0
	Dimethyl benzyl cerbinyl acetate*	0	0	0	0	0	0
	Diphenylamine	0	0	0	0	0	0
	Diphenyl ether	0	0	0	0	0	0
	Diphenylmethane	0	0	1	0	0	0
	Geraniol#	0	0	0	0	0	0
	D.L.B.-Phenyl-Ethyl-Methyl-Ethyl-Carbinol	1	1	2	0	0	0
	Indole	0	0	0	0	0	0
	α-Ionone	0	1	4	0	0	0
	β-Ionone	0	0	1	0	0	0
	Linalol*	5	5	5	0	0	0
	Methyl anthranilate	0	0	0	0	0	0
	Methyl benzoate	0	0	1	0	0	0
	Methyl disulfide	0	0	2	0	0	0
	Phenyl ethyl ether	1	2	3	0	0	0
	Phenyl ethyl dimethyl carbinol*	1	1	1	0	0	0
	Phenyl ethyl methylethyl carbinol	0	0	0	0	0	0
	Piperonal	2	2	3	0	0	0
	Salicylaldehyde	1	1	2	0	0	0
	Triethylphosphine	1	2	3	0	0	0
	α-Terpineol*	0	0	2	0	0	0
Pepperminty	2-sec-Butyl cyclohexanone*	2	2	3	0	0	0
	Cyclohexanone	1	1	2	0	0	0
	Cyclopentanone	1	1	1	0	0	0
	Eugenol#	0	0	1	0	0	1
	Menthone	2	3	3	0	0	0
	Pinacoline	2	3	3	0	0	0
	Pulezone	1	1	1	0	0	0

TABLE I (*continued*)

Class	Chemicals	Negative EOG 1/36	1/6	1/1	Positive EOG 1/36	1/6	1/1
Musky	Musk 89*	0	0	0	0	0	0
	Thibetolide	0	0	1	0	0	0
	Tonalid*	0	0	0	0	0	0
Putrid	Indole*	0	0	0	0	0	0
	Methylamine	4	1	0	0	1	3
	α-Naphthylamine	0	0	0	0	0	0
	Trimethylamine	4	4	5	0	0	0
Almond	Cyclohexanone	1	1	2	0	0	0
	Cyclopentanone	1	1	2	0	0	0
	2,4-Dichlorobenzaldehyde	0	0	0	0	0	0
	3,4-Dichlorobenzaldehyde	0	0	0	0	0	0
Aromatic	γ-Butyrolactone	1	2	2	0	0	0
	Phenylhydrazine	1	1	1	0	0	0
	Quinoline	1	1	1	0	0	0
Aniseed	Benzothiazole	1	1	1	0	0	0
	Isoquinoline	0	0	0	0	0	0
	(Quinoline	1	1	1	0	0	0)
Rancid	n-Capric acid	3	3	2	0	0	0
	n-Caproic acid	0	0	1	0	0	0
	Methyl n-nonyl ketone	1	1	2	0	0	0
	n-Valeric acid	1	1	1	0	0	0
Amber	Fixateur 404	1	1	1	0	0	0
Fruity	p-Hydroxy benzyl acetone*	0	0	0	0	0	0
Lemon	Limonene	1	1	2	0	0	0
Citrus	Methyl nonyl acetaldehyde*	0	0	0	0	0	0
Cedar	Cedrol	0	0	0	0	0	0
Garlic	Allylsulfide	1	2	2	0	0	0
Miscellaneous	iso-Amyl alcohol	0	0	1	0	0	0
	iso-Amyl acetate	2	3	5	0	0	0
	Butyric acid	1	1	2	0	0	0
	n-Butyl alcohol	2	2	3	0	0	2
	iso-Butyl alcohol	1	1	1	1	2	3
	sec-Butyl alcohol	2	2	3	2	2	3
	Citronellol	0	0	0	0	0	0
	2,6-Dichlorobenzaldehyde	0	0	0	0	0	0
	Ethyl alcohol	1	1	1	0	0	0
	2-Ethyl aminoethanol	0	0	0	0	0	0
	Ethyl ether	3	4	5	0	0	0
	Methanol	1	1	1	0	0	0
	Mono-chlorobenzene	1	1	2	0	0	0
	Petroleum ether	4	4	4	0	0	0
	Pyridine	2	2	1	0	0	0

disappear, as was shown in previous papers.[33,34] The 16 odorant vapors found to elicit the electropositive EOGs are put on the degenerating epithelium at three different concentrations, and the EOGs produced are compared with those in the normal epithelium.

RESULTS

Odors that elicit negative and positive EOGs

Among the 122 odors, 106 (87 per cent) elicited only negative EOGs. They are classified as Group I odors. With increasing concentrations (from 1/36 to 1/1), they simply increased the amplitudes of the negative EOGs. However, 16 other odors elicited the positive EOGs (Table I and II). They can be classified into Groups II, III, and IV (Table II). Six odorous vapors in Group II elicited only the positive EOGs. With increasing concentrations (from 1/36 to 1/1) they only increased in amplitude (Figure 1A).

Seven odorous chemicals belong to Group III (Table II). Except for eugenol, the negative-on EOGs were always elicited at a low concentration (1/36), and were accompanied by positive after-potentials, except for n- and sec-butyl alcohols (Table II

TABLE II
LIST OF ODORANTS WHICH ELICIT THE POSITIVE EOGs IN THE EXCISED NORMAL AND DEGENERATED OLFACTORY EPITHELIA
They are divided into three groups according to their shapes. These results were obtained from excised olfactory epithelia (cf. Discussion).

		Normal						Degenerated					
		Neg EOG			Pos EOG			Neg EOG			Pos EOG		
Group	Odors	1/36	1/6	1/1	1/36	1/6	1/1	1/36	1/6	1/1	1/36	1/6	1/1
	Acetonitrile (E)	0	0	0	1	2	3	0	0	0	1	2	3
	Carbon disulphide (E)	0	0	0	1	1	1	0	0	0	1	1	1
II	Chloroform (E)	0	0	0	2	3	4	0	0	0	2	3	5
	Dichloromethane (E)	0	0	0	1	2	4	0	0	0	1	2	4
	1,2-Dichloroethane (E)*	0	0	0	1	2	3	0	0	0	1	2	3
	Trichloroethylene (E)	0	0	0	1	1	1	0	0	0	1	1	1
	n-Butyl alcohol	2	2	3	0	0	2	0	0	1	0	0	2
	iso-Butyl alcohol	1	1	1	1	2	3	0	0	0	1	2	3
	sec-Butyl alcohol	2	2	3	0	0	2	0	0	1	0	0	2
III	tert-Butyl alcohol (Cam)	1	1	1	1	2	5	1	1	1	1	2	4
	Ethyl acetate (E)	2	4	4	0	0	1	0	0	0	0	0	1
	Eugenol (M)	0	0	1	0	0	1	0	0	0	0	0	0
	Tetrahydrofuran (E)	1	3	4	1	2	4	1	1	1	1	2	4
	Formic acid (Pun)	2	0	0	0	2	3	0	0	0	0	2	3
IV	Methylamine (Put)	4	1	0	0	1	3	0	0	0	0	1	3
	Ethylene chloride (E)*	3	0	0	0	1	2	0	0	0	0	1	4
	Ethyl ether (Takagi et al.[29])												

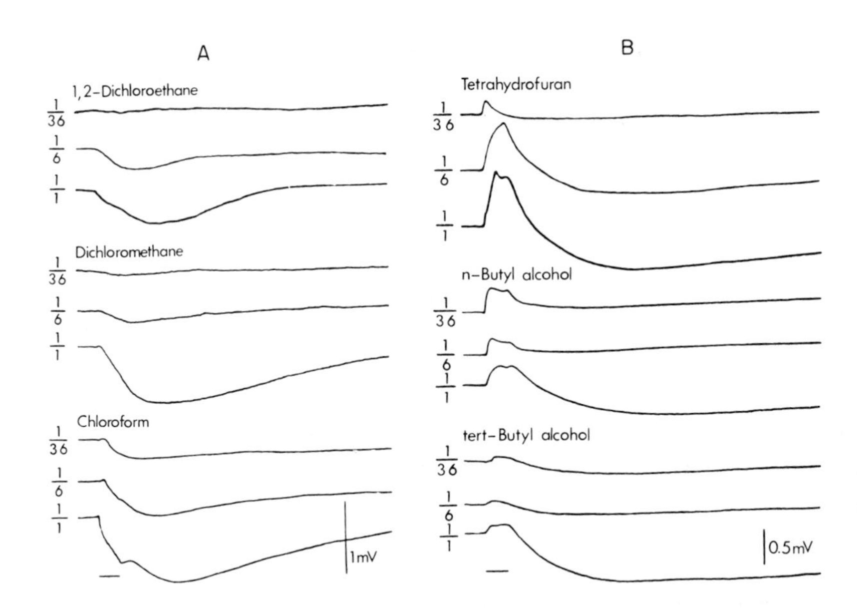

Figure 1 A: EOGs elicited by Group II odorants. Positive EOGs increase in amplitude with increasing odor concentrations. B: EOGs elicited by Group III odorants. The initial negative potentials do not always increase in amplitude with increasing odor concentrations, while the positive after-potentials do increase.

and Figure 1B). With increasing concentrations, the positive after-potentials appeared in all cases and increased in amplitude, while the negative EOGs increased, decreased, or did not change. EOGs of this type have been found in the rabbit by the application of benzene and ethyl acetoacetate.[11] Three chemicals belong to Group IV. Formic acid, methyl amine, and ethylene chloride (Kokusan Chemicals Co., Tokyo) elicited only negative EOGs at a low concentration (1/36), and only positive EOGs at high concentrations (1/6 and/or 1/1) (Table II and Figure 2A). It is noteworthy that the negative EOGs completely disappeared when the positive EOGs appeared. 1, 2-dichloroethane (from the Matheson Coleman & Dell Co.), although equivalent to ethylene chloride, elicited only positive EOGs and consequently was included in the Group II (indicated by asterisks in Table II).

In a previous paper[27] it was shown that negative-on EOGs were produced by ethyl ether at low concentrations, but that with increasing concentrations they decreased in amplitude; in exchange, the negative-off EOGs appeared and increased in amplitude, and eventually the negative-on EOGs disappeared or often reversed their polarity. In other words, the positive-on, negative-off EOGs frequently appeared. From these data, it is clear that ethyl ether should be placed in the Group IV. However, in the present experiment, ethyl ether produced only the negative on-off EOGs at the saturated concentration. It is presumed that the positive-on negative-off EOGs occurred in the previous experiment because the saturated ether vapour was applied at higher velocities than in the present experiment.

Ionic mechanisms of the positive EOGs. The positive EOGs elicited by the odors of the Group II decreased remarkably in amplitude, as did the positive EOGs elicited by chloroform vapor, when Cl^- in the bathing Ringer's solution was replaced by SO_4^{--}. Thus, the principal role of Cl^- was proven. The once-decreased positive EOGs recovered considerably in the modified Ringer's solutions in which Cl^- was replaced by Br^- (Figure 3) or F^-, but not in the Cl^--free solutions, in which Cl^- was replaced by one of the other anions shown in the Method section. Similar relations were found in the positive EOGs elicited by the odors of Groups III and IV (Figure 3). Thus, it is clear that the ionic mechanism of the positive EOGs elicited by the odors of the above three groups is practically the same as that of the positive EOGs elicited by chloroform vapor (cf. reference 32 and Table III, this paper).

Positive EOGs in the eminential and ceiling epithelia. Considerable differences in shape of the positive EOGs recorded in the eminential and ceiling epithelia were sometimes found (Figure 4). The positive components in the EOGs elicited by chloroform and other vapors in the eminential epithelium are often not so striking as those found in the ceiling epithelium. When the positive component was small in amplitude, the negative component was large, and it seems that there is

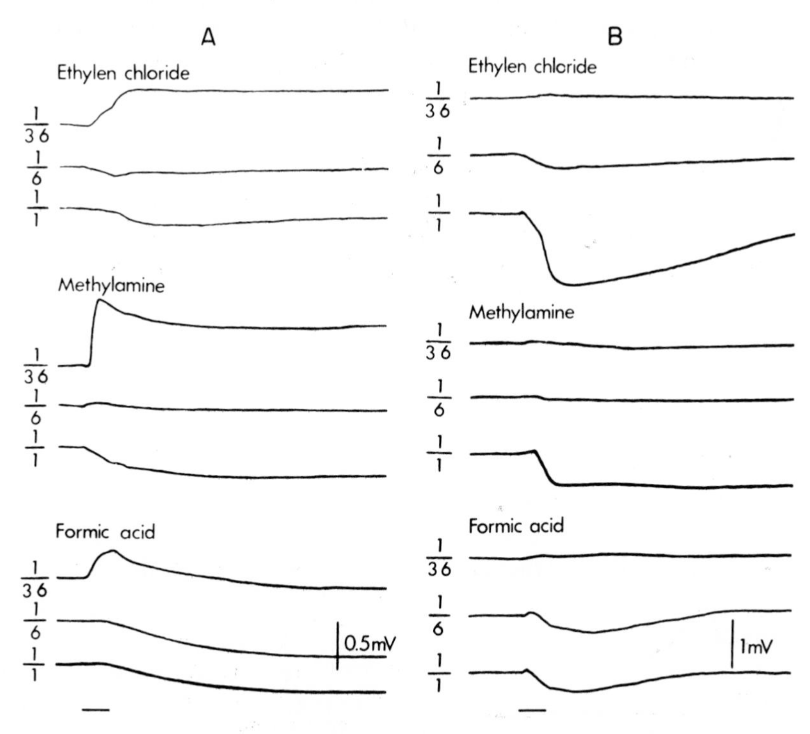

Figure 2 EOGs elicited by Group IV odorants. A: Only the negative EOGs are elicited by ethylene chloride, methylamine, and formic acid at low concentrations (1/36), while only the positive EOGs are elicited at high concentrations. B: In the degenerated olfactory epithelia, the same odorants elicit the positive EOGs, but not the negative EOGs.

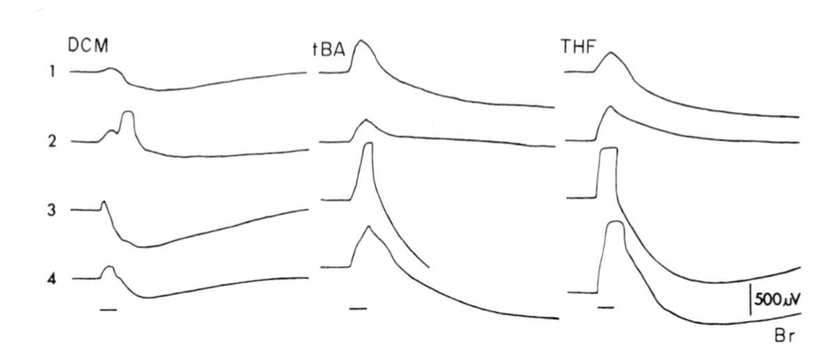

Figure 3 Replacement of Cl^- by Br^-. The positive EOGs elicited by dichloromethane (DCM), tertiary butyl alcohol (tBA) and tetrahydrofuran (THF), shown in 1, decrease in amplitude when Cl^- is replaced by SO_4^{--} in the Ringer's solution (2). The positive EOGs recover only when SO_4^{--} is replaced by Br^-, F^- or HCO_2^-. Recoveries in the Br^- solution (3 and 4) are remarkable.

competition between the processes that generate the two EOGs.[31] In some cases, the positive components were minor in the beginning of the experiments, but after several stimulations they increased in amplitude and became very similar in shape to the EOGs elicited in the ceiling epithelium by the same odors. Thus, the typical chloroform EOGs usually found in the ceiling epithelium were also obtained in the eminential epithelium in many cases. It can be said, therefore, that there is not a fundamental difference between the EOGs recorded in these two epithelia.

EOGs recorded with the electrodes of different types. When the EOGs elicited by chloroform vapor were recorded by means of the zinc electrode, the positive components clearly appeared. However, the amplitudes of the negative and positive EOGs themselves were not large unless saturated vapors of chloroform and

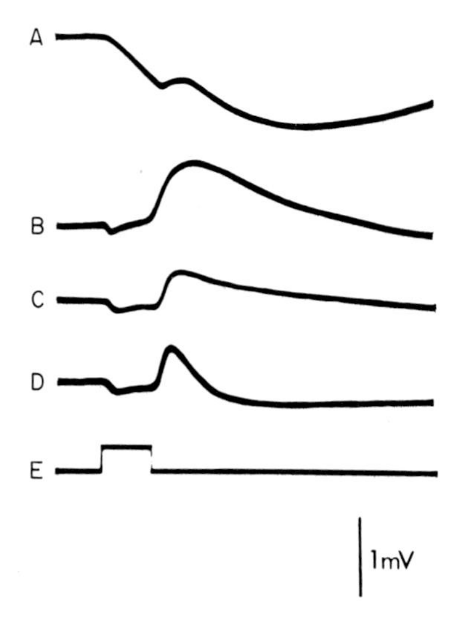

Figure 4 Positive EOGs of various types.

A: Typical EOG elicited by chloroform vapor. This shape is always observed in the ceiling olfactory epithelium of the nasal cavity.

B: Small on-potential is followed by a large negative off-potential. This is often followed by a small positive after-potential.

C and D: Intermediate shapes between A and B. The EOGs in B, C, and D are most often observed in the olfactory eminentia.

Time of stimulation is indicated at bottom right.

other odorants were used. When the vapors were diluted to 1/36 by purified air, chloroform and other odorants elicited very small EOGs. When electrodes of the other types were used, the EOGs could be recorded by the same odorants of much lower concentrations. This may be attributed to the smaller diameters of these electrodes. The shape of the EOGs in these cases more often was a small negative- or positive- on EOG followed by a large negative off- and a small positive- on after-potential (Figure 4B, C, D), although the typical chloroform EOGs (Figure 4A) were also recorded.

The most important finding is that these positive EOGs were recorded by means of the calomel half-cell electrodes, because Ottoson and Shepherd[19] claimed that no positive EOGs could be so recorded. Except for the amplitudes of the EOGs, it can be reiterated that there is not a fundamental difference among EOGs recorded with electrodes of different types.

Positive EOGs in vivo. When the EOGs elicited by chloroform and other vapors in Groups II to IV were recorded in the olfactory eminentia of bullfrogs immobilized with d-tubocurarine, the negative on- and off-potentials appeared strikingly, and the positive after-potentials, if they occurred, were relatively small in amplitude (cf. Figure 4B). Furthermore, methanol and ethanol elicited the positive EOGs, although the negative EOGs were recorded from decapitated animals.[26] Thus EOGs appeared very differently in vivo. It was our impression that the positive EOGs appeared with smaller amplitudes and the negative EOGs were predominant when circulation of blood with d-tubocurarine was good, but vice versa when the circulation was not good. This may be an important finding for future EOG investigations.

Positive EOGs in degenerating epithelium. In previous papers,[33,34] we have shown that EOGs elicited by amyl acetate, ethyl ether, and chloroform vapors disappeared in the olfactory epithelium several days after the olfactory nerve was sectioned. In those experiments, the EOGs were recorded in the olfactory eminentia by means of silver-wire electrodes. Consequently, the positive EOGs were not so obvious and were overlooked (see Figure 4B and C). In the present experiments, recording was performed on the ceiling epithelium by means of Zn-$ZnSO_4$-Ringer-gelatin electrodes. Consequently, the remarkable positive EOGs could always be recorded in the normal epithelium. It is worthy of note that the negative-off EOGs always disappeared, but the positive EOGs survived after the degeneration of the olfactory cells (Figure 2B). The odors of Groups II, III, and IV were applied to the degenerating or degenerated epithelia. In response to the odors of Group II, the positive EOGs appeared but the negative EOGs did not. In response to the odors of the Group III, the positive EOGs appeared, but the negative ones did not, or appeared only with much decreased amplitudes. The behaviors of the EOGs elicited by the odors of the Group IV are more worthy of note (Figure 2B). The negative EOGs usually elicited in the normal epithelium by these odors at low concentration (1/36) did not appear, but the positive EOGs elicited at higher concentrations (1/6 and 1/1) continued to appear in the degenerating epithelium. Thus, it is clear that the positive EOGs are elicited in the degenerating olfactory epithelia. The origin of the rest of the small negative EOGs which are rarely found in the degenerating epithelia, and then only when the odors of the Group III are applied, may be the remaining olfactory cells.

Ionic mechanism of the positive EOGs in the degenerated olfactory epithelia. Using the same method as before, we examined the ionic mechanisms of the positive EOGs in the degenerating epithelia. When Cl^- was replaced by SO_4^{--} in the Ringer's solution, the positive EOGs decreased in amplitude. The once-decreased EOGs recovered to various degrees in Cl^--free Ringer's solutions in which Cl^- was replaced by Br^- or F^-. The decreased EOGs never recovered in the solutions in which Cl^- was replaced by NO_3^-, I^-. ClO_3^-, BrO_3^-, BF_4^- or ClO_4^-. The results are summarized in Table III.

DISCUSSION

The positive EOGs

Ottoson[16] (1956) found that a small positive potential often preceded the negative EOG when the volume of the stimulating air exceeded 0.5 cc or when the tip of the pipette was kept close to the surface of the epithelium. Because the amplitude of the positive EOG is related not to the intensity of the stimulus, but to the humidity of the

TABLE III

PERMEABLE AND NONPERMEABLE ANIONS THROUGH THE POSITIVE EOG-GENERATING MEMBRANE IN THE DEGENERATED OLFACTORY EPITHELIUM

Anions in the left column are arranged from top to bottom according to naked ion size, beginning with the smaller sizes. In the second column from the left the number and results of the experiments are shown. Plus (+) and minus (−) indicate whether an anion can recover the once-decreased positive EOGs; in other words, whether the anion is permeable. Results are given in the third column. Fraction denominators indicate the number of experiments performed; the numerator indicates the number of cases in which the positive EOGs recovered. Compare with Table II in the preceding paper, this volume.

Ion	Exp. No. 1	2	3	4	5	Total	
F	+	+	+	+	+	7/8	
Br	+	+	+	+	+	7/7	Permeable
HCO_2	−	±	±	±		3/4	
NO_3	−	−	−			0/3	
I	−	±	−	−		1/4	
ClO_3	−	−	−			0/3	Nonpermeable
BrO_3	−	−	−			0/3	
BF_4	−	−	−			0/3	
ClO_4	−	−	−			0/3	

stimulating air, he concluded that the initial positive potential is caused by positively charged water particles. Consequently, he concluded that the initial positive potential is a nonbiological potential. Since then, positive potentials have been found by several workers. Ottoson and Shepherd[19] assumed that all the positive potentials found by these workers are elicited by the complex potential field in the olfactory mucosa during excitation, particularly when stimulation is unevenly applied on the mucosa. As an example they cited a negative EOG in the cat's olfactory mucosa that sometimes was preceded by a positive potential that distributes over a complex system of turbinals.[12] It is likely that a positive potential appears artifactually in such a manner, but it is also true that real positive EOGs exist, as will be stated below.

In our experiments during the past four years, mainly nonpolarizable zinc electrodes were used. Electrodes of this type were essential when the effects of polarizing currents on the EOGs were studied.[6] Moreover, this type has been recommended by Kimura[10] and Mozell[14] because nonbiological artifactual responses were either nonexistent or so small that they could not be measured. However, Ottoson and Shepherd[19] indicated the probable effect upon the EOGs of polarizing currents that might arise in the electrode-agar or gelatin interfaces of these electrodes. The effect of polarizing currents had been clearly demonstrated by Higashino and Takagi.[6] Consequently, Ottoson and Shepherd recommended the calomel half-cell electrode, and claimed that the initial positive EOGs could not be recorded with it. Thus, they doubted the real existence of the positive EOGs.

In the present experiment, we demonstrated that the positive EOGs can be recorded by calomel half-cell electrodes even when the negative EOGs disappeared in the degenerating or degenerated olfactory epithelia.

The positive-on potentials were always accompanied by the negative off-potentials when ethyl ether, chloroform, and some other odorants were applied at high concentrations, but they were not observed with these stimulants at low concentrations.[27] It is interesting that in the newt, transferred to land, the positive-on EOGs began to appear, accompanied by the negative-off potentials, when the negative-on EOGs began to appear.[22,23] These findings indicated that this positive potential is different from the initial positive potential found by Ottoson.[16] Further, it was found that this potential appears only in the living epithelium and in response to a small number of odorants but not to the others (Table I). The effects of polarizing currents on the EOGs and the ionic mechanisms of the EOGs disclosed the difference of the positive potential from the artifactual potential and from the negative EOGs, and indicated that the negative and positive EOGs are generated by two different underlying processes.[6,31,32] Gesteland et al.[4] reached the same conclusion, based on different experiments. Consequently, it is now clear that true biological positive EOGs in the olfactory epithelium occur in response to some kinds of odors.

Concentrations of odors

In most of our previous experiments, saturated vapors of amyl acetate, menthone, ethyl ether, and chloroform were used. The saturated vapors are apparently very strong—perhaps too strong in some cases—as olfactory stimulants. Consequently, the results derived therefrom have been doubted by other investigators.[13] It is true that single neurons recorded in the olfactory epithelium and bulb by a microelectrode change their response types or become unresponsive to subsequent stimuli after the application of these strong odors[8] (also Aoki and Takagi, unpublished). Similar phenomena have been reported in the turtle (Tucker and Shibuya, personal communication; Schneider, personal communication). However, it is also a fact that these saturated vapors can elicit negative and positive EOGs repeatedly in the excised preparations for more than two or three hours without significant change in magnitude.[4,31,32] Since these strong odors are found to be destructive to the impulse-generating mechanism, it is assumed that the EOG-generating mechanism is more resistant than the impulse-generating mechanism to irritation from external agents.

Here, it should be made clear why we have used saturated vapors of odorants in most cases. With the zinc electrodes, the EOGs elicited by odorant vapors of low concentrations (1/36 or lower) appeared very small. On the other hand, the study of the ionic mechanisms required large EOGs. As stated above, large EOGs appeared consistently for more than two or three hours in response to the saturated vapors. Moreover, it was far more convenient to apply saturated vapors than those of lower concentrations.

When the electrodes of the other types were used, the EOGs could be recorded at the same sites with much larger amplitudes than with the zinc electrodes. Thus, it is clear that, with the other electrodes, the EOGs can be studied by applying odors of much lower concentrations. It is believed, however, that the EOGs elicited by these very strong odors and recorded with the zinc electrodes are not essentially different from those elicited by much weaker odors and recorded with the other electrodes.

Four kinds of odors

In the present experiment, it was found that most (87 per cent) of the odorous chemicals elicit only the negative EOGs at the three concentrations, with a simple increase in amplitude with increasing concentration. We have called these odors the "general odors group." The six odors (Group II) that produce only the positive potentials may be called the "chloroform odor group." They all belong to the ethereal group, according to Amoore.[1] The seven chemicals in Group III (Table II) produce the negative-positive type of potential. It is interesting that the four isomers of butyl alcohols in this group have the same molecular weight and elicit similar potentials, but all have different odor qualities.

As mentioned before, ethyl ether produces a negative-on EOG at low concentrations (1/16), but smaller negative-on or even positive-on EOGs at higher concentrations (1/4 and 1/1).[27] The latter potentials were recorded by silver-wire electrodes when saturated ethyl ether vapor was applied at higher velocities than in the present experiment. The same phenomena were found with the zinc electrode in cases of ethylene chloride, formic acid, and methyl amine (Table II). It should be added, however, that once the saturated vapors of formic acid and methyl amine were applied, the EOGs usually were not maintained. It is clear that these vapors have destructive effects upon the EOG-generating mechanism.

Shapes of the positive EOG

The positive EOG discovered in the olfactory eminentia by Takagi et al.[27] had a shape of the positive-on potential followed by the negative-off potential (Figure 4C). The positive-on potential continued to appear to the end of the stimulation. EOGs of this type were often observed in the olfactory eminentia when chloroform vapor was applied and when recorded with silver-wire electrode. In these conditions the positive components of the EOGs are very often smaller in amplitude, while the negative ones are bigger (Figure 4B, C). The shapes of these EOGs are in contrast with the big positive-on, small negative-off, and big positive-after potentials elicited in the ceiling epithelia by the same vapors (Figure 4A). In the ceiling epithelium it seems that the positive components surpass the negative ones in the EOGs; in the eminential epithelium the opposite situation obtains. However, the latter type of EOG (Figure 4D) also appeared often in the olfactory eminentia.

EOGs recorded in vivo

Some of the EOGs from the present experiments are very different in shape from those elicited by the same odors in Gesteland's experiment on frogs immobilized with d-tubocurarine.[3] Above all, the effects of methanol and ethanol are conspicuous in that in his work only the positive EOGs were elicited, while in ours only the negative EOGs occurred (Table I). Since most of our experiments were performed on decapitated frogs, we suspect that the difference may originate in the lack of either blood circulation or d-tubocurarine.

In order to examine this possibility, odors in Group II, III, IV, and some others were applied in vivo. It was found that chloroform, dichloroethane, and dichloromethane elicited negative- or positive-on, negative-off and positive-after potentials, that the other odors elicited only the negative EOGs, although they were small in amplitude. Moreover, methanol and ethanol elicited small positive EOGs.[26] Thus, the results of our experiment on the immobilized frog were very similar to those of Gesteland. It is remarkable that the EOGs change their shapes and potentials according to the experimental conditions.

Hosoya and Yoshida,[9] and Yoshida[36] studied the EOGs in the excised olfactory epithelium of the dog; and Ottoson[16] and Takagi with his collaborators[6,7,22–34] studied them in decapitated frogs; Ottoson,[17,18] Kimura,[10] Mozell,[14] Gesteland et al.,[3,4] and Tucker and Shibuya[35] studied the EOGs in frogs or turtles immobilized with d-tubocurarine or other anesthetics. However, no one has yet indicated that such a large difference in the EOGs resulted from the difference in the condition of the animal. In other receptor organs, including the retina, receptor potentials and ERGs have been studied primarily in excised states (in vitro), and probable differences in the potentials recorded in vitro and in vivo have not been considered. In the olfactory epithelium of decapitated frogs, the positive potentials appeared predominantly in response to the odors in Groups II to IV, while the negative potentials appeared for nearly all odors in Table I in vivo. Therefore, it is assumed that generative processes for the negative potentials are more dependent upon blood circulation and that the positive potentials which are obscure in immobilized frogs become apparent for some odors in the decapitated state. From these results one may suppose that the olfactory epithelium devoid of blood circulation is far from normal. The authors, however, do not support such a supposition, because in the decapitated state spike discharges can be recorded well by microelectrode and the negative EOGs, especially, can be elicited consistently for more than two or three hours by most odors shown in Table I.[6,7,16,23–24] Although it is important to study in vivo the responses of single olfactory cells stimulated with various odors, olfactory epithelia in the decapitated frog can be used for the study of the negative EOGs and may be more important and useful for the analysis of the positive EOGs, which are generally hidden in the immobilized frogs.[32]

Origin of the positive EOG

In the degenerating or degenerated olfactory epithelia, it was shown that obvious positive, but not negative, EOGs appeared when chloroform and other vapors were applied. It was presumed, therefore, that the positive EOGs originate in elements other than the receptor cell. In these degenerated epithelia there remain the supporting cells, basal cells, and the Bowman's glands. Since the basal cells are situated at the bottom of the olfactory epithelium, they probably do not take part in generating the positive potential, so that either or both of the other two elements may be the origin of the positive EOGs.

Electron- and light-microscopy disclosed that the supporting cell secretes when chloroform vapor is applied to either normal or degenerated olfactory epithelia.[15] The present research showed that the ionic mechanisms are not different between the positive EOGs in the normal and degenerated epithelia. Consequently, it is likely that the supporting cell produces a positive potential. On the other hand, it is probable that some positive potential is also elicited by the olfactory receptor cells, because inhibitory phenomena have been found via microelectrode in the spike discharges[27,28] (also Aoki and Takagi, unpublished). Thus, it may be concluded that the positive EOGs recorded in the normal olfactory epithelium are composites of the positive potentials elicited in both the olfactory and the supporting cells.

Full details of the EOGs of the opposite polarities elicited by methanol and ethanol in vivo and in the decapitated state will be published elsewhere.[26]

SUMMARY

One hundred twenty two odorous chemicals were applied to olfactory epithelia excised from the ceilings of the olfactory cavities of bullfrogs. Three different concentrations (1/36, 1/6, 1/1) were used and the results were classified into the following groups.

Group I. 106 chemicals (87 per cent) elicited only negative EOGs, which increased in amplitude with increasing concentrations (Table I).

Group II. Six chemicals, including chloroform (Table II), elicited only positive EOGs, which increased in amplitude with increasing concentrations.

Group III. Seven chemicals produced negative accompanied by the positive EOGs (Table II).

Group IV. Three chemicals produced only negative EOGs at the 1/36 concentration, and only positive EOGs at the high concentration of 1/1.

In the degenerated olfactory epithelia, the olfactory nerves of which had been sectioned previously, positive EOGs appeared in response to the chemicals of Groups II, III, and IV; negative EOGs were not elicited by any of the four groups.

The ionic mechanisms of the positive EOGs elicited by the odors of Groups II, III, and IV were examined in both normal and degenerated olfactory epithelia. It was found that these EOGs have practically the same ionic mechanisms as the positive EOGs elicited by chloroform.[32]

For comparison, the positive EOGs were recorded in the eminential epithelium in normal and degenerated states by various types of recording electrodes, including the calomel half-cell electrodes. In spite of the difference in shape, no fundamental differences appeared among the positive EOGs recorded in the eminential and ceiling epithelia by the various types.

A remarkable difference in shape and polarity was found between the EOGs recorded in decapitated frogs and those in frogs immobilized with d-tubocurarine (in vivo). In vivo, most odors in Groups II, III, and IV produced the negative EOGs sometimes accompanied by the positive afterpotentials, while the same odors produced clear positive EOGs in decapitated frogs.

The origin of the positive EOGs was discussed. Very probably, they are composites of the positive potentials elicited at least in the olfactory and supporting cells.

ACKNOWLEDGMENTS

This work was supported by a grant for scientific research from the Ministry of Education of Japan and by an Air Force Office of Scientific Research Grant through DA-CRD-AFE-S92-544-67-G67 of the United States Army Research and Development Group (Far East), Department of the Army.

REFERENCES

1. Amoore, J. E. 1962. The stereochemical theory of olfaction. 1. Identification of the seven primary odours. *Proc. Sci. Sect. Toilet Goods Assoc.* Suppl. to No. 37, pp. 1–12.
2. Döving, K. B. 1966. Analysis of odour similarities from electrophysiological data. *Acta Physiol. Scand.* **68**, pp. 404–418.
3. Gesteland, R. C. 1964. Initial events of the electro-olfactogram. *Ann. N.Y. Acad. Sci.* **116**, pp. 440–447.
4. Gesteland, R. C., J. Y. Lettvin, and W. H. Pitts. 1965. Chemical transmission in the nose of the frog. *J. Physiol.* (*London*) **181**, pp. 525–559.
5. Gesteland, R. C., J. Y. Lettvin, W. H. Pitts, and A. Rojas. 1963. Odor specificities of the frog's olfactory receptors. *In* Olfaction and Taste I (Y. Zotterman, editor). Pergamon Press, Oxford, etc., pp. 19–34.
6. Higashino, S., and S. F. Takagi. 1964. The effect of electrotonus on the olfactory epithelium. *J. Gen. Physiol.* **48**, pp. 323–335.
7. Higashino, S., S. F. Takagi, and M. Yajima. 1961. The olfactory stimulating effectiveness of homologous series of substances studied in the frog. *Japan. J. Physiol.* **11**, pp. 530–543.

8. Higashino, S., H. Takeuchi, and J. E. Amoore. 1969. Mechanism of olfactory discrimination in the olfactory bulb of the bullfrog. *In* Olfaction and Taste III (C. Pfaffmann, editor). Rockefeller Univ. Press, New York, pp. 192–211.
9. Hosoya, Y., and H. Yoshida. 1937. Ueber die bioelektrische Erscheinungen an der Riechschleimhaut. *Japan. J. Med. Sci. III Biophys.* **5**, pp. 22–23.
10. Kimura, K. 1961. Olfactory nerve response of the frog. *Kumamoto J.* **14**, pp. 37–46.
11. MacLeod, P. 1959. Première données sur l'électro-olfactogramme du lapin. *J. Physiol.* (*Paris*) **51**, pp. 85–92.
12. MacLeod, P. 1965. Variations de l'électro-olfactogramme et du potentiellent glomérulaire en fonction du stimulus olfactif. *Rev. Laryngol.-Otol.-Rhinol.* 86^{e}année, Suppl., pp. 855–859.
13. Moulton, D. G., and L. M. Beidler. 1967. Structure and function in the peripheral olfactory system. *Physiol. Rev.* **47**, pp. 1–52.
14. Mozell, M. M. 1962. Olfactory mucosal and neural responses in the frog. *Amer. J. Physiol.* **203**, pp. 353–358.
15. Okano, M., and S. F. Takagi. 1969. (In preparation)
16. Ottoson, D. 1956. Analysis of the electrical activity of the olfactory epithelium. *Acta Physiol. Scand.* **35**, Suppl. 122, pp. 1–83.
17. Ottoson, D. 1959. Studies on slow potentials in the rabbit's olfactory bulb and nasal mucosa. *Acta Physiol. Scand.* **47**, pp. 136–148.
18. Ottoson, D. 1959. Comparison of slow potentials evoked in the frog's nasal mucosa and olfactory bulb by natural stimulation. *Acta Physiol. Scand.* **47**, pp. 149–159.
19. Ottoson, D., and G. M. Shepherd. 1967. Experiments and concepts in olfactory physiology. *In* Progress in Brain Research, Vol. 23 (Y. Zotterman, editor). Elsevier Publishing Co., Amsterdam, pp. 83–138.
20. Shibuya, T. 1960. The electrical responses of the olfactory epithelium of some fishes. *Japan. J. Physiol.* **10**, 317–326.
21. Shibuya, T. 1963. (Unpublished data)
22. Shibuya, T., and S. F. Takagi. 1963. A note on the on- and off-responses observed in the olfactory epithelium of a newt. *Gunma J. Med. Sci.* **11**, pp. 63–68.
23. Shibuya, T., and S. F. Takagi. 1963. Electrical response and growth of olfactory cilia of the olfactory epithelium of the newt in water and on land. *J. Gen. Physiol.* **47**, pp. 71–82.
24. Takagi, S. F. 1968. Ionic basis of olfactory receptor activity. *In* Theories of Odors and Odor Measurement (N. Tanyolac, editor). Robert College Research Center, Istanbul, Turkey, pp. 509–521.
25. Takagi, S. F. 1969. EOG problems. *In* Olfaction and Taste III (C. Pffaffmann, editor). Rockefeller Univ. Press, New York, pp. 71–91.
26. Takagi, S. F., and M. Iino. 1969. (In preparation)
27. Takagi, S. F., and K. Omura. 1960. Micro-electrode study on the electrical activity of the olfactory epithelium. *J. Physiol. Soc. Japan.* **22**, p. 768.
28. Takagi, S. F., and K. Omura. 1963. Responses of the olfactory receptor cells to odours. *Proc. Japan Acad.* **39**, pp. 253–255.
29. Takagi, S. F., T. Shibuya, S. Higashino, and T. Arai. 1960. The stimulative and anaesthetic actions of ether on the olfactory epithelium of the frog and the toad. *Japan. J. Physiol.* **10**, pp. 571–584.
30. Takagi, S. F., and G. A. Wyse. 1965. The ionic mechanism of olfactory receptor potentials. Abstract of paper presented at XXIII International Congress of Physiological Sciences, Tokyo, p. 379.
31. Takagi, S. F., G. A. Wyse, H. Kitamura, and K. Ito. 1968. The roles of sodium and potassium ions in the generation of the electro-olfactogram. *J. Gen. Physiol.* **51**, pp. 552–578.
32. Takagi, S. F., G. A. Wyse, and T. Yajima. 1966. Anion permeability of the olfactory receptive membrane. *J. Gen. Physiol.* **50**, pp. 473–489.
33. Takagi, S. F., and T. Yajima. 1964. Electrical responses to odours of degenerating olfactory epithelium. *Nature* **202**, p. 1220.
34. Takagi, S. F., and T. Yajima. 1965. Electrical activity and histological change of the degenerating olfactory epithelium. *J. Gen. Physiol.* **48**, pp. 559–569.
35. Tucker, D., and T. Shibuya. 1965. A physiologic and pharmacologic study of olfactory receptors. *Cold Spring Harbor Symp. Quant. Biol.* **30**, pp. 207–215.
36. Yoshida, H. 1950. Bioelectrical phenomena in the olfactory epithelium. *Hokkaido J. Med.* **25**, pp. 454–458. (In Japanese)

ACTIVITIES OF SINGLE OLFACTORY RECEPTOR CELLS

TATSUAKI SHIBUYA
Zoological Institute, Faculty of Science, Tokyo Kyoiku University, Tokyo, Japan

INTRODUCTION

It is well known that single activities of olfactory receptor cells in response to odors have been studied by several investigators for several recent years with microelectrodes. Gesteland et al.[1,2] recorded spike discharges from the layer of the olfactory nerve fibers in frog with their metal microelectrode and reported on odor specificity of the olfactory receptors. Takagi and Omura[9] obtained spike discharges of the olfactory receptor cell in decapitated frog with ordinary glass microelectrodes. They mentioned seven types of responses in olfactory cells from their results. Shibuya and Shibuya,[7] in gopher tortoise (Reptilia), recorded simultaneously the unitary responses and the negative slow potential (EOG) in response to odors in the olfactory mucosa with glass microelectrodes filled with potassium ferricyanide. Shibuya and Tucker[8] were able to record unitary spike discharges of the olfactory mucosa in response to several odors in two kinds of vulture, a warm-blooded animal. The unitary spikes appeared in response to various concentrations of odors and the spike heights were 1–2 mV. The unitary spikes indicated various patterns according to change in odor concentration.

Ottoson has studied and analyzed the negative slow potential recorded from the surface of the olfactory mucosa in frog.[4] However, the origin and role of the EOG are not yet clear. Then studies on the EOG were continued by Takagi and Shibuya,[10] and they found the "off" negative slow potentials. Electropositive slow potentials with long duration, caused by chloroform vapor, were recently reported by Takagi et al.[11,12] They have observed effects of various ions on negative and positive EOGs in the isolated roof olfactory mucosa of frog and have discussed the possible ionic mechanisms for them.

On the other hand, Shibuya's paper experiment has shown that the negative EOG could be dissociated from the olfactory nerve-twig discharges.[6] This raised some question whether the negative EOG is a true generator potential which elicits afferent impulses in the olfactory nerve. But the origins of the olfactory mucosal potentials are still unknown. If the EOG may not be a generator potential, a true generator potential should be recorded separately from the EOG.

This work was undertaken to discover if it is possible to record a true generator potential in the extremely small olfactory cells with fine microelectrodes, and then to examine the problem of impulse generation in the olfactory cell in addition to the relation between the EOG and activities of the single olfactory cell. Part of the results shown in this paper were obtained in Dr. L. M. Beidler's laboratory at Florida State University.

MATERIALS AND METHODS

The animals used were the gopher tortoise, *Gopherus polyphemus*, in Florida, and the stone turtle, *Geoclemys reevesii*, in West Japan. In these studies the animal was anesthetized with ethyl urethane (2.5 g/kg intraperitoneally) and its head was fixed in a holder. The skin, bone, and cartilage over the olfactory cavity were cut to make a small window into the cavity. To record the olfactory nerve discharges, a small twig (20–40 μ in diameter) of the olfactory nerve was exposed inside the cranial cavity by the methods described by Tucker[14,15] and in our previous papers.[6,17]

Using a micromanipulator a microelectrode was inserted into the olfactory mucosa on the septal wall through the window to the cavity. In addition, a glass capillary bridge to a calomel electrode was placed close to the microelectrode (within 1 mm distance) on the same wall to record EOG. Microelectrodes were filled with 10–15 per cent potassium ferricyanide or sometimes with 1 M KCl. The tip diameter was about 0.5 μ and resistance was 80–150 megohms.

To record spike discharges of the olfactory cell, the signal was led through a simple cathode follower to a d-c setting of Grass P6 amplifier. Also, the EOG was led to another P6 amplifier, and the olfactory twig discharges to a Grass P5 amplifier. Then these signals were recorded with a kymograph camera (Grass) or a long recording camera (Nihon Koden).

Tucker's type of olfactometer was used for the stimulation of odors.[14] The odor concentrations were varied by 10^{-3} to 10^{0} per saturated odors. The odor vapor was blown through the nostrils, and cleaned air was flushed through after every stimulation. The flow rate was about 25–50 cc/sec in all experiments. Odors used were high-grade amyl acetate, amyl alcohol, butyric acid, etc.

RESULTS

Spike discharges of single olfactory receptor cells

The olfactory mucosa of the gopher tortoise was about 1.0 mm in whole thickness, slightly thicker than that of the stone turtle. Histological structure of the olfactory mucosa of both animals was approximately the same, except that that of the stone turtle has more mucus.

When the microelectrode was penetrated very slowly into the olfactory mucosa from the surface, spike discharges in response to odors from the layer of the olfactory cells were recorded at about 200–250 μ. Spike discharges were always initially positive relative to the EOG (Figures 1, 2, and 5) and were monophasic or diphasic in shape. In stable recordings, the height of the spike was 1–2 mV and the duration was 3–4 msec; also it was held for about 60–90 min. The spike heights decreased in succession. This had already been reported on the decrement of the spike in the unitary response of the olfactory mucosa of the gopher tortoise and the vulture.[7,8] The decrement was particularly noted when the odor concentration was high. Also, spike discharges were easily damaged by repeated short-interval stimuli at high concentration. It is suggested that the decrement may be derived from strong depolarization in the olfactory cell.

Figure 1 shows simultaneous recordings of EOG, spike discharges of a single olfactory cell, and olfactory nerve-twig discharges in response to various concentrations of amyl acetate. Usually, these three responses occurred in parallel, although each single olfactory cell had a threshold value at some odor concentration. In many experiments with amyl acetate the thresholds were 10^{-3} to $10^{-1.5}$ of saturation. Spontaneous spike discharges were sometimes observed that were depressed or did not respond to an odor stimulus even if the EOG produced normally. As shown in Figures 1 and 3, "off" responses appeared with stimulus at $10^{-0.5}$ and 10^{0}, but mechanisms of the "off" response in the olfactory cells are still not clear, and this subject needs further investigation.

D-c-positive shift with spike discharges

Ottoson has reported[4] that the EOG produced by olfactory stimulation decreased gradually without reversal as the tip of the microelectrode was lowered from the sur-

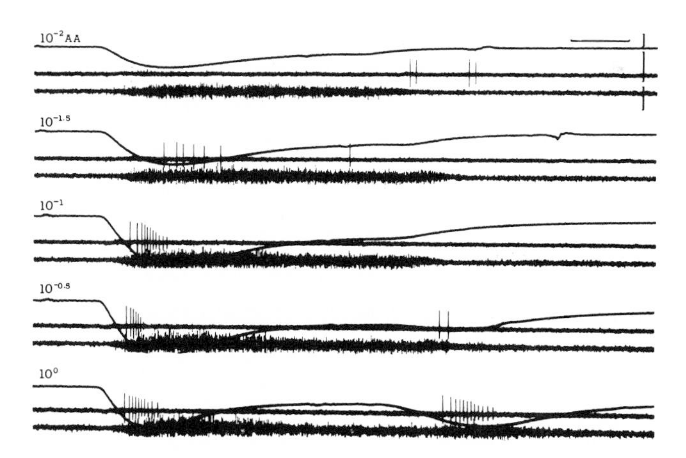

Figure 1 Simultaneous recording of EOG, spike discharges, and olfactory nerve-twig discharges in response to various concentrations of amyl acetate. Usually, the three kinds of responses were in parallel. The duration of the stimulation was for 5 sec. Time: 1 sec. Calibration: 2 mV; 1 mV; 100 μV from above.

Figure 2 EOG and activity of an olfactory cell recorded simultaneously through d-c amplifiers. Odor: 10^{-2} of amyl acetate. Time: 1 sec. Calibration: 2 mV; 1 mV from above.

face to the basal membrane. The same result was observed in the olfactory mucosa of both the gopher tortoise and the stone turtle.

When the microelectrode was carefully advanced from the surface of mucus by steps of 5–10 μ, resting potentials of 10–20 mV were first obtained when it passed through the outer limiting membrane of the olfactory mucosa. Then potential changes of about 10–25 mV in magnitude were observed. Presumably the tip of the microelectrode passed through the membranes of numerous cells that composed the olfactory mucosa. It was difficult to determine a true resting potential of the olfactory receptor cell.

However, a d-c–positive shift together with spike discharges in response to odor was recorded close to the layer where a number of the bodies of the olfactory receptor cell were packed. These signals were observed to be stable for a long time even though the EOG was not recorded in the deep layer. But only spike trains, not these responses, were obtained in the deeper regions below the inner limiting membrane.

The simultaneous recordings of the EOG and the response of an olfactory cell at 10^{-2} of amyl acetate are shown in Figure 2. The EOG at the surface was about 2 mV, the d-c–positive shift about 270 μV, and the spikes about 2 mV, but the d-c–positive shift could not always be dissociated from spike discharges. The d-c–positive shift and spikes in response to amyl acetate in the same olfactory cell are shown in Figure 3. When four concentrations of amyl acetate ($10^{-1.5}$, 10^{-1}, $10^{-0.5}$, and 10^{0}) were applied to the olfactory cell, the heights and the rates of rise of the d-c–positive shifts increased with the increment of concentration (Figures 3 and 4). The responses of the olfactory cell to butyric acid scarcely appeared, although this stimulus elicits discharges in the receptor cells of the vomeronasal organ.

Relation between the EOG and the d-c–positive shift with spike discharges

Usually, the EOG appeared in parallel with the single spike discharges. Simultaneous recordings of EOG and activities of an olfactory cell to amyl acetate with a series of concentrations are shown in Figure 5. The vertical line on the left indicates the onset of the EOG presumably elicited by contact of odor molecules. The olfactory cell had a threshold of between 10^{-3} and $10^{-2.5}$ amyl acetate. In spite of a constant stimulus of 3 sec in all records, the d-c–positive shifts lengthened in duration in proportion to concentration as the height increased. However, the d-c–positive shifts appeared with some delay from the start of the EOG, as follows: 400 msec at 10^{-2}, 360 msec at $10^{-1.5}$, 240 msec at 10^{-1}, 120 msec at $10^{-0.5}$, and 80 msec at 10^{0}, respectively (Figure 6).

In the preparation shown in Figure 7, the magnitudes of the EOG and the d-c–positive shift increased gradually, but only the magnitude of the EOG at 10^{0} (saturated vapor of amyl acetate) curved down, in spite of augmentation of the d-c–positive shift. These results suggest that the origin of the EOG may be different from that of the d-c–positive shift.

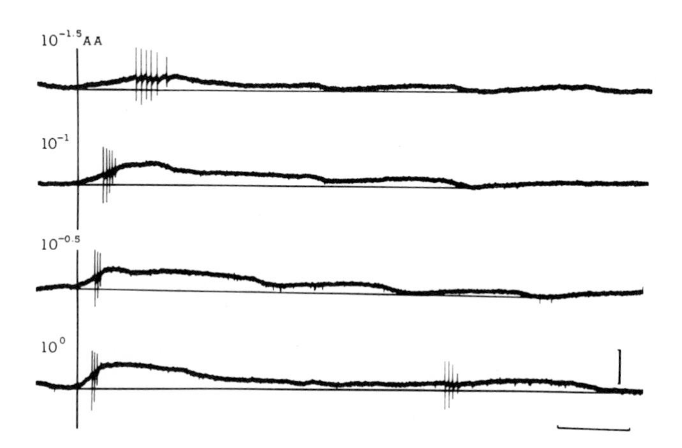

Figure 3 The d-c recordings of the activity of an olfactory cell to amyl acetate. The vertical line on the left indicates beginning of the d-c shifts. Time: 1 sec. Calibration: 1 mV.

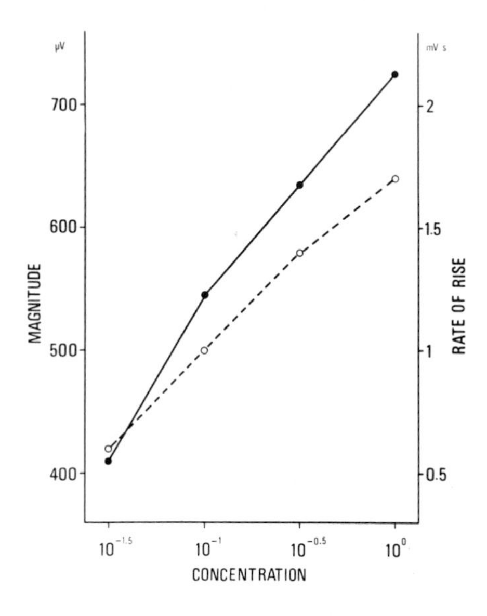

Figure 4 Magnitudes and rates of rise of the d-c positive shift in response to amyl acetate.

CONCLUSION

Gesteland et al.[2] and Takagi and Omura[9] recorded spike discharges from the layer of the olfactory nerve fibers in frog, but they have not tried d-c recordings. Shibuya and Shibuya[7] tried d-c recordings of unitary spikes from the olfactory mucosa in tortoise, but have not discussed the phenomena in detail.

Single-spike discharges in response to odor from the layer of the olfactory cell body gave stable recordings and the spikes always appeared to combine a d-c–positive shift. However, it is still not established that these are activities of intracellular recordings in the olfactory receptor cell, because the resting potentials of the olfactory cell were not measured exactly. In addition, spikes were only 2 mV, although the cells are extremely small, and the d-c shift, which seems to have initiated with impulses at the axon of the olfactory cell, was up to 700 μV in height (Figure 3).

In view of the relation between signals, it seems possible that the d-c–positive shifts and spikes may arise at the olfactory receptor cell. In particular, the d-c shift has appeared with some delay (400 msec at 10^{-2} amyl acetate) from the start of EOG and the delay was shortened at high concentration. The height of EOG usually decreases with the depth,[4] but it does not reverse with mirror image at any layer. It is suggested that the d-c–positive shifts may not be a reversal potential of EOG. Moreover, the d-c shift could be obtained independently in depths at which EOG could not be observed. It is presumed that the d-c–positive shifts with spikes may be activities of the single olfactory receptor cell that were recorded extracellularly or semi-intracellularly with microelectrodes.

Recently, Gesteland et al.[1] reported on the pre-positive potential for part of EOG, and Takagi et al.[11,12] obtained the positive potential with long duration aroused by chloroform vapor in the isolated olfactory mucosa of frog. It is inferred that the d-c–

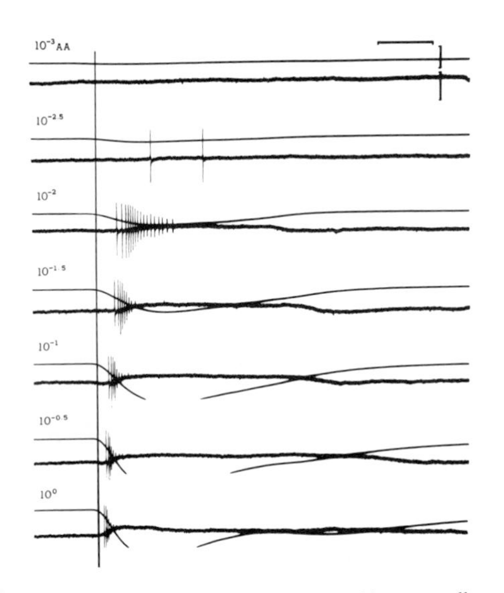

Figure 5 EOGs and activities of an olfactory cell to amyl acetate recorded simultaneously. The vertical line on the left indicates beginnings of EOG (see text). Time: 1 sec. Calibration: 2 mV, 1 mV from above.

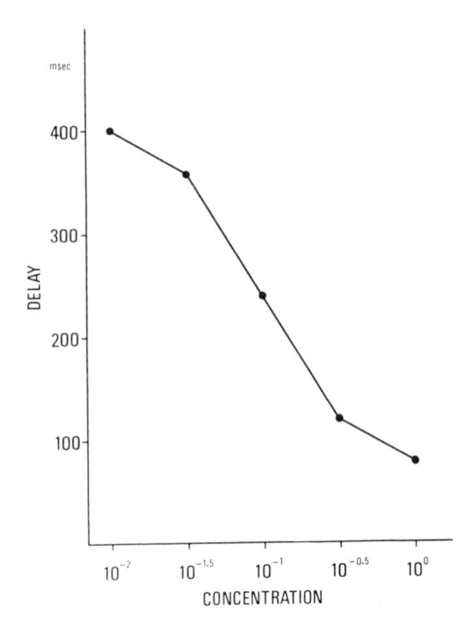

Figure 6 Time delay between EOG and d-c–positive shift to amyl acetate.

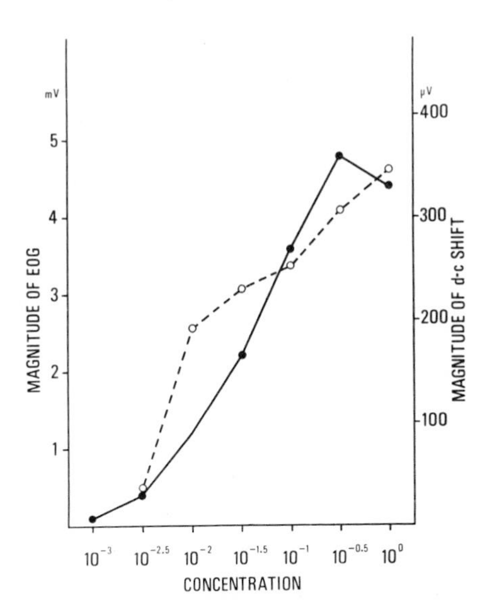

Figure 7 Comparison of magnitudes of EOG and d-c–positive shift to amyl acetate.

positive shifts from the olfactory cell may be entirely different from the positive potentials recorded at the surface. Takagi et al.[11] used saturated vapors of odors for all stimuli; the single activities of the olfactory cells were soon damaged at high concentration of odors and the responses were difficult to recover.

Shibuya[6] dissociated EOG and olfactory nerve discharges by his simple paper experiment. Tucker and Shibuya[17] suggested that there may be electrical coupling between the receptors and the supporting cells, and that the supporting cells may be stimulated directly by odorants. Reese[5] and Okano[3] hypothesized the release of secretion droplets from the supporting cells in frog by electronmicroscopy. The magnitude of EOG produced at 10^{0} of amyl acetate was smaller than that at $10^{-0.5}$, which was contrary to the d-c–positive shift. Production of the EOG may be suppressed by some function, for example, secretion of the supporting or other cells. The magnitude of the d-c–positive shifts was only about one-tenth of that of the EOG. Accordingly, it is strongly suggested that the larger EOG may be produced from nonreceptor cells, although a generator potential may be embedded in it at some time. Thus, one possibility is that the recorded d-c–positive shift is separate from EOG and may be a true generator potential which initiates afferent impulses in the olfactory nerve. We will try to get complete intracellular recordings of the olfactory receptor cell in our further experiments to clarify impulse-generation mechanisms in the olfactory receptor cells.

Recently, Tucker[16] succeeded in removing olfactory cilia from the surface of the olfactory mucosa. However, he observed that the olfactory nerve-twig discharges soon recovered to normal magnitude. Also, Takagi and Shibuya[13] obtained negative slow potentials from the surface of the vomeronasal mucosa in Japanese snakes. They observed with electronmicroscopy that the vomeronasal mucosa has no cilia on the receptor cells. These results seem to be very important for generation mechanisms of EOG.

ACKNOWLEDGMENTS

The author wishes to thank Dr. L. M. Beidler and Dr. D. Tucker, Department of Biological Science, Florida State University, for their encouragement and support of this work. Part of this work was supported by U.S. National Institutes of Health Research Grant NB 07725-01 and U.S. Public Health Service Grant TI NB 5258.

REFERENCES

1. Gesteland, R. C., J. Y. Lettvin, and W. H. Pitts. 1965. Chemical transmission in the nose of the frog. *J. Physiol.* (*London*) **181**, pp. 525–559.
2. Gesteland, R. C., J. Y. Lettvin, W. H. Pitts, and A. Rojas. 1963. Odor specificities of the frog's olfactory receptors. *In* Olfaction and Taste I (Y. Zotterman, editor). Pergamon Press, Oxford, etc., pp. 19–34.
3. Okano, M. 1967. (Personal communication)
4. Ottoson, D. 1956. Analysis of the electrical activity of the olfactory epithelium. *Acta Physiol. Scand.* **35** (suppl. 122), pp. 1–83.
5. Reese, T. S. 1965. Olfactory cilia in the frog. *J. Cell Biol.* **25**, pp. 209–230.
6. Shibuya, T. 1964. Dissociation of olfactory neural response and mucosal potential. *Science* **143**, pp. 1338–1339.
7. Shibuya, T., and S. Shibuya. 1963. Olfactory epithelium: Unitary responses in the tortoise. *Science* **140**, pp. 495–496.
8. Shibuya, T., and D. Tucker. 1967. Single unit responses of olfactory receptors in vultures. *In* Olfaction and Taste II (T. Hayashi, editor). Pergamon Press, Oxford, etc., pp. 219–233.
9. Takagi, S. F., and K. Omura, 1963. Responses of the olfactory receptor cells to odours. *Proc. Jap. Acad.* **39**, pp. 253–255.
10. Takagi, S. F., and T. Shibuya. 1960. The "on" and "off" responses observed in the lower olfactory pathway. *Jap. J. Physiol.* **10**, pp. 99–105.
11. Takagi, S. F., G. A. Wyse, H. Kitamura, and K. Ito. 1968. The role of sodium and potassium ions in the generation of the electro-olfactogram. *J. Gen. Physiol.* **51**, pp. 552–578.
12. Takagi, S. F., G. A. Wyse, and T. Yajima. 1966. Anion permeability of the olfactory receptive membrane. *J. Gen. Physiol.* **50**, pp. 473–489.
13. Takagi, T., and T. Shibuya. 1967. Fine structure and olfactory response in the Jacobson's organ in Japanese snake. *Jap. J. Exp. Morph.* **21**, p. 483. (In Japanese)
14. Tucker, D. 1963. Physical variables in the olfactory stimulation process. *J. Gen. Physiol.* **46**, pp. 453–489.
15. Tucker, D. 1963. Olfactory, vomeronasal and trigeminal receptor responses to odorants. *In* Olfaction and Taste I (Y. Zotterman, editor). Pergamon Press, Oxford, etc., pp. 45–69.
16. Tucker, D. 1967. Olfactory cilia are not required for receptor function. *Fed. Proc.* **26**, p. 544.
17. Tucker, D., and T. Shibuya. 1965. A physiologic and pharmacologic study of olfactory receptors. *Cold Spring Harbor Symp. Quant. Biol.* **30**, pp. 207–215.

THE INTERACTION OF THE PERIPHERAL OLFACTORY SYSTEM WITH NONODOROUS STIMULI

THOMAS V. GETCHELL
Biological Sciences, Northwestern University, Evanston, Illinois

INTRODUCTION

The prime function of chemoreceptors is to encode and convey to higher centers of the central nervous system information concerning changes in the chemical composition of the internal and external environment. As a result of this sensory imput the organism alters its behavorial patterns or adjusts internal physiological or biochemical mechanisms to meet the changing environment. Psychophysical perception or awareness of the impinging stimulant need not be an integral part of the response. In terrestrial vertebrates the olfactory and gustatory receptor organs have become anatomically orientated to detect chemical changes in a gaseous or fluid environment. Thus the interaction of the stimulant molecule and the receptor membrane may be more limited by the physical parameters of the stimulant than by the receptor membrane.

In terrestrial vertebrates the olfactory receptor cells are sensitive to many volatile chemicals. We call substances odors physiologically if, when they are presented to the nose at or above some threshold level, they cause change in activity of the receptor cells and cells higher in the olfactory pathway. Psychophysically we call substances odors if we perceive them as an odor, or if we can demonstrate behavorial responses to the substance when they act on the olfactory receptors. As the interaction of the stimulant molecule and the chemoreceptive membrane cannot be monitored on the molecular level, knowledge of the extramental odor world is necessarily based on introspective psychophysical judgments or behavorial responses of experimental animals. Thus from the human point of view a substance has an odor if one can "smell" it. The frog stuck with an electrode "smells" an odor if a particular olfactory receptor cell's firing frequency increases or decreases, whereas a beagle's yelp upon sniffing the air informs me that the rabbit is near. Clearly, then, for an effector substance three classes of response exist, each having its own threshold which cannot be adequately equated with either of the other two.

The experiment described in this paper demonstrates that, in the frog, common chemical substances which are nonodorous to humans evoke slow voltage changes across the olfactory receptor-cell layer indistinguishable from those voltages produced by more conventional stimuli. In many cases nonodorous taste stimuli have been utilized in an attempt to determine if the receptor mechanisms of gustatory and olfactory receptors have underlying similarities.

When an odorous substance interacts with the olfactory tissue of the frog a voltage change can be recorded; this is the result of the receptor membrane current change dropping across the extracellular resistance of the cell layer. We will call any substance applied to this chemosensitive tissue that can produce a voltage change an "adequate effector substance," if we can demonstrate that it is caused by the action of the substance on the receptor cells and is not an electrochemical artifact. We do not call all of these substances odors, because olfactory sensations need not and generally are not produced when solutions of them are atomized into a human nose. Thus the question asked of the frog's nose is: what substances stimulate the receptor cells either by excitation or inhibition?

Decorporate preparations of the leopard frog *Rana pipiens* were used for all experiments.[4] Although I generally avoid this preparation, initial investigations demonstrated the necessity for a larger exposed working area over the sheet of receptor cells than I could easily prepare using a live pithed frog. Decapitation avoids the spillage of any blood onto the intact mucosal surface. When a stream of clean moist air plays on the mucosa surface[2] normal reproducible EOGs are evoked for at least an hour by a puff of an odor if sufficient time elapses between stimuli.

MATERIALS AND METHODS

The oxygenated carrier medium is delivered to the dorsomedial aspect of the exposed olfactory cavity through polyethylene tubing (Intramedic P.E. 90, I.D. 0.034″) by gravity flow at 8 cc/min. The carrier medium then flows anteriorly over the ventral surface of the olfactory cavity, including the eminentia. The possibility of contaminants leaching from the tubing poses no problem in a given experiment, for we are concerned with voltage changes from the steady state due to a change in concentration of a particular chemical substance. On the average, 10^{-4} ml of the stimulus is drawn into the tip of a micropipette pulled from Kimax 0.7–1.0 mm melting point capillary with an open tip diameter ranging from 20 to 50 μ. All stimuli were made up in the carrier medium as the solvent. Reagent grade stimuli were used without further purification. Initial stimulus concentrations in the delivery micropipette were from 10^{-1} to 10^{-5} M. The pressure of the flow system causes the carrier medium to be forced up into the delivery pipette as it touches the flow stream, and a further dilution of the stimulus and concomitant change in volume may occur. In the final analysis, taking this dilution into account, approximately 3×10^{-4} ml of the substance in concentration ranges of 3×10^{-2} to 3×10^{-6} M is injected into the flow stream in the area of the receptor cells in approximately 3 sec. The stimulus aliquot injection rate does not significantly alter the carrier flow rate. If we assume uniform mixing, the concentration of the stimulus substance in the flow stream is between 2×10^{-4} to 2×10^{-9} M.

The electronic and recording apparatus are similar to those employed by Gesteland et al.[3]

Initial physical experiments with electrode systems demonstrated the expected electrochemical potentials at the silver–silver chloride solution interface in response to changes in ionic concentration as well as to changes in pH. By careful control of the configuration of the flow stream used for liquid stimulation and the air stream used for vapor stimulation, these were prevented from causing electrochemical voltage changes at the reference electrode. Silver–silver chloride electrodes, with either KCl or Ringer-gelatin pipettes as bridges, were used as the recording electrodes; a small silver–silver chloride plate wrapped in moist gauze and placed against the back of the animal's skull served as the reference electrode.

As in the work of Tucker and Shibuya,[7] it was necessary to determine experimentally the proper chemical composition of the perfusing carrier medium. One criterion employed was the minimal attenuation of the EOGs when the nose was soaked in the oxygenated test solution. The olfactory cavity was filled with the test solution for 5 min and then drained out; a puff of odor was given and the EOG recorded. This procedure was repeated at 5 min intervals for 1 hr. Three solutions were finally selected for testing in the flow system: Bradbury Ringer,[1] 245 mM sucrose and 245 mM glycerol. Iso-osmotic sucrose was commonly used, as it could be employed in the perfusion system for approximately 1 hr with only a 15 to 20 per cent decrease in the amplitude of the EOG generated by an odorous puff. Evidently the local environment surrounding the free olfactory receptor cell endings is not seriously changed by sucrose solution flowing for an hour. The mechanism of retention or replenishment of ions in extracellular fluid near the receptors must be a matter for speculation at the present. Another criterion was that the conductivity of the perfusing fluid be low enough so that the EOG evoked in the fluid not be attenuated to below system noise level. The usual difficulties of extracellular recordings in a conductive medium are compounded in this case because of electrochemical potentials which occur as the chemical composition of the flow stream is changed. Therefore, sucrose and glycerol, which are nonionic, consistently worked better than the Ringer solution or any ratio of sucrose and Ringer.

A small enhancement of the negative EOG in the frog by sucrose solutions as reported by Takagi et al.[6] was observed during the initial time intervals, but the extraordinarily large EOG reported by Tucker and Shibuya for the box turtle[7] was not observed.

By using the perfusion system, electrochemical potentials can easily be separated from the biological receptor response. If a Ringer-gelatin electrode is touching the mucosal surface as an aliquot of the stimulus, 2×10^{-5} M caffeine/245 mM sucrose is injected into the carrier stream (Figure 1 a) and a purely negative slow voltage change or a voltage change with positive and negative components occurs. It looks like the EOG waveform evoked by many odorous puffs in air. If the electrode is then withdrawn from the mucosal surface but remains in the carrier medium and an identical

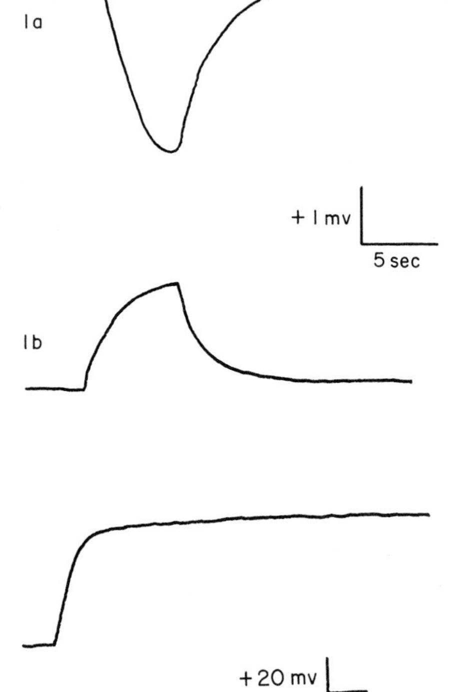

Figure 1 (a) Tracing of an EOG evoked by a "fluid puff" of 2×10^{-5} M caffeine/245 mM sucrose with the electrode touching the olfactory mucosal surface. (b) Tracing of a purely positive electrochemical potential evoked by a "fluid puff" of 2×10^{-5} M caffeine/245 mM sucrose when the electrode is raised up from the mucosal surface into the carrier stream.

Figure 2 Tracing that shows the establishment of the steady state as 245 mM sucrose begins to flow over the frog's olfactory mucosa.

aliquot is delivered, a purely positive electrochemical potential which arises at the electrode tip is recorded (Figure 1 b). If distilled water is substituted for the aliquot, the polarity of the electrochemical potential is reversed to purely negative. By injecting a more concentrated solution of sucrose, hence increasing the concentration of sucrose at the electrode tip, a purely positive electrochemical voltage is again seen (Figure 4 b).

When an attempt is made to draw current from the voltage generator, another significant difference between cellular activity and electrochemical potentials is seen. When the electrode is shunted by a resistor of appropriate value, the voltage changes recorded in the fluid drop to near zero, while there is little attenuation of the voltage changes recorded from the mucosal surface. The electrochemical potential is thus a high impedance source, and the excited receptor cells have a much lower source impedance.

Salts and acids were not used as stimuli, because of their well-known indiscriminate action on proteins and their effects on membrane potentials. Urea, caffeine, sucrose, and dextrose effects are reported here.

As the sucrose carrier solution begins to flow over the receptor tissue (Figure 2) a positive voltage change occurs, and a steady state is reached in approximately 10 sec. This response is undoubtedly the result of many simultaneous voltage changes, both biological and electrochemical. If the sucrose is serially diluted with Ringer, a new steady state, less positive than the previous one, is reached until a reversal of the polarity to purely negative is reached with 100 per cent Ringer. Thus, when a 100 per cent Ringer solution is placed on the nose, it elicits a purely negative voltage change. As the Ringer solution is serially diluted with distilled water, a biphasic response appears with an enhancement of the initial positive component at each successive decrease in ionic concentration. These voltages are probably evoked by receptor cell activity, changes in other cells of the mucosal layer, and electrochemical events at the interfaces.

By using 245 mM sucrose as the carrier and injecting a "fluid puff" of *n*-butanol (Figure 3 a) or isoamyl alcohol (Figure 3 b) as examples of odorous stimuli, an EOG is recorded that is most difficult to distinguish in waveform, amplitude, and duration from those produced by a moderately strong puff of the same odorous substances in air. For instance, with *n*-butanol the small initial positive phase (I, Figure 3 a), the large negative-going "on" phase (II, Figure 3 a) and the negative-going "off" deflection (III, Figure 3 a) are clearly seen. There is a considerable difference in the threshold for identical "fluid puffs" of three stimuli in the same preparation. For example, an aliquot of 2×10^{-7} M ethyl-*n*-butyrate will evoke a negative EOG of about -3 mV, whereas an identical "fluid puff" of *n*-butanol evokes no response. Benzaldehyde evokes an intermediate response. Thus the postulated receptor cell mechanisms responsible for generating the EOG in air are apparently operational when a smell is delivered in a fluid stream.

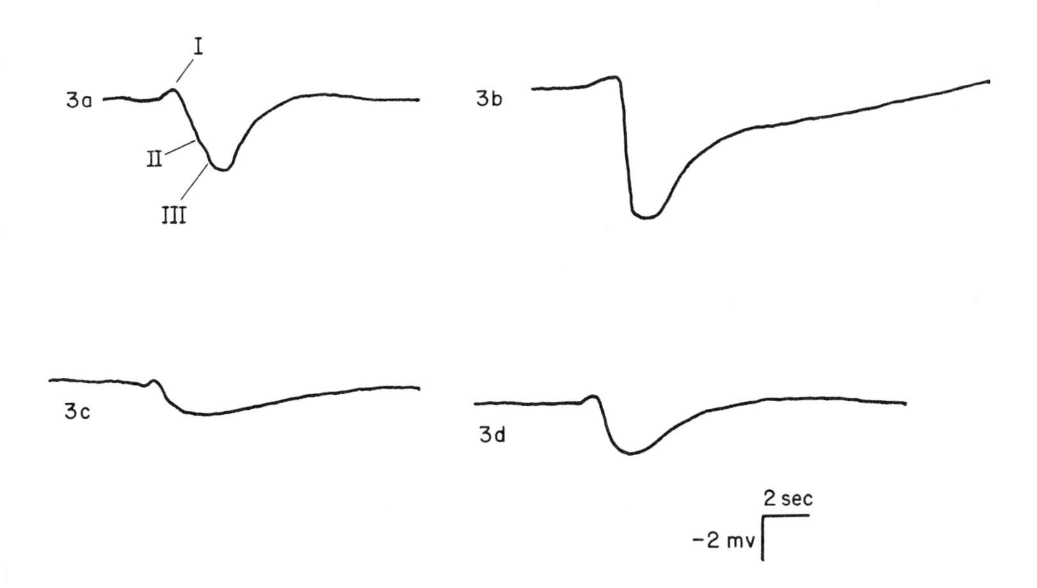

Figure 3 A comparison of the EOGs evoked by "fluid puffs" of odorous substances: (a) 2×10^{-5} M *n*-butanol, (b) 2×10^{-6} M isoamyl alcohol; and nonodorous substances: (c) 2×10^{-7} M caffeine, (d) 2×10^{-4} M urea. The carrier medium in each case is 245 mM sucrose. The record is a tracing.

An aliquot of 2×10^{-7} M caffeine and 2×10^{-4} M urea, as examples of non-odorous stimuli, also stimulate slow voltage changes across the olfactory receptor sheet of cells, as seen in Figure 3 c and 3 d. In each response the negative component is preceded by a small initial deflection. In one preparation urea had a threshold three log units higher than caffeine. Although the sensitivity of the olfactory receptors is more acute than those psychophysical thresholds reported for urea taste[5] in man, we found that urea always had a higher threshold than caffeine in the frog's nose. As the concentration of a stimulus increases, the duration and magnitude of the negative component of the EOG also increases. There is also an apparent increase in the rise time for an increase in stimulus concentration, but this is most likely due to an increase in the rate of stimulus-concentration change.

By substituting 245 mM glycerol for sucrose as the carrier, the biological receptor response of sucrose was determined. A 2×10^{-5} M aliquot of sucrose will elicit a purely negative EOG between -2 and -4 mV as seen in Figure 4 a. It may be

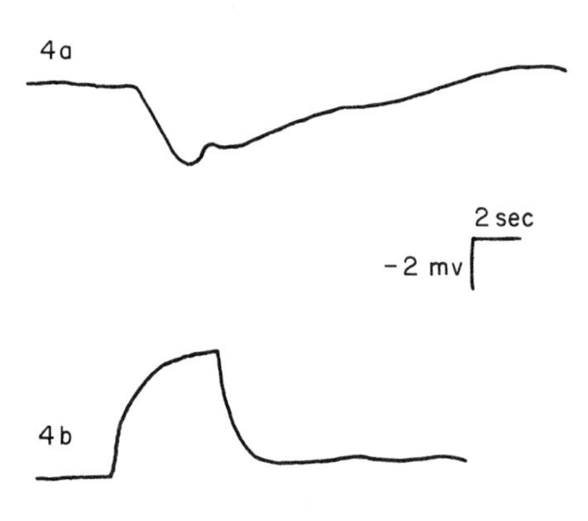

Figure 4 (a) With the electrode touching the mucosal surface a purely negative EOG is evoked by 2×10^{-5} M sucrose with 245 mM glycerol as the carrier medium (tracing). (b) When the electrode is raised up into the carrier medium, 245 mM glycerol, a purely positive electrochemical potential is evoked by 2×10^{-5} M sucrose (tracing).

argued that this is a negative-going electrochemical artifact caused by the dilution of the carrier medium, yet, if the electrode is raised from the mucosal surface but remains in the carrier medium, a purely positive voltage change is observed (Figure 4 b). Although the initial rise time of the negative component is rapid, the return to base line is variable, ranging from 10 to 20 sec. The response of the nose to a 2×10^{-5} M aliquot of dextrose is more similar in waveform and amplitude to the response of sucrose than to that of urea or caffeine. The thresholds of the nose are variable, as indicated by the EOG to sucrose and dextrose. It is not clear why this should be so. It is interesting that the EOGs produced by an aliquot of sucrose or dextrose more closely resemble those evoked by a weak organic acid such as acetic or butyric than they do caffeine or urea.

If it is accepted that the waveform of the EOG or any component thereof is a manifestation of receptor events, it can logically be concluded that the receptor layer of tissue responds by permeability changes to these nonodorous stimuli and produces the characteristic generator response.

The literature on peripheral vertebrate olfactory physiology is filled with problems of interpretation of the EOG, usually based on its origin as a multi-unit receptor system or on the ability to record slow voltage changes from nonliving systems and nonolfactory tissue. In my own experience I have never observed in any purely electrochemical voltage change the complexity in waveform so characteristic of the EOG. Because, at the present, routine intracellular recordings from olfactory bipolar neurons are impractical, we are forced to interpret the physiological activity of the peripheral olfactory system by using indirect methods. We must interpret the answer the frog's nose has given us to the question posed in light of the techniques used. What lines of evidence can be presented to demonstrate that the responses observed are signs of olfactory receptor cell activity?

1. The cellular receptor response can clearly be separated and identified as distinct from nonbiological electrochemical potentials in the flow stream.
2. The EOGs produced by a "fluid puff" of odor mimic those produced by a vapor puff of the same odor.
3. The EOGs evoked by a "fluid puff" of nonodorous substance mirror those produced by a "fluid puff" of odorous substance.
4. Current work in progress demonstrates that the time course of the EOG parallels a change of impedance across the receptor cell layer.
5. Urea, caffeine, sucrose, and dextrose, as well as salts and acids, evoke responses from axons of single olfactory receptor cells and from single cells of the olfactory bulb. This activity mimics the effects of odorous stimulation.

It thus seems likely that common chemical substances generally considered to be nonodorous by humans do act as stimulating molecules for the frog's olfactory receptor membrane, provided that the substance can get to the tissue. Would the frog, using its olfactory apparatus as its sole afferent imput, respond behaviorally to these "nonodorous" stimuli? If so, it would be possible to attempt to correlate molecular properties of chemical stimuli with the frog's psychophysical responses. At least it seems that considerable caution is necessary before physiological evidence from frogs and other experimental animals can be applied to support odor theories based on human psychophysical responses.

ACKNOWLEDGMENTS

The encouragement, fruitful discussions and constructive criticism of Professor Robert C. Gesteland during the preparation of this manuscript is respectfully acknowledged.

The work was supported in part by U.S. Air Force Contract No. F 33615-67-C1497, N.I.H. Grant No. 1 R01 NB06063-02, and U.S. Army Grant No. D.A. AR0-D-31-124-G991.

REFERENCES

1. Bradbury, M. W. B., H. Bagdoyan, A. Berberian, and C. R. Kleeman. 1968. Effect of osmolarity on cell water and electrolytes in isolated frog brain. *Amer. J. Physiol.* **215**, pp. 730–735.
2. Gesteland, R. C. 1964. Initial events of the electro-olfactogram. *Ann. N.Y. Acad. Sci.* **116**, pp. 440–447.

3. Gesteland, R. C., J. Y. Lettvin, and W. H. Pitts. 1965. Chemical transmission in the nose of the frog. *J. Physiol.* (*London*) **181**, pp. 525–559.
4. Ottoson, D. 1956. Analysis of the electrical activity of the olfactory epithelium. *Acta Physiol. Scand.* **35**, Suppl. 122, pp. 1–83.
5. Pfaffmann, C. 1959. The sense of taste. *In* Handbook of Physiology—Neurophysiology I (J. Field, editor). American Physiological Society, Washington, D.C., pp. 507–533.
6. Takagi, S. F., G. A. Wyse, H. Kitamura, and K. Ito. 1968. The roles of sodium and potassium ions in the generation of the electro-olfactogram. *J. Gen. Physiol.* **51**, pp. 552–578.
7. Tucker, D., and T. Shibuya. 1965. A physiologic and pharmacologic study of olfactory receptors. *Cold Spring Harbor Symp. Quant. Biol.* **30**, pp. 207–215.

II PSYCHOPHYSICS AND SENSORY CODING

II · PSYCHOPHYSICS AND SENSORY CODING

OLFACTORY ADAPTATION AND THE SCALING OF ODOR INTENSITY

WILLIAM S. CAIN *and* TRYGG ENGEN

Brown University, Providence, Rhode Island. Dr. Cain's present address is John B. Pierce Foundation Laboratory, New Haven, Conn.

THE PSYCHOPHYSICAL FUNCTION FOR ODOR INTENSITY

Many sensory continua, including olfactory intensity,[9] obey the psychophysical power law proposed by S. S. Stevens.[18] This law states that sensory magnitude ψ is related to physical magnitude φ by a power function of the form

$$\psi = k\varphi^{\beta},$$

where the size of β depends on the particular sensory continuum and certain conditions of stimulation. Research in the Brown University laboratory has shown that the size of the exponent of the psychophysical power function for aliphatic alcohols decreases as chain length increases.[6,7] The exponents were small, ranging from 0.13 to 0.52 and averaging about 0.26, when stimulus intensity was specified as the mole fraction of odorant in liquid solution. In similar experiments with a variety of odorants[8–10,14] the average exponent was greater than 0.50.

The original purpose of the present experiments was to learn whether certain procedural differences might account for the difference between the exponents obtained in these earlier studies and in the studies on the aliphatic alcohols. This led to the study of olfactory adaptation.

Intertrial Interval

In the experiments on the alcohols, the observers waited as long as one minute between exposures to stimuli in order to prevent adaptation. Delays this long could easily strain the observer's memory. Reese and Stevens,[14] who reported an exponent of 0.55 for coffee odor, allowed the observers to proceed at their own pace; this resulted in intertrial intervals of 20 to 30 seconds. Such intervals would presumably favor good memory, but at the same time might introduce adaptation effects. In order to check this notion, the subjective intensity of propanol and butanol was scaled for three different groups of 24 observers and with three intervals between stimulus presentations: 45 sec (Group I), 20 sec (Group II), and 7 sec (Group III). It was expected that, as the interval between stimulus presentations was decreased, the concomitant effects of memory or adaptation would increase the exponents of the psychophysical functions. This was not confirmed; the three groups yielded very similar exponents. As in previous experiments, the exponent for propanol was larger than for butanol (on the average, 0.5 and 0.3, respectively).

Effect of an Interpolated Standard

In some of the earlier experiments[8–10] a standard concentration, typically one from the middle of the stimulus series, was presented to the observer immediately before each stimulus trial in order to give him a constant reference stimulus against which to judge other concentrations. In addition to serving as a reference, however, the standard could have served as an adapting stimulus and thereby increased the exponents of the psychophysical functions. This possibility was tested in the present case with n-amyl acetate, the odorant used by Engen and Lindström.[8]

Figure 1 shows the effect of a reference concentration, the middle concentration, smelled immediately before the test concentrations of amyl acetate ($N = 17$). Without a standard, the exponent was 0.30; with the standard, the exponent increased to 0.52, a value almost identical to the exponent obtained by Engen and Lindström.[8] Thus, sniffs of the standard appreciably altered the relation between suprathreshold intensities, increasing the exponent of the psychophysical function by more than 70 per cent.

These experiments show that, because of adaptation effects, the exponent for odor may have been overestimated in some experiments. Since recovery from adaptation is so rapid, however, inflation of the exponent could depend on small differences in recovery time.

In all but one of the experiments discussed above, the concentration series was prepared by dissolving the odorants in a liquid diluent. Reese and Stevens,[14] the sole

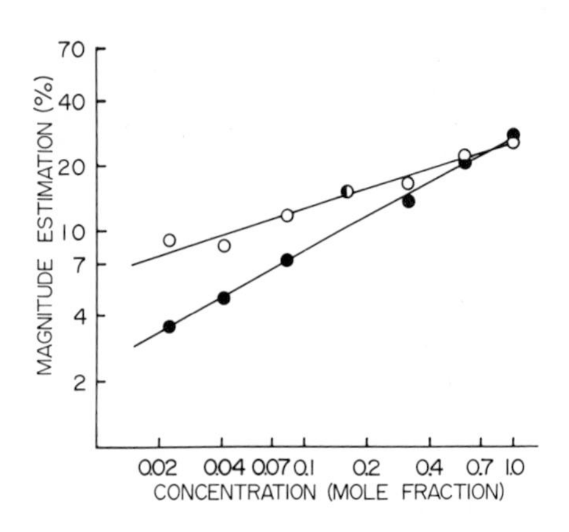

Figure 1 The psychophysical power functions for amyl acetate obtained when the middle concentration was presented only at the beginning of the session (circles) and when it was presented immediately before each of 6 test concentrations (dots). The half-filled circle represents the judgment of the standard concentration in both conditions.

exceptions, prepared concentrations by air-dilution. The large exponent for coffee odor that they obtained suggests that the diluent is another factor that deserves consideration. Odors diluted with air may yield larger exponents than those diluted with liquid solvents.

The procedure used to obtain the data in Figure 1 had been applied earlier to problems of adaptation in vision[17] and in taste.[11] It seemed particularly appropriate to extend the use of such a procedure to olfaction, where more traditional approaches had failed to clarify the principles underlying the adapting effectiveness of odors. The next section focuses on some issues in olfactory adaptation in which the need to expand beyond traditional approaches seems critical.

PREVIOUS RESEARCH ON OLFACTORY ADAPTATION

The degree to which absolute sensitivity to odor changes under adaptation conditions depends on both the duration and the concentration of the adapting stimulus. For the small group of odors that has been investigated, threshold concentration* grows at approximately a 0.7 power of adapting concentration when the adapting stimulus is presented for one-breath exposures.[2,3,13] A similar negatively accelerated relationship holds when the observer is exposed to the adapting stimuli for longer periods of time.[20]

Sensitivity, as measured by the threshold, decreases rapidly upon exposure to a constant adapting stimulus. The change over time is also negatively accelerated and its rate and magnitude are directly related to adapting concentration.[20]

For suprathreshold stimuli the effects of adaptation have typically been studied by "time-course" methods, in which observers judge at intervals the intensity of a constant adapting concentration. The results parallel those obtained at the threshold level; perceived intensity is a negatively decelerated function of adapting time.[4,19] Woodrow and Karpman[21] found that the duration necessary for complete adaptation, i.e., for an odor to become imperceptible, was linearly related to the concentration of the adapting stimulus. How adapting concentration might affect the parameters of time-course functions needs further specification.

Because of the similarity of the time-course procedure to everyday experience with odors, observers tend to come to the experimental situation with pronounced biases. For example, they usually expect the intensity to decrease during the adaptation period and may even make assumptions about the form of the adaptation function. Moreover, they can be influenced by the expectations of the experimenters. For example, Woodrow and Karpman[21] assumed that all of the odors they presented would adapt completely if only the observers could learn to ignore the "irrelevant sensations" attending the odorous stimulus. Ekman et al.[4] and Stone[19] did not make

* The term "threshold" is used in an operational sense, for there is real doubt about the existence of such a value on the sensory continuum. See Engen, T. Man's ability to perceive odors. *In* Advances in Chemoreception: Vol. 1, Communication by Chemical Signals (J. W. Johnston, Jr., D. C. Moulton, and A. Turk, editors). (In press)

this assumption and, unlike Woodrow and Karpman,[21] found that over many minutes of stimulation complete adaptation occurred for only some observers.

Comparisons Between Odors

Most of the basic data on olfactory adaptation seem straightforward, if incomplete, until comparisons between odors are made. Perhaps the only generalization that can be made across odorants is that the relation between adapting concentration and threshold is described by a power function with an exponent of 0.7. It is generally accepted that odors differ from each other in their rates of adaptation (both at threshold and at suprathreshold levels), in their adapting effectiveness, and in the time necessary for disappearance. The principles that would relate these differences have not been discovered. Yet the situation may be simpler than it seems, for it is even possible that the apparent differences among odors stem as much from procedural differences among experiments and from failure to consider certain psychological variables as from real chemical or physical differences among the odorants themselves.

Olfactory research has typically focused on small segments or points on the dynamic range of intensity. Threshold experiments give a limited and sometimes a misleading picture of intensity. An odorant can have a low threshold but never become very intense at any concentration,[7,13,22] or an odorant can have a high threshold but become very intense even at moderate concentrations. It is therefore perilous to rely on thresholds alone or on single suprathreshold concentrations for comparisons between the adapting efficiencies of different odorants.

The threat that response bias might affect the traditional time-course measurements looms larger when the results for various odors are compared. This is especially true when observers rate intensity on a limited category scale, say a 7-point scale where 1 stands for the weakest intensity and 7 for the strongest. As Schutz, Overbeck, and Laymon[16] pointed out, observers may tend to assign inappropriately high ratings to weak odors at the beginning of adaptation in order to leave a sufficient numerical range in which to work during the remainder of the exposure. Thus, an observer may feel uncomfortable assigning 2 and thereby forcing himself to respond only 2 or 1 later in the adapting period. When this response bias operates, comparisons of time-course functions for different odorants or for different concentrations of the same odorant would become almost meaningless.

Ekman et al.[4] are apparently the only investigators to scale the time course of olfactory adaptation by a method that does not restrict the observer to a limited number of response categories. They used a two-step scaling procedure. In one session the observers matched the perceived span between thumb and forefinger to the apparent intensity of hydrogen sulfide at specified intervals in the adapting period. In a second session they scaled finger span by magnitude production, a method similar to magnitude estimation. The experimenters then were able to convert matched finger span into numerical estimates of the subjective intensity of odor.

Two advantages derive from this procedure. First, because perceived finger span is a nonlinear function of physical finger span, observers cannot easily utilize an hypothesis about the form of the adaptation function. Second, the observers are not faced with the problem of "using up" the available responses, because no matter what span they match to the initial olfactory magnitude, they can continue indefinitely to narrow the finger span.

Ekman et al. reported that subjective intensity is a negative exponential function of duration for four different adapting concentrations. They did not, however, compare the time course of different odorants.

Psychophysically Equivalent Adapting Stimuli

Perhaps the foremost reason various odorants appear to adapt differently from one another is that comparisons of adapting effectiveness have not been made with psychophysically equivalent adapting stimuli. Great advantage might derive from specifying the subjective rather than the physical magnitude of the adapting stimulus. Two reasons for this follow.

(1) Rarely do odorants have equal subjective intensity when matched in concentration. Thus, adapting stimuli that are matched in concentration usually are psychophysically different to begin with, and there is no reason to expect them to be psychophysically equivalent as adapting stimuli. Schutz et al.[16] (see also Schutz and Laymon[15]) tried to correlate the rates of adaptation of 30 suprathreshold odors with each of several physical variables such as hemolytic accelerating activity, ultraviolet absorption, etc. The low correlations they found might have been much higher had the subjective intensity of the various odorants been the same at the onset of adaptation. Differences in subjective intensity alone could easily have caused different apparent rates of adaptation.

(2) The rate of growth of subjective intensity as a function of concentration differs among odorants. Two stimuli matched for apparent intensity may not match at a multiple of the original concentrations. Differences in growth of adapting effectiveness as a function of concentration could reflect differences in the growth of the suprathreshold intensity of the adapting stimuli. In other words, odorants may appear to have different adaptation effects because of differences in their psychophysical functions.

Moncrieff[13] maintained that the "strength" of an odor could be defined in terms of threshold concentration and the extent to which adaptation to the same odor (i.e., self-adaptation) raises the threshold. He used these two factors to rate the intensity of undiluted odorants and assigned to each a scale value that combined the threshold under adaptation and the threshold under nonadaptation conditions. Implicit in his approach was the assumption that odors matched for apparent intensity have equivalent adapting effectiveness. For example, if two odors had identical threshold concentrations and smelled equally strong in undiluted form it was expected that they would yield the same scale value. Moncrieff did not test this implicit assumption, but did note that effectiveness in raising the threshold correlated with his own subjective impressions of the suprathreshold intensity of the adapting stimuli.

The following study was partially designed to test the hypothesis that odors have equal self-adapting effectiveness when equated for apparent intensity. If this is so, the relative effectiveness of adapting stimuli could be predicted from knowledge of their relative subjective intensities.

EXPERIMENT I: SELF-ADAPTATION

The purpose here was to determine, in light of the preceding discussion, how both the intensity and the duration of adapting stimuli affect the relative subjective intensity produced by concentrations that span a large segment of the dynamic range. One specific question was whether weak adapting stimuli, smelled over a long period of time, produced adapting effects similar to effects of strong adapting stimuli smelled for a short period. Another, mentioned above, was whether subjectively matched adapting stimuli have equal adapting effectiveness. In approaching the second question it appeared, for the time being, more fruitful to work with two odorants under a large number of adaptation conditions than with many odorants under one or two adaptation conditions.

Procedure

The apparatus for controlling and delivering the stimulus was a two-channel, air-dilution olfactometer. One channel controlled the adapting concentrations and the other the test concentrations. The principle of operation was to saturate dry, odorless air with the odorant, and to control concentration by mixing saturated air with pure air in various proportions. Materials used in the construction of the olfactometer were all normally odorless and easily cleaned.

There were two stimulus delivery tubes of 1.5 inches in diameter mounted on the front wall of the experimental chamber. Lights signaled the observer when to place his nose over one of the tubes and when to begin breathing the stream of odorous air. Whenever more than one inspiration was required, a metronome was used to pace breathing rate at 30 cycles per minute. A series of relays and variable timers controlled the opening and closing of the solenoid valves used to deliver the stimuli. These allowed precise control of the temporal conditions of the experiment.

n-Propanol (C_3) and *n*-pentanol (C_5) were the odorous stimuli. The over-all flow-rate for propanol was 4 l/min and for pentanol 6 l/min.

Nineteen observers were tested. The adaptation data from one of them were very erratic and were therefore not counted. In the first part of the experiment the method of magnitude estimation was used to scale the apparent intensity produced by a series of propanol and pentanol. Adaptation per se was not studied in this part. The observers made two judgments of each of seven concentrations. In each of these two scaling sessions they also judged a concentration of ethyl acetate, so that later the psychophysical functions could be tied together at a common point. The observers were instructed to assign to the apparent intensity of the first stimulus (middle concentration of the series) any number deemed appropriate, and to let their judgments of subsequent concentrations reflect the ratio relations among the perceived intensities. For example, if one concentration smelled five times as strong as another its intensity was to be assigned a number five times as large. If it smelled one-half as strong it was to be assigned a number one-half as large, etc.

The second part of the experiment used this same procedure under sixteen different adaptation conditions. After the first trial the observers smelled an adapting stimulus before judging each of the test concentrations. Throughout any given test session the adapting stimulus was constant in concentration and duration.

Adaptation conditions for both odorants are presented in Table I. Subjectively

TABLE I

SELF-ADAPTATION CONDITIONS FOR PROPANOL (C_3) AND PENTANOL (C_5)

Adapting Intensity	Matched Concentrations (mg/l) C_3	C_5	Exposure (no. of breaths)
Low	2.0	0.5	3
Low	2.0	0.5	8
Middle	6.3	2.1	3
Middle	6.3	2.1	5
Middle	6.3	2.1	8
Middle	6.3	2.1	8 plus 3 recovery
Middle	6.3	2.1	15
High	21.6	9.2	3

matched levels of the adapting stimuli had already been measured in a previous experiment.[1] Within the limits of experimental error the present group confirmed the earlier result. The subjective ratio between adjacent adapting levels was 2.25 for both odorants.

Exposure durations were chosen to cover a range that could be comfortably handled in the context of a scaling experiment. Longer exposures would have necessitated long intervals between stimuli and would have taxed the observers. In all but one condition the observers judged each of the seven test concentrations twice and waited 1.5 min between trials. In the 15-breath conditions they judged each concentration once and waited 2.5 min between trials.

The number of breaths the observers took at the output of the adapting channel was only the number required by the condition being run at the time. In the non-adaptation condition, for example, the observers simply took one breath from the test channel and none from the adapting channel. This procedure was used because a control experiment indicated that the psychophysical function was not affected by exposing the observers ($N = 15$) to a stream of pure air from the adapting channel before presenting each test concentration. Each observer judged the intensity of pentanol after 0, 3, and 8 breaths of this stream of pure air. The range of the exponent for the three conditions was 0.02.

Results

The data for each experimental condition were normalized to a standard of 10. For example, if an observer used a modulus of 100 (i.e., called the first stimulus 100) his estimates at each concentration were multiplied by 0.1, if he used a modulus of 5 his estimates were multiplied by 2, etc. When the observers made two estimates at each concentration the geometric mean of these estimates was then computed. The median was used as the measure of central tendency across observers.

Functions Under No Adaptation. As expected, the exponent for pentanol, 0.58, was lower than that for propanol, 0.74 (Figure 2). A previous experiment[1] using magnitude estimation yielded an exponent of 0.70 for propanol with smaller deviations of the points from the line of best fit.

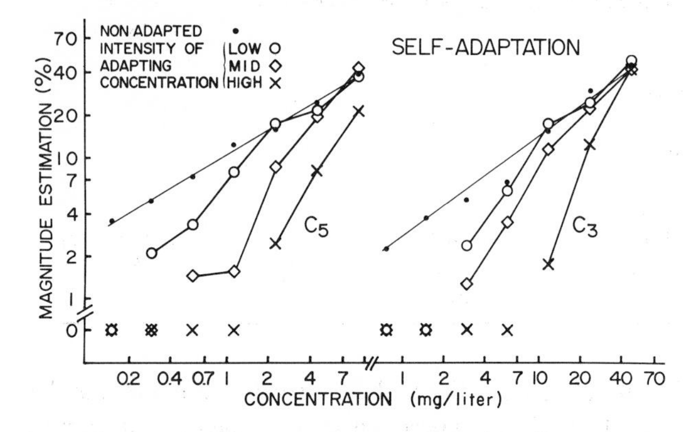

Figure 2 Psychophysical functions of pentanol (C_5) and propanol (C_3) obtained under different levels of adaptation. Each adapting concentration was presented for 3 breaths. The straight lines are the power functions fitted to the magnitude estimates obtained under non-adaptation conditions.

These exponents for propanol are just about twice as large as that obtained by Engen et al.[7] This confirmed the notion, mentioned in reference to the results of Reese and Stevens,[14] that odorants presented in a stream of air yield higher exponents than odorants presented in liquid solutions. At this moment, no simple explanation for these differences is obvious. The present data and the data obtained earlier by Cain[1] illustrate that observers are much more reliable in judging stimuli presented in air. This may have to do as much with ease of judgment as with precision of stimulus control.

Effect of Adapting Intensity. Under conditions of adaptation, the psychophysical functions became steeper, indicating that the dynamic relation between subjective intensity and physical concentration undergoes profound change as a result of adaptation (Figure 2). Every increase in adapting concentration further steepened the function and further elevated the threshold.

The functions for different adapting concentrations converge at the highest test concentration for each odorant. An exception to this general rule occurred for pentanol when its adapting concentration equaled the concentration of the strongest test stimulus.

Effect of Adapting Duration. Time-course experiments indicate that the most dramatic effects are in the early stages of adaptation. Any large effect of duration of exposure should be seen in the psychophysical functions obtained under the durations employed in this experiment, i.e., three to 15 cycles of respiration or six to 30 seconds.

Figure 3 shows that duration had only minor influence for low-intensity adaptation

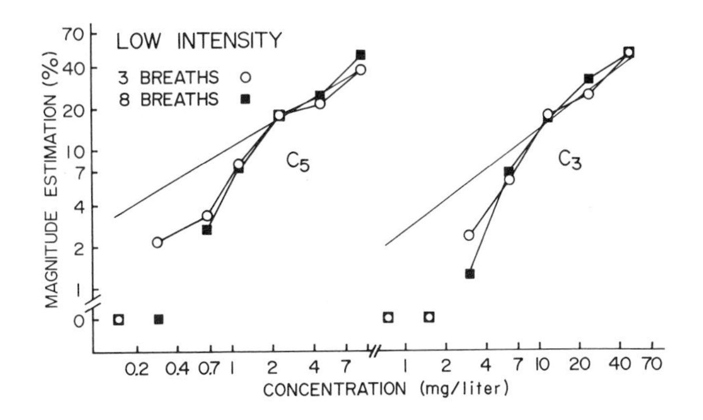

Figure 3 Low-intensity adaptation under exposures of 3 breaths and 8 breaths. The straight lines, provided for reference, are the psychophysical functions obtained under nonadaptation conditions.

conditions, although small differences in the expected direction showed up between the functions for three and for eight breaths. For adaptation with the middle intensity the effect was also rather small (Figure 4). Three breaths seemed to produce slightly less adaptation than longer exposures, but the longer exposures produced very similar functions. A small divergence can be seen in the 5- through 15-breath conditions with propanol. This divergence is absent from the pentanol functions, but the variability could have concealed small effects.

To summarize, Figures 3 and 4 show that duration of exposure had only a small effect on the psychophysical functions for odor. A large increase in adapting time

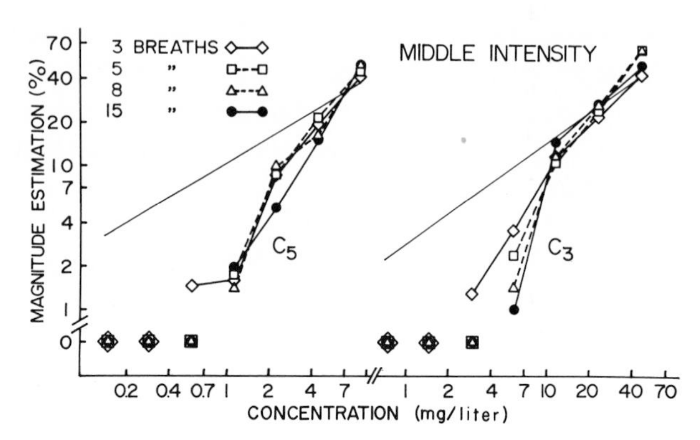

Figure 4 Middle-intensity adaptation under exposures ranging from 3 to 15 breaths.

produced a much less dramatic change than a moderate difference in the subjective intensity of adapting stimuli. On the other hand, a recovery period of only three breaths sizeably reduced the effects of adaptation (Figure 5). This is consistent with earlier findings that recovery from adaptation is quite rapid.[4,19,20]

Form of the Function Under Adaptation. Stevens and Stevens[17] found that brightness functions obtained under adaptation could be fitted by a generalized form of the psychophysical power law, $\psi = k(\varphi - \varphi_0)^\beta$, where φ_0 approximates the absolute threshold. This equation describes a straight line in log-log coordinates when subjective intensity ψ is plotted against physical intensity above threshold $(\varphi - \varphi_0)$. When ψ is plotted against simple physical intensity φ, the function is relatively straight at high values of the independent variable, but becomes concave downward at lower values, eventually reaching an asymptote at the value of φ_0. As the brightness of an adapting field increases both β and φ_0 increase. For taste functions obtained under adaptation, φ_0 increases with adapting level but β remains relatively constant.[11]

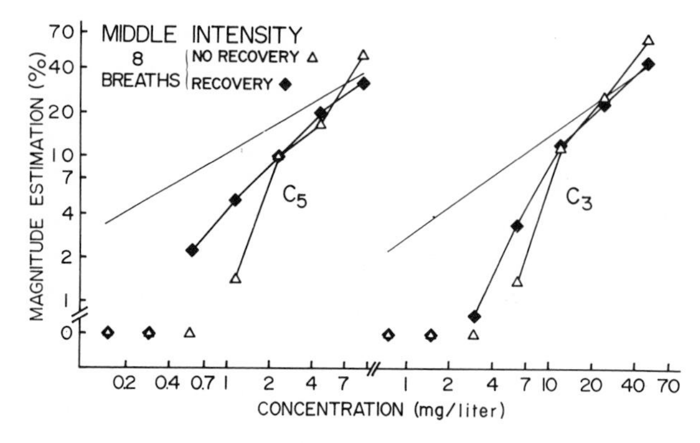

Figure 5 Psychophysical functions obtained under 8 breaths' exposure to the middle intensity adapting stimulus with 3 breaths' recovery and with no recovery.

The adaptation functions obtained in the present experiment did not all have the same form. No single equation could describe all of the results. Those functions that were concave downward could be well fitted by the generalized power function but those that were concave upward in the lower range could not. Figure 6 illustrates some of the adaptation data fitted by the generalized power equation. These functions were determined by plotting, in log-log coordinates, the magnitude estimates against the concentration re threshold, i.e., against ($\varphi - \varphi_0$), and choosing for φ_0 the value that maximized the fit of the data to a straight line.

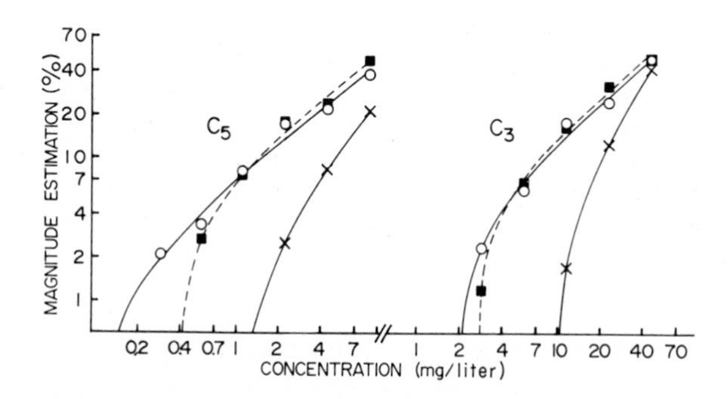

Figure 6 Adaptation data fitted by the generalized form of the psychophysical power law. The adaptation conditions represented are low intensity, 3 breaths (circles); low intensity, 8 breaths (squares); and high intensity, 3 breaths (crosses).

For each odorant, the functions for low-intensity adaptation have practically the same exponent, but the eight-breath conditions yielded higher values of φ_0 than did the three-breath conditions. High intensity adaptation yielded both larger values of φ_0 and larger exponents.

One possible reason why the functions obtained in some of the conditions were not concave downward over the lower concentrations may be that some observers took deeper inhalations of weak stimuli in order to obtain a clearer sensation. The observers were warned against doing this, but there was no way to check whether they actually followed the instruction. The exact form of the psychophysical function under adaptation is, therefore, still open to question. Future experiments could well employ pneumographic monitoring to insure uniform breathing of the stimuli. It may yet turn out that refinements of method will reveal that the generalized form of the psychophysical power law, which has worked so well in vision and taste, will also adequately describe olfactory adaptation.

Equivalence. Figure 7 compares the various adaptation functions for propanol and pentanol. The close agreement between the functions demonstrates that propanol and pentanol have very similar self-adapting efficiencies under a large number of adaptation conditions. These two odorants are physically and chemically

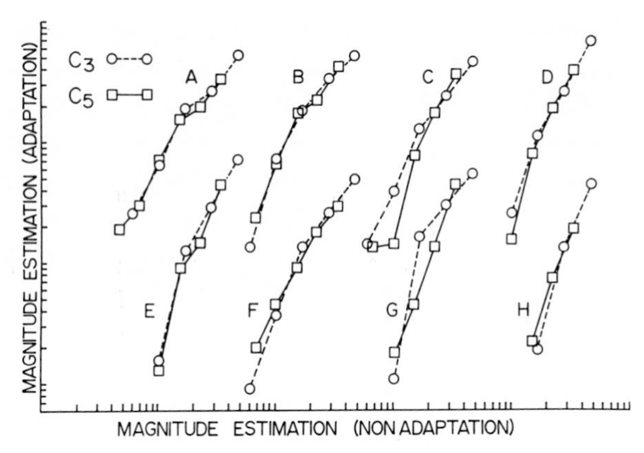

Figure 7 Magnitude estimation under eight conditions of adaptation as a function of magnitude estimation under nonadaptation conditions. The coordinates are relative. Values of zero, included in previous figures, were excluded from this figure for clarity. The conditions are designated as follows: A, low intensity, 3 breaths; B, low intensity, 8 breaths; C, middle intensity, 3 breaths; D, middle intensity, 5 breaths; E, middle intensity, 8 breaths; F, middle intensity, recovery; G, middle intensity, 15 breaths; H, high intensity, 3 breaths.

rather similar, and it is not yet possible to generalize this finding to dissimilar odorants. However, the psychophysical power functions for the two odorants had different exponents, indicating that under nonadaptation conditions the physical differences between them were psychophysically significant. Consequently, the equal self-adapting efficiencies of these odorants could not have been predicted simply from the psychophysical results obtained under nonadaptation conditions. This lends support to the possibility that the same general relation may obtain between subjective intensity and adapting efficiency for all odorants. If so, we would be in the excellent position of being able to predict the relative adapting efficiency of concentrations of any two odorants simply from their relative subjective intensities.

EXPERIMENT II: CROSS-ADAPTATION

The usual purpose of cross-adaptation experiments is to "throw light on the usefulness of classifying odors by adaptation, on the existence or otherwise of fundamental odours, and perhaps on the mechanisms whereby olfactory adaptation comes about" (Moncrieff,[12] p. 303). It has generally been assumed that cross-adaptation and self-adaptation are two aspects of the same phenomenon. Schutz et al.[16] essentially challenged this notion by claiming that cross-adaptation does not occur on the suprathreshold level. However, Engen[5] found that observers' ability to identify odors of aliphatic alcohols was impaired after exposure to an alcohol of a different quality from the test stimulus. This finding supported the existence of cross-adaptation between suprathreshold stimuli.

The following experiments examined cross-adaptation between propanol and pentanol in order to answer two questions: (1) Do cross-adaptation and self-adaptation have the same effect on the form of the psychophysical function? And (2) to what degree will pentanol and propanol cross-adapt each other when the cross-adapting concentrations are matched for subjective intensity?

These experiments used the same procedure and adapting concentrations used in the self-adaptation experiments. Nineteen observers estimated the intensity of propanol and pentanol under nonadaptation conditions and under six conditions of cross-adaptation. Propanol was the cross-adapting stimulus in three of the conditions and pentanol in the other three. In the low and middle intensity conditions observers were given an eight-breath exposure to the adapting stimuli and in the high intensity condition a three-breath exposure.

The function for pentanol obtained under nonadaptation conditions in the present experiment had a smaller exponent, 0.43, than the function obtained in the self-adaptation experiment. This apparently means that the adapting concentrations were not so closely matched for the observers in the cross-adaptation as in the self-adaptation experiment. This is not of major consequence, however, given the grossly unequal cross-adapting effects of propanol and pentanol. As Figure 8 shows, propanol had a

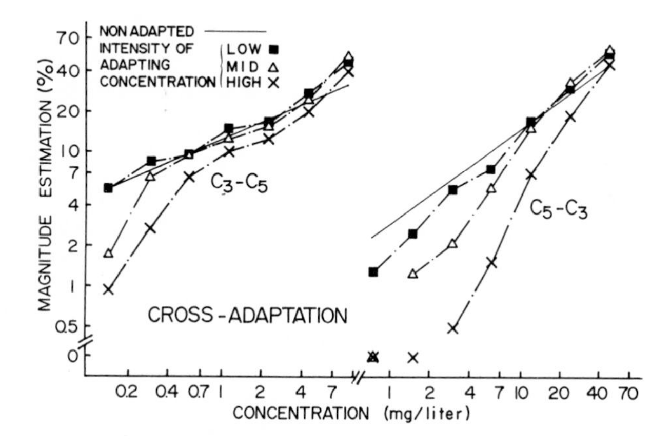

Figure 8 Psychophysical functions obtained under different levels of cross-adaptation. The straight lines are the power functions fitted to the magnitude estimates obtained under non-adaptation conditions. The data points from the nonadaptation conditions have been omitted for clarity.

very small cross-adapting effect on pentanol. The high-intensity, cross-adapting stimulus was not even as effective as the low-intensity, self-adapting stimulus for pentanol. Pentanol, on the other hand, had a sizeable cross-adapting effect on propanol.

Figure 9 compares C_5–C_3 cross-adaptation with C_3 self-adaptation. The cross-adapting stimulus of middle intensity was almost as effective as the self-adapting stimulus at low intensity. The basic similarity between the functions obtained under cross-adaptation and self-adaptation conditions also reveals itself in the figure.

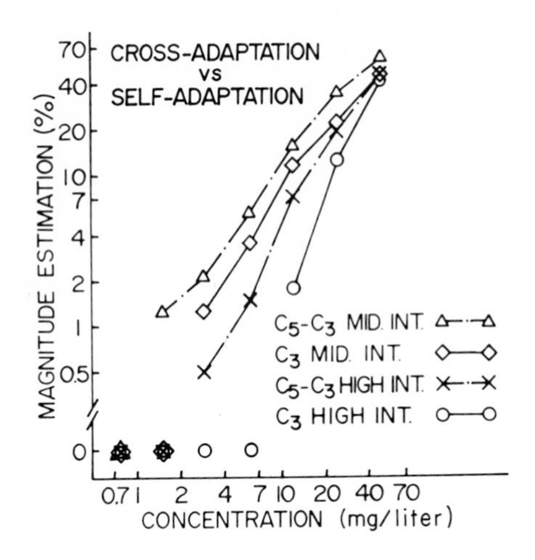

Figure 9 Comparison between self-adaptation of propanol and cross-adaptation of propanol by pentanol.

CONCLUSIONS

(1) If observers smell a reference concentration of moderate intensity immediately before smelling test concentrations, the psychophysical function steepens. However, interstimulus intervals as short as 7 sec permit adaptation effects to dissipate.

(2) Presentation of odors in a stream of air leads to steeper psychophysical functions than sniffing odors presented in liquid form.

(3) Increasing the intensity of adapting concentrations by subjective ratios of 2.25 leads to large changes in psychophysical functions. However, the mathematical form of the functions obtained under adaptation could not be specified with certainty.

(4) The psychophysical functions change only slightly with increases in adapting time. Presumably, only very long exposures would affect the functions as dramatically as moderate increases in the subjective intensity of adapting concentration.

(5) *n*-Propanol and *n*-pentanol have almost identical self-adapting effectiveness when their adapting stimuli are matched for subjective intensity.

(6) Closely matched concentrations of propanol and pentanol have unequal cross-adapting effectiveness. This asymmetrical effect of cross-adaptation within pairs of odorants must be considered in any attempt to use cross-adaptation as a basis for the classification of odor quality.

(7) Cross-adaptation and self-adaptation seem to produce psychophysical functions of the same shape, suggesting that they are two aspects of the same phenomenon.

The rules of olfactory self-adaptation may be simpler than previously thought. We have here the results for only two odorants, but these results strongly suggest that comparisons between odors are likely to be far more fruitful when adapting stimuli are specified in terms of subjective intensity.

ACKNOWLEDGMENT

This work was supported by National Science Foundation Research Grant (GB-724) to Trygg Engen. The authors wish to acknowledge the assistance of Mr. Geoffrey Loftus, Mrs. Tiina Corbit, and Mrs. Georgia Hayden in the execution of the experiments.

REFERENCES

1. Cain, W. S. 1968. Olfactory adaptation and direct scaling of odor intensity. Unpublished doctoral dissertation. Brown University.
2. Cheesman, G. H., and S. Mayne. 1953. The influence of adaptation on absolute threshold measurements of olfactory stimuli. *Quart. J. Exp. Psychol.* **5**, pp. 22–30.
3. Cheesman, G. H., and M. J. Townsend. 1956. Further experiments on the olfactory thresholds of pure chemical substances, using the "sniff-bottle method." *Quart. J. Exp. Psychol.* **8**, pp. 8–14.
4. Ekman, G., B. Berglund, U. Berglund, and T. Lindvall. 1967. Perceived intensity of odor as a function of time of adaptation. *Scand. J. Psychol.* **8**, pp. 177–186.
5. Engen, T. 1963. Cross-adaptation to the aliphatic alcohols. *Amer. J. Psychol.* **76**, pp. 96–102.
6. Engen, T. 1965. Psychophysical analysis of the odor intensity of homologous alcohols. *J. Exp. Psychol.* **70**, pp. 611–616.
7. Engen, T., W. S. Cain, and C. K. Rovee. 1968. Direct scaling of olfaction in the newborn infant and the adult human observer. *In* Theories of Odors and Odor Measurement (N. Tanyolac, editor). Robert College Research Center, Istanbul, Turkey, pp. 271–294.
8. Engen, T., and C. O. Lindström. 1963. Psychophysical scales of the odor intensity of amyl acetate. *Scand. J. Psychol.* **4**, pp. 23–28.
9. Jones, F. N. 1958. Scales of subjective intensity for odors of diverse chemical nature. *Amer. J. Psychol.* **71**, pp. 305–310.
10. Jones, F. N. 1958. Subjective scales of intensity for three odors. *Amer. J. Psychol.* **71**, pp. 423–425.
11. McBurney, D. H. 1966. Magnitude estimation of the taste of sodium chloride after adaptation to sodium chloride. *J. Exp. Psychol.* **72**, pp. 869–873.
12. Moncrieff, R. W. 1956. Olfactory adaptation and odour likeness. *J. Physiol.* (*London*) **133**, pp. 301–316.
13. Moncrieff, R. W. 1957. Olfactory adaptation and odor-intensity. *Amer. J. Psychol.* **70**, pp. 1–20.
14. Reese, T. S., and S. S. Stevens. 1960. Subjective intensity of coffee odor. *Amer. J. Psychol.* **73**, pp. 424–428.
15. Schutz, H. G., and R. S. Laymon. 1959. Investigation of olfactory adaptations. *Amer. Psychol.* **14**, p. 429. (Abstract)
16. Schutz, H. G., R. C. Overbeck, and R. S. Laymon. 1958. Relationship between flavor and physicochemical properties of compounds. Final Report, Battelle Memorial Institute, Contract No. DA-19-129-QM-1141, Quartermaster Food and Container Institute for the Armed Forces.
17. Stevens, J. C., and S. S. Stevens. 1963. Brightness function: effects of adaptation. *J. Opt. Soc. Amer.* **53**, pp. 375–385.
18. Stevens, S. S. 1957. On the psychophysical law. *Psychol. Rev.* **64**, pp. 153–181.
19. Stone, H. 1966. Factors influencing behavioral responses to odor discrimination: A review. *J. Food Sci.* **31**, pp. 784–790.
20. Stuiver, M. 1958. Biophysics of the Sense of Smell. Excelsior, 'S-Gravenhage (Netherlands).
21. Woodrow, H., and B. Karpman. 1917. A new olfactometric technique and some results. *J. Exp. Psychol.* **2**, pp. 431–447.
22. Wright, R. H. 1964. The Science of Smell. Allen and Unwin, London.

INTENSITY IN MIXTURES OF ODOROUS SUBSTANCES

E. P. KÖSTER
Psychologisch Laboratorium der Rijksuniversiteit, Utrecht, The Netherlands

INTRODUCTION

In earlier work with binary mixtures of odorous substances at suprathreshold level[5] we were interested in three phenomena that were described in the literature. Zwaardemaker[7] and Backman[1] showed that when two substances of given concentration are mixed in a given ratio, the resulting odor of the mixture may be far less intense than that of each of the separate components, and that eventually it may not be perceptible at all. This phenomenon, which is essentially a decrease of olfactory intensity, was called *compensation*. Another phenomenon, called *synergy*, which consists of an enhancement of the odorous intensity of a substance by the addition of another substance, was never reported by the authors just mentioned, but Kendall and Neilsen[4] claimed to have found synergy often. Complete *additivity*, or the summation of the two intensities of the components of the mixtures, has been reported by a number of authors[2,3,6] who found the occurrence of each of the mentioned phenomena to be dependent on or concurrent with a number of different circumstances.

The main conclusions of our own work, in which we investigated the relative intensities of mixtures of a considerable number of pairs of odorous substances, were that:

(1) Synergy, or enhancement of odorous intensity, is very rare and could not be found in a reproducible way in the 18 different pairs we investigated.

(2) Compensation, or decrease in intensity, occurred frequently, especially in mixtures in which the mixing ratios of the components were more extreme, such as 1:6 or 1:8.

(3) Additivity was more likely to occur with less extreme mixing ratios, such as 1:1 or 1:2.

(4) Intensity seemed to influence the occurrence of additivity. In the two pairs of odorous substances that were mixed at different levels of intensity, additivity was more frequent at high concentrations.

In the present study we are interested in the relationship between the cross-adaptational influences that substances exert on the sensitivity to each other and the phenomena they show in mixing. If, for instance, two substances do influence the sensitivity to each other when presented successively as in a cross-adaptation experiment, it will be interesting to see whether they also influence each other when they are presented simultaneously in a mixture. Therefore the intensity of mixtures has been investigated in two groups of three odorous substances. The groups were chosen on the basis of the results of a series of cross-adaptation experiments in such a way that within each group:

(1) Two substances showed no cross-adaptational influence (i.e., response reduction as a result of cross-adaptation). These substances were perceived just as well when they were presented one after the other as when they were presented after pure air. (In group I: cyclohexanone and dioxan; in group II: citral and benzyl acetate.)

(2) Two substances showed a positive and almost equal amount of cross-adaptational influence upon each other. Both substances were perceived less well when they were presented after each other than when they were presented after pure air. (In group I: cyclohexanone and cyclopentanone; in group II: citral and safrole.)

(3) Two substances showed cross-adaptational influence upon each other to a very different degree. The sensitivity to one of the substances was much less when it was presented after the other than when it was presented after pure air, but the sensitivity to the other was not or was barely affected by previous presentation of the first odor. (In group I: dioxan [positive influence] and cyclopentanone; in group II: benzyl acetate [positive influence] and safrole.)

It was hoped that the pairs of substances showing the same cross-adaptational relationships in both groups would also show the same characteristic phenomena in the mixing experiments.

METHODS

Four experiments were carried out, including one matching experiment to determine the approximate ranges of concentrations of the mixture components and three experiments with mixtures. In each of these last experiments the intensities of five dilutions of seven different mixtures of each of two pairs of components were compared with the intensity of a standard stimulus of 1 M *m*-xylene. The 70 comparisons (35 mixtures for each pair of components) were made by the same 96 subjects (48 male, 48 female) in all experiments. These subjects had also been used in the matching experiment.

In all experiments presentation order within and between comparison pairs was carefully balanced. The subjects merely indicated whether they judged the mixture to be stronger or weaker than the standard stimulus.

Data Treatment

The data were treated in a way that has been described in detail elsewhere.[5] This treatment consisted of the following steps.

(1) The percentage "stronger than standard" judgments was calculated for each mixture.

(2) For each mixture of a given mixing ratio the percentages "stronger" obtained at the five different concentrations at which the mixture was presented were plotted against the concentrations.

(3) The resulting sigmoidal curves (differential sensitivity functions) were cut off at three levels of equal subjective intensity, 70 per cent, 50 per cent, and 30 per cent stronger than standard judgments.

(4) The exact concentrations of both components were calculated for each cut-off point of each mixture of given ratio.

(5) The concentrations thus found for each mixture at a given level of subjective intensity were divided by the concentration found for the pure (unmixed) substance at the same level of subjective intensity. The resulting ratio is called the "relative concentration" of a given substance in the mixture.

Thus, a relative concentration of cyclopentanone of 1.26 in a mixture of cyclopentanone and dioxan means that, in order to get an impression of equal intensity, 1.26 times as much cyclopentanone is needed as when cyclopentanone is presented alone. Therefore, adding dioxan results in reduction of the perceived intensity and makes it necessary to add more cyclopentanone to restore the original intensity. In such a mixture the amount of dioxan can be expressed in the same way.

(6) The relative concentrations calculated for the different mixtures were plotted in a graph like that shown in Figure 1, where the relative concentration of one substance is represented along the abscissa and that of the other is given at the ordinate.

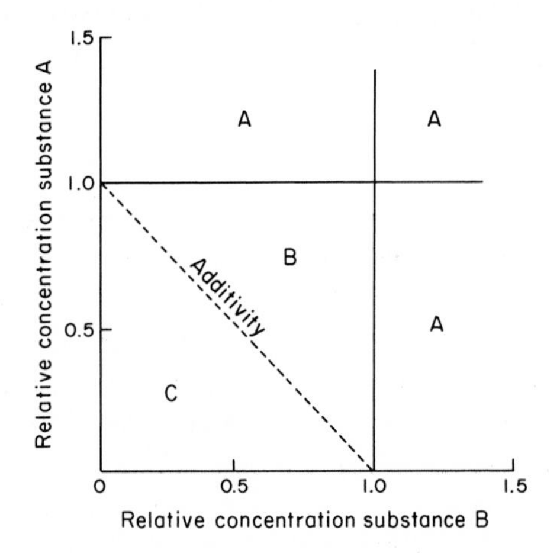

Figure 1 Schematic curves of isointensity of substance odors. Compensation is represented in A areas. Solid horizontal and vertical lines represent independence of subjects in the mixture. Partial additivity occurs in area B. Complete additivity is reached when points lie on dotted line.

In this graph the different areas marked A, B, and C represent the different phenomena described earlier. Compensation is represented in the areas marked A. Whenever a point falls in one of these areas, it means that in order to reach equal subjective intensity at least one of the two components must be added to the mixture in a higher concentration than when it is presented alone. The solid horizontal and vertical lines represent independence of the substances in the mixture. Whenever a point lies on one of these lines it indicates that at least exactly as much of one substance is needed in the mixture as would be needed of the substance alone in order to get an impression of the given intensity.

In area B, partial additivity takes place. Whenever a point lies in this area both substances are needed in a lower concentration in the mixture than when they are presented alone. Nevertheless, the additivity is incomplete because the summated relative concentrations of the two substances needed to get an impression of the given intensity are larger than 1. Complete additivity is reached when the points lie on the dotted line. Here, for instance, half of the concentration of substance A can be replaced by half of the concentration of substance B to obtain an impression of equal intensity to that of the pure substances.

Below the line in area C, true synergy or odor enhancement takes place. For each point in this area the sum of the relative concentrations of the two substances is less than 1, indicating that the loss of a given amount of substance A can be replaced by a smaller amount of substance B.

RESULTS

The results obtained in the two groups of three substances will be discussed successively. In each of the figures the points obtained at the three different levels of subjective intensity (30 per cent = low sensitivity; 50 per cent = medium sensitivity; 70 per cent = high sensitivity) are given. The points of the same level are connected by lines, but the lines have no other meaning than to serve as a visual aid.

In Figure 2, the results of cyclohexanone and dioxan are given. These two substances do not influence each other in cross-adaptation experiments. Partial additivity and even compensation are found, especially in the curves for the higher intensities

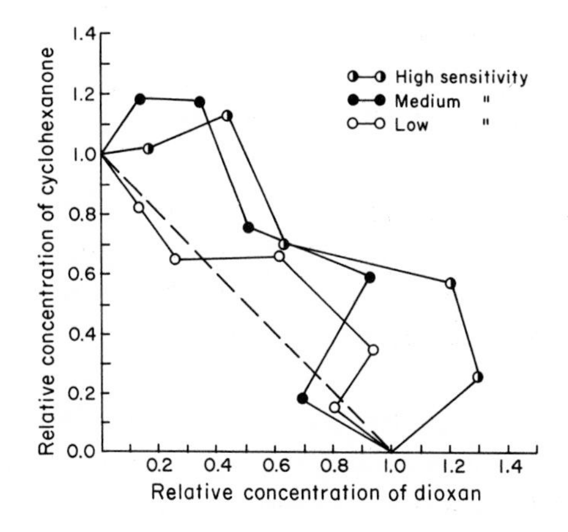

Figure 2 Isointensity curves for mixtures of cyclohexanone and dioxan. To reach the 70 per cent (high sensitivity) criterion, 1.29 times more dioxan is needed when a bit (0.27 times) of cyclohexanone is added.

and when the mixing ratios are more extreme. It is clear that considerably more dioxan (1.29 times) is needed to reach the 70 per cent "stronger than standard" criterion when a bit of cyclohexanone (0.27 times) is added than when no cyclohexanone is present.

The fact that the shape of the curves is dependent upon the level of sensitivity or subjective intensity is in accordance with the old data of Zwaardemaker[7] and Backman,[1] who claim that the perceived reduction of intensity (compensation) is dependent upon two factors: the mixing ratio of the components; and the intensity of the components. Compensation takes place only within specific limits of intensity for each given mixing ratio, according to these authors. In Figure 2 no true signs of synergy are found, and even complete additivity seems to be an exception.

The results obtained with cyclopentanone and cyclohexanone, which influence each other positively and to an almost equal degree in cross-adaptation experiments, are given in Figure 3. Here, addition of a small amount of cyclohexanone results in com-

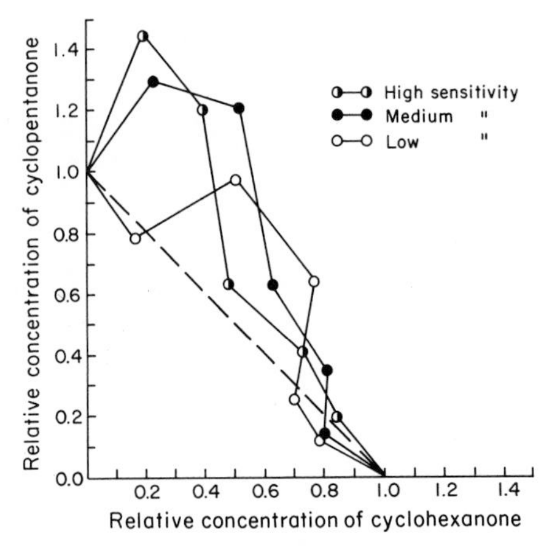

Figure 3 Isointensity curves for mixtures of cyclopentanone and cyclohexanone. The components influence each other positively to an almost equal degree in cross-adaptation experiments.

pensation in at least two of the curves (70 per cent and 50 per cent), but addition of small amounts of cyclopentanone does not decrease the odorous intensity and leads to almost complete additivity. Again, it can be seen that the relationships differ somewhat according to the level of subjective intensity chosen. No enhancement or synergy is found.

In Figure 4 the results for dioxan and cyclopentanone are given. Dioxan has a much stronger cross-adaptational influence on the sensitivity to cyclopentanone than cyclopentanone has on the sensitivity to dioxan. Compensation is provoked when dioxan is added in small quantities, but cyclopentanone does not give rise to compensation when added in small amounts. Here, additivity increases with the subjective intensity in a regular way, as can be seen from the order of the three curves.

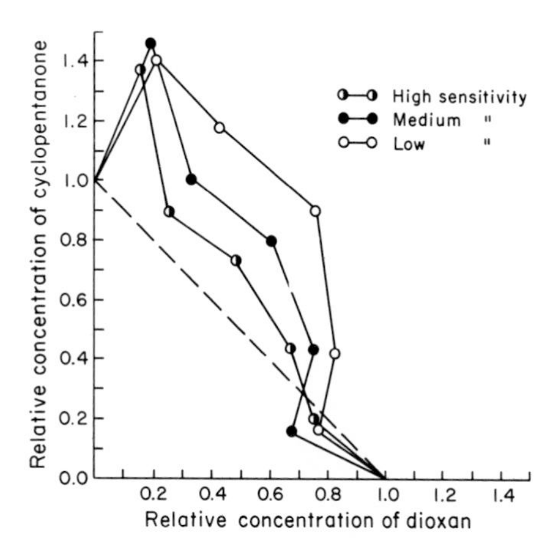

Figure 4 Isointensity curves for mixtures of cyclopentanone and dioxan. Curves show that additivity increases with the subjective intensity in a regular way.

In the second group citral and benzyl acetate are the substances that exert no cross-adaptational influence on each other. The lines in Figure 5 indicate that additivity is prevailing in the mixtures. Only when citral is present in small quantities is there some tendency to compensation. On the other hand, there are some clear indications that synergy occurs in mixtures with a mixing ratio of about 1:1.

The results of citral and safrole, which influence each other to the same degree in cross-adaptation experiments, are shown in Figure 6. Both substances reduce the intensity of the mixture when introduced in small quantities. At low subjective intensity, however, a small amount of safrole has the opposite effect, and synergy is found. Additivity prevails again when the ratios are less extreme; in these cases there

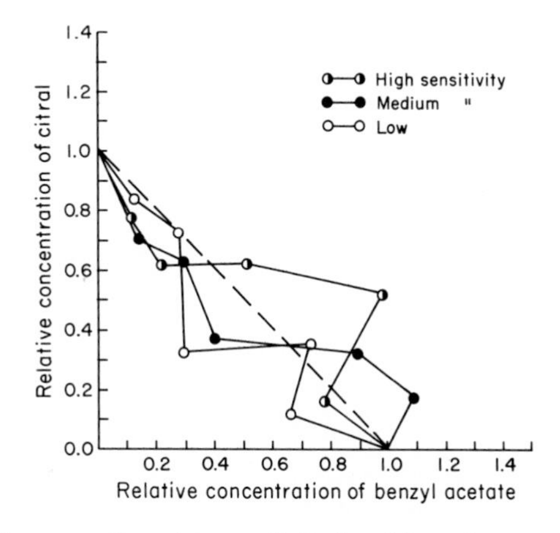

Figure 5 Isointensity curves for mixtures of citral and benzyl acetate. Additivity prevails except when citral is present in small quantities. Synergy probably occurs in mixing ratios of about 1:1.

is even a tendency towards synergy, although only the point at the lowest subjective intensity level lies significantly below the additivity line.

In Figure 7 the results obtained for benzyl acetate and safrole are illustrated. Benzyl acetate exerts a cross-adaptational influence on safrole but safrole does not affect the sensitivity to benzyl acetate. As can be seen in Figure 7, both substances cause compensation at the low-intensity level when the mixing ratios are extreme, but when the mixing ratios are less extreme, or when the intensities are higher, additivity becomes prevalent. When the ratio is 1:1 there is evidence of synergy.

CONCLUSIONS

The results of these experiments show that intensity reduction or compensation occurs in the majority of cases in which two substances are mixed in uneven proportions. Especially when the mixing ratios are in the order of 1:8 or 1:6 true compensation tends to occur, but when the mixing ratio is about 1:1 there is a tendency towards additivity in many pairs and in some cases even a clear indication of synergy, or intensity enhancement. Except for the last finding, these results are in good agreement with the data obtained in an earlier study.[5] Synergy, however, although definitely found here, remains rare. In the experiments described, there was no clear relation-

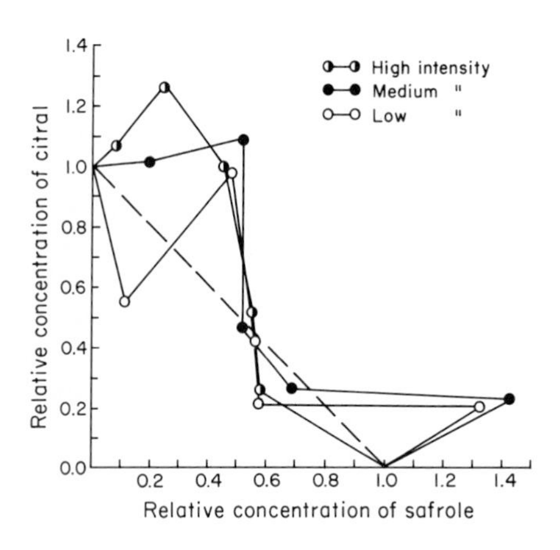

Figure 6 Isointensity curves for mixtures of citral and safrole. In small quantities, both reduce the mixture intensity; at low subjective intensity, a small amount of safrole causes synergy.

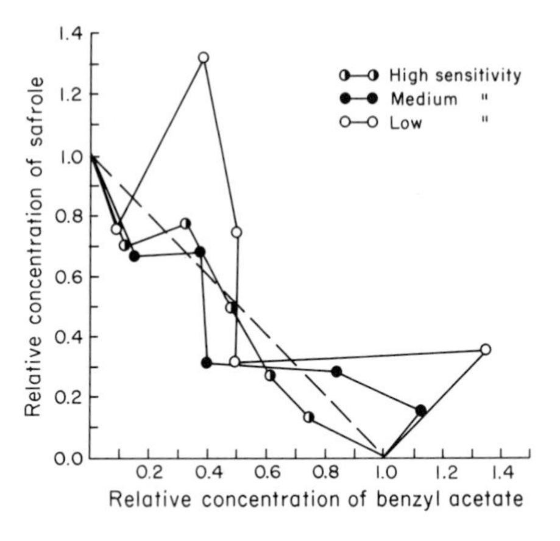

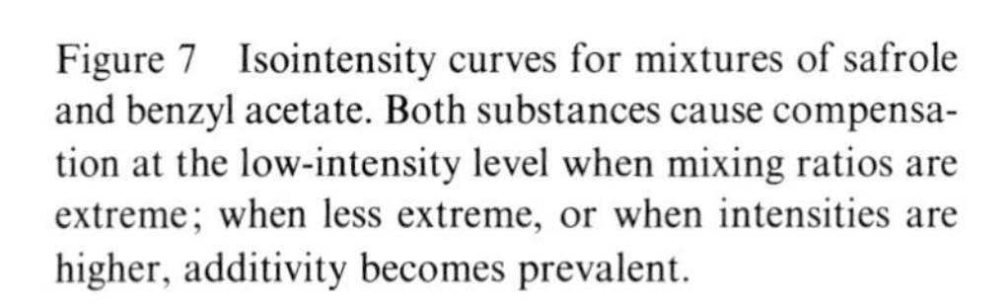

Figure 7 Isointensity curves for mixtures of safrole and benzyl acetate. Both substances cause compensation at the low-intensity level when mixing ratios are extreme; when less extreme, or when intensities are higher, additivity becomes prevalent.

ship between high subjective intensity of the stimuli and additivity. Only the data in Figure 4 and in Figure 7 show such a tendency towards increasing additivity with increasing subjective intensity. Instances of the inverse relationship can be found in the other figures.

The fact that small amounts of added substance show a more pronounced compensation effect than do large amounts almost certainly excludes explanations of the intensity reducing effect in terms of overlap in the occupation of receptor sites. In that case, compensation would be expected to grow with the amount of added substance.

A comparison of the mixture data (simultaneous stimulation) with the cross-adaptation data (successive stimulation) shows no clear relationship. Sometimes a substance that has a strong cross-adaptational influence on another does not show any influence when presented in a mixture; sometimes a substance that does not decrease the sensitivity to another substance in cross-adaptation does influence the intensity of the mixtures strongly when added in small quantities. Inverse relationships also occur.

The lack of specific relationship may be due to differences in the levels of intensity at which the two sets of data were obtained. On the other hand, the occurrence of compensation seems to be dependent to a large degree on the substances used. Cyclopentanone and benzyl acetate do not seem to give rise to compensation when added in small amounts, whereas addition of the other four substances does influence the intensity in a negative way.

REFERENCES

1. Backman, E. L. 1917. Experimentella undersökningar över luktsinnets fysiologi. *Upsal. Lakareforenings Forhandl.* **22**, pp. 319–464.
2. Baker, R. A. 1964. Response parameters including synergism-antagonism in aqueous odor measurement. *Ann. N.Y. Acad. Sci.* **116**, pp. 495–503.
3. Jones, F. N., and M. H. Woskow. 1964. On the intensity of odor mixtures. *Ann. N.Y. Acad. Sci.* **116**, pp. 484–494.
4. Kendall, D. A., and A. J. Neilsen. 1966. Sensory and chromatographic analysis of mixtures formulated from pure odorants. *J. Food Sci.* **31**, pp. 268–274.
5. Köster, E. P. 1968. Relative intensity of odor mixtures at supra-threshold level. *Olfactologia* **1**, pp. 29–41. (Suppl., Cahiers d'Oto-Rhino-Laryngologie **3**, No. 5.)
6. Rosen, A. A., J. B. Peter, and F. M. Middleton. 1962. Odor thresholds of mixed organic chemicals. *J. Water Pollut. Contr. Fed.* **35**, pp. 7–14.
7. Zwaardemaker, H. 1907. Über die Proportionen der Geruchs Kompensation. *Arch. Anat. Physiol.* (*Leipzig*) **31**, Suppl. Bd. **59**, p. 70.

A LINEAR RELATIONSHIP BETWEEN OLFACTORY EFFECTIVENESS AND IDENTIFIED MOLECULAR CHARACTERISTICS, EXTENDED TO FIFTY PURE SUBSTANCES

PAUL LAFFORT

Collège de France, Paris, France

INTRODUCTION

Since the work of Passy,[11] it has been known that the relation between human olfactory threshold values and chain length in various homologous series provides dissimilar complex curves. For example, Figure 1 shows these relationships in four homologous series: saturated hydrocarbons, primary alcohols, ethanol esters, and carboxylic acids, according to a table of standard thresholds we established in 1963.[6]

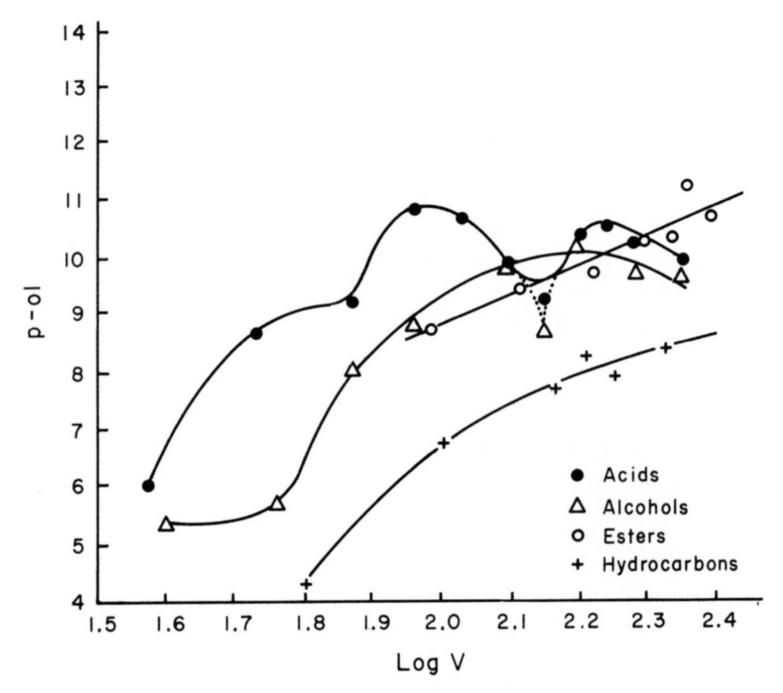

Figure 1 Relations of four homologous series between the cologarithm of threshold molar concentration in air (p-ol) and the logarithm of molecular volume.

That same year we were able to show[7] that, by taking the air-water partition coefficient into account, these values calculated in the aqueous phase provided linear relationships in log-log coordinates (Figure 2). This result was evidence of the important role played by the aqueous mucus in which olfactory cilia are immersed. This role is confirmed by the analogy between the straight lines calculated in aqueous phase in mammals and those determined in gaseous phase in insects, in which receptors are deprived of a superficial aqueous layer.[3,8]

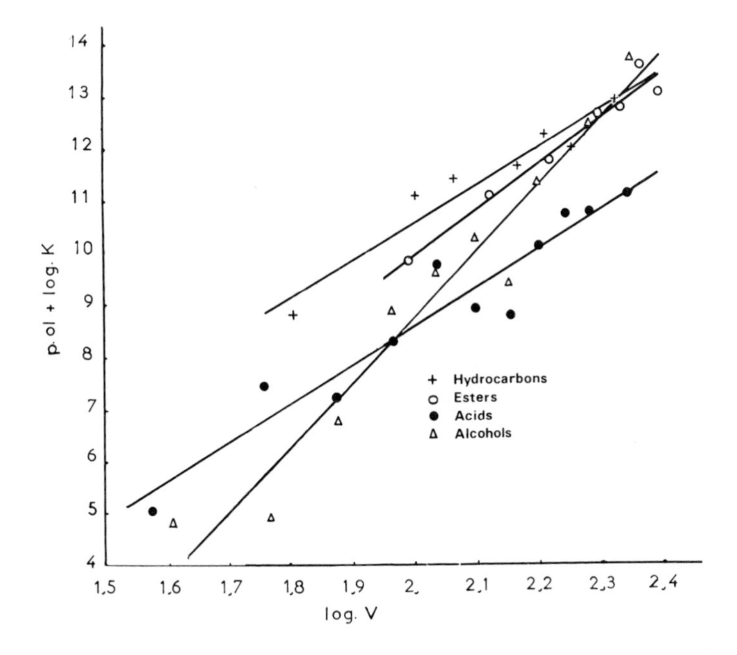

Figure 2 Relations of the same substances as in Figure 1 between p-ol in aqueous phase (p-ol + log K) and the logarithm of molecular volume.

The findings opened up the possibility of estimating the olfactory effectiveness of molecules. By taking two experimental values into account (or even only one, because of the convergence of the straight lines), it was possible to infer all other terms.

To obtain a generalized relationship between molecular properties and olfactory effectiveness, it was necessary to determine the physicochemical parameters responsible for the differences of slopes and origins in ordinates. We have been led to determine three parameters, which we have designated A, H, and P, respectively. Their combination enables us to account not only for the activity of the 35 compounds in the preceding four homologous series, but also for 15 additional compounds, including poly- and monosulfide, poly- and monochloride, bromide, and nitrogenous substances of ethylenic, benzenic, and paraffinic types.

THE APOLAR TERM: A

The apolar parameter A is dependent on the molar volume at the boiling point (V_b), which is more representative of the actual molar volume than is that measured at 20°C. V_b is an additive molecular property. This allows us to calculate A by use of increments. We have adopted Kopp's method (quoted by Partington,[10] pp. 17–28), which has already been used by Davies.[2] In some cases, the measured value can be determined by the volumic expansion coefficient. A is expressed by the following relation:

$$A = \frac{15.7\ V_b}{48.6 + V_b}$$

Numerical coefficients have been empirically determined, based upon the best expression of the paraffinic hydrocarbon series. When the co-log of molar concentration in water at threshold (p-ol + log K) is plotted against A for the four homologous series (those shown in Figure 1), the graph obtained is almost identical to the one in Figure 2. It is striking that, in these figures, the higher the slopes of the four series, the higher the tendency to form hydrogen bonds. This could be interpreted by assuming that among odor molecules dissolved in aqueous mucus, those associated with water by hydrogen bonding are inactivated or unavailable for olfactory receptors. If so, it was necessary to look for a quantitative expression of that molecular property.

HYDROGEN BONDING INDEX: H

Presumably, it is possible to determine the level of hydrogen bonding of various compounds dissolved in heavy water by measuring the shift of the infrared optical density band. Gordy and Stanford[5] used this method for heavy methanol.

First, we can assume that the tendency of a substance to form a hydrogen bond with water is comparable to its tendency to form a hydrogen bond with itself in its pure form. A variety of indexes designed to account for this latter property has been proposed by various authors.[4,10] All the indexes are abnormally high, when we take into account the molecular size, boiling point, viscosity, heat of evaporation, or surface tension due to stronger cohesive forces. None of those proposed indexes appeared suitable for our purposes, so we looked for a linear relationship in saturated and ethylenic hydrocarbons between boiling point, on the one hand, and molecular refraction and molecular volume, on the other. The straight line of Figure 3 allows us to propose the following relationship:

$$H = \frac{T}{48} + 5.1\left(1 - \sqrt{\frac{R_m}{\rho}}\right)$$

in which: T = boiling point in °K; R_m = molar refraction; ρ proportional to radius of mole = $\sqrt[3]{V_b}$.

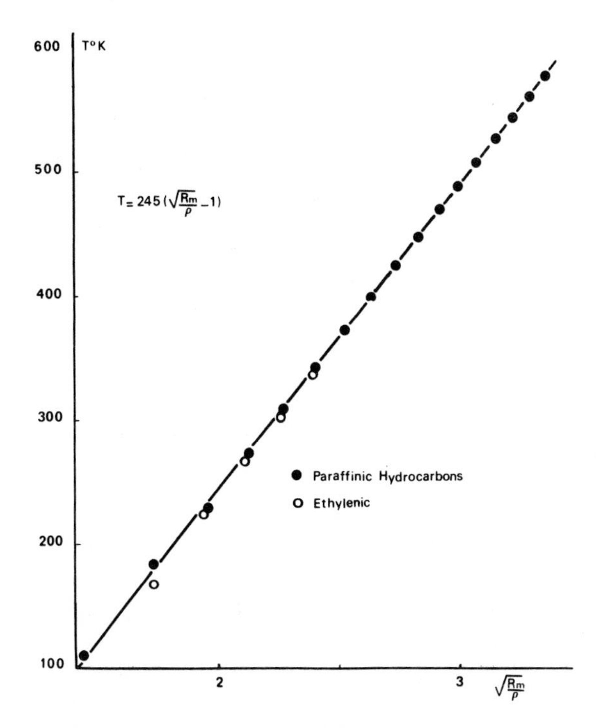

Figure 3 Linear relation for paraffinic and ethylenic hydrocarbons, permitting an estimate of a hydrogen bonding index, H.

In all the substances studied, H is positive or zero. Moreover, this parameter seems to give a good measure of the knowledge concerning hydrogen bonding: high values for small hydroxyl molecules, mean values for esters and ketones, etc.

The co-log of thresholds in the aqueous phase (p-ol + log K) plotted against A–H for the four homologous series (saturated hydrocarbons, primary alcohols, ethanol esters, carboxylic acids) displays a single straight line with an acceptable deviation ($1 - r^2 = 0.09$). This result is comparable to that of Davies, for example. Only the first two acid terms, shown in Figure 4, differ from the theoretical value by more than two log units. The A–H expression could be compared to the oil-water partition coefficient that has been used by various authors.

As shown in Figure 4, the consideration of the air-water partition coefficient and of A–H does not account for all cases. Among the 15 additional substances chosen, all of which have a small molecular volume (less than 125 ml), seven do not fall on the theoretical straight line; the highest deviations are for NH_3 and sulfurous compounds. Davies had already noted that the seven deviant compounds were exceptions in his proposed general relationship between molecular properties and olfactory effectiveness, and he suggested that they had a distinct stimulatory mechanism. However, as we shall see, the strong olfactory effectiveness of these small molecules may be explained by the consideration of their "polarizability."

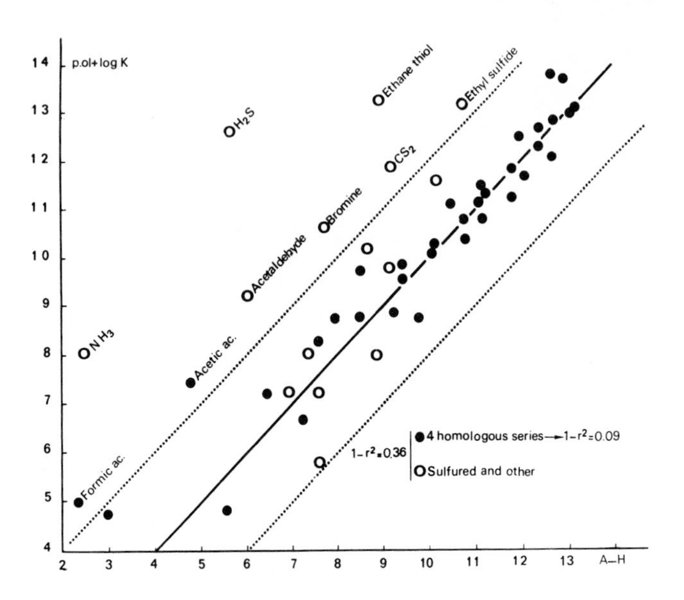

Figure 4 p-ol + log K varies linearly in terms of A–H for four homologous series, but does not account for the activity of some small molecules.

ATOMIC VOLUMIC POLARIZABILITY: P

When an electric field is applied to a molecule, the centers of gravity of positive and negative charges move. As a result, an electrical moment proportional to the field E is produced. The proportionality factor is called "molecular polarizability" $\mu = \alpha E$. The Claussius-Mosotti-Debye relation allows us to calculate α:

$$\frac{4\,N\alpha}{3} = \frac{n^2 - 1}{n^2 + 2}\frac{M}{d}$$

in which N is the Avogadro number and n the refractory index for the D ray of sodium. The second term of this relation is named "molecular refraction" R_m.

The molecular refraction is an additive property, which may, like V_b, be calculated by increments. Thus, the volumic polarizability, or density of the polarizability, may be calculated for each of the constitutive atoms of the molecules by the ratio of molecular refraction increment r to molecular volume increment v. The highest r/v ratio of each molecule has proved to be the exclusive fourth molecular factor of olfactory effectiveness. Particularly, it accounts for the high efficacy of sulfurous compounds and for the fact that this efficacy is not dependent on the number of sulfur atoms in the molecule.

We have used the methods of Vogel and his colleagues.[12] The parameter P is given in the relation:

$$P = 38\ r/v$$

The polarity of the cellular receptor membrane may explain the efficacy of this polarizability of the stimulus.

CONCLUSION

The final relation is:

$$\text{co-log O.T. M or p-ol} = \text{co-log K}\ {}^{\text{air}}_{\text{water}} + (\text{A or P}) - \text{H}$$

Only the higher value, whether A or P, is considered. Figure 5, in which p-ol + log K is plotted against (A or P)—H, shows that all the values are on a single straight

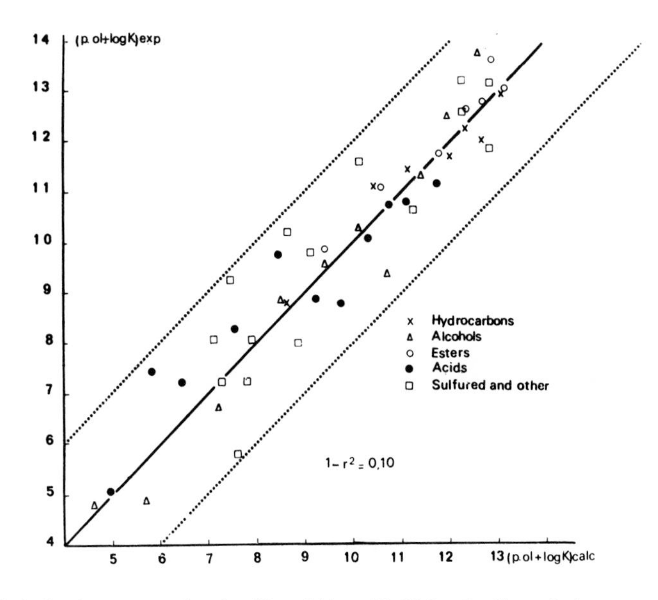

Figure 5 Relation between p-ol + log K and (A or P)-H for the 50 studied compounds.

line with a deviation ($1 - r^2$) equal to 0.10. Finally, Figure 6 shows the correlation between calculated and experimental values of thresholds. The deviation between these two series of values is slightly higher because of a narrower range of values. But for the 50 substances studied, the difference between calculated and experimental thresholds is below one log unit for 42 compounds and between one and two log units for the remaining eight.

All numerical data used for the composition of Figures 4, 5, and 6 are presented in Table I.

In further investigations it should be possible to determine experimentally the parameters log K, A, P, and H by gas liquid chromatography, the first by means of a static dosage from an equilibrium chamber and the other three by the combination of retention volumes of apolar columns and polar columns with and without hydroxyl groups.

TABLE I

Compound	log K_{37}	P	A	H	p-ol exper	p-ol calc
Ethane	4,601	8,951	8,263	0,317	4,20	4,03
Butane	4,431	8,951	10,423	−0,001	6,66	5,99
Pentane	4,346	8,951	11,095	−0,021	7,05	6,77
Heptane	4,045	8,951	12,025	−0,009	7,63	7,99
Octane	4,152	8,951	12,362	0,001	8,11	8,21
Nonane	4,270	8,951	12,643	0,004	7,75	8,37
Hendecane	4,560	8,951	13,083	0,012	8,30	8,51
Methanol	−0,594	8,951	7,287	4,340	5,37	5,20
Ethanol	−0,821	8,951	8,814	3,257	5,68	6,51
Propanol	−1,218	8,951	9,826	2,600	7,96	8,44
Butanol	0,079	8,951	10,693	2,200	8,75	8,41
Pentanol	0,356	8,951	11,300	1,882	9,20	9,06
Hexanol	0,563	8,951	11,779	1,649	9,69	9,57
Heptanol	0,730	8,951	12,158	1,448	8,63	9,98
Octanol	1,228	8,951	12,472	1,288	10,06	9,96
Decanol	2,869	8,951	12,959	1,011	9,59	9,08
Dodecanol	4,187	8,951	13,319	0,729	9,54	8,40
Ethyl acetate	1,204	9,883	10,828	1,412	8,64	8,21
Ethyl butyrate	1,623	9,883	11,915	0,835	9,47	9,46
Ethyl hexanoate	2,090	9,883	12,602	0,824	9,67	9,69
Ethyl octanoate	2,423	9,883	13,083	0,731	10,21	9,93
Ethyl nonanoate	2,483	9,883	13,271	0,563	10,29	10,23
Ethyl decanoate	2,458	9,883	13,433	0,562	11,16	10,41
Ethyl hendecylate	2,408	9,883	13,575	0,419	10,64	10,75
Formic acid	−1,011	9,883	7,222	4,907	6,06	5,99
Acetic acid	−1,416	9,883	8,906	4,092	8,86	7,21
Propionic acid	−1,951	9,883	10,006	3,571	9,15	8,39
Butyric acid	−2,492	9,883	10,799	3,226	10,78	10,06
Pentanoic acid	−0,870	9,883	11,444	2,973	10,62	9,34
Hexanoic acid	−0,712	9,883	11,896	2,709	9,61	9,90
Heptanoic acid	−0,434	9,883	12,272	2,516	9,20	10,19
Octanoic acid	−0,127	9,883	12,581	2,258	10,24	10,45
Nonanoic acid	0,281	9,883	12,838	2,100	10,46	10,46
Decanoic acid	0,670	9,883	13,056	1,943	10,10	10,44
Dodecanoic acid	1,320	9,883	13,406	1,661	9,86	10,43
Acetaldehyde	0,318	9,883	8,440	2,417	8,90	7,15
Acetone	−0,337	9,883	9,649	2,070	7,57	8,15
Allylic alcohol	−1,186	9,883	9,486	2,597	8,43	8,47
Ammonia	0,413	10,208	5,567	3,092	7,67	6,70
Bromine	2,079	11,946	8,377	0,689	8,54	9,18
Carbon disulfide	2,362	13,054	9,384	0,220	9,48	10,47
Chloroform	2,549	9,733	9,984	0,856	7,25	6,58
Dichloro-1-2-ethane	2,052	9,733	10,179	1,537	8,12	6,59
Isopropyl alcohol	−0,947	8,951	9,893	2,287	6,74	8,55
Propenal	−0,038	9,883	9,331	1,990	8,09	7,93
Pyridine	−1,301	9,883	10,315	1,476	9,29	10,14
Hydrogen sulfide	2,847	13,001	6,418	0,763	9,72	9,39
Ethane thiol	2,526	13,001	9,654	0,742	10,68	9,73
Ethyl sulfide	2,528	13,317	11,217	0,500	10,62	10,29
Benzene	3,387	9,883	10,492	0,334	8,18	6,77

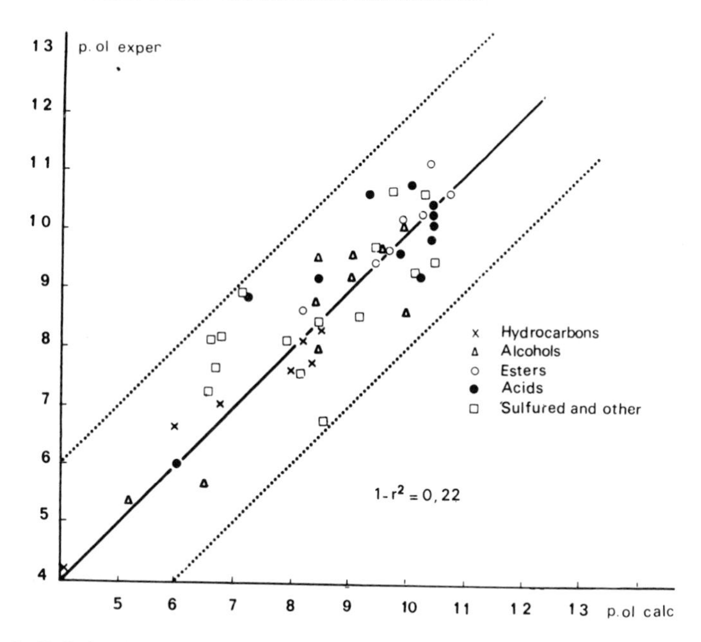

Figure 6 Relation between experimental thresholds and thresholds calculated from the proposed formula, expressed by p-ol.

It should be noted that the flexibility or the rigidity of the molecule, as well as its form, which has been referred to by several authors,[1,9] does not seem to be important. As I have pointed out, the determining role of the active molecular parameters discussed in this study could be a guide to further research on the fine mechanisms of the excitatory process of receptor cells.

REFERENCES

1. Beck, L. H. 1964. A quantitative theory of the olfactory threshold based upon the amount of the sense cell covered by an adsorbed film. *Ann. N.Y. Acad. Sci.*, **116**, pp. 448–456.
2. Davies, J. T., and F. H. Taylor. 1959. The role of adsorption and molecular morphology in olfaction: the calculation of olfactory thresholds. *Biol. Bull.* **117**, pp. 222–238.
3. Dethier, V. G. 1954. Olfactory responses of blowflies to aliphatic aldehydes. *J. Gen. Physiol.* **37**, pp. 743–751.
4. Duclaux, J. 1934. Traité de chimie physique appliquée à la biologie. Hermann, Paris.
5. Gordy, W., and S. C. Stanford. 1941. Spectroscopic evidence for hydrogen bonds: comparison of proton-attracting properties of liquids. *J. Chem. Phys.* **9**, pp. 204–223.
6. Laffort, P. 1963. Essai de standardisation des seuils olfactifs humains pour 192 corps purs. *Arch. Sci. Physiol.* **17**, pp. 75–105.
7. Laffort, P. 1963. Mise en évidence de relations linéaires entre l'activité odorante des molécules et certaines de leurs caractéristiques physicochimiques. *C. R. Acad. Sci.* (*Paris*) **256**, pp. 5618–5621.
8. Laffort, P. 1965. Efficacité odorante et activité thermodynamique. *Rev. Laryngol.* (*Bordeaux*) (Suppl., Oct.), pp. 860–879.
9. Mullins, L. J. 1955. Olfaction. *Ann. N.Y. Acad. Sci.* **62**, pp. 247–276.
10. Partington, J. R. 1951. An Advanced Treatise on Physical Chemistry. Vol. II: The Properties of Liquids. Longmans, Green and Co., London.
11. Passy, J. 1895. Revue générale sur les sensations olfactives. *Année Psychol.* **2**, pp. 363–410.
12. Vogel, A. I., W. T. Cresswell, G. H. Jeffrey, and J. Leicester. 1952. Physical properties and chemical constitution. Part XXIV. *J. Chem. Soc.*, Part I, pp. 514–549.

A PLAN TO IDENTIFY MOST OF THE PRIMARY ODORS

JOHN E. AMOORE

Western Regional Research Laboratory, Agricultural Research Service, U.S. Department of Agriculture, Albany, California

INTRODUCTION

Is the human sense of smell composed of primary odors or not? After decades of discussion, the debate goes on. A recently revived approach to this problem consists of studying certain naturally occurring defects in the olfactory system, known informally as "odor-blindness." Each type of odor-blindness may correspond with the absence or loss of one of the primary odors. Accordingly, I have begun a systematic search for different examples. I have hunted through the meager literature; I have combed out my own laboratory notebooks; I have badgered my professional colleagues; I have made appeals in my lectures; and I have published requests in my papers. Not long ago the editors of two journals kindly printed the following letter, which will serve to introduce the problem.[5,6]

Are You "Odor-Blind"?

To the Editor:

It is well known that human color-vision is based on a system of three primary colors: red, green, and blue. This makes possible the reproduction of any color by means of three inks in color printing, three emulsions in color photography, or three phosphors in color television. Corresponding defects in components of the human visual system result in three main types of color-blindness: red-blindness which afflicts about 2% of men; green-blindness, also affecting 2% of men; and a rare blue-blindness (1 in 50,000).

However, it is a very little-known fact that a similar situation applies to odor. Some persons, while having an apparently normal sense of smell, find that they are unable to detect one particular odor. About 10% of people cannot smell the poisonous hydrogen cyanide. Two per cent cannot smell the sweaty odor of isovaleric acid. One person in a thousand cannot smell the mercaptan odor of the skunk. This phenomenon has been called "odor-blindness." It is known scientifically as "specific anosmia," or specific loss of one component of the sense of smell.

It had been suspected for centuries that color-vision works on just three primaries, because of the common knowledge among artists that three pigments suffice to mix any color. But no such practical rule has ever been found for odor, which leads to the suspicion that odor, if indeed it is based on primaries, must be far more complex than color.

This is where odor-blindness comes in. It is a very reasonable hypothesis that each type of odor-blindness is due to a malfunction of one of the primary odors of the sense of smell. Hence, if we knew just how many different types of odor-blindness exist, we would have a clear indication of how many primary odors there are. This idea was put forward twenty years ago by Marcel Guillot,[29] who listed 8 types of odor-blindness as a beginning. My own recent investigation of the newly discovered isovaleric acid anosmia has firmly established the corresponding "sweaty" primary odor.[4] But these observations and others have made it obvious that the list of known specific anosmias is far from complete.

The scientific community can contribute quite effortlessly toward clarifying an ancient problem. Many readers must be able to recall personal experiences which, on looking back, may have been due to odor-blindness. You know that you have a perfectly good sense of smell, that faithfully registers the aroma of food, the scent of flowers, the allure of perfume, the stench of decay, or the warning of fire. Yet inexplicably you encounter some particular odor which is obvious to your colleagues or family, but which is a complete blank for you. It might be a perfume, a flower, a foodstuff, or a specific chemical compound. Or you may perceive some unconventional reaction, like a fruity smell in isobutyric acid, or a sulfury smell in pineapple. You probably dismissed your discovery with a joke, and thought no more of it. But the chances are that you happen to be specifically anosmic to that one particular odor, or class of odors.

If so, do please write and tell me about it, because the information you provide may furnish one more missing piece in the jig-saw puzzle of the sense of smell.

(signed) John E. Amoore

Western Regional Research Laboratory
(postal address given)

RESULTS

Specific Anosmias

The search for specific anosmias has been quite productive. It has raised the number of known chemical examples of odor-blindness from the eight described by Guillot in 1948 to a grand total of 62 at the time of writing. Lest the reader despair, it should be explained immediately that this does not necessarily imply the existence of 62+ primary odors. It is very likely that the list contains many "duplicates," i.e., two or more different chemical examples which are normally perceived by the same deficient primary odor receptor.

The data are consolidated in Table I. I have called them "*Reputed* specific anosmias" because the observations were often of an informal or fragmentary nature. Nevertheless, the specific defect was sufficiently striking to the person concerned that he made a spontaneous comment, or even felt motivated to sit down and write me a letter. It is worth recalling that a single casual observation of equally modest scientific validity[22] led, after experimental refinement, to the first thoroughly established primary odor.[4,7,12]

The listings are restricted to single named chemical compounds of known structure, and it was established that the deficient observer has a substantially normal sense of smell toward other odors. Where known, the concentration of test odorant that the subject failed to smell is specified, and if normal threshold data are available, the minimum ratio of the anosmic to normal threshold concentrations is given. The experimental test methods varied from the simplest spontaneous comment on an odor or lack thereof, all the way up to elaborate, statistically controlled, threshold measurements, sometimes with an olfactometric instrument. A rough idea of the frequency of occurrence of the specific anosmia in the tested group is often available; and here the encouraging feature, from my point of view, is that interesting anosmias

TABLE I REPUTED SPECIFIC ANOSMIAS

The compounds are listed alphabetically, ignoring prefixes.

Compound[a]	Concentration[b]		
	ppm	Diluent	Intensity
Acetic acid	976	water	32 x Th
Adamantane	10,000, w/v	mineral oil	Ma
Allicin	(garlic)[g]	--	--
Allyl isothiocyanate	saturated	air	10^5 x Th
$\Delta^{4,16}$-Androstadien-3-one	saturated	air	--
Δ^{16}-Androsten-3-ol	saturated	air	--
Anisic aldehyde	30	water	Ma
Anisole	6	water	Ma
Benzene	1000	water	5 x Ma
Benzyl alcohol	40,000	water	Ma
Benzyl salicylate	saturated	air	--
tert.Butyl carbinol	80	water	Ma
n-Butyl mercaptan	750	methanol	--
tert.Butyl mercaptan	0.0004	air	5 x Th
Caproic acid	86	water	9 x Th
Cedryl acetate	20, w/v	water	Ma
1,8-Cineole	10	water	830 x Th
Cinnamaldehyde	0.1	water	2 x Th
Cyclotene	10	water	5 x Th
p-Dichlorobenzene	3, w/v	water	Ma
Dimethyl disulfide	0.1	water	83 x Th
Ethylene brassylate	saturated	air	--
Ethylenedichloride	800	water	28 x Th
Eugenol	45	water	Ma
Farnesol	saturated	air	--
Formic acid	45,000	water	33 x Th
Geranial	--	--	--
Geraniol	10,000	DNP	M.E.P.
2-Heptanone	--	--	--
Δ^{9}-Hexadecenolactone	2	water	Ma
Hexylamine	500	DNP	M.E.P.
Hibiscolide	saturated	air	--
Hydrogen cyanide	(1% KCN)	water	--
Indole	10,000, w/v	DNP	M.E.P.
Iodoform	20,000, w/v	DNP	M.E.P.
Ionone	saturated	air	--
Isobutyraldehyde	0.1	water	11 x Th
Isobutyric acid	61	water	7 x Th
Isovaleric acid	3.1	water	16 x Th

For further information see text and footnotes.

Type of test[c]	Specific anosmics[d]			References[e]		Table II line no.[f]
	Found	Tested	%	Main	Supple-mentary	
Threshold, 2/5	2	420	0.5	W		33
Similarity	1	27	4	W		4
Informal	1	--	--	23		20
Olfactometer	1	6	17	51		20
Informal	2	20	10	24		44
Informal	1	12	8	30		19
Similarity	1	27	4	W		17
Similarity	1	25	4	W		5
Similarity	1	28	4	W		3
Similarity	1	27	4	W		5
Informal	1	12	8	28		17
Similarity	1	24	4	W		8
Informal	4	4030	0.1	47		32
Olfactometer	7	35	20	54		32
Threshold, 2/5	9	420	2	12	15	25
Similarity	2	26	8	W		42
Screening, 2/5	6	420	1.4	W		4
Threshold, 1/2	2	20	10	W		6
Threshold series	1	20	5	53		17
Similarity	1	28	4	W		4
Screening, 2/5	15	420	3.6	W		32
Informal	1	--	--	37		19
Screening, 2/5	5	420	1.2	W		3
Similarity	1	27	4	W		5
Informal	1	12	8	29		16
Screening, 2/5	10	420	2.4	W		33
Informal	1	--	--	49		11
Descriptive	11	38	29	31	39,49,36	10
Informal	3	--	--	50		38
Similarity	3	23	13	W		19
Descriptive	several	--	--	33		21
Informal	1	--	--	37		19
Threshold series	198	2885	6.9	17	43,38,26	12
Descriptive	--	--	26	33		30
Descriptive	--	--	20	33		22
Informal	1	12	8	28		16
Threshold, 2/5	29	89	32	W		12
Threshold, 2/5	10	420	2.4	12	22	25
Threshold, 2/5	8	443	1.8	8	12	25

TABLE I (*continued*) REPUTED SPECIFIC ANOSMIAS
The compounds are listed alphabetically, ignoring prefixes.

Compound[a]	Concentration[b]		
	ppm	Diluent	Intensity
Menthol	20,000, w/v	DNP	M.E.P.
d,l-Menthone	6	water	35 x Th
Methional	--	--	>200 x Th
Methyl cyclopropyl ketone	500	water	Ma
Methyl mercaptan	(asparagus)	urine[g]	--
Musk xylol	saturated	air	--
Naphthalene	1, w/v	water	Ma
PEME carbinol	300	water	47 x Th
Pentadecanolactone	1, w/v	water	1430 x Th
Phenylacetic acid	saturated	air	--
Phenylethyl alcohol	5000	DNP	M.E.P.
Phenyl isocyanide	saturated	air	--
Phenyl isothiocyanate	saturated	air	10^5 x Th
1-Propenylsulfenic acid	(onions)[g]	--	--
Putrescine	--	--	--
Salicylaldehyde	10	water	Ma
Skatole	1000, w/v	DNP	M.E.P.
Thiophane	3	methane	3000 x Th
Thymol	saturated	air	--
Trichloroethylene	200	water	Ma
γ-Undecalactone	20	water	Ma
Vanillin	--	--	--
Versalide	2, w/v	water	Ma

[a] Trade or trivial names requiring amplification are as follows: Cyclotene is 1-methylcyclopentenol-2-one-3; Hibiscolide is cyclodioxa-(1,6)-pentadecacarbanone-(17); Methional is 3-methylthiopropanal; Musk xylol is 2,4,6-trinitro-3,5-dimethyl*tert.*butylbenzene; PEME carbinol is phenylethylmethylethyl carbinol; Thiophane is tetrahydrothiophene; Versalide is 1,1,4,4-tetramethyl-6-ethyl-7-acetyl-1,2,3,4-tetrahydronaphthalene.

[b] The concentration of each compound is in parts per million (v/v) except where indicated otherwise. "Saturated in air" means that the straight undiluted chemical was not perceived. DNP = dinonyl phthalate. Wherever possible, some indication of odor intensity is provided. Th = mean threshold concentration for normal observers. Ma = "matching" concentration, usually about 32 × Th[9] but sometimes much higher. M.E.P. = moderately easily perceptible.[33]

[c] Various types of smell test were employed. "Threshold, 2/5" means that a personal threshold was found, using a concentration series of 2 in 5 sample bottle sorting tests.[12] "Threshold, 1/2" is similar, but with a 1 in 2 sorting test.[27] "Threshold series" used a concentration series without sorting.[17] "Olfactometer" shows that a personal threshold was found, using an olfactometric instrument.[45] "Screening 2/5" indicates that only a single concentration 2 in 5 sorting test was used.[12] "Similarity" shows that an observer spontaneously commented that this odor was

For further information see text and footnotes.

Type of test[c]	Specific anosmics[d]			References[e]		Table II line no.[f]
	Found	Tested	%	Main	Supple-mentary	
Descriptive	--	--	26	33		8
Screening, 2/5	6	420	1.4	W		8
Informal	1	--	--	36		41
Similarity	1	24	4	W		3
Informal	some	--	--	40		32
Descriptive	17	53	32	32		19
Similarity	1	20	5	W		4
Screening, 2/5	1	420	0.2	W		13
Screening, 2/5	28	420	6.7	W	30,33	19
Informal	1	4	25	46	34	25
Descriptive	5	70	7	32		10
Informal	1	--	--	44	14	27
Olfactometer	1	6	17	51		20
Informal	1	--	--	23		20
Descriptive	2	24	8	13		29
Similarity	1	24	4	W		6
Descriptive	4	40	10	32		30
Informal	1	--	--	52		32
Informal	1	12	8	28		9
Similarity	2	25	8	W		3
Similarity	1	25	4	W		1
Informal	--	--	--	25		17
Similarity	2	25	8	W		19

imperceptible when it was presented as an unknown in the matching standards method of odor analysis.[9,10] "Descriptive" means that an observer had difficulty perceiving this odor when it was offered for descriptive analysis.[33] "Informal" includes a variety of spontaneous incidental observations, usually relating to the undiluted chemical, and reported by a single observer who was not directly involved in experimental olfactometry.

[d] Percentage frequencies of occurrence for different anosmias are not usually comparable, because the test odorant concentrations are seldom a constant multiple of the threshold.

[e] References given as "W" are observations recorded by the author at the Western Regional Research Laboratory (in the course of other experiments) but not previously published. Other references are listed by number at the end of the paper.

[f] This column gives the tentative assignment of the compound to the indicated horizontal line on the right side of Table II. (Chemical names may sometimes be abbreviated.)

[g] The chief active principles of garlic and of onion are probably allicin and 1-propenylsulfenic acid respectively.[18] The belief that the characteristic odor of the urine after eating asparagus is due to methyl mercaptan may be somewhat apocryphal.[42]

are common enough in the population to provide an accessible basis for systematic experiments.

In the references at the right of the Table, I have cited first the most detailed source of information, but I have also tried to give supplementary credit to all other informants relating to the same compound. I would like to take this opportunity to thank publicly my many colleagues, collaborators, and correspondents who have generously furnished much of this information, here published for the first time.

Odor Classifications

The variety of smells represented by the total gamut of odors is proverbial. There have been at least eight major schemes proposed for dividing odors into limited numbers of classes. In the left half of Table II these eight classifications are compared, and I have made an attempt to match up, on the same horizontal line, those classes that appear to be equivalent or at least very similar. I started with Zwaardemaker's classification of 1895, which, though not the earliest, was the first detailed comprehensive system.[57] His nine classes, partly based on the results of cross-adaptation experiments, were further divided into 30 subclasses. The great classifier Linnaeus had turned his attention to odors in 1756, but his seven classes are mostly exemplified by plants with exhalations of unknown, and probably mixed, chemical composition.[41] In 1915 Henning suggested six classes arranged in an "olfactory prism," and he gave chemical examples.[35] Further compression to four classes was suggested by Crocker and Henderson in 1927,[19] but the system requires expert handling to obtain consistent results.

My own theory of seven primary odors was put forward in 1952. It was based on considerations of molecular rigidity and the frequency of occurrence of odor descriptions in the chemical literature.[1–3] The year 1964 saw two new classifications of odors, independently arrived at by computerized factor analyses of multidimensional odor similarity comparisons among large groups of odorants by panels of observers; Schutz suggested nine classes[48] and Wright and Michels settled on eight.[55] A swing away from the simplifying trend (four to nine classes) of the earlier twentieth century has been suggested recently by the 44-class system of Harper, Bate-Smith, Land, and Griffiths.[33] In the miscellaneous column are added a few extra odor descriptions that do not seem to fit comfortably into any of the major classifications. (Sensations which are probably not true odors at all are relegated from consideration to the foot of the Table.)

Each of these classifications of odors is the work of one or more investigators who devoted a great deal of knowledge, thought, and often experimental endeavor, to the problem. Despite the apparent contradictions among them, it would be prudent to assume that each classification contains at least a germ of the truth, and that together they may very well contain most of the truth. Although a total of 118 odor classes were suggested by the various authors, it required very little arbitrariness on my part to reduce them to 44 lines.

TABLE II
CLUES TO THE OLFACTORY CODE

		GENERAL ODOR CLASSIFICATIONS									SPECIFIC ANOSMIA ANALYSES					
LINE NO.		ZWAARDEMAKER 1895 30 (SUB) CLASSES	LINNAEUS 1756 7 CLASSES	HENNING 1915 6 CLASSES	CROCKER & HENDERSON 1927 4 CLASSES	AMOORE 1952 7 CLASSES	SCHUTZ 1964 9 CLASSES	WRIGHT & MICHELS 1964 8 CLASSES	HARPER ET AL. 1968 44 CLASSES	MISCEL-LANEOUS ADDITIONAL	REPUTED SPECIFIC ANOSMIAS				ESTABLISHED PRIMARY ODORANT	PROBABLE PRIMARY ODOR
1		FRUITY						HEXYL ACETATE	FRUITY		γ-UNDECA-LACTONE					
2		WAXY							SOAPY							
3		ETHEREAL				ETHEREAL	ETHERISH		ETHERISH; SOLVENT		ETHYLENE-DICHLORIDE	TRICHLORO-ETHYLENE	BENZENE	METHYL-CYCLOPROPYL KETONE		
4		CAMPHOR				CAMPHOR			CAMPHOR; MOTHBALLS		1.8-CINEOLE	NAPHTHALENE	p-DICHLORO-BENZENE	ADAMANTANE		
5		CLOVE	AROMATIC						AROMATIC		EUGENOL	BENZYL ALCOHOL	ANISOLE			
6		CINNAMON		SPICY				SPICE	SPICY		CINNAM-ALDEHYDE	SALICYL-ALDEHYDE				
7		ANISEED						BENZO-THIAZOLE								
8		MINTY				MINTY			MINTY		MENTHONE	MENTHOL	TERT-BUTYL CARBINOL			
9		THYME									THYMOL					
10		ROSY									GERANIOL	PHENYL-ETHANOL				
11		CITROUS		FRUITY				CITRAL	CITROUS		GERANIAL					
12		ALMOND					SPICY		ALMOND		HYDROGEN CYANIDE	ISOBUTYR-ALDEHYDE				
13		JASMINE		FLOWERY		FLORAL			FLORAL		PEME CARBINOL					
14		ORANGE-BLOSSOM	FRAGRANT		FRAGRANT		FRAGRANT		FRAGRANT							
15		LILY														
16		VIOLET									IONONE	FARNESOL				
17		VANILLA					SWEET		VANILLA; SWEET		VANILLIN	BENZYL SALICYLATE	ANISIC ALDEHYDE	CYCLOTENE		
18		AMBER							ANIMAL							
19		MUSKY	AMBROSIAL			MUSKY			MUSK		MACROCYCLIC MUSKS (4)	ANDROSTENOL	MUSK XYLOL	VERSALIDE		
20		LEEK	ALLIACEOUS						GARLIC		ALLYL ISO-THIOCYANATE	PHENYL ISO-THIOCYANATE	ALLICIN	PROPENYL-SULFENIC ACID		
21		FISHY							AMMONIA; FISHY		HEXYLAMINE					
22		BROMINE									IODOFORM					
23		BURNT		BURNT	BURNT		BURNT	AFFECTIVE	BURNT							
24		PHENOLIC							CARBOLIC							
25	*	CAPROIC	HIRCINE		CAPRYLIC				SWEATY		ISOBUTYRIC ACID	PHENYLACETIC ACID	CAPROIC ACID		ISOVALERIC ACID	SWEATY
26	—	CAT-URINE													=25?	—
27		NARCOTIC	REPULSIVE								PHENYL-ISOCYANIDE					
28		BED-BUG														
29		CARRION	NAUSEOUS						SICKLY		PUTRESCINE					
30		FECAL							FECAL		SKATOLE	INDOLE				
31				RESINOUS				RESINOUS	RESINOUS; PAINT							
32				FOUL		PUTRID	SULFUROUS	UNPLEASANT	PUTRID; SULFUROUS		MERCAPTANS (3)	DIMETHYL DISULFIDE	THIOPHANE			
33					ACID				ACID		FORMIC ACID	ACETIC ACID				
34							OILY		OILY							
35	—						RANCID		RANCID						=25?	—
36							METALLIC		METALLIC							
37									MEATY							
38									MOLDY		2-HEPTANONE					
39									GRASSY							
40									BLOODY							
41									COOKED-VEGETABLE		METHIONAL					
42										SANDAL	CEDRYL ACETATE					
43										WATERY						
44										URINOUS	ANDROSTA-DIENONE					
NON-OLFACTORY						(PUNGENT)		(TRIGEMINAL)	(PUNGENT; & 5 OTHERS)							

Possible Assignment of Specific Anosmias to Odor Classes

Following Guillot's suggestion,[29] the list of observed specific anosmias itself constitutes an odor classification, and each chemical example of specific anosmia may represent a primary odor. However, my enthusiasm for tracking down specific anosmias has apparently led to "l'embarras de la richesse." Some of the examples are undoubtedly redundant, especially among the macrocyclic musks, where Guillot noted that the same observer (himself) could not smell any of them.[30] Many of the lower fatty acids have likewise been proved to belong to a common primary odor.[4]

In the right side of Table II, I have entered all known chemicals reputedly exhibiting specific anosmia. I have done my best to enter each one on that horizontal line to which it seems most naturally to belong. One way this was done was by considering the qualitative odor character of the chemical, for a normal observer, and matching it with the descriptive odor names in the classifications at the left. An alternative way was by taking into account any chemical and/or stereochemical similarity between the anosmia chemical and any chemical substances cited by the classifiers as typical representatives of their classes.

It should be emphasized that the parallels drawn in Table II are provisional, and are open to alteration and refinement. Nevertheless, it is interesting that it was necessary to use only 27 lines to accommodate with reasonableness all the 62 known anosmia-exhibiting chemicals. The unoccupied lines admittedly may sometimes represent rare specific anosmias, or scarce organic chemicals. However, there is a fair likelihood that certain of the lines will prove truly superfluous when each primary odor is eventually worked out in detail. This seems quite plausible for line 26 (cat-urine) and line 35 (rancid) which may well be variants of the thoroughly established line 25 (sweaty). Hence there may be between 20 and 30 primary odors in the final analysis.

DISCUSSION

Clues to the Olfactory Code

The information assembled in Table II is gathered from a multitude of sources, and it is often superficial, but I am convinced that it encompasses a very valuable collection of clues to the "olfactory code".[4] By this I mean the stereochemical code, consisting of a limited number of chemically specific primary odor detectors, capable between them of transducing the information content of any mixed odorous stimulus. This is to be contrasted with the electrophysiological code, by which the primary detector signals are transmitted to and processed in the central nervous system.[56] It is likely to be an extremely difficult undertaking to unravel the secondary or electrophysiological code that integrates the information, before the primary stereochemical code which generates that information has been analyzed. In fact, this approach from specific anosmia may offer the most promising avenue of attack on olfactory coding that is available to existing methodology.

As discussed previously,[7] it is very reasonable to assume that each primary odor is represented by a specific receptor protein, whose absence or defectiveness is responsible for a specific anosmia. Shortly after that speculation of mine on odor receptor proteins was made (Sept. 1966), it was announced by Dastoli that specific receptor proteins for the sweet and bitter tastes can be isolated from tissues of cow or pig tongue by conventional biochemical techniques.[20,21] In the wisdom of hind-sight, it could have been surmised that specific taste receptor proteins should exist, because of the long-known inherited taste-blindness to phenylthiocarbamide (PTC, see reference 16). Chemical defects in living organisms are nearly always traceable to a defective protein.

A Plan of Attack on the Olfactory Code

I shall now describe the sequence of events which to my mind has the best chance of success in achieving a solution to the stereochemical code of olfaction. Part of the work, possibly the first three phases and the fifth, I can hope to accomplish myself with the aid of collaborators; but the fourth phase of the campaign will demand either dividing up the tasks among cooperating laboratories, or establishing a special organization to assume the responsibility.

1. *Genetic analysis.* First there should be a thorough genetic study of specific anosmia. Only the cyanide-smelling deficiency has been investigated from this point of view, and there is lack of agreement as to whether the defect is inherited. Kirk and Stenhouse[38] gave evidence, almost exactly corroborated by Fukumoto et al.[26] that the nonsmelling character is inherited as a sex-linked recessive, but Brown and Robinette[17] could find no support at all for sex linkage, and concluded that there was very little influence of inheritance compared with the total variation in sensitivity. I am planning to conduct a new investigation of this question, but working with an established primary odorant, purified isovaleric acid, which should give the clearest attainable indication of the inheritability or otherwise of the corresponding specific anosmia to the sweaty odor.

2. *Olfactory receptor proteins.* If the genetic studies confirm the suspicion that this is an inheritable biochemical defect, it will provide a strong incentive to attempt the isolation of a corresponding olfactory receptor protein. Ample raw material is available, albeit rather inaccessibly, in the olfactory epithelium of slaughter-house animals. Radioautography with labeled isovaleric acid might give an indication of which parts of the receptor cells (rod or cilia) contain the selective receptor protein. If it is the cilia, and their notable surface area is suggestive in this context, an easy first-stage purification might be the isolation of olfactory cilia. Many subtle techniques are available in modern preparative and analytical biochemistry, so there should be a good prospect for an olfactory success following the gustatory lead provided by Dastoli.[20,21]

3. *Pilot survey of specific anosmias.* If the biochemical work yields specific olfactory receptor proteins, it will provide a cogent rationale for the basic concept of primary odors. This will make it imperative to glean the fullest information possible

from the fragmentary data on different types of specific anosmia. Here a pilot screening program would probably save work in the long run. The multifarious reputed specific anosmias should be confirmed and compared by quantitative tests run on the same large group of observers. I would envisage finding average human threshold values for each compound (in the right half of Table II) that has been cited as demonstrating a specific anosmia. Then a standard screening test concentration should be selected, perhaps four binary steps or 16 times the average threshold (i.e., about two standard deviations above the mean[8]). The screening presentation would take the form of a 2/5 sample sorting test,[12] to prove any failure to detect the odor. The tests might be run on the entire personnel of a large laboratory (we can muster about 450 here). For good measure, it would be sensible to include one chemical representative of each odor class of all the existing odor classifications (left half of Table II), to take advantage of every scrap of relevant information input. This would require presentation of 118 odor class compounds plus 62 specific anosmia compounds, or a total of 180 test odorants. A computer analysis of the odor detection failures in the matrix of 180 odorants and 450 subjects would show which compounds most likely belong to the same primary odor, and which subjects exhibit the corresponding specific anosmia.

4. *Systematic mapping of primary odors.* The preceding pilot survey should reveal which areas of the olfactory gamut probably contain suspected primary odors. The fourth phase of the campaign is a commitment an order-of-magnitude greater. It is the detailed analysis of each and every type of specific anosmia, to map out the corresponding primary odor. It will demand thorough chemical purification of each test odorant, together with individual threshold measurements for each subject in the panel of normal observers, as well as each of the specifically anosmic subjects in the test panel of observers having the relevant olfactory defect. Potential panels of specific anosmics for each primary odor would have been identified in the pilot screening program outlined above. This scheme should result in the identification of most of the primary odors in the human sense of smell. However, a few primaries may yet slip through the net, either because no classifier has thought of them, or because the corresponding specific anosmia has never been noticed.

The only primary odor so far worked out in detail required 18 carefully chosen test odorants to achieve the minimum acceptable mapping of the stereochemical height and boundaries of the primary odor sensitivity.[4,12] This demanded about one man-year of intensive professional-level work, plus the far-from-negligible time required for the essential purification of the test chemicals. Considering that there may be between 20 and 30 primary odors, the total task is clearly beyond the capabilities of an individual researcher. Possible solutions to this problem lie outside the scope of a scientific paper.

5. *Stereochemical assessment of primary odor specificities.* Certain applications of this basic research in odor technology (flavor, perfume, pollutants, etc.) will demand an expansion of the survey to include hundreds or even thousands of addi-

tional compounds. This is because most naturally occurring odors encountered in practice have been proved by gas chromatography to be composed of a hundred or more individual odorous compounds. It would be extremely tedious to assay all these experimentally by presentation of each compound to all the different panels of specific anosmics in order to estimate the contribution of each primary odor to the total odor. A short cut might be achieved by submitting the compounds to a computer analysis of the molecular configurations and affinities, such as we have begun with the Probabalistic Automatic Pattern Analyzer (PAPA), a pattern recognition machine.[11] In this way the stereochemical requirements for each primary odor may be worked out. The generalizations could then be applied to the theoretical assessment of any given odorant molecule, by a programmed computer technique, in terms of its expected primary odor characteristics.

I do not believe that any practically useful analysis of the discriminating power of the sense of smell will be achieved by any plan less ambitious than that I have outlined above. If the demands of applied science become sufficiently pressing, and there are indications from market, famine, pollution, and infestation that they will, we may enjoy the benefits of an organized attack on the olfactory code.

SUMMARY

The results of a survey are presented, showing that specific anosmias have been noted for 62 distinct chemical compounds. When these specific anosmias are compared with the 118 odor classes listed in the eight major classifications of odors, there are indications that between 20 and 30 primary odors may be represented.

A general plan of attack on the olfactory code would include several phases. These are a genetic analysis of specific anosmia, a search for olfactory receptor proteins, a pilot screening of all reputed specific anosmias, and a systematic, detailed analysis of each and every type of specific anosmia in order to map out each corresponding primary odor.

REFERENCES

1. Amoore, J. E. 1952. The stereochemical specificities of human olfactory receptors. *Perfumery Essent. Oil Rec.* **43**, pp. 321–323; 330.
2. Amoore, J. E. 1962. The stereochemical theory of olfaction. 1. Identification of the seven primary odors. *Proc. Sci. Sect. Toilet Goods Assoc.* **37**, Suppl., pp. 1–12.
3. Amoore, J. E. 1962. The stereochemical theory of olfaction. 2. Elucidation of the stereochemical properties of the olfactory receptor sites. *Proc. Sci. Sect., Toilet Goods Assoc.* **37**, Suppl., pp. 13–23.
4. Amoore, J. E. 1967. Specific anosmia: A clue to the olfactory code. *Nature* **214**, pp. 1095–1098.
5. Amoore, J. E. 1968. Are you "odor-blind"? *Drug and Cosmetic Ind.* **102**, (2), p. 128.
6. Amoore, J. E. 1968. Are you "odor-blind"? *J. Chem. Educ.* **45**, p. 209.
7. Amoore, J. E. 1968. Specific anosmias and primary odors. *In* Theories of Odor and Odor Measurement (N. Tanyolac, editor). Robert College Research Center, Istanbul, Turkey, pp. 71–81.
8. Amoore, J. E. 1968. Odor-blindness as a problem in odorization. *Proc. Operating Sect., Amer. Gas Assoc. Distribution Conf.* pp. 242–247.
9. Amoore, J. E., and D. Venstrom. 1966. Sensory analysis of odor qualities in terms of the stereochemical theory. *J. Food Sci.* **31**, pp. 118–128.

10. Amoore, J. E., and D. Venstrom. 1967. Correlations between stereochemical assessments and organoleptic analysis of odorous compounds. *In* Olfaction and Taste II (T. Hayashi, editor). Pergamon Press, Oxford, etc., pp. 3–17.
11. Amoore, J. E., G. Palmieri, and E. Wanke. 1967. Molecular shape and odor: Pattern analysis by PAPA. *Nature* **216**, pp. 1084–1087.
12. Amoore, J. E., D. Venstrom, and A. R. Davis. 1968. Measurement of specific anosmia. *Percept. Mot. Skills* **26**, pp. 143–164.
13. Anderson, J. A. Telephone, July 1, 1968.
14. Aylett, K. C. Letter of March 29, 1968.
15. Baker, E. G. S. Letter of March 12, 1968.
16. Blakeslee, A. F. 1932. Genetics of sensory thresholds. Taste for phenyl thio carbamide. *Proc. Nat. Acad. Sci. U.S.* **18**, pp. 120–130.
17. Brown, K. S., and R. R. Robinette. 1967. No simple pattern of inheritance in ability to smell solutions of cyanide. *Nature* **215**, pp. 406–408.
18. Carson, J. F. 1967. Onion flavor. *In* Symposium on Foods: The Chemistry and Physiology of Flavors (H. W. Schultz, E. A. Day, and L. M. Libbey, editors). Avi Publishing Co., Westport, Conn., pp. 390–405.
19. Crocker, E. C., and L. F. Henderson. 1927. Analysis and classification of odors. An effort to develop a workable method. *Amer. Perfumer* **22**, pp. 325–327; 356.
20. Dastoli, F. R., and S. Price. 1966. Sweet sensitive protein from bovine taste buds: Isolation and assay. *Science* **154**, pp. 905–907.
21. Dastoli, F. R., D. V. Lopiekes, and A. R. Doig. 1968. Bitter-sensitive protein from porcine taste buds. *Nature* **218**, pp. 884–885.
22. Davis, A. R. Letter of February 19, 1964.
23. Draper, A. L. Letter of March 22, 1968.
24. Edwards, B. E. Letter of July 12, 1968.
25. Fullman, B. 1963. Stereochemical theory of olfaction. *Nature* **199**, p. 912.
26. Fukumoto, Y., H. Nakajima, M. Uetake, A. Matsuyama, and T. Yoshida. 1957. Smell ability to solutions of potassium cyanide and its inheritance. *Japan. J. Human Genet.* **2**, (1), pp. 7–16.
27. Guadagni, D. G., R. G. Buttery, and S. Okano. 1963. Odour thresholds of some organic compounds associated with food flavours. *J. Sci. Food Agr.* **14**, pp. 761–765.
28. Guillot, M. 1948. Sur quelques caractères des phénomènes d'anosmie partielle. *Compt. Rend. Soc. Biol., Paris* **142**, pp. 161–162.
29. Guillot, M. 1948. Anosmies partielles et odeurs fondamentales. *Compt. Rend. Acad. Sci.* **226**, pp. 1307–1309.
30. Guillot, M. 1956. Aspect pharmacodynamique de quelques problèmes liés à l'olfaction. *Actualités Pharmacol.* **9**, pp. 21–34.
31. Harper, R. Orally, September 1, 1966.
32. Harper, R. 1968. *In* Theories of Odor and Odor Measurement (N. Tanyolac, editor) Robert College Research Center, Istanbul, Turkey, p. 194.
33. Harper, R., E. C. Bate-Smith, D. G. Land, and N. M. Griffiths. 1968. A glossary of odour stimuli and their qualities. *Perfumery Essent. Oil Rec.* **59**, pp. 22–37.
34. Harris, G. Orally, July 25, 1966.
35. Henning, H. 1915. Der Geruch, I. *Zeit. Psychol.* **73**, pp. 161–257.
36. Henning, G. J. Letter of August 2, 1967.
37. Keefe, E. F. Letter of March 25, 1968.
38. Kirk, R. L., and N. S. Stenhouse. 1953. Ability to smell solutions of potassium cyanide. *Nature* **171**, pp. 698–699.
39. Köster, E. P. Orally, September 1, 1966.
40. Lettvin, J. Y., and R. C. Gesteland. 1965. Speculations on smell. *Cold Spring Harbor Symp. Quant. Biol.* **30**, pp. 217–225.
41. Linnaeus, C. 1756. Odores medicamentorum. *Amoenitates Academicae* **3**, pp. 183–201.
42. Loebisch, Harnuntersuchung, p. 325; quoted by H. Zwaardemaker. 1895. Die Physiologie des Geruchs, p. 230. Wilhelm Engelmann, Leipzig, Germany.

43. Meyer, J. 1935. Wie riecht Blausäure? Gasmaske **7**, p. 112.
44. Noebels, H. J. Letter of January 9, 1967.
45. Ough, C. S., and H. Stone. 1961. An olfactometer for rapid and critical odor measurement. *J. Food Sci.* **26**, pp. 452–456.
46. Owens, F. H. Letter of April 15, 1968.
47. Patterson, P. M., and B. A. Lauder. 1948. The incidence and probable inheritance of "smell blindness." *J. Heredity* **39**, pp. 295–297.
48. Schutz, H. G. 1964. A matching-standards method for characterizing odor qualities. *Ann. N.Y. Acad. Sci.* **116**, pp. 517–526.
49. Shearer, D. A. Letter of July 27, 1967.
50. Shearer, D. A. Letter of May 22, 1968.
51. Stone, H., G. Pryor, and J. Colwell. 1967. Olfactory detection thresholds in man under conditions of rest and exercise. *Percept. Psychophys.* **2**, pp. 167–170.
52. Vlasek, H. C. Letter of June 24, 1968.
53. Wasserman, A. E. Letter of July 9, 1968.
54. Wilby, F. V. 1968. Variation in recognition odor threshold of a panel. Paper presented at Air Pollution Control Annual Meeting, St. Paul, Minnesota.
55. Wright, R. H., and K. M. Michels. 1964. Evaluation of far infrared relations to odor by a standards similarity method. *Ann. N.Y. Acad. Sci.* **116**, pp. 535–551.
56. Wright, R. H., J. R. Hughes, and D. E. Hendrix. 1967. Olfactory coding. *Nature* **216**, pp. 404–406.
57. Zwaardemaker, H. 1895. Die Physiologie des Geruchs. Wilhelm Engelmann, Leipzig, Germany.

CORRELATIONS BETWEEN ELECTRO-PHYSIOLOGICAL ACTIVITY FROM THE HUMAN OLFACTORY BULB AND THE SUBJECTIVE RESPONSE TO ODORIFEROUS STIMULI

JOHN R. HUGHES, D. EUGENE HENDRIX, NICHOLAS WETZEL, *and* JAMES W. JOHNSTON, JR.

Department of Neurology and Psychiatry, Department of Surgery, Northwestern University Medical Center, Chicago, Illinois, and Department of Physiology and Biophysics, Georgetown University Schools of Medicine and Dentistry, Washington, D.C.

INTRODUCTION

The ways in which messages are transmitted from one portion of the brain to another represents one of the greatest challenges to neurophysiology. The classical approach to this problem is to delineate some aspect of a response which correlates with some particular aspect of the stimulus producing the response. The significance of this type of correlation remains questionable as long as there is no indication of the total response of the organism. An example is that an increased amplitude of response may be considered the coding mechanism for an increased intensity of stimuli, but such a correlation may be only a meaningless epiphenomenon until the subject gives an indication of perceiving an increase in stimuli intensities. Animals can sometimes be taught to communicate their perceptions grossly with certain indexes of behavior, but man is an ideal subject, since he can describe these perceptions orally. Throughout the past four years we have had the opportunity of studying this crucial problem of coding in man by correlating certain aspects of olfactory stimuli, their electrophysiological responses in the olfactory bulb, and the psychological or subjective descriptions of these responses. This particular report will feature the coding of quality and intensity at the level of the olfactory bulb.

METHODS

This study included 11 patients (six men and five women) who received therapeutic neurosurgical operations involving the necessary exposure of the olfactory bulb and tract. Tumors were found in three (two pituitary and one thalamic); aneurysms in four; three had hypophysectomies; and one patient had an arteriovenous malformation. Electrodes were implanted within the olfactory bulbs of these patients and were of Type 316 Nilstain (120 μ diameter) supplied by W. B. Driver Co. (Newark, N.J.) and insulated with Teflon by the Warren Wire Co. (Pownal, Vt.). In most patients, the olfactory tract was sectioned behind the bulb, but in some, electrodes were implanted in the intact bulb. Four electrodes, each with a 1 mm exposed tip, were placed 1 mm apart and sealed into one stalk, which also included a reference electrode (at least 3.5 cm from the tip of the stalk) and a ground lead. The stalk of electrodes was anchored to the dura with silk sutures, but was easily removed within a week or so after the implantation and produced no difficulties.

Odoriferous stimuli were at first presented to the patients by means of an odorless receptacle which contained a piece of absorbent material soaked with 0.5 ml of the liquid stimulus. This cup was held 1 in from the tip of the nose and the patient was instructed to breathe deeply in order to perceive the stimulus. Recently, stimuli have been presented by a portable two-channel olfactometer (see Figure 1). Compressed air was passed through differential pressure regulators and flow valves to 18 feet of circular glass tubing within two separate Dewar flasks for each of the two channels. (The flasks contained liquid nitrogen, but we now use dry ice and isopropanol for air purification, because the former cooled the air to approximately −200° C and resulted in freezing within the glass tubing.) The tubing then led to a stream splitter with stop-cocks containing variable flow Teflon plugs and an entry into an enclosed lucite chamber, representing a constant temperature (thermostatically-controlled) air bath at 25° C. Both lines led to flow-raters; one of the two lines led to a mixing chamber and the other to a sparging vessel containing the odorant and then to the mixing chamber. Beyond the mixing chamber there were three-way valve controls (solenoids) which could be directed to the exhaust system of the room through a total flow-rater or to a nose-cone for presentation of the olfactory stimulus to the patient. Figure 2 shows this apparatus. The total flow-rate varied between 3000–3500 cc/min with a usual sparging rate of 420 cc/min or slightly less if foaming occurred. The dilution ratio of odorless air to odorant, however, varied between 5:1 and 15:1, but most often was approximately 8:1. For compounds freely soluble in water, distilled water was used as the diluent; for most other compounds diethyl phthalate was used. However, standards were diluted to the desired concentration with mineral oil except for formic acid, with which distilled water was used. The other standards[4] were d,l-B-phenylethylmethylethyl carbinol-25% (floral); 1,2-dichloroethane-100% (ethereal); 1,8-cineole-10% (camphoraceous); d,l-menthone-25% (minty); formic acid-20% (pungent); and dimethyl disulfide-5% (putrid). The musk standard was cyclopentadecanone (500 mg/ml)—different from that used by Amoore and Venstrom.[4] These concentrations were suprathreshold for most of our subjects.

Responses from the implanted electrodes were recorded on an electroencephalographic machine (Model VI, Grass Instrument Co.), but for analysis these responses have been stored on a four-channel FM Mnemotron tape system (Type 204) at the speed of 1 7/8 in per sec.

Figure 3 shows the system of analysis of these responses, which tend to be rhythmical in character. Since this system has been previously described,[8] only a brief description will be presented here. Tapes of 30 sec of responses are looped for continuous playback at a speed of 15 inches/sec. The responses were analyzed by a Heterodyne Wave Analyzer (Type 1900A, General Radio Co.) with amplitudes integrated by Operational Amplifiers (Type O, Tektronix Co.) and recorded on the Y-axis of an X-Y recorder (Type 2D-2A, Moseley Co.). The X-axis is a measure of frequency and is driven by the output from a Helipot, driven by a slowly revolving synchronous motor, which also drives the frequency dial of the analyzer. The resulting write-out from the X-Y recorder is a type of frequency, or spectral, analysis.

RESULTS AND DISCUSSION

The background rhythm of the olfactory bulb, similar to Adrian's "intrinsic waves,"[1] was recorded without presentation of any olfactory stimuli while the patient was quietly at rest. Figure 4 shows an example of this background activity, consisting of desynchronized and irregular waveforms, with little organization into rhythmical bursts. When an odorant is presented, rhythmical bursts, similar to Adrian's "induced waves,"[1] are usually seen synchronously with each inspiration. Figure 5 shows a long-lasting rhythmical burst to the very complex stimulus, cigarette smoke, usually the most effective olfactory stimulus in man. The rhythmical activity shows faster frequencies at the beginning of the burst and a progressive showing in the later portions of the burst. This rhythmical activity is longer in duration than most bursts, and the Figure is presented to show how extraordinarily distinctive can be some rhythmical bursts from the human olfactory bulb. The stimulus was especially effective on the electrode #2, the same electrode that shows the highest amplitude fast activity in the background rhythm. Figure 6 shows an example of a more typical burst with a duration between 1 to 2 sec, and is a response to *n*-heptanol from a different patient (D.M.). This Figure demonstrates again the differences between the various electrode placements within the bulb.

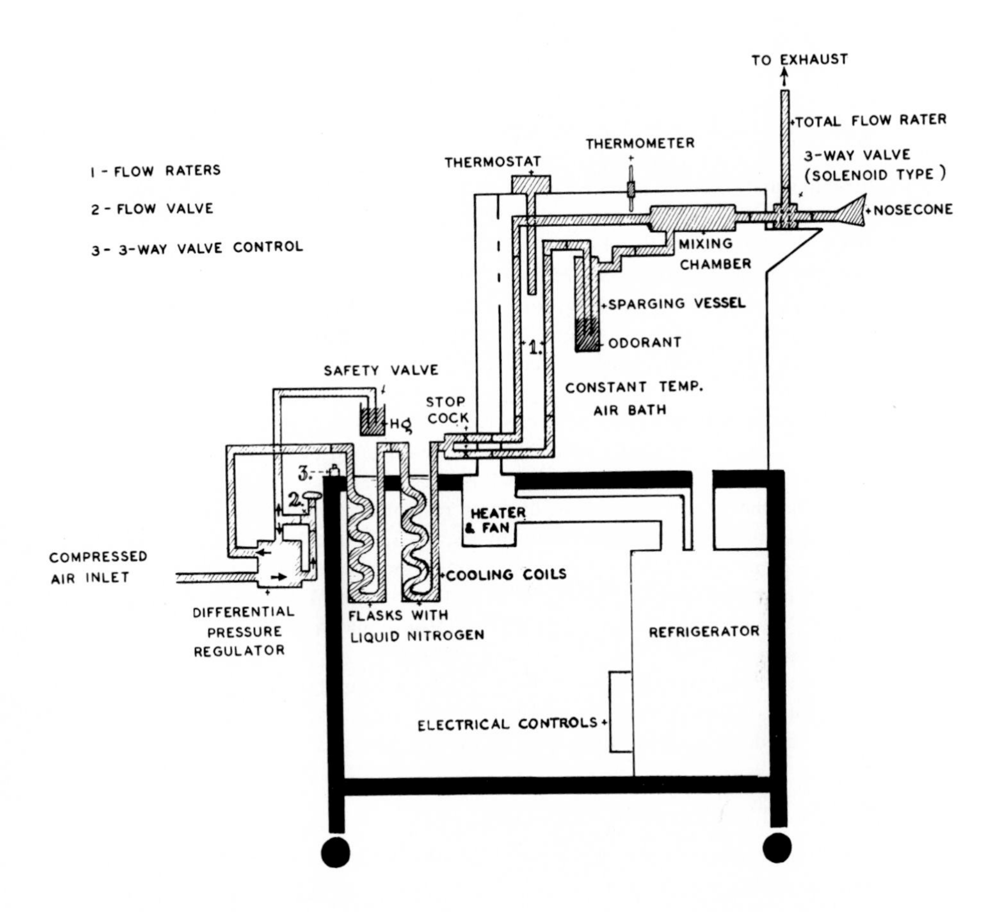

Figure 1 Schematic diagram of portable two-channel olfactometer, designed by Dr. James W. Johnston, Jr. (Washington, D.C.).

An analysis of these responses to odorants is exemplified in Figure 7, showing the analyzed response to d-limonene. The peaks of activity, seen especially between 20 to 50/sec and also up to 70/sec, represent different frequency components, which provide the major dependent variable for this study. It is these frequency components, represented by the peaks in the analyzed response, that will be used to characterize the responses to be discussed. An analysis of the background activity with the patient at rest without any olfactory stimulus is similar to that seen in response to an odorant (Figure 7), but without the peaks of activity; this type of curve without either high peaks or prominent frequency components is consistent with the desynchronized character of this "intrinsic" activity. The background level is then subtracted from the analyzed response, since it has previously been shown that the olfactory bulb

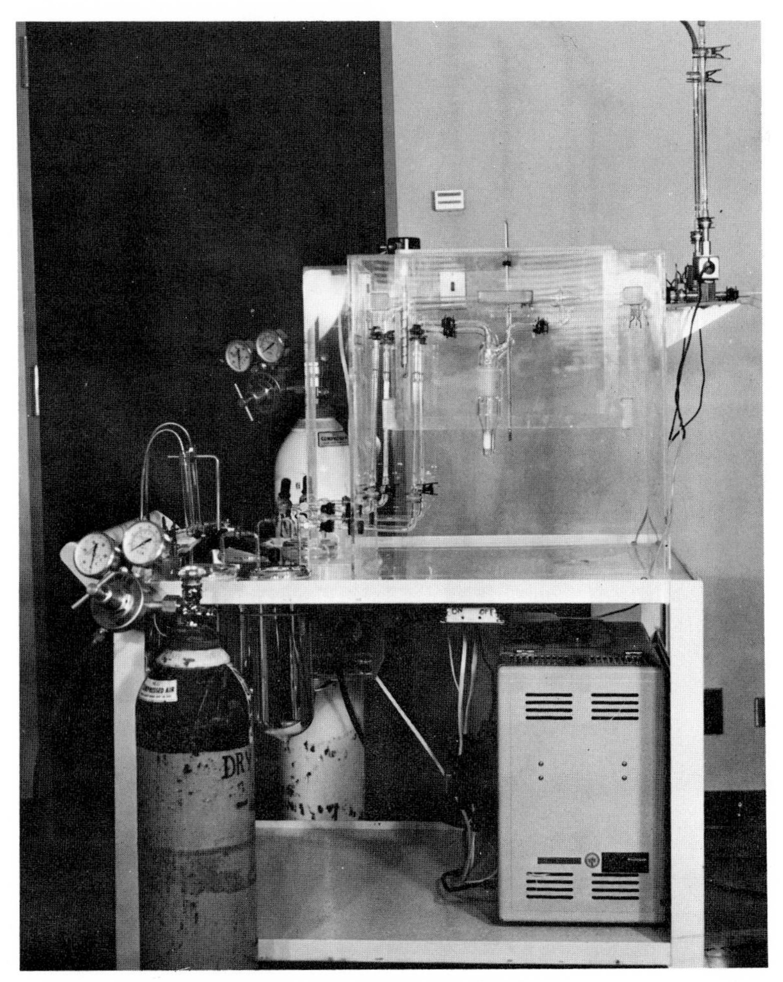

Figure 2 The two-channel olfactometer.

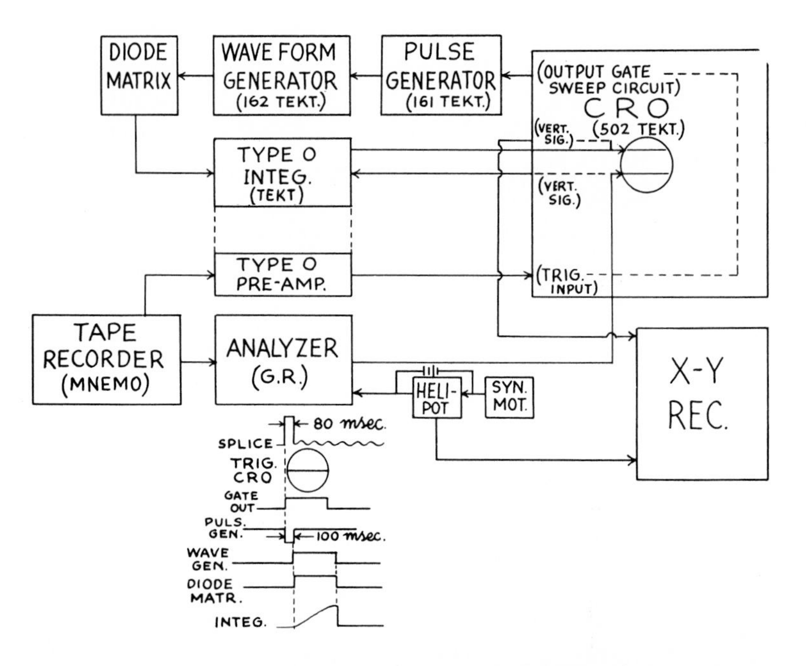

Figure 3 System of analysis of responses. See text for details. Note that the bottom of the Figure shows how the timing of integration was precisely controlled by the pulse generator, waveform generator, and diode matrix.

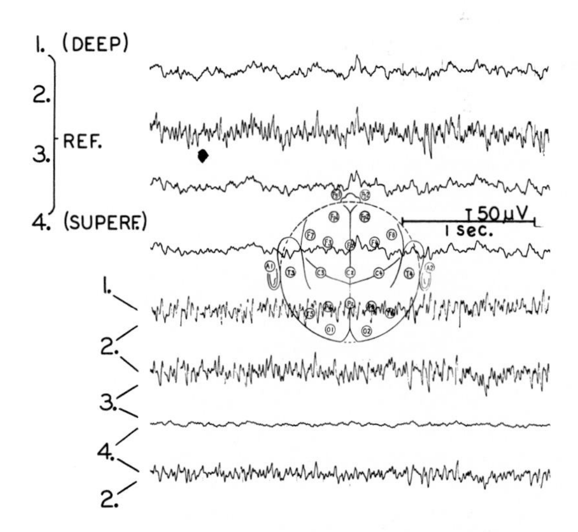

Figure 4 Background activity in the human olfactory bulb (patient B.W.). The first four channels show a referential recording from the four different electrodes with electrode #1 deep in the bulb and #4 superficial. The last four channels represent a bipolar recording from these electrodes. Note that electrode #2 shows more fast activity than the other electrodes. Calibrations in this and the next two figures of 50 μV and 1 second are shown.

seems to deal with the *change* in activity from the background that has occurred with the presentation of the stimulus rather than with the absolute level of the total response.[5]

Previous studies on the Frequency Component Hypothesis[5,7] have presented evidence from infrahumans that these components represent the essential coding mechanism for the *quality* of the odor in the olfactory bulb. Figure 8 shows an example of the different frequency components found in response to stimuli in seven different stereochemical categories.[3] For this patient (H.L.), the stimuli which were *perceived* and *characterized* as olfactory were:

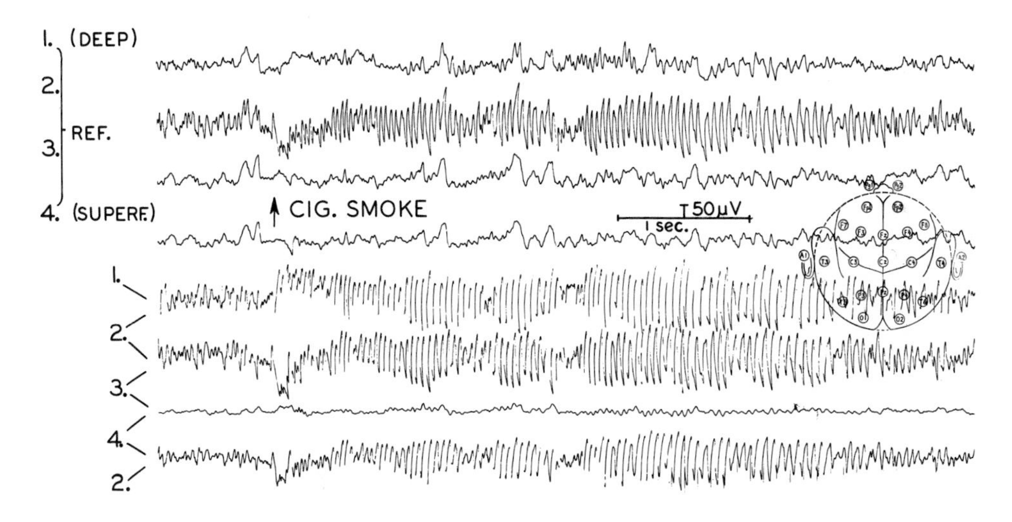

Figure 5 Rhythmical bursts of activity in response to cigarette smoke (patient B.W.). Electrode montage same as in Figure 4. Note long-lasting rhythms, especially on electrode #2. Arrow indicates onset of inspiration.

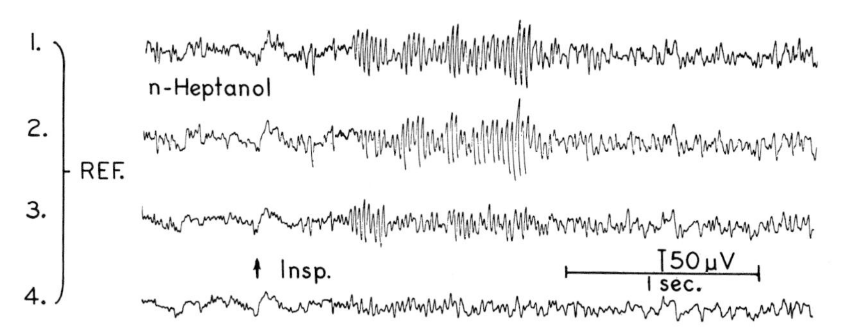

Figure 6 Rhythmical burst in response to n-heptanol. The duration of this burst is more typical from man's olfactory bulb than that seen in Figure 5. Arrow indicates onset of inspiration. Electrodes #1–4 are represented in this referential recording on channels #1–4.

Floral	*Camphoraceous*	*Pepperminty*	*Ethereal*
phenylethylmethylethyl carbinol	pinene	menthone	1-2 dichloroethane (a)
anethole	cineole (a)		1-2 dichloroethane (b)
anise oil	cineole (b)		1-2 dichloroethane (c)
safrole	fenchone		ethanol
sassafras oil			

Musk	*Pungent*	*Putrid* (?)
cyclopentadecanone (a)	formic acid	indole
cyclopentadecanone (b)	acetic acid	

As the Figure shows, the components for electrode #1 range mainly from 23.9 to 56.3/sec but are also found up to 93.8/sec. The major point in the figure is that each stimulus category is associated with a different combination of components. The components for a given stereochemical category were chosen because they were prominently represented in all of the responses within that given stimulus category. The predictive value of these data would be, for example, that any odor which produces a response with components at 23.9, 34.8, 45.8, 48.0, 50.4 and 71.8/sec should be a floral odor or that any floral odor should produce a response with those same components—for that patient and for that given electrode placement.

Our studies on rabbits and monkeys[6,8] have shown that different responses are seen with different electrode placements within the bulb. Figure 9 shows all of the frequency components for all four electrode placements, rather than for only one electrode, as in the previous figure. The same frequency components can be found within a given stereochemical category for different electrode placements and also

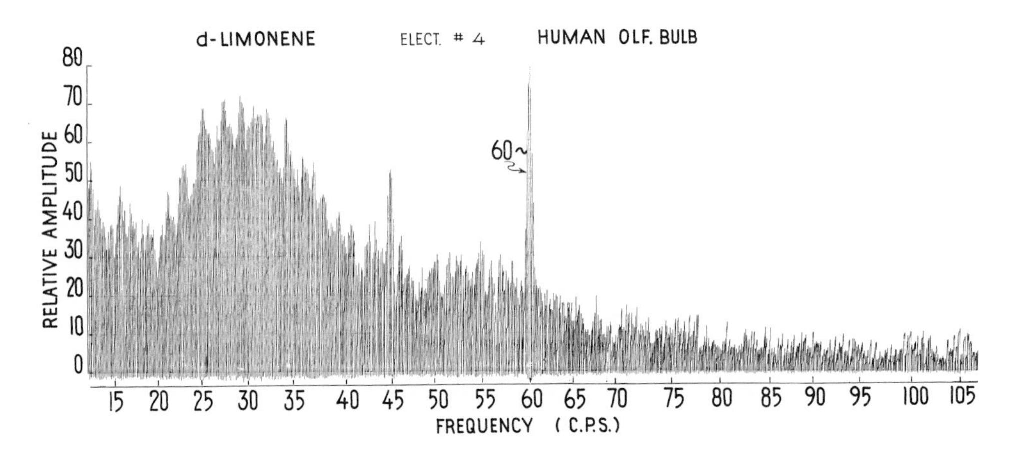

Figure 7 Analyzed response to an odorant (d-limonene). Note the separate peaks, especially between 20–50/sec, representing different frequency components, in patient H.M. Frequency is represented along the horizontal axis and relative amplitude along the vertical axis. The peak at 60/sec represents the presence of some 60-cycle line-current in the recording, but these peaks helped to confirm the designation of the precise frequencies along the X-axis.

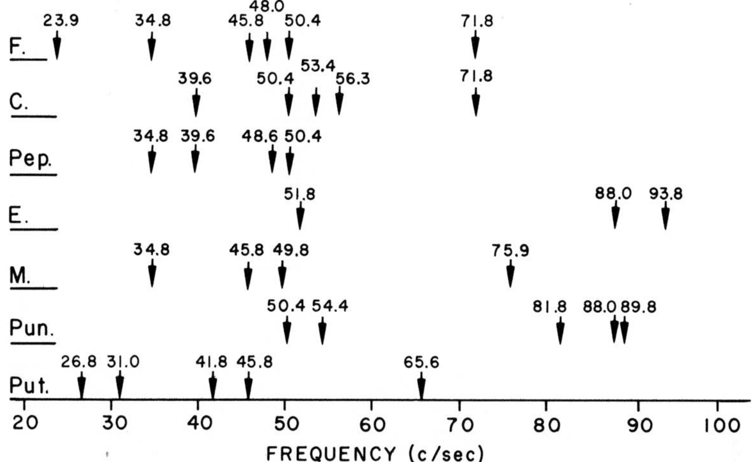

Figure 8 Different frequency components in the responses of the human olfactory bulb to different categories of odorants (electrode #1; patient H.L.). The stereochemical categories of stimuli are listed vertically (F. = floral, C. = camphoraceous, Pep. = pepperminty, E. = ethereal, M = musky, Pun. = pungent, Put. = putrid). Frequency is indicated along the horizontal axis, and the arrows with the frequencies above them represent the different frequency components.

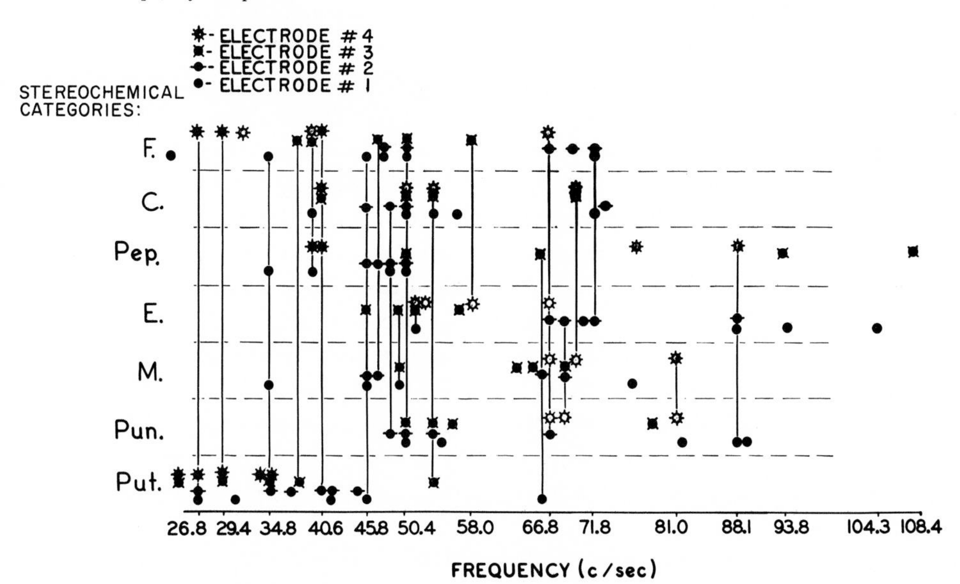

Figure 9 Different frequency components for different (stereochemical) stimulus categories for all four electrode placements (patient H.L.). Symbols for each electrode on horizontal axis within a given stimulus category show that each area within a bulb has its own distinctive response or its own group of frequency components. The 20 vertical lines connect those frequency components which are exactly the same (±0.25 sec) among the different stimulus categories. Under the numbers are listed the stimuli and their stereochemical categories.

within a given electrode placement and for different stimuli. The 20 frequency components (vertical lines) seem to be the most prominent in all of the responses recorded from the four different electrode positions in this patient. It is, at least, interesting to speculate whether these 20 components are necessary and sufficient to characterize the responses to all of the odorants presented to the patient. Also, it is tempting to speculate whether the 20 components could possibly be related to the number of olfactory "primaries" in that patient. Table I shows these 20 components listed for the different stimulus categories. Note that 12 of them are found in the floral odorants and as a few as six are found in the ethereals. The underlined numbers are those that are the highest in amplitude among the different stereochemical categories. In the floral class, only one of its 12 components was the highest among all the stimuli, while four of those found in camphoraceous odors were the highest. The numbers 4, 5, 8, 13, 14, and 15 are not underlined, since equal amplitudes were found in the different categories of stimuli. The underlined components then reduce the number from 20 to 14. However, the larger battery of odorants presented to this patient may have resulted in those six components finding a maximal representation in the response to some other given stimulus.

Figure 9 shows that at times the same frequency components are found in different stereochemical categories. Table II shows the incidence of common components found for all combinations of stimulus categories; combinations which resulted in the maximum of five frequency components that were seen in both stimulus groups are underlined. Those stereochemical categories were the combinations of floral, camphoraceous, and pepperminty, in addition to floral combined with putrid. The latter combination is probably related to the fact that some investigators have also categorized indole (the only stimulus called putrid presented to patient H.L.) as a floral, especially at a low stimulus intensity.[2] On the left of Figure 9 are those stimuli that are considered *primarily* floral (phenylethylmethylethyl carbinol, anethole, anise oil, safrole, and sassafras oil), and then those mainly camphoraceous (pinene, cineole, and fenchone), and pepperminty (menthone). Also listed are the other stereochemical categories that are usually considered to be represented in those stimuli.[2] The many instances in which floral, camphoraceous, and pepperminty odors are represented within the same stimulus would seem to be consistent with the large number of times in which the same frequency component is found among those three stimulus categories. Thus, if combinations of frequency components represent the coding for quality, some similarity of components should be found among the responses to odorants which share common stimulus categories. Table II has provided evidence on this point.

Previous studies on infrahumans[5,7–9] have given evidence for odor localization, referring to the idea that different odorants produce maximal responses in different areas within the bulb. Table III shows evidence for odor localization in the human. For indole (putrid-floral?), electrode position #3 showed the maximal representation, as was the case in the florals. Camphoraceous stimuli were equally represented

TABLE I
PRESENCE OF 20 PROMINENT FREQUENCY COMPONENTS IN DIFFERENT STIMULUS CATEGORIES
(Patient H.L.—Nov. 16–18/66)
Combinations producing minimum frequency components in each stimulus group are underlined.

Floral	1	2	3	4	5	6	7	8			11		13		15			18			= 12	1
Camphoraceous					5	6	7		9		11	12					17	18			= 8	4
Pepperminty			3		5	6	7	8	9		11			14						20	= 9	1
Ethereal							7			10			13		15	16				20	= 6	1
Musk			3				7	8		10				14	15	16	17		19		= 9	3
Pungent									9		11	12			15	16			19		= 6	1
Putrid (?)	1	2	3	4		6	7					12		14							= 8	3
																				Total	=	14

TABLE II
COMMON COMPONENTS FOUND FOR ALL COMBINATIONS OF STIMULUS CATEGORIES
(Patient H.L.)
Combinations producing maximum of five frequency components in both stimulus groups are underlined.

	F.	C.	Pep.	E.	M.	Pun.	Put.
Floral	–	5	5	3	4	2	5
Camphoraceous	5	–	5	2	2	3	3
PEPperminty	5	5	–	2	4	3	4
Ethereal	3	2	2	–	4	3	1
Musk	4	2	4	4	–	3	3
PUNgent	2	3	3	3	3	–	1
PUTrid	5	3	4	1	3	1	–

Phenylethylmethylethyl carbinol = F. PEP.
Anethole = F. PEP. C.
Anise oil = F. C. PEP.
Safrole = F. C. PEP.
Sassafras oil = F. C. PEP.
Pinene = C. E.
Cineole = C.
Fenchone = C. E.
Menthone = PEP. F. C.
Indole = PUT. F.

on electrodes #3 and #4; pepperminty was found best on #1, while musky, ethereal, and pungent showed the highest amplitude responses on electrode #2. In each case, the highest amplitude component was used for each response for all four electrode positions, then the percentage of amplitude was determined for each electrode. The total numbers of responses (10–16) are given parenthetically.

Other differences are seen at different locations within the bulb. The prominent frequency components emphasized in Figure 9 are listed in Table IV according to

TABLE III
PROMINENT COMPONENTS FROM PERCEIVED ODORANTS
(Patient H.L.—Nov. 16–18/66)
Incidence of given electrode with maximum amplitude.

	Category						
Electrode	Put.? (Indole)	Fl.	Pep.	Musk	Camp.	Eth.	Pun.
#1	16%	25%	<u>37%</u>	18%	22%	0%	18%
#2	25	19	18	<u>40</u>	22	<u>40</u>	<u>37</u>
#3	<u>43</u>	<u>31</u>	18	20	<u>28</u>	30	18
#4	16	25	27	30	28	30	27
	100% (12)	100% (16)	100% (11)	100% (10)	100% (14)	100% (10)	100% (11)

TABLE IV
PROMINENT FREQUENCY COMPONENTS ON SAME ELECTRODE FROM DIFFERENT STIMULUS CATEGORIES
(Patient H.L.—Nov. 16–18/66)

	Electrode										
(Deep)	#1)		34.8	39.6	45.8	50.4			71.8		88.0
	#2)	←——	———	———	45.8, 47.1, 48.5	50.4	66.8, 68.1		71.8		
	#3)	←——	38.1			49.8, 50.4, 53.4	———	———	———	———	——→
(Superficial)	#4)	26.8, 29.4		39.6, 40.6 ←	———	———	→66.8	69.5		81.0	

whether they were best represented on electrodes #1, 2, 3 or 4, with #1 as the deepest in the bulb and #4 the most superficial. The most superficial electrode (#4) best represented the low- and the high-frequency components, while the adjacent electrodes (#2, #3) showed the mid-frequencies, especially #3 (38.1–53.4/sec). The deepest electrode (#1) showed a wide range of frequencies from 34.8 to 88/sec. These data suggest that each area within the bulb has its own repertoire of frequency components. Further evidence for this idea is seen in the following analysis. In 89 instances (patient H.L.), the same frequency component could be found in the responses to stimuli within two different stereochemical categories. In 68 per cent of the instances, this component was found on the same electrode and in 32 per cent on one of the other three electrodes. With a random distribution of components among all four electrodes, 25 per cent would be expected on the same electrode by chance. The value of 68 per cent is significantly different from that expected by chance ($p = 0.02$). The conclusion from these data is that a given area within the olfactory bulb has a given repertoire of frequency components for its response to an odorant; however, there seems to be a relatively large number of components available for each area, so that each odorant is associated with different combinations of these components within each electrode position.

The previous figures and tables have presented evidence that the odor quality is coded in terms of frequency components. The next question is the coding of odor intensity. Our first attempt was to determine the differences in the analyzed responses between those odorants perceived or detected by the patient and those that were not detected. Figure 10 shows the results from 31 different odorants that were detected, contrasted to 11 that were not. This Figure shows that the amplitude of the detected responses was, on the average, higher for the perceived odorants. Also, a larger number of peaks (frequency components) was found for the perceived odorants.

An example of a given odorant from the 42 represented in the previous figure is seen in Figure 11. The responses to pinene are contrasted with regard both to amplitude and numbers of peaks. During one presentation, the pinene was perceived (and characterized appropriately) by the patient, and in two instances was undetected. As in the previous figure, the amplitude and the number of peaks were higher when the stimulus was detected than when it was not. Electrode positions #3 and #4 showed the maximal difference for both measures. Figure 12 gives another example in which the amplitude differences are seen best on certain electrodes (#1 and 2), while the difference between the number of peaks is seen best on another area within the bulb (electrode #3). Generally, the most consistent correlations with odor detection were found for the measure of amplitude of peaks than for the number of peaks.

The results on another patient were similar, and Figure 13 shows an example (patient C.S.) of the higher amplitudes and larger numbers of peaks on 33 detected responses contrasted with 44 nondetected responses. Some detected responses were characterized in such a way that it was likely that a V nerve activation had occurred.

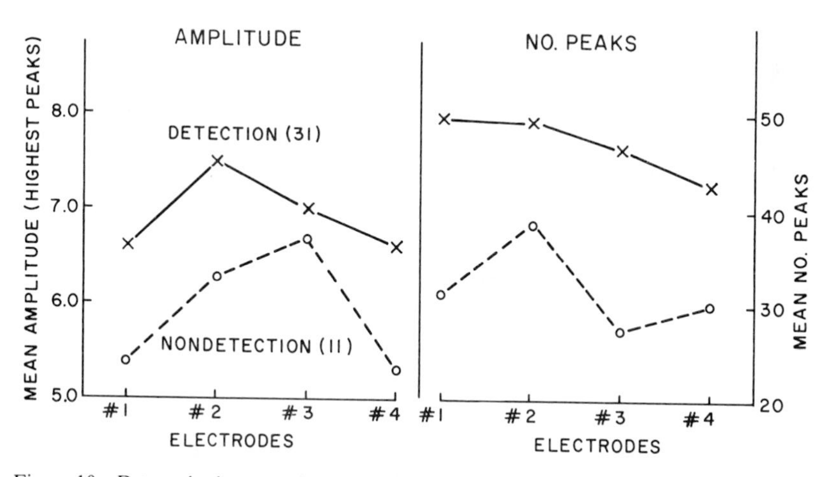

Figure 10 Detected odors gave larger amplitudes and greater number of peaks than did non-detected responses (patient H.L.). Values for amplitude (left) were determined by averaging (mean) and highest peaks of 31 detected responses (solid lines) and 11 undetected responses (broken lines); mean values were also used for the number of peaks (right). Electrode positions #1-4 are shown on the abscissa. Note the higher values in the detected responses.

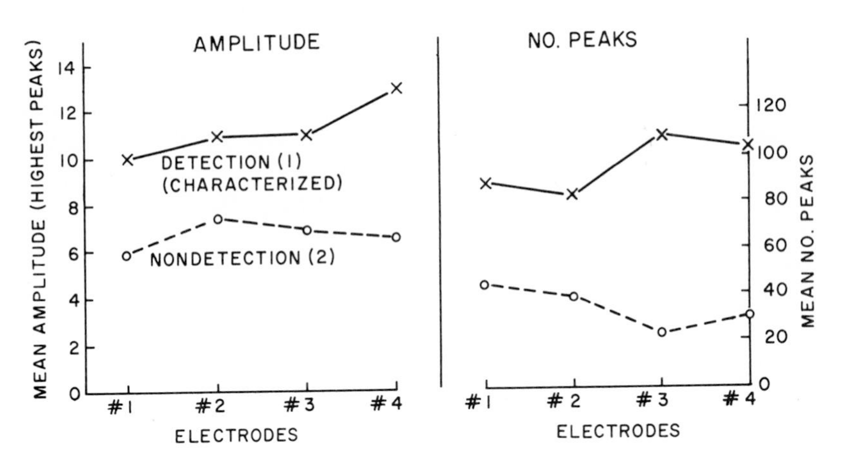

Figure 11 An example of an odorant (pinene) with a greater response during olfactory detection than during nondetection. Note the higher amplitudes and numbers of peaks for the one response that was detected and characterized (solid lines) than for the two responses not detected (broken lines).

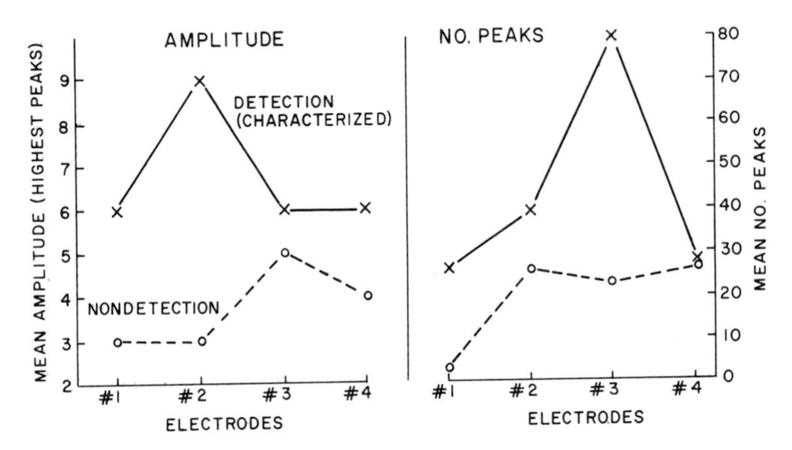

Figure 12 An example of responses during detection and nondetection (phenylethylmethylethyl carbinol). Note the greatest difference in amplitude on electrodes #1 and #2, but the most pronounced difference in number of peaks on #3. No difference in the numbers of peaks is seen on electrode #4.

The detected responses were then separated into those with strictly olfactory characterization and those which had at least a V nerve component and probably an olfactory one, as well. In the case of amplitude, electrodes #3 and #4 showed higher values for the olfactory responses than did those with V nerve activation, but no consistent differences could be found on other electrodes for either amplitude or number of peaks.

An example (Figure 14) is shown in which the amplitudes and number of peaks in

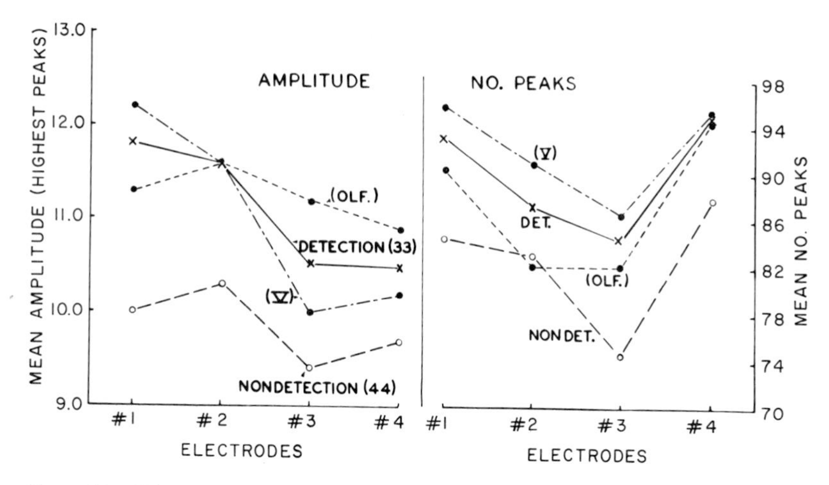

Figure 13 Olfactory detection vs. nondetection (patient C.S.). Note that the amplitudes and numbers of peaks of detected responses (solid lines) were higher than in the non-detected responses (long broken lines). The 33 detected responses were divided into those with olfactory characterizations (dotted lines) and those with V nerve activation (dotted and broken lines).

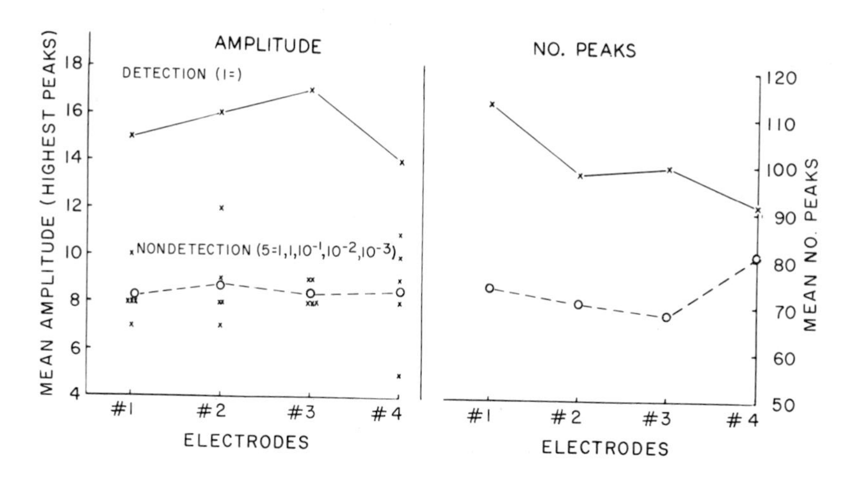

Figure 14 Example of detection and nondetection (phenylethylmethylethyl carbinol; patient C.S.). The relative stimulus intensity in the detected response was 1; in the nondetected responses it was 1, 1, 10^{-1}, 10^{-2} and 10^{-3}. An example of range of values is given for the five nondetected responses by the Xs around the broken line, representing their mean value.

a detected response are higher than in the five nondetected responses. All four electrode positions show this difference. However, Figure 15 shows an example of another odorant in which these differences were seen only on certain electrode positions (#1 and #2). This localized effect within the deeper electrodes was noted for both amplitude and number of peaks. The Figure emphasizes that any failure to find some of these described effects may be related to an inappropriate position of the recording electrode within the bulb, because of the localization of function within the structure.

An attempt was made to determine if differences could be found between responses that were not detected, those that were only perceived but not characterized, and those that were both perceived and appropriately characterized. Figure 16 shows an

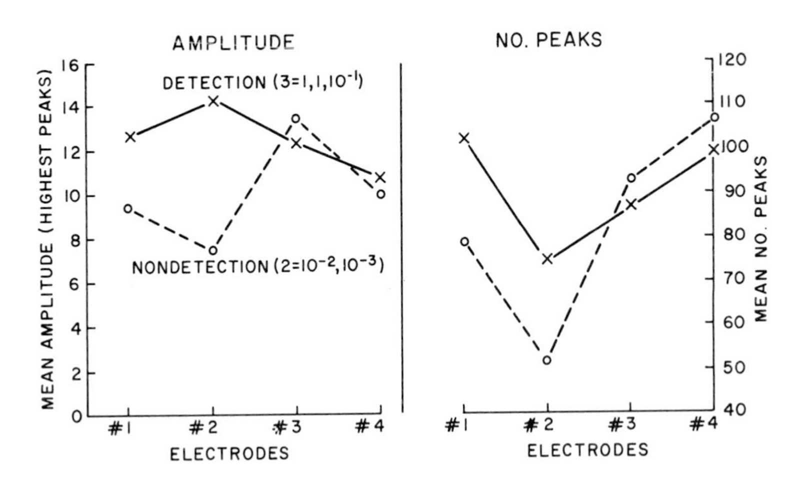

Figure 15 Example of a localized difference in the responses to detected and nondetected responses. Note that the three detected responses to cineole at intensity values of 1, 1, and 10^{-1} showed higher amplitudes and higher numbers of peaks only on electrode positions #1 and #2, contrasted to the two nondetected responses from the relative stimulus intensities of 10^{-2} and 10^{-3}.

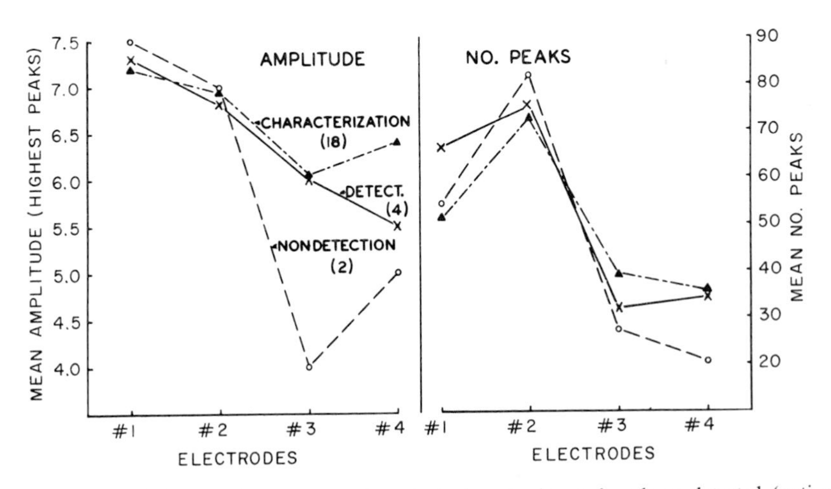

Figure 16 Differences in responses characterized, only detected and nondetected (patient B.W.). Note on electrode #4 the highest amplitude for the characterized (18) responses (dot and dash), less high for detected (14) responses (solid line), and lowest for nondetected (2) responses (broken lines).

example from another patient (B.W.), in which the highest amplitudes were seen from the 18 characterized responses with lower values from the four detected but uncharacterized responses, and the lowest values from the two nondetected responses. Electrode position #4 showed these differences, while #3 showed a great difference only between the nondetected and the detected responses. Similar, but less convincing, changes were seen with the numbers of peaks. For both measures of activity, electrodes #1 and #2 showed similar values for the characterized, detected, and nondetected responses. This Figure again demonstrates an effect seen only within a given area within the olfactory bulb, or represents an example of odor localization. However, the most important point in the Figure is that higher amplitudes of frequency components are required to change an undetected response to a perceived one, and still higher amplitudes are required to produce a characterization of the stimulus.

Figures 10 to 16 have provided evidence that the coding for intensity in the olfactory bulb is likely to be represented by the amplitude of the frequency components. In the previous figures, only the frequency components of highest amplitude were considered, and the question arises whether only those prominent components are higher in the detected responses or whether many other components are involved. Figure 17

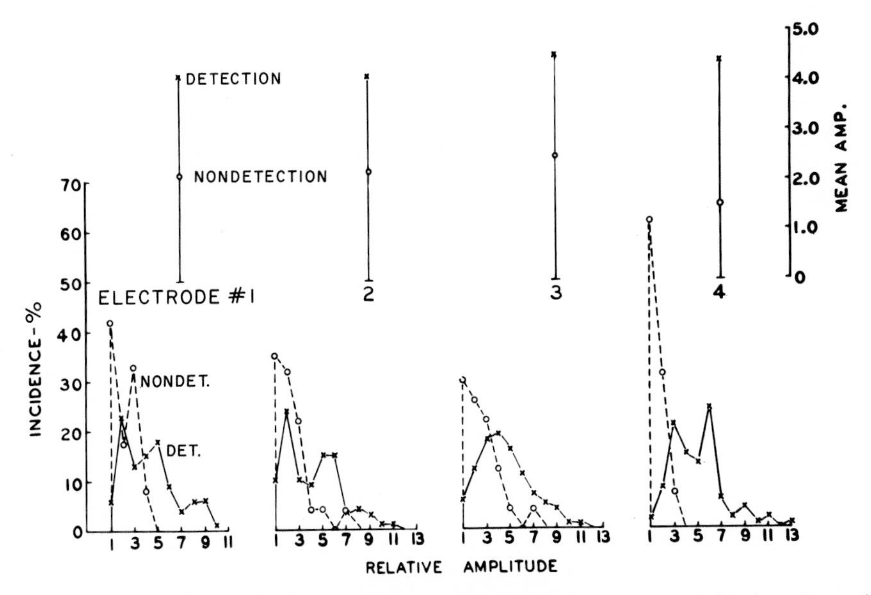

Figure 17 Comparison of the amplitudes of all components seen in detected and nondetected responses (pinene; patient H.L.). On the top, mean amplitudes for all components are shown for electrodes #1 (left) to #4 (right) showing that the mean amplitude in detected responses (Xs) was higher than in nondetected responses (circles). On the bottom, amplitudes of all components are shown with relative amplitude on the abscissa, and incidence (%) of those amplitude values on the ordinate. The extent to which the solid lines (detected responses) are shifted to the right of the dashed lines (nondetected) is a measure of how many components in the detected responses were higher in amplitude than in those not perceived.

shows that the detected response (patient H.L.) includes many components, not only the highest or the most prominent, that are greater in amplitude than in the nondetected response. Figure 18 provides another example of the same point in another patient (C.S.). In this patient the responses to phenylethylmethylethyl carbinol included many components which were higher in amplitude when the odorant was detected than when it was not. Thus, it is not only the most prominent components that show higher amplitudes in the detected responses; many components tend to be higher when an individual detects an odorant than when he does not.

We considered other possibilities for the neurophysiological differentiation between a detected and an undetected odorant, including whether the presence of certain frequency components determined detection or nondetection. In Figure 19, the frequency components common to each of three detected responses were determined; then the presence or absence of these components was checked in the two different nondetected responses. For electrode #1, two out of three components in all of the detected responses were also found in one or the other nondetected response. For electrode #2, it was four of the seven; for electrode #3, six of the seven; and in electrode #4, all 14 common components in the detected responses were seen in one or the other of the nondetected responses. The conclusion from these data is that perception of an odorant probably does not depend upon the *presence* or *absence* of any

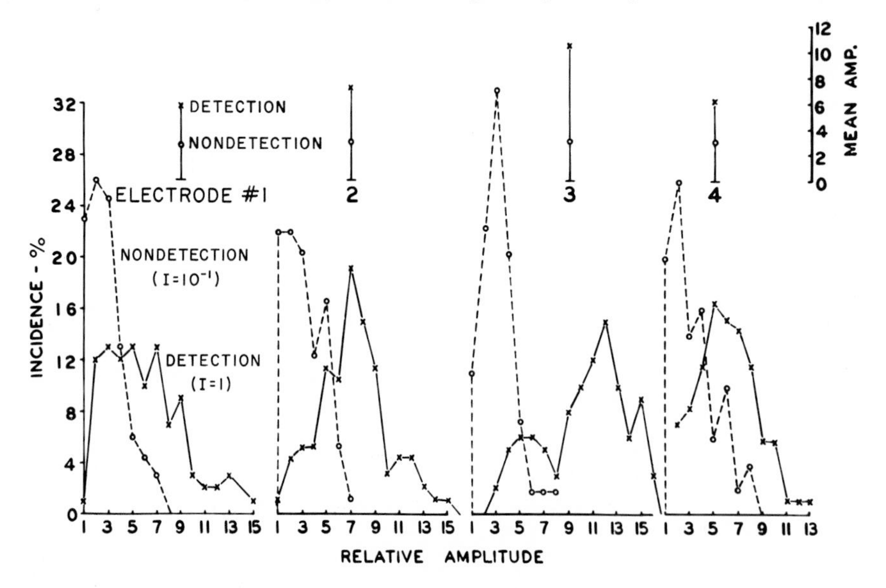

Figure 18 Comparison of amplitudes of components in detected and nondetected responses (phenylethylmethylethyl carbinol; patient C.S.). As in Fig. 17, note that the mean amplitudes for the detected responses (intensity = 1) was higher for the four electrodes than in the nondetected response (intensity = 10^{-1}). Note that on the bottom the solid line for the detected responses are shifted markedly to the right of the broken lines (nondetected response), indicating that many components in the detected response were higher in amplitude than in the response that was not perceived.

given frequency components, but rather, as previously noted, from the *amplitude* of those components.

We considered another possibility for the coding of detection or perception of an odorant. The data were analyzed to determine if an odorant was perceived only if given frequency components produced an inter-areal synchrony. In this case, the same component could be seen on more than one electrode position at the same time, so that a synchronization could take place between different areas within the bulb. Table V shows an example of the response to cyclopentadecanone (patient H.L.). For three given frequency components found in both the perceived and undetected responses, only two electrodes showed these components when the odorant was detected, while an even higher number of electrodes (three or four) showed the same components with nondetection. In each case, the amplitude of these components was higher when the odorant was detected. For other odorants, 13 more components were analyzed in both the detected and nondetected responses. The average (mean) number of electrodes showing the same component in the detected responses was 2.6, compared to the even higher value of 3.1 for the nondetected responses. The conclusion from these data is that the perception of an odorant does not depend upon an inter-areal synchrony, i.e., whether common frequency components can be found on a number of different electrode positions at the same time within the olfactory bulb.

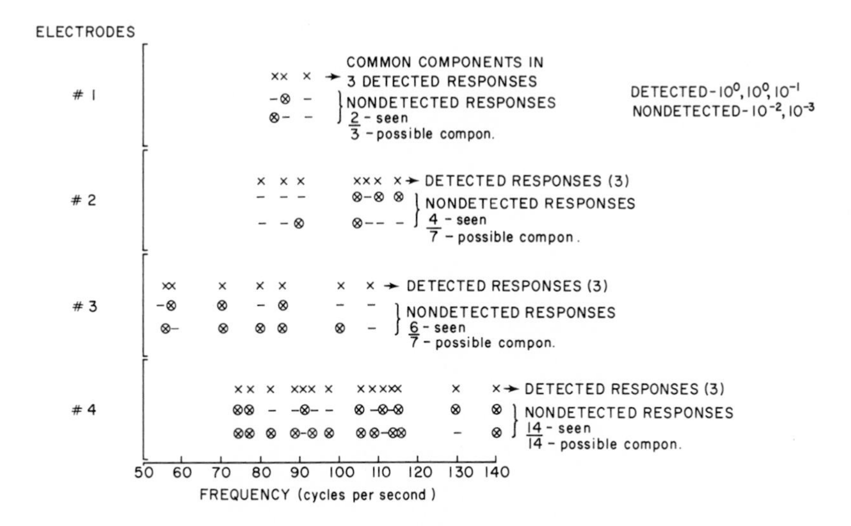

Figure 19 Common components in detected and nondetected responses (cineole; patient C.S.). Frequency is represented along the horizontal axis and the common frequency components are shown by Xs in the detected responses (intensities = 10^0 or 1, 1, 10^{-1}). Under each X a circled X indicates that the same component was seen in one or the other nondetected response (intensities = 10^{-2}, 10^{-3}) and a dash indicates the absence of that component. Note that the majority of common components seen in the three detected responses were also seen in one or the other nondetected response.

TABLE V
APPEARANCE OF FREQUENCY COMPONENTS ON DIFFERENT ELECTRODE POSITIONS
Example: Cyclopentadecanone (Patient H.L.)

	Detected			Nondetected		
		Electrode	(amp)		Electrode	(amp)
Component:	45.3/sec	#1	(6)	45.3/sec	#2	(7)
		#2	(8)		#3	(5)
					#4	(2)
	49.6/sec	#1	(6)			
		#3	(2)	49.6/sec	#2	(3)
					#3	(4)
	66.3/sec	#2	(8)		#4	(1)
		#4	(6)			
				66.3/sec	#1	(1)
					#2	(6)
					#3	(5)
					#4	+(5)

The detailed analysis of the responses of the olfactory bulb often shows the presence of many frequency components. The question of replication of results arises, especially in dealing with conscious subjects in whom variability of response is often noted. Previous studies of infrahumans[5] have shown that similar frequency components were found in two responses to the same odorant, if stable conditions existed at the time that the responses were recorded. Figure 20 shows how often *exactly the same* frequency components can be found in two different perceived responses to the same stimulus in man. For all four electrodes, more than one half of the many components

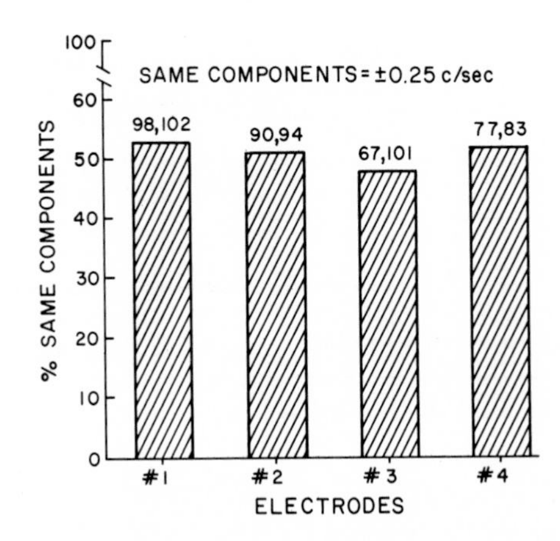

Figure 20 Incidence of the same frequency components found in two different responses to the same stimulus (cineole). Electrodes #1–#4 are represented on the horizontal axis and the incidence (%) of the same components (±0.25/sec) are seen on the vertical axis. Above each column is the number of components seen in the first and second response. Note that 50–53% of all components are seen in two different responses for all four electrode positions.

seen in the responses can be found in a second response. Which of these peaks represent "noise" or insignificant information is probably found within the portion of the components uncommon to the two responses. The higher amplitude components were found in both analyses, consistent with one of the major themes of this paper—whether an odorant is perceived seems dependent upon the amplitude of these components.

SUMMARY

Electrodes were implanted into the human olfactory bulb in 11 patients requiring therapeutic neurosurgical operations involving a necessary exposure to the anterior fossa. Responses to many different odorants were tape-recorded and analyzed for frequency components. Also, patient descriptions of the odorant were carefully noted.

Evidence is presented for the Frequency Component Hypothesis in man, maintaining that the coding of *quality* of an odorant is related to the different combinations of frequency components found in the analyzed response.

Evidence is also presented that nondetection or detection of an odorant is determined mainly by the amplitude of these frequency components. Also, characterization of an odorant seems associated with higher amplitudes than does simple detection.

ACKNOWLEDGMENT

* Supported by Grant NB-05377 from NINDB (USPHS), by the Warner-Lambert Pharmaceutical Company, Morris Plains, N.J., and the Givaudan Corporation, Delawanna, N.J.

REFERENCES

1. Adrian, E. D. 1950. The electrical activity of the mammalian olfactory bulb. *Electroencephalog. Clin. Neurophysiol.* **2**, pp. 377–388.
2. Amoore, J. E. 1962. The stereochemical theory of olfaction. *Proc. Sci. Sect. Toilet Goods Assoc.*, Spec. Suppl. to No. 37, pp. 1–12.
3. Amoore, J. E. 1963. Stereochemical theory of olfaction. *Nature.* **198**, pp. 271–272.
4. Amoore, J. E., and D. Venstrom. 1967. Correlations between stereochemical assessments and organoleptic analysis of odorous compounds. *In* Olfaction and Taste II (T. Hayashi, editor). Pergamon Press, Oxford, etc., pp. 3–17.
5. Hughes, J. R., and D. E. Hendrix. 1967. The frequency component hypothesis in relation to the coding mechanism in the olfactory bulb. *In* Olfaction and Taste II (T. Hayashi, editor). Pergamon Press, Oxford, etc., pp. 51–87.
6. Hughes, J. R., and J. A. Mazurowski. 1962. Studies on the supracallosal mesial cortex of unanesthetized conscious mammals. II. Monkey. B. Responses from the olfactory bulb. *Electroencephalog. Clin. Neurophysiol.* **14**, pp. 635–645.
7. Hughes, J. R., D. E. Hendrix, and N. Wetzel. 1968. Evidence from the human olfactory bulb for the frequency component hypothesis. *In* Theories of Odors and Odor Measurement (N. Tanyolac, editor). Robert College Research Center, Istanbul, Turkey, pp. 84–114.
8. Hughes, J. R., and J. A. Mazurowski. 1964. Comparative studies on the frequency analysis of responses from the olfactory bulb of unanesthetized monkeys and rabbits. *In* Proceedings of the Conference on Data Acquisition and Processing in Biology and Medicine. Pergamon Press, Oxford, etc., pp. 243–257.
9. Mozell, M. M., and C. Pfaffmann. 1954. The afferent neural processes in odor perception. *Ann. N.Y. Acad. Sci.* **58**, pp. 96–108.

MECHANISM OF OLFACTORY DISCRIMINATION IN THE OLFACTORY BULB OF THE BULLFROG

SHOJI HIGASHINO, HIDEO TAKEUCHI, *and* JOHN E. AMOORE

Department of Physiology, School of Medicine, Gunma University, Maebashi, Japan. Dr. Higashino's present address is Department of Physiology, Columbia University, New York. Dr. Takeuchi's present address is Department of Internal Medicine, Gunma University, Maebashi, Japan. Dr. Amoore's address is Western Regional Research Laboratory, U.S. Department of Agriculture, Albany, California.

INTRODUCTION

The mechanism of odor discrimination has been a central topic for research among investigators in various fields who have an interest in olfaction, but it is still far from clear. By means of the neurophysiological technique, the problem has been studied in the olfactory bulb by a few workers. Adrian[1-4] recorded the pulses induced by stimulation in the olfactory bulb and suggested the importance in odor discrimination of the observed spatio-temporal differences in the induced spike discharges. Shibuya, Ai, and Takagi[25] examined the various response patterns in the olfactory bulb of the frog and classified them into seven types: excitatory on, excitatory on-off, excitatory off, postexcitatory inhibition, postinhibitory excitation, inhibition, and nonresponding. Mancia, Green, and v. Baumgarten[19] also found some of these response patterns in the olfactory bulb of the rabbit. Döving[11-14] classified the spike discharges in the olfactory bulb into three types (excitation, nonresponding, and inhibition) and he applied a statistical analysis to clarify the mechanism of odor discrimination. As Döving himself stated in his papers, classification into three types is a relatively coarse treatment of the true responses in the olfactory bulb.

Following Moncrieff's suggestion,[20,21] correlation between the quality of an odor and the size and outer configuration of the odorous molecule has been affirmed by other workers.[10,26] One of the authors of this paper (Amoore[5]) has indicated that the stereochemical arrangement of a molecule is the most important factor in the determination of odor quality, and he classified odorous substances into seven classes on the basis of their descriptions in the literature. Amoore and Venstrom[9] subsequently provided statistical proof of a correlation between the similarity of odors and the similarity of molecular configuration.

The present experiment was designed to investigate the neurophysiological responses of the frog olfactory bulb to eight odors selected from the camphoraceous and the pepperminty classes of Amoore's classification.[5] The experiment was repeated with four isomers of butyl alcohol and ethyl ether. For each pair of odorants the correlation coefficient (r) provided a measure of the similarity between the odors as demonstrated by neurophysiological responses. The similarities within the same sets of odorants were also assessed stereochemically by measurements of molecular models[6] and organoleptically by a panel of human observers.[8] From the three sets of data, correlations were calculated between the similarities among the same groups of odorants, experimentally assessed in three different ways—neurophysiologically, stereochemically, and organoleptically.

METHODS

Preparation

Two hundred bullfrogs (*Rana catesbiana*) of 200 to 400 g were used. They were immobilized with *d*-tubocurarine chloride (1 mg/100 g body weight). Under a binocular microscope, the olfactory eminentia on one side and both olfactory bulbs were exposed with minimal bleeding. The pia mater over the bulbs was carefully removed.

Recording

In order to record spike discharges of single neurons, a superfine microelectrode filled with 3M of KCl was inserted into the bulb on the same side as the exposed eminentia. Microelectrodes with resistances of 100 to 200 megohms were found best for long recording of single-cell discharges. A silver wire was inserted into a glass pipette with a tip diameter of about 200 μ, which was filled with Ringer's solution. This electrode was used to record the slow potential change (electro-olfactogram, EOG) in the olfactory epithelium. A Ringer-soaked cotton-wool pad was put into the frog's mouth and grounded.

In Figure 1 the general set-up of the recording apparatus is shown. The EOGs were amplified with a dc amplifier and were recorded with an ink-writing pen-recorder (Nihon-Kohden Co., Tokyo). Spike discharges of single neurons were amplified with an RC coupled amplifier (time const. = 0.03 sec). Unwanted slow components, such as heart beats and induced waves, were removed with a band-pass filter of 70 c/sec to 40 kc/sec. After the positive phases of the spike discharges were cut with a slicer, the monophasic spike discharges were recorded with the pen-recorder and/or a dual-beam oscilloscope (Tektronix 502). When necessary, the data were stored by tape-recorder (TEAC R400, Tokyo) for future use.

Stimulants

In the first series of experiments, four odorants were selected (by S.H.) from each of the camphoraceous and pepperminty classes, according to Amoore's classification.[5] The odorants are shown in Table I. In the second series of experiments, four isomers of butyl alcohol and ethyl ether were used (Table I). These five odorants (also the choice of S.H.) have the same molecular weight.

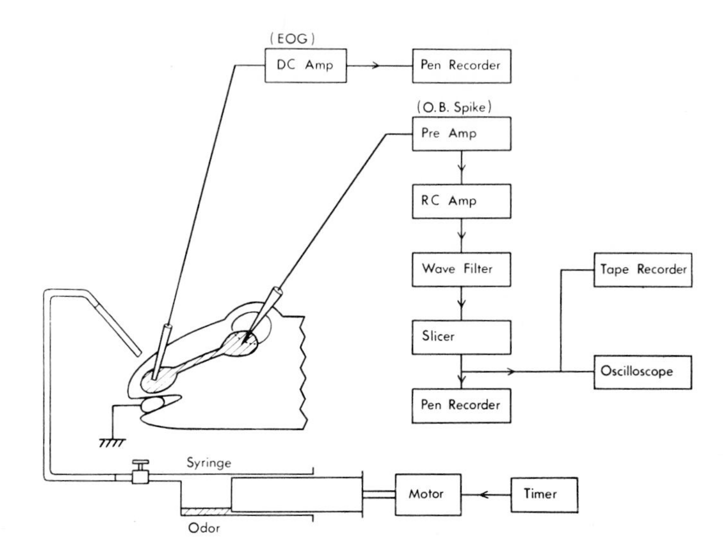

Figure 1 A diagram of the experimental procedure. The EOG was fed through a dc amplifier to one channel of a pen recorder or to an oscilloscope. Spike discharges in the olfactory bulb were fed through pre- and rc-coupled amplifiers to another channel of the recorder or to the oscilloscope. Interfering slow components in the discharges were removed with a wave filter, and one phase of the biphasic spike potentials was cut with a slicer. The plungers of the 20 ml syringes containing various diluted odorants were driven by a dc motor.

TABLE I
ODORANTS USED AS STIMULANTS
Concentrations were chosen to equalize intensity of stimulus.

Group	Compound	Abbr.	Concentration (%)	
			Frog	Human
Camphoraceous	dl-Camphor	CM	1*	0.004
	d-Borneol	BL	2*	0.0015
	Methyl isobutyl ketone	MK	0.1*	0.006
	1,8-Cineole	CL	0.1*	0.001
Minty	d,l-Menthone	MN	0.1*	0.0006
	l-Menthol	ML	20*	0.0075
	Cyclopentanone	CP	1*	0.18
	Cyclohexanone	CH	1*	0.015
Butyl Alcohol	Normal	NOR	8	0.05
	Isomeric	ISO	10	0.25
	Secondary	SEC	10	0.25
	Tertiary	TER	40	5
Ether	Ethyl ether	EET	5	0.02

* In propylene glycol. The remaining dilutions were made in water.

When saturated vapors of the odorants were applied, it was observed that spike discharges of single cells continued for several minutes after the stimulus was stopped. When the vapors were applied repeatedly, the responses of single neurons became variable and did not show the same patterns at each stimulation. Consequently, all the saturated vapors were diluted so that the after-effect of a stimulation would not influence the responses to succeeding stimuli.

Odorants of the camphoraceous and pepperminty classes were diluted with odor-free propylene glycol, which by itself elicited no spike discharges and did not give rise to any change in the spontaneously appearing spike discharges.[23] Since menthol elicited the smallest EOG among the eight odorants, 20 per cent menthol was used as a standard stimulus, and the remaining seven odorants were diluted so that the EOGs elicited were the same size as the EOG produced by 20 per cent menthol. In this manner, comparable concentrations of the eight odors were established (Table I). In the second series of experiments, the butyl alcohols and ethyl ether were diluted with deodorized distilled water, and appropriate concentrations were decided in the same way. Diluted odorant (1 ml) was placed in a 20 ml syringe. A battery of separate syringes and delivery tubes was prepared, one for each odorant included in the test series. To apply a stimulus, the vapor was expelled by a dc motor at the rate of 1.5 ml/sec through the respective teflon tube (0.8 mm diameter). A short glass capillary with a diameter of 0.8 mm was inserted into the end of each tube and directed onto the olfactory epithelium from a distance of about 1 cm.

When a neuron of the excitatory type was encountered, the vapors of all eight odorants were applied for one second each, with intervals of 30 seconds between applications. After each stimulation, deodorized air was applied for three seconds at a rate of 100 ml/min to remove any possible after-effect of the odors used previously.

The purity of all the odorants used in the neurophysiological experiments was examined by means of a gas chromatograph at the Takasago Perfumery Company in Tokyo. It was found that the menthone contained 20 per cent isomenthone and the isobutyl alcohol contained 3 per cent impurity, but that all the other odorants were pure.

Data Processing

In order to compare the effects of the eight odor stimuli, we counted, for the initial second of the response, the number of spike discharges which appeared corresponding to the rising phase of the EOG. The sequence of odor application was changed several times in order to determine whether the spike discharges appeared with the same pattern to the same odor. The spike discharges usually continued somewhat longer than one second. However, when odors were applied for as long as 2 sec, the spike discharges ceased before the 2-sec stimulus was complete. In order to minimize error caused by uncertainty as to the exact duration of the response, the odors were henceforth applied for one sec only and the elicited spike discharges were counted for that first second.

The method of data processing employed in the present experiment was based on that described by Erickson[16] for his study of gustatory neural response patterns. The similarity between any given pair of odorants, with respect to their patterns of neurophysiological response, was estimated as follows. For each active neuron examined, the number of spike discharges elicited by one odorant was plotted against the number of discharges elicited by the other odorant from the same neuron. Each pair of odorants was compared by at least 20 and sometimes as many as 50 discrete neurons. Examples of the resulting graphs are shown in Figure 4, which illustrates some good correlations, and Figure 5, in which correlations are absent. It was considered that the coefficient of correlation (r) between the sets of responses for each odorant would provide an experimental measure of the degree of similarity between those odorants, as indicated by the neurophysiological responses of the frog.

A separate graph was prepared for each of the 28 possible pairs among the eight camphoraceous and minty odorants of the first series. Similarly, the numbers of spike discharges were plotted for the 10 pairs among the four butyl alcohols and ethyl ether in the second series. The correlation coefficient (r) was calculated for each graph. Considerable use was made of these correlation coefficients in subsequent comparisons between the different methods of assessing the similarities between odorants. However, the statistical significance of the neurophysiological criterion of similarity was also established by calculating Student's "t" from r and N (number of neurons examined) by the formula:

$$t = r\sqrt{N - 2} / \sqrt{1 - r^2}.$$

Then, from the t-distribution table, the probability P was obtained, and recorded in Tables II and III.

Stereochemical Measurements

The similarities in molecular shape between each pair of odorants were estimated by comparing radial measurements on photographs of scale molecular models, using the "shadow-matching" method described elsewhere.[6] The molecular similarities are designated Sm.

Organoleptic Assessments

The general procedure for obtaining human judgments of the similarities in odor quality for each pair of odorants has been published by Amoore and Venstrom.[8] However, the wording of the instructions to the judges was changed to: "Compare the qualities of the two odors, and rate the degree of similarity between them" (on a scale from 0 to 8). The odorants were diluted in water, and the matching intensity concentrations are shown in Table I. The chemicals were different samples from those used in the frog work. Five sample pairs were presented per session. The human organoleptic similarities are designated Sh.

RESULTS

When a microelectrode was advanced to a depth of about 250 to 400 μ below the surface of the olfactory bulb, neurons were frequently encountered that showed little spontaneous discharge but responded well to some odors. Most neurons had no spontaneous discharges, but did produce spike discharges when certain odors were applied. In many cases, however, these neurons did not respond to other odors. With controlled odorous stimuli of the above limited concentrations, no inhibition occurred in the spontaneously appearing spike discharges, although inhibition was observed when odors of higher concentrations were applied. Thus, only the excitatory-on and the nonresponding types of response were found in the present experiments.

A single neuron could exhibit spike discharges in response to odorants for anything from several to twenty minutes before they decreased in amplitude and disappeared. To each neuron the eight (or five) odors were applied as often as possible. In one instance, they were applied no fewer than eight times without appreciable changes in the respective responses. However, in many other cases, the eight odors could be applied only once or twice before the spike discharges disappeared.

When the microelectrode was advanced deeper, neurons were often encountered that exhibited spontaneous spike discharges of greater amplitude at higher frequencies. Some of them disappeared in several seconds, but others appeared spontaneously for as long as 20 to 30 minutes. The latter neurons did not show any change in frequency when odors were applied. Consequently, when this kind of spike discharge appeared, advancement of the microelectrode was stopped and then resumed anew at a different locus.

Camphoraceous and Pepperminty Odors

When these eight odorous vapors were applied, some neurons responded to all of them. However, even in such instances the patterns of spike discharges were different for some odors than for others. Figure 2 shows an example of one such neuron. Spike discharges appeared more readily to methyl isobutyl ketone, cyclopentanone, and cyclohexanone than to the other odorants. When the number of spike discharges in the initial second was counted, there was good resemblance among the above three odorants. In other instances, spike discharges appeared only to a few odors. For the neuron illustrated in Figure 3, spike discharges appeared only in response to *d,l*-camphor, *d*-borneol, and cineole, but not to the other odors. From these results, it was concluded that neurons exist in the olfactory bulb that are selectively responsive in their discharge patterns to several odors.

When the numbers of spike discharges per initial second were plotted between paired odorants, good correlations were found in some cases (Figure 4) but not in others (Figure 5). The corresponding values of r, N and P are shown for each pair of odorants. The results for all 28 odor pairs are summarized in Table II. According to the values of P the odor pairs are separated into four groups by horizontal dotted lines.

The similarity in steric structure of the molecules was examined between all odorant pairs by the shadow-matching method.[6] The molecular similarity values (Sm) are shown in the second column from the right in Table II. When r, P and Sm are compared, the best parallel relation is to be found in the four odor pairs shown in the top group. From the top group down, the values of r and P gradually decrease in parallel with those of Sm. In the bottom group, however, there are some exceptions. Although

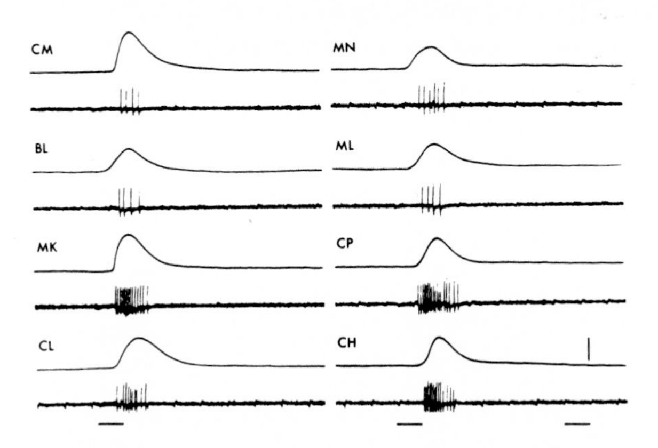

Figure 2 EOGs and spike discharges. This neuron responded to all eight odors in the camphor and minty classes. CM, camphor; BL, borneol; MK, methyl isobutyl ketone; CL, cineole; MN, menthone; ML, menthol; CP, cyclopentanone; CH, cyclohexanone. The responses to MK, CP and CH are conspicuously stronger, similar to each other, and different from those to the other odors. The horizontal bars near the left of the lower tracings indicate time and duration of odor applications (1 sec).

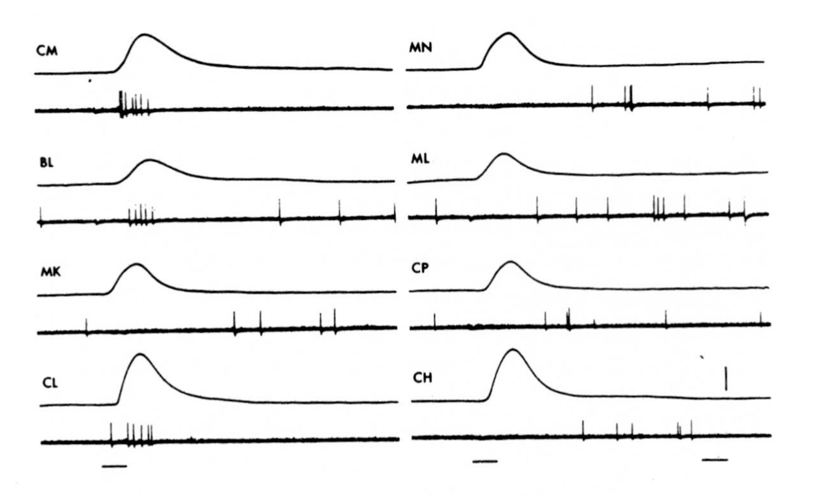

Figure 3 EOGs and spike discharges. This neuron responded only to CM, BL and CL.

the Sm values are greater than 0.6 in the pairs MK–MN, MK–ML, CL–MN, CL–ML, CL–CH and MN–ML (indicated by asterisks in Table II), the values of r were small or even negative. Especially in the pair menthone–menthol (MN–ML) the similarity of the molecular shapes is very close (Sm = 0.693; second highest value among the odor pairs used), while the value of r is −0.15 (most negative value among the odor pairs used in the present experiment).

According to the values of P, the degrees of correlation among the eight odorants have been drawn schematically (Figure 6). The odorants above the horizontal broken line belong to the camphoraceous class and those below the line to the pepperminty class. It is interesting to find that, in addition to some expected good correlations between odorants within the same odor class, there were also good correlations between odorants belonging to different odor classes, while there were several poor correlations between odorants in the same odor class. These results will be discussed later.

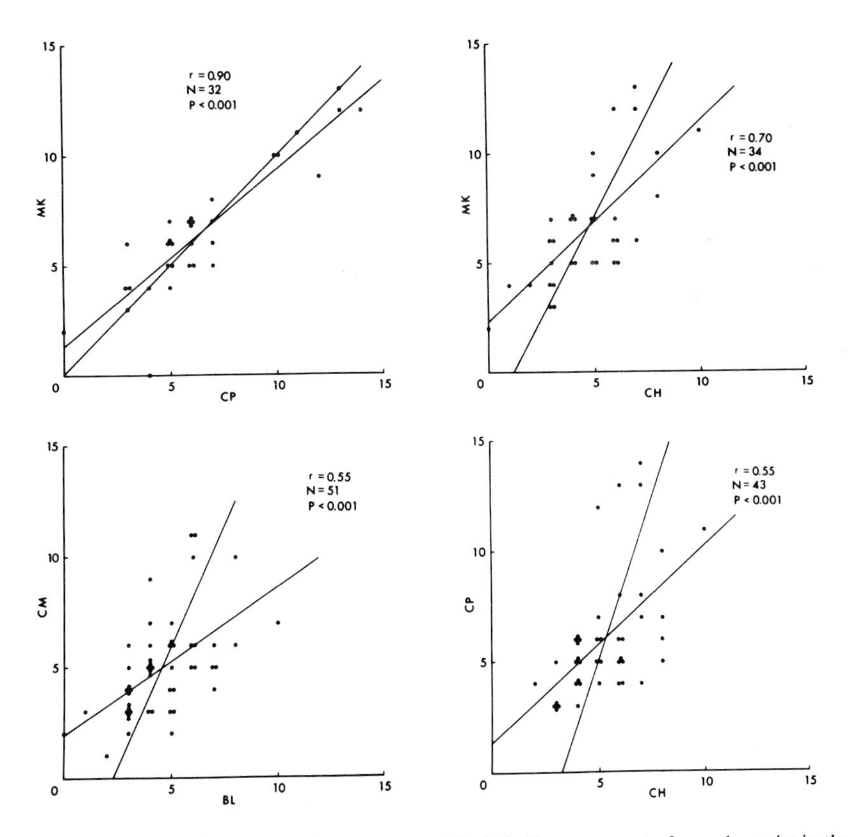

Figure 4 Correlations between the responses of single bulbar neurons to four odor pairs in the camphor and mint classes. These are examples of good correlations, as indicated by the reasonably convergent regression lines.

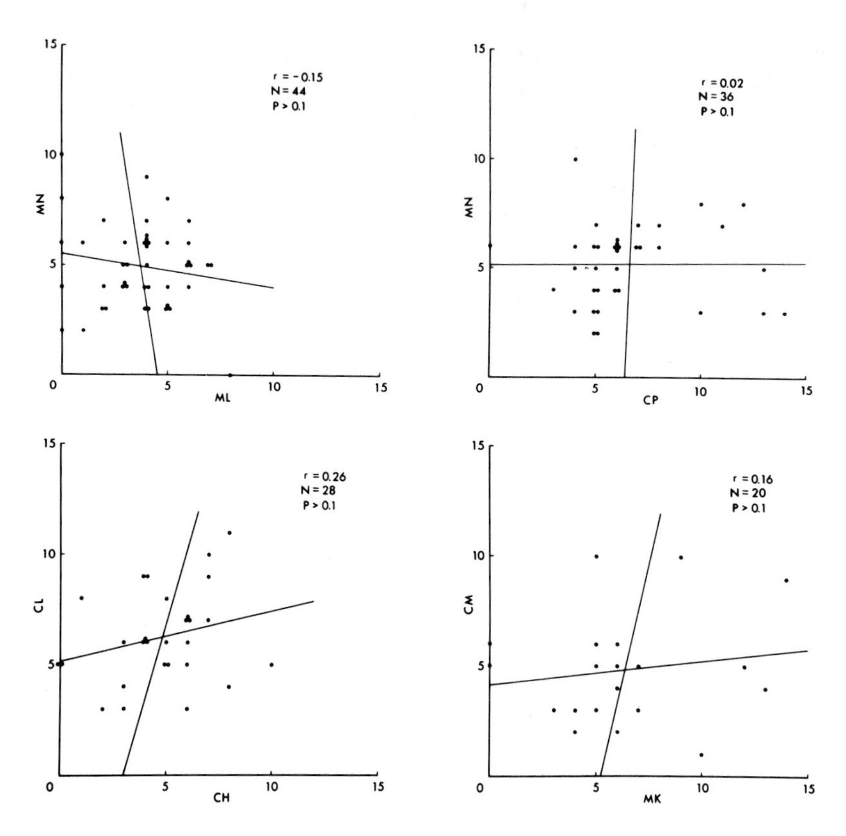

Figure 5 Correlations between the responses of single bulbar neurons to odor pairs; examples of poor correlations, with regression lines wide apart.

Butyl Alcohols and Ethyl Ether

The spike discharges were recorded after applying these five odorants. Although spike discharges appeared in response to all five odorants in certain neurons, there was most often a selective response to only a few odorants. The numbers of spike discharges for the initial second were compared between each pair of odorants. Examples of good and bad correlations were observed. The correlation coefficients r, numbers of neurons recorded N, and probabilities P are summarized in Table III. The odor pairs are separated into four groups according to the values of P. The best r of 0.75 was obtained between *sec*-butyl alcohol and ethyl ether. The next best r of 0.57 was found in two cases. In these three odor pairs, P was less than 0.001 as shown in the top group of Table III. From the top down, the values of r and P gradually decrease. According to the values of P, correlations among the five odorants are schematically drawn in Figure 7. Although the correlation is good between the isobutyl and the *sec*-butyl alcohols, none was found between the *sec*- and the *tert*-butyl alcohols, or between the normal and the *tert*-butyl alcohols.

TABLE II
CORRELATIONS AMONG CAMPHORACEOUS AND PEPPERMINTY ODORS

For key to odorant abbreviations, see Table I. r, correlation coefficient; N, number of neurons recorded; t, Student's "t"; P, probability; Sm, similarity of shapes of odorous molecules by the shadow-matching method; Sh, odor similarity assessed by human judges.

Odor pairs	r	N	t	Probability	Sm	Sh
CM -- BL	0.55	51	4.609	$P < 0.001$	0.684	5.07
MK -- CP	0.90	32	11.300		0.617	3.47
MK -- CH	0.70	34	5.544		0.637	4.08
CP -- CH	0.55	43	4.216		0.695	4.80
CM -- CL	0.51	25	2.843	$0.001 < P < 0.02$	0.663	4.93
MK -- CL	0.43	35	2.736		0.604	3.96
BL -- CL	0.44	26	2.400	$0.02 < P < 0.1$	0.641	3.49
BL -- ML	0.36	26	1.890		0.577	2.65
CM -- MN	0.44	26	2.400		0.554	3.87
MN -- CH	0.34	33	2.012		0.566	3.13
CL -- CP	0.37	27	1.991		0.549	3,40
CM -- MK	0.16	20	0.687	$0.1 < P$	0.548	3.59
CM -- ML	0.34	24	1.695		0.562	4.92
CM -- CP	0.35	21	1.628		0.511	3.96
CM -- CH	0.14	25	0.678		0.580	2.00
BL -- MK	−0.14	23	−0.648		0.539	4.07
BL -- MN	0.16	28	0.826		0.574	2.53
BL -- CP	−0.04	20	−0.170		0.483	2.71
BL -- CH	−0.11	24	−0.519		0.540	3.05
MK -- MN	0.15	34	0.858		0.605*	3.87
MK -- ML	0.14	25	0.678		0.602*	4.43
CL -- MN	−0.02	36	−0.117		0.631*	5.16
CL -- ML	0.09	30	0.478		0.634*	4.68
CL -- CH	0.26	28	1.372		0.614*	3.05
MN -- ML	−0.15	44	−0.983		0.693*	5.03
MN -- CP	0.02	36	0.117		0.519	3.75
ML -- CP	0.27	32	1.535		0.528	2.96
ML -- CH	0.08	31	0.432		0.583	3.99

* Indicates lack of correspondence between r and Sm.

The four isomeric butanols and their structural isomer ethyl ether were also examined as regards similarity of molecular shape. The results for Sm are given in Table III, in the second column from the right. There is a fairly definite parallel trend between the r values and those for Sm, indicating some relation between molecular shape and neurophysiological response in the frog.

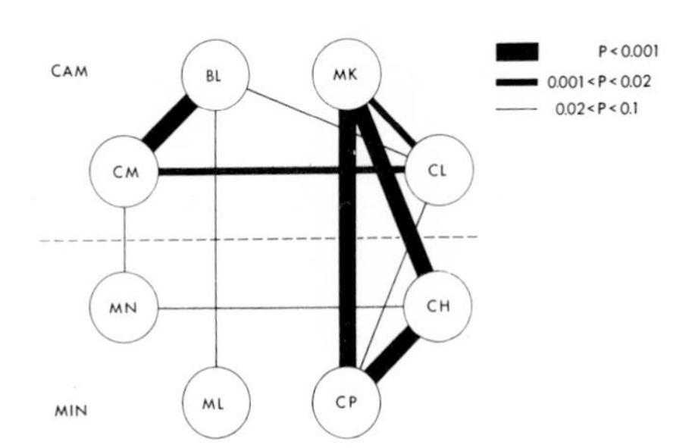

Figure 6 Degrees of correlation among the eight odors from the camphor and minty classes, as schematized by the values of P. The camphoraceous odors (CAM) and the pepperminty odors (MIN) are separated by a horizontal dotted line.

DISCUSSION

EOG as a Criterion of Olfactory Stimulus Intensity

In the present experiment, spike discharges of single neurons in the olfactory bulb were recorded together with the EOG. It was supposed that the EOG indicates the beginning and continuation of the response of the olfactory receptor cells.[22,24] Later Döving[12–14] assumed that the intensities of olfactory stimulating power of various odors can be made equal by equalizing the magnitude of the EOG elicited by each odor. However, the origin and generative mechanism of the EOG have not been determined,[27] and it is well known that negative as well as positive EOGs appear in the olfactory epithelium, depending upon the kinds of odors applied.[28,30–32] In addition, the EOG is a mass potential, very probably elicited by the excitation of a large number of olfactory receptor cells. Consequently, it is doubtful that the magnitude of the EOG really parallels the stimulating power of an odor. When recorded by a

TABLE III
CORRELATIONS AMONG BUTYL ALCOHOLS AND ETHYL ETHER
(For key to odorant abbreviations, see Table I.)

Odor pairs	r	N	Probability	Sm	Sh
NOR --- EET	0.57	34	$P < 0.001$	0.842	2.17
ISO --- SEC	0.57	42		0.681	3.75
SEC--- EET	0.75	32		0.671	4.02
NOR --- ISO	0.38	41	$0.001 < P < 0.02$	0.624	4.18
NOR --- SEC	0.42	46		0.691	3.25
ISO --- TER	0.44	41		0.656	4.52
ISO --- EET	0.36	35	$0.02 < P < 0.1$	0.611	4.38
TER --- EET	0.31	33		0.582	4.25
NOR --- TER	0.18	40	$0.1 < P$	0.591	2.53
SEC --- TER	0.28	45		0.651	3.78

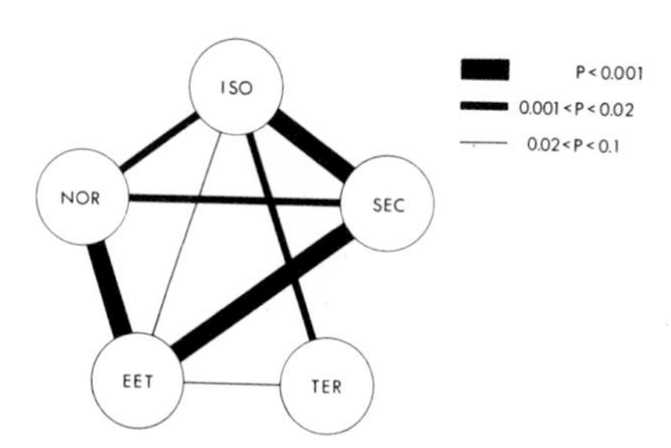

Figure 7 Degrees of correlation among the five odors of the butyl alcohols and ethyl ether, as schematized by the P values.

microelectrode in the olfactory epithelium, it was found that spike discharges are elicited by odor stimulation, but that they often disappear before the end of the stimulus. Thus, the beginning of the EOG may be used to indicate an average beginning of the responses of many olfactory receptor cells, but continuation of the EOG does not always indicate continuation of spike discharges of the olfactory cells. Spike discharges disappear earlier or they persist much longer, depending on the kinds and concentrations of odors.

It is well known that, for the human observer, odor quality can vary when the concentration of an odorous vapor is changed. Yet in the frog there is no other known objective criterion, apart from the EOG, by which the stimulating powers of odors can be decided or equalized among many different odorants. The EOG is one of the objective expressions of olfaction. Furthermore, it is probably more than a mere coincidence that in the butyl alcohol series the test odorant concentrations, independently determined by two different methods (EOG and intensity matching), are in exactly the same rank order, even though the species are different (Table I: the camphors and mints are not comparable, due to change of solvent). Consequently, if the EOG is used as a standard to decide the concentrations of various odors, it is believed that the data obtained with these odorants at these concentrations may survive analysis in the future, even when the real generative mechanism of the EOG is clarified. In this sense, the authors (S. H. and H. T.) support Döving's method and use the magnitude of the EOG to decide the concentrations of the various odors. Because menthol had the weakest ability to elicit the EOG, a 20 per cent solution of menthol in propylene glycol was adopted, and the magnitude of the EOG elicited was made a standard for the other odors. In case of the butyl alcohol series, a 40 per cent solution of *tert*-butyl alcohol in water was chosen as an independent standard, for the same reason.

Effect of Concentration of Odor on Neurophysiological Response

At the concentrations of odors used in the present experiments, spike discharges either appeared for all odors or failed to appear for some of them. Thus, responses of the excitatory and nonresponding types were obtained, but responses of the inhibitory type were not observed.

When the concentrations of the odors were increased, the excitatory response was replaced by the inhibitory response in the same neurons. In his experiments, Döving[12–14] classified the responses in the olfactory bulb into three types: excitatory (+), inhibitory (—) and nonresponding (0), and he presented a statistical analysis. Considering that a transition in the type of response was found in our preliminary experiments, depending on the concentrations of the odors applied, it is very probable that Döving's inhibitory results could disappear if he were to use the same odors at lower concentrations. In his experiments, odorous puffs of less than 0.5 ml were delivered to the olfactory epithelium by pushing the plungers of 10 ml glass syringes by hand. The odorous puffs were adjusted to give EOGs of nearly equal amplitudes. It is apparent that it was always the saturated vapors of the odorants that were being applied, although at different velocities. Hence, the concentrations of odors used in his experiments were much higher than those of the odors in our experiments. Conjecturally, our results might have been different from his, even if we were to use the same odorants that he used. This may be one of the most difficult problems in reconciling the various researches on olfactory discrimination.

In Table IV, differences in the response patterns between Döving's work and ours are shown. In the present experiment, some neurons were found that did not respond to any of the odorants shown in Table I, but to other odors. Such neurons are omitted from Table IV.

Shibuya, Ai, and Takagi[25] classified the bulbar spike discharges elicited by odorous stimulation into seven types: on, on-off, off, simple inhibition, postinhibitory excitation, postexcitatory inhibition, and nonresponding. Similar response types were found in the olfactory epithelium.[29] These types had been obtained when saturated vapors were applied. From the above findings, it is presumed that a classification of

TABLE IV

COMPARISON OF DÖVING'S DATA WITH THE PRESENT RESULTS

The values are percentages of neurons responding with excitation (+) or inhibition (—), or not responding (0).

Odorants	Döving	Higashino, Takeuchi & Amoore	
Camphoraceous odors	+ 16% 0 11% (1966b) — 73%	+ 45–65% 0 55–35% — 0%	
Pepperminty odors		+ 45–65% 0 55–35% — 0%	
Normal butyl alcohol	+ 25% 0 10% (1964) — 65%	+ 73% 0 27% — 0%	For the other butyl alcohols and ethyl ether: + 56–93% 0 44–7% — 0%

the responses into three (+, −, 0) categories may not be accurate, as Döving himself realized.[11] The role of the three inhibition types in the afferent olfactory discharges[25] may be very important at high odor intensities, but they remain a problem for future studies of olfactory information processing.

Grouping of Odors

In the present experiment, it was found that neurons which responded well to camphor responded well to borneol and cineole (Figure 3). By continuing this method of analysis, the odorants were found to fall into two groups: camphor, borneol, and cineole belong to one group; methyl isobutyl ketone, cyclopentanone, and cyclohexanone belong to another group. Neurons which responded well to one member of a given group of odors (Figure 6) were very likely to respond well to other odors of the same group. Gesteland, Lettvin, Pitts, and Rojas[18] studied the responses of the olfactory receptor cells and classified the cells into five groups. Upon finding that some cells responded in similar ways to several odors, they suggested that odorants may be classified into groups. In the first series of the present experiments, interrelations were studied among only eight odors—four from the camphoraceous class and four from the pepperminty class. In the second series the relationships were examined among only five odors—four butyl alcohols and ethyl ether. A considerable degree of relatedness was proved among certain groupings of odorants. If our research could be continued in this way, increasing the number of odorants applied to the olfactory organ of the frog, we might eventually be able to classify all odorants into a limited number of groups, according to their neurophysiological relationships. If so, those groups may represent primary odors for the frog.

Odor Similarity in the Frog and Shape Similarity of Molecules

When the values of r are compared with the values of similarity of odorous molecular silhouettes (Sm) in Table II, fairly good correlations are found between the two values in the upper three groups of the Table. In the bottom group, there are also good correlations between r and Sm in most odor pairs; but there are some exceptions. As indicated by the asterisks beside the values of Sm, these six odor pairs have very small values of r, whereas the values for Sm are quite large. However, it was found that the values of stereochemical similarity are always large for the odorant pairs in which the numbers of spike discharges are closely correlated. Consequently, the Moncrieff-Amoore supposition—that the stereochemical structures play important roles in the discrimination of odor qualities—can be supported by our data. In the exceptional cases indicated by asterisks in Table II, it may be inferred that some factors other than the molecular profiles may exert influences upon the odor quality.

Frog Olfaction and Human Olfaction

It was shown that there is a very good correlation among methyl isobutyl ketone, which belongs to the camphoraceous class, and cyclopentanone and cyclohexanone, both of which belong to the pepperminty class. On the other hand, no correlation was found between cyclopentanone or cyclohexanone and menthone, although these three belong to the pepperminty class. Thus, the descriptive classification of odors by Amoore[5] did not seem applicable to the data in the present experiment on frogs. However, Amoore's earlier classification was based on descriptions of odors of chemicals assembled in the literature survey; the descriptions of different chemicals are not so uniform as could be desired. Therefore, the relatedness of odors was experimentally re-examined, using a panel of 29 human judges and providing standard odorant solutions for reference purposes.[9] According to the new results from human judges, cyclohexanone is stronger in camphoraceous component than in pepperminty component, and cyclopentanone is stronger in camphoraceous or ethereal component than in pepperminty. If so, the results of our experiments on the frog coincide better with the results of the improved assessments by human judges. Close similarity had already been observed among the profiles of these odorous molecules by the shadow-matching method. In camphor, borneol, and cineole, all of which belong to the camphoraceous group, a fairly close correlation was also found among the electrophysiological data (r), the results of human olfactory tests (Sh), and the similarities of the profiles of the molecules (Sm).

However, other data did not indicate close correlation between frog and human olfaction. There was no correlation between the odors of menthone and menthol in the frog, although both compounds belong to the pepperminty class, according to the similarity in profiles of these molecules. Nevertheless, according to the matching standards tests on human olfaction, menthol is rather more camphoraceous than minty.[9] Consequently, it may be said that the result on the frog is not entirely inconsistent with the result on human olfaction. Döving,[14] dividing the response patterns of the olfactory bulb neurons into the three categories, found that menthone is closely akin to menthol. However, Gesteland, Lettvin, and Pitts[17] showed that receptor cells in the olfactory epithelium of the frog responded to menthol but not to menthone. This finding is in complete agreement with the result in the present experiment. It was later found that the menthone used in the present experiment contained 20 per cent isomenthone. It is not clear how much influence this impurity may have had upon the results in the present experiments. When disagreement occurs between the results on the frog and the measurements on the odorous molecules, it may indicate that some factors other than the profile of an odorous molecule contribute to odor discrimination.

Döving[13] used the same homologous alcohols in his frog experiments as those used by Engen[15] in odor similarity tests on human judges. It was found that the rank correlation coefficient between Engen's psychological data and Döving's physiological chi-square values was 0.868, which is significant at $P = 0.001$. Consequently, he assumed that the discriminatory mechanisms of the olfactory system are similar in frogs and humans. In our research, however, there is a total lack of correlation between the frog similarity values and the human similarity values (Sh) in both experiments (r is 0.07 and 0.04 respectively). In contrast, the correlation r between frog similarity value and shape similarity value is 0.34, which is barely significant at P about 0.05 when N is 28. A better correlation exists between the similarity of odor for the human judges and the similarity of molecular shape (r is 0.58 at P about 0.001). In the present stage of our experiments, the reason for results in our experiment differing from those in Döving's experiment are open to question. Quite possibly the difference originates in the different statistical criteria of bulbar response similarity employed in the two experiments. In addition, this joint experiment shows reasonable correlations between the molecular shape Sm on the one hand and both human odor judgments Sh and frog nervous responses r on the other, whereas the frog and the human results do not correlate at all. It is possible that the natural groupings of the odors (e.g., the primary odor specificities) for the two species may be very different, so that although both species show some dependence on molecular shape, there may be practically no correspondence between the species.

Butyl Alcohols and Ethyl Ether

Among the four isomers of butyl alcohol, good correlation was found only between iso- and *sec*-butyl alcohols. Equally good correlation was found between the two pairs of ethyl ether and normal butyl alcohol, and ethyl ether and *sec*-butyl alcohol. In contrast, no correlation at all was found between *sec*- and *tert*-butyl alcohols or between normal and *tert*-butyl alcohols. Remembering that all these five odorants have the same molecular weight, it may be said that molecular weight per se cannot be related with the qualities of odors.

Correlations Between Neurophysiological, Stereochemical, and Organoleptic Assessments of Odor Similarity

Within each discipline of fundamental odor research, whether it be analytical chemistry, odor theories, electrophysiology, or psychometrics, there has been little attempt in the past to reconcile the results obtained by different investigators. Regarding direct experimental contact between the disciplines, the situation is worse; it may occassionally have been contemplated, but it has very seldom been enacted.

As mentioned in the body of this paper, some tentative but statistically significant connections have been found between electrophysiology and psychometrics,[13] between stereochemistry and psychometrics,[9] and now between electrophysiology and stereochemistry. An important bonus of the present investigation is that, for the first time, direct correlations among all three disciplines have been obtained on the same set of odorous compounds.

The results are summarized in Table V. The correlation between molecular shape and frog nervous response is weak, but just significant, for the mints and camphors, and for the butyl alcohol group it is stronger, but due to smaller N, not more significant. Tenuous as these results may be, because the compounds examined were barely numerous enough, the result is consistently in the direction of indicating that stereochemical principles play an important part in olfactory discrimination by the nervous system of the frog. With the mints and camphors, the correlation coefficient between

TABLE V
STATISTICAL SIGNIFICANCE OF INTERDISCIPLINARY CORRELATIONS

Odorants	Correlation	r	~P
8 Camphors & mints (N = 28 pairs)	Molecule/Frog	0.34	0.05
	Molecule/Human	0.58	0.001
	Frog/Human	0.07	1
5 Butanols & ether (N = 10 pairs)	Molecule/Frog	0.59	0.05
	Molecule/Human	−0.57	0.05
	Frog/Human	0.04	1

molecular shape and human judgment has the substantial value and statistical significance we have come to expect from earlier studies of this question.[6,9] However, stereochemistry met another of its occasional failures with the butyl alcohol group, which had a barely significant negative correlation. Amoore[7] had encountered a similarly disappointing result with the eight isomers of amyl acetate. The interpretation advanced then still stands; within a group of isomers of a small molecule the variations in odor and in molecular shape are probably much the same as the standard errors of the measurements themselves.

The gross statistical result that shows no correlation whatsoever between frog and human assessments of odor similarity is, unfortunately, misleading, and deserves closer scrutiny, especially in respect to the camphors and mints of the first series. We have rank-ordered the 28 odorant pair comparisons for each variable (molecule, frog, and human). From the purely stereochemical point of view, the eight compounds fall rather obviously into three sub-groups. Camphor, borneol, and cineole are bicyclic terpenes with fairly large spherical molecules. There are three possible paired comparisons among them. The group, which smells camphoraceous to the human, showed the best agreement among molecular shape, frog responses, and human judgments, with the sole exception of the human judgment of the pair BL–CL. A second group (of two) is menthone and menthol, monocyclic terpenes with molecules about the same size as those of camphors, but wedge-shaped or triangular. Only one paired comparison is possible; it shows excellent agreement for human odor judgment (probably mainly minty) and molecular shape, but the frog finds it totally distinct. The third chemical group is methyl isobutyl ketone, cyclohexanone, and cyclopentanone, which are small ketones with molecular volumes roughly half those of the camphors and menthols. Obviously, the frog finds these three pairs very similar indeed, and there is also high similarity in molecular shape; but in the human judgments the odors are not notably similar. The human odor impression seems to be a mixture of ethereal, camphor, and minty, according to the matching standards method of scaling analysis.

Our tentative conclusion from these observations is to suggest that there is evidence that both the frog and the human make use of primarily stereochemical criteria in recognizing the odor of the camphors. Humans, but not frogs, also use stereochemical assessments in recognizing the menthols, but the frog uses either some obscure, minor, stereochemical feature that does not show up in the gross molecular silhouette analysis, or completely nonsteric information, such as the presence of a particular chemical functional group (e.g., alcohol versus ketone). Finally, the frog, but apparently not the human, uses steric factors in recognizing the small ketones and, possibly, the butyl alcohols as well. It may be premature to offer the following interpretation, but perhaps both frog and man possess a camphoraceous primary odor based on stereochemical principles. However, man, but not frog, possesses a minty primary odor based on stereochemistry, and frog, but not man, has a small-ketone primary and an alcoholic primary for which we have no right to assign any anthropomorphic names. Of course, it is disappointing to find apparently only one out of four possible primary odors common to both species, because the visual pigments seem to have much greater phylogenetic constancy. However, chemical detection may require far more interspecies variety and specificity than color appreciation.

On the whole, we feel encouraged with the results of this collaboration, particularly in respect to the camphors, considering that we have been working with different species, chemical samples, and experimental methods, and that each method is open to general or particular intrusions of experimental error (standard deviation, biological variation, chemical impurities, and conformational uncertainty). We hope that an increasing number of such interdisciplinary comparisons will take place in the future.

SUMMARY

Spike discharges of single neurons were recorded by microelectrodes from the olfactory bulb of the bullfrog. Eight odorants, four from the camphoraceous and four from the pepperminty odor class, were used as stimulants. Five other odorants, four isomers of butyl alcohol and ethyl ether, were also used. Each of the odorants was diluted so the EOGs could be equalized in amplitude. In this way, the concentrations of these odorants were just strong enough to elicit responses of the excitatory type, but not so strong as to elicit responses of the inhibitory type. For a given electrode position, the number of spikes occurring in the initial second of the discharge was counted for different odorants, and the correlation coefficient (r) was calculated for each pair of odorants. When the r values were compared with the similarities between the molecular profiles of the odorants (Sm), a fairly good correlation was found, although there were some exceptions. Thus, the importance of molecular shape in odor discrimination was generally indicated, and the Moncrieff-Amoore hypothesis received support from experiments in the frog. The exceptions may indicate the contributions of some additional factors to odor discrimination. Although good correlation between human olfaction (Sh) and molecular similarity is well established, there was no correlation between frog olfaction and human olfaction.

ACKNOWLEDGMENTS

We are indebted to Prof. Sadayuki F. Takagi for suggesting this joint investigation, for his advice and help throughout the experiments, and for his aid in drafting this manuscript. We also thank Miss T. Yajima and other members of the laboratory for their aid in making figures and tables. Our sincere thanks are due to Mr. M. Kainosho, Mr. H. Yoshigi and Miss N. Karasuda in the Takasago Perfumery Company, Ltd. for valuable suggestions, a supply of odorous chemicals, and gas chromatographic analysis. The neurophysiological experiments were performed at Gunma University in Maebashi, Japan, and supported by a grant for scientific research from the Ministry of Education and by the U.S. Air Force Office of Scientific Research through Grant No. DA-CRD-AFE-S92-544-68-G110 of the U.S. Army Research and Development Group (Far East). The molecular silhouette measurements and organoleptic assessments were made at Albany, California with the assistance of Mrs. Delpha Venstrom.

Reference to a company or product name does not imply approval or recommendation of the product by the U.S. Department of Agriculture to the exclusion of others that may be suitable.

REFERENCES

1. Adrian, E. D. 1950. Sensory discrimination: with some recent evidence from the olfactory organ. *Brit. Med. Bull.* **6**, pp. 330–332.
2. Adrian, E. D. 1952. The discrimination of odours by the nose. *Schweiz. Med. Wschr.* **82**, pp. 973–974.
3. Adrian, E. D. 1953. Sensory messages and sensation. The response of the olfactory organ to different smells. *Acta Physiol. Scand.* **29**, pp. 5–14.
4. Adrian, E. D. 1954. The basis of sensation. Some recent studies on olfaction. *Brit. Med. J.* **1**, pp. 287–290.
5. Amoore, J. E. 1962. The stereochemical theory of olfaction. 1. Identification of the seven primary odors. *Proc. Sci. Sect. Toilet Goods Assoc.* Suppl. No. 37, pp. 1–12.
6. Amoore, J. E. 1965. Psychophysics of odor. *Cold Spring Harbor Symp. Quant. Biol.* **30**, pp. 623–637.
7. Amoore, J. E. 1967. Stereochemical theory of olfaction. *In* Symposium on Foods: The Chemistry and Physiology of Flavors (H. W. Schultz, E. A. Day and L. M. Libbey, editors). Avi Publishing Co., Westport, Conn., pp. 119–147.
8. Amoore, J. E., and D. Venstrom. 1966. Sensory analysis of odor qualities in terms of the stereochemical theory. *J. Food Sci.* **31**, pp. 118–128.
9. Amoore, J. E., and D. Venstrom. 1967. Correlations between stereochemical assessments and organoleptic analysis of odorous compounds. *In* Olfaction and Taste II (T. Hayashi, editor). Pergamon Press, Oxford, etc., pp. 3–17.
10. Beets, M. G. J. 1957. Structure and odour. *In* Molecular Structure and Organoleptic Quality. Monograph No. 1. Society of Chemical Industry, London, pp. 54–90.
11. Döving, K. 1964. Studies of the relation between the frog's electro-olfactogram (EOG) and single unit activity in the olfactory bulb. *Acta Physiol. Scand.* **60**, pp. 150–163.
12. Döving, K. 1965. Studies on the responses of bulbar neurons of frog to different odour stimuli. *Rev. Laryngol. (Bordeaux)* **86**, pp. 845–854.
13. Döving, K. 1966. An electrophysiological study of odour similarities of homologous substances. *J. Physiol. (London)*, **186**, pp. 97–109.
14. Döving, K. 1966. Analysis of odour similarities from electrophysiological data. *Acta Physiol. Scand.* **68**, pp. 404–418.
15. Engen, T. 1962. The psychological similarity of the odors of aliphatic alcohols. *Rep. Psychol. Lab. Univ. Stockholm* **127**, pp. 1–10.
16. Erickson, R. P. 1963. Sensory neural patterns and gustation. *In* Olfaction and Taste I (Y. Zotterman, editor). Pergamon Press, Oxford, etc., pp. 205–213.

17. Gesteland, R. C., J. Y. Lettvin, and W. H. Pitts. 1965. Chemical transmission in the nose of the frog. *J. Physiol.* (*London*), **181**, 525–559.
18. Gesteland, R. C., J. Y. Lettvin, W. H. Pitts, and A. Rojas. 1963. Odor specificities of the frog's olfactory receptors. *In* Olfaction and Taste I (Y. Zotterman, editor). Pergamon Press, Oxford, etc., pp. 19–34.
19. Mancia, M., J. D. Green, and R. v. Baumgarten. 1962. Response patterns of olfactory bulb neurons. *Arch. ital. Biol.* **100**, pp. 449–462.
20. Moncrieff, R. W. 1951. The Chemical Senses (2nd ed.). Leonard Hill Ltd., London.
21. Moncrieff, R. W. 1956. Olfactory adaptation and odour likeness. *J. Physiol.* (*London*), **133**, pp. 301–316.
22. Ottoson, D. 1956. Analysis of the electrical activity of the olfactory epithelium. *Acta Physiol. Scand.* **35**, pp. 1–83.
23. Ottoson, D. 1958. Studies on the relationship between olfactory stimulating effectiveness and physico-chemical properties of odorous compounds. *Acta Physiol. Scand.* **43**, pp. 167–181.
24. Ottoson, D. 1959. Comparison of slow potentials evoked in the frog's nasal mucosa and olfactory bulb by natural stimulation. *Acta Physiol. Scand.* **47**, pp. 149–159.
25. Shibuya, T., N. Ai, and S. F. Takagi. 1962. Response types of single cells in the olfactory bulb. *Proc. Jap. Acad.* **38**, pp. 231–233.
26. Stoll, M. 1957. Facts old and new concerning relationships between molecular structure and odour. *In* Molecular Structure and Organoleptic Quality. Monograph No. 1. Society of Chemical Industry, London, pp. 1–12.
27. Takagi, S. F. 1967. Are EOG's generator potentials? *In* Olfaction and Taste II (T. Hayashi, editor). Pergamon Press, Oxford, etc., pp. 167–179.
28. Takagi, S. F., K. Aoki, M. Iino, and T. Yajima. 1969. The electropositive potential in the normal and degenerating olfactory epithelium. *In* Olfaction and Taste III (C. Pfaffmann, editor). The Rockefeller Univ. Press, New York, pp. 92–108.
29. Takagi, S. F., and K. Omura. 1963. Responses of the olfactory receptor cells to odours. *Proc. Jap. Acad.* **39**, pp. 253–255.
30. Takagi, S. F., T. Shibuya, S. Higashino, and T. Arai. 1960. The stimulative and anaesthetic actions of ether on the olfactory epithelium of the frog and the toad. *Japan J. Physiol.* **10**, pp. 571–584.
31. Takagi, S. F., and G. A. Wyse. 1965. Ionic mechanisms of olfactory receptor potentials. *Proc. 23rd Intern. Congr. Physiol. Sci.*, Tokyo, p. 379.
32. Takagi, S. F., G. A. Wyse, and T. Yajima. 1966. Anion permeability of the olfactory receptive membrane. *J. Gen. Physiol.* **50**, pp. 473–489.

RECIPROCAL INHIBITION AT GLOMERULAR LEVEL DURING BILATERAL OLFACTORY STIMULATION

J. LEVETEAU *and* P. MACLEOD

Laboratoire de Physiologie Générale, 9 Quai Saint-Bernard, Paris 5e, and Laboratoire de Neurophysiologie, Paris, France

INTRODUCTION

Paired sense organs of vertebrates are centrally interconnected in such a way that they contribute to spatial orientation of the subject. Stereoscopic vision, still incompletely understood, has been known for a long time. More recently, stereophonic hearing has been demonstrated and a fairly well-developed bilateral reciprocal inhibition was discovered by von Békésy[2] in the sense organs of taste and smell in man.

A very small difference in time (a few tenths of a millisecond) or in concentration (10 per cent) are enough, during birhinal stimulation, to completely suppress the smell perception of either the later or the weaker stimulated side. Obviously, this effect should be efficient in subserving the mechanism of head orientation toward a proximal olfactory source.

The present experiment was undertaken in order to ascertain whether any electrophysiological support could be found, at the glomerular level, for the reciprocal inhibition suggested by these psychophysical findings.

MATERIALS AND METHODS

The experiments were performed on young rabbits. Both olfactory bulbs and the dorsal aspects of olfactory sacs were uncovered under light pentothal anesthesia. Since it is known that barbiturates have a long-lasting suppressive action on bulbar slow potentials,[1,7] the animal was then curarized and artificially ventilated for at least 6 hr before we started to record.

We made two types of recordings: (1) layer recording of glomerular activity,[8] for which we used a monopolar glass micropipette filled with KCl 3 M; and (2) single glomerulus recording, for which we used a specially designed bipolar glass micropipette.[3] Two conventional micropipettes were cemented side by side with a longitudinal shift that separated their tips by about 150 μ, approximately the diameter of a glomerulus.

Olfactory stimulations were supplied by a simple two-channel device (Figure 1): a low-pressure nitrogen supply was divided into two streams separately controlled by two needle valves in series with two electromagnetic valves and bubbled into simple odor flasks. The electromagnetic valves were driven by an ordinary two-channel, square-wave, electronic stimulator. The apparatus was connected to the animal's nostrils by short lengths of polyethylene tubing of 1.5 mm outer diameter.

An over-all control of the actual delay between the two-odor channel was readily achieved by simply putting two microphones along the lateral faces of the rabbit's nose and displaying the acoustic signals on a dual-beam oscilloscope. The precision of the setting was in the range of 0.1 msec. Within experimental limits, the quantitative action of the needle valves did not interfere with the delay, which depended only on the electronics and was therefore easily monitored.

The chemical used as odorant was almost always propanol in 0.01 M aqueous solution.

The intensity of stimulation (needle-valve setting) was adjusted in respect to the amplitude of simultaneously recorded EOGs[6]: equal intensities were supposed to elicit equal fractions of maximal response on corresponding sides.

RESULTS

Layer Recording, Time Effect

When both stimulations are adjusted for equal intensities and separated by an interval of a few milliseconds, a sharp minimum appears in the homolateral response, which is roughly halved when the heterolateral stimulation is given about 3 msec before the

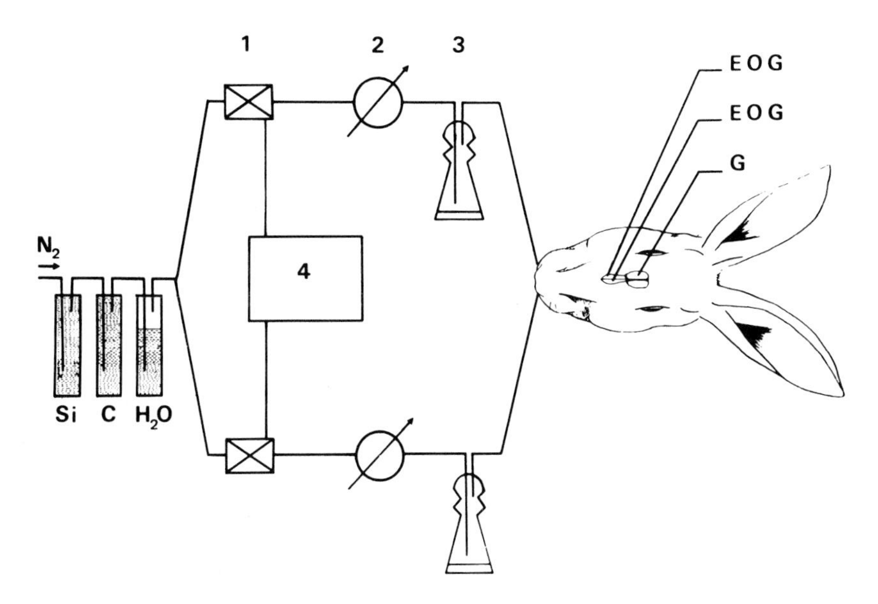

Figure 1 Stimulating apparatus diagram. 1, electromagnetic valve; 2, needle valve; 3, odor flask; 4, electronic stimulator.

homolateral one (Figure 2). The inhibition completely disappears within ± 10 msec around this value.

Layer Recording, Concentration Effect

When the delay is adjusted for maximum time effect and heterolateral concentration varies, it is consistently found that homolateral response amplitude decreases as heterolateral stimulus strength increases, until both stimuli become equal; as heterolateral stimulus becomes greater, the homolateral response amplitude remains constant (Figure 3).

Single Glomerulus Recording, Concentration Effect

When the homolateral stimulus is weak, a weaker heterolateral stimulus produces inhibition, but a stronger one produces an unexpected facilitation (Figure 4).

When the homolateral stimulus is strong, the reverse is observed: a weaker heterolateral stimulus produces facilitation and a stronger one has no effect, possibly because of heterolateral glomerular saturation.

When the homolateral stimulation is medium, nothing appears but a hardly noticeable facilitation.

Exactly the same occurs if the heterolateral side is stimulated by a different chemical—pyridine instead of propanol.

DISCUSSION

The difference observed in concentration effect on layer response as compared to that of a single glomerulus is puzzling. Nothing can presently help us decide whether

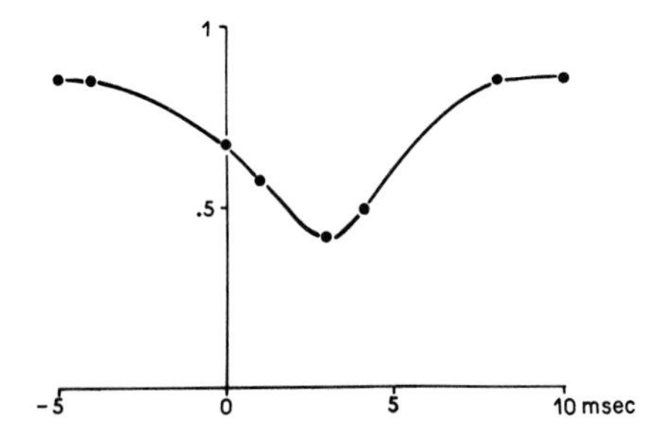

Figure 2 Time effect on the layer response. Abscissae: time elapsed between heterolateral and homolateral stimulations; Ordinates: relative magnitude of the homolateral response; magnitude 1 refers to unilateral stimulation.

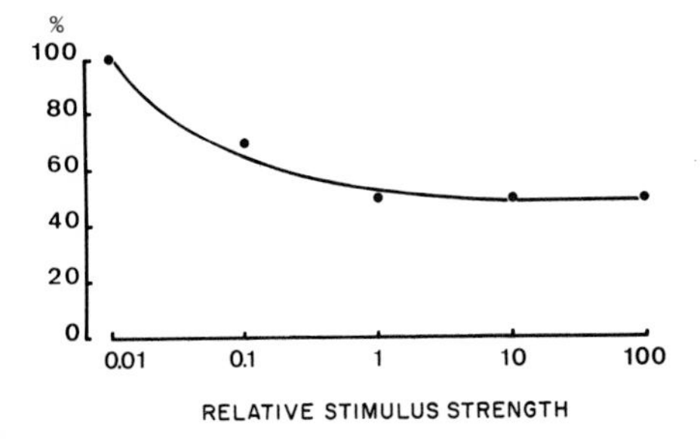

Figure 3 Concentration effect on the layer response. Abscissae: the relative stimulus strength is the ratio heterolateral:homolateral concentration. The "concentration" is measured by previous EOG scaling of the needle-valve setting, using known dilutions as references. Ordinates: magnitude of the response in per cent of the response to unilateral stimulation.

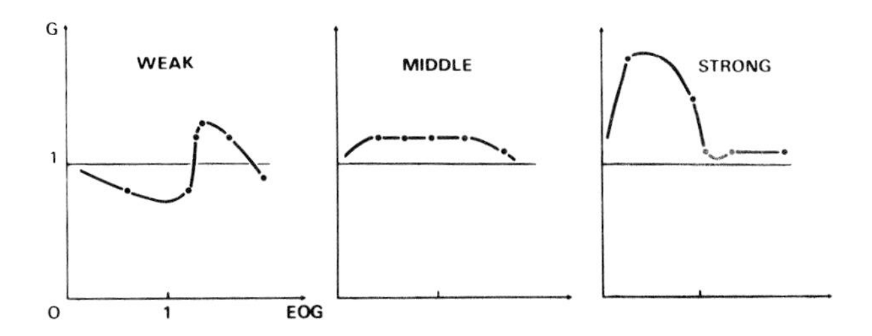

Figure 4 Concentration effect on single glomerulus response. The abscissae are the ratio heterolateral EOG amplitude: homolateral EOG amplitude. The ordinates are the ratio actual homolateral response: response to unilateral stimulation.

it is caused by the summation of many different individual responses or by the addition of some other components from adjacent interneurons.

The facilitation observed in the response of a single glomerulus was a quite unexpected finding. It leads to the assumption that interglomerular interaction might be twofold: inhibitory with a low threshold, and facilitory with a higher threshold, the facilitation becoming predominant at higher stimulus strengths. The pathway and latency of this facilitation remain unknown; higher centers, such as the limbic system and/or the reticular formation, might be involved. The time effect on the layer response is not maximum when stimulations are simultaneous, but rather when heterolateral stimulation is given 3 msec before homolateral. This indicates that two synapses are probably involved in the interbulbar pathway, a fact consistent with anatomical data from Lohman.[4,5] This further suggests that more centrally located structures, such as anterior olfactory nuclei, must be concerned in the birhinal intensity balance: rather close to each other and almost equidistant from peripheral inputs, they should interact maximally when these inputs are simultaneous. Hence, the observed glomerular inhibition would be a diffusion effect, parallel and secondary to olfactory nucleus inhibition, and merely reinforcing and lengthening it by reverberation.

REFERENCES

1. Adrian, E. D. 1950. The electrical activity of the mammalian olfactory bulb. *Electroencephalogr. Clin. Neurophysiol.* **2**, pp. 377–388.
2. Békésy, G., von. 1964. Olfactory analogue to directional hearing. *J. Appl. Physiol.* **19**, pp. 369–373.
3. Leveteau, J., and P. MacLeod. 1966. La discrimination des odeurs par les glomérules olfactifs du lapin (étude électrophysiologique). *J. Physiol.* (*Paris*) **58**, pp. 717–729.
4. Lohman, A. H. M. 1963. The anterior olfactory lobe of the guinea pig. *Acta Anat.* **53**, Suppl. 49, pp. 1–109.
5. Lohman, A. H. M., and H. J. Lammers. 1967. On the structure and fiber connections of the olfactory centers in mammals. *In* Sensory Mechanisms: Progress in Brain Research, Vol. 23 (Y. Zotterman, editor). Elsevier Publishing Co., Amsterdam, The Netherlands, pp. 65–82.
6. MacLeod, P. 1959. Premières données sur l'électro-olfactogramme du lapin. *J. Physiol.* (*Paris*) **51**, pp. 85–92.
7. Ottoson, D. 1959. Studies on slow potentials in the rabbit's olfactory bulb and nasal mucosa. *Acta Physiol. Scand.* **47**, pp. 136–148.
8. Yamamoto, C. 1961. Olfactory bulb potentials to electrical stimulation of the olfactory mucosa. *Jap. J. Physiol.* **11**, pp. 545–554.

EFFECT OF ETHMOIDAL NERVE STIMULATION ON OLFACTORY BULBAR ELECTRICAL ACTIVITY

HERBERT STONE

Life Sciences Research, Stanford Research Institute, Menlo Park, California

INTRODUCTION

In two previous investigations[11,12] data were presented which suggested that the response of an animal to an odor stimulus could be altered by blocking or interrupting the trigeminal pathways. These experiments suggested to us the concept of an "olfactory-trigeminal" response.

The observation that some odor stimuli are capable of stimulating trigeminal receptors is not new.[1,6,10] Early studies showed that highly concentrated irritants and pungent compounds were capable of stimulating trigeminal acceptors. More recently Tucker[13,14] and others (Beidler[2]) reported that some odorants were capable of stimulating the trigeminal free nerve endings below concentrations required to stimulate olfactory receptors. These observations raised questions as to a possible role the trigeminal system might have in olfaction in addition to its primary role in controlling the well-being of the animal.

In our experiments, rabbits were implanted with bipolar recording macro-electrodes (stainless steel) in both olfactory bulbs, and with a cannula in each gasserian ganglion, permitting reversible blocking of trigeminal inputs. Results from one of these experiments are shown in Figure 1. In the trigeminally blocked animal, bulbar-induced sinusoidal wave activity was significantly increased in frequency and amplitude compared to the activity in the unblocked animal.[12] Experiments with animals that had lesions in the gasserian ganglia showed similar results. These experimental observations led us to conclude that the response to an odor stimulus is mediated in part by the trigeminal system. How this is effected, i.e., whether there are specific pathways or the reaction is a general one, remains unknown.

From these experiments, assuming the behavioral changes reflected neural changes, it was hypothesized that a better understanding of the events taking place could be achieved through study of the neural changes. Therefore, a series of experiments were initiated in which the ethmoidal nerve was stimulated and changes in bulbar activity were monitored. The rationale for this series of experiments was the knowledge that the ethmoidal nerve, a branch of the ophthalmic division of the trigeminal nerve, innervates the olfactory region and adjacent nasal tissue and would be a primary contributor to the observed phenomena. It is recognized that there are numerous other branches of the trigeminal that are stimulated by odorants and might provide additional inputs to the system.[4] For purposes of simplicity, present efforts were confined to the ethmoidal branch.

EXPERIMENTAL

The animals (young, male, adult, New Zealand white rabbits, 2.5 to 3.0 kg) were anesthetized with sodium pentobarbital (30 mg/kg intravenously [iv]) and a tracheal tube was inserted. Subsequent doses of pentobarbital were given intraperitoneally (ip) to maintain the level of anesthesia. Heart rate was monitored with electrodes placed on the skin and was maintained at about 210 to 230 beats/min. Before retraction of the eye, the animal was immobilized with gallamine triethiodide (0.25 cc, iv) and a respirometer used to control respiration.

Retraction of the eye and its associated tissue was followed by careful dissection

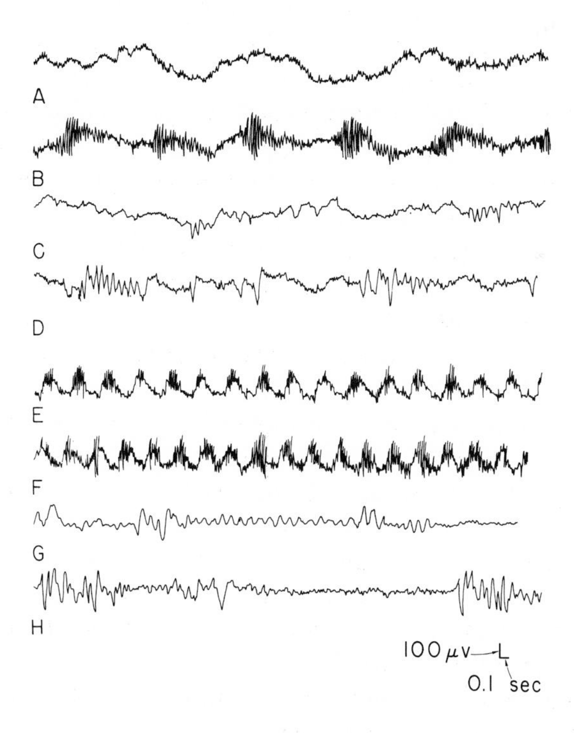

Figure 1 Olfactory bulbar response following stimulation by n-propyl alcohol (100% saturation). A, awake animal before administration of Xylocaine (lidocaine hydrochlorid); B, awake after Xylocaine (2% solution, 0.05 cc in each cannula); C, anesthetized (pentobarbital, 35 mg/kg, intravenously), before Xylocaine; D, anesthetized, after Xylocaine; E, awake, before lesions of the ganglia; F, awake 3 days after lesions; G, anesthetized, before lesions; H, asleep, 3 days after lesions. The upper records are part of the total record from one of the three Xylocaine-blocked animals, and are typical of the responses obtained from them. The lower records are from one of the animals with lesions.

of the ethmoidal branch at that point where it enters the ethmoidal foramen. The cut end of the nerve was placed on a bipolar, platinum, stimulating electrode. The olfactory bulbs were exposed and a monopolar needle electrode lowered into position. For reference, a small screw was placed in the skull above the sensory motor cortex region.

Stimulation of the nerve was accomplished with a Grass Stimulator (Model S4) using single pulses of brief duration (0.1 msec), from 0.2 to 12 volts, AC, and a frequency of not more than one every 5 sec. In some experiments, trains of impulses (200 msec duration) were employed. Data output was led through a cathode follower into a Grass P5A.C. pre-amplifier with amplitude frequency settings of 7 and 10,000 cps. The output from the pre-amplifier was led to a tape recorder for data storage and to an oscilloscope for display purposes.

In the first experiment, the effects of ethmoidal nerve stimulation on total bulbar activity (both ipsilateral and contralateral bulbs) and, by means of appropriate filters, the slow activity (less than 100 cps) and the fast activity (greater than 100 cps) were studied. Multi-unit fast activity was initially depressed in frequency and amplitude while changes in slow activity were unclear. The changes in fast activity seemed most evident at moderate to high voltages (4 to 8); however, the threshold could not be established in this series of experiments. On closer examination, it appeared that the changes were occurring within the first 100 msec stimulation, and more precise analysis would be required before any definitive explanations might be presented. Such analyses are now in progress.

A second experimental approach now under consideration is the generation of evoked potentials in the olfactory bulb by stimulation of the ethmoidal nerve, and the characterization of these potentials as a function of varying stimulus conditions.

In these experiments, data were summated by means of a computer of average transients (CAT Model 400C). Stimulation consisted of single pulses of 0.1 msec duration, presented every 5 sec. Approximately 50 such sweeps were made and summated for observation. Electrode placement began at the surface of the anterior portion of the bulb followed by stimulation. If a response did not occur, the electrode was lowered and the process was repeated. Based on preliminary results, evoked potentials in the olfactory bulb exhibited a 0.15 to 0.25 msec duration, and an amplitude that was voltage-dependent. Verification of a response was usually accomplished by polarity reversal on the stimulating electrodes and, at the conclusion of the experiments, by passing a current through the animal after expiration. The most successful experiments occurred when the recording electrode was at least 2mm into the bulb and in the anterior segment but not at the very tip of the bulb. Present experiments are designed to characterize this response better, to delineate in which areas and hopefully what structures are involved.

DISCUSSION

Kerr and Hagbarth,[7] Moulton,[8] Carreras et al.,[3] and others have shown that there are centrifugal connections between the olfactory bulbs, as well as other brain structures, that appear to exert an influence on bulbar activity. The reports by Hernandez-Peón et al.,[5] and Moyano[9] provide further information on structures that are influenced by olfactory stimulation or are capable of influencing bulbar electrical activity. However, the fundamental question of whether these observations do, in fact, affect olfactory bulbar activity remains to be answered. If one assumes, for example, that the trigeminal input does have an effect, it is worthwhile to consider how this is achieved. It is hypothesized that the trigeminal input, via the ethmoidal nerves, could exert its influence on such structures as the reticular formation and prepyriform cortex. This influence could then take the form of a capacitor capable of inhibition, facilitation, or no response, depending on the stimulus and its concentration. So far, only inhibitory effects have been observed; however, more experiments need to be carried out before one can expect adequately to test such a hypothesis.

SUMMARY

Acute preparations were employed as a means of studying the effect of trigeminal stimulation on olfactory bulbar activity of the rabbit. The ethmoidal branch of the ophthalmic division of the trigeminal nerve was dissected from the surrounding tissue, severed at the ethmoidal foramen and the cut end placed on a hooked, bipolar, platinum, stimulating electrode.

Stimulation of the ethmoidal branch of the trigeminal nerve with single pulses (0.2 to 12 volts) and with trains of pulses (same voltages) of brief duration appeared to influence olfactory bulbar electrical activity in the anesthetized, immobilized rabbit. Slow activity (between 7 and 100 cps) and fast activity (between 100 cps and 10 kc) in the ipsilateral and contralateral bulbs showed changes that support the concept of interaction between the olfactory and trigeminal systems. Multi-unit fast activity was initially depressed in amplitude and frequency while slow activity changes were not as clearly defined; both inhibitory and contralateral bulb changes were less than those in the ipsilateral bulb.

It is hypothesized that a pathway exists, possibly involving the reticular formation and the prepyriform cortex, which is capable of influencing bulbar activity. Such a system would add greater informational capacity to the olfactory system by inhibiting, facilitating, or not affecting the afferent inputs from the bulbs to the higher centers of the brain.

ACKNOWLEDGMENT

Research supported by Grant No. R01 NB07866 from the Division of Neurological Diseases and Blindness, National Institutes of Health. The author acknowledges the excellent technical assistance of Mrs. Donna Oman and the helpful advice of Drs. E. J. A. Carregal and Lawrence R. Pinneo.

REFERENCES

1. Allen, W. F. 1937. Olfactory and trigeminal conditioned reflexes in dogs. *Amer. J. Physiol.* **118**, pp. 532–540.
2. Beidler, L. M. 1965. Comparison of gustatory receptors, olfactory receptors, and free nerve endings. *Cold Spring Harbor Symp. Quant. Biol.* **30**, pp. 191–200.

3. Carreras, M., D. Mancia, and M. Mancia. 1967. Slow potential changes induced in the olfactory bulb by central and peripheral stimuli. *In* Olfaction and Taste II (T. Hayashi, editor). Pergamon Press, Oxford, etc., pp. 181–191.
4. Dawson, W. W. 1962. Chemical stimulation of the peripheral trigeminal nerve. *Nature.* **196**, pp. 341–345.
5. Hernández-Peón, R., A. Lavin, C. Alcocer-Cuarón, and J. P. Marcelin. 1960. Electrical activity of the olfactory bulb during wakefulness and sleep. *Electroencephalog. Clin. Neurophysiol.* **12**, pp. 41–58.
6. Katz, S. H., and E. J. Talbert. 1930. Intensities of odors and irritating effects of warning agents for inflammable and poisonous gases. *U.S. Bur. Mines, Tech. Papers.* Doc. **480**.
7. Kerr, D. I. B., and K. E. Hagbarth. 1955. An investigation of olfactory centrifugal system. *J. Neurophysiol.* **18**, pp. 362–374.
8. Moulton, D. G. 1963. Electrical activity in the olfactory system of rabbits with indwelling electrodes. *In* Olfaction and Taste I (Y. Zotterman, editor). Pergamon Press, Oxford, etc., pp. 71–84.
9. Moyano, H. F. 1966. Interaction patterns of preoptic and olfactory bulb evoked responses in the rat. *Experientia* **22**, pp. 397–399.
10. Parker, G. H., and E. M. Stabler. 1913. On certain distinctions between taste and smell. *Amer. J. Physiol.* **32**, pp. 230–240.
11. Stone, H., E. J. Carregal, and B. Williams. 1966. The olfactory trigeminal response to odorants. *Life Sci.* **5**, pp. 2195–2201.
12. Stone, H., E. J. Carregal, and B. Williams. 1968. The role of the trigeminal nerve in olfaction. *Exp. Neurol.* **21**, pp. 11–19.
13. Tucker, D. 1963. Physical variables in the olfactory stimulation process. *J. Gen. Physiol.* **46**, pp. 453–489.
14. Tucker, D. 1963. Olfactory vomeronasal and trigeminal receptor responses to odorants. *In* Olfaction and Taste I (Y. Zotterman, editor). Pergamon Press, Oxford, etc., pp. 45–69.

EVIDENCE FOR THE DIFFERENTIAL MIGRATION OF ODORANT MOLECULES ACROSS THE OLFACTORY MUCOSA

MAXWELL M. MOZELL

Physiology Department, Upstate Medical Center, State University of New York, Syracuse, New York

We generally think of sensory coding in terms of the selective sensitivity of the receptors themselves, i.e., that the receptors per se are tuned to particular types of stimuli. This is not always true. There are no firm data, for instance, suggesting that the receptors in the auditory system are themselves selectively tuned to frequency. Instead, the receptors merely reflect or encode the activity of an altogether different analyzing agent, the basilar membrane, which operates according to physical principles not at all peculiar to the receptors themselves.

For the past several years, we have been looking at data from the level of the olfactory mucosa. These suggest that one of the mechanisms—and I stress that phrase, one of the mechanisms—by which odors are encoded may be not based upon the tuning of the receptors themselves, but rather upon physical processes which are involved in the movement or migration of molecules across the olfactory receptor sheet. This, of course, is not a new concept. Similar suggestions have been made by Adrian,[1] Moncrieff,[3] and Beidler.[2] I will try here to arouse a greater interest in differential molecular migration as a model for olfactory vapor analysis. I hope to foster this interest by presenting electrophysiological evidence which demonstrates not only that molecules of different odorants migrate differentially across the olfactory receptor sheet, but, in addition, that these differences across the mucosa are subsequently encoded in the discharge of the olfactory nerve.

We sampled different regions of the olfactory mucosa by recording the activity from two widely separated olfactory nerve branches.[4] One, the most medial, supplied a region of the olfactory mucosa near the external naris, and the other, the most lateral, supplied an area near the internal naris. The odorized air stream flowed from the external to the internal naris, i.e., from the region supplied by the medial nerve branch to that supplied by the lateral nerve branch. The neural discharges were processed by an electronic summator, a process which yielded traces like those shown in Figure 1. The area under each trace is proportional to the activity in the nerve branch producing it, and this, in turn, should be proportional to the activity occurring at the region of the olfactory mucosa supplied by that nerve branch. The important point of the Figure is that the activity of the lateral nerve relative to that of the medial nerve (the LB/MB ratio) is different for the different chemicals; d-limonene is about 1, citral less than 1, octane greater than 1, and geraniol considerably less than 1. Here we have a possible coding system, apparently dependent upon a gradient of activity across the olfactory mucosa. Correlated with such a spatial code is a temporal code. We found that those chemicals giving the largest ratios also displayed the shortest time differ-

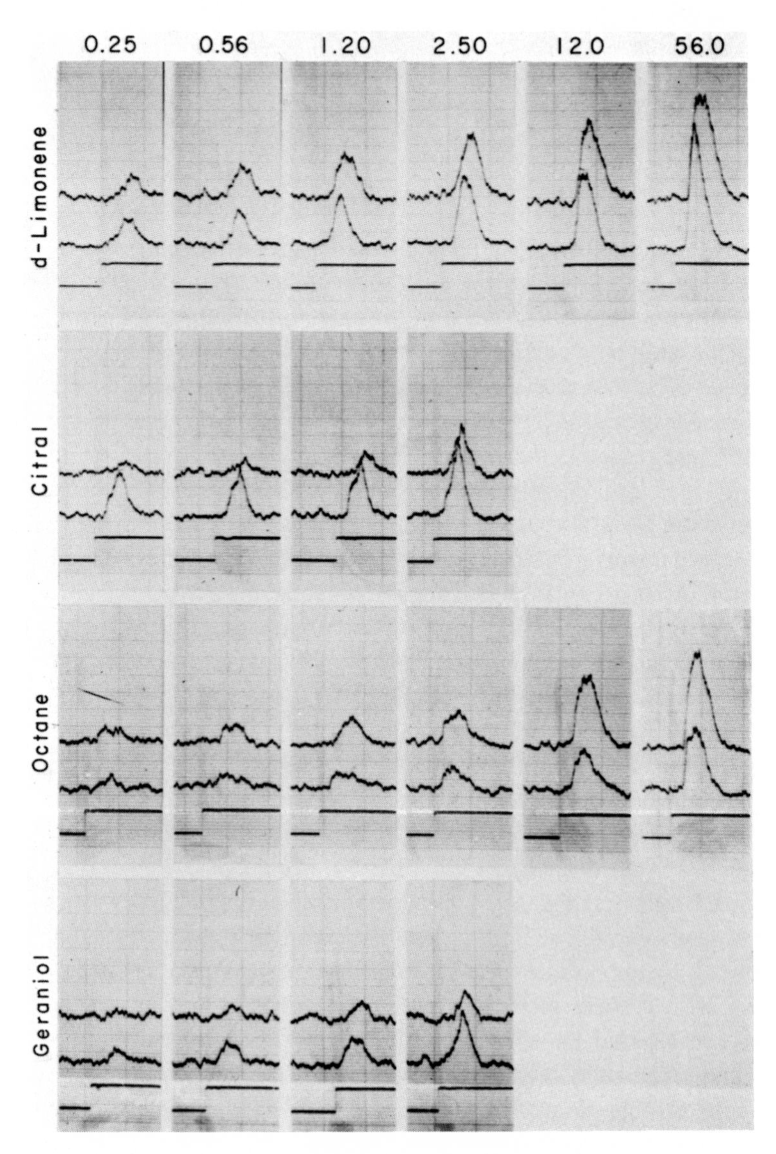

Figure 1 Visicorder records of summated neural discharges, showing the responses of one frog to a single presentation of every stimulus. Partial pressures are given along the top in terms of $\times 10^{-2}$ mm Hg. The flow rate was 8.24 cc/min. The upper response of each pair is recorded from the lateral nerve branch and the lower from the medial nerve branch. The stimulus marker shows only the onset of the stimulus. Vertical time lines occur once every 10 sec. (From Mozell[4])

ences between the onset of the two responses. Thus, d-limonene gave a shorter time difference than did citral. We believe that these two measures are two sides of the same coin.

What mechanism might be responsible for this spatiotemporal pattern of activity across the mucosa? There are at least two possibilities. First, there may be a gradient of selective receptor sensitivities. For instance, there may be more receptors selectively sensitive to geraniol near the external than the internal naris. This would account for the comparatively small LB/MB ratios produced by geraniol. On the other hand, the receptors across the whole mucosa may be more equally sensitive to d-limonene, thus accounting for d-limonene's characteristic LB/MB ratio of about unity. Second, perhaps the propensity of the molecules to migrate across the mucosa differs for different chemicals. For instance, perhaps comparatively few geraniol molecules are able to cross the mucosa and those that do cross take a comparatively long time. This would also account both for geraniol's small LB/MB ratio and for the long time difference between the responses it produces on the two nerve branches. The larger ratios and shorter time differences produced by d-limonene could then be explained by the greater ability of its molecules to migrate across the mucosa.

To differentiate between these two possibilities, we reversed the direction of the odorized airflow, i.e., we introduced the odorant through the internal naris and allowed it to pass through the external naris. When we compared the results of this reversed flow to those produced by the usual flow direction, we found that the relative magnitudes of the responses on the two nerve branches also reversed. For instance, for geraniol, the lateral nerve now gave the larger response and the medial nerve the smaller.

This reversal is difficult to explain in terms of a gradient of receptor sensitivities across the mucosa, because such a gradient should remain more or less constant, regardless of the airflow direction. However, the reversal is consistent with the molecular migration hypothesis. For instance, geraniol molecules, still unable to cross the mucosa with great facility, will now be delayed around the region of the internal naris rather than around the external naris, yielding a larger response on the lateral nerve branch than on the medial nerve branch. We think this suggests differential migration of molecules across the mucosa. If this is the case, olfaction could be likened to a nonbiological system, which, also in order to separate or analyze chemicals, uses the same basic phenomenon of differential molecular migration through a medium. That is, olfaction could be likened to chromatography.

So far, data based upon only four chemicals have been presented. In order to support this chromatographic model of olfaction, we would have to show that it holds for more than just four chemicals. We have recently expanded this number to 16, each presented at four different concentrations (as measured by partial pressure) and at two different flow rates. Figure 2, based upon the data from 12 animals, shows the median LB/MB ratio produced by 13 of these chemicals at one of the flow rates and one of the partial pressures. (At the ambient room temperature, the vapor pressures of three of the chemicals were too low to be included.) Remember that the larger the ratio, the greater the response in the lateral nerve branch as compared with the medial nerve branch. The ratio produced depends upon the chemical presented. In other words, each chemical sets up a characteristic gradient of activity across the mucosa.

The chemicals might have been plotted in any order, but we have chosen, in Figure 2, to plot them in the order of retention times in a gas chromatograph fitted with a 20 M Carbowax column supported by Chromosorb P (a combination that yields a moderately polar column). Such a plot shows a rather clear relation between the LB/MB ratio and the chromatographic retention time. This is further established by the resulting correlation coefficient (−0.72) and its probability (<0.01). Indeed, if it were not for butyl alcohol, the only chemical that deviates markedly from the general trend, the correlation coefficient would be −0.85. It appears, then, that the ability of these chemicals (as measured by their LB/MB ratios) to move across the

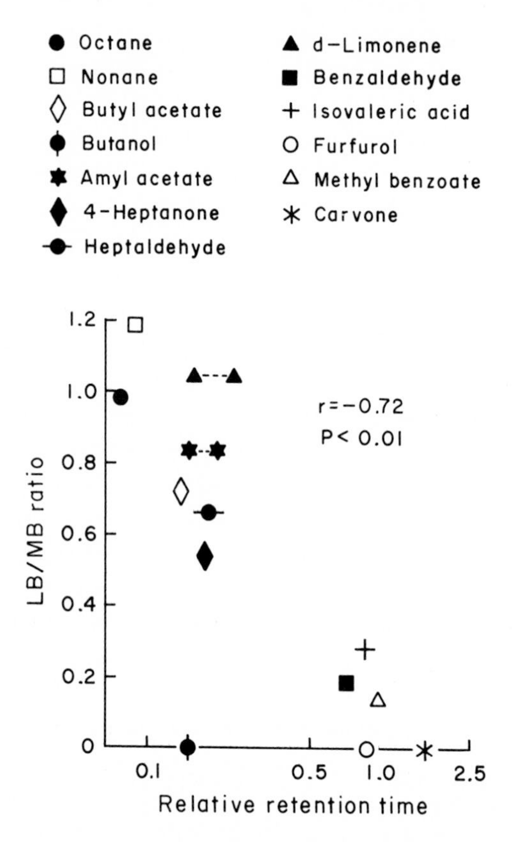

Figure 2 LB/MB ratios as a function of the relative retention times of 13 chemicals. Partial pressure of the chemicals: 12×10^{-2} mm Hg. Flow rate: 20.6 cc/min. Symbols linked together by broken lines are shown for those chemicals with more than one chromatographic peak, the two symbols representing the shortest and longest retention times. r = correlation coefficient; P = probability.

olfactory mucosa is related to their ability (as measured by their retention times) to pass through the chromatographic column. This suggests that similar processes are at work in the two systems.

It must be emphasized, however, that the LB/MB ratios produced by at least some of the chemicals could be altered somewhat by varying the concentration or flow rate at which they were presented. This is more true of some chemicals than of others. Of course, such variations could distort the concept which suggests that the coding of odorants depends upon the gradient of activity produced across the mucosa. Obviously, the central nervous system of the animal could not use such an activity gradient to identify odorants if the gradient were greatly confounded by flow rates and concentrations and did not carry information predominantly referable to odor quality per se. However, quantitative tests, including the analysis of variance, showed that although concentration and flow rate play a role in the determination of the LB/MB ratios, their effects are relatively minor when compared with the effect engendered by changing the chemicals. In comparison to the variations in the median ratio produced by changing the chemicals shown in Figure 2, each tenfold increase in concentration increased the median ratio by only 0.09 and a fivefold increase in the flow rate increased that ratio by only 0.08. (Flow rates much greater than this, although possible experimentally, would have gone beyond the normal physiological limits.) In passing, it may be suggested that perhaps this slight increase in ratio, which occurs with an increase in the concentration of a chemical, may be just enough to explain why some chemicals do smell somewhat different at higher concentrations than at lower concentrations. The basic point, however, is that the odorant itself is primary in the determination of the activity gradient across the mucosa.

In conclusion, these data lead us to entertain the possibility that one of the mechanisms for coding in the olfactory system does not involve the selective tuning of the receptors themselves, but depends instead upon the propensity of the molecules of different odorants to migrate across the mucosa.

ACKNOWLEDGMENT

This work has been supported by National Institutes of Health Grant 5-R01-NB03904.

REFERENCES

1. Adrian, E. D. 1950. Sensory discrimination: With some recent evidence from the olfactory organ. *Brit. Med. Bull.* **6**, pp. 330–333.
2. Beidler, L. M. 1957. Facts and theory on the mechanism of taste and odor perception. *In* Chemistry of Natural Food Flavors (J. H. Mitchell, Jr., editor). Quartermaster Food and Container Institute for the Armed Forces, Chicago, Ill., pp. 7–47.
3. Moncrieff, R. W. 1955. The sorptive properties of the olfactory membrane. *J. Physiol.* (*London*) **130**, pp. 543–558.
4. Mozell, M. M. 1966. The spatiotemporal analysis of odorants at the level of the olfactory receptor sheet. *J. Gen. Physiol.* **50**, pp. 25–41.

SUMMARY OF OLFACTION ROUNDTABLE

Prepared by CARL PFAFFMANN

Further discussion of olfactory coding was carried out at a roundtable chaired by Dr. Dietrich Schneider. He called attention to the several levels on which this problem can be approached. The most peripheral is that of the donor-acceptor or stimulant-molecule-receptor-site specificity. Although highly specific molecule-receptor site relations are known, there are also instances of a broad side spectrum of reactivity. In either case, the neuron acts as an integrator. Excitation of its dendritic processes is summed, leading to the arousal of a train of impulses traveling centrally up the axon. The impulse train is a type of frequency modulation more or less proportional to the degree of depolarization. At the synaptic junctions in the central nervous system, there is still another coding process or interpretation of the frequency modulation. Neural activity may occur as an increase or decrease above or below some resting level of discharge. Because there may be no response, the code has three states, +, 0, −. Excitation at the periphery of the receptor is associated with receptor membrane depolarization, inhibition or decrement, with hyperpolarization. In the glomerulus of the olfactory bulb of vertebrates there is a great central convergence of the order of 1000 to 1, from afferent fibers to second-order cells. Similar convergence may occur in insects. Many recurrent fibers, and specifically inhibitory afferent inputs to the olfactory bulb, add to the inhibition seen at this level. Any mechanism for lateral inhibition of the type seen in visual or auditory systems has not yet been demonstrated at the peripheral level, but possibly it exists.

In odor perception, the most recurring question is the "Geruchs problem," that is, the nature of the olfactory neural code by which such a wide range of odors can be differentiated. Does electrophysiological evidence give any information concerning this problem?

Insect research indicates that there are two classes of olfactory receptor. First are the highly specialized receptors for pheromones, making up perhaps 50 per cent of the receptors in any one insect. These are remarkably specific and may react with extreme sensitivity to one odorant, while being much less sensitive to others. In such an animal as the honeybee, there may be a handful of pheromone "specialist" receptors, some with overlapping, others with no overlapping, spectra. This class mediates the specific signal to the animal, and this usually leads to a specific behavioral response.

Second, there are the "generalist" receptors, located on special sensilla. However, they do not respond to the pheromones, but to a wide range of odorous substances, with an increase or decrease in activity. Each receptor cell has its own individual, but consistent, spectrum of response; no cells have identical spectra, so that the responses of many cells to different odorants provide a unique response matrix for any one odorant. Such a matrix could be read by a programed computer which could discriminate and identify different odors. But we do not know whether the nervous system uses the information in such a manner.

Although behavioral data indicate pheromone-like effects in mammals, there is no electrophysiological data showing the existence of mammalian odor specialists. However, not enough work has been done using natural animal scents as odor stimuli. Dr. Mathews presented single-unit responses from the curarized rat olfactory bulb (second-order cells).[7] These data can also be plotted as a matrix, showing responses of a series of olfactory bulb neurons to a battery of different odorants (Figure 1). The neurons in the olfactory bulb of the rat fall into three classes: (1) purely facilitory; (2) purely inhibitory; and (3) those which respond with facilitation to some odorants and with inhibition to others. The units are neither highly specific nor highly non-

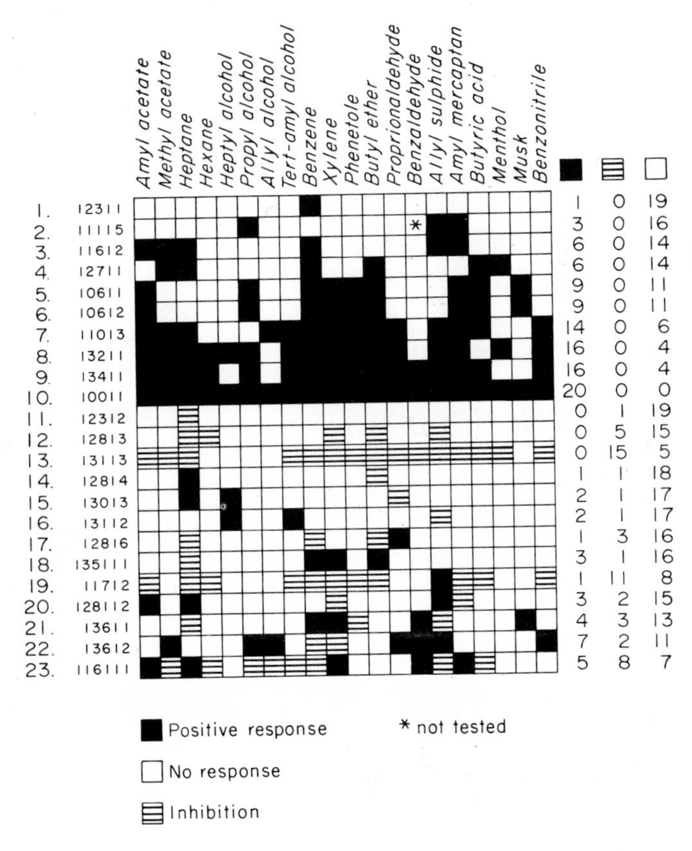

Figure 1 Responses of olfactory bulb neurons to different odorants.

specific, but are distributed between these two extremes. Each odorant generates an individual pattern of response that usually is not duplicated by any other odorant, in a manner that very closely resembles the "generalists" in insects described by Dr. Schneider and his group. It is interesting that the bulbar olfactory system of a mammal so closely resembles the peripheral system of the insect in this regard.

Similar evidence was presented for the all-or-none responses of single glomeruli of the rabbit, studies carried out by Dr. J. Leveteau and Dr. P. Mac Leod. Figure 2 shows their matrix pattern. In this case, there was not much evidence of inhibition. The left part of the Figure shows the matrix for 96 different glomeruli and 12 odors (A through L) in approximately 30 different animals. There is no way to arrange them in a systematic way, and only two or three glomeruli give exactly the same pattern. The right side of the Figure shows the distribution of glomeruli with respect to selectivity. Any amount of selectivity can be found, with some glomeruli responding to all stimuli, but there is some grouping around 4 or 5 and 7 or 8. The same investigation was repeated at a 100-times-lower stimulus concentration to avoid a possible artifact from too high a concentration. The results are more or less the same, except that the nondiscriminating class is smaller. Each point in Figure 3 presents the anatomical location of the glomerulus. It can be seen that the coordinates of glomeruli responding to each chemical are not arranged systematically in the olfactory bulb. There is no rostral-caudal distribution, only a distribution among chemicals that stimulate very

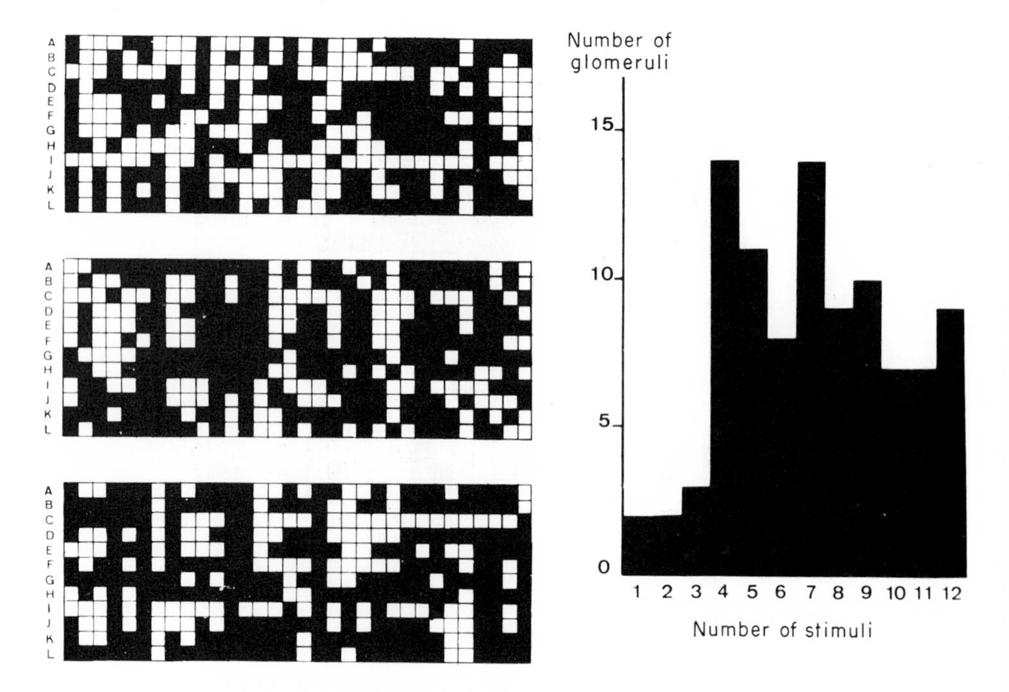

Figure 2 Glomerular response to high-level stimulation.

poorly and others that stimulate strongly. Stimulus intensities had been adjusted to give the same EOG amplitude.

Dr. D. G. Moulton raised certain questions concerning spontaneous activity, inhibition, and the intensity of stimulation often employed in electrophysiological studies. For example, he took issue with the assumption that spontaneous activity is a normal occurrence, noting that the rate of spontaneous activity may result from the presence of the microelectrode close to the receptor cell. This, in turn, could determine the extent to which inhibition is observed. Single-unit recordings from the periphery of the vertebrate olfactory system have often shown inhibition, but, in a recent study by O'Connell and Mozell,[9] only one case of inhibition was seen in 101 units studied. Recent studies at the level of the olfactory bulb show inhibition, but, as Dr. R. von Baumgarten pointed out, many other factors must be taken into account, such as centrifugal control from the reticular formation or other limbic structures. Moulton

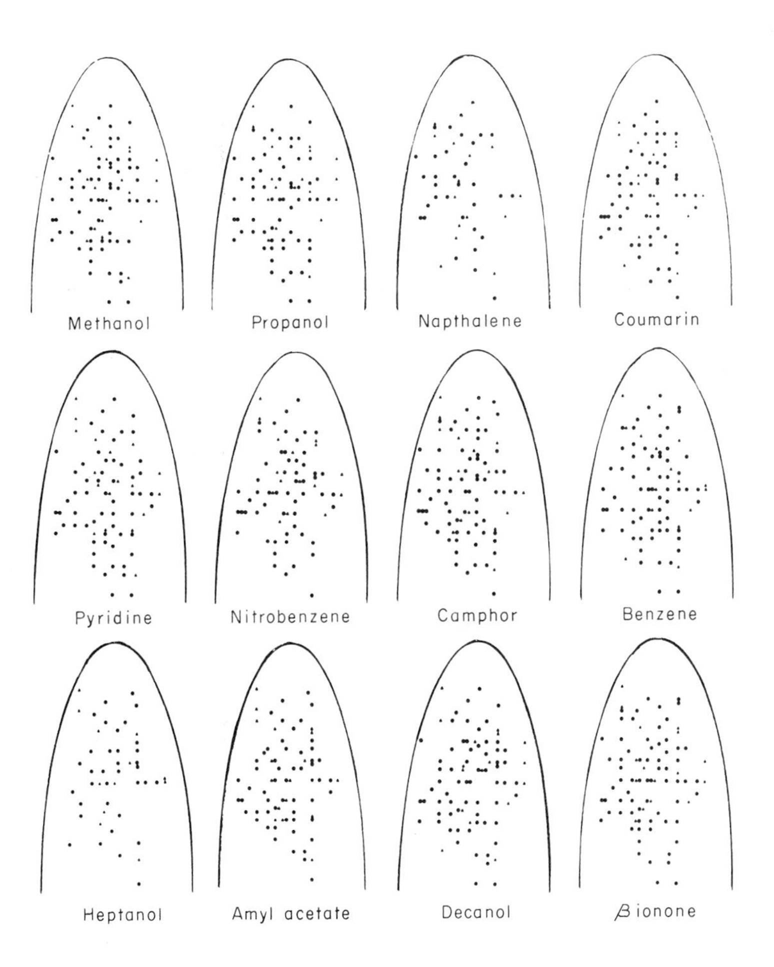

Figure 3 Topography of glomeruli activated by each of 12 different stimuli.

also thought that interactions analogous to lateral inhibition in the limulus eye were possible. When electrodes are chronically implanted in the rabbit olfactory bulb, one is particularly impressed by the lack of inhibition (except at very high concentrations or for certain odorants) when the centrifugal fiber input is cut. As noted above, Dr. Mac Leod found little inhibition of the glomerular response.

Table I indicates the single-fiber activity from the goldfish olfactory bulb to a

TABLE I

RESPONSES OF SINGLE BULBAR UNITS IN THE GOLDFISH (*Crassius auratus* L.). (Ai[1])

	Conc. (S)	No. of units tested	Response (%) +		0
Food Extract	$(10^{-2} + 10^{-1})$*	244	46	0	54
Skin Extract	$(10^{-2} + 10^{-1})$*	210	15	4	81
L-Alanine	$5 + 10^{-2}$ M	95	41	15	44

* Expressed as a dilution of a standard solution (10 gm solid/100 ml distilled water).

complex food extract odor.[1] For the food extract stimulus, there is no evidence whatsoever for inhibition in 244 units studied, in marked contrast to other recent studies in which the odorants tested often caused quite marked inhibition. Until further work has been done, it cannot be assumed that inhibition plays an important role in the coding mechanism. Activity in some units may be related more to the function of the olfactory system as an activator for other sensory systems than as part of the discriminatory mechanism.

Moulton further noted that many studies employ compounds not normally present in the environment of the animals being studied. Even if present, the concentrations tested are sometimes relatively high. "It's rather like studying vision by using lightning flashes," he said. Behavior patterns critical for survival are often triggered by odorous substances in extremely low concentration. For example, the amino acid L-serine is thought to be the constituent of human hand rinses that triggers the alarm reaction in the migrating adult coho salmon.[5] Kleerekoper and Mogensen[6] have worked on the amine in water rinses from the brook trout that elicits locomotor activity in the sea lamprey. The alarm reaction of fish substances released from the skin of other injured fish, and the discrimination of odors characteristic of the home stream which controls homing in salmon are other examples of great sensitivity to chemical stimuli present at very low concentrations.[2–4,10] Such patterns and their coding mechanisms must be very efficient to recognize odors in low dilutions. In recording, one can often find units that are sensitive to only one of several compounds and not to others. This might be described as a fairly high degree of specificity. Moulton suggested that it might be worth looking at compounds responsible for triggering behavior mechanisms at low levels.

The absence of any good evidence on vertebrate odor "specialists" may largely reflect the tremendous population of receptors in the vertebrates. For example, the rabbit has roughly a hundred million. Presumably, no more than a few of these units need to be occupied with signaling highly specific information, but the likelihood of sampling one of these units is small.

Moulton concluded with Figure 4, which showed recordings from the olfactory bulb of a freely moving rabbit with chronically implanted electrodes. Complex patterning of excitation might be of two types, he said. One is that which Lord Adrian first reported and Dr. Mozell discusses in this volume. This is a relatively gross distribution between the caudal and the rostral, found by averaging. Far more impressive is the detailed and complex continuously evolving pattern shown in the Figure. It seems to lack any systematic spatial distribution of excitation in terms of lateral-medial or rostral-caudal loci, etc. In other words, he concluded that the one pattern may be related to the way in which molecules are distributed across the receptor sheet; the other may reflect the partial specificity of the receptors themselves, which, presumably, is basic to the olfactory code.

REFERENCES

1. Ai, N. 1969. Unpublished observations. (With permission of the investigator)
2. Frisch, K. von. 1941. Über einen Schreckstoff der Fischhaut und seine biologische Bedeutung. *Z. vergleich. Physiol.* **29**, pp. 1–45.
3. Hara, T. J., K. Veda, and A. Gorbman. 1965. Electroencephalographic studies of homing salmon. *Science* **149**, pp. 884–885.
4. Hasler, A. D., and W. J. Wisby. 1951. Discrimination of stream odors by fishes and its relation to parent stream behavior. *Amer. Natur.* **85**, pp. 223–238.
5. Idler, D. R., U. Fagerlund, and H. Mayoh. 1956. Olfactory perception in migrating salmon 1. L-serine, a salmon repellent in mammalian skin. *J. Gen. Physiol.* **39**, pp. 889–892.
6. Kleerekoper, H., and J. Mogensen. 1963. Role of olfaction in the orientation of *Petromyzon marinus* I. Response to a single amine in prey's body odor. *Physiol. Zool.* **36**, pp. 347–360.
7. Mathews, D. F. 1966. Response patterns of single units in the olfactory bulb of the unanesthetized, curarized rat to air and odor. Unpublished Ph.D. thesis. Brown Univ., Providence, R.I.
8. Moulton, D. G. 1967. Spatio-temporal patterning of response in the olfactory system. *In* Olfaction and Taste II (T. Hayashi, editor). Pergamon Press, Oxford, etc., pp. 109–116.
9. O'Connell, R. J., and M. M. Mozell. 1968. Quantitative stimulation of frog's olfactory receptors. *J. Neurophysiol.* **32**, pp. 51–63.
10. Skinner, W. A., R. O. Mathews, and R. M. Pankhurst. 1962. Alarm reaction of the top smelt, *Atherinops affinis* (Ayres). *Science* **138**, pp. 681–682.

Figure 4 *appears on page* 232.

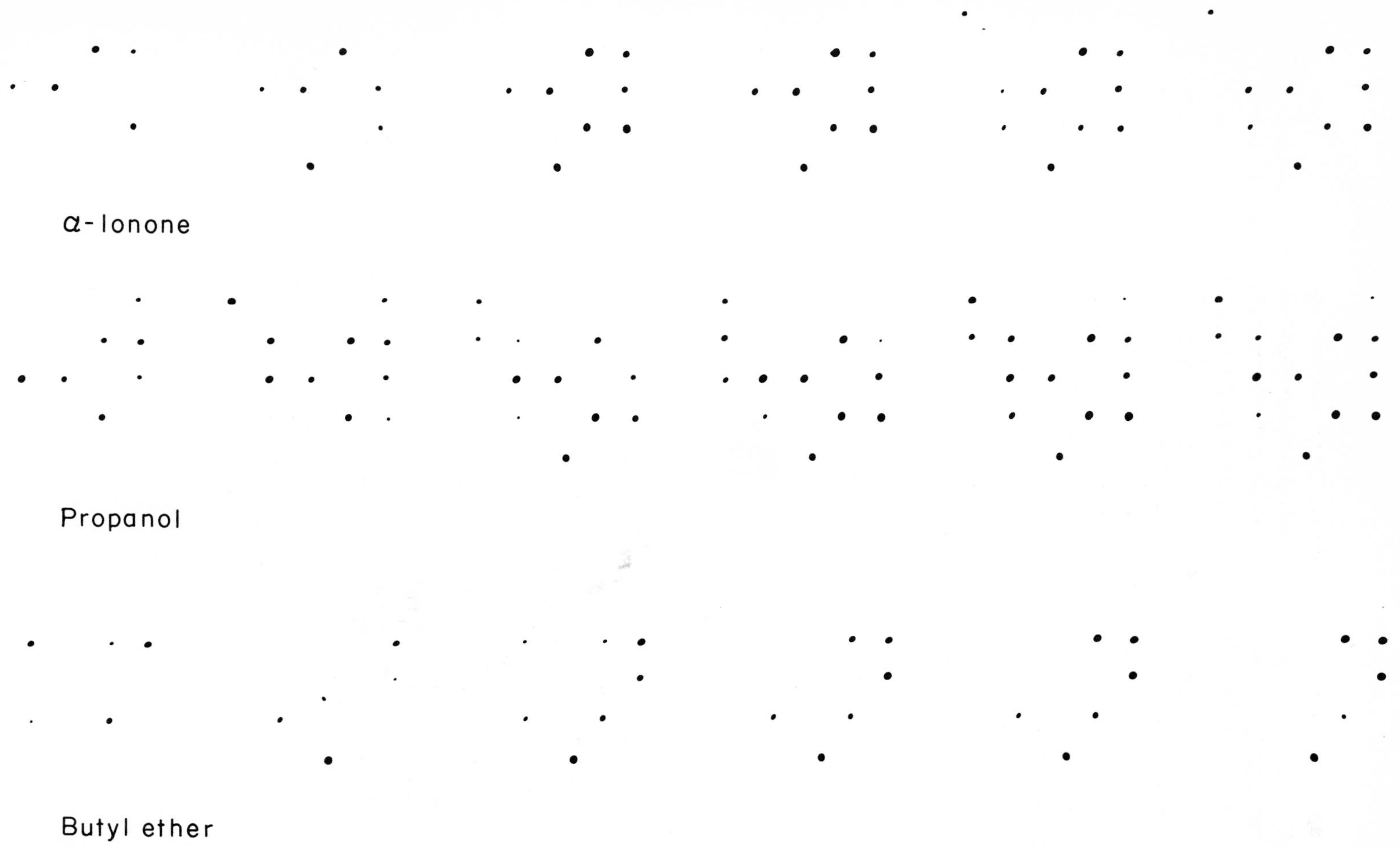

Figure 4 Average multiunit activity from the olfactory bulb of an unanesthetized rabbit displayed as a mosaic of points on an oscilloscope face. The first frame in each series represents the resting activity. Subsequent frames were taken at one-sec intervals after the odorant had been introduced into the rabbit's face mask. (Modified from Moulton[8])

III ROLE OF OLFACTION IN BEHAVIOR

III · ROLE OF OLFACTION IN BEHAVIOR

A NEW APPROACH TO INSECT PHEROMONE SPECIFICITY

ERNST PRIESNER

Max-Planck-Institut für Verhaltensphysiologie, Seewiesen, Germany

Substances produced by one individual that convey information to other individuals of the same species have been called pheromones.[12] A distinction between releaser pheromones (release of a specific behavior response) and primer pheromones (trigger of endocrine, morphogenetic, or metabolic activities) has been proposed.[6,19,30–32]

Of the main groups of insect releaser pheromones—female sex attractants, male sex attractants and aphrodisiacs, aggregating pheromones, trail substances, alarm substances, and other olfactory markers in social insects—the alarm pheromones of social insects are chemically the best-known group (26 compounds identified). The sex attractants produced by the female to attract the male are the most abundant ones (known for 12 insect orders).

Reaction thresholds, intensity ranges, and reaction spectra of pheromone receptor cells in insects have been determined with electrophysiological techniques.[3,4,11,18–20,22,24,29] These techniques are also adequate for the determination of interspecific pheromone effects.[13,15,16,19] Some aspects of the application of these techniques are shown here for the female sex pheromones of Lepidoptera.

Until now, chemical analysis has not contributed to the knowledge of their specificity. Of the formulae published on the chemical composition of female sex pheromones of six lepidopteran species (Table I), those for *Bombyx mori* have been fully accepted by all other authors, but those for *Lymantria dispar* and *Pectinophora gossypiella* have been proved to be incorrect.[7,19,27]

About 100,000 lepidopteran species are likely to produce female sex pheromones. Many cases are known from field observations as well as from laboratory behavior tests of interspecific pheromone effects between related species.[13] All known female sex pheromone glands of Lepidoptera[1] are organs homologous to the sacculi

TABLE I
CHEMICAL COMPOSITION OF FEMALE SEX PHEROMONES OF LEPIDOPTERA

	Formula	Authors	Further proof
Bombycidae:			
Bombyx mori	(trans-10,cis-12)hexadecadien(1)ol "bombykol"	Butenandt, Beckmann, stamm, and Hecker[5]	fully accepted by all other authors
Lymantriidae:			
Lymantria dispar	D(10)acetoxy(cis-7)hexadecen(1)ol "gyptol"	Jacobson, Beroza, and Jones[9]	incorrect? (see text)
Gelechiidae:			
Pectinophora gossypiella	(10)propyl(trans-5,9)tridecadien(1)ol acetate "propylure"	Jones, Jacobson, and Martin[10]	incorrect? (see text)
Noctuidae:			
Trichoplusia ni	(cis-7)dodecen(1)ol acetate	Berger[2]	Green, Jacobson, Henneberry, and Kishaba[8]
Spodoptera frugiperda	(cis-9)tetradecen(1)ol acetate	Sekul and Sparks[26]	—
Tortricidae:			
Argyroploce leucetreta	(trans-7)dodecen(l)ol acetate	Read, Warren, and Hewitt[13a]	—
Argyrotaenia velutinana	(cis-11)tetradecen(l)ol acetate "riblure"	Roelofs and Arn[14]	—

laterales[28] of the *Bombyx mori* female. By single cell recordings, the sensilla trichodea of the male antenna[25] have been identified to be the pheromone receptors.[16,17,19,24] Schneider[16] first used their summated receptor potentials for the analysis of pheromone interspecificity.

With improved techniques,[13] the interspecific effects of the chemically unknown female sex pheromones have been determined quantitatively among 600 species from 27 moth families (Saturniidae, Sphingidae, Lasiocampidae, Bombycidae, Endromididae, Lemoniidae, Brahmaeidae, Notodontidae, Thaumatopoeidae, Noctuidae, Lymantriidae, Arctiidae, Endrosidae, Syntomididae, Nolidae, Cymbidae, Brephidae, Geometridae, Drepanidae, Cymatophoridae, Zygaenidae, Cochlidiidae, Pyralidae, Tortricidae, Psychidae, Sesiidae, and Cossidae). Some families have been investigated worldwide, including species from all subfamilies and all main systematic groups. In the Saturniidae, the interspecific pheromone effects among 104 species representing more than half of the existing genera have been determined in 1900 species combinations.[13] These effects among 25 species of one single saturniid subfamily are shown in Figure 1A. (For the full pattern, including all saturniid subfamilies, see Figures 3 and 4 of Priesner[13].) Figure 1B shows relations among 25 species of four sphingid subfamilies.

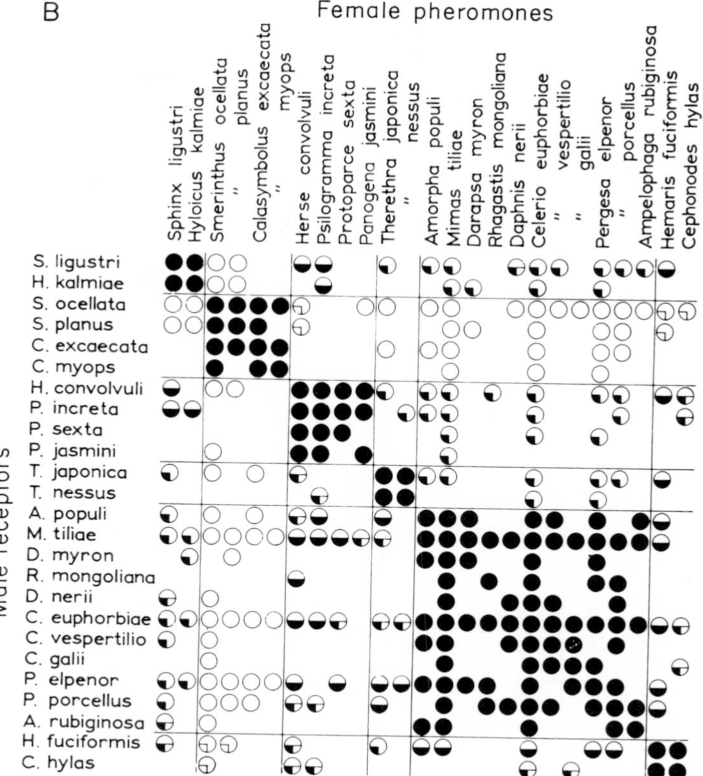

Figure 1 Interspecific effectiveness of female sex pheromones of Lepidoptera, as determined by EAG recordings from the male pheromone receptor cells. Solid circles = full interspecific effects which cannot be distinguished from intraspecific ones. Empty circles = no interspecific effect up to the highest gland concentrations. Other symbols = differing degrees of partial reactions. A. Saturniidae: 25 species of the subfamily Saturniinae. (From Priesner[13]) Section of a chart showing the entire pattern among 60 species of five subfamilies. B. Sphingidae: 25 species of four subfamilies. (From Priesner, unpublished)

Species reacting fully in both directions (♂A♀B, ♂B♀A) have been summarized as a reaction group (RG). Figure 1A shows six RGs; the 104 saturniid species investigated are distributed into 19 RGs, the 600 species of the 27 families into 102 RGs. No conclusion will be made about the chemical identity: the species of the same RG may have the same female sex pheromone molecule, or they may have different ones which are fully intereffective.

Compared with the morphological system of a family, an RG is a genus or a group of very closely related genera. In a few cases it is a distinct species group within a genus. In none of the 27 families have two of the closest related species fallen into two different RGs. All the species of the same RG are directly related and can be reduced to a common origin.[13] The pattern of pheromone interspecificity has been used in some families to correct the present morphological system. From this pattern it has also been shown that the female sex pheromones are not the sexual isolating mechanisms between very closely related species. Species of the same genus are separated not by their female sex pheromones, but by a scale of spatial, seasonal, temporal, ecological, ethological, and other isolating mechanisms that safeguard the integrity of the species.

In single-cell recordings from the sensilla trichodea of the male antenna, all sex pheromone receptor cells of the same species showed the same spectrum of reaction. All the pheromones effective on one species—their own and that of other species—acted on the same cell but elicited different degrees of excitation in the same concentration. In the silkmoth *Bombyx mori*, more than 500 individual pheromone receptor cells investigated had the identical reaction spectrum. This spectrum has been analyzed quantitatively with chemically known molecules. The actual molecular concentration on the receptor has been calculated from experiments with tritium-labeled bombykol.[21,23] By changing the number and position of the double bonds, and by introducing other functional groups into the molecule, reaction thresholds and intensity reaction curves have been determined comparatively for the pheromone receptor cells of *Bombyx mori* and species of thirteen other families (Priesner, unpublished). Of the primary unsaturated aliphatic alcohols tested, some examples are shown in Table II to demonstrate effects of chain length, configuration, position, and the number of double bonds in four species. About the same concentration threshold as in *Bombyx* to bombykol has been found for seven different compounds in species of other families.

All substances activating the bombykol receptor cells of the male *Bombyx mori* (their own pheromone; other pheromones and bombykol derivatives in higher concentrations; certain nonpheromone substances in much higher concentrations) elicited the sexual behavior response. No case has been found in which a substance caused sexual behavior in the male but did not excite this type of cell in the respective concentration. The difference between the threshold of a substance and the bombykol threshold was the same in the electrophysiological and in the behavioral experiments.[22]

TABLE II

SEX PHEROMONE RECEPTOR THRESHOLDS FOR *BOMBYX MORI* AND THREE SPECIES OF CLOSELY RELATED FAMILIES TO 12 HIGHLY PURIFIED PRIMARY UNSATURATED ALIPHATIC ALCOHOLS, AS OBTAINED BY EAG AND SINGLE-CELL RECORDINGS.

The threshold values are stated in μg of stimulus source concentration for identical stimulation. They were determined by increasing concentration steps logarithmically from 10^{-6} to 10^{3} μg.

	BOMBYCIDAE *Bombyx mori*	SATURNIIDAE *Aglia tau*	ENDROMIDIDAE *Endromis versicolora*	SPHINGIDAE *Deilephila euphorbiae*
(trans-10,cis-12)hexadecadien(1)ol	0.001	0.1	0.01	0.1
(cis-10,trans-12)hexadecadien(1)ol	0.1	0.1	0.1	0.1
(cis-10,cis-12)hexadecadien(1)ol	1	1	1	1
(cis-10,cis-12)heptadecadien(1)ol	10	0.1	10	10
(trans-10,trans-12)hexadecadien(1)ol	1	0.01	0.1	0.1
(trans-10)hexadecen(12)in(1)ol	100	0.01	100	
(trans-10)hexadecen(1)ol	1	0.1	0.1	0.1
(cis-9)hexadecen(1)ol	100	10	1000	10
(cis-9)octadecen(1)ol	1000	1	>1000	100
(cis-9,cis-12)octadecadien(1)ol	1000	1	>1000	100
(trans-9)octadecen(1)ol	>1000	100	>1000	1000
(trans-11)octadecen(1)ol	>1000	0.1	>1000	100

ACKNOWLEDGMENT

Some of the substances listed in Table II were kindly provided by Professor E. Hecker (Institut für Biochemie der Universität Heidelberg) and by Dr. G. Kasang (Max-Planck-Institut für Biochemie, München).

REFERENCES

1. Barth, R. 1960. Orgaos odoriferos dos Lepidópteros. Rio de Janeiro: *Parque Nac. do Itatiaia Bol.* **7**, pp. 1–159.
2. Berger, R. S. 1966. Isolation, identification, and synthesis of the sex attractant of the cabbage looper, *Trichoplusia ni. Ann. Entomol. Soc. Am.* **59**, pp. 767–771.
3. Boeckh, J., K.-E. Kaissling, and D. Schneider. 1965. Insect olfactory receptors. *Cold Spring Harbor Symp. Quant. Biol.* **30**, pp. 263–280.
4. Boeckh, J., E. Priesner, D. Schneider, and M. Jacobson. 1963. Olfactory receptor response to the cockroach sexual attractant. *Science* **141**, pp. 716–717.
5. Butenandt, A., R. Beckmann, D. Stamm, and E. Hecker. 1959. Über den Sexual-Lockstoff des Seidenspinners *Bombyx mori.* Reindarstellung und Konstitution. *Z. Naturforsch.* **14** b, pp. 283–284.
6. Butler, C. G. 1967. Insect pheromones. *Biol. Rev.* **42**, pp. 42–87.
7. Eiter, K., E. Truscheit, and M. Boness. 1967. Neuere Ergebnisse der Chemie von Insektensexuallockstoffen. Synthesen von D,L-10-Acetoxy-hexadecen-(7-cis)-ol-(1), 12-Acetoxy-octadecen-(9-cis)-ol-(1) ("Gyplure") und 1-Acetoxy-10-propyl-tridecadien-(5-trans.9). *Justus Liebigs Ann. Chem.* **709**, pp. 29–45.
8. Green, N., M. Jacobson, T. J. Henneberry, and A. N. Kishaba. 1967. Insect sex attractants. VI. 7-Dodecen-1-ol acetates and congeners. *J. Med. Chem.* **10**, pp. 533–535.
9. Jacobson, M., M. Beroza, and W. A. Jones. 1960. Isolation, identification, and synthesis of the sex attractant of gypsy moth. *Science* **132**, pp. 1011–1012.

10. Jones, W. A., M. Jacobson, and D. F. Martin. 1966. Sex attractant of the pink bollworm moth: isolation, identification, and synthesis. *Science* **152**, pp. 1516–1517.
11. Kaissling, K.-E., and M. Renner. 1968. Antennale Rezeptoren für Queen Substance und Sterzelduft bei der Honigbiene. *Z. Vergleich. Physiol.* **59**, pp. 357–361.
12. Karlson, P., and M. Lüscher. 1959. "Pheromones": a new term for a class of biologically active substances. *Nature* **183**, pp. 55–56.
13. Priesner, E. 1968. Die interspezifischen Wirkungen der Sexuallockstoffe der Saturniidae (Lepidoptera). *Z. Vergleich. Physiol.* **61**, pp. 263–297.
13a. Read, J. S., F. L. Warren, and P. H. Hewitt. 1968. Identification of the sex pheromone of the false codling moth (*Argyroploce leucetreta*). *Chem. Commun.* **14**, pp. 792–793.
14. Roelofs, W. L., and H. Arn. 1968. Sex attractant of the red-banded leaf roller moth. *Nature* **219**, p. 513.
15. Ruttner, F., and K.-E. Kaissling. 1968. Über die interspezifische Wirkung des Sexuallockstoffes von *Apis mellifica* und *Apis cerana*. *Z. Vergleich. Physiol.* **59**, pp. 362–370.
16. Schneider, D. 1962. Electrophysiological investigation on the olfactory specificity of sexual attracting substances in different species of moths. *J. Insect Physiol.* **8**, pp. 15–30.
17. Schneider, D. 1963. Electrophysiological investigation of insect olfaction. *In* Olfaction and Taste I (Y. Zotterman, editor). Pergamon Press, Oxford, etc., pp. 85–103.
18. Schneider, D. 1963. Vergleichende Rezeptorphysiologie am Beispiel der Riechorgane von Insekten. *Jahrb. Max-Planck-Ges. Göttingen.* pp. 150–177.
19. Schneider, D. 1966. Chemical sense communication in insects. *Symp. Soc. Exp. Biol.* **20**, pp. 273–297.
20. Schneider, D. 1967. Wie arbeitet der Geruchssinn bei Mensch und Tier? *Jahrb. Max-Planck-Ges. Göttingen.* pp. 294–314.
21. Schneider, D. Insect olfaction: Deciphering system for chemical messages. *Science* (In press)
22. Schneider, D., B. C. Block, J. Boeckh, and E. Priesner. 1967. Die Reaktion der männlichen Seidenspinner auf Bombykol und seine Isomeren: Elektroantennogramm und Verhalten. *Z. Vergleich. Physiol.* **54**, pp. 192–209.
23. Schneider, D., G. Kasang, and K.-E. Kaissling. 1968. Bestimmung der Reichschwelle von *Bombyx mori* mit Tritium-markiertem Bombykol. *Naturwissenschaften* (In press)
24. Schneider, D., V. Lacher, and K.-E. Kaissling. 1964. Die Reaktionsweise und das Reaktionsspektrum von Reichzellen bei *Antheraea pernyi* (Lepidoptera, Saturniidae). *Z. Vergleich. Physiol.* **48**, pp. 632–662.
25. Schneider, D., and R. A. Steinbrecht. 1968. Checklist of insect olfactory sensilla. *Symp. Zool. Soc. London.* pp. 279–297.
26. Sekul, A. A., and A. N. Sparks. 1967. Sex pheromone of the fall armyworm moth: isolation, identification, and synthesis. *J. Econ. Entomol.* **60**, pp. 1270–1272.
27. Stefanović, D. K., B. Grujić-Injać, and D. Mićić. 1963. Result of the activity of the synthesis of the sexual odour of the gypsy moth female (Gyptol and its higher homologue Gyplure). Plant Protection (Beograd) **73**, pp. 235–249.
28. Steinbrecht, R. A. 1964. Feinstruktur und Histochemie der Sexualduftdrüse des Seidenspinners *Bombyx mori* L. *Histochemie* **64**, pp. 227–261.
29. Stürckow, B. 1965. The electroantennogram (EAG) as an assay for the reception of odours by the gypsy moth. *J. Insect Physiol.* **11**, pp. 1573–1584.
30. Wilson, E. O. 1963. Pheromones. *Sci. Amer.* **208**, No. 5 pp. 100–114.
31. Wilson, E. O. 1965. Chemical communication in the social insects. *Science* **149**, pp. 1064–1071.
32. Wilson, E. O., and W. H. Bossert. 1963. Chemical communication in animals. *Recent Progr. Hormone Res.* **19**, pp. 673–716.

OLFACTION AND BEHAVIORAL SOPHISTICATION IN FISH

J. ATEMA, J. H. TODD, *and* J. E. BARDACH

School of Natural Resources, The University of Michigan, Ann Arbor, Michigan.
Dr. Todd's present address is Department of Zoology, San Diego State College, San Diego, California

THE CLASSICAL CONCEPT OF THE FISH AS A REFLEX-DOMINATED VERTEBRATE

Fishes are considered primitive compared to mammals, which implies that their behavior is far simpler and that their nervous system is at a lower level of development. The peripheral sensory systems of fishes are certainly as well developed as those of mammals, but their central nervous system appears simpler. Even in the adult animal, it presents itself as an unfolded neural tube with local thickenings in the wall that correspond to collections of incoming neurons from various peripheral sensors. Many of the brain centers so formed are dominated by one sensory modality each, and connect input with output in a rather direct way. For example, touch and taste steer the motor output (i.e., locomotion) of the fish (Figure 1). The scheme is based on behavioral experiments and neuroanatomy.[9] This neurological pattern led to the classical concept of fish behavior as one of reflexes and stereotyped action patterns even before the ethologists appeared on the scene. There is little doubt that this concept is correct for a number of behavior patterns with their underlying nervous connections. But, as we will see, there is more to it.

NEUROANATOMY

Herrick[9,10] has shown that higher nonspecific correlation centers exist in the fish brain. The primary cerebral centers of the more important sensory systems, or groups

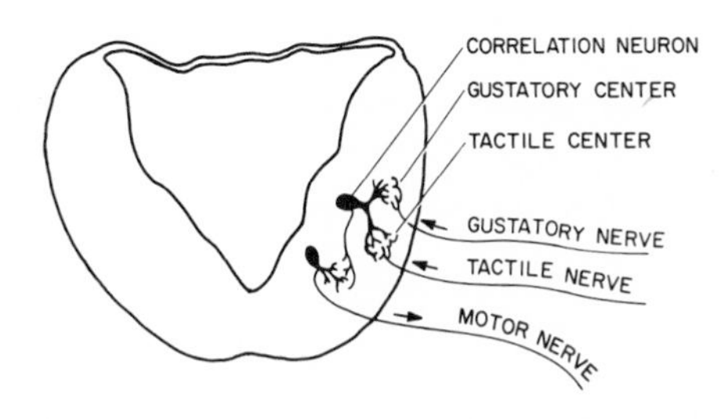

Figure 1 Diagram of some of the connections of tactile and gustatory nerves in the medulla oblongata of larval Amblystoma as seen in cross section. (Herrick[9])

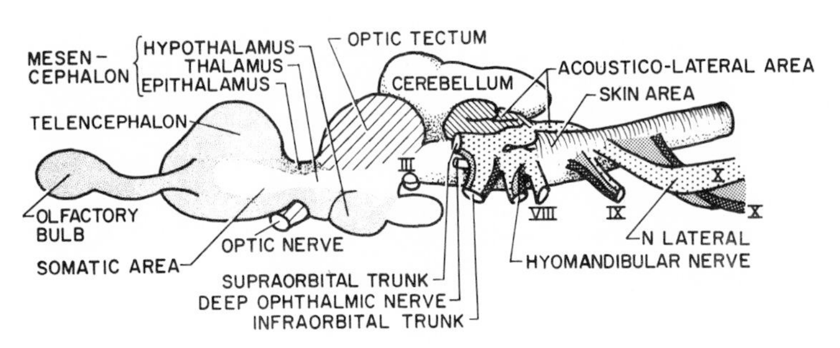

Figure 2 Dogfish brain seen from the left side. The primary (specific) centers are shaded. The stippled area is olfactory. Nonspecific areas are left blank. (After Herrick[9])

of related systems, in the dogfish are shown in Figure 2. The nonspecific areas are in the telencephalon and diencephalon. Parts of the telencephalic portion are homologous with the amygdala and striatum in higher vertebrates, and the diencephalic portion is the thalamic area common to all vertebrates. Thalamus and hypothalamus receive indirect inputs (mainly secondary and tertiary paths) from almost all brain centers. This organization suggests that some of the forebrain centers serve a higher correlative function, evaluating inputs from most of the sensory systems. It is also evident that the olfactory input is connected with all the centers in both the telencephalon and the thalamic region, and that it even dominates some of them: dorsal, lateral, and ventral olfactory areas have strong direct connections with epithalamus (habenula), thalamus, and hypothalamus respectively. In general, all the brain centers that can be expected to serve a "higher" correlative function are in physical contact with the olfactory input.

The cerebellum is a correlation center in a more limited sense: it organizes vestibular responses and is considered an outgrowth of the acoustico-lateral area in the dorsal wall of the medulla. For a detailed description of the fish brain, see Ariëns Kappers.[1]

BEHAVIOR AND THE OLFACTORY SYSTEM

A disproportionate number of fish behavior studies have been done on one laboratory animal, the goldfish, for which both visual and chemosensory information are important. However, most experiments have involved the visual aspect, and for this reason, as we will see, the goldfish and its experimenters have become, to a high degree, responsible for the classical concept that the fish is predominantly a reflex-oriented vertebrate.

We find this reflected in Maier and Schneirla,[14] for instance, who wrote about the class Pisces: "The fishes are subject to limitations prescribed mainly by a virtual independence in the control exerted by the major exteroceptive systems over behavior, since there appears to be a lack of any important intersensory correlation in the qualitative control of adjustments to the environment. For this reason in particular, the behavior of the class is characterized by directly determined activities and by a lack of plasticity."

The concept deserves re-examination, which is best initiated by a few examples relating olfactory development to behavior. We are restricted in our choice to fish, the behavior of which is described in sufficient detail.

The first is the almost anosmic, highly visual stickleback*, with behavior mainly dependent on visual sign stimuli. Summarizing his ample experience with this fish, Tinbergen[19] writes: "A stickleback is different from a rat. Its behavior is much more purely innate and much more rigid. Because of its relative simplicity, it shows some phenomena more clearly than the behavior of any mammal can. The dependence on sign stimuli, the specificity of motivation, the interaction between two types of motivation with the resulting displacement activities are some of these phenomena." And further: ". . . mammals are in many ways a rather exceptional group, specializing in 'plastic' behavior. The simpler and more rigid behavior found in our fish seems to be the rule in most of the animal kingdom."

In short, the stickleback has very stereotyped behavior and conforms to the concept of the reflex-dominated fishes. It will act predictably in a given situation and its behavior depends directly on stimuli in a quantitative way.

Observers of the behavior of cichlid fishes† generally express similar ideas. Behavior sequences are dependent on sign stimuli and develop in predictable steps. The sequence stops when the necessary stimulus is not present. Individual learning does not seem to play an important role.[2] However, the cichlids recognize their own young and care for them preferentially. Myrberg[16] summarizes his findings as follows: "Chemoreception apparently plays the most important role in the parental preference of wiggling young, while vision has this role in the parental preference of free-swimming young. This highly adaptive change in the primary modality used by a parent to distinguish between young apparently occurs by the association of parental care with certain visual cues of the young in the presence of some chemical factor(s) of these same young." In other words, the chemosensory (presumably olfactory) input not only mediates individual learning in the cichlid parents; it also facilitates learning based on other (visual) stimuli from the environment (their young).

Both stickleback and cichlid behavior have been studied and described to near perfection by excellent ethologists (Tinbergen and Baerends, loc. cit). In both cases emphasis was on stereotypy, on "reflex behavior" that reinforced the classical concept.

The next two examples are more recent and less well known. Behavior and morphology of the glandulocaudine fishes‡ (the name indicates they possess glands on their tails) was studied by Nelson,[17] who demonstrated that "consistent differences in [one animal's] behavior toward different animals lasted over periods of several

* Gasterosteiformes, Gasterosteidae.
† Perciformes, Cichlidae.
‡ Cypriniformes, Characidae.

months." The observations led him to suggest that a true pair bond might exist in certain glandulocaudine species. The existence of individual recognition is clearly indicated, and with it the fact that these fish, in their social behavior, operate on a basis of individually learned experiences, a sign of behavioral sophistication. Nelson (loc. cit.) also discusses basic contrasts in behavior patterns between sticklebacks and glandulocaudines. In the latter "... no activity is seen to occupy a period of time exclusively, with the possible exception of spawning. Thus, within the course of a day, a *Corynopoma* male may court, feed, fight, partake in repeated aggregation and dispersal, and change its dominance relations several times." In the male stickleback, on the other hand "... long periods of time may be occupied successively by nest-building, territorial defense, courtship, and finally parental care, each to the near exclusion of other activities."

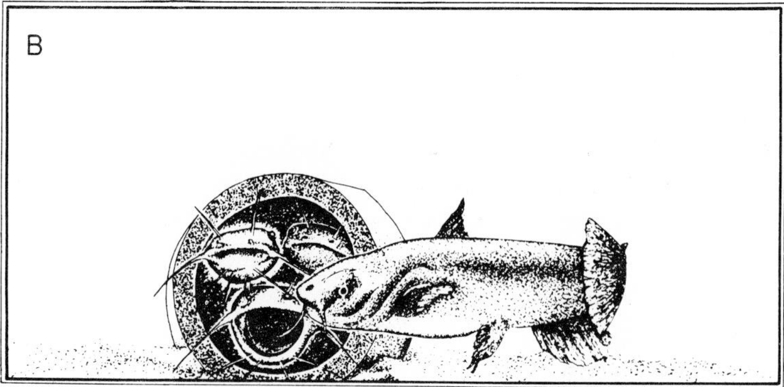

Figure 3 A strange bullhead is introduced into a community consisting of three others in a stable dominance hierarchy:

A. His chemical presence has caused the β and γ fish to leave their own territories and to seek shelter in the tile, which represents the center of the α territory. Such cooperative behavior occurs only during times of stress.

B. After some seconds, the strange fish begins to explore the new environment. The α fish displays aggression against the stranger when he approaches. A little later he will initiate a fight. When the stranger has disappeared he will no longer allow his acquaintances in his territory and the original territorial situation will be reestablished. (Todd[20])

That there is a chemosensory basis for their behavior seems obvious from the anatomy of the fish. Structures around the gland opening and many movements during courtship behavior are well adapted to direct the secretion from the gland to the partner's head. It is more than tempting to postulate a pheromone role for the gland secretion and to claim the nose as the receptive channel.

Although Nelson's work was not geared to disclosing an olfactory theory of this higher behavior, there is ample evidence that the glandulocaudine fishes exhibit behavior patterns similar to those of the catfishes that we studied, which will be described below.

The last example in the series is the catfish order (Siluriformes). The members of this order are chiefly nocturnal, and the chemosensory system dominates the brain. The catfishes, particularly the bullheads (family Ictaluridae), have been studied for a century and are thus better known than the glandulocaudines. Herrick[6,7,8] studied the histology of bullhead brains and nerves. Parker,[18] von Frisch,[5] and others used this fish in physiological experiments that were mainly concerned with the highly developed sense of taste. It was recently established that blind bullheads localize and ingest food even when deprived of their sense of smell.[3] The well-developed olfactory organ of bullheads mediates individual recognition, the capacity for which disappears when the olfactory epithelium is destroyed.[21] From a behavioral point of view there then exist two fairly distinct chemosensory channels in this fish: taste, which dominates food search and ingestion, and smell, which governs social behavior.

A systematic study of bullhead behavior[20] revealed that they live in communities, where each fish recognizes the others as individuals. Individual recognition appears to be the basis for territoriality and hierarchies within a community. The fish have good memories and can recall individuals after separation of up to two months' duration. Aggressive behavior between strange bullheads is very different from that between residents of a community. Cooperative behavior occurs when residents of a community are threatened (Figures 3 and 4). The social organization collapses when the fish are deprived of their olfactory sense. In summary, bullheads display sophisticated behavior in which smell plays a crucial role. (We have chosen the phrase sophisticated behavior to indicate its higher vertebrate character. The term is practically synonymous with plastic or versatile behavior.) Similar results have been obtained from work on behavior in other catfishes (Pimelodidae, McLarney, in preparation).

An interesting comparison can be made if we look at the brains and noses of the four fish just described (Figures 5, 6, 7, and 8). Some of the dorsolateral outgrowths of the brainstem represent the primary centers that collect the nerves of one sensory system. The development of these centers is an indication of the development of the system, which in turn indicates the type of behavior and habitat we can expect for the fish.[4]

The visually dominated stickleback *Eucalia inconstans* (Figure 5), shows an exaggerated development of the optic lobes, which push the cerebellum backwards so it overlays the medulla. The telencephalon is even less well developed than the dorsal view indicates, because of its dorsoventral compression. The olfactory bulbs are very small, the nerves hardly distinguishable, and there are no more than two olfactory lamellae in the nose. One small opening connects it with the surrounding water.

The cichlid *Cichlasoma nigrofasciatum* (Figure 6) has large optic lobes, but its telencephalon is far better developed than the stickleback's, as are the olfactory bulbs and nerves. The nose has 13 hill-shaped lamellae and one opening to the external environment.

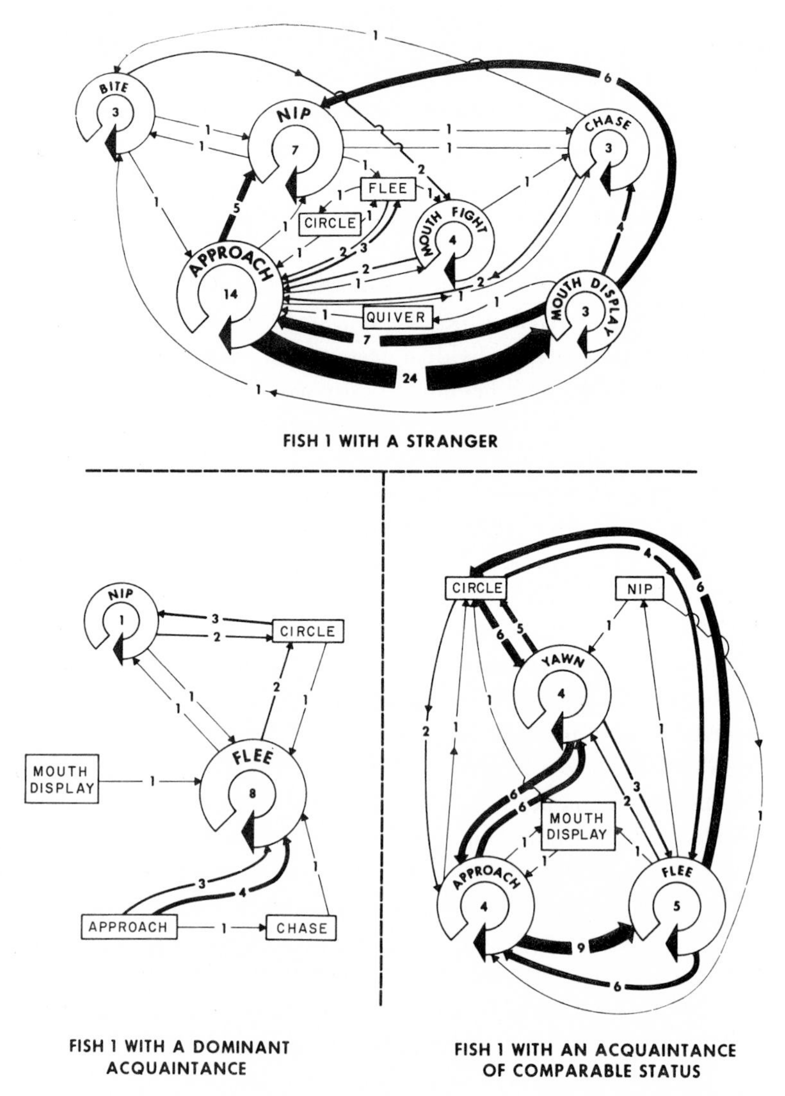

Figure 4 Ethograms showing the interactions of a test fish with three others of different status on different occasions. The behavior with a stranger is complicated and unpredictable; with an acquaintance of comparable status, the behavior pattern becomes more ritualized; with a dominant acquaintance there is a formal relationship. The chemical "composition" of the different fish is smelled and evaluated by the test fish. Memory plays an important role.

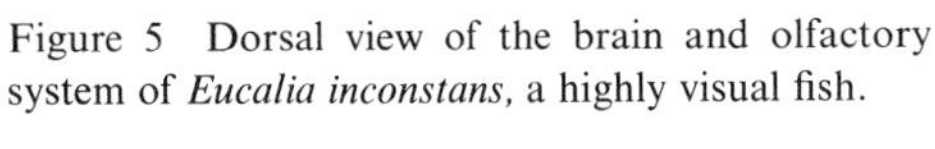

Figure 5 Dorsal view of the brain and olfactory system of *Eucalia inconstans*, a highly visual fish.

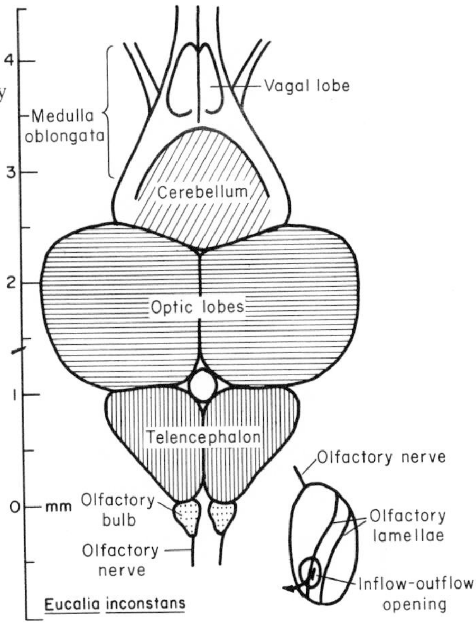

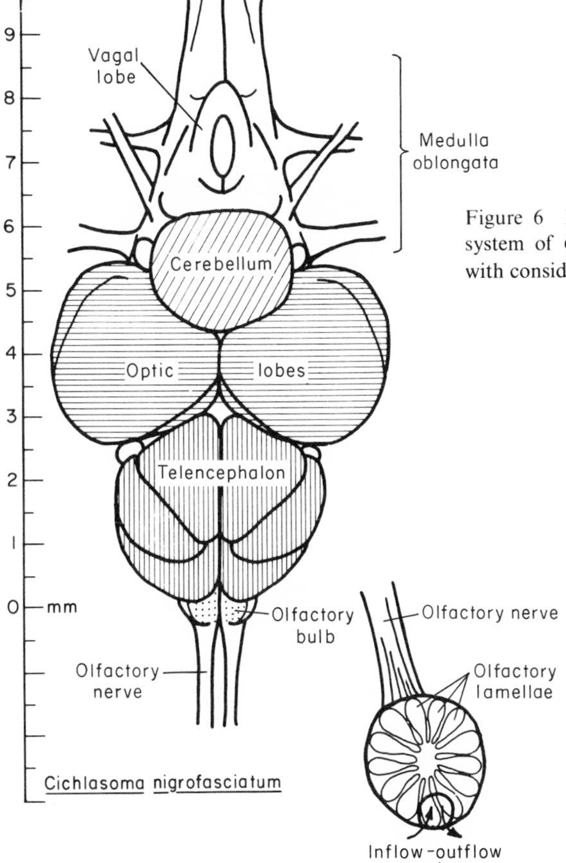

Figure 6 Dorsal view of the brain and olfactory system of *Cichlasoma nigrofasciatum*, a visual fish with considerable chemosensory acuity.

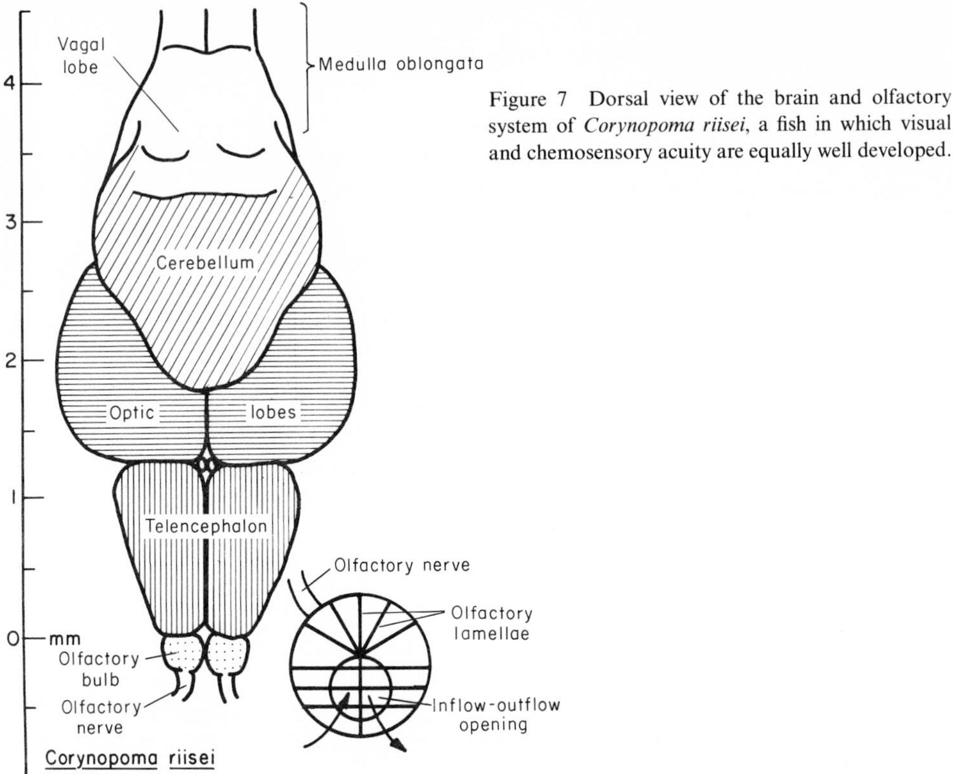

Figure 7 Dorsal view of the brain and olfactory system of *Corynopoma riisei*, a fish in which visual and chemosensory acuity are equally well developed.

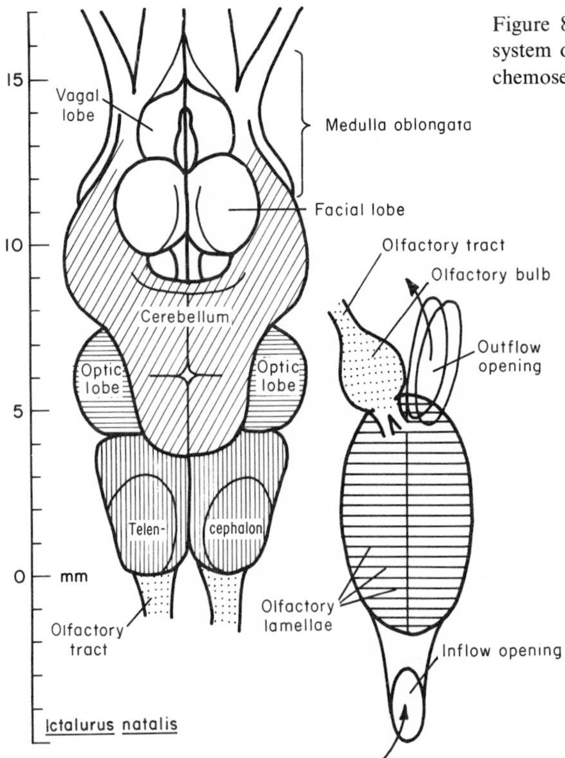

Figure 8 Dorsal view of the brain and olfactory system of *Ictalurus natalis*, a fish dominated by the chemosensory system.

The trend continues with the glandulocaudine *Corynopoma riisei* (Figure 7). It has smaller optic lobes and a larger cerebellum. The olfactory bulbs and nerves are of the same relative size. The 11 bookleaf-shaped lamellae in the nose are flushed with water through one large opening. Olfactory development is similar to that in the cichlid, but the latter has a larger optic system, indicating the greater dependence on vision.

The bullhead *Ictalurus natalis* (Figure 8) shows a tremendous development of the facial and vagal lobes, which collect the taste nerves from body and pharynx respectively. The cerebellum overgrows the small optic lobes and even the caudal part of the telencephalon. The olfactory bulb is located directly behind and under the nose, connected with the brain by a long tract. The olfactory nerve is consequently short. The large nose has separate in- and out-flow openings. Cilia on the olfactory epithelium maintain a variable water flow over 46 densely packed lamellae.

As indicated, the nonspecific centers in the forebrains of fish are in direct and indirect relation with the olfactory input, and it is therefore attractive to conclude that they are responsible for plasticity in fish behavior. Preliminary work indicates that a bullhead with the telencephalon removed responds in a much more stereotyped manner to visual and tactile stimuli than does a normal animal. Individual recognition seems to be lost; so is exploratory behavior, another remarkable phenomenon in normal bullhead behavior.

Bullheads without telencephalons turn into fighting robots, unable to evaluate an opponent's strength. The conclusive stage in a normal bullhead fight (the mouth fight) lasts three seconds at most. In one case, a fish without telencephalon had a mouth fight that lasted over half an hour with an opponent twice its size. Even then the smaller fish continued to act aggressively toward the other.

Similar results can be obtained by destroying the primary input (olfactory epithelium) and leaving the telencephalon intact. Bullheads will then relate only aggressively to others.

PARALLELS IN MAMMALS

Prior to this report, behavioral sophistication and plasticity as described here have been claimed mainly for mammals. To quote Maier and Schneirla[14] again: "In general, the visual determinant appears more rigid and specific in its control than does the olfactory." They further refer to Herrick[9]: "It is through the evolution of the forebrain, and hence through mechanisms primitively under the dominance of olfaction, that higher vertebrates become less subject to the specific enforcement of stereotyped behavior."

We will present some striking similarities between the situation described for fish above and the one described for mammals. We will not deal here with the other vertebrate classes. However, the differentiation between plastic and stereotyped behavior seems to occur within all the vertebrate classes, and olfaction might well be correlated with behavioral plasticity in general.

Although there is no hard experimental evidence directly correlating olfaction and social behavior in mammals, skin glands have been correlated with behavior, and it seems clear that skin gland products are smelled. (A number of papers in this volume indicate the phenomenon.) A striking example is found in the social behavior of the rabbit (*Oryctolagus cuniculus*) vs. that of the hare (*Lepus europaeus*).

A fully grown hare is four times as heavy as a rabbit and most of its glands are larger, but its anal and chin glands are only one tenth the size of those in rabbits. The hare is a solitary animal, whereas rabbits live in small groups. In rabbit groups both males and females establish a dominance hierarchy. The most dominant male and female rule over the group and the territory, which is defended by its occupants and recognized by rabbits of other groups. The rabbit hierarchy is constantly reinforced by chasing, sniffing, being sniffed at, and submission.[15]

A similar correlation between smell and sophisticated social behavior is found in prairie-dog coteries. The importance of the sense of smell in their social communication is shown not only in their friendly kissing but also by the fact that strangers turn their tails toward each other and expose their anal glands for olfactory exploration.[12]

Telencephalon ablations in fish correspond in a very gross way to temporal lobe ablations in mammals. (The temporal lobe includes hippocampus and amygdala, structures that developed from dorsal and lateral olfactory areas in the fish telencephalon.) After lobectomy, rhesus monkeys became stereotyped and responded in a compulsive way to all visual stimuli, a phenomenon called hypermetamorphosis. They also showed "psychic blindness," that is, they indicated no visual recognition and detection of the meaning of objects, although vision itself was not impaired. Their emotional behavior was changed in such a way that they showed no fear of animate or inanimate objects, even those that would have frightened them under normal conditions.[13] Iwata and Ando[11] repeated these experiments in an intraspecific context. They also found inhibited emotional behavior in general and especially the lack of normal fear and caution. However, two lobectomized monkeys, brought together seven months after the operation, fought so viciously that they had to be separated after 30 minutes. Such intense fighting never occurs in normal social relationships. The authors describe this as being due to a conspicuous lack of rank relationship, called social agnosia. This all appears identical to our findings in forebrain-ablated bullheads.

CONCLUSION

Each succeeding step in vertebrate evolution has been one toward more behavioral plasticity, as reflected in the forebrain development that finally produced the human neocortex. However, considerable differences in behavior exist within each vertebrate class and within smaller taxa; many groups include both species that have developed stereotyped behavior and others with plastic behavior, correlated respectively with visual or olfactory dominance in their central nervous system. This dualism can be shown clearly in the fishes, where stickleback and bullhead can be taken as extreme examples.

ACKNOWLEDGMENT

We wish to thank Dr. W. O. McLarney of the School of Natural Resources and Dr. J. Olds of the Brain Research Laboratory of the University of Michigan for their invaluable criticisms during the composition of this paper.

REFERENCES

1. Ariëns Kappers, C. U., G. C. Huber, and E. C. Crosby. 1936. The Comparative Anatomy of the Nervous System of Vertebrates, Including Man. Macmillan Co., New York. (Reprinted by Hafner Publishing Co., New York, 1965.)
2. Baerends, G. P., and J. M. Baerends-van Roon. 1950. An introduction to the study of the ethology of cichlid fishes. *Behaviour*, Suppl. 1, pp. 1–243.
3. Bardach, J. E., J. H. Todd, and R. Crickmer. 1967. Orientation by taste in fish of the genus *Ictalurus*. *Science* **155**, pp. 1276–1278.
4. Davis, B. J., and R. J. Miller. 1967. Brain patterns in minnows of the genus *Hybopsis* in relation to feeding habits and habitat. *Copeia* **1**, pp. 1–39.
5. Frisch, K. von. 1926. Vergleichende Physiologie des Geruchs und Geschmackssinnes. *Handb. Norm. Pathol. Physiol.* **11**, pp. 203–239.
6. Herrick, C. J. 1903. The organ and sense of taste in fishes. *U.S. Fish Comm. Bull.* **22**, pp. 237–272.
7. Herrick, C. J. 1905. The central gustatory paths in the brains of bony fishes. *J. Comp. Neurol.* **15**, pp. 375–456.
8. Herrick, C. J. 1906. On the centers for taste and touch in the medulla oblongata of fishes. *J. Comp. Neurol.* **16**, pp. 403–439.
9. Herrick, C. J. 1924. Neurological Foundations of Animal Behavior. Hafner Publishing Co., New York and London (Reprinted 1962), pp. 155–169.
10. Herrick, C. J. 1933. The functions of the olfactory parts of the cerebral cortex. *Proc. Natl. Acad. Sci. U.S.* **19**, pp. 7–14.
11. Iwata, K., and Y. Ando. 1968. Socio-agnostic behavior in temporal lobectomized monkeys. *In* Japan-U.S. Joint Seminar on Neurophysiological Basis of Learning and Behavior. (Org.: N. Yoshii and N. A. Buchwald.)
12. King, J. A. 1964. *In* Social Behavior and Organization Among Vertebrates (W. Etkin, editor). Univ. of Chicago Press, Chicago, p. 281.
13. Kluever, H., and P. C. Bucy. 1939. Preliminary analysis of functions of the temporal lobes in monkeys. *A. M. A. Arch. Neurol. Psychiat.* **42**, pp. 979–1000.
14. Maier, N. R. F., and T. C. Schneirla. 1935. Principles of Animal Psychology. McGraw-Hill, New York. (New ed., Dover Publications, Inc., New York, 1964), pp. 176–198.
15. Mykytowycz, R. 1968. Territorial marking by rabbits. *Sci. Amer.* **218**, pp. 116–126.
16. Myrberg, A. A. 1966. Parental recognition of young in cichlid fishes. *Anim. Behav.* **14**, pp. 565–571.
17. Nelson, K. 1964. Behavior and morphology in the glandulocaudine fishes (Ostariophysi, Characidae). Univ. of Cal. Publ. in Zool. **75**, **2**, pp. 59–152. Univ. of Calif. Press, Berkeley and Los Angeles.
18. Parker, G. H. 1908. The sense of taste in fishes. *Science* **27**, p. 453.
19. Tinbergen, N. 1952. The curious behavior of the stickleback. *In* Psychobiology. W. H. Freeman & Co., San Francisco and London, 1966, pp. 5–9.
20. Todd, J. H. 1968. The social behavior of the yellow bullhead (*Ictalurus natalis*). Ph.D. thesis. Univ. of Michigan, Ann Arbor, Mich.
21. Todd, J. H., J. Atema, and J. E. Bardach. 1967. Chemical communication in social behavior of a fish, the yellow bullhead (*Ictalurus natalis*). *Science* **158**, pp. 672–673.

MAMMALIAN PHEROMONES

W. K. WHITTEN

The Jackson Laboratory, Bar Harbor, Maine

INTRODUCTION

Pheromones have been clearly demonstrated in three families of rodents,[6,11,17] in rhesus monkeys,[12] and in cats,[12] and there is a growing body of evidence to suggest that they may function generally for mammals. The evidence has been derived from the study of odor glands and secretions, the use of these secretions for trail laying and territorial demarcation, and the behavioral and physiological responses evoked by them.

The mammalian pheromones may not be essential for successful breeding, particularly in captivity, but it appears that they may confer an advantage in some environments and so facilitate adaptation. Variation in the pheromones, or the responses to them, could occur and provide the mechanism for isolation and subsequent speciation. Some evidence from genetic studies of mouse pheromones, which may support this hypothesis, will be discussed after a brief introduction to the subject of mammalian pheromones in general.

SCENT GLANDS

Mykytowycz[14] has reviewed the distribution and structure of mammalian scent glands and discussed the use of their secretions to lay trails and to mark territory, and Kingston[9] has described the chemistry of their products. However, few studies have reported the behavioral responses to the known chemical constituents of these secretions, and as yet the chemical identity has not been established for any mammalian pheromone. These problems are by no means simple, and one may well ask what information a substance like civetone, which is produced by civets of both sexes, conveys from one civet to another. The substance may serve to identify members of the species and the amount produced may reflect the gonadal activity of an individual. However, other factors such as temperature, surface area, air movement, time, and distance would confound the assessment of gonadal activity if based on the intensity of the odor which an animal produces. Civetone, or the crude secretion, has been used as a fixative by perfumers and serves to extend more valued components of the perfume. Perhaps civetone and similar substances also function in animals as fixatives for other more ephemeral odors by which individuals are identified and status within a hierarchy recognized.[22] These odors are no doubt determined by genetic and environmental factors and may vary with the endocrine status.

BEHAVIORAL RESPONSES

Few adequately controlled experiments that demonstrate behavioral responses of males to odors from females have been described. The elegant studies of Michael and Keverne[12] demonstrated that male rhesus monkeys respond with increased bar pressing to a pheromone produced by spayed females that had received intravaginal doses of estrogen, too small to elicit a systemic response. The increased bar pressing did not occur when the olfactory area of the male was closed by suitable surgical packing, but returned after the packing was removed. These findings indicate that the pheromone is produced by the estrogen-stimulated vagina and that it stimulates the olfactory epithelium of the males.

Likewise, there have been very few reports of behavioral responses by females to odors of males. Signoret and Buisson[15] have shown that the odor of a boar, along with other stimuli, elicits the immobilization reflex of estrous sows that is a necessary prelude to successful mating. Michael and Keverne[12] have reported the crouching and treading response of estrous cats when they were placed in a cage recently occupied and marked by an active male. However, the response did not occur if the cage was washed after the male's occupancy.

The degree to which these behavioral responses are innate or dependent on learning and imprinting has not been determined. Some evidence has been provided by Todd[16] that the treading reaction of estrous cats is determined genetically.

RESPONSES TO PRIMER PHEROMONES

Most of the information on the endocrinological responses to pheromones has been derived from studies with laboratory mice.[20] Males of this species secrete a pheromone in their urine that stimulates the release of gonadotrophin from the anterior pituitary of diestrous females so that the occurrence of follicle development, ovulation, and estrus is advanced. This results in a degree of synchronization of estrus, on the third night after pairing, if the females are kept remote from males beforehand. A greater degree of synchrony occurs, in some strains, if the females are also grouped together prior to pairing with the males. The greater synchrony occurs because grouping of females prolongs the diestrous phase of the estrous cycle. This prolongation of the cycle may itself be caused by a pheromone, but it has yet to be demonstrated that it can be produced by an exterior product.

One of the interesting aspects of the estrus acceleration is that significant shortening of the cycle has only been demonstrated after 48 hours' exposure to the pheromone. This time lag corresponds to that required for animals to respond to injected gonadotrophins, and suggests that the pheromone acts during both follicle development and ovulation. It also suggests that adaptation to the stimulus does not occur. In addition, there does not appear to be any learned component in the response to the pheromone, because virgins respond as well as do retired breeders.[17] However, the role of imprinting by the male parent during development needs to be investigated with adequate environmental control.

Recently mated females, of a few susceptible strains, respond to a pheromone in the urine of some male mice and ovulate and mate again on the third night after exposure to the new male.[3] This response only occurs if the new male is from a strain which is sufficiently different from that of the original sire. In this case, like that described earlier, the pheromone has been demonstrated in the urine and found to be androgen-dependent. These similarities and the fact that both cause ovulation—usually on the third night after the females are exposed—indicate that the pheromones are very similar. They may be closely related substances or the same substance, the action of which is modified by some factor, such as that whereby the strangeness of the new male is recognized.

The Receptors

The pheromone from male mice most probably stimulates receptors in the peripheral olfactory system, but there is little direct evidence to support this assumption. Results observed after removal of the olfactory bulbs should be interpreted with caution. This operation prevents pregnancy block, so the embryos are carried to term,[4] whereas ovarian atrophy occurs in nonpregnant animals of some strains.[13,18] Experiments designed to determine whether the receptors are located in the vomeronasal organ have been inconclusive, and it has not been possible to obtain a mutant in which this structure is defective. Electrophysiological studies are eagerly awaited, and it is hoped that the alteration of the electrical pattern is not too small to detect, as might be inferred because of the prolonged stimulation which is required to shorten an estrous cycle significantly.[19]

Some indirect evidence which supports the role of olfactory receptors in the response to the pheromone has recently been obtained by Whitten, Bronson, and Greenstein,[21] who observed that mice housed two meters downwind from a cage containing males exhibited estrus synchrony, whereas those housed upwind did not.

The Pathway

Bronson and Desjardins (in press) have examined the pituitaries of spayed mice exposed to the pheromone and have observed a significant decrease in the level of follicle-stimulating hormone (FSH) and a corresponding increase in the level of FSH in the blood. The values for luteinizing hormone (LH) were too variable to arrive at any conclusions concerning the release of that hormone.

Dominic[8] has shown that pregnancy block can be prevented by reserpine administration, but the effects of more specific inhibitors of transmitter substances have yet to be determined.

STUDIES WITH MALE MOUSE URINE

The pheromone has been shown to be present in the urine of males; it disappears soon after castration, but is restored by testosterone treatment. Such treatment can also induce females to secrete the pheromone. It is present in urine collected directly from the bladder, after the proximal end of the urethra has been ligated. This observation strongly suggests that it is secreted by the kidney and not by any of the accessory sex glands, including the preputial glands and the urethral diverticulum. There is one report that the pheromone is produced by the preputial glands,[7] but we have not been able to confirm the findings.

Male mouse urine possesses a strong mousy odor, and I have seen it used to mark a fresh cage. Contrary to the report by Lane-Petter,[10] the odor is present in bladder urine and not in the secretions of the preputial glands. It is probable that the substance responsible for this odor is also the pheromone. If this assumption is correct, one could follow the extraction of the pheromone by its odor. If it is incorrect, we may derive some interesting information about mouse odors. To follow the extraction of the odorous substance, it is essential to have urine which is entirely free from feces, and this precludes the use of metabolism cages for collecting urine. The following procedure is followed. Mature mice from which the preputial glands have been surgically removed are housed individually and permitted to mark their cages adequately for two or three days. They are frequently handled, so that they do not void urine immediately upon being caught. Urine is collected by holding the animals over a test tube and squeezing the bladder through the abdominal wall. Care is taken not to express the accessory gland secretions; about 0.5 ml of clean urine can be obtained in this manner.

Three fractions were obtained when male mouse urine was passed through a sephadex column. The first was odorless and opalescent and contained the mouse urinary protein. The second was yellow and possessed the strong characteristic odor. The third was clear and odorless. When these fractions were assayed by the Bronson and Whitten method[2] only the odorous fraction contained detectable quantities of the pheromone.

Samples of the gaseous phase from containers partly-filled with male or castrated mouse urine have been seen through a gas chromatograph. A peak has been observed with the sample from the males, and this is associated with the characteristic mouse odor from the shunt portal. These problems are being actively explored and will be reported on later.

GENETIC OBSERVATIONS

Male mice from all strains so far examined induce estrus synchrony. However, it has been claimed that males heterozygous for the lethal yellow allele fail to do so, even though they do cause pregnancy block.[1] This is the only report of pregnancy block without estrus synchrony, and because of the theoretical importance of the findings the problem was reinvestigated. "Yellow" males are predisposed to obesity and reduced fertility, therefore only young, nonobese animals were used. Two strains of males with the yellow genotype were examined and compared with "nonyellow" males from the same strain. Peaks of mating were observed on the third night and it was concluded that the earlier workers had confounded production of the pheromone with ability to copulate. They used bigamous matings, which further reduced the ability of the yellow males to copulate with all estrous females.

Female mice normally show a peak frequency of mating on the third night after pairing, but those from the BALB/c inbred strain do not. There is no difference between the proportions of these females that mate on the first three nights, but significantly reduced numbers mate on the fourth (Table I). This reduction suggests that

TABLE I

Number of vaginal plugs observed in SJL/J, BALB/cDg and BALB/c × SJL/J F_1 hybrids on the first 5 days after mating

Female Type	No.	Day 1	Day 2	Day 3	Day 4	Day 5	No. unmated	P*
SJL/J	270	39	37	133	13	5	37	<0.001
BALB/cDg	502	117	134	137	31	11	63	<0.001
BALB/c × SJL/J F_1	282	50	65	110	6	3	50	<0.001

* Heterogeneity: days (1–4).

the mice do respond to the pheromone but in some unusual manner. BALB/c females also respond to a strange male with pregnancy block, which indicates that they are sensitive to that pheromone. They are, however, unusual because they continue to show normal fertile estrous cycles even after removal of the olfactory bulbs.

With one exception, the pheromone from males induces estrus synchrony on the third night after pairing. The exception occurs when SJL/J × BALB/cDg F_1 males are used. The two parental strains produce normal third-night peaks and the males from each can block pregnancies initiated by the other, but when the hybrid is used the peak often occurs on the second night (Table II). The earlier-than-usual mating

TABLE II

Number of vaginal plugs observed in SJL/J females on the first five days after pairing with SJL/J, BALB/c and hybrid males

Type of male	No. of females	Day 1	Day 2	Day 3	Day 4	Day 5	No. unmated	P*
SJL/J	112	17	19	55	4	2	15	<0.001
BALB/cDg	120	15	22	50	19	1	16	<0.001
SJL/J × BALB/c hybrid	115	33	42	35	2	0	3	<0.001

* Heterogeneity: days (1–4).

is interpreted as the result of a high pheromone content and hybrid vigor. These animals have been used to supply urine for the study of chemistry of the pheromone.

Pregnancy block has been demonstrated only in a few inbred strains of mice. Hybrids from two strains, one of which responds while the other is resistant, also fail to respond.[5] When this resistant hybrid is backcrossed to the susceptible strain, approximately half of the offspring are susceptible whereas the remainder are not. The data are not inconsistent with inheritance of susceptibility to pregnancy block as a simple recessive. In addition, preliminary observations suggest that the ability of males to induce pregnancy block may be similarly determined.

ACKNOWLEDGMENT

This work was supported in part by Public Health Research Grant HD-00473 from the National Institute of Child Health and Human Development.

REFERENCES

1. Bartke, A., and G. L. Wolff. 1966. Influence of the lethal yellow (A^y) gene on estrous synchrony in mice. *Science* **153**, pp. 79–80.
2. Bronson, F. H., and W. K. Whitten. Oestrus-accelerating pheromone of mice: assay, androgen-dependency and presence in bladder urine. *J. Reprod. Fertil.* **15**, pp. 131–134.
3. Bruce, H. M. 1959. An exteroreceptive block to pregnancy in the mouse. *Nature* **184**, p. 105.
4. Bruce, H. M., and D. M. V. Parrott. 1960. Role of olfactory sense in pregnancy block by strange males. *Science* **131**, p. 1526.
5. Chapman, V., and W. K. Whitten. Inheritance of pregnancy block in inbred mice. (In preparation)
6. Clulow, F. V., and J. R. Clark. 1968. Pregnancy-block in *Microtus agrestis* and induced ovulation. *Nature* **219**, p. 511.
7. Gaunt, S. L. L. 1967. Classification and effect of the preputial pheromone in the mouse. *Amer. Zool.* **7**, p. 713. (Abstract 22)
8. Dominic, C. J. 1966. Observations on the reproductive pheromones of mice. II. Neuro-endocrine mechanisms involved in the olfactory block to pregnancy. *J. Reprod. Fertil.* **11**, pp. 415–421.
9. Kingston, B. H. 1964. The chemistry and olfactory properties of musk, civet and castoreum. Proc. 2nd Int. Cong. Endocrinol., London. Excerpta Medica Foundation, Amsterdam, etc., pp. 209–214.
10. Lane-Petter, W. 1967. Odour in mice. *Nature* **216**, p. 794.
11. Marsden, H. M., and F. H. Bronson. 1964. Estrous synchrony in mice: alteration by exposure to male urine. *Science* **144**, p. 1469.
12. Michael, R. P., and E. B. Keverne. 1968. Pheromones in the communication of sexual status in primates. *Nature* **218**, pp. 746–749.
13. Mody, J. K. 1963. Structural changes in the ovaries of IF mice due to age and various other states: demonstration of spontaneous pseudopregnancy in grouped virgins. *Anat. Rec.* **145**, pp. 439–447.
14. Mykytowycz, R. The role of skin glands in mammal communication. *In* Advances in Chemoreception (J. W. Johnston, D. G. Moulton, and A. Turk, editors). Appleton-Century-Crofts, New York. (In press)
15. Signoret, J. P., and F. du Mesnil du Buisson. 1961. Étude du comportement de la truie en oestrus. Fourth Int. Cong. Anim. Reprod. Artif. Insem. (The Hague), Vol. 2, pp. 171–175.
16. Todd, N. B. 1963. The catnip response. Doctoral dissertation, Harvard Univ.
17. Whitten, W. K. 1956. Modification of the oestrous cycle of the mouse by external stimuli associated with the male. *J. Endocrinol.* **13**, pp. 399–404.
18. Whitten, W. K. 1956. The effect of removal of the olfactory bulbs on the gonads of mice. *J. Endocrinol.* **14**, pp. 160–163.
19. Whitten, W. K. 1958. Modification of the oestrous cycle of the mouse by external stimuli associated with the male. Changes in the oestrous cycle determined by vaginal smears. *J. Endocrinol.* **17**, pp. 307–313.
20. Whitten, W. K. 1966. Pheromones and mammalian reproduction. *In* Advances in Reproductive Physiology, Vol. 1 (A. McLaren, editor). Logus and Academic Press, London, pp. 155–177.
21. Whitten, W. K., F. H. Bronson, and J. A. Greenstein. 1968. Estrus-inducing pheromone of male mice: transport by movement of air. *Science* **161**, pp. 584–585.
22. Whitten, W. K., and F. H. Bronson. Role of pheromones in mammalian reproduction. *In* Advances in Chemoreception (J. W. Johnston, D. C. Moulton, and A. Turk, editors). Appleton-Century Crofts, New York. (In press)

BEHAVIORAL AND ELECTROPHYSIOLOGICAL RESPONSES OF MALE RATS TO FEMALE RAT URINE ODORS

D. PFAFF *and* C. PFAFFMANN
The Rockefeller University, New York

INTRODUCTION

LeMagnen[13] and Carr and his coworkers[3,4] have reported that normal male rats approach or investigate the odors emanating from receptive female rats more intensively than they approach odors from nonreceptive females. The sources of the odors which cause this phenomenon have not been identified. Odors from urine communicate information about hormonal states among individuals in groups of mice[22] and in dogs.[1] In the present investigation we used urine odors from estrous and ovariectomized female rats for behavioral and electrophysiological studies. The purpose of the behavioral study was to determine the relative preferences of normal, sexually experienced male rats and castrated, inexperienced male rats for urine odors from estrous and ovariectomized female rats. The purpose of the electrophysiological study was to search for nerve cells which responded to either or both of the female urine odors in a distinctive way.

MATERIALS AND METHODS

Behavioral

The subjects were 19 male Sprague-Dawley rats weighing between 400 and 500 gm., and the sources of urine were female Sprague-Dawley rats, weighing between 250 and 300 gm. All rats were obtained from Charles River Breeding Laboratories, Inc. Ten of the male rats were intact and had been used as breeders at Charles River, and thus had extensive sexual experience. The other nine males had been castrated six weeks before testing and were sexually inexperienced. Among the females, some were intact and were rendered receptive by subcutaneous injections of 0.2 mg estradiol benzoate in 0.2 cc sesame oil 48 hr before the beginning of urine collection, and 0.5 mg progesterone in 0.5 cc sesame oil at the beginning of urine collection. Other females had been ovariectomized six weeks before the experiment and received only injections of sesame oil. The behavioral receptivity of the intact hormone-injected females was confirmed by behavioral testing, and these females are referred to as "estrous" or "receptive." The absence of behavioral receptivity in the ovariectomized, oil-injected females was also confirmed, and these females are referred to as "ovariectomized" or "nonreceptive." Urine was collected over a 24-hr period in standard metabolism cages. It was used when fresh or after storage in a frozen state for 24 hrs.

The situation used for testing responses by males to female urine odors was similar to that of Carr et al.[3] The males were tested in their home cages, $11\frac{1}{2} \times 11\frac{1}{2} \times 17\frac{1}{2}$ in. stainless-steel cages, the floors of which were covered with Sanicel cellulose bedding (Paxton Processing Co., Paxton, Ill.). Males, housed in pairs in the home cages, were removed just before testing, and two glass bottles, $2\frac{1}{2}$ in. diameter and 5 in. long, were placed in the cage. The metal screw top ($2\frac{1}{2}$ in. diameter) of each bottle had been drilled with 16 $\frac{1}{4}$-in. holes, and a Teflon sleeve extended 1 in. out from the top, forming an antechamber into which the male rat could put his nose. Each bottle contained a small amount of clean Sanicel bedding, which had been left clean or soaked with 10 drops (0.5 cc) of urine from ovariectomized or estrous females. The bottles were placed on their sides, with their bottoms in the two front corners of the cage, so that their tops and sleeves extended diagonally toward the middle of the cage.

Each test was 2 min long, starting at the time a male was put back in his home cage with the two bottles in position. Tests as long as 5 min tended to give poorer results. The behavioral response measured in most of the tests was the amount of time the male rat extended his nose into the antechamber formed by the sleeve, *plus* the amount of time that he sniffed with his nose against the sides of the bottle *after* he had first sniffed inside the antechamber. In one series of tests, only the amount of time the male rat extended his nose into the antechamber was measured. Most of the tests were conducted "blind."

On a given day, each male was tested only once, and all males were tested on that day with the same pair of odorants. All tests were done during the middle of the dark phase of an artificially reversed day-night cycle. For the ten days of testing using a given pair of odorants, each of the odorants was located on the two sides of the cage equally often, to eliminate the possible effect of position preferences. All males were tested on all of the odorant pairs used. During tests in which the two bottles contained only clean Sanicel bedding, one was labeled and treated as distinct from the other, to determine whether two otherwise identical bottles would yield average times of investigation which were comparable.

For a given pair of odorants, results for each male were averaged over the ten tests, giving a matched pair of average measurements. The statistical test used to determine if a set of males investigated one of the odorants in the pair significantly more than the other was the Walsh test for two related samples.[20]

Electrophysiological

The subjects were castrated male Sprague-Dawley rats weighing between 200 and 350 gm., obtained from Charles River Breeding Laboratories. The rats were anesthetized with urethane (1.5 g/kg i.p.) and were usually immobilized with Flaxedil (50 mg/kg i.p.). A routine tracheotomy was performed to permit artificial respiration (Rodent Respirator, Harvard Apparatus Co.). Air was drawn through the nose by intermittent application of a vacuum to a polyethylene tube which reached into the back of the nose from the site of tracheotomy. The head of the rat was stabilized in a stereotaxic instrument. After incision and retraction of the skin, the bone dorsal to olfactory bulb or the appropriate cortical region was thinned by drilling. The deepest layer of bone and the dura were then removed carefully with forceps.

Single units were recorded with micropipettes filled with 3 M sodium chloride. Tip diameters were between 1 and 3 μ, and d.c. resistances were between 0.5 and 5 megohms. The reference electrode was clipped to the retracted skin. The micropipette was positioned by a micromanipulator (La Précision Cinématographique), according to the stereotaxic atlas of König and Klippel,[12] except for the olfactory bulb, for which an atlas was not required.

Conventional recording procedures were used. Input from the micropipette was led to a Grass P6 preamplifier equipped with a cathode follower, and then to a Tektronix 502 oscilloscope and an audio monitor. Permanent records of spike discharges were made on magnetic tape (Magnecord model 1048) or on film (Grass Kymograph Camera). The frequency of spike discharges was quantitatively evaluated by use of either an integrating circuit or a spike-frequency histogram generator equipped with a Schmitt trigger and a counting circuit. The outputs of these devices were displayed with an Esterline Angus penwriter or a Grass Model 7 Polygraph.

Air was cleaned with silica gel and activated charcoal and was passed by the nose of the rat at a rate measured with a flowmeter. A Sage infusion pump was used to add measured amounts of odor-saturated air from a syringe to the stream of cleaned air. In each experiment, the concentrations of all the odor stimuli were the same, and were between 1/100 and 1/10 of saturation. Urine samples for use as odor sources were collected from female rats prepared in the way we have described for the behavioral study.

Electrode localization was confirmed by lesioning the tissue at the tip of the pipette (50 microamps d.c., for 50 sec) or by marking with a dye[21] and locating the lesion or mark in stained frozen sections.

The statistical significance of the differences between types of responses in the olfactory bulb and the preoptic area was determined by using the test for the difference between two proportions.[7] Units that did not respond to odors were included in the statistics only when they were recorded from preparations in which at least one other unit did respond.

RESULTS

Behavioral

The results of the behavioral study are summarized in Table I. When the bedding in one of the bottles was soaked with receptive female urine and in the other bottle it was left clean, normal, experienced males investigated the urine odor bottle significantly longer than the clean bottle ($p. < 0.005$). The castrated males did not show a significant difference. The urine odor from nonreceptive female rats did not elicit a differential response from either the normal males or the castrated males. When the odor of urine from receptive females was paired with that from nonreceptive females, the normal males again significantly preferred the odors from the receptive females ($p. < 0.005$) and the castrated males did not. Another series of tests was done with this pair of odors in which only the amount of time the male's nose was inside the antechamber was measured. With this more restricted measure, very low times resulted, and there were no preferences exhibited. The times of sniffing at the top and sides of two clean bottles by the normal males and the castrates did not differ significantly.

TABLE I

MEAN TIME MALE RATS SPENT INVESTIGATING ODOR BOTTLES

(Total sec/rat, during ten 2 min tests)

Odor Bottles Paired A vs. B	Normal Males N = 10		Castrated Males N = 9	
	Time at Bottle A (sec)	Time at Bottle B (sec)	Time at Bottle A (sec)	Time at Bottle B (sec)
Receptive vs. Clean Bottle	222**	131	201	174
Nonreceptive vs. Clean Bottle	119	117	166	140
Receptive vs. Nonreceptive	231**	158	147	175
Receptive vs. Nonreceptive (Antechamber only)	37	36	39	38
Clean vs. Clean	109	104	121	114

** Larger than B, $p < 0.005$.

Two other comparisons were drawn from the data in Table I. The average number of seconds which the normal males spent sniffing at the bottle containing the non-receptive female urine was significantly larger when the other bottle contained urine from receptive females (average = 158 sec) than when the other bottle was clean (average = 119 sec) ($p. < 0.05$). That is, the presence of receptive female urine enhanced sniffing at the nonreceptive female urine odor by the normal males. Moreover, the time the normal males spent sniffing at the clean bottle was greater when the other bottle contained receptive female urine (average = 131) than when the other bottle was also clean (average = 104), suggesting enhancement of sniffing at the clean bottle by the presence of receptive female urine, but this trend did not reach statistical significance ($p. < 0.1$).

The total time normal males sniffed at the tops and sides of the bottles was significantly greater for tests which included the odor of receptive female urine than for tests which did not ($p. < 0.01$).

For the castrated males, the odor of receptive female urine did not enhance sniffing either at the nonreceptive female urine odor or at the clean bottle. Also, the odor of urine from nonreceptive females did not enhance sniffing, either for the normal or for the castrated males.

Finally, when pairs of males were put back in their home cages, one tried to copulate with the other, by mounting and thrusting, after an average of 20 per cent of the tests that included both the odors of estrous and ovariectomized female urine. This occurred significantly less often (4%) after tests in which either of these odors was not present ($p. < 0.01$).

Electrophysiological

The odors of estrous or ovariectomized female rat urine were used during microelectrode recording from 41 olfactory bulb units and 71 preoptic area units in male rats (Table II). Fifty-seven (88%) of the preoptic area units and 28 (68%) of the

TABLE II

RESPONSES BY UNITS IN THE PREOPTIC AREA AND OLFACTORY BULB OF MALE RATS TO RAT URINE AND OTHER ODORS

ODORANT	PREOPTIC AREA				OLFACTORY BULB			
	No. of Units	No. Responsive (%)	Excited	Inhibited	No. of Units	No. Responsive (%)	Excited	Inhibited
Estrous female urine odor	65	57 (88%)	37	20	41	28 (68%)	23	5
Ovariectomized female urine odor	38	33 (87%)	21	12	34	27 (80%)	22	5
Amyl acetate	56	52 (93%)	37	15	37	36 (97%)	32	4
Cineole	33	29 (88%)	20	9	27	26 (96%)	20	6

olfactory bulb units responded to estrous female urine odor ($p. < 0.05$). Thirty-three (87%) of the preoptic area units and 27 (80%) of the olfactory bulb units responded to ovariectomized female urine odor. The over-all percentages of units which responded and the percentages of responses which were excitatory were not significantly different between the two urine odors or between the urine and the nonurine odors. However, there was a tendency toward more frequent responses to the nonurine odors than to the urine odors in olfactory bulb units and no difference in preoptic area units. No units have been found which respond exclusively to either or both urine odors.

Figure 1 shows responses from a preoptic area unit and Figure 2 an olfactory bulb unit which illustrate a difference in responses between these two sites. Units in the preoptic area tended to respond differently to estrous than to ovariectomized female urine odor, but tended not to respond differentially among the nonurine odors. The reverse was true for the responses of units in the olfactory bulb. The data are summarized in Table III. Responses of a unit to two odors were counted as different from each other on the basis of any one of these three criteria: (a) presence vs. absence of

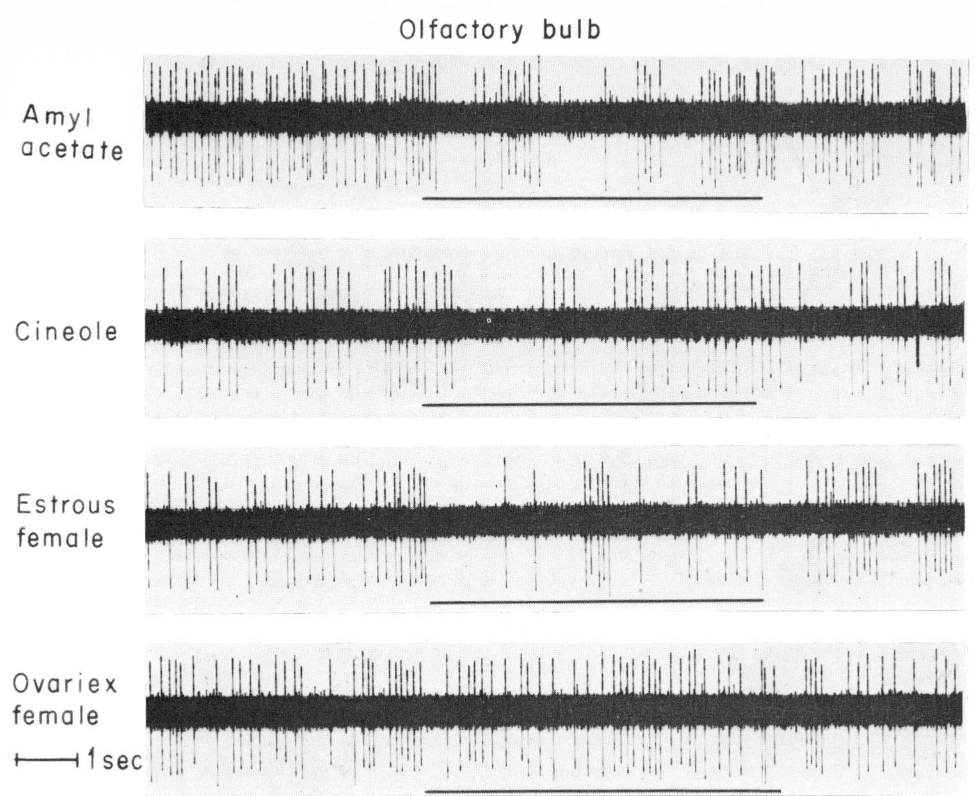

Figure 1 Responses of a unit in the preoptic area to the odors of amyl acetate and cineole were similar to each other, but its response to the odor of estrous female urine was different from its response to urine odor from ovariectomized females.

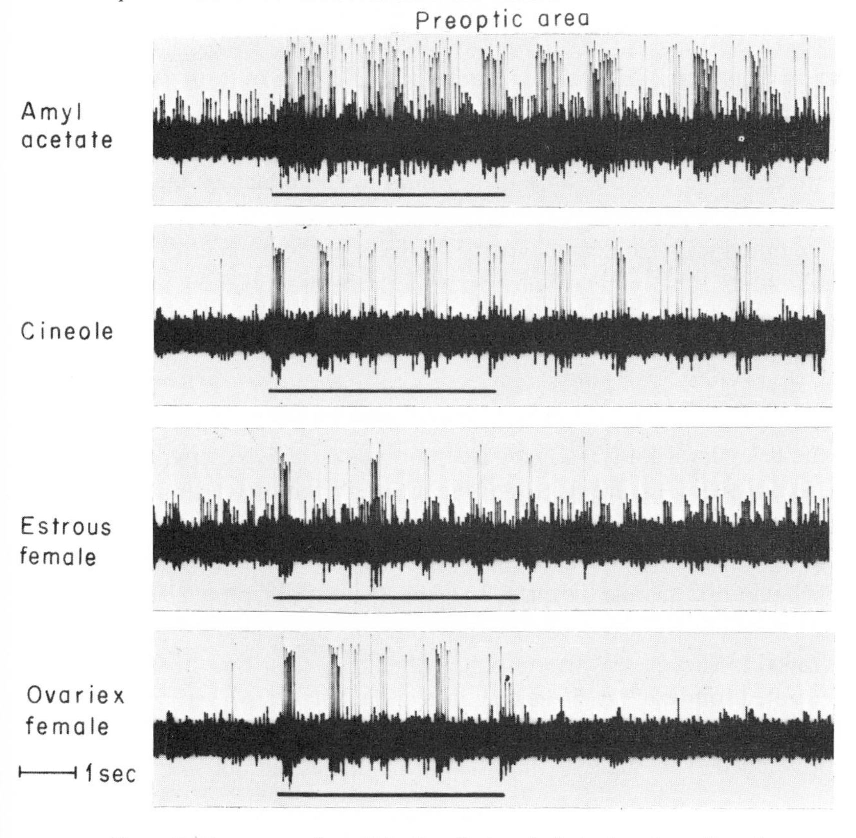

Figure 2 Responses of a unit in the olfactory bulb to the odors of amyl acetate and cineole were different from each other, although responses to urine odors from estrous and ovariectomized females were similar to each other.

TABLE III
PROPORTION OF UNITS RESPONDING DIFFERENTLY TO THE TWO MEMBERS OF AN ODOR PAIR

ODOR PAIR	PREOPTIC AREA Mode of differential response			OLFACTORY BULB Mode of differential response		
	Presence vs. Absence	Excitation vs. Inhibition	Difference in Magnitude	Presence vs. Absence	Excitation vs. Inhibition	Difference in Magnitude
Urine Odors Estrous vs. ovariectomized female rat urine	7 units	3	9	3 units	1	1
	19 total, responding differentially / 33 units tested = (57%)			5 total, responding differentially / 29 units tested = (17%)		
Nonurine odors Amyl acetate vs. Cineole, Benzaldehyde vs. Cineole and Amyl acetate vs. Benzaldehyde	9 units	4	4	9 units	3	20
	17 total, responding differentially / 42 units tested = (40%)			32 total, responding differentially / 42 units tested = (76%)		
Urine vs. Nonurine odors Estrous or ovariectomized female rat urine vs. Amyl acetate, cineole or benzaldehyde	23 units	24	50	28 units	15	43
	97 total, responding differentially / 137 units tested = (71%)			86 total, responding differentially / 115 units tested = (75%)		

* Different from preoptic area, $p < 0.01$.

response; (b) excitation vs. inhibition; or (c) consistent difference of response magnitude, quantitatively determined. Using these criteria, we found that the proportion of units responding differentially to estrous vs. ovariectomized female urine odor was significantly greater in the preoptic area (57%) than in the olfactory bulb (17%) (p. < 0.01), but that the proportion of units responding differently to the members of pairs of nonurine odors was significantly greater in the olfactory bulb (76%) than in the preoptic area (40%) (p. < 0.01) (Figure 3). In both the preoptic area and in the olfactory bulb, a high percentage of units discriminated between the urine odors and the nonurine odors.

DISCUSSION

The results of the behavioral study, showing that normal, sexually experienced male rats selectively investigate odors from receptive female rats, but that castrated males do not, confirm the previous reports of Le Magnen[13] and Carret al.[3,4] The present investigation shows for the first time that urine odors are sufficient to give this effect.

Some of the behavioral data raise the question of whether the differential behavioral responses by the normal males were more exploratory or more hedonic in nature. Normal males failed to show a selective response when only their time of sniffing at the most concentrated source of odor (the openings in the bottle tops) was measured, although they did respond selectively when time of sniffing along the sides of the bottle was also included. This suggests that the normal males did not simply prefer

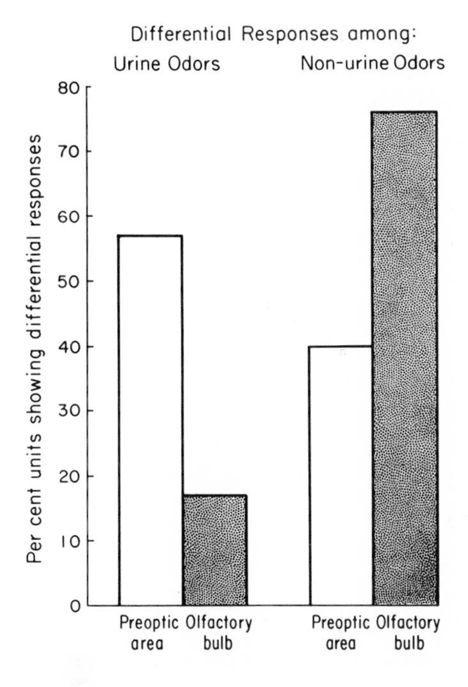

Figure 3 Bar graph showing greater differential responses among urine odors by units in preoptic area than in olfactory bulb, compared with responses to nonurine odorants.

the odor of receptive female urine, in the hedonic sense, but instead that the odor of receptive female urine enhanced the response of exploration with sniffing along the sides of the bottle. The total time of sniffing by the normal males was greater in tests which included receptive female urine than in tests which did not. Moreover, the presence of odor from receptive female urine enhanced exploration with sniffing at the other bottle in the cage, whether it contained odor from nonreceptive female urine or just clean Sanicel bedding. Finally, the presumed goal of an exploration triggered by odor from receptive female urine—the performance of the copulatory response pattern—was activated by the combined presence of receptive and non-receptive female urine odors during the test.

The results of LeMagnen,[13] Carr et al., [3,4] and of the present behavioral study indicate that odors of urine from receptive and nonreceptive female rats are coded differently in the nervous system of the normal male rat. The electrophysiological results reported above show that cells in the olfactory bulb and in the preoptic area respond to the two urine odors, and that differential responses to the two odors occur more frequently in the preoptic area. Comparison of the behavior of castrated and normal males in LeMagnen's, Carr's, and this behavioral study suggests that testosterone can affect differential electrophysiological responses to the two urine odors by nervous-system mechanisms concerned with tendencies to approach or avoid, although previous studies have shown that mechanisms concerned with detection and discrimination of the two odors are apparently not sensitive to testosterone.[2,5] Intraperitoneal injections of testosterone during the present electrophysiological experiments frequently affected the responses of units in both the olfactory bulb and in the preoptic area to the urine odors,[15a] but the testosterone injections did not consistently cause an increased (or decreased) distinction between unit responses to the receptive and nonreceptive odors. Thus, the mechanism of the testosterone effect on the selective approach-avoidance tendencies by the male rat in response to female urine odors has not yet been elucidated.

Although many responses to the odors of estrous and ovariectomized female urine were recorded from units in the olfactory bulb and preoptic area, no units were found which responded exclusively to either or both urine stimuli. Failure to find cells with such specific responses to the odors of natural products could be due to an unexplained difficulty in isolating such cells with the recording techniques used, to an unfortunate selection of recording sites or stimuli, or to poor preservation of the natural products used. Nevertheless, the results gathered thus far suggest that sex-related stimuli are coded in a less specific, all-or-none manner in the rat brain than in the insect nervous system, and that in the rat brain the specificity which does exist becomes apparent only after more processing than is required in the insect nervous system, in which marked specificity may be demonstrated at the receptor level. Instead of exclusive responses to sex attractants, as Schneider[19] has recorded from olfactory receptors in moths, units in the rat brain seem to distinguish urine from nonurine odors (in the olfactory bulb and preoptic area) and between the two urine odors themselves (more in the preoptic area than in the olfactory bulb) by subtler variations in the direction and magnitude of response. Thus, the differential aspects of both behavioral responses (shown here and in previous studies) and electrophysiological responses of the male to odors from the female seem to have a more probabilistic character in rats than insects.

This phylogenetic comparison has parallels in taste and in the visual system. In taste, recording from receptors in the blowfly[6] has revealed greater specificity than has been discovered at any stage of the taste pathways in the rat or the cat.[16,17,18] In vision, recording from retinal ganglion cells of frogs[14] and pigeons[15] has shown highly constrained stimulus requirements at this early stage of visual information processing, whereas recording from higher animals such as cats[8,10] and monkeys[9,11] has shown that in these animals such specific stimulus requirements do not appear until later stages in the visual pathways.

SUMMARY

1. Normal, sexually experienced male rats spent significantly more time investigating bottles containing the urine odor of estrous female rats than bottles containing the urine odor of ovariectomized rats. Castrated male rats did not show this differential response. The odor of estrous female rat urine enhanced exploration with sniffing primarily, but not exclusively, in the region of the estrous urine odor by the sexually experienced male rats.

2. A high percentage of units in the olfactory bulb and in the specific area responded to urine odors from estrous or ovariectomized female rats. No units have been found which responded exclusively to either or both urine odors. The proportion of units

responding differentially to estrous vs. ovariectomized female urine odor was significantly higher in the preoptic area than in the olfactory bulb, but the proportion of units responding differently to the members of pairs of nonurine odors was significantly greater in the olfactory bulb than in the preoptic area.

ACKNOWLEDGMENTS

This work was supported by NSF grant GB4198X to Carl Pfaffmann and was performed during Donald Pfaff's tenure as an NSF post-doctoral fellow. Mrs. Louise Musser provided valuable technical assistance.

REFERENCES

1. Beach, F. A., and R. W. Gilmore. 1949. Response of male dogs to urine from females in heat. *J. Mammal.* **30**, pp. 391–392.
2. Carr, W. J., and W. F. Caul. 1962. The effect of castration in the rat upon the discrimination of sex odors. *Animal Behav.* **10**, pp. 20–27.
3. Carr, W. J., L. S. Loeb, and M. L. Dissinger. 1965. Responses of rats to sex odors. *J. Comp. Physiol. Psychol.* **59**, pp. 370–377.
4. Carr, W. J., L. S. Loeb, and N. R. Wylie. 1966. Responses to feminine odors in normal and castrated male rats. *J. Comp. Physiol. Psychol.* **62**, pp. 336–338.
5. Carr, W. J., B. Solberg, and C. Pfaffmann. 1962. The olfactory threshold for estrous female urine in normal and castrated male rats. *J. Comp. Physiol. Psychol.* **55**, pp. 415–417.
6. Dethier, V. G. 1968. Chemosensory input and taste discrimination in the blowfly. *Science* **161**, pp. 389–391.
7. Freund, J. E. 1952. Modern Elementary Statistics. Prentice-Hall, Englewood Cliffs, N.J., p. 208.
8. Hubel, D. H. 1960. Single unit activity in lateral geniculate body and optic tract of unrestrained cats. *J. Physiol. (London)* **150**, pp. 91–104.
9. Hubel, D. H., and T. N. Wiesel. 1960. Receptive fields of optic nerve fibres in the spider monkey. *J. Physiol. (London)* **154**, pp. 572–580.
10. Hubel, D. H., and T. N. Wiesel. 1962. Receptive fields, binocular interaction and functional architecture in the cat's visual cortex. *J. Physiol. (London)* **160**, pp. 106–154.
11. Hubel, D. H., and T. N. Wiesel. 1968. Receptive fields and functional architecture of monkey striate cortex. *J. Physiol. (London)* **195**, pp. 215–243.
12. König, J., and R. A. Klippel. 1963. The Rat Brain. Williams and Wilkins, Baltimore, Md.
13. LeMagnen, J. 1952. Les phénomènes olfacto-sexuels chez le rat blanc. *Arch. Sci. Physiol.* **6**, pp. 295–331.
14. Lettvin, J. Y., H. R. Maturana, W. S. McCulloch, and W. H. Pitts. 1959. What the frog's eye tells the frog's brain. *Proc. Inst. Radio Engrs.* **47**, pp. 1940–1951.
15. Maturana, H. R., and S. Frenk. 1963. Directional movement and horizontal edge detectors in the pigeon retina. *Science* **142**, pp. 977–979.

15a. Pfaff, D. W., and C. Pfaffmann. 1969. Olfactory and hormonal influences on the basal forebrain of the male rat. *Brain Research.* (In press)

16. Pfaffmann, C. 1941. Gustatory afferent impulses. *J. Cell. Comp. Physiol.* **17**, pp. 243–258.
17. Pfaffmann, C. 1955. Gustatory nerve impulses in rat, cat, and rabbit. *J. Neurophysiol.* **18**, pp. 429–440.
18. Pfaffmann, C., R. P. Erickson, G. P. Frommer, and B. P. Halpern. 1961. Gustatory discharges in the rat medulla and thalamus. *In* Sensory Communication (W. A. Rosenblith, editor). Wiley, New York, pp. 455–473.
19. Schneider, D. 1963. Electrophysiological investigation of insect olfaction. *In* Olfaction and Taste I (Y Zotterman, editor). Pergamon Press, Oxford, etc., pp. 85–103.
20. Siegel, S. 1956. Nonparametric Statistics for the Behavioral Sciences. McGraw-Hill, New York.
21. Thomas, R. C., and V. J. Wilson. 1965. Precise localization of Renshaw cells with a new marking technique. *Nature* **206**, pp. 211–213.
22. Whitten, W. K. 1969. Mammalian pheromones. *In* Olfaction and Taste III (C. Pfaffmann, editor). Rockefeller Univ. Press, New York, pp. 252–257.

SOCIAL COMMUNICATION BY CHEMICAL SIGNALS IN FLYING PHALANGERS

(*Petaurus breviceps papuanus*)

THOMAS G. SCHULTZE-WESTRUM

Zoologisches Institut der Universität München, Munich, Germany

INTRODUCTION

The causal relations between intraspecific communication, the social system as a whole, and the population structure of flying phalangers (*Petaurus breviceps papuanus*) were studied experimentally both in the laboratory and in the animal's natural habitat in tropical New Guinea.[1,3,4] The phalangers are nocturnal, arboreal, and live in social nesting communities consisting of up to six adult males and females together with their young. Intraspecific agonistic behavior is displayed only between communities. One to two dominant males per community perform most of the social activities, especially odor distribution, territory maintenance, territory patroling, aggression against foreign community members, and mating.

Petaurus communicates by chemical, acoustic, visual, and tactile signals. Experiments demonstrate, however, that only through odors is an individual- and community-specific communication achieved. Vocalizations and visual signals are only used to transmit anonymously the physiological states of an individual, not its individuality (unpublished). The chemical communication of *Petaurus* can also be anonymous, in the same way.

Characteristics of the *individual-* and *community-specific* means of chemical communication in *Petaurus* are: (1) Each male sender supplies at least three, and females at least two, individually differentiated odors, which have a great range of functions. (2) The specificity of function of the three male odors in the receiver does not depend on the originating odor-producing organ, but on the *social attitude* of the receiver to the sender; all these odors have the same range of functions. (3) This social relationship is community-specific and depends upon former chemical communication and the (central) processing of odor signals by learning or habituation. (4) Chemical communication in *Petaurus* can have both a releaser and a primer effect, which means that the chemical signals not only elicit immediate motor reactions, but cause relatively slow physiological and anatomical changes as well, e.g., in size and secretory activity of scent glands, body weight, rank position, and aggressiveness. (5) A last feature is the frequent occurrence of multiple social ties between conspecifics by means of odor signals. For instance, a young animal receives information that its mother is a conspecific, a mother, and its own mother. Another example is that an adult member of the community is known by odor signals as a conspecific, a group member, and an individual. All these functions were demonstrated separately by experiments.

THE ODORS

Adult males possess at least three body regions that permanently produce odors: a frontal gland (Figure 1), a sternal gland, and the urogenital-anal region. In the latter are three sources of odor: urine, secretions of proctodaeal glands, and those of paraproctal glands, which reach the body surface through the joint urogenital-anal outlet characteristic for marsupials. (The paraproctal glands produce a secretion that apparently is used only in interspecific communication.) The saliva also contains scent substances. Females do not possess frontal and sternal gland organs, but have scent glands in the inner skin of their pouches, which are active from the time prior to giving birth until the offspring leaves the pouch. Females also have an individual body odor which is permanently present. (For the histology of the glands, see[1])

The frontal and sternal glands, and the urogenital-anal region of a male produce different odors. This was demonstrated by experiments in which an animal sniffed at filter paper scented with frontal or sternal secretion, or urine of a completely un-

Figure 1 Male *Petaurus*, one year old, and female, more than five years old. Note the frontal gland organ on the male's forehead.

familiar male. When the frontal secretion was offered several times it became known to the sniffing animal; this resulted in habituation and a reduction of sniffing activity per unit time. Then, instead of the frontal secretion, the odor from the sternal gland was offered. This caused a rise in sniffing activity, and proved that the animal made a distinction between frontal and sternal odor. Anal odor was tested in the same manner (Figure 2). To exclude the possibility that, for instance, urine was transferred to the

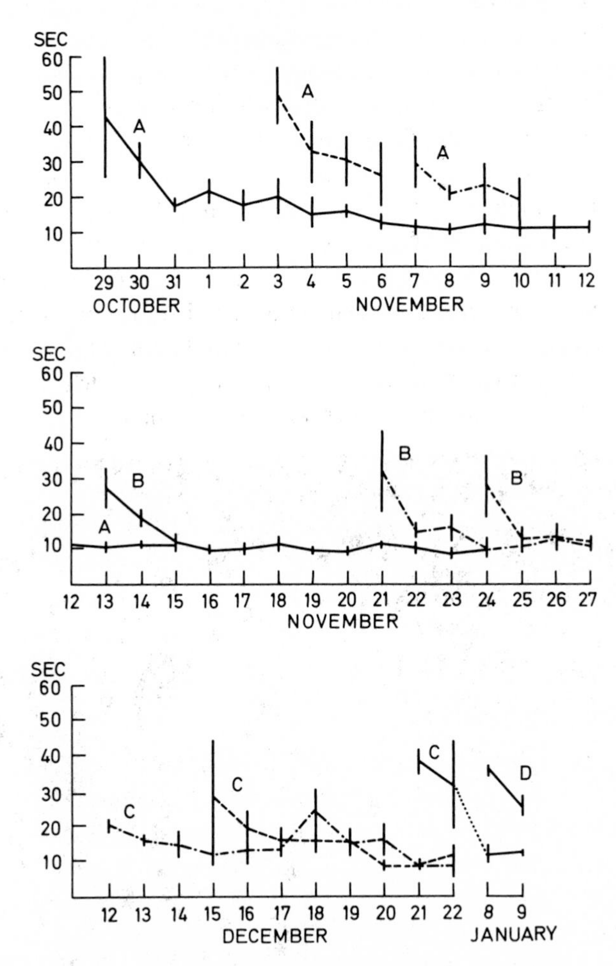

Figure 2 Registration of sniffing activity on scented filter paper, to demonstrate the odor differentiation between frontal, sternal, and urogenital-anal region. The scent was taken from foreign males which were completely unknown to the sniffing animal prior to the first offer. Note the high sniffing activity when an odor was offered for the first time. The curves indicate the average sniffing time of two experimental males per day (in sec/5 min experimental time unit); N = number of experiments per day with odor from the same body area of one male. The sniffing times of both males per day were added up and the sum divided by the number of experiments, which were: when one odor was offered, $N = 2$ (on the first day of offering), or $N = 4$ (on the following days); when two odors were offered by turns, $N = 6$ (on the first day), or $N = 8$ (on the following days). A, B, C, D = individual male donors of odor; ——— odor from frontal gland; — — — odor from sternal gland; —·—·—· odor from urine.

sternal region but not to the frontal region and thus caused an odor differentiation between the two, an experimental animal was habituated to frontal odor and urine odor before sternal odor was offered.

In addition to distinctly different body odors of one male there exists an odor differentiation between the same scent-producing body regions of different animals. This is the basis of individual- and community-specific chemical communication. It was proved by an extensive series of experiments for the frontal and sternal glands, the urogenital-anal region, and the pouch glands. The odor of saliva is at least community-specific. Parameters used in the experiments were sniffing activity, marking behavior, intraspecific fighting behavior, and social attractiveness. All these odor differentiations are independent of the food consumed (although, generally speaking, food may possibly influence scent quality) or odors of the local environment, and do not change in their composition, at least for several days.

Nothing is known about the chemical nature of the scent substances. One speculative explanation of the individual differentiation is that the secretion of the frontal gland, for example, consists of several components which are present in all normal males, but whose quantitative combination is individual-specific. If there are, to give a theoretical example, ten distinct scent substances in the frontal gland secretion, these ten are present in each male and indicate species specificity and other anonymous information (see next section). The quantitative ratio of these substances, however, is possibly individual-specific, and might provide means for the creation of an individually differentiated odor quality.

The scent substances are distributed in the territory by six distinct patterns of active marking behavior. The body surface of conspecifics is also marked actively (Figure 3). By this behavior and additional passive scent transfer in the nest, the community-specific odors are created.

THE INTRASPECIFIC FUNCTIONS OF ODORS

Odors in *Petaurus* were classified according to their functions. There are three main groups. (1) *Species odors*, which act anonymously and have, according to separate experiments, the following functions: indication of species membership; sexual distinction; sexual stimulation; and anonymous mutual mother-young attraction. (2) *Individual odors*, which allow mutual recognition of adult individuals, of their own young, and of the mother by the young. (3) *Community odors*, distributed both in the territory and on the body surface of community members and originating mainly from the dominant males. Their functions will be discussed later (Table I).

Both individual and community odors are based on the individual differentiation of odors, whereby there is no function difference between odors from the frontal, sternal, and urogenital-anal regions of an individual male. This is striking when compared with insect pheromones, where as a rule odors of distinct glandular secretions have distinct functions. In *Petaurus* it is primarily the central processing in the receiver that determines the significance of an individually differentiated odor signal by allowing recognition of the social relationship with the sender. In this context it

TABLE I

THE SOCIAL FUNCTIONS OF COMMUNITY ODORS

(produced mainly by the dominant males) in male receivers of the producer's community (= associated), or of a foreign community (= not associated). After Schultze-Westrum,[1,3] Schultze-Westrum and Braun,[4] and unpublished material.

Induced motor pattern or influenced physiological disposition	A in case of association[(a)] to the community odor	B in case of no association[(a)] to the community odor
1. intraspecific agonistic behavior	not released	released (if signal originates directly from an individual)
2. potential aggressiveness[(b)]	*kept low* (primer effect)	
3. effective aggressiveness[(c)]	raised (immediate effect of territorial scent marks)	lowered
4. rank position	kept low	
5. mating activity	kept low	apparently not influenced
6. scent production	kept low	increased
7. marking behavior	kept low	stimulated
8. patrol activity in territory	kept low	stimulated
9. increase in body weight	kept low	?

(a) See p. 274.

(b) Physiological disposition for intraspecific fighting behavior depending upon individual general constitution, age, *rank*, and sex (4, above).

(c) Based on 2, above, but additionally determined by the momentary general physiological condition, and the momentary *territorial maintenance* (odor-dependent).

should be stressed that recognition of membership in the animal's own species by species odor or another signal is essential for individual and community-specific communication (groups 2 and 3, above). Individual differentiation of the odors without simultaneous indication of species membership has no social significance.

One experiment to demonstrate that frontal-, sternal-, and urine-differentiated odors of one male elicit the same responses was the following. Nest young of about 95 days of age can only distinguish communities, but do not yet know the members of their own community individually, except for their own mothers. They display defense behavior against conspecifics bearing the odor of a foreign community. If a female of the young animal's community was artificially scented with foreign frontal odor, it was identified by the young as a member of a foreign community, and the young displayed intense defensive motor reactions. If urine or sternal odor from the same foreign male was applied instead of frontal odor, exactly the same response was elicited. In controls, when the female was not scented with foreign odor, the young displayed no defense.

A series of physiological processes that determines the nature of a response undoubtedly takes place from the time an individually differentiated odor is produced in a gland until the response occurs.

1. Production of the scent substances in the sender; or scent produced in one individual and transferred to another, who now becomes a sender, which changes the significance of the scent. Individual odors of dominant males, for instance, become community odors after transfer to the body surface of conspecifics or into the territory.
2. Perception by the receiver.
3. Central learning processes and *habituation.* Former learning of a signal results in *recognition.* Central processes after repeated perception of a signal may also lead to "*association*" to the signal (see next section and Table I), which results in a change of the releaser and primer effects.
4. Influencing physiological dispositions, e.g., aggressiveness, and primer effects on growth, glandular activity, and other processes.
5. Release of motor reactions (e.g., aggression).

Physiological disposition is regarded as a quantitative measure of specific excitation. A change of its level (increase or decrease) causes a qualitative change in the motor pattern which is peculiar for the disposition, e.g., in intraspecific fighting behavior. Influencing a physiological disposition by odor signals, however, does not necessarily lead to a motor reaction. Territorial scent marks of the receiver's own community, for instance, cause an increase in aggressiveness, and foreign scent marks a decrease. But motor reactions are released only if additional signals have been perceived indicating the presence of an individual. They can be, according to the level of aggressiveness, defensive, aggressive, or of intermediate stage (threat).[4]

Figure 3 Active marking of a conspecific. The male (on the right) holds the female's chest region with his forepaws and rubs his frontal gland on the fur of the female's sternal region. The same behavior is displayed by females on males. In that case, sternal secretion is transferred from the male's sternal gland to the female's forehead.

Levels of dispositions are influenced by a complex system of social factors (in addition to hormones and other effects). For instance, experiments show that the effects of odor stimuli that cause intraspecific fighting can be counteracted by those that cause sexual excitation. A male that has received signals indicating the presence of both a foreign male and an estrous female first copulates with the female and only afterwards displays aggression against the rival.

The responses to an odor signal may be quite different, depending on the degree of habituation and the kind of recognition (point 3, above). Experiments show that other social factors determining the kind of response to an odor signal are the receiver's age, sex, and rank position. The range of functions of community odor (frontal, sternal, or anal) in male receivers is demonstrated in Table I. This Table also shows that the type of response depends upon the social attitude of the receiver to the sender.

The results in Table I are compiled from several series of experiments and observations both in the laboratory and in the field: elimination of acoustic, visual, and tactile signals as possible releasers; artificial removal or transfer of community odor in the territory or on the body surface of conspecifics; artificial overpopulation, resulting in an excessive supply of community and individual odors (to study their inhibiting effects upon community members); removal and reintroduction of dominant males, which produce most of the community odors at normal population density; other changes in group composition; field observations on rank position in relation to glandular activity, body weight, and aggressiveness; registration of marking activity in own, foreign, and neutral cages; and observation of patrol and mating activities of dominant and subordinate males in groups of different social composition.

ASSOCIATION

Animal species that form closed communities distinguish basically two classes of conspecifics of the same age and sex—individuals that belong to the animal's own community, and foreigners. The process leading to a social attitude characteristic for intracommunity relations is here termed *association.* In *Petaurus*, association takes place during postembryonic development and is based on the processing of odor signals. The young associates to the community odor of its mother's community and the adults associate to the odors of the new young community member. Association is completed noticeably on the 92nd day, when nest young distinguish between communities. Between adults it can only be induced artificially by forcing two conspecifics (two males or two females) of different communities to use one common nest synchronously over a period of several days. Continuous stay in a common cage (not nest) does not lead to association. The repeated central processing of a community odor that causes association does not effect a gradual variation in the response to the signal in such a way that the releaser value increases or decreases; rather it results in a qualitative change of intraspecific function, as shown in Table I (A, B).

Association was demonstrated quantitatively by registering sniffing activity on filter paper scented with frontal gland secretion. Figure 4 shows schematically the decrease of sniffing activity with habituation and association. It should be noted

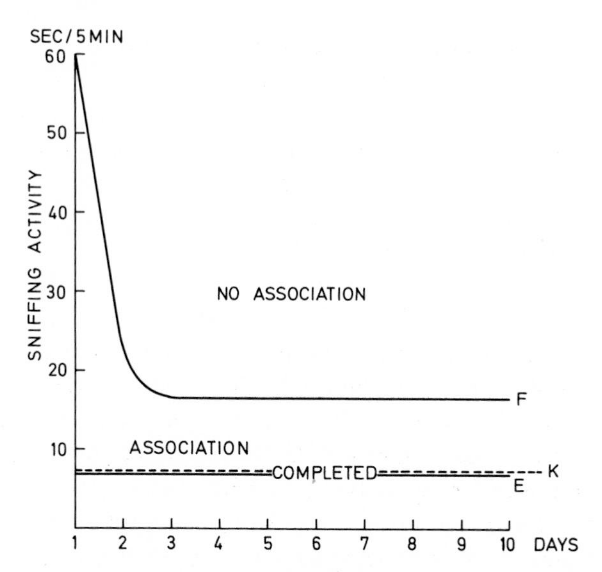

Figure 4 Dependence of sniffing activity on scented filter paper upon social relation of receiver to donor of odor is shown schematically. For further explanation, see section on "association." The number of days is chosen arbitrarily.

that the levels do not indicate the time necessary for odor identification (two conspecifics identify each other within 2 sec), but rather the degree of "excitation." When a completely unknown odor from a foreign community is offered, the sniffing activity is very high (up to 60 sec, see Figure 2, October 29). After about six experiments the sniffing activity reaches a level (F) that remains very stable in further experiments. Examples: F 1 (average for four male donors of odor) 17.4 $\pm$ 1.0 sec/5 min ($N = 44$ experiments); F 2 (average for two other male donors, otherwise the same experimental conditions as in F 1) 15.5 $\pm$ 0.7 sec/5 min ($N = 66$ experiments). This horizontal level F indicates the degree of excitation characteristic of a *known* male odor from a *foreign* community. Even continuous offering of the foreign odor to the experimental animal for several weeks by keeping both donor and receiver in a common cage, but with no common nest, does not lead to a lowering of sniffing activity below the level F; there is no further decrease of excitation and there is no association. Odor from the sniffing animal's own community, however, causes a sniffing activity of only 6.8 $\pm$ 0.5 sec/5 min ($N = 44$ experiments). This is the level characteristic of completed association (E). Between F and E there is a strong barrier, which can only be overcome, as was mentioned previously, in the natural course of postembryonic development, or compulsorily by the "common-nest procedure."

This barrier between F and E on the one hand and the qualitative change of odor function accompanying the transition from F to E (see Table I, nonassociated, associated) indicates that association is not merely a high degree of habituation. Thus one cannot describe association as the result of gradual progression from no habituation to medium habituation (level F) to full habituation (level E). Central processing of an unknown nature must be involved for transition from F to E. Certainly, however, association is correlated with full habituation, because sniffing activity on unscented filter paper (level K: 7.0 ± 1.4 sec/5 min, $N = 24$ experiments) is almost identical with level E. Despite the full habituation to odor from the animal's own community, the signal still is recognized (for instance, see the inhibiting effect in Table I). At the present stage of knowledge it is impossible to define the terms habituation and recognition satisfactorily.

INHIBITION BY COMMUNITY ODORS

The inhibiting effect of chemical conspecific signals upon community members (Table I) needs special attention. Normally the community odor originates from the dominant male, and keeps the social activities of subordinate, associated males at a low level. Its absence after removal of the dominant male (or males) results in an increase of marking and patrol activity in the territory, scent production, mating activity, potential aggressiveness, and body weight in other males of the community. These processes are reversible by reintroducing the former dominant male (males). Rank order is established by the distribution of community odors without individual encounters or fights. Usually the older males are the dominant ones.

In case of artificial overpopulation (in captivity), the total quantity of odor distributed by "subordinate" males is great enough to have an inhibiting effect upon "dominant" males, also, so that inhibition is present in all males of the community. Therefore overpopulation results in a general quantitative decrease of social activities, *including potential aggressiveness* (Schultze-Westrum, unpublished, see also[2]).

In *Petaurus* the main social factors that induce social pressure and may lead to stress are known; they are the individually differentiated odors (individual and community odor signals). The excitation caused after perception of such odors can be determined quantitatively by recording sniffing activity (see previous section). But there is no simple correlation between the level of excitation and the degree of social pressure: those odors to which the individual is associated inhibit social activities, although these odors cause the lowest level of excitation (level E).

In animals such as *Petaurus* that form communities, there is no linear correlation between population density and intensity of social pressure which may cause stress; members of the animal's own community exert quite a different kind of influence than do other conspecifics (see Table I).

The effectiveness of odor signals as stressors is modified not only by the social relation of the receiver to the sender, but also by other social factors, e.g., the maintenance of a territory. There is no breakdown of the population structure as long as territories exist.

ONTOGENY OF CHEMICAL COMMUNICATION

The pouch odor of a mother acts as an attractant for newborn *Petaurus*. Communication at this stage is anonymous—that is, the young do not distinguish between their own and other mothers of the same physiological state, and mothers will adopt any young of the same age. Only when the young begin to leave the pouch temporarily (on about the 74th day) do mutual individually differentiated relations by odor signals become noticeable. Then young are more attracted by their own mothers, and an interchange of young becomes impossible. However, other mothers still remain more attractive than do other conspecifics. The young at this age make no distinction between community members and other conspecifics, and are attracted by species odor of all conspecifics. The species odor of the young prevents it from being regarded as prey or otherwise foreign to species. Foreign adult individuals are normally not aggressive towards young up to an age of about 92 days. At this important stage of development, the young are able to distinguish communities and display defense against conspecifics bearing an odor of a foreign community. Only later do the young phalangers learn to distinguish individuals (other than their own mother). Behavioral differences depending on sex do not occur before the young become independent of the mother (on about the 110th day). Intraspecific fighting behavior, for instance, is displayed towards members of same and opposite sex. But after sexual maturation of the receiver, species odors indicating that a (foreign) donor is of the opposite sex prevent the release of fighting. The earliest visible secretion of the frontal gland was found in a male 153 days old.

From the 62nd day onwards young *Petaurus* produce a strong odor from the secretion of the paraproctal glands in the anal region. This odor is also produced by adults of both sexes in case of danger. No intraspecific function for this odor was found, but there is evidence that it is used in interspecific communication.

ACKNOWLEDGMENTS

We wish to thank Professor Dr. D. Schneider, Dr. Katherine Ralls, Mrs. Helene J. Jordan, and Dr. R. Loftus for their valuable comments on the manuscript.

REFERENCES

1. Schultze-Westrum, Th. G. 1965. Innerartliche Verständigung durch Düfte beim Gleitbeutler *Petaurus breviceps papuanus* Thomas (*Marsupialia, Phalangeridae*). *Z. Vergleich. Physiol.* **50**, pp. 151–220.
2. Schultze-Westrum, Th. G. 1967. Biologische Grundlagen zur Populationsphysiologie der Wirbeltiere. *Naturwissenschaften* **54**, pp. 576–579.
3. Schultze-Westrum, Th. G. The population structure of the flying phalanger *Petaurus breviceps papuanus* (*Marsupialia*). (In preparation)
4. Schultze-Westrum, Th. G., and B. Braun. Social dependence of intraspecific aggression in *Petaurus* (*Marsupialia*). (In preparation)

BEHAVIORAL CHANGES IN PIGEONS AFTER OLFACTORY NERVE SECTION OR BULB ABLATION

B. M. WENZEL, P. F. ALBRITTON, A. SALZMAN, *and* T. E. OBERJAT

Department of Physiology and Brain Research Institute, UCLA School of Medicine, Los Angeles, California

Mrs. Albritton's present address is F.P.O. Box No. 2, Seattle, Washington

Dr. Salzman's present address is Hospital Pirovano, Sala 9, Monroe 3555, Buenos Aires, Argentina

INTRODUCTION

The possibility that the primary olfactory system serves a wider function than simply the processing of information about odorous stimuli has been in the literature at least since Herrick's speculations in 1933.[5] Supporting evidence has been scattered and scanty, but it has been suggested that the olfactory bulbs may be involved in such functions as osmoregulation,[8] aggression,[4] and certain aspects of visual discrimination learning.[11]

In the last case, we found that the main effect of removing the olfactory bulbs from pigeons, or of sectioning their olfactory nerves, could be seen during those phases of training that might be termed orientation, or adaptation, rather than during acquisition of visual discrimination behavior. Thus, the experimental pigeons were very much slower than control birds in approaching and eating from the food hopper during the adaptation period before training began, and they were also relatively slow in initiating pecking on the response key during response shaping. Once these adjustments to the situation had been completed, they performed the discrimination tasks with no more errors than the birds in the two control groups, which had sustained sham lesions and bilateral lesions in the hyperstriatum, respectively.

Further experiments were then undertaken to extend these observations. The results presented in this paper were collected in experiments that were designed to examine various aspects of arousal behavior of pigeons with lesions in the primary olfactory system. An automatic shaping procedure for the key-pecking response was substituted for the hand-shaping procedure we had used earlier, gross motor orientation toward the hopper was studied, and heart rate was recorded during repeated presentations of a visual stimulus.

All of the data reported here were obtained on the same four groups of pigeons. Eleven control subjects were divided into two groups, one of which received only a sham lesion, while the other sustained the unilateral loss of an amount of tissue from the surface of the hyperstriatum that was approximately equivalent to the total size of the olfactory bulbs. Nine experimental pigeons were divided into two groups according to whether their olfactory bulbs had been removed or their olfactory nerves sectioned. Some birds in each group had been born and raised in a breeding colony of racing homing pigeons on the roof of the UCLA vivarium; the rest were "public square" pigeons obtained from a local supplier. The birds were housed individually in adjacent cages in the same room.

Surgery was done under Equithesin anesthesia (2 cc/kg of body weight) and local application of procaine. A small jeweler's bit was used to section the portion of the olfactory nerves that runs in a thin bony sheath between the orbits just anterior to the olfactory bulbs, and also to destroy the olfactory bulbs by drilling rapidly through the overlying venous sinus and through the bulbs. Hemòstasis was induced immediately with Gelfoam. The hyperstriatal ablations were made by suction under direct vision. The sham-lesioned group experienced anesthetization and cutting and suturing of the scalp, but the skull was not opened. At the end of the experiments, the pigeons were anesthetized and perfused through the heart with normal saline followed by a 10 per cent buffered formalin solution. After fixation, the brains of the birds with olfactory-system lesions were sectioned and stained. Damage to the olfactory nerve was verified by degeneration in the olfactory bulb, as described previously.[11]

ORIENTING BEHAVIOR

The first experiment was concerned with an analysis of overt motor reactions to each of the separate components of food-hopper presentation, viz., the sound of the solenoid, the onset of a light over the hopper, and the presence of grain in the hopper. The general procedure consisted of placing each bird individually in the same Grason-Stadler test chamber we had used before. This was located inside a sound-shielded Industrial Acoustics enclosure, and we observed the birds' activity by closed-circuit television. Descriptions of behavior were recorded according to a standardized checklist.

On the first three days, each bird was put into the box for 5 min and allowed to explore with no stimuli presented. On the fourth day, 5 min after the bird was put into the box, the washed hopper was presented 11 times with no light coming on above it and with no grain inside—i.e., the only stimulus was the sound of the hopper. The reactions of the control and experimental pigeons to the first of these auditory stimuli were sharply distinguished. None of the experimental birds approached the end of the box where the hopper was located, and eight of the nine either turned away from the source of the sound if they had been facing in that direction or remained stationary. Seven of the 11 control birds, on the other hand, either looked toward the sound or actually moved toward it, and only three moved away. These differences disappeared with repeated presentations of the hopper noise, so that the four groups met a criterion of adaptation to the sound (no visible reaction on six out of 10 successive presentations) in the same number of sessions, on the average. The contrasting effects of the initial presentation may well have been manifestations of opposing central processes that are involved in the control of eventual approach and eating from the hopper under the standard experimental procedure, and that were differentially emphasized as a result of the olfactory lesions.

AUTOMATIC RESPONSE SHAPING

In the second experiment, begun about one month after completion of the first, key-pecking for grain was initiated by means of an automatic shaping procedure similar to that described by Brown and Jenkins.[3] The same test chamber was used and all programing and recording was automatic. One member of the sham-lesioned group died between the two experiments. The birds, at 80 per cent of free-feeding body weight, were first trained individually by a standardized procedure to eat promptly from the hopper. On their first day in the box, a small dish of grain was placed in front of the opening to the food hopper. A piece of masking tape with several pieces of grain fastened to it led from the dish to the hopper opening. On the second day, the external dish was not present, but the grain-dotted masking tape hung from the hopper opening and the hopper was continuously available. This phase was repeated until the bird ate the grain from the tape, after which the hopper alone was presented until the bird ate steadily from the hopper for at least 2 min. To this point, each of the sessions lasted until the behavioral criterion was met or until 30 min had elapsed. In the next two sessions, the hopper was presented 10 times at intervals of from 10 to 30 sec, by which time all birds were eating within 2 to 3 sec of hopper presentation. At this point, hopper training was considered complete and the automatic shaping procedure was begun.

In the automatic procedure, a 30-sec intertrial interval was terminated by 5 sec of green or red light on one of two response keys, which was followed by a 3-sec presentation of the hopper. Each daily session consisted of 60 trials. In the first nine sessions, pecking the lighted key had no effect on hopper presentation; in the next five sessions, each peck on the lighted key was reinforced with immediate access to the hopper. The light on the key went out when the hopper was presented and the intertrial interval began when the hopper was withdrawn. If the key was not pecked, the hopper became available at the end of the 5-sec key light as before. Position of the response key and color of the light on it were balanced within each group so that all four combinations of left, right, red, and green were represented as equally as possible. After completion of the sessions with reinforced pecking on the first key or after meeting a criterion of pecking on at least 56 trials in one session, two further sessions were held on the next two days in which only the second key was lighted and the second color was used. Each peck was reinforced immediately in these sessions. Fifty days elapsed between the first nine sessions and the remaining ones, during which time the birds had grain continuously available. Except for this break, sessions were held on consecutive days.

Presumably because of the experience with the training box and aspects of hopper presentation already acquired in the first experiment, differences throughout hopper training in this experiment were not as striking as those seen before. On the first day only 22 per cent of the experimental birds ate immediately from the small open dish of grain, whereas 60 per cent of the control birds ate at once. Similarly, the cumulative mean time for onset of eating from the open dish, the tape, and the hopper in the early phases of hopper training was 565 sec for the experimental birds and 291 sec for the control birds.

The groups behaved very differently with regard to pecking on the first key. Figure 1 shows group performance in terms of the mean number of pecks on the lighted key during the first nine sessions when pecking was not regularly followed by immediate presentation; in Figure 2 are the group curves for the mean number of

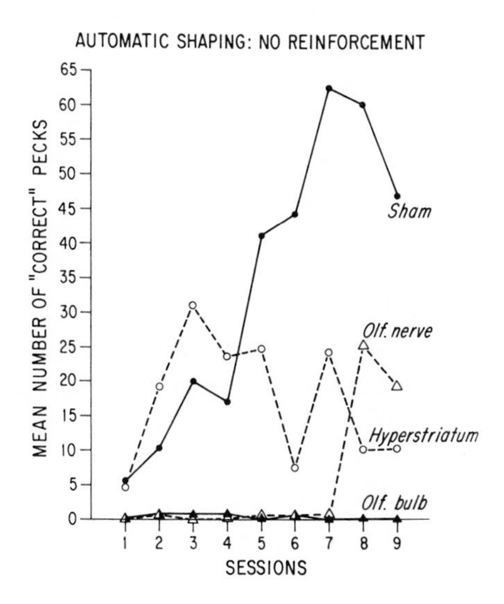

Figure 1 Group mean numbers of pecks on the lighted key during sessions in which hopper presentation did not immediately follow the first peck.

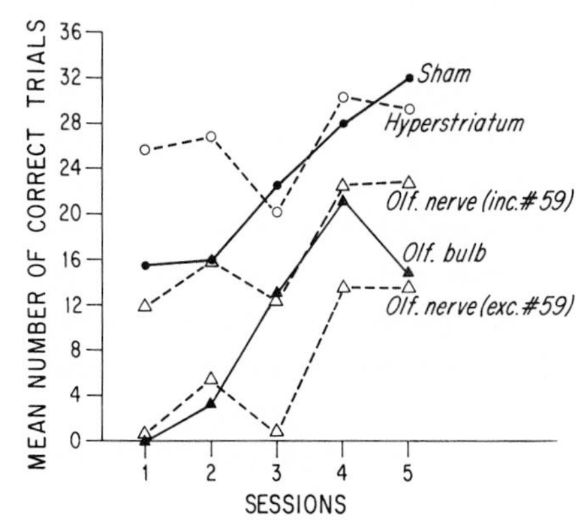

Figure 2 Group mean numbers of trials on which the lighted key was pecked during sessions when hopper presentation immediately followed the first peck. Two curves are shown for the nerve-sectioned group, with and without one atypically active bird.

correct trials during the five sessions when each peck was reinforced at once and terminated the trial. One bird (#59) accounted for all of the activity in the nerve-sectioned group in Figure 1 and the major portion of it in Figure 2. The extent of this bird's contribution is shown explicitly in Figure 2, where two curves have been presented for the nerve-sectioned group, one with and one without Pigeon #59. Obviously, this bird's behavior differed from the others with olfactory pathway damage, but we have no ready explanation because it did not perform atypically at other times. It was one of the pigeons of heterogeneous stock, all members of which showed a tendency to peck at a higher rate than did the racing homing pigeons from our own colony. It may be that #59 was simply an extreme case of that characteristic. Some birds in each group never pecked the key during the entire experiment, although all but five birds, two with bulb ablations and three with sham lesions, ate from the hopper in all sessions virtually every time it was presented. The others either failed to eat on early trials in some sessions or showed erratic patterns.

The per cent of birds that pecked the key at least once during the two parts of the shaping procedure can be seen in Table I. For the pigeons that pecked, the mean trial

TABLE I

PER CENT OF BIRDS IN EACH GROUP THAT PECKED THE LIGHTED KEY AT LEAST ONCE DURING TWO TYPES OF SESSIONS.

In one type of session the first peck was not reinforced immediately by hopper presentation; in the other type of session it was.

	Per Cent Responding							
	Sham		Hyperstriatum		Olfactory nerve		Olfactory bulb	
Session	No reinf.	Reinf.	No reinf.	Reinf.	No reinf.	Reinf.	No reinf.	Reinf.
1	40	60	40	80	0	40	0	0
2	40	80	60	80	20	40	25	25
3	40	80	60	80	0	40	25	50
4	40	80	60	80	20	60	25	50
5	60	80	60	80	20	60	0	50
6	60		60		20		25	
7	60		60		40		0	
8	60		60		40		0	
9	60		60		20		0	

number on which they made their first response for the sessions in which they responded at all was 11.8 for the sham-lesioned group, 11.5 for the hyperstriatal-lesioned group, 22.3 for the olfactory bulb-ablated group, and 19.4 for the olfactory nerve-sectioned group. Only four birds met the criterion of pecking on at least 56 trials in one session—two with sham lesions, one with a hyperstriatal lesion, and one with olfactory nerve section (#59). No reliable group differences appeared during the two sessions in which only the second key was lighted and the second color presented.

The groups might have separated if more sessions had been given under this condition, for at least one bird in each group had not begun to peck by the end of the second session.

It is apparent that a smaller number of the birds with olfactory lesions began to peck the key and that fewer pecks were made by those that did respond. The omission of the reinforcement during the first sessions amplified the group differences, because multiple pecks could be made during a single trial due to the absence of a reinforcing event that would have terminated the trial after a single peck. The birds that never began to peck the key typically stood quietly near the panel containing the keys and the opening to the food hopper. They did not appear apprehensive or unobservant, but were simply inactive until the hopper was presented.

HEART RATE CHANGES TO VISUAL STIMULI

Three to four months after the preceding experiment, a third experiment was conducted on rate of habituation to the onset of light. The purpose was to measure the heart-rate response to light onset as a function of repeated experience with the stimulus. Two birds had died in the interval, one with a sham lesion and one with olfactory nerve section. Each pigeon, at free-feeding body weight, was tested in nine daily sessions, consecutive except for a two-day break between sessions 5 and 6. Each session consisted of 30 presentations of a 75-watt light for 10 sec, with 2 min between stimuli. The bird was wrapped lightly in stretch bandage and placed on a stand in a dark box exactly like the test enclosure used in the two preceding experiments. The stimulus light was mounted on the wall in front of the bird. Heart rate was recorded through needle electrodes inserted into the left pectoral muscle and the left thigh at the beginning of each session and connected to a portable Offner electroencephalograph. Recording was done continuously throughout the session. A delay of 1 to 2 min occurred before presenting the first stimulus, so that the heart rate at that time would not reflect the activities involved in preparing the bird for the session. At the end of the ninth and final session, after 270 visual stimuli had been presented, two auditory stimuli were given—two 10-sec tones of 390 cps and 76 db.

The effect of each stimulus on heart rate was measured by calculating the difference in rate between the 10-sec period immediately preceding light onset and the 10-sec period during light. All groups had essentially the same mean rate in the prestimulus periods. Figure 3 gives the mean changes for each group in each session and also shows the responses to the two tones that were presented at the end of the experiment. A clear difference in trend appears between the two experimental and two control groups with regard to heart-rate changes in the later sessions. Although none of the groups habituated completely to the light stimulus within the limits of this experiment, the two control groups did show a reduction in the heart-rate response by the last session. At the same time, however, the two experimental groups were showing an increased reaction to light onset. It is notable that all groups showed about the same reaction in the first session as well as to the two tones at the end of the experiment.

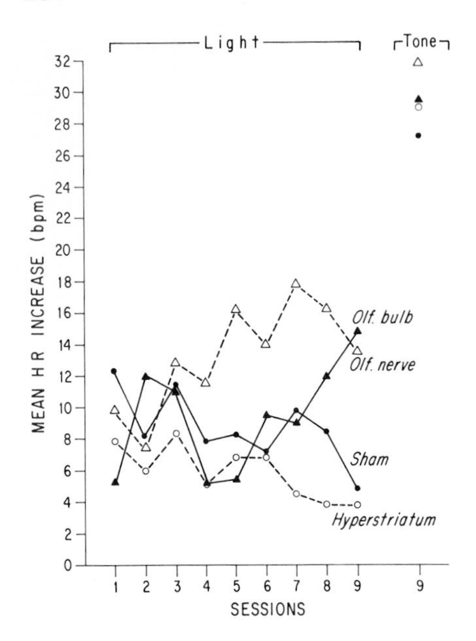

Figure 3 Group mean increases in heart rate to 30 presentations of light in each session. The responses to two tones at the end of the last session are also shown.

Thus, the group differences appear to lie in their reactions to repeated stimuli rather than to the initial stimuli.

DISCUSSION

In the interpretation of these combined results certain points are clear. First, the effect is not limited to specific responses, but seems rather to be a general one that results in altered behavior in a variety of situations. Second, the lesioned control group is more similar in behavior to the unlesioned group than it is to the two groups with olfactory system damage, indicating that the effects of the latter lesions are in some degree specific to locus. Third, the close correspondence between the bulb-ablated and nerve-sectioned groups further strengthens this concept of significance specifically for the olfactory system. Fourth, there seems to be no basis for suspecting that the loss of odor perception can account for the changed behavior of the experimental pigeons either in their reaction to the sound of the empty food hopper or in their heart-rate response to a repeated light stimulus. It becomes increasingly likely, therefore, that the integrity of the primary olfactory system, or some crucial aspect of it, is important for the maintenance of certain features of normal, nonolfactory behavior. The experiments described here and the one previously published converge on such an implication.

The various situations we have used undoubtedly have elicited both startle and orienting reactions. Neurophysiologists have implicated a number of central neural structures in the mediation of these arousal reactions (cf. references 6 and 10), including the hippocampus, fornix, and entorhinal cortex, all of which have connections with the olfactory bulb via relays.

It would be premature to try to relate our results to proposed neurophysiological mechanisms at this time, inasmuch as we cannot yet state exactly what types of behavior are affected. On the basis of our original experiment, as well as the first two of the present three, the descriptions that come readily to mind are inactive, less alert, harder to arouse, etc. On the other hand, the heart-rate experiment shows an increased excitability as the result of a procedure that normally decreases arousal. From this combination, it could be deduced that new and sudden stimuli result in a greater alerting response than normal, and that its increased intensity may produce repulsion or distraction, either of which would interfere with appropriate responses to the stimuli. The experimental groups' reaction to the first hopper sound in the orienting experiment above fits such an interpretation especially well, and their behavior in the learning experiments is consistent with it. If this interpretation is correct, it implies a more active arousal system as a result of these lesions. Alternatively, the effect may be a dual one of producing both more and less activity in different areas so that the usual balance is shifted. According to this idea, the experimental birds' failure to respond as actively as the controls in the early phases of the learning experiments reflects lowered activity in certain central neural areas at that time; their increased cardiac reaction to the later visual stimuli in the habituation experiment suggests an eventual increase in activity with repeated stimulation. More research is needed to clarify the critical features, and the importance of diversified behavioral measures is emphasized by the apparently opposite results obtained here.

One neurophysiological experiment has special relevance for our results. The work of Beteleva and Novikova on changes in electrical activity of various brain regions after destruction of the olfactory membrane in rabbits showed that both rate and amplitude decreased in the hippocampus and sensorimotor and visual areas, while increases occurred in the reticular formation. They interpreted their data as indicative of "general neurodynamic changes" including both a "reduction in the level of [cortical] functional mobility and the presence of diffuse inhibition" and "a raising of the level of excitability" in the reticular formation (reference 1, p. 554).

Much more needs to be known about downstream influences from the olfactory bulbs. The common experience of ready arousal of memories and of visceral reactions by odors has indicated that the bulbar efferents are distributed to some important control areas, but the connections and their activity have not been worked out even in mammals, let alone in birds. The major outflow is to the prepyriform cortex, where the electrical activity has been found to be closely related to that of the olfactory bulb and to be as sensitive to changes in behavioral state as are the bulbs themselves.[2] Similar relationships have also been reported for the amygdalar area.[7] Centrifugal influences on the bulb are well documented (cf. reference 12). Therefore, the necessary ingredients for interaction between olfactory bulb mechanisms and other central processes are known to exist, at least in mammals. The results presented here and in previous reports[11,12] point to the likelihood of analogous circuitry in birds, and may help to explain the preservation of a functional olfactory system in avian forms for which it has seemed to have little importance.

The possibility must be considered that the birds with olfactory damage are showing merely the effect of reduced environmental stimulation due to the loss of input from the olfactory receptors. This notion cannot be completely ignored, although it represents a major shift in emphasis from the long-held opinion that most birds, and especially seed-eaters, experience little if any odor sensation. In work with macrosmatic forms, it would be especially important. The most suitable test procedure is to maintain the subjects in an odor-free environment. This is virtually impossible to achieve, however, if for no other reason than that food must be provided and the drastic alternatives of intragastric or intravenous feeding could hardly be carried out in an odorless manner. Another type of test could be done by eliminating an equivalent amount of input over a different modality. Establishing equivalence is again a severe challenge, but on the reasonable assumption that pigeons do not have a rich smell life, section of a cutaneous nerve might be appropriate, such as the ophthalmic branch of the trigeminal, which is relatively easy to reach and is not essential for normal maintenance of healthy birds.

Should this phenomenon be valid and widely applicable across species, it would raise the need for some caution in interpreting the outcome of experiments in which a lesion is placed in the primary olfactory system as a control condition in studies of olfactory perception. According to our results, certain types of failure to respond might be attributable more to a generalized effect than to a specific one involving the receipt of olfactory information, and should at least be checked with control stimuli in another modality.

SUMMARY

Three experiments were conducted in further study of our earlier finding that pigeons with ablated olfactory bulbs or sectioned olfactory nerves were much slower than controls in approaching the food magazine, beginning to peck the first response key, and transferring pecking to a second key during training in visual discrimination. Four groups of pigeons have been studied, those with olfactory bulbs ablated, olfactory nerves sectioned, hyperstriatum unilaterally lesioned, and sham-operated The previously noted slowness of the olfactory-damaged birds to peck the key in operant conditioning through hand-shaping was seen again with an automatic shaping procedure in which periodic illumination of the response key was always followed by presentation of the grain hopper. Under these conditions, the two control groups spontaneously pecked the lighted key more often than the experimental birds, whether each peck was followed by hopper presentation or not. In one of two experiments on orienting behavior, it was found that the control birds reacted to the sound of the empty grain hopper by looking or moving toward it, while the experimental birds typically did not. Measurements of heart-rate increases to a repeated light stimulus showed that the olfactory-damaged groups became more reactive in later sessions when the control birds were responding less. At least part of the effect thus appears to reflect an altered pattern of reactivity to nonolfactory stimuli.

ACKNOWLEDGMENT

Supported by grant GB 4596 from the National Science Foundation to B. M. Wenzel. Mr. Oberjat held a predoctoral traineeship under grant MH 6415 from the National Institute of Mental Health to UCLA.

REFERENCES

1. Beteleva, T. G., and L. A. Novikova. 1961. Electrical activity in various cortical regions and in the reticular formation after elimination of the olfactory analyzer. *Pavlov J. Higher Nerv. Activ.* **11**, pp. 547–555.
2. Boudreau, J. C., and W. F. Freeman. 1963. Spectral analysis of electrical activity in the prepyriform cortex of the cat. *Exp. Neurol.* **8**, pp. 423–439.
3. Brown, P. L., and H. M. Jenkins. 1968. Auto-shaping of the pigeon's key-peck. *J. Exp. Anal. Behav.* **11**, pp. 1–8.
4. Didiergeorges, F., M. Vergnes, and P. Karli. 1966. Privation des afférences olfactives et aggressivité interspécifique du rat. *C. R. Séances. Biol.* **160**, pp. 866–868.
5. Herrick, C. J. 1933. The functions of the olfactory parts of the cerebral cortex. *Proc. Nat. Acad. Sci. U.S.* **19**, pp. 7–14.
6. Lynn, R. 1966. Attention, Arousal and the Orientation Reaction. Pergamon Press, Oxford, etc.
7. McLennan, H., and P. Graystone. 1965. The electrical activity of the amygdala and its relationship to that of the olfactory bulb. *Can. J. Physiol. Pharmacol.* **43**, pp. 1009–1017.
8. Novakova, V., and H. Dlouha. 1960. Effect of severing the olfactory bulbs on the intake and excretion of water in the rat. *Nature* **186**, pp. 638–639.
9. Sieck, M. H., and B. M. Wenzel. 1969. Electrical activity of the olfactory bulb of the pigeon. *Electroencephalogr. Clin. Neurophysiol.* **26**, pp. 62–69.
10. Sokolov, E. N. 1963. Higher nervous functions: the orienting reflex. *Ann. Rev. Physiol.* **25**, pp. 545–580.
11. Wenzel, B. M., and A. Salzman. 1968. Olfactory bulb ablation or nerve section and pigeons' behavior in non-olfactory learning. *Exp. Neurol.* **22**, pp. 472–479.
12. Wenzel, B. M., and M. H. Sieck. 1966. Olfaction. *Ann. Rev. Physiol.* **28**, pp. 381–434.

OLFACTORY MEDIATION OF IMMEDIATE X-RAY DETECTION

JAMES C. SMITH *and* DON TUCKER
Departments of Psychology and Biological Science, Florida State University, Tallahassee, Florida

During the past decade it has been shown repeatedly that a variety of animals react immediately to the onset of an X-ray beam (e.g., Kimeldorf and Hunt[15]). In some cases this detection of ionizing radiation apparently is mediated by the visual system (e.g., Smith and Kimeldorf[19]). In several mammalian species and in the pigeon, however, the sensory modality most prominently involved in X-ray detection is olfaction.[2,3,4,6,7,8,14,20]

The purpose of this paper is to review the evidence obtained in this laboratory that an intact olfactory system is necessary for immediate detection of X-rays. In addition, a description is given of how the behavioral technique used in the X-ray-detection problems can be applied in measuring odor intensity and quality discriminations in both pigeons and monkeys.

The behavioral technique in question is a modified version of conditioned suppression. This technique was first used by Hendricks[9] to measure flicker-fusion thresholds in pigeons and by Morris[16] to measure X-ray dose-rate thresholds in albino rats. Essentially, it is a classical conditioning paradigm (i.e., a warning stimulus terminated with an unavoidable electric shock) superimposed on an ongoing operant baseline.

Figure 1 illustrates the procedure used. The top line represents time in 10 sec units. The animal is deprived of food and trained to manipulate a device for food reward. The rat is trained to press a bar for sucrose, the monkey learns to press a lever for food pellets, and the pigeon is trained to peck a key for mixed grain reinforcement. Typical responding early in training is seen on line A. By carefully scheduling the presentation of reinforcement on a variable time base, a smooth rate of responding can be obtained, as is shown in line B. When the animal is given a brief electric shock, the smooth responding can be suppressed, as is seen in line C. The duration and intensity of the shock can be adjusted so that the suppression is marked, but not too long in duration. A warning stimulus is now presented just prior to electric shock. Initially, one may see some suppression to the warning stimulus alone (line D), but it is not very marked. Later, however, after several pairings of warning stimulus and shock, the animal becomes conditioned to the shock and complete or almost complete suppression occurs, as can be seen in line E. If one samples the behavior of the animal for a comparable period of time prior to the onset of the warning stimulus, baseline responding can be obtained to compare with responding during the warning stimulus.

The suppression can then be quantified in the manner described by Hoffman et al.[13]

$$\text{Suppression Ratio (S.R.)} = \frac{\text{Pre Rs} - \text{WSRs.}}{\text{Pre Rs}}$$

One can see from this equation that if complete suppression occurs during the warning stimulus the ratio is 1.00, and if no suppression occurs the ratio is 0.00 (assuming an absolutely uniform rate of responding).

We have successfully used an X-ray beam as the warning stimulus in such a design for rats, monkeys, and pigeons. In order to insure that the animal was responding to the beam and not to some artifact associated with its onset, control trials were always run. These trials consisted of operating the X-ray machine in the same fashion as used in the X-ray trials, but the beam was either directed away from the animal or attenuated by the addition of a lead filter. To insure that suppression of responding in the presence of X-rays could be distinguished from random pausing in the rate of lever pressing, baseline trials were run. These consisted of sampling responses during two consecutive periods of time equal in duration to the warning stimulus period. When the suppression ratio formula is applied to control trials and baseline trials, if there are no artifacts, these trials should yield ratios near zero in value. Neither baseline nor control trials were ever terminated with shock.

Rats learn extremely rapidly to suppress lever pressing in the presence of the X-ray beam. Of the 14 animals trained in the study by Dinc and Smith,[6] eight rats reached a suppression ratio of 0.90 or higher on the second day of training. Five more animals had suppression ratios of 0.93 or higher on the third day, and it took only one animal four days of training to surpass a ratio of 0.90. Only five to 10 trials with X-rays were presented each day at a dose rate of 0.5 r/sec for 15 sec stimulus duration. The rhesus monkey learns to detect X-rays as rapidly as the rat. Table I illustrates significant acquisition of suppression to X-rays in fewer than 20 trials for four subjects. Median

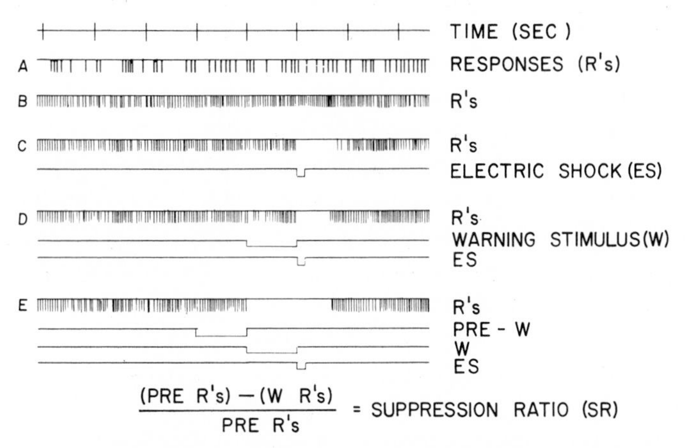

Figure 1 A model of the conditioned suppression technique used. The responses are not actual data, but represent typical responding in rats, pigeons, or monkeys. See text for detailed description.

TABLE I
MEDIAN SUPPRESSION RATIOS OF MONKEYS TO X-RAYS (IN BLOCKS OF FIVE TRIALS)

		Suppression Ratio		
Subject	Block	X-ray	Baseline	Control
M391	1	0.29	0.14	0.12
	2	0.60	0.09	−0.05
	3	0.60	−0.03	0.26
	4	1.00	−0.07	0.02
M283	1	0.08	0.12	0.11
	2	0.26	−0.05	0.19
	3	0.56	0.59	0.05
	4	0.81	0.06	−0.11
M389	1	0.45	0.00	0.03
	2	0.12	0.00	0.03
	3	0.45	0.12	0.03
	4	0.78	−0.06	0.00
M390	1	0.74	0.08	0.33

suppression ratios for blocks of five trials are presented for X-ray, baseline, and control trials. Most of the suppression ratios for the baseline and control trials are near zero, but the ratios for X-ray trials approach 1.00. Subject M-390 began suppressing to the X-ray on the second trial.

Garcia et al.[7] and Hull et al.[14] did partial body-exposure experiments to show that the head of the rat was most sensitive for the immediate detection of X-rays. By using a collimated X-ray beam, they showed that the area of the olfactory bulbs, which is closely enveloped anteroventrally by the olfactory organs, was the most sensitive. Cooper and Kimeldorf,[3] Cooper et al.,[5] and Cooper and Kimeldorf[4] demonstrated that X-rays to the head of the animal evoked responses in single olfactory bulb neurons in rats, rabbits, dogs, and cats.

Dinc and Smith[6] demonstrated that animals which were taught to suppress to X-ray stimulation lost this sensitivity when the olfactory bulbs were removed. Table II shows the mean suppression ratio on the day before surgery and the first day of training two weeks later. A marked difference can be noted between the rats with complete bulb removals and animals with sham operations, frontal lobe ablations, or incomplete bulb removals.

Taylor et al.[20] reported similar results after sectioning the olfactory tracts in rhesus monkeys. Table III shows that monkeys which had the tracts sectioned underwent a drastic reduction in suppression ratios. Sham-operated animals, in which the tracts were exposed but not sectioned, maintained a high level of suppression to X-rays. The one animal with a partially intact tract on one side demonstrated a marked rise

TABLE II

EFFECTS OF SURGICAL PROCEDURES ON MEAN SUPPRESSION RATIOS OF RATS TO X-RAYS

Subject	Suppression Ratio Pre-op	Suppression Ratio Post-op	Post-mortem Examination
4B	0.92	0.24	Complete Bulb Removal
11B	1.00	0.02	Complete Bulb Removal
8A	1.00	0.06	Complete Bulb Removal
3B	1.00	0.97	Incomplete Bulb Removal
6B	1.00	1.00	Incomplete Bulb Removal
10B	1.00	0.94	Incomplete Bulb Removal
3A	1.00	1.00	Frontal Lobe Removal
9A	0.95	1.00	Frontal Lobe Removal
10A	0.96	0.86	Sham Operation
11A	0.93	0.96	Sham Operation

in dose-rate threshold, but it remained able to detect X-rays. The behavioral technician who tested the monkeys before and after surgery and the pathologist who did the post-mortem examinations were unaware of the identities of experimental and control animals.

Additional monkeys trained in this laboratory demonstrate that sectioning of the olfactory tract does not inhibit learning of a suppression problem in the rhesus. Seven animals were trained to suppress to X-rays and given training on other problems after the olfactory tract was sectioned. Data from one operated and one sham-operated animal are illustrated in Figure 2.

The first panel demonstrates the acquired conditioned suppression to X-rays at a dose rate of 10 r/min delivered as a 15 sec warning stimulus. Each datum point represents the mean of 10 trials. It can be seen that suppression to the X-ray beam was high.

TABLE III

EFFECTS OF SURGICAL PROCEDURES ON MEAN SUPPRESSION RATIOS OF MONKEYS TO X-RAYS

Subject	Suppression Ratio Pre-op	Suppression Ratio Post-op	Olfactory Tracts After Surgery *Left*	Olfactory Tracts After Surgery *Right*
M283	0.88	0.07	opened	opened
M391	0.86	0.14	opened	opened
M595	0.85	0.03	opened	opened
M46G	0.89	0.03	opened	opened
M390	0.75	0.72	partially interrupted	opened
M389	0.80	0.89	intact	intact
M600	0.81	0.82	intact	intact

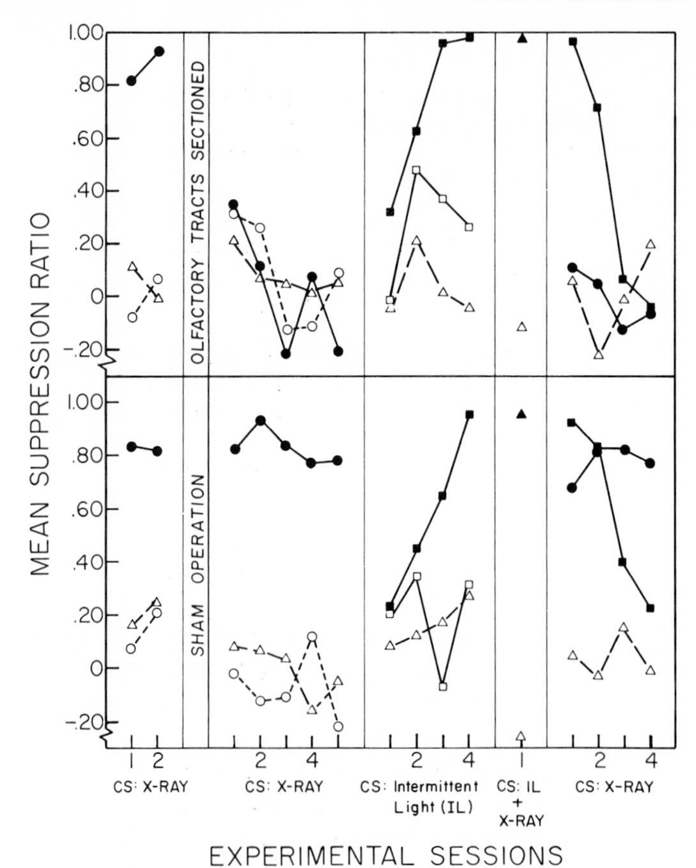

Figure 2 Mean suppression ratios as a function of trials and different experimental conditions are presented for two animals. The solid circles represent trials with X-ray as the warning stimulus, the open circles represent sham X-ray exposures, and the open triangles represent baseline trials. The solid squares represent suppression to flickering light; the open squares show the mean suppression ratios to a fused light. The solid triangles represent trials in which X-rays and flickering light were presented concurrently. The conditioned (or warning) stimulus for each panel is indicated at the bottom of the graph.

Baseline and control trials yielded S.R.s of near zero. In the second panel it can be seen that suppression to X-rays was lost in the animal with the olfactory tracts sectioned and suppression was maintained in the sham-operated subject. The animals were then conditioned to suppress when the warning stimulus was a flickering light (approx. 2/sec). Intermittent trials were given with the same light not flickering and these trials were never followed by shock. As is seen in the third panel, suppression also occurred at first to the fused light, but soon the monkeys discriminated, and by the fourth session they suppressed markedly to flicker and not to fused light. In the fourth panel the warning stimulus was flickering light administered simultaneously with X-ray. Here, almost complete suppression occurred. In the last condition, the

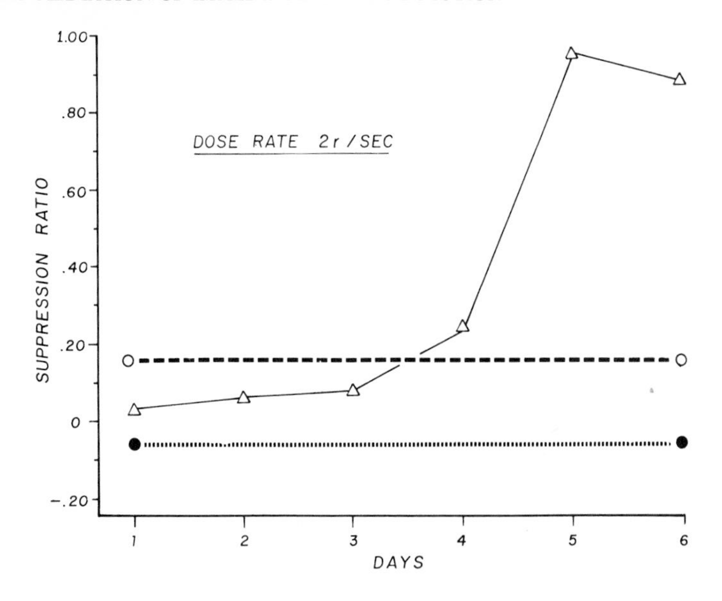

Figure 3 Mean suppression ratios for each of six days are presented for one pigeon during X-ray trials (solid line), baseline trials (dotted line), and sham exposure trials (dashed line).

flicker was presented but was no longer paired with shock, and extinction of the suppression took place. The animals continued to receive intermittent trials in which X-ray was paired with shock and the beam was detected only by the control animal. Sham-control X-ray trials always resulted in mean suppression ratios near zero.

Because of the difference in the olfactory anatomy of birds, we wanted to train pigeons to detect X-rays. The olfactory organ and bulb in the pigeon are spaced several millimeters apart, permitting sectioning of the primary olfactory nerves without damage to the other structures. Training the pigeon to detect X-rays has proved to require much more time and higher dose rates than does mammal training.

The acquisition curve of one of the faster-learning birds is illustrated in Figure 3. This example shows acquisition only after five days and approximately 100 pairings of the X-ray with shock. We have attempted to train approximately 50 birds and have seen the X-ray detection in only about half of the subjects. The dose rate was higher than with the mammals, i.e., about 2 r/sec. The birds sustained an extremely large cumulative dose in the training process.

Birds that do detect, however, lose this detection after the olfactory nerves are sectioned. Several of the pigeons were treated in the manner described in Figure 4. The pigeon demonstrated X-ray detection and a sham operation was performed, i.e., the bird was anesthetized and the olfactory nerves were exposed, but not sectioned. The wound was packed with Gelfoam and the bird was run the following 2 days in the test chamber. It can be seen that X-ray detection was still apparent. However, when the nerves were sectioned in a subsequent operation, the detection was lost.

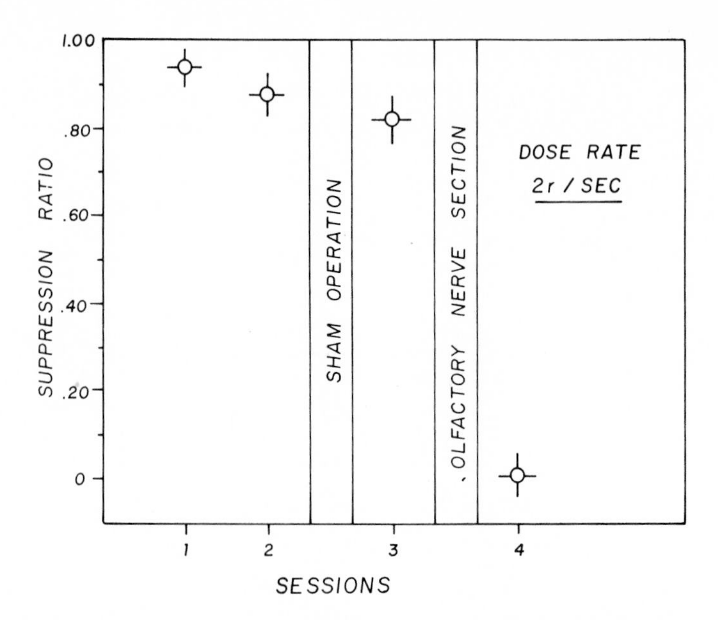

Figure 4 Mean suppression ratios of one pigeon to X-ray stimulations. Suppression ratios for two sessions are presented prior to a sham operation. The mean suppression is presented for the third session, which was followed by an operation in which both olfactory nerves were sectioned. The mean suppression ratio after the olfactory surgery is shown.

These experiments give evidence that an intact olfactory system is necessary for this immediate detection of X-rays. We still do not know if the detection is the result of an odor formed in the vicinity of the receptors by action of the ionizing radiation or if it is the result of some direct action on elements of the peripheral or central nervous system. The presumed odor sensation due to X-irradiation may be weak, since Gasteiger and Helling[8] obtained evidence that various odorants masked the X-ray odor sensation. The spotty results obtained with pigeon X-ray detection would fit in with the belief that they orient visually. Perhaps it would be profitable to reduce odor backgrounds in X-ray-detection experimentation. The pigeon's loss of detection after the olfactory nerves are severed is presumptive evidence for mediation of the effect at the peripheral level, but one can hardly rule out the possibility that a bulbar mechanism might be deranged by loss of the normal afferent inflow.

One outgrowth of these experiments with X-ray detection has been the development of an animal psychophysical technique to study odor discriminations in a behaving animal. We have conducted extensive experiments with pigeon olfaction because of the canard extant in the literature that birds cannot smell, and have begun to apply the technique to the study of olfaction in the rhesus monkey.

We constructed a glass olfactory breathing chamber and placed it in the pigeon test chamber. The pecking key and the opening for the food hopper were located in the breathing chamber. A more complete description of the apparatus can be found elsewhere.[10,11,12,18] The bird was trained to peck in a flow of clean air. The warning stimulus was an odor (amyl acetate) introduced into the breathing chamber by a flow director. Figure 5 shows the acquisition of olfactory detection in a typical pigeon when

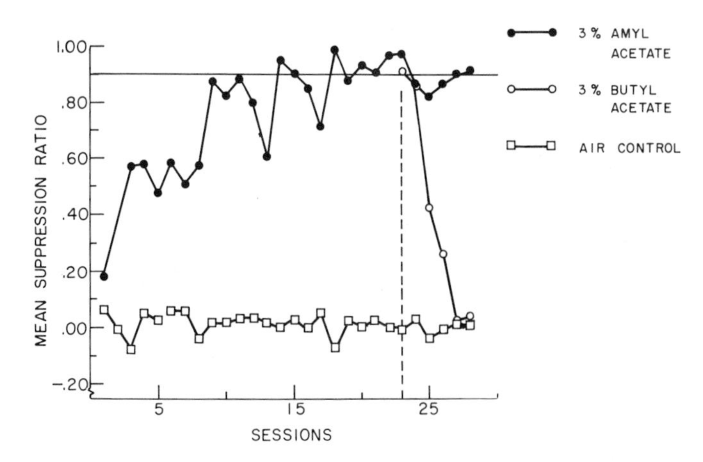

Figure 5 Mean suppression ratios are presented as a function of sessions for one pigeon. The warning stimulus was the odor of amyl acetate (solid circles) and control trials were run when the clean air was presented (open squares). During the 23rd session some trials were initiated using butyl acetate as a stimulus. Butyl acetate was never paired with shock.

the air is 3 per cent saturated with amyl acetate vapor.[10] To insure that the bird was not responding to a transient, trials were run in which the intensity of the odor was reduced to zero and the flow director was merely changed from air to air. It can be seen from the open squares in Figure 5 that no suppression was evident on these trials. After learning to amyl acetate was evident, trials were administered with butyl acetate (at 3 per cent of vapor saturation). When the butyl acetate was presented, it was never followed by shock, but the amyl acetate trials were always followed by shock. It can be seen from Figure 5 that at first the bird generalized to the amyl and suppressed on butyl acetate trials. However, after two such sessions suppression to the butyl trials diminished, and by the fifth session there was no more suppression than on the air control trials. Each session included 10 shocked amyl acetate, five nonshocked butyl acetate, and five nonshocked air control trials. These data indicate that the pigeon can learn to discriminate between amyl and butyl acetates at the concentrations described. Approximately 20 additional pigeons have been trained in this laboratory to make odor discriminations.

In a similar design, Shumake et al.[18] have demonstrated that the pigeon can also make odor intensity discriminations. In this study, pigeons were trained to suppress to 7 per cent of vapor saturation amyl acetate; 7 per cent trials were always followed by shock. In other trials we administered, the concentration of the amyl acetate was lower and the trials never followed by shock. Figure 6 shows that suppression was high

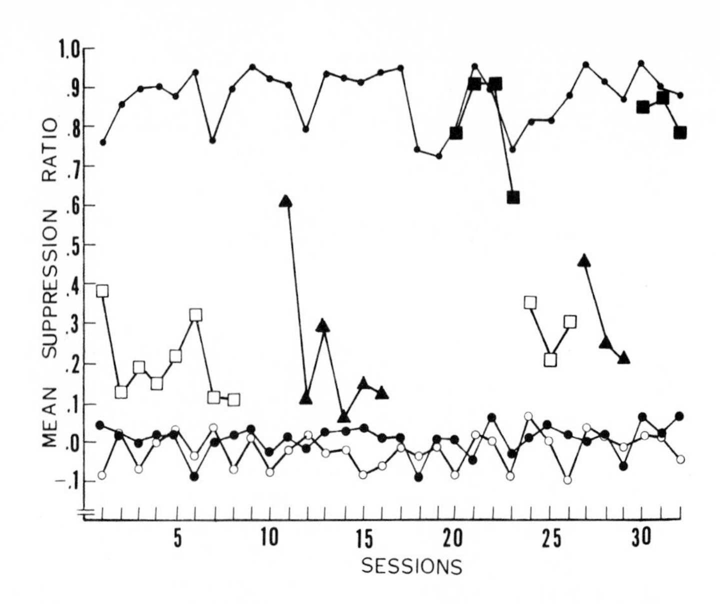

Figure 6 Mean suppression ratios as a function of sessions for one pigeon are shown. The small solid circles represent suppression to amyl acetate of 7% vapor saturation; trials of amyl acetate of 1% (open squares), 2% (closed triangles), and 3% (closed squares) vapor saturation are also presented. The air control trials (closed circles) and baseline trials (open circles) are presented for all 32 sessions. When amyl acetate of 7% vapor saturation was used as the warning stimulus, it was terminated by electric shock. All other trials were not shocked.

on the 7 per cent trials, but much lower when the concentration was 1 or 2 per cent (illustrated by the open squares and closed triangles, respectively). When the concentration was 3 per cent, however, (closed squares), the pigeon could not discriminate this concentration from the 7 per cent. In sessions 24–32, the experiment was replicated with similar results. Air control trials (closed circles) and baseline trials (open circles) yielded mean suppression ratios of approximately zero throughout all of the sessions. Similar results were obtained for five additional pigeons.

To give further evidence that the suppression is the result of olfaction, the primary olfactory nerves of several birds were sectioned after they had demonstrated suppression to the amyl acetate trials. The results of such an experiment are presented in Figure 7.[17] Prior to the surgery, the suppression to amyl acetate averaged higher than 0.90. Following surgery, however, the mean suppression ratios for amyl acetate trials could not be distinguished from baseline or air control trials. Various types of experimentation by Wenzel,[21] especially with pigeons, and anatomical studies by Bang and Cobb[1] are other recent evidences of avian olfactory ability.

We have recently constructed an olfactory chamber for the rhesus monkey and have successfully trained two monkeys to suppress lever pressing in the presence of amyl acetate. Results similar to those described above with pigeons have been obtained in demonstrating detection and discrimination.

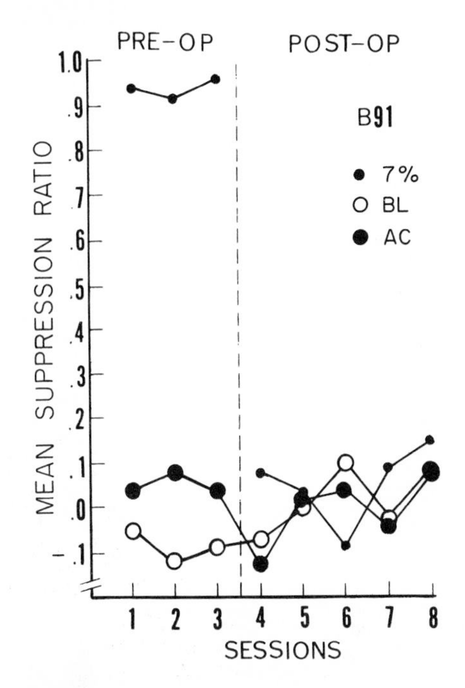

Figure 7 Mean suppression ratios as a function of pre-operation and post-operation sessions are presented. It can be seen that suppression to amyl acetate of 7% vapor saturation was lost following sectioning of the primary olfactory nerves.

ACKNOWLEDGMENT

This research was supported by United States Air Force Contract Number F 29600-67-0012 under Project 6893, the United States Atomic Energy Commission Contracts AT-(40-1)-2903 and AT-(40-1)-2690 (Division of Biology and Medicine), and NIH Research Grant 5 R01 NB01083-12 CMS.

REFERENCES

1. Bang, B. G., and S. Cobb. 1968. The size of the olfactory bulb in 108 species of birds. *Auk* **85**, pp. 55-61.
2. Brust-Carmona, H., H. Kasprzak, and E. L. Gasteiger. 1966. Role of the olfactory bulbs in X-ray detection. *Radiation Res.* **29**, pp. 354-361.
3. Cooper, G. P., and D. J. Kimeldorf. 1966. The effects of X-rays on the activity of neurons in the rat olfactory bulb. *Radiation Res.* **27**, pp. 75-86.
4. Cooper, G. P., and D. J. Kimeldorf. 1967. Responses of single neurons in the olfactory bulbs of rabbits, dogs, and cats to X-rays. *Experientia* **23**, pp. 137-138.
5. Cooper, G. P., D. J. Kimeldorf, and G. C. McCorley. 1966. The effects of various gases within the nasal cavities of rats on the response of olfactory bulb neurons to X-irradiation. *Radiation Res.* **29**, pp. 395-402.
6. Dinc, H. I., and J. C. Smith. 1966. Role of the olfactory bulbs in the detection of ionizing radiation by the rat. *Physiol. Behav.* **1**, pp. 139-144.
7. Garcia, J., N. A. Buchwald, B. H. Feder, R. A. Koelling, and L. Tedrow. 1964. Sensitivity of the head to X-ray. *Science* **144**, pp. 1470-1472.
8. Gasteiger, E. L., and S. A. Helling. 1966. X-ray detection by the olfactory system: ozone as a masking odorant. *Science* **154**, pp. 1038-1041.
9. Hendricks, J. 1966. Flicker thresholds as determined by a modified conditioned suppression procedure. *J. Exp. Anal. Behav.* **9**, pp. 501-506.

10. Henton, W. W. 1966. Suppression behavior to odorous stimuli in the pigeon. Doctoral dissertation, Florida State University.
11. Henton, W. W. 1969. Conditioned suppression to odorous stimuli in pigeons. *J. Exp. Anal. Behav.* **11**, pp. 175–185.
12. Henton, W. W., J. C. Smith, and D. Tucker. 1966. Odor discrimination in pigeons. *Science* **153**, pp. 1138–1139.
13. Hoffman, H. S., M. Fleshler, and P. Jensen. 1963. Stimulus aspects of aversive controls: The retention of conditioned suppression. *J. Exp. Anal. Behav.* **6**, pp. 575–583.
14. Hull, C. D., J. Garcia, N. A. Buchwald, B. Dubrowsky, and B. H. Feder. 1965. Role of the olfactory system in arousal to X-ray. *Nature* **205**, pp. 627–628.
15. Kimeldorf, D. J., and E. L. Hunt. 1965. Ionizing Radiation: Neural Function and Behavior. Academic Press, Inc., New York.
16. Morris, D. D. 1966. Threshold for conditioned suppression using X-rays as the pre-aversive stimulus. *J. Exp. Anal. Behav.* **9**, pp. 29–34.
17. Shumake, S. A. 1967. Olfactory intensity difference thresholds in the pigeon. Masters thesis, Florida State University.
18. Shumake, S. A., J. C. Smith, and D. Tucker. 1969. Olfactory intensity differences thresholds in the pigeon. *J. Comp. Physiol. Psychol.* (In press)
19. Smith, J. C., and D. J. Kimeldorf. 1964. The bioelectrical response of the insect eye to beta-radiation. *J. Insect Physiol.* **10**, pp. 839–847.
20. Taylor, H. L., J. C. Smith, A. H. Wall, and B. Chaddock. 1969. Role of the olfactory system in the detection of X-rays by the rhesus monkey. *Physiol. Behav.* (In press)
21. Wenzel, B. M. 1967. Olfactory perception in birds. *In* Olfaction and Taste II (T. Hayashi, editor). Pergamon Press, Oxford, etc., pp. 203–217.

X-RAY AS AN OLFACTORY STIMULUS

JOHN GARCIA, KENNETH F. GREEN, *and* BRENDA K. McGOWAN

State University of New York at Stony Brook; California State College at Long Beach; Barrow Neurological Research Institute, Phoenix, Arizona

INTRODUCTION

Ionizing radiation has long been considered a form of energy that penetrates living mammalian species without stimulating them. Exposures that might prove fatal later were assumed to be colorless, odorless, and tasteless to the victim. Recent evidence indicates that this assumption is probably wrong on all three counts, even for relatively low doses. It is probably more accurate to say that ionizing rays glow yellow-green, smell pungent, and taste sour. The perceived effect depends, quite naturally, upon the specific receptors which have absorbed the radiant energy. The diverse effects of radiation upon receptors of vertebrates and invertebrates have been reviewed recently elsewhere.[28,29] In this volume Smith and Tucker[40] present evidence demonstrating that radiation stimulates a variety of laboratory animals via the olfactory system.

We will limit our discussion specifically to the following propositions.

A. Radiation stimulates the olfactory receptors directly, not through an airborne odorant; and as such has some unique properties.

B. By completely different mechanisms, radiation exposure produces delayed noxious stimulation via internal receptors. This internal stimulation is integrated with chemoreceptive stimuli at the first synaptic relay station in the brain stem.

IMMEDIATE EFFECTS UPON THE OLFACTORY SYSTEM

Several questions are inevitably raised by assertions that animals can detect ionizing radiation. First, since radiation sources necessarily involve either apparatus to generate the rays, such as an X-ray machine, or to expose an isotope emitter, such as cobalt 60, what assurances do we have that the animal is not actually responding to some visual or auditory cue associated with operation of the machinery? Second, how do we know that radiation is not producing an effect such as fluorescence of nearby materials or ozone in the atmosphere, which in turn is perceived by the animal by means of some conventional receptor process? Since either of these possibilities would render our phenomena trivial, let us discuss the precautions and the data which bear on these two questions.

Arousal to Radiation and the Artifact Problem

The initial studies carried out here and in Russia demonstrated the immediate stimulus properties that ionizing rays exerted on sleepy laboratory mammals. These animals were habituated to the ambient noise of radiation machinery in constant operation until they became drowsy; then a shutter was opened silently, allowing the rays to impinge upon the animals. Within seconds, the animals displayed characteristic signs of arousal, electroencephalographic records revealed waves of low amplitude and high frequency,[9,42] heart rate was increased, and motor movements were observed.[26,27] Operation of the radiation machinery while delivering zero dose did not arouse the animals, indicating that the responses were not false ones elicited by coincidental stimuli associated with exposure. Animals that previously had been bilaterally ophthalectomized displayed similar arousal responses, indicating that the retina was not an essential site of action.

Subsequent behavioral studies on the immediate stimulating effect used radiation as a conditional stimulus or signal as described by Smith and Tucker[40] in this volume. For example, the first study employed brief X-ray exposures to signal a subsequent electric shock to the paws of rats.[11] Shock tends to disrupt sequential performance of animals, such as pressing a lever for food or licking at a spout for water, and after a number of training sessions in which discrete radiation exposures are immediately followed by shock, radiation also acquires the capacity to suppress performance. This response measure is functionally related to the log of the stimulus intensity (X-ray dose rate), which indicates that this response is not the result of stimulus artifacts.[10]

Since ionizing radiation is a pervasive form of energy that penetrates the living organism, it was postulated that arousal responses might be due to a direct effect upon the neural tissue. Conceivably, this stimulus might be operating diffusely upon ganglia and synapses distributed throughout the nervous system. This host of minute changes, amplified by the reticular activating system, could produce an arousal similar to the spontaneous arousal from sleep.[26,27] However, subsequent studies indicated that radiation was producing a specific effect via the olfactory system, and the notion of a direct activating effect upon the nervous system was abandoned, at least for lower levels of stimulus intensity (i.e., dose rate).

Radiation as a Signal and the Ozone Question

A narrow, collimated X-ray beam (0.45 cm) directed at the olfactory area of rats is an effective conditional stimulus. The beam loses its effectiveness as it is moved away from the olfactory area.[12] If the olfactory bulbs are removed, the animals are no longer responsive to the radiation stimulus.[4,6,8,24] Furthermore, single neurons in the olfactory bulbs of laboratory animals respond to the onset of exposure,[7] clearly establishing olfactory mediation of this stimulus.

The possibility that animals might be detecting an odor produced in irradiated air was raised by studies that used ozone as a masking odorant.[22] Ozone disrupted the rat's conditioned responses to X-ray, but other odorants, such as wintergreen, did not. Furthermore, ozone did not disrupt the animal's ability to smell out bits of apple, hence it was not producing a general impairment of the animal's ability to smell. It is known that X-rays produce active molecules, such as ozone and oxides of nitrogen, so one might conclude that the animals were sniffing an airborne product of irradiated air.

Other observations do not support this conclusion. Neither irradiation of the nasal inhaling air passages anterior to the olfactory receptors, nor irradiation of the lungs and the exhaling air passages posterior to olfactory receptors, produce effective stimulation.[12] Yet rats with chronic tracheotomies that prevent nasal inhalation of ozone, or any other radiation by-products, readily arouse to X-ray.[24]

Discrimination of Flickering X-rays

Rats appear to suppress conditioned responses within 200 milliseconds of the onset of exposure, and often begin responding again within one second after the end of exposure. This appears to be a discrimination too fine to be dependent upon airborne products in an uncontrolled atmosphere. Let us present some new evidence bearing upon this particular point.

Our apparatus has already been described in detail.[21] Essentially, it consists of a 280 kv X-ray-therapy machine, which is continuously operating during an experimental session to eliminate "off-on" noises which could serve as cues. Between exposures the animal is shielded from X-rays by a lead shutter which can be operated silently. The precise duration of the exposure is determined by a lead disk with an adjustable open section that is continuously revolving at constant speed between the animal's compartment and the X-ray tube. The shutter is opened and the open section in the disk is allowed to pass over the animal once, exposing it briefly to 280 kv filtered X-rays (half-value layer 1.4 mm copper).

Our training procedures have been described previously and were summarized above. Thirsty animals were trained with a 1.0-sec exposure at approximately 0.4 r/sec signaling a subsequent shock to the paws within two seconds. They soon learned to suppress licking at the water spout during this X-ray signal. Then they were tested with a flickering X-ray exposure consisting of two 0.25-sec pulses separated by an 0.50-sec interval. They suppressed their licking responses to the flickering X-ray, also indicating that they detected it but did not differentiate it from the steady-state exposure on which they had been trained.

These rats were then given discrimination training in which the steady exposure was followed by shock, but the flickering exposure was presented without shock. They soon learned to discriminate between the two signals; that is, they continued to suppress responses to the steady (shock) signal but no longer suppressed responses to the flickering (nonshock) signal. Figure 1 summarizes the results obtained before and after discrimination training with four animals. Other rats in our laboratory learned to discriminate between a four-second steady-state and a train of five pulses

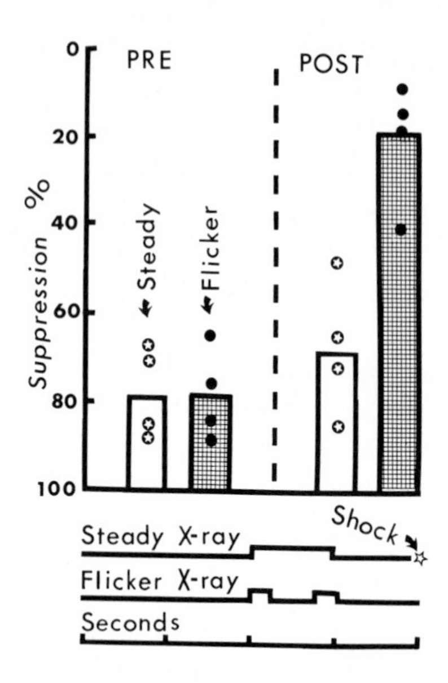

Figure 1 Discrimination of X-ray flicker by rats. Animals were first trained to suppress drinking by administration of steady X-ray only, followed by shock; they failed to distinguish between steady X-ray and flicker X-ray in the pretest. When given further training with steady followed by shock and flicker without shock they made the discrimination in the post-test.

of the same duration. Examination of the individual records indicates that many animals were apparently discriminating the end of the first pulse.

It seems to us that air conduction would "smear" the pulses so that this kind of discrimination would be impossible. The alternative hypothesis is that X-rays produce a sharp change in the mucous sheath or on the surface of the olfactory receptors. This latter notion is supported by recent investigations of unit activity in the plexiform layer of the olfactory bulb in response to beta irradiation of the olfactory epithelium.[5] Beta irradiation produced neuronal activity only if it was focused directly upon the olfactory epithelium. The firing rate was a function of the logarithm of the dose rate, and the threshold dose rate was 10 mR/sec. This corresponds closely with previous behavioral data.[10,31]

The most reasonable interpretation is that radiation generates short-lived chemical substances, perhaps ozones or peroxides, at the surface of the receptor. These act as an adequate stimulus. Only relatively few bulb neurons respond to X-ray.[5,7] Thus, ozone in the atmosphere may produce its specific masking effect by occupying the same few sites which are sensitive to radiation exposure.

DELAYED EFFECTS VIA INTERNAL RECEPTORS

After this initial period of olfactory stimulation that coincides with the exposure period, there appears to be a symptom-free period for sublethal doses that lasts for 20 minutes to several hours, followed by an internal malaise. Human patients complain of nausea, monkeys vomit, and rats become inactive.[28] Let us make a clear distinction between the immediate and the delayed effects by the use of a rather laborious analogy. If a person walks into a room filled with a poison gas, he first smells the gas and later, as he accumulates a critical dose in his system, he becomes ill. Both the olfactory sensation and the illness may be dependent upon the same molecule. Or, in a room with a mixture of gases, the olfactory sensation may be dependent upon one molecule and the illness upon another. But in both cases, inasmuch as the molecule, or molecules, are operating via different afferent systems, the olfactory sensation is psychologically distinct from the sensation of illness, and any association between the two is dependent upon whether their juxtaposition is favorable for learning.

Radiation operates in a similar fashion; it produces a host of chemical changes. Some, operating upon the olfactory epithelium, produce smells, while the same chemical changes, or perhaps others, produce illness by operating on internal homeostatic receptors. Any association between these two sensations depends upon learning. Let us consider some empirical findings.

If an animal drinks a distinctively flavored fluid prior to the radiation-induced malaise, he will develop a conditioned aversion for that fluid because of the association of the flavor and the subsequent illness. The aversion is proportional to the total radiation dose.[17] Similar aversions can be established when ingestion of a distinctive substance is followed by injections of toxins and emetic drugs,[19] or with transfusions of whole blood or serum from irradiated donor animals.[15,25] Apparently the radiation malaise is due to circulating by-products of energy absorption that produce gastric disturbances.

The Specificity of Radiation-Induced Aversions

The conditioned aversion is of interest to us on several accounts. Animals develop aversive reactions specifically to gustatory and olfactory stimuli presented prior to the malaise, but not to auditory and visual stimuli so presented. Conversely, avoidance reactions induced by peripheral pain transfer readily to auditory and visual signals but not to gustatory and olfactory stimulation.[18,20]

Gustatory aversions can be established even though the radiation exposure or the drug injection is delayed for many hours after the animal has consumed the distinctive substance.[14,33,39] We are concerned with the nature of the gustatory stimulus and the related capacity of the animal to form associations over such long time periods. It seems clear that a distinctive chemical attribute of the substance ingested is required for this kind of learning. Nonchemical attributes, such as size of pellet, do not serve as cues. Let us examine some new studies in this regard.

Chemical Cues in Radiation-Induced Aversions

All the experiments discussed below were performed on young male white laboratory (Sprague-Dawley) rats. The general procedure was, first, to habituate the animals to one 10-minute drinking period at the same time each day in their home cage, then to present them with a distinctively flavored water every third day. The animals were removed from their home cage sometime after drinking flavored water and exposed

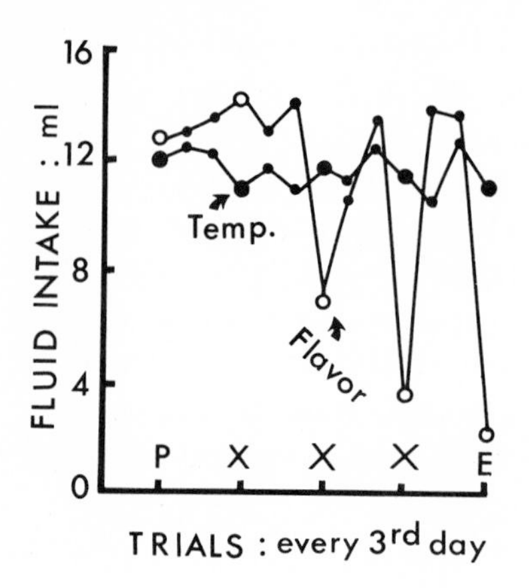

Figure 2 The development of a conditioned aversion for saccharin-flavored water repeatedly associated with X-ray exposure. Animals that had a temperature difference of 23° C associated with the same X-ray treatment did not develop an aversion.

to 280 kv filtered X-radiation (half value layer 1.4 mm copper) at the rate of approximately 15 R/min in another room. On days between flavor and X-ray trials the animals were given their usual 10-min drinking period.

Figure 2 summarizes the results from two groups treated in this way. The flavor group (N = 10) was given water flavored with one gram of saccharin per liter of water, then a radiation exposure (100 R), which began five minutes after drinking. The temperature group (N = 13) was given distilled water which differed by 23° C on the X-ray days. The water was either warm (43° C) or cool (20° C). Seven of the animals had warm water paired with X-rays and cool water during their usual drinking period. The remaining animals were treated conversely.

The first presentation of flavor or temperature difference served as a pretest. The next three presentations were followed by X-ray, while the last presentation served as a post-test.

The animals which were given a flavor difference developed a significant aversion ($P < 0.01$ by ranks test) after a single exposure. This aversion became progressively greater with each exposure. In contrast, the temperature group never developed an aversion despite the obvious temperature difference, which was repeatedly paired with radiation. Others have shown that rats will not acquire a conditioned aversion to a combination of temperature and consistency of a gelatinous nutrient,[36] nor to the position of the food cup[38] when these cues are followed by illness. Hence it seems clear that a chemical cue is required for the development of a radiation-induced aversion.

Transient Gustatory Cues and Delayed Visceral Feedback

Apparently a chemical substance must be ingested to provide an effective cue when the noxious effects of ingestion are delayed. Since a relatively large amount (13 ml) of extremely sweet saccharin-water is consumed, it is entirely possible that some traces of this substance linger in the system and are thus physically present when the animal experiences the delayed noxious effects of radiation or injected drugs. On the other hand, the temperature differences would have dissipated, making this association impossible. Therefore we repeated the above study with a chemical substance that provides a flavor but does not linger in the system.

The transient taste cue selected was water flavored with 0.5 ml hydrochloric acid per liter of water. Rats prefer unflavored water but will accept this solution when the former is not available. Humans report that this solution has a mildly sour taste which does not seem to linger in the mouth. Litmus paper tests on rats and men indicate that the mouth acidity returns to normal within two minutes. Furthermore, the sour water that is swallowed is soon mixed with gastric juices, which contain this acid in higher concentration. Our animals were exposed 10 minutes after eating and became ill probably no sooner than 30 minutes after drinking this solution.

The results are summarized in Figure 3, where only the HCl test days are plotted;

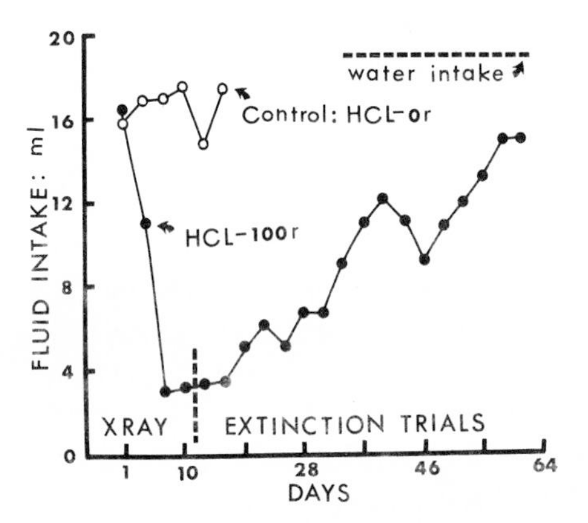

Figure 3 The development of an aversion for a transient flavor (0.5 ml hydrochloric acid per liter of water) when paired with X-ray exposure, and its persistence for more than a month following radiation treatment.

the scores on the interpolated water days were eliminated for clarity. Consumption of unflavored distilled water on the days between test days tended to be higher, as indicated by the water intake level line in Figure 3. The irradiated animals ($N = 10$) developed a profound aversion, which was significant compared to the controls ($N = 10$) after a single exposure ($P < 0.01$ by ranks test). The aversion persisted for more than a month after the final X-ray exposure although the animals were thirsty (due to 24-hour water deprivation) whenever the solution was presented.

In a second study we compared the effectiveness of the same acid solution with another chemical cue—0.025 g of quinine per liter of water. Rats prefer distilled water to this slightly bitter solution but accept it on equal terms with the hydrochloric acid solution when given a choice between the two.

This study differed in method from the two previous studies. Thirsty animals were given either bitter (quinine) water or sour (hydrochloric acid) water to drink for 10 minutes in the home cage. All animals drank 7.0 ml or more. The animals were then given dry laboratory food to eat for one hour. All animals ate more than 5.0 g and the average amount was approximately 9.0 g. Immediately after eating, the animals were exposed either to a 300-R dose of X-rays or sham exposed to a zero dose.

A two-bottle preference test for the two solutions (sour vs. bitter) was conducted in the home cages for a 24-hour period beginning two days after exposure. The results, summarized in Figure 4, illustrate the amount of hydrochloric acid solution con-

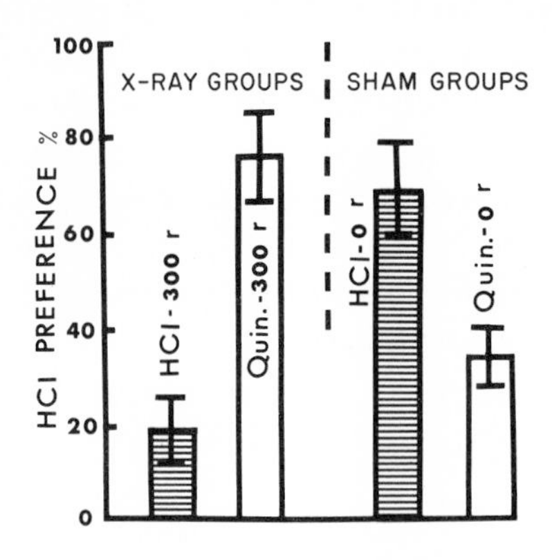

Figure 4 Mean preference (with standard deviations) for a weak acid solution when presented with a weak quinine solution for four groups of rats which had one of the two fluids previously paired with either 300 R or X-rays as indicated on the bars. Bars indicate consumption of the acid solution as a per cent of total fluid consumed from both bottles.

sumed as a per cent of the total fluid consumed from both bottles. Animals ($N = 7$) that had the sour taste paired with the noxious X-ray effects preferred the bitter taste, drinking less than 20 per cent from the hydrochloric bottle; their controls ($N = 6$), which had previously consumed the sour water without ill effects, preferred the sour taste, drinking over 70 per cent from the hydrochloric bottle. Conversely, animals ($N = 7$) that had the bitter taste associated with the X-ray effects preferred the sour taste; they drank over 75 per cent of their fluid from the hydrochloric bottle. Their control animals ($N = 6$), which drank the bitter water without ill effects, preferred the bitter water, drinking less than 35 per cent from the hydrochloric bottle. Both effects were statistically significant ($P < 0.01$ by ranks test). If a flavor is followed by no ill effects (or by beneficial effects, as we shall see below) the animal will increase its preference for that fluid.

Hydrochloric acid again proved to be an effective stimulus, even though, on the average, an animal consumed only a minute physical amount (0.005 in a 10 ml drink) and then gnawed and swallowed a dry meal of 9.0 g. Thus it seems that the physical stimulus would be lost, as food and water are mixed with gastric acids in the stomach for an hour before exposure begins.

Aversions had been established when sucrose consumption was followed by X-ray exposure seven hours later,[33] when milk consumption was followed by X-rays one hour later,[34] and when the smell of perfume was followed by a subsequent radiation illness.[19] In all these cases it would seem that the original stimulus is completely absorbed or altered before illness sets in. The most reasonable hypothesis is that the gustatory information is somehow coded and stored in the neural mechanism of memory until the delayed noxious visceral feedback arrives to complete the association.

A conditioned aversion in a natural setting has been observed in birds. The blue jay, which preys on butterflies, does not eat the monarch butterfly. This insect feeds on a toxic plant and apparently retains the toxin in its system. Consumption of the butterfly by the bird causes emesis and as a result the bird develops a conditioned aversion for the insect.[3] Numerous observations of this kind of behavior, described as "bait-shyness," have been noted in field tests of poisons on wild rodent populations.[1]

Animals also exhibit increased preferences for flavors associated with beneficial after-effects. Thus thiamine-deprived animals will increase their consumption of sweetened water if it is followed by thiamine injections.[16] It is well known that animals can learn to select beneficial diets on the basis of flavor even though the meal must be absorbed—thus destroying or radically altering the flavor cues of the required nutrient—before the onset of beneficial effects.[35,37]

CONCLUSIONS

Analysis of the Afferent Chemoreceptive Inputs

Since a gustatory aversion can be established with a single trial and persists for weeks, an excellent opportunity for tests of generalization to other gustatory stimuli is available. In an early study Nachman[32] provided us with a model. He gave rats a toxic solution of lithium chloride and studied generalization of the induced aversion to other salt solutions of equal molar concentration. He demonstrated that behavioral measures of generalization in drinking preferences correlated with similarity of afferent volleys in the chorda tympani. More recently Tapper and Halpern[41] have initiated a similar program. They produced an aversion for one sweet solution and tested its generalization to other sweet substances in an effort to quantify the similarity of sweet stimuli. They are now surgically interfering with gustatory afferents to study the effects of lesions upon generalization.

Braun[2] has begun to test the stereochemical hypotheses of olfaction with conditioned aversions for odorants. Using a rat drinking in a wind tunnel, he creates an aversion for one odor by following it with an injection of a noxious drug. Then he selects other odorants of similar molecular structure and tests for generalization of the aversion by noting its effect upon drinking.

Neural Convergence of Conditionally Paired Afferents

It has been argued elsewhere[13] that the rapid and specific association of gustatory stimuli with internal visceral feedback reflects the neural convergence of gustatory and visceral inputs. It appears that the commonsense notion "it must have been something I ate," often voiced by patients and displayed behaviorally by our animals, is a "hypothesis" which is neurologically programed in a wide variety of animals. Rozin[36] has demonstrated a more elaborate hypothesis testing behavior in thiamine-deficient rats. These animals tend to accept new foods more readily under deficiency conditions; they limit their eating to short bouts of one food at a time and then wait several hours before sampling another food. This "scientific" behavior maximizes their chances of finding the adequate vitamin supply in the experimental cafeteria.

C. Judson Herrick[23] described the convergence of the gustatory and visceral afferents in medulla oblongata of the tiger salamander. The visceral sensory neuropil receives the fibers both from the gustatory analyzers and the visceral receptors, providing an excellent opportunity for their specific integration. In mammalian species, gustatory afferents and visceral afferents converge via the vagus on the nucleus of the fasciculus solitarius. In addition, the area postrema, with its internal monitors, discharges afferents to this same nucleus.[30] It appears that the neural circuits involved in gustatory-visceral conditioning may require relatively few neuronal elements and synaptic junctions, thus providing an excellent locus from which to study the process of associative learning within these modalities.

ACKNOWLEDGMENT

Much of this research was carried out in the Behavioral Laboratories at Massachusetts General Hospital under the direction of William H. Sweet and Frank R. Ervin, and was supported by NIH Grants RH00467, R01 NHI 4380-01A1 NAD, and AEC contract #AT (30-1)-3698.

REFERENCES

1. Barnett, S. A. 1963. The Rat: A Study in Behavior. Aldine Press, Chicago.
2. Braun, J. J. Yale University. (Personal communication)
3. Brower, L. P., W. N. Ryerson, L. L. Coppinger, and S. C. Glazier. 1968. Ecological chemistry and the palatability spectrum. *Science* **161**, pp. 1349–1350.
4. Brust-Carmona, H., H. Kasprzak, and E. L. Gasteiger. 1966. Role of the olfactory bulbs in x-ray detection. *Radiation Res.* **29**, 354–361.
5. Cooper, G. P. 1968. Receptor origin of the olfactory bulb response to ionizing radiation. *Amer. J. Physiol.* **215**, pp. 803–806.
6. Cooper, G. P. and D. J. Kimeldorf. 1965. Effects of brain lesions on electroencephalographic activation by 35 kvp and 100 kvp x-rays. *Intern. J. Radiation Biol.* **9**, pp. 101–105.
7. Cooper, G. P., and D. J. Kimeldorf. 1966. The effect of x-rays on the activity of neurons in the rat olfactory bulb. *Radiation Res.* **27**, pp. 75–86.
8. Dinc, H. I., and J. C. Smith. 1966. Role of the olfactory bulbs in the detection of ionizing radiation by the rat. *Physiol. Behav.* **1**, pp. 139–144.
9. Garcia, J., N. A. Buchwald, G. Bach-y-Rita, B. H. Feder, and R. A. Koelling. 1963. Electroencephalographic responses to ionizing radiation. *Science* **140**, pp. 289–290.
10. Garcia, J., N. A. Buchwald, B. H. Feder, and R. A. Koelling. 1962. Immediate detection of x-rays by the rat. *Nature* **196**, pp. 1014–1015.
11. Garcia, J., N. A. Buchwald, B. H. Feder, R. A. Koelling, and L. F. Tedrow. 1964. Sensitivity of the head to x-ray. *Science* **144**, pp. 1470–1472.
12. Garcia, J., N. A. Buchwald, C. D. Hull, and R. A. Koelling. 1964. Adaptive responses to ionizing radiations. *Bol. Inst. Estud. Med. Biol.* (*Mex.*) **22**, pp. 101–113.
13. Garcia, J., and F. R. Ervin. 1968. Gustatory-visceral and telereceptor-cutaneous conditioning—adaptation in internal and external milieus. *Commun. Behav. Biol.* Part A, **1**, pp. 389–415.
14. Garcia, J., F. R. Ervin, and R. A. Koelling. 1966. Learning with prolonged delay of reinforcement. *Psychonomic Sci.* **5**, pp. 121–122.
15. Garcia, J., F. R. Ervin, and R. A. Koelling. 1967. Toxicity of serum from irradiated donors. *Nature* **213**, pp. 682–683.

16. Garcia, J., F. R. Ervin, C. H. Yorke, and R. A. Koelling. 1967. Conditioning with delayed vitamin injections. *Science* **155**, pp. 716–718.
17. Garcia, J., D. J. Kimeldorf, and R. A. Koelling. 1955. A conditioned aversion towards saccharin resulting from exposure to gamma radiation. *Science* **122**, p. 157.
18. Garcia, J., and R. A. Koelling. 1966. The relation of cue to consequence in avoidance learning. *Psychonomic Sci.* **4**, pp. 123–124.
19. Garcia, J., and R. A. Koelling. 1967. A comparison of aversions induced by X rays, toxins, and drugs in the rat. *Radiation Res.* Suppl. **7**, pp. 439–450.
20. Garcia, J., B. K. McGowan, F. R. Ervin, and R. A. Koelling. 1968. Cues—Their relative effectiveness as a function of the reinforcer. *Science* **160**, pp. 794–795.
21. Garcia, J., J. R. Schofield, and D. Oper. 1966. Tachistoscope for x-rays. *Amer. J. Psychol.* **79**, pp. 318–320.
22. Gasteiger, E. L., and S. A. Helling. 1966. X-ray detection by the olfactory system: ozone as a masking odorant. *Science* **154**, pp. 1038–1041.
23. Herrick, C. J. 1948. The Brain of the Tiger Salamander-*Ambystoma tigrinum*. Univ. of Chicago Press, Chicago.
24. Hull, C. D., J. Garcia, N. A. Buchwald, B. Dubrowsky, and B. H. Feder. 1965. Role of the olfactory system in arousal to x-ray. *Nature* **205**, pp. 627–628.
25. Hunt, E. L., H. W. Carroll, and D. J. Kimeldorf. 1965. Humoral mediation of radiation-induced motivation in parabiont rats. *Science* **150**, pp. 1747–1748.
26. Hunt, E. L., and D. J. Kimeldorf. 1962. Evidence for direct stimulation of the mammalian nervous system with ionizing radiation. *Science* **137**, pp. 857–859.
27. Hunt, E. L., and D. J. Kimeldorf. 1963. Arousal reactions with a brief partial- and whole-body x-ray exposure. *Nature* **200**, pp. 536–538.
28. Kimeldorf, D. J., and E. L. Hunt. 1965. Ionizing Radiation: Neural Function and Behavior. Academic Press, New York.
29. Levy, C. K. 1967. Immediate transient responses to ionizing radiation. *Current Topics in Radiation Res.* **3**, pp. 97–137.
30. Morest, D. K. 1967. Experimental study of the projections of the nucleus of the tractus solitarius and the area postrema in the cat. *J. Comp. Neurol.* **130**, pp. 277–293.
31. Morris, D. D. 1966. Threshold for conditioned suppression using x-rays as the pre-aversive stimulus. *J. Exp. Anal. Behav.* **9**, pp. 29–34.
32. Nachman, M. 1963. Learned aversion to the taste of lithium chloride and generalization to other salts. *J. Comp. Physiol. Psychol.* **56**, pp. 343–349.
33. Revusky, S. H. 1968. Aversion to sucrose produced by contingent x-irradiation—temporal and dosage parameters. *J. Comp. Physiol. Psychol.* **65**, pp. 17–22.
34. Revusky, S. H., and E. W. Bedarf. 1967. Association of illness with prior ingestion of novel foods. *Science* **155**, pp. 219–220.
35. Richter, C. P. 1956. Salt appetite in mammals: Its dependence on instinct and metabolism. *In* L'Instinct Dans Le Comportement Des Animaux Et De L'Homme (M. Autuori, editor). Masson et Cie, Paris, pp. 577–632.
36. Rozin, P. University of Pennsylvania. (Personal communication)
37. Rozin, P. 1967. Thiamine specific hunger. *In* Handbook of Physiology. Vol. I, Section 6 (C. F. Code, editor). American Physiological Society, Washington, D.C., pp. 411–431.
38. Rozin, P. Specific aversions as a component of specific hunger. 1967. *J. Comp. Physiol. Psychol.* **64**, pp. 237–242.
39. Smith, J. C., and D. L. Roll. 1967. Trace conditioning with x-rays as the aversive stimulus. *Psychonomic Sci.* **9**, pp. 11–12.
40. Smith, J. C., and D. Tucker. 1969. Olfactory mediation of immediate x-ray detection. *In* Olfaction and Taste III (C. Pfaffmann, editor). Rockefeller Univ. Press, New York, pp. 288–298.
41. Tapper, D. N., and B. P. Halpern. Cornell University. (Personal communication)
42. Tsypin, A. B., and Yu. G. Grigor'ev. 1960. Quantitative measurements of the sensitivity of the central nervous system to ionizing radiation. *Bull. Exp. Biol. Med.* **49**, pp. 21–23.

TASTE

IV RECEPTOR MECHANISMS

IV · RECEPTOR MECHANISMS

THE ULTRASTRUCTURE OF VERTEBRATE TASTE BUDS

P. P. C. GRAZIADEI
Department of Biological Science, Florida State University, Tallahassee, Florida

INTRODUCTION

Some hundred years ago Leydig[28] provided the first histological description of taste buds in fish and soon afterwards Schwalbe[41] and Loven[29] described similar organs in the lingual papillae of mammals. The structure of these peripheral apparatuses was commonly described as little buds, inside the covering epithelium, made of slender, fusiform cells among which the sensory fibers provided a delicate plexus. Their localization in the oral cavity was found typical of terrestrial vertebrates, while aquatic forms were found to have taste organs localized on the exterior of the mouth as well as on the head or, more caudally, on the surface of the body. After these pioneering studies, a series of papers appeared between the end of the last century and the first twenty years of the present one dealing with the fine morphology of taste buds in several animal species. The broad comparative approach of these studies supported the idea that taste buds in all vertebrate classes maintain the basic arrangement first described by the above-mentioned students. Agreement was not reached among authors regarding the "types" of cells forming the taste buds. According to Schwalbe,[42] Merkel,[31] Hermann,[20,21] Lenhossék,[27] Gråberg,[14] and others, at least two different cell types were present in taste buds. Only one of the types was presumed to be the true sense cell, while the others were supporting elements. However, Hermann[21] described mitotic figures at the base of taste buds, and the possibility that morphological differences of the taste bud cells were related to the aging processes of a unique type and not to specific and distinct types of cells was supported by Kolmer,[25] Retzius,[39] Heidenhain,[19] Olmsted,[36] and others.

Parallel to the cytological observations of taste buds were studies made on the nervous supply of these sense organs. It was clear that in vertebrates there is no discrete gustatory nerve as there are olfactory and optic nerves. Gustatory fibers run in the V, VII, IX, X cranial nerves, as was proved by the impairment of the gustatory sensation after severing branches belonging to these nerves (see among others the works of Vintschgau and Hönigschmied,[47] Zander,[48] Herrick,[22] and May[30]). It was also clear that cutting the nervous supply induced taste bud degeneration, and subsequent regeneration of the nerves induced reappearance of taste buds.[32,37,38]

The fine innervation of taste buds was clearly investigated around the end of the last century, contemporary with the appearance of specific staining methods for the nervous system such as those of Golgi, Herlich, and Cajal. It was observed that nerve fibers penetrated among taste cells, and that these contacts were extended to all epithelial components of the taste buds (taste cells proper and so-called supporting cells). Fusari and Panasci[13] were the first to employ specific methods for the demonstration of nerves in taste buds. Subsequently, Retzius,[39] Arnstein,[1] and others were able to demonstrate a fine interlacing of fibers among taste cells. These researches showed that, besides gustatory fibers ending among taste cells (*intragemmal fibers*), others (*perigemmal fibers*) ended freely in the epithelium surrounding the buds. A third category was also described in the epithelium of the tongue; these, which do not immediately surround the taste organs, were termed *intergemmal fibers*. The exact nature of the perigemmal and intergemmal fibers is still to be determined.

The status of our knowledge on the histology of taste buds was reviewed by Kolmer,[26] after which histologists' interest in studying these organs decreased, possibly because of the limitations imposed by the technique. The electron microscope has provided, fairly recently, new ways of observing biological material, and quite recently a series of morphological papers appeared dealing with the ultrastructure of taste buds. Little attempt, however, has been made by recent investigators to study the taste organs from an anatomocomparative point of view, and the bulk of information so far available derives from studies on rabbit and rat, while studies on lower vertebrates are still largely incomplete.

The first study on the ultrastructure of taste buds was performed on rabbit by Engström and Rytzner in 1956.[9,10] These authors provided a great deal of information that has been confirmed by other students. It was clear from their observations that taste cells lack cilia but bear microvilli. Cells with different morphological characteristics were observed in the taste buds, but were cautiously interpreted as stages of a possibly unique type of cell. Direct contacts between taste cells and nerve fibers were also observed, but discrete synaptic areas were not mentioned.

Several papers subsequently appeared on taste bud ultrastructure in rabbit,[4–6,24,34,35,40,43] rat,[11,15,17] monkey,[33] frog,[45,46] and fish.[7,8,23,44] All these researches show that, as previously observed in older papers, a common organization pattern exists in the structure of taste buds of different species. Slender epithelial cells showing microvilli or microvilli-like structures at their free surface are consistently found. A belt-like system of tight junctions is situated at the apical pole of the taste cells, thus sealing off the environment from the intercellular spaces of the taste bud. Nerve fibers in direct contact with taste cells have been found regularly; however, authors have made no specific mention of whether these contacts are more common or are even present in all the cell types of the taste buds or are only in the ones described as "taste cells proper." The nature of these contacts appears somewhat obscure because, particularly in mammals, they lack the typical details of synapses (see Gray and Watkins[15] for a survey). Only recently have synaptic contacts been described in humans.[18] In lower vertebrates,[18,46] nerve to taste-cell contacts show morphological details similar to the ones described in CNS synapses. The problem that has most puzzled the authors, however—the classification of types of cells in the taste buds—has so far received little help from recent ultrastructural observations. Several classifications of cell types have been proposed for rabbit and rat taste buds,[11,34,35,40] but there is not yet complete agreement among authors as to the functional role of these types and their specific part in taste transduction. In lower vertebrates, taste and supporting cells are recognized as discrete, morphologically recognizable elements.[7,8,24,44]

Using H^3 thymidine label and light microscope radioautography, Beidler and co-workers[2,3] were able to show that in rat the cells from the surrounding epithelium enter the buds at a rate of one every 10 hr, with a renewal time of approximately 11 days. This observation has given support to the hypothesis put forward by many authors from Kolmer[26] on that morphological differences in taste cells may be due to aging processes. It does not rule out, however, the possibility that cell types other than the ones marked could be present in the buds.

In my opinion, a comparative ultrastructural study of taste buds in different animal species could yield information with regard to essential morphological details. These details, consistent in all the species and possibly more easily recognizable (morphologically) in some of them, should allow the student to discriminate them from not relevant, although impressive, structures that are not necessary for taste transduction. A comparative ultrastructural study is now being carried on in our laboratory on taste buds of several animals. The preliminary results of this study will be presented below.

METHODS

Together with orientative histological preparations (Figure 1) and specific silver methods such as Cajal, Bodian, Holmes, and Bielschowsky, the data were obtained from material prepared for ultrastructural observations both with the scanning electron microscope and the transmission electron microscope. Methods of preparation of these specimens will be mentioned in the picture captions. The following species have been investigated so far: adult humans, rat, frog, catfish, and guppy.

RESULTS

The first interaction between taste compounds and sensory structures is presumed to take place at the pore region of the taste bud (Figure 1). The ultrastructural details of this part of the sensory organ may then yield valuable information on taste mechan-

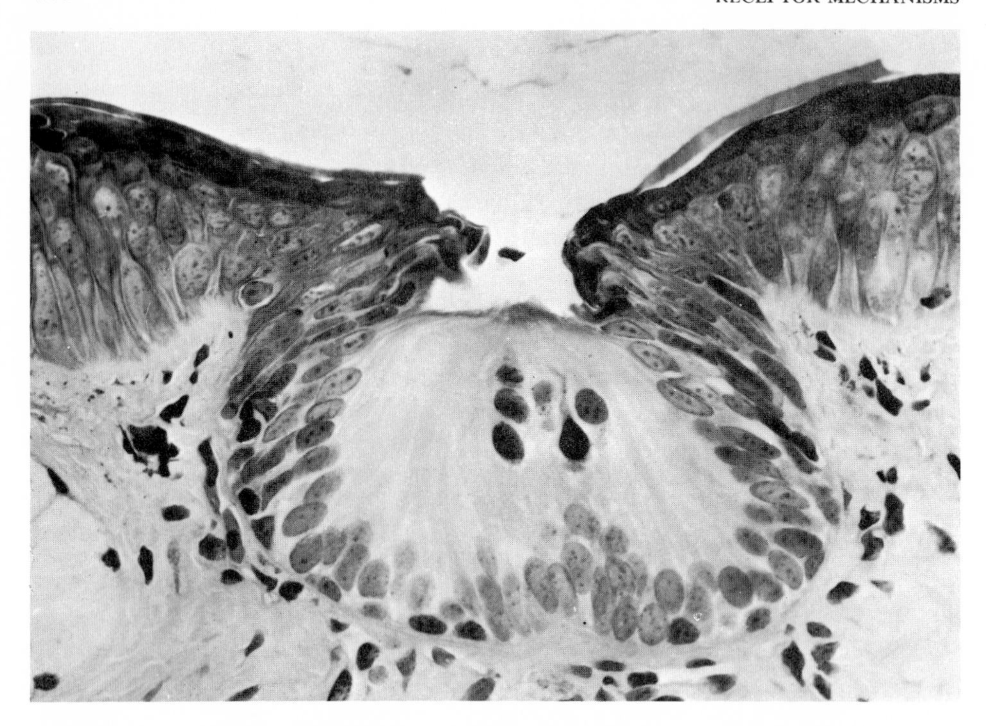

Figure 1 The histological section through the lip of amphiuma shows a typical taste bud. Slender cells compose the sense organ and their nuclei have a quite regular basal position. Cell nuclei darkly stained at the center possibly belong to mature cells in the process of degeneration. In this micrograph, which is representative of the structure of taste buds in all vertebrates, cell types are not obvious. Fixation: Bouin. Stain: iron hematoxylin. ×249.

ism. From the observations of Engström and Rytzner[9,10] in rabbit, it was clear that, contrary to olfactory and other vertebrate receptors, taste cells lack cilia. This has been consistently confirmed in all the species so far examined. The conspicuous microvillar apparatus, demonstrated in rabbit by Trujillo-Cenóz[43] and many other authors after him, has not been observed in this study to have the same extension in other species. Taste cells in human and rat fungiform papillae (Figure 2) are provided with rather short microvilli, some 0.5 μ long and with a diameter of some 0.1 μ or less. Their internal core maintains the low homogeneous electron density of the cytoplasm of the cell and does not show the filamentous structures reported by other authors[34,35,40] in rabbit. In frog, where taste and sustentacular cells are morphologically discernible, both have irregular microvilli some 0.5 μ in length. The pattern of these microvilli is clearly seen in surface views of the buds, as in Figures 3 and 4. In both guppy and catfish, while the cells presumed to be supporting have fine microvilli some 900 Å in diameter with a core of densely packed tubular structures some 90 Å in width, the taste cells do not have microvilli, but have a unique finger-like protrusion 1500 Å to 2000 Å in diameter and 0.4 μ long in which tubular structures are observed as well (Figure 5). Microvilli as such seem then not to be consistent structures in all taste cells, while protrusions of different size and shape are (Figures 6 and 7). Their function is still to be precisely determined, but these cellular processes have in common the property of increasing the free receptor surface, sometimes considerably.[16]

A junctional complex system similar to the one previously described by Farquhar and Palade[12] in several epithelia has been consistently described in the apical portion of the taste bud cells in all the species examined.

Several attempts have been made by previous authors to provide morphological criteria to substantiate a classification of types of cells in taste buds of different animal species. Two animals in particular have been studied from this point of view—the rat and rabbit.[4–6,24,34,35,40] In spite of some divergence, there seems to be agreement among students that the morphological variations of taste cells, at least in mammals, are due to aging processes more than to different categories of cells. However, so far, quantitative and histochemical data are lacking, and more observations are needed to give to the data of Beidler and co-workers[2,3] a precise morphological base. In humans, morphological differences in taste cells have been noticed

Figure 2 Surface view of the tongue of rat showing several filiform papillae and one fungiform papilla. At the center of the fungiform papilla a little region marked with an arrow represents the pore of the taste bud. The reduced length of microvilli does not allow them to be visualized in this surface view, contrary to the way they can be seen in the taste pore of fish (see Figure 6). Stereoscan scanning electron micrograph. Fixation in 10% formalin and specimen coated after dehydration with gold. ×612.

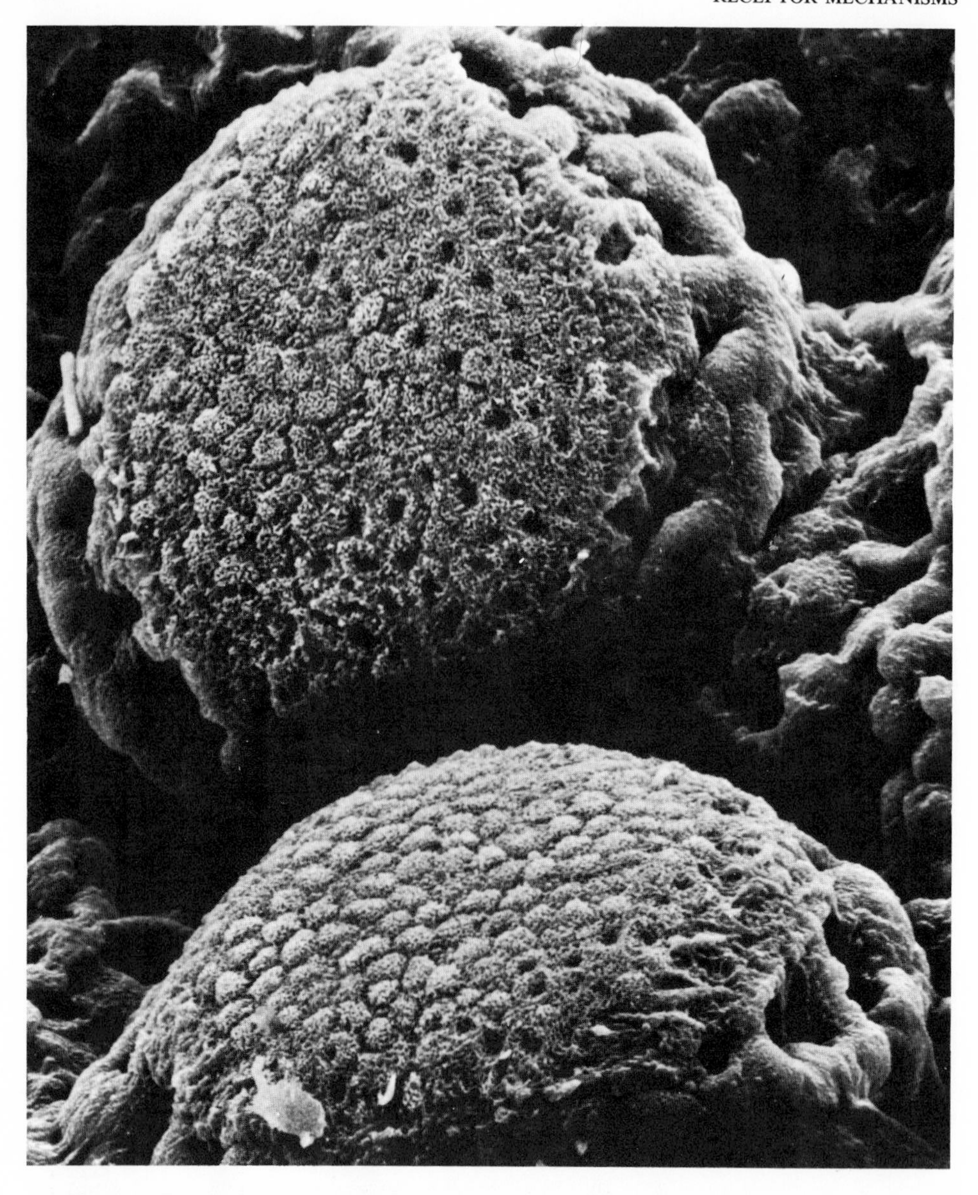

Figure 3 Tongue of a frog showing two taste buds in different projections. The pattern on the surface of these taste buds is made by the discrete arrangement of microvilli on top of every taste cell. Same method of preparation as in Figure 2. × 1050.

in the course of these observations, but no attempt has been made to classify them quantitatively because of the difficulty in establishing reliable criteria. These morphological differences are no greater or more specific than those observed in the cells of the basal layer of the common epithelium surrounding the taste buds.[17] In the study of partially reconstructed taste buds through ultra-thin serial sections, moreover, it was observed that one cell may vary considerably in organelle content

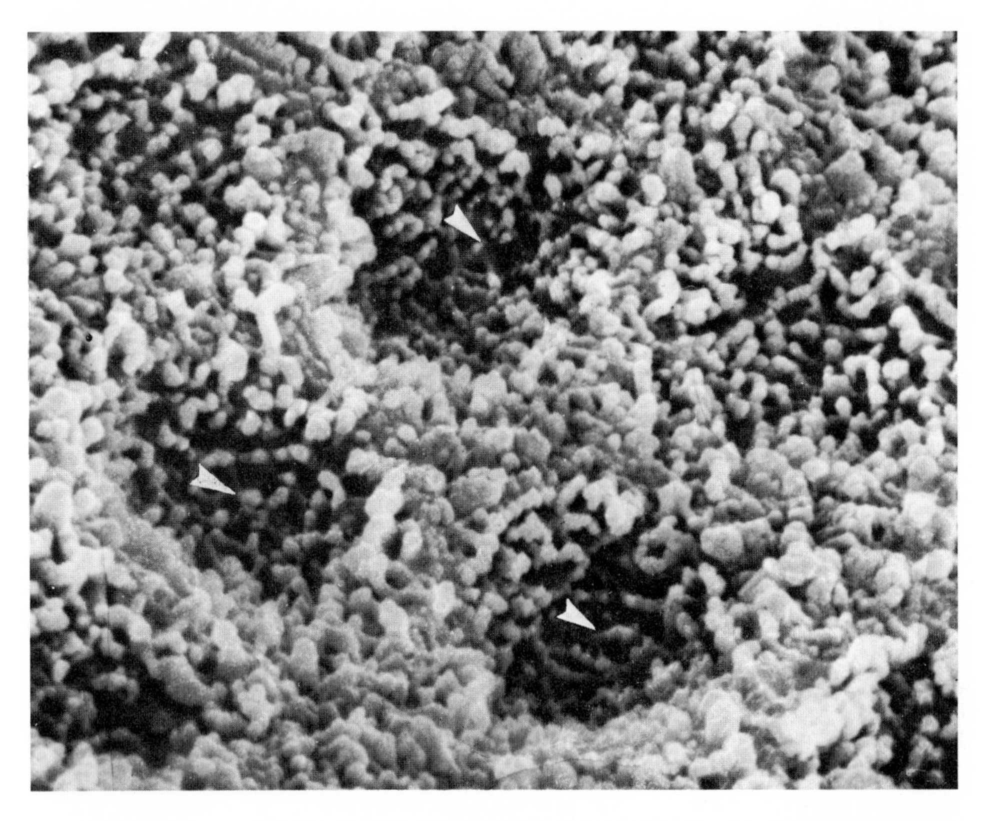

Figure 4 Detail of the surface of a frog taste bud to show the different arrangements of microvilli. The zones marked by arrows represent the presumed sensory cells while the other areas are occupied by the so-called supporting cells. Same preparation as in Figure 2. $\times$ 10,150.

from one region to the other, so that criteria for their morphological classification must rely on complete reconstructions of entire taste buds. The cells show a variable number of mitochondria, which are more numerous in the supranuclear region, sparse profiles of the endoplasmic reticulum both of the smooth and rough type, and free ribosomes clustered in rosettes. Tonofilaments, common in the epithelial cells surrounding the buds, are very scanty in the taste cells and desmosome attachments are reduced in size if compared with those of the common epithelium. Tubular structures, already reported by Murray and Murray[34] in rabbit, are also observed in the apical portion of taste cells of rats and humans. These organelles consist of tubules 200 Å in diameter, longitudinally arranged along the main axis of the cell. Similar tubules are observed in the supranuclear region of taste cells of the lower vertebrates, and closely resemble the tubules described in the olfactory receptors of many vertebrates.[16]

In lower vertebrates, taste bud cells are usually classified in at least two groups, one of which is the taste cell, while the other presumably is the supporting cell. In the present observations, the differences are more obvious in frog, where our results are

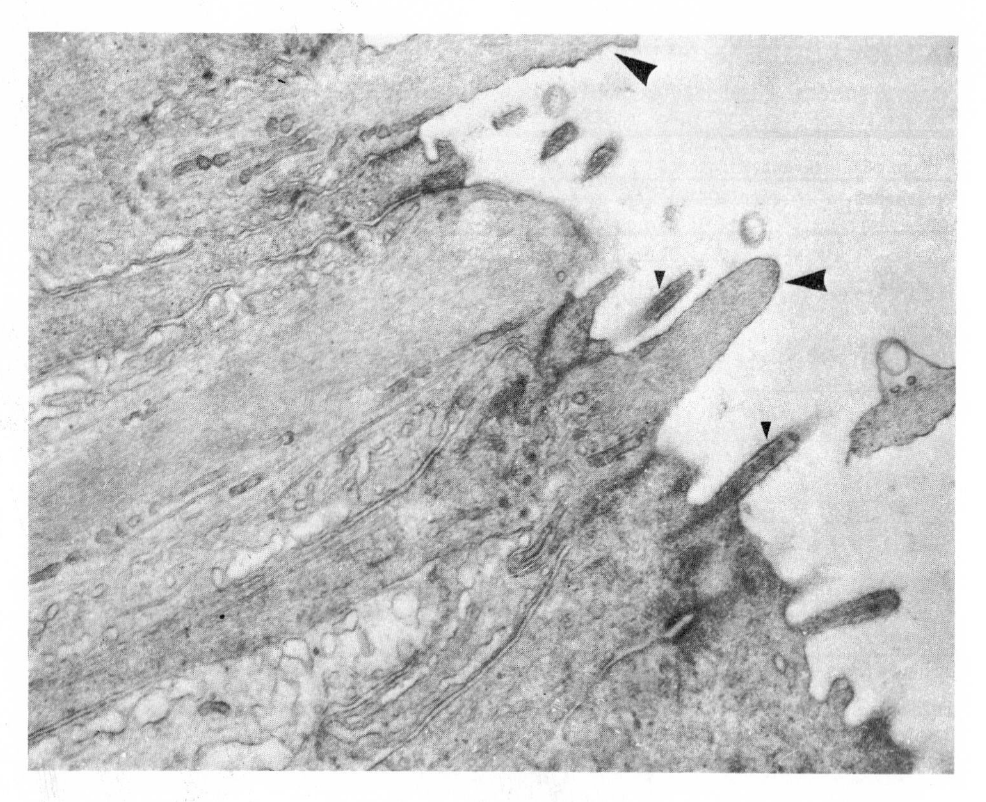

Figure 5 The apical portion of taste cells in fish shows two types of protrusions. Some cells have long, unique, finger-like appendages (large arrows) while all others have more conventional microvilli structure (small arrows). The structure of these two cell types shows few morphological differences (see text). Taste bud from a catfish barbel. Fixation: 2% OsO_4, uranyl acetate + lead citrate stain. ×62,500.

similar to those described by Uga.[45] In fish, apart from the differences observed at the free part of the cell where microvilli and cellular protrusions characterize the two types of cells, criteria for differentiating the cell types are less obvious. The presumed taste cells have their supranuclear portion filled with tubular structures 200 Å in diameter and few canalicular and vesicular profiles belonging to the endoplasmic reticulum system. Few mitochondria are observed in this part of the cell. The presumed supporting cells do not have specific cellular components to differentiate them from the first category of cells, apart from the before-mentioned microvilli, but usually they present fewer tubules and a more irregularly scattered system of endoplasmic reticulum profiles. The critical analysis of the morphological differences, at least in the material so far observed, does not provide clear-cut criteria for differentiating two cell types. There seem to be some differences, not in the quality of organelles but in their relative number. One can say that a clear distinction of cell types is not yet possible as it is, for instance, in the lateral line organs of fish, in which the receptors are clearly

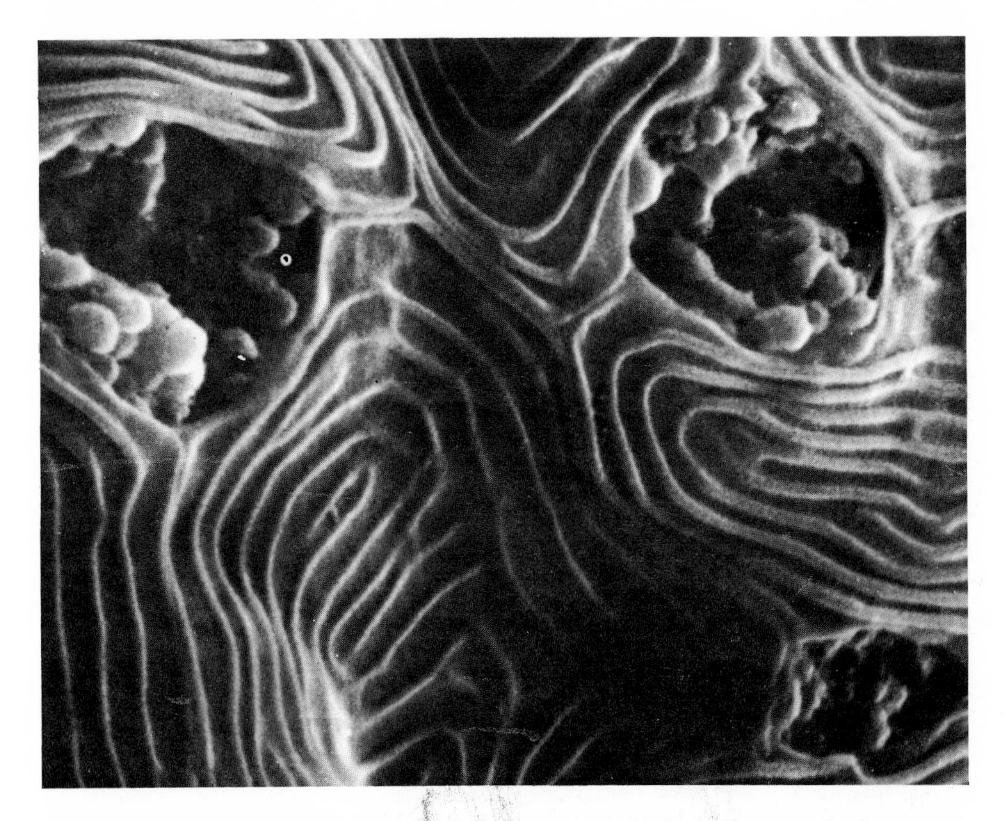

Figure 6 In the lip of fish, several taste buds are recognizable by their pore opening between scales. Large ball-like structures represent the apex of the projections observed in taste cells. See Figure 5. Same preparation as in Figure 2. × 8600.

different from the supporting cells both in specific morphological details and relations with nerves. From our observations, we can conclude that in lower vertebrates some differences do exist among at least two categories of taste cells, and that they are more obvious than in higher vertebrates. Comparative histochemical, ultrastructural, and radioautographic observations, however, should be completed in order to discover positively if substantially different types of cells exist or if only metabolic differences characterize the cells of the vertebrate taste buds.

In sense organs such as cochlea, semicircular canals, and lateral line of fish, sensory nerve fibers establish discrete contacts in the form of typical synaptic apparatuses as can be seen clearly with the electron microscope. In taste buds, nerve to taste-cell relations are rather atypical.[15] Discrete synaptic areas have seldom been described in mammals. In the course of these observations, it has been established that in rat these contacts are extremely sparse. The nerve fibers penetrate the buds and run among taste cells, often invaginated in grooves, but no clear synaptic apparatuses have been detected. In few instances (five cases during extensive examination

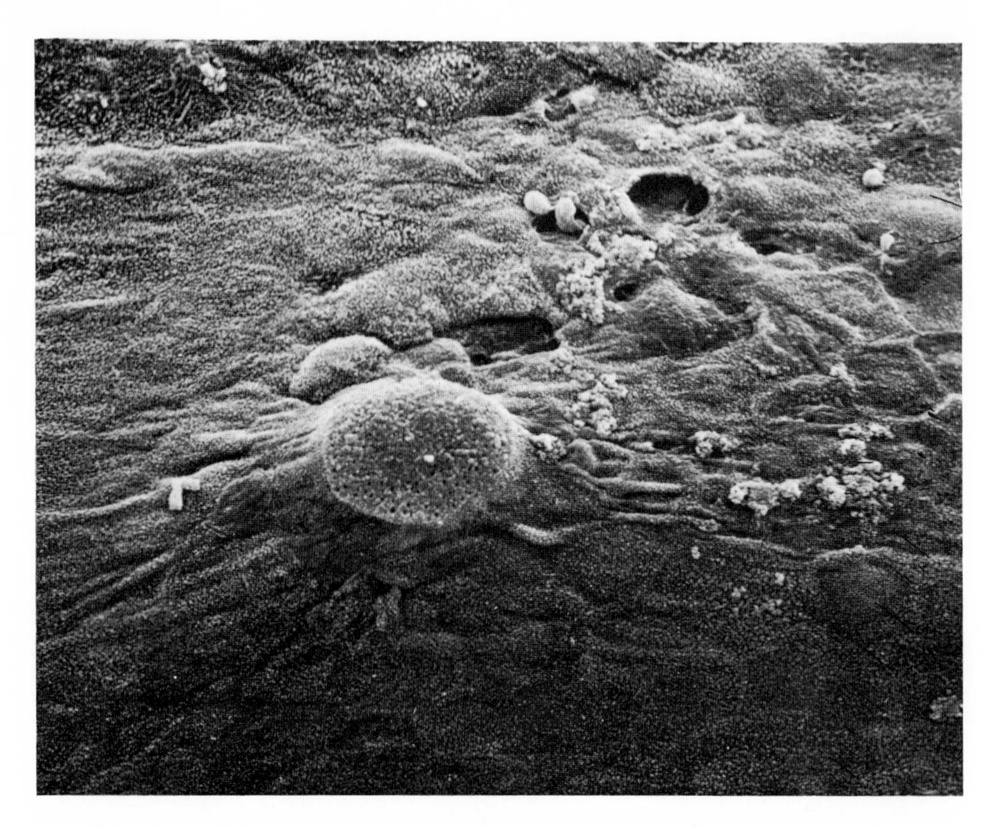

Figure 7 On the barbels of catfish, several taste buds can be recognized. One of them appears in the center of the figure like a little hill. Same preparation as in Figure 2. ×1700.

of some 40 taste buds) discrete areas of increased density of membranes with closely clumped vesicles have been observed (Figure 8). In humans discrete zones with the morphological characteristics of synapses have been observed. I have described these contacts[18] as having a double polarity, judged by the localization of the vesicles. In one case, vesicles are clumped on the nerve fiber membrane side and in the other on the taste cell membrane side (Figures 9 and 10). One could postulate on the basis of this morphological evidence that both afferent and efferent synapses occur. It is, however, premature at present to establish the nature of the two types of fibers and their precise role in taste mechanism.

In frog and fish, synaptic contacts have been observed by a number of authors (in frog, Uga;[45] in fish, Desgranges.[7,8]). Discrete synaptic contacts have been confirmed in frog, catfish, and guppy, and in all these species they are localized at the basal portion of the taste cells where a core of nerve fibers establishes contacts with processes of the taste cells (Figures 11 and 12).

The presence of muscle fibers in the core of the papillae bearing taste buds is not commonly mentioned in the ultrastructural studies so far performed. The observation, however, is rather common in frog, and will be dealt with in detail in a separate paper. These muscle fibers are striated, and have been observed running in the core of the papillae peripheral to the trunk of nerves and vessels directed toward the taste bud. Contraction of these fibers alters the length of the papilla, as they are longitudinally arranged in its core, but their precise function has not yet been established.

CONCLUSIONS

From the first observations of taste buds in the last century until the present, attempts have been made to classify morphologically the cell types constituting the taste bud. So far, these attempts have not been completely successful. A pattern common to all vertebrate species cannot be found, and even in mammals the classifications proposed by different authors vary considerably. The recent observations of Beidler and co-workers[3] proving that, at least in rat, taste bud cells turn over and that every cell has a life span of approximately 11 days, lead the morphologists to indicate one or more "stages" instead of "types" of cells in the taste buds. However, "types" could be

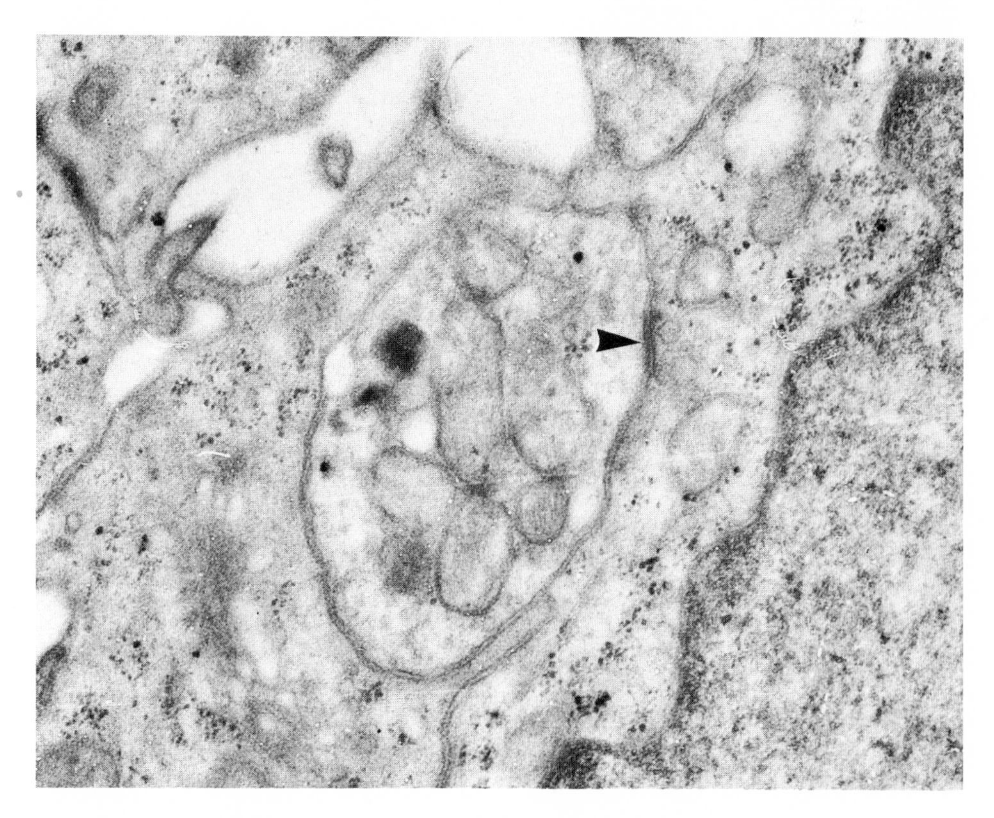

Figure 8 Nerve fibers often run in grooves of taste cells. One such fiber in a rat taste bud is shown contacting two cells. At the arrow, increased density of the membranes is observed and few vesicles lie close in the taste cell cytoplasm. In rat, these contacts are sparse, but represent the only morphological specialization so far observed between these taste cells and nerve fibers. Rat taste bud from fungiform papillae. Fixation: 2% OsO_4 in veronal acetate buffer. Lead citrate stain. ×51,350.

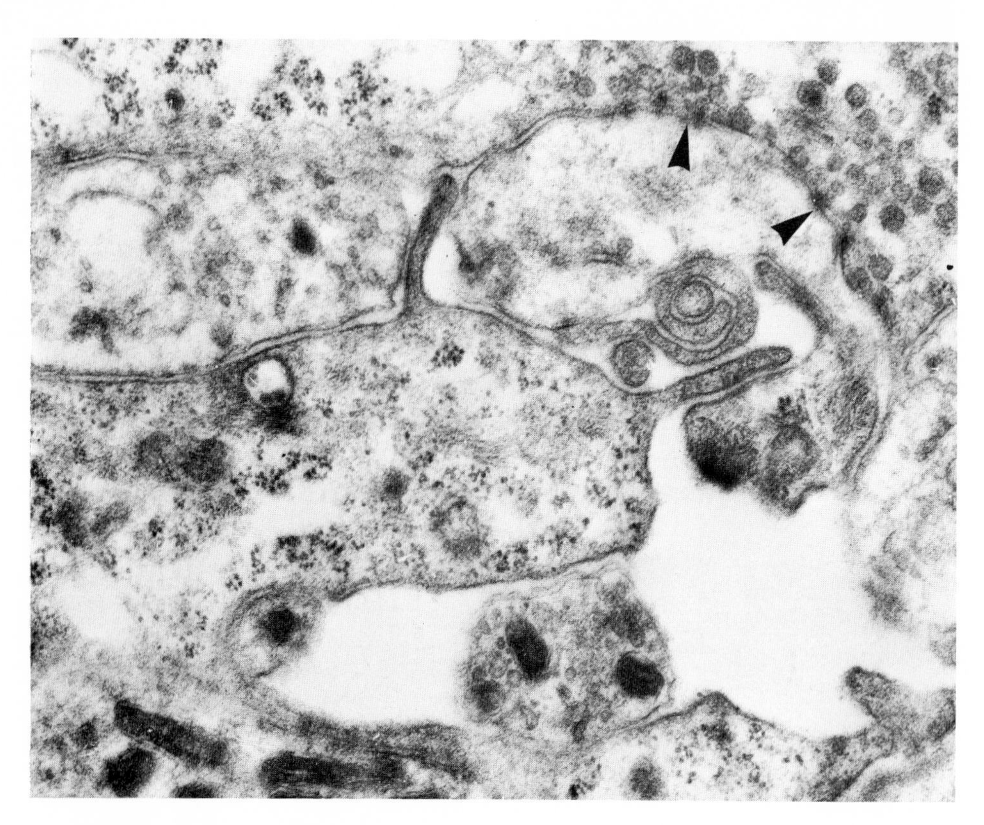

Figure 9 Human taste bud. Several vesicles in the cytoplasm of a taste cell clump in the region of contact with a nerve fiber (arrows). In humans, these contacts have vesicles either on the taste cell side or on the nerve fiber side (see Figure 10). Fixation: 2% OsO_4 in phosphate buffer. Uranyl acetate + lead citrate stain. ×30,750.

present, as the results of Beidler do not rule out this possibility but state only that cells penetrate inside the buds from the surrounding epithelium at a given rate. The possibility need not be overlooked that different types of cells do really exist with different turnover rates. Definitive proof has still to be provided, and the data so far available do not indicate specific morphological characteristics of any taste cells. The problem, then, is to establish the turnover rate, if any, of every cell type, the relations of each cell to the nervous supply, and, eventually, the role of each cell in taste mechanism. It is worth mentioning at this point that some differences exist between the conditions observed in mammals and those in lower vertebrates. In the first, a clear distinction between taste and supporting cells does not seem possible on the basis of ultrastructural findings only.[11,17,34,40] In amphibians and fish, taste cells are often described as opposed to supporting cells,[7,8,45] and in frog these differences are seen consistently.

There seems to be a marked difference in the nerve to taste-cell contacts in different animal species. In mammals, nerve fibers penetrate the bud and are singularly in-

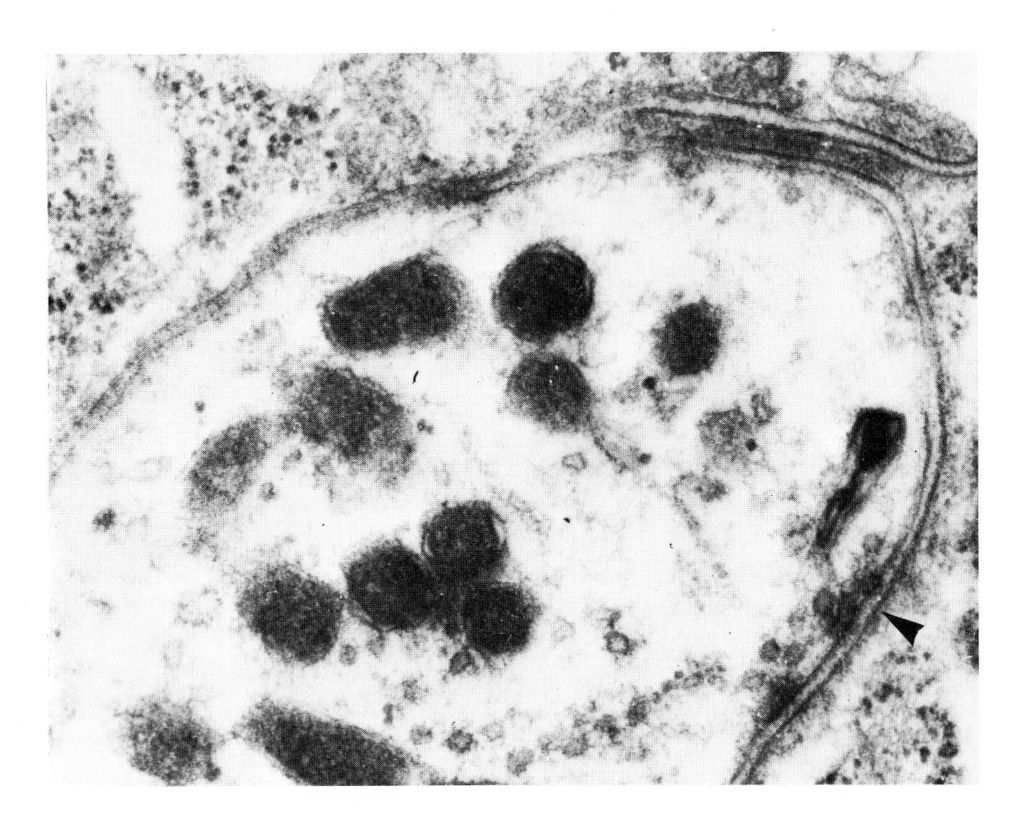

Figure 10 A nerve profile from a human taste bud shows vesicles clumping near a zone where increased density of membranes occurs (arrow). The polarity of this contact as judged from the position of vesicles is reversed if compared with the one of Figure 9. Same material, fixation, and stain as in Figure 9. $\times$ 108,930.

vaginated in pockets of the cells, reaching levels not far from the pore. In the lower vertebrates, nerve fibers make contacts only at the base of the taste cells. Classic synaptic apparatuses have been described only in humans.[18] In rat, these specializations have not been clearly observed. In the course of this symposium, Murray (p. 000) has provided evidence of membrane specializations with clumping of vesicles along the course of nerves in the taste buds of rabbit. In lower vertebrates, previous observations[8,45] as well as the present ones point out that clear synaptic apparatuses do exist. This variable morphology of the nerve to taste-cell contact is not clear. The fact that a clear turnover exists, as has been proved by Beidler in rat, indicates that this contact is not permanent, but is limited in time. It may be postulated that, during this relatively short period, morphological specializations such as those which we know happen in the CNS have no time to develop, in spite of adequate functional properties of the contacts between nerve and taste cells. Clearly, the relations of nerves to taste cells must be further investigated to explain the discrepancies of our observations thus far.

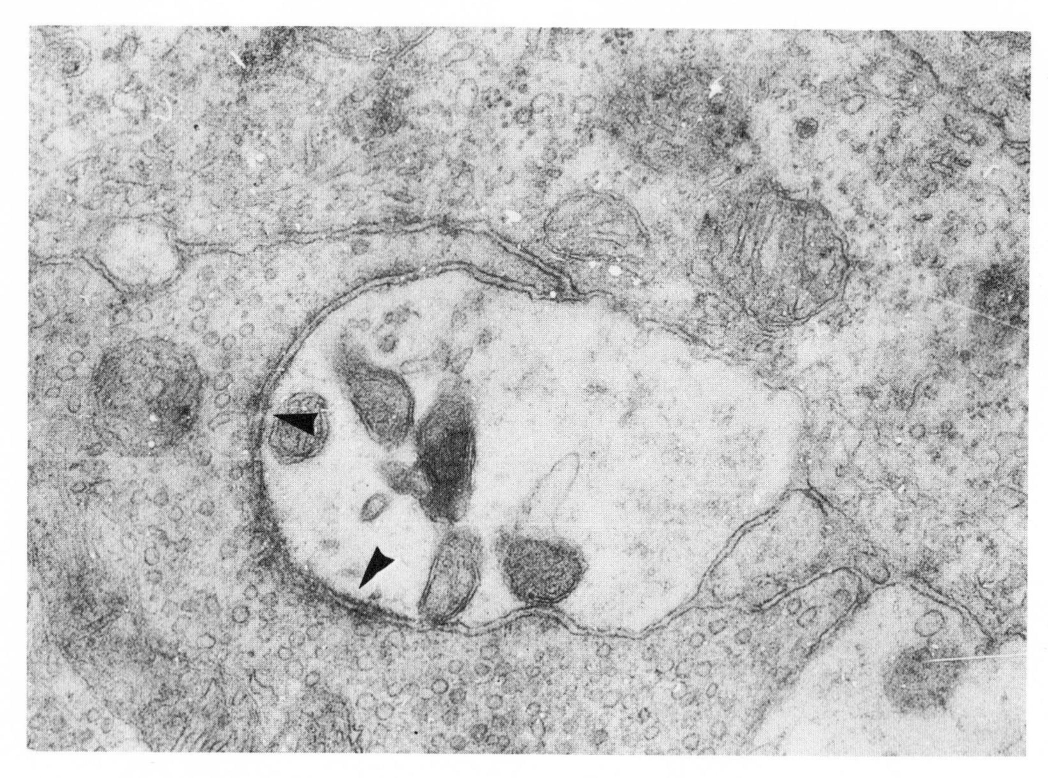

Figure 11 Detail of the synaptic zone in the taste bud of catfish. Notice large number of vesicles in the cytoplasm of the taste cell. These vesicles clump in discrete regions of contact with a nerve profile (arrows). Same material, fixation, and stain as in Figure 5. × 104,000.

ACKNOWLEDGMENT

The author wishes to acknowledge the photographic assistance of Mr. D. Fleming and the technical help of Mr. R. Parker and Mrs. N. Kemp in preparing the material.

This work was supported in part by the following grants: The Nutrition Foundation, Inc. #400; U.S. Atomic Energy Commission, between the Division of Biology and Medicine, Public Health Service grant SPO 1 NBO7468-02 NBP, and the Florida State University.

REFERENCES

1. Arnstein, C. 1893. Die Nervenendigungen in den Schmeckbechern der Säuger. *Arch. Mikroskop. Anat.* **41**, pp. 195–218.
2. Beidler, L. M., M. S. Nejad, R. L. Smallman, and H. Tateda. 1960. Rat taste cell proliferation. *Fed. Proc.* **19**, p. 302.
3. Beidler, L. M., and R. L. Smallman. 1965. Renewal of cells within taste buds. *J. Cell Biol.* **27**, pp. 263–272.
4. De Lorenzo, A. J. 1958. Electron microscopic observations on the taste buds of the rabbit. *J. Biophys. Biochem. Cytol.* **4**, pp. 143–150.
5. De Lorenzo, A. J. 1960. Electron microscopy of the olfactory and gustatory pathways. *Ann. Otol. Rhinol. Laryngol.* **69**, pp. 410–420.
6. De Lorenzo, A. J. D. 1963. Studies on the ultrastructure and histophysiology of cell membranes, nerve fibers and synaptic junctions in chemoreceptors. *In* Olfaction and Taste I (Y. Zotterman, editor). Pergamon Press, Oxford, etc., pp. 5–17.

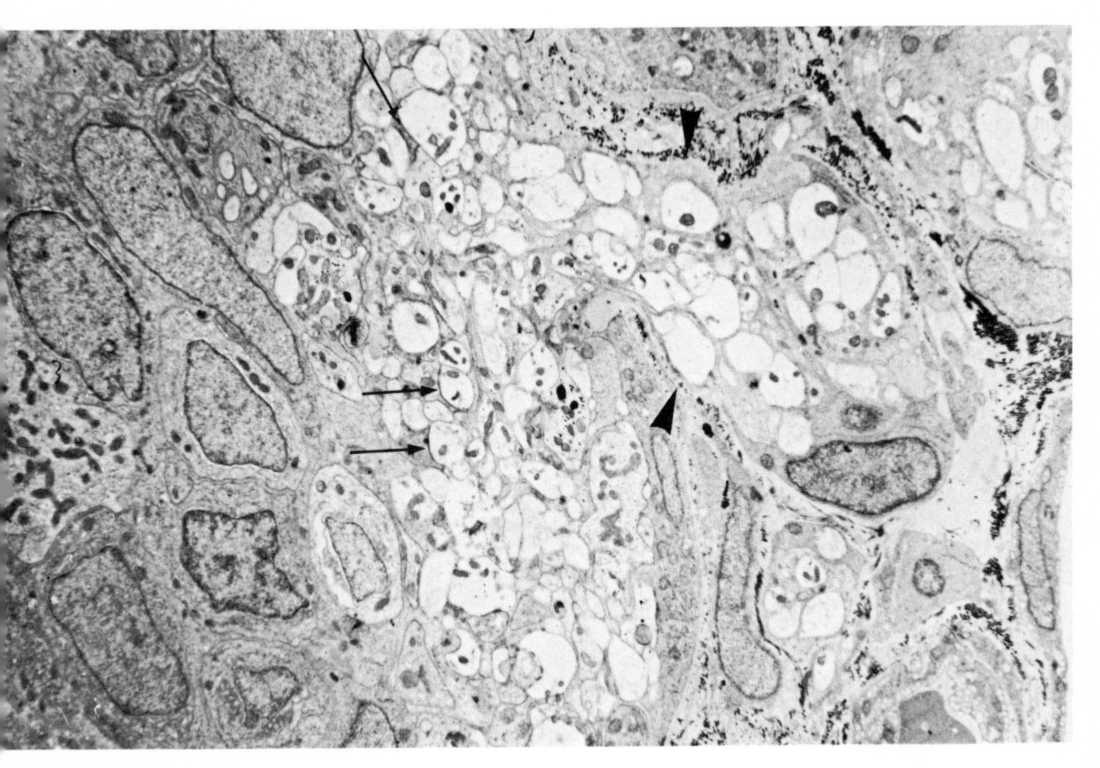

Figure 12 In the taste bud of fish, nerve to taste-cell contacts occur at the base of the sense organ. The sensory nerve (between large arrows) penetrates the basal membrane and gives rise to a neuropil-like structure where synaptic contacts occur between nerve fibers and the basal portion of the taste cells (small arrows). Nerve profiles are generally clear and sparsely provided with mitochondria. Same material, fixation, and stain as in Figure 5. × 14,500.

7. Desgranges, J. C. 1965. Sur l'existence de plusieurs types de cellules sensorielles dans les bourgeons du goût des barbillons du Poisson-chat. *C. R. Hebd. Séances Acad. Sci. Paris* **261**, pp. 1095–1098.
8. Desgranges, J. C. 1966. Sur la double innervation des cellules sensorielles dans des bourgeons du goût des barbillons du Poisson-chat. *C. R. Hebd. Séances Acad. Sci. Paris* **263**, pp. 1103–1106.
9. Engström, H., and C. Rytzner. 1956. The structure of taste buds. *Ann. Oto-Laryngol.* **46**, pp. 361–367.
10. Engström, H., and C. Rytzner. 1956. The fine structure of taste buds and taste fibres. *Ann. Oto. Rhinol. Laryngol.* **65**, pp. 361–375.
11. Farbman, A. I. 1965. Fine structure of the taste bud. *J. Ultrastruct. Res.* **12**, pp. 328–350.
12. Farquhar, M. G., and G. E. Palade. 1963. Junctional complexes in various epithelia. *J. Cell Biol.* **17**, pp. 375–412.
13. Fusari, R., and A. Panasci. 1891. Les terminaisons des nerfs dans la muqueuse et dans les glands sereuses de la langue des mammifères. *Arch. Ital. Biol.* **14**, pp. 240–246.
14. Gråberg, J. 1899. Zur Kenntnis des zellulären Baues der Geschmacksknospen beim Menschen. *Anat. Hefte* **12**, pp. 337–368.
15. Gray, E. G., and K. C. Watkins. 1965. Electron microscopy of taste buds of the rat. *Histochemie* **66**, pp. 583–595.
16. Graziadei, P., and L. H. Bannister. 1967. Some observations on the fine structure of the olfactory epithelium in the domestic duck. *Histochemie* **80**, pp. 220–228.
17. Graziadei, P. P. C. 1969. The ultrastructure of taste buds in mammals. *In* Second Symposium on Oral Sensation and Perception. Charles C Thomas, Springfield, Ill. (In press)

18. Graziadei, P. P. C. 1968. Synaptic organization in vertebrate taste buds. Proceedings XXIV International Congress of Physiological Sciences **7**, p. 167.
19. Heidenhain, M. 1914. Ueber die Sinnesfelder und die Geschmacksknospen der Papilla foliata des Kaninchens. Beiträge zur Teilkörpertheorie. III. *Arch. Mikroskop. Anat. Entwickl.* **85**, pp. 365–479.
20. Hermann, F. 1885. Beitrag zur Entwicklungsgeschichte des Geschmacksorgans beim Kaninchen. *Arch. Mikroskop. Anat.* **24**, pp. 216–229.
21. Hermann, F. 1888. Studien ueber den feineren Bau des Geschmacksorgans. *Sitzungsber. Math.-Naturwiss. Kl. Bayer. Akad. Wiss. Muenchen.* **18**, pp. 277–318.
22. Herrick, C. J. 1902. The organ and sense of taste in fishes. *Bull. U.S. Fish. Comm.* **22**, pp. 239–272.
23. Hirata, Y. 1966. Fine structure of the terminal buds on the barbels of some fishes. *Arch. Histol. Jap.* **26**, pp. 507–523.
24. Iriki, T. 1960. Electron microscopic observation on the taste buds of the rabbit. *Acta Med. Univ. Kagoshima.* **2**, pp. 78–94.
25. Kolmer, W. 1910. Ueber Strukturen im Epithel der Sinnesorgane. *Anat. Anz.* **36**, pp. 281–299.
26. Kolmer, W. 1927. Geschmacksorgan. *In* Handbuch der mikroskopischen Anatomie des Menschen, Haut und Sinnesorgane, Vol. III (W. v. Moellendorff, editor). Springer Verlag, Berlin, pp. 154–191.
27. Lenhossék, M. von. 1893. Die Geschmacksknospen in den blattförmigen Papillen der Kaninchens. *Verh. Phys.-Med. Ges. Wurzburg* **27**, pp. 191–266.
28. Leydig, F. 1851. Ueber die Haut einiger Süsswasserfische. *Z. Wiss. Zool.* **3**, pp. 1–12.
29. Loven, C. 1868. Beiträge zur Kenntnis vom Bau der Geschmackswärzchen der Zunge. *Arch. Mikroskop. Anat.* **4**, pp. 96–110.
30. May, R. M. 1925. The relation of nerves to degenerating and regenerating taste buds. *J. Exp. Zool.* **42**, pp. 371–410.
31. Merkel, F. 1880. Über die Endigungen der sensiblen Nerven in der Haut der Wirbelthiere. H. Schmidt, Rostock, p. 214.
32. Meyer, S. 1897. Durchschneidungsversuche am Nervus Glossopharyngeus. *Arch. Mikroskop. Anat. Entwickl.* **48**, pp. 143–145.
33. Murray, R. G., and A. Murray. 1960. The fine structure of the taste buds of Rhesus and Cynomalgus monkeys. *Anat. Rec.* **138**, pp. 211–233.
34. Murray, R. G., and A. Murray. 1967. Fine structure of taste buds of rabbit foliate papillae. *J. Ultrastruct. Res.* **19**, pp. 327–353.
35. Nemetschek-Gansler, H., and H. Ferner. 1964. Über die Ultrastruktur der Geschmacksknospen. *Histochemie* **63**, pp. 155–178.
36. Olmsted, J. M. D. 1920. The nerve as a formative influence in the development of taste-buds. *J. Comp. Neurol.* **31**, pp. 465–468.
37. Olmsted, J. M. D. 1920. The results of cutting the seventh cranial nerve in *Amiurus nebulosus* (Lesuer). *J. Exp. Zool.* **12**, pp. 368–401.
38. Ranvier, L. A. 1889. Traité Technique d'Histologie. 2nd edition. F. Savy, Paris, pp. 724–733.
39. Retzius. G. 1892. Die Nervenendigungen in dem Geschmacksorgan der Säugetiere und Amphibien. *Biol. Unters.* **4**, pp. 19–32.
40. Scalzi, H. A. 1967. The cytoarchitecture of gustatory receptors from the rabbit foliate papillae. *Histochemie* **80**, pp. 413–435.
41. Schwalbe, G. 1867. Das Epithel der Papillae vallatae. *Arch. Mikroskop. Anat. Entwickl.* **3**, pp. 504–508.
42. Schwalbe, G. 1868. Ueber die Geschmacksorgane der Säugethiere und des Menschen. *Arch. Mikroskop. Anat. Entwickl.* **4**, pp. 154–187.
43. Trujillo-Cenóz, O. 1957. Electron microscope study of the rabbit gustatory bud. *Z. Zellforsch. Mikroskop. Anat.* **46**, pp. 272–280.
44. Trujillo-Cenóz, O. 1961. Electron microscope observations on chemo- and mechanoreceptor cells of fishes. *Histochemie* **54**, pp. 654–676.
45. Uga, S. 1966. The fine structure of gustatory receptors and their synapses in frog's tongue. *Symp. Cell Chem.* **16**, pp. 75–86.
46. Uga, S., and K. Hama. 1967. Electron microscopic studies on the synaptic region of the taste organ of carps and frogs. *J. Electron Microscop.* (*Japan*) **16**, pp. 269–277.
47. Vintschgau, M. von, and J. Hönigschmied. 1877. Nervus glossopharyngeus und Schmeckbecher. *Arch. Gesamte. Physiol. Menschen Tiere* (*Pfluegers*) **14**, pp. 443–448.
48. Zander, R. 1897. Ueber das Verbreitungsgebiet der Gefühls- und Geschmacksnerven in der Zungenschleimhaut. *Anat. Anz.* **14**, pp. 131–145.

CELL TYPES IN RABBIT TASTE BUDS

RAYMOND G. MURRAY
Department of Anatomy and Physiology, Indiana University, Bloomington, Indiana

The considerable variation in appearance of the several cells comprising mammalian taste buds has been interpreted in various ways. There is general agreement that the differences are not entirely artifacts of preparation, but at least three other possibilities must be considered: (1) a single cell type with variations reflecting functional change; (2) a single functional cell type, but with transitional stages from a less differentiated precursor cell; (3) more than one functional cell type. Since presumably we are dealing here with neuroepithelial cells, and marked morphological change with function is not characteristic of this group, the first possibility probably will not be sufficient to account for the variations noted. On the other hand, it has been demonstrated[1] that the taste bud cells are constantly being renewed from less differentiated cells at their margins. Stages of transition in this process could account for a wide range of variation, and the functionally significant cells may be of one type only. However, the less differentiated cells, in addition to supplying precursors for new gustatory receptor cells, might act in support of the latter and thus constitute a second cell type. That would describe the most widely held conception of cell-type relations in taste buds.[2]

OBSERVATIONS ON RABBIT TASTE BUDS

There has, however, been confusion regarding which cell should be designated as supporting, and which as gustatory. In a recent paper on taste buds of rabbit foliate papillae,[6] we have developed the idea that the primary gustatory cell is a large, relatively lightly-staining cell with variable but distinct microvilli projecting into the taste pit and extensive, but largely unspecialized, contact with nerve endings (Figures 1, 2, and 7). These cells are almost entirely surrounded by darker cells with more slender nuclei, longer apical parts with more numerous microvilli, apical granules which are precursors of the dense substance of the pit, and a relation to the nerves much like that of Schwann cells. We have designated these dark cells as supporting cells, but with the reservation that they might also play a role in gustation. This analysis is at variance with the more frequently stated opinion[3,8,10] that the dark cells are the gustatory cells and the light cells are supportive. Further observations in our laboratory on fungiform papillae of rabbits (Figures 3 and 4) indicate that a similar distinction can be made in the taste buds of this location, although there are no secretory granules in the dark cells to make the distinction obvious.

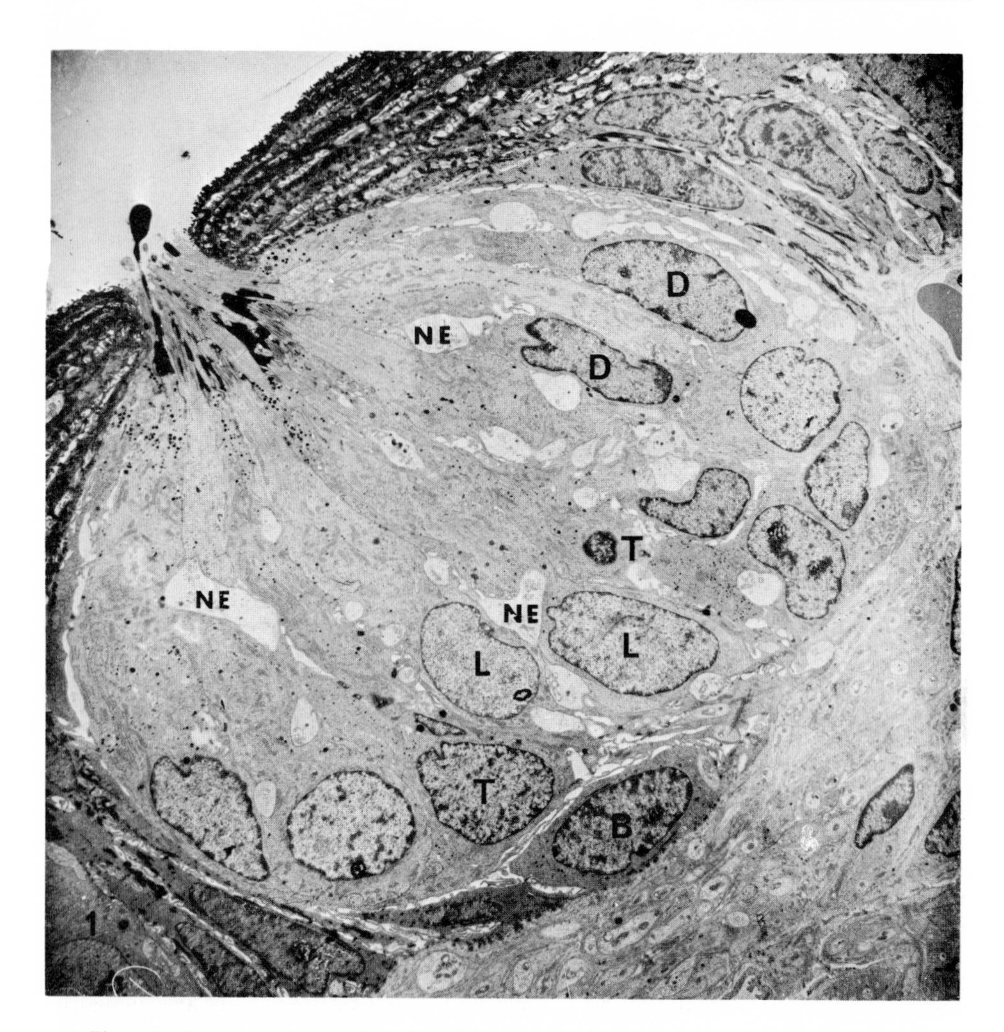

Figure 1 Low-power survey of a rabbit foliate taste bud from one of the near-serial sections. In addition to light cells (L) and dark cells (D), two type III cells (T) are present, as well as a basal cell (B). Numerous pale nerve endings (NE) are present throughout. Uranyl acetate stain only. × 2,200. (Unless otherwise indicated, each illustration in this chapter is of material fixed by perfusion with 2.5 per cent glutaraldehyde buffered to pH 7.3 with phosphate buffer, post-fixed in veronal-buffered 1 per cent OsO_4, embedded in Epon, sectioned on an LKB microtome, stained with uranyl acetate and lead citrate, and examined in a Hitachi HU 11 microscope.)

Returning to buds of the rabbit foliate papillae, it must be noted that there is a range of variation within each cell group. Many of the dark cells at the periphery closely resemble the adjacent perigemmal cells. Among the light cells are some with little vacuolization, and others that are highly vacuolated (Figure 2). In addition to these differences, however, we must note more fundamental differences in a few instances, if the full range of variation is to be accounted for. Among cells that do not precisely fit the two categories are the following:

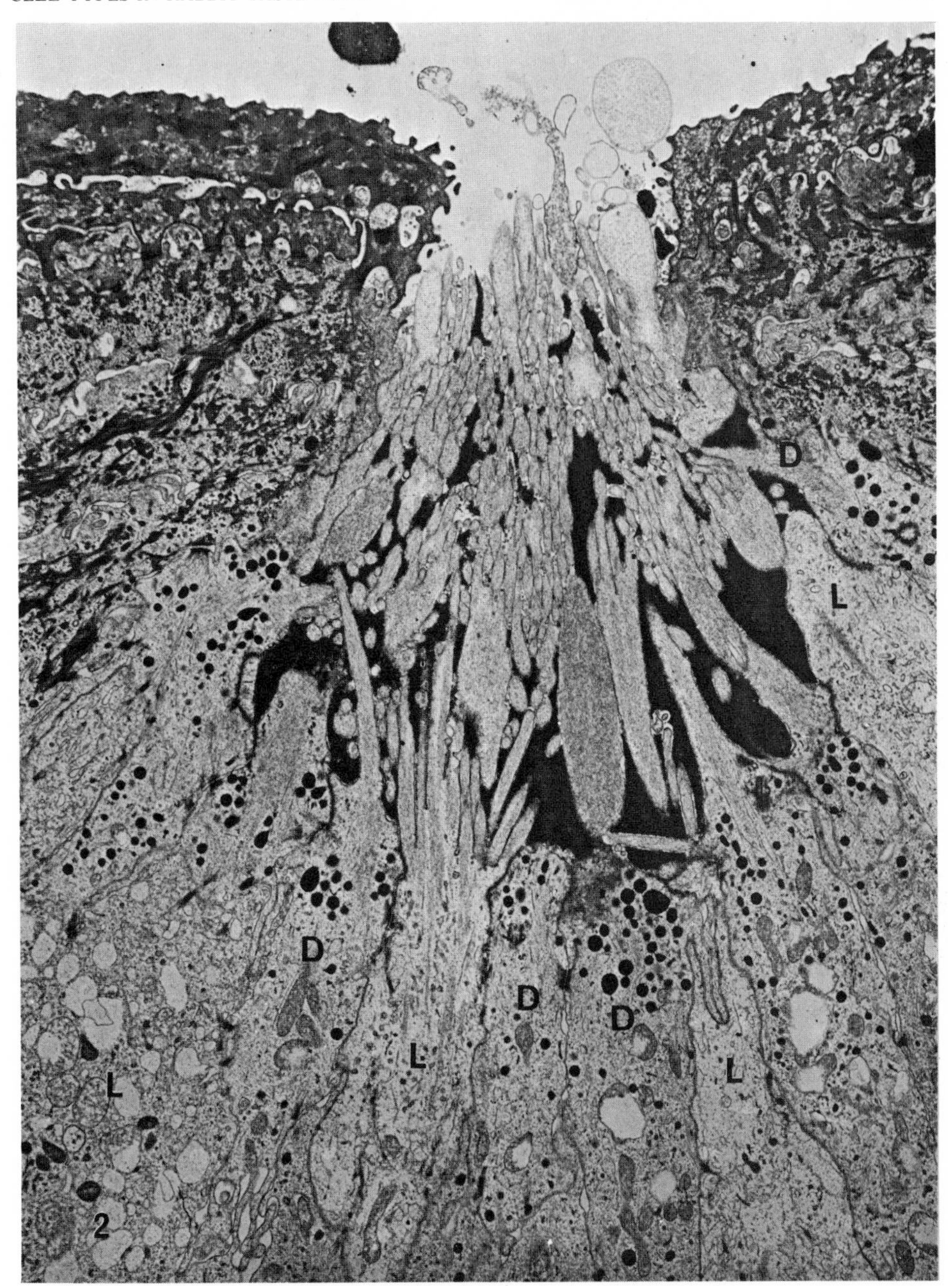

Figure 2 Detail of pit and pore from another taste bud similar to that of Figure 1. Light cells (L) may have many microvilli or none. Dark cells (D) have a variety of configurations at their apices. Dense granules in the dark cells resemble the dense substance of the pit. Note heavily vacuolated light cell far left. × 14,000.

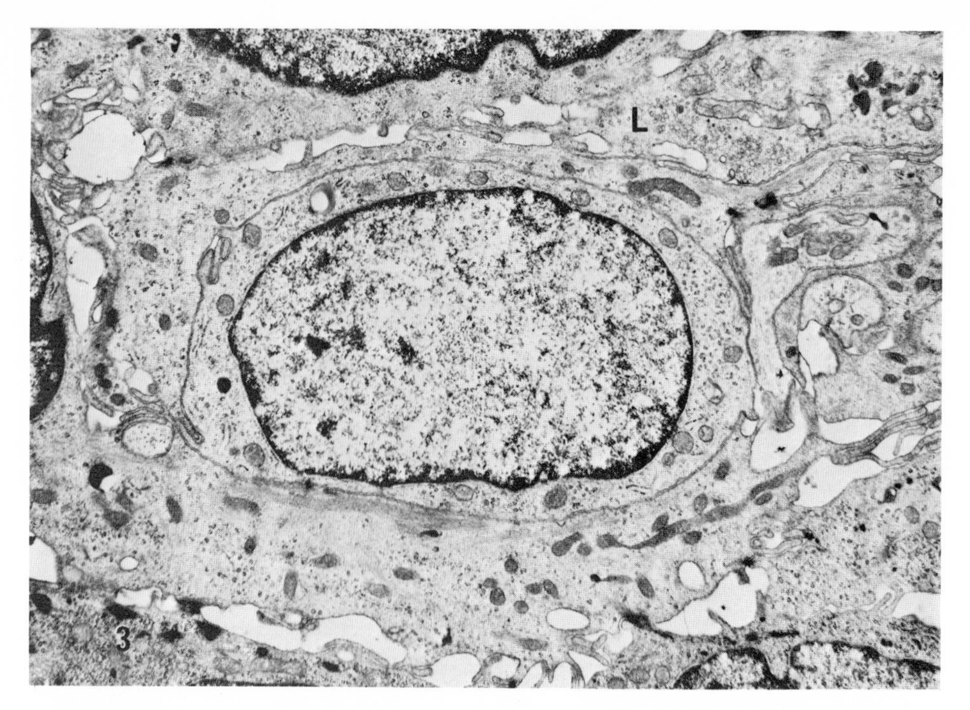

Figure 3 Portion of taste bud from a rabbit fungiform papilla. Central cell is a light cell surrounded by several dark cells. A portion of another light cell at L contains numerous small vesicles. × 9,000.

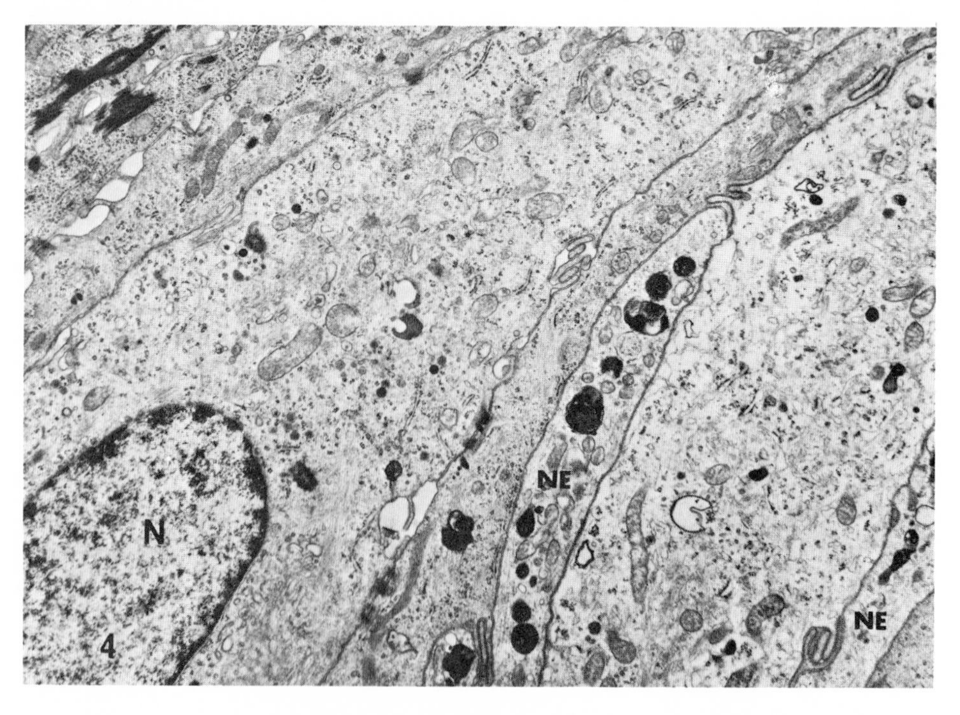

Figure 4 Another region of the same taste bud as Figure 3. A few dark-cored vesicles and a centriole lie to the right of the nucleus (N) of a light cell. This cell is bounded on each side by dark cells. At far right is another light cell in extensive contact with a nerve ending (NE) on each side. × 9,000.

(1) Cells in the lateral-basal region of the bud (Figures 1 and 5) that do not appear to make contact with the pit, but otherwise resemble the dark cells of the most undifferentiated type, and may, in rare instances, be in mitosis. These would correspond to the basal cells described by several authors,[4,5,8,9] both in rabbit foliate taste buds and in other species and locations.

(2) Cells which otherwise resemble the dark cells, but do not have the long necks from which the microvilli typically originate (Figure 9).

(3) Cells with a neck and villus structure like the dark cells, but without granules in the apical part (see Figure 17 of reference 6).

(4) Cells that reach the pit, but are without microvilli (Figures 2 and 9).

(5) Cells with irregular, apparently degenerating, cytoplasmic contents (Figure 5).

(6) Cells that contain in their basal cytoplasm numerous dark-cored vesicles which are smaller than, and otherwise clearly distinguishable from, the granules of the dark cells (Figures 1 and 5).

The foregoing observations are based on a relatively random sampling, which does not permit complete analysis of the cells represented by given cell profiles. From such sampling, for example, one cannot be sure that a "basal cell" does not extend into other sections to reach the pore. Variations in structures at the pit region could not always be related to the structure of the remainder of the cell in question. We have found it useful, therefore, to employ near-serial sections to answer some of these objections. By this we mean that a taste bud is sectioned continuously from one side to the other, but not all sections are photographed. The intervals vary with the region under study and usually the gaps between samples are of less than one micron; however, at times, because of technical difficulty, they may be several microns. Low-magnification micrographs of entire buds are prepared and examined in a series, which may include as many as 100 or more micrographs. From such observations it is possible to establish several points which otherwise are uncertain:

(1) Virtually all cells reach the pore, and come close to or reach the underlying connective tissue. The basal cell can be positively identified as an exception to this rule.

(2) The very extensive, apparently functionally significant, contact of nerve endings with the light cells is verified, including the fact that nerves may wrap in a spiral around the body of a light cell (Figures 6 and 7).

(3) Nerve endings may terminate within indentations of the nuclei of dark cells (Figure 8). In random sections the nerves seem only to be passing through the dark cells on the way to contact with light cells, so it had been assumed that these fibers were only indenting the nuclei of the dark cells as they passed.

(4) It can be verified that the same nerve fiber makes apparently functional contact with more than one cell of different types.

(5) Numbers of cells of the two main types can be accurately determined. Dark cells comprise from 55 to 75 per cent, and light cells from 15 to 30 per cent.

(6) Cells with dark-cored vesicles in their basal parts are found to be present consistently, and unique enough to be classified separately as cell type III. They constitute from 5 to 15 per cent.

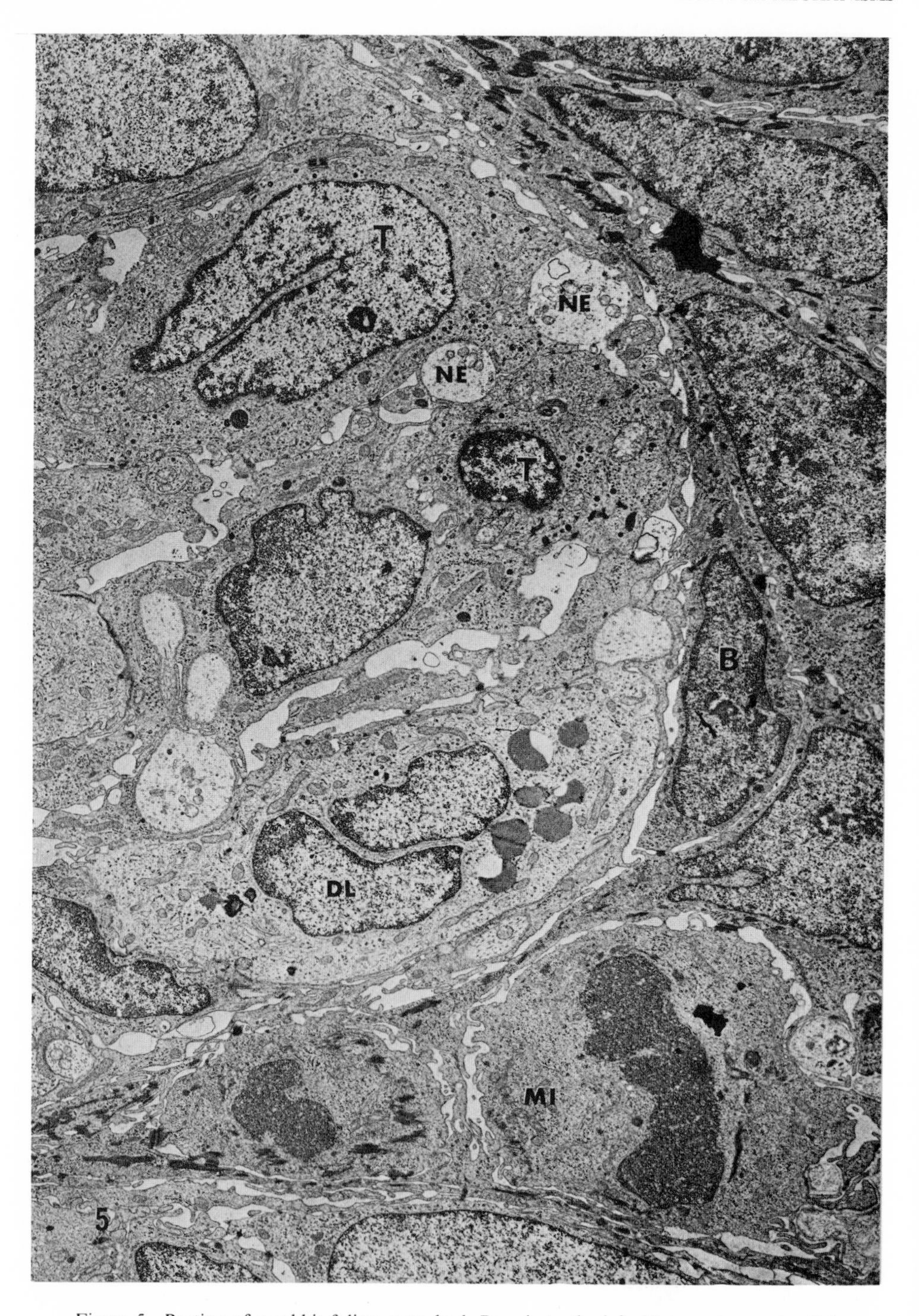

Figure 5 Portion of a rabbit foliate taste bud. Pore is to the left. Note perigemmal cell in mitosis (MI), degenerating light cell (DL), and two type III cells (T) with dark-cored vesicles, which are in contact with each other and with two large nerve endings. The cell at base (B) is either a basal cell, or about to become one. × 5,700.

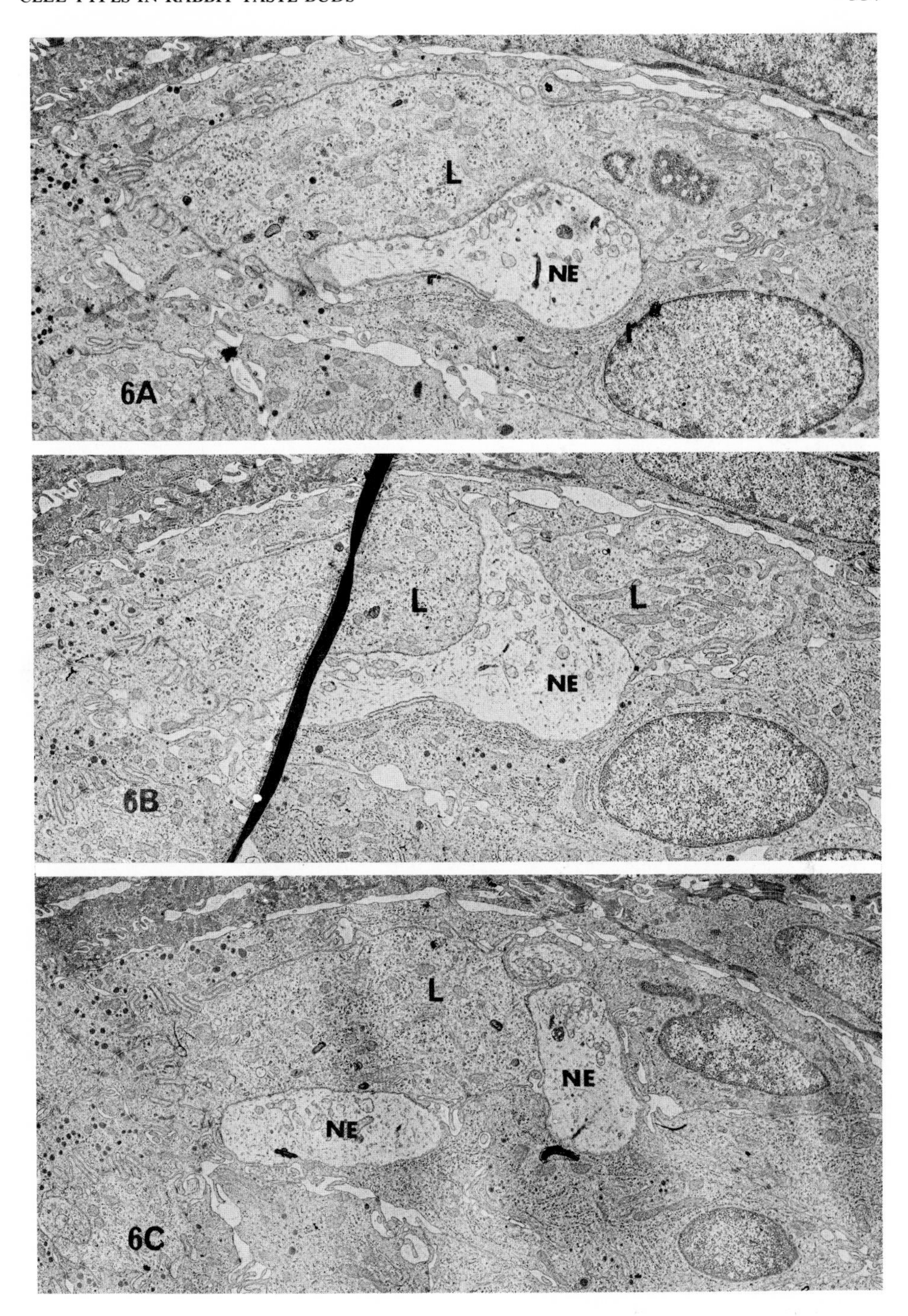

Figure 6 Three sections from a near-serial sampling of a rabbit foliate taste bud. The pore is to the left. In 6A two small parts of the tangentially sectioned nucleus of a light cell (L) appear to the right of the nerve ending (NE) which is in extensive contact with the cell. In 6B the nucleus has disappeared and the nerve ending extends across to the other side of the cell as well as along the same side. In 6C the right portion of the cell has disappeared, the nerve ending has divided but remains in extensive contact with the cell. × 5,600.

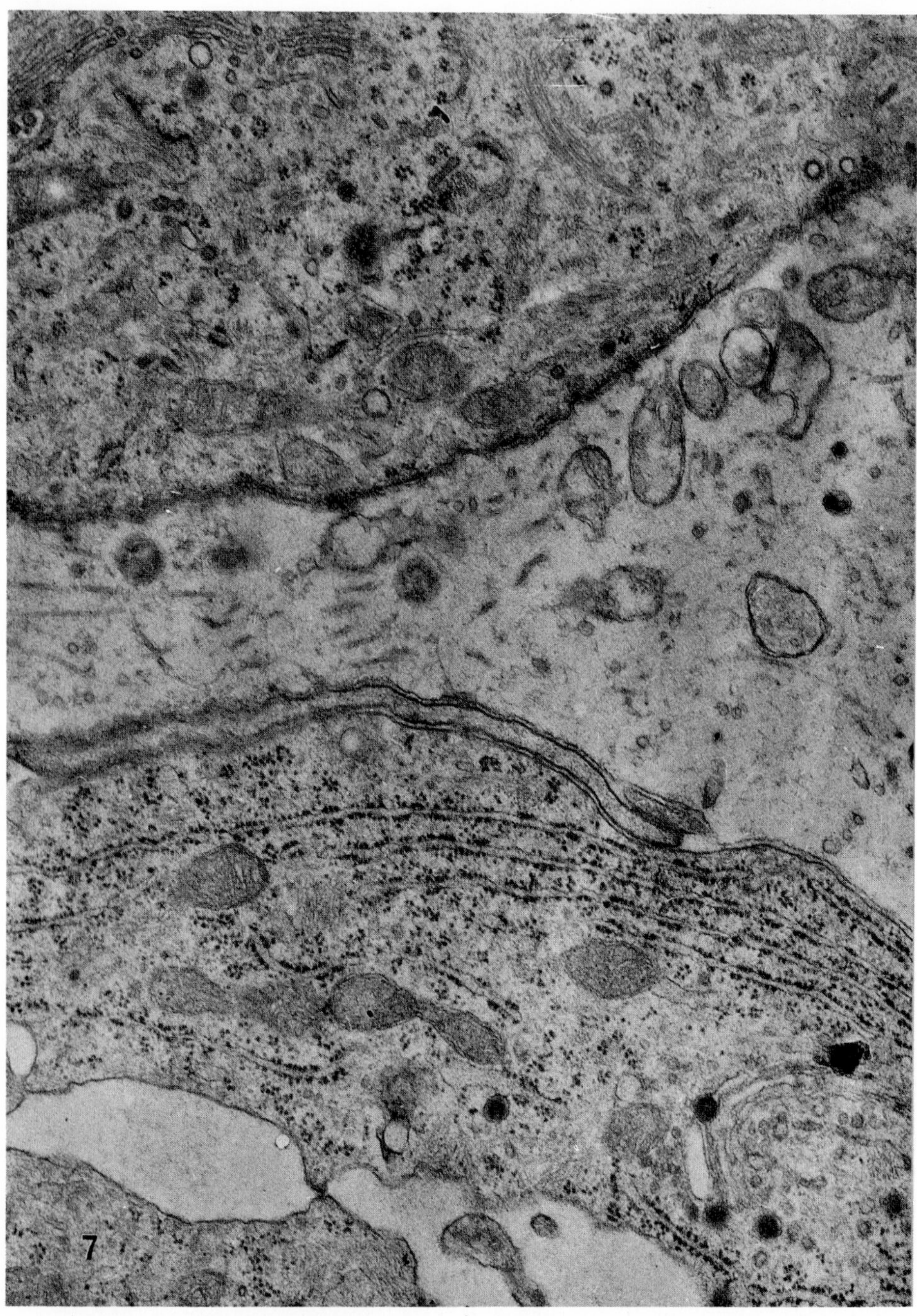

Figure 7 Detail from Figure 6A. The large nerve ending crossing the field separates a light cell above from the dark cell below. The extensive, rough-surfaced endoplasmic reticulum is characteristic of dark cells. Golgi zone of dark cell, lower right, shows presence of dense granules, while Golgi of light cell, upper right, shows a few empty vesicles. Massing of mitochondria, in both light cell and nerve, and densities in the light cell against the plasma membrane, suggest a synaptic region. $\times$ 37,000.

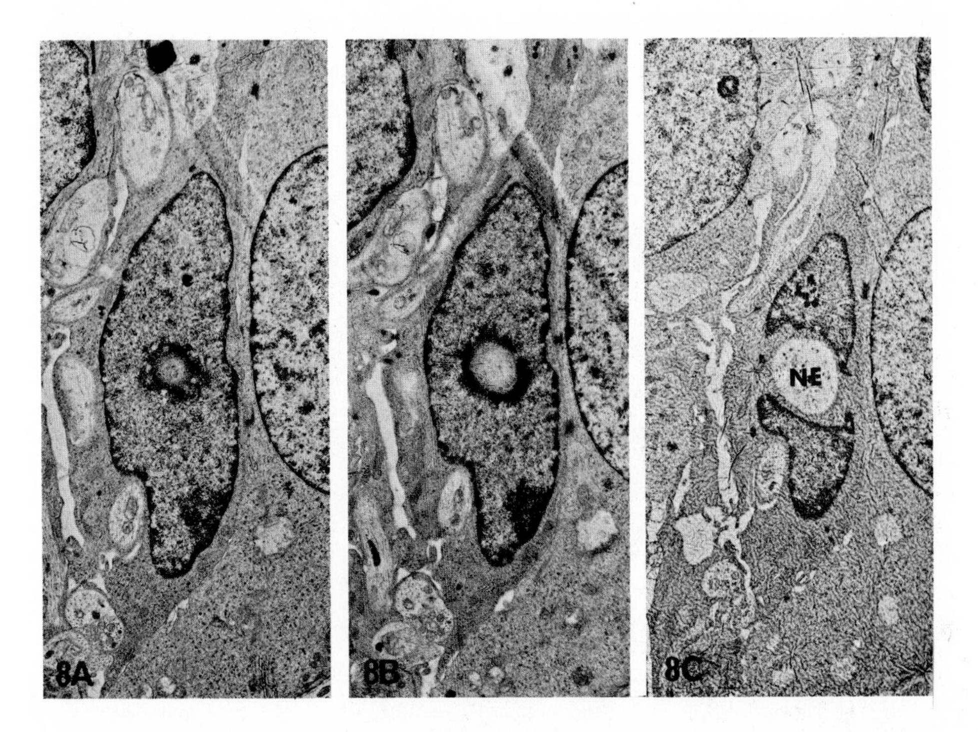

Figure 8 Portions of three near-serial sections of a rabbit foliate taste bud to illustrate the termination of a nerve in a deep indentation of a dark cell nucleus. In 8C the pale nerve ending (NE) is seen between portions of the nucleus, in 8A and 8B the nerve terminates as the nucleus assumes its full profile. Uranyl acetate stain only. × 5,000.

CHARACTERISTICS OF CELL TYPE III

Systematic examination of type III cells in near-serial sections reveals a nucleus of a density much like that of the central dark cells, and often irregular in shape (Figure 5). The cytoplasm has a density between that of dark and light cells and never contains the swollen cisternae of the endoplasmic reticulum. Type III cells do not surround other cells, but are surrounded by dark cells. Their isolation is not as complete as that of the light cells, for extensive contact between type III cells is characteristic (Figure 5). They can almost always be traced to the taste pit above, but do not usually make direct contact with the underlying connective tissue. Their apical ends vary from narrow necks to blunt or irregular surfaces in contact with the dense substance of the pit. One or two centrioles may be seen, the location varying from near the apex (Figure 9) to near the nucleus. In addition to the dark-cored vesicles, there may rarely be a few dense granules in the apical region. In the deeper parts, also, one finds large numbers of smaller, apparently empty vesicles of the type usually called synaptic vesicles. The most striking and significant feature is the aggregation of these latter vesicles at sites adjacent to nerve endings and in association with cytoplasmic densities, the complex resembling, in all important respects, the classic picture of the

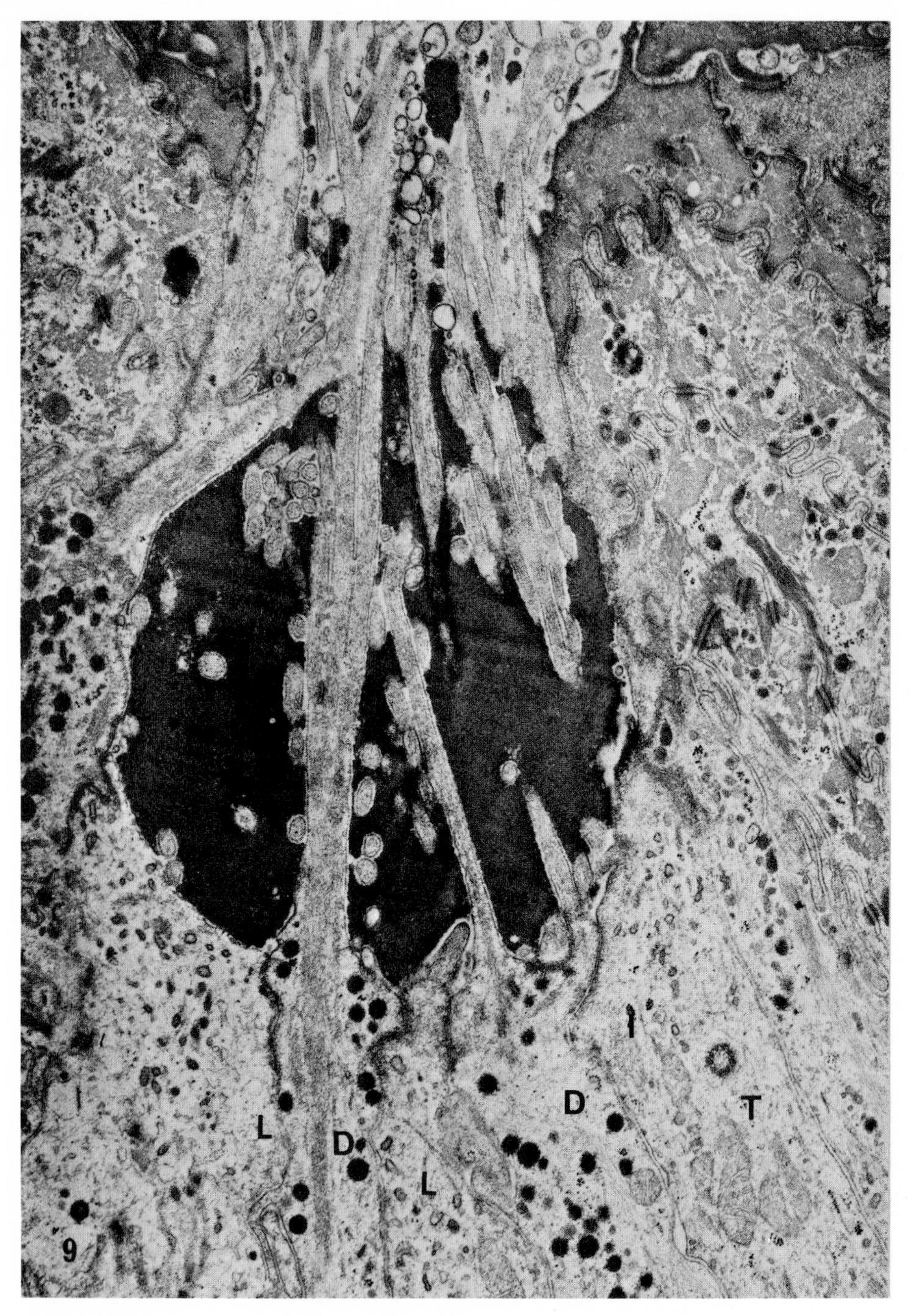

Figure 9 Pit and pore of a rabbit foliate taste bud. Note particularly the light cells (L) with no microvilli, dark cells (D) with and without necks, and a type III cell (T) which reaches the pit and has a well defined centriole. × 20,000.

chemical synapse (Figure 10). From the location of the vesicles and of the cytoplasmic densities, it seems clear that such synapses are sensory in function.

DISCUSSION

What is the significance of these cell types for the function of taste reception? We have earlier[6] stressed the probability that the region of the taste pore and pit is the zone in which the first processes of transduction occur. This is apparently guaranteed by the tight junctions which seal off the pit from the intercellular spaces below. Since each of the three main cell types makes contact with the pit, their apical surfaces may be involved in this first process. The nerves are never in contact with the pit, hence a cell of the bud must act as an intermediary, and it is possible that any or all of these three main types may do so. The most likely one, on first examination, is cell type III, since it has a typical synaptic contact with nerve endings. However, the relation of the light cells to the nerves is so extensive, and peculiar in a different way, that it is difficult to ignore the probability that these also have a function in transduction. Lastly, the dark cells are in perhaps the most extensive contact with the nerves, although this contact is largely with fibers rather than "endings," and in the manner of Schwann

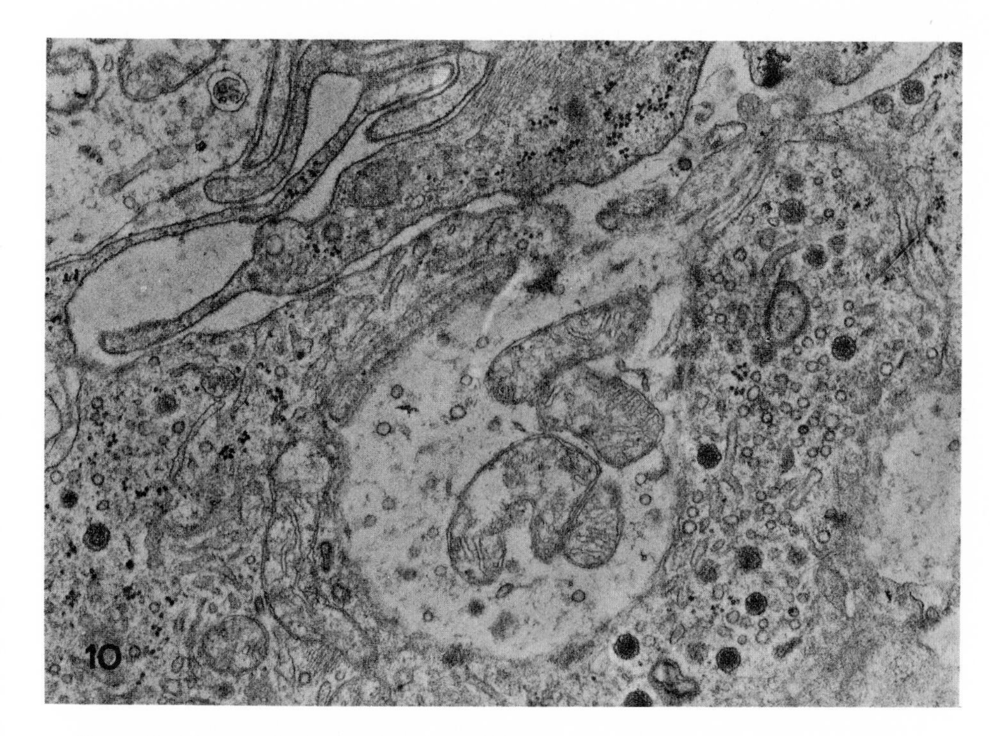

Figure 10 Nerve ending in type III cell of rabbit foliate taste bud. Synaptic vesicles and dark-cored vesicles crowd against the nerve, and relate to presynaptic densities at the upper right portion of the nerve ending. × 31,000.

cells. The relation is not entirely of this form, for profiles of nerves within dark cells are often broad, and contain many mitochondria. Moreover, terminations of bulging nerve endings in indentations of dark cell nuclei (Figure 8) are strongly suggestive of more than supportive function for this cell.

It is possible that three different paths for transduction exist within the taste buds of rabbit foliate papillae. This suggestion may, for the moment at least, confuse rather than clarify. However, if single taste buds can respond to several or all the taste qualities, such morphologic complexity may play a role in this complex response. This need not take the form of specific cells for the several taste qualities, but might involve different patterns of response among the several paths, which patterns could code for various taste substances in unique ways.

How far this division into three types can be verified in taste buds of the fungiform papillae, or in other species and locations, remains to be investigated. We have seen cells in the taste buds of rabbit fungiform papillae similar to type III cells of the foliate regions, with occasional dark-cored vesicles (Figure 4), groups of synaptic vesicles and presynaptic densities (Figure 11). Other cells of these buds make extensive contact with nerve endings in a manner similar to the light cells in the foliate buds (Figure 4). Farbman[4] described small vesicles in the taste bud cells of rat fungiform papillae. In his "basal cell" these vesicles were aggregated next to a nerve ending in a manner suggestive of a synapse. He did not interpret this as a synapse, for he considered the basal cells to be relatively undifferentiated, and to lack contact with the pore. Without serial sections, however, it is not absolutely certain if profiles which are identified as basal cells might not extend to the pore. If this were the case, the relation of vesicles and nerves suggests the presence of a type III cell in the rat fungiform taste buds.

On the other hand, it is possible that the buds of the fungiform papillae may represent a distinctly different class from those of the foliate, since they are supplied by a different nerve. Preliminary study of the circumvallate papillae of rabbits, which have a nerve supply similar to that of the foliate region, indicates greater similarity to the foliate than to the fungiform buds, including the presence of typical type III cells in the circumvallate location.[7]

The complexity of possibilities is not completely described in the existence of three cell types, each capable of transduction. Also, nerves make contact with several cells and with cells of different types in succession. To this must be added the probability that fibers may branch to supply cells in two different buds. (This last point would not be a factor in the case of the fungiform papillae, where usually only one bud is present.) One further observation must be added: none of the thousands of nerve endings examined can be fitted confidently into the category of efferent ending. This is surprising in view of the presence of such endings in other sensory transducers. Whether this simplifies or further complicates the already complex picture must be left to those better able to judge.

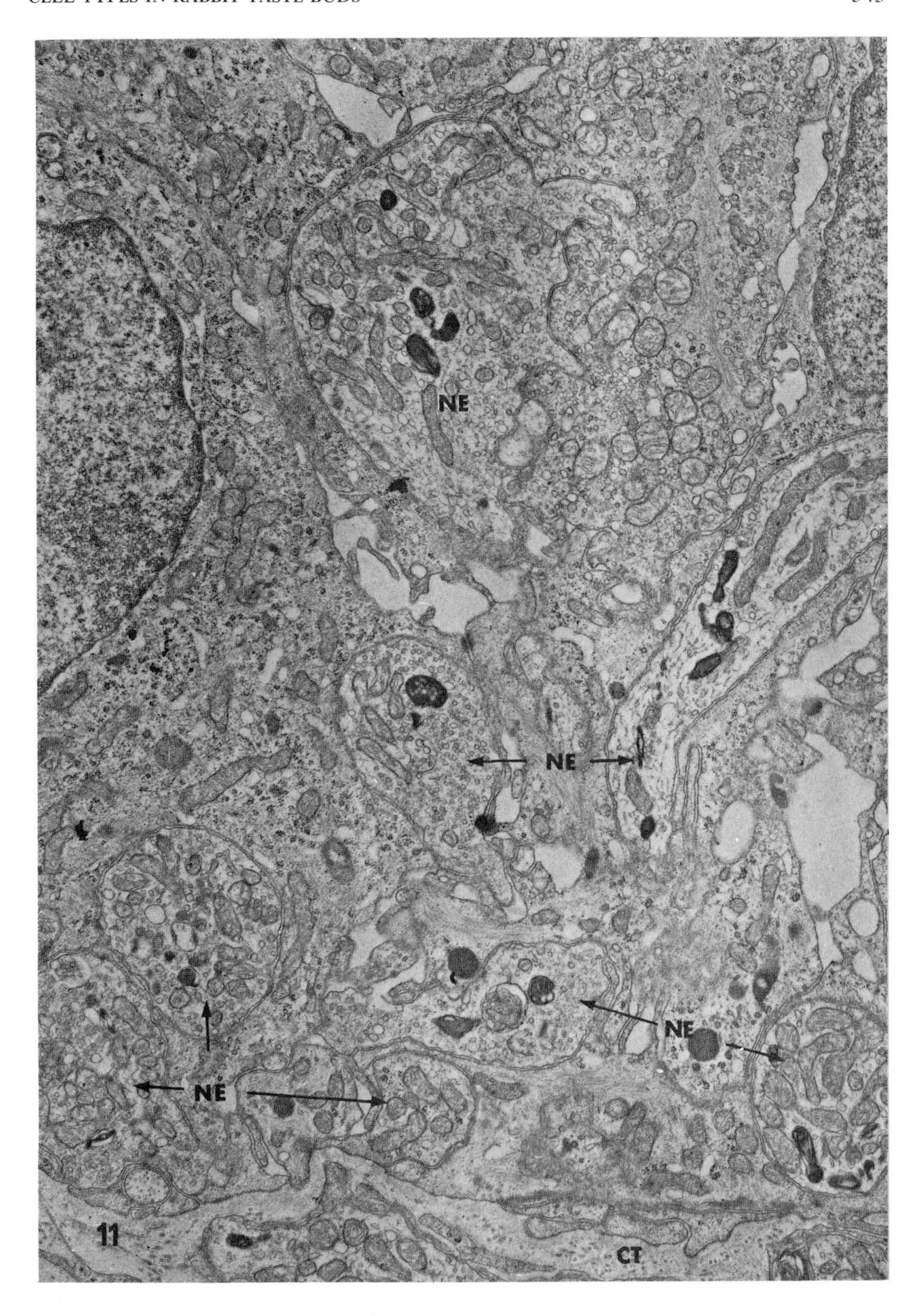

Figure 11 Region at base of a rabbit fungiform taste bud, with connective tissue (CT) below. Eight nerve endings (NE) are seen, the large one above being related to a mass of small vesicles and many mitochondria. Several nerve endings, particularly the one in the center of the picture, contain numerous "synaptic vesicles." Fixed directly in 1 per cent OsO_4. $\times$ 14,000.

ACKNOWLEDGMENT

This work was supported in part by a grant, PHS NB 07472, from the National Institutes of Health, United States Public Health Service.

REFERENCES

1. Beidler, L. M., and R. L. Smallman. 1965. Renewal of cells within taste buds. *J. Cell Biol.* **27**, pp. 263–272.
2. Bloom, W., and D. W. Fawcett. 1968. A Textbook of Histology. 9th edition. W. B. Saunders Co., Philadelphia, Pa., pp. 513–515.
3. De Lorenzo, A. J. 1963. Studies on the ultrastructure and histophysiology of cell membranes, nerve fibers and synaptic junctions in chemoreceptors. *In* Olfaction and Taste I (Y. Zotterman, editor). Pergamon Press, Oxford, etc., pp. 5–18.
4. Farbman, A. I. 1965. Fine structure of the taste bud. *J. Ultrastruct. Res.* **12**, pp. 328–350.
5. Kolmer, W. 1927. Geschmacksorgan. *In* Handbuch der mikroskopischen Anatomie des Menschen, Haut und Sinnesorgane, Vol. III (W. v. Moellendorff, editor). Springer Verlag, Berlin, pp. 154–191.
6. Murray, R. G., and A. Murray. 1967. Fine structure of taste buds of rabbit foliate papillae. *J. Ultrastruct. Res.* **19**, pp. 327–353.
7. Murray, R. G., A. Murray, and S. Fujimoto. 1969. Fine structure of gustatory cells in rabbit taste buds. *J. Ultrastruct. Res.* (In press)
8. Nemetschek-Gansler, H., and H. Ferner. 1964. Über die Ultrastruktur der Geschmacksknospen. *Histochemie* **63**, pp. 155–178.
9. Scalzi, H. A. 1967. The cytoarchitecture of gustatory receptors from the rabbit foliate papillae. *Histochemie* **80**, pp. 413–435.
10. Trujillo-Cenóz, O. 1957. Electron microscope study of the rabbit gustatory bud. *Z. Zellforsch. Mikroskop. Anat.* **46**, pp. 272–280.

LOCAL POTENTIAL CHANGES AT SENSORY NERVE FIBER TERMINALS OF THE FROG TONGUE

HIROMICHI NOMURA *and* SANYA SAKADA
Department of Physiology, Tokyo Dental College, Tokyo, Japan

INTRODUCTION

A fungiform papilla and its connected nerve fascicle (1–2 mm in length) dissected from the frog tongue provides a good preparation for studies of gustatory receptors. Because the nerve fascicle of this preparation contains only a small number of nerve fibers, one can easily recognize the impulses of individual sensory units by their shape and amplitude when afferent nerve impulses are induced.

Another advantage of this preparation is that the distance between the sensory nerve fiber terminal and the recording site is only a few hundred microns. Thereby, the local potential change at the sensory nerve fiber terminals is often recorded, at least to a certain extent.

The present paper deals especially with the local potential changes that take place in conjunction with the water response of the taste organ.

MATERIALS AND METHODS

The animals used in the present work were two kinds of common Japanese frog, *Rana nigromaculata* and *Rana japonica*. The experiments were carried out mainly from late September to January at temperatures of 20°–25°C.

The schematic diagram of the set-up is shown in Figure 1. The essential parts of the equipment are three pieces of glass plate. The fungiform papilla was placed on the narrow middle piece of glass plate and the nerve fascicle was bridged across an air gap. Ringer's solution and the test solution were applied through two small glass tubes (G_1 and G_2, respectively), each of which was connected to a small syringe. The solutions applied through the glass tube were sucked along a cotton wick and finally into a large cotton wad.

Tactile stimulation was applied to an individual fungiform papilla by means of a thin glass stylet. A crystal of Rochelle salt was used to move the stylet. This was driven by a rectangular current from a pulse generator.

Instead of tap water, a taste solution with a high calcium concentration was used. It consists of 5 mM $CaCl_2$ and 100 mM NaCl. By using this solution, we could reduce the junction potential change at the electrodes.

RESULTS

Mechanoreceptors in the Fungiform Papilla

Each fungiform papilla in the frog is thought to be a functional analogue of a single taste bud in mammals, but it has mechano- and other receptors in addition to taste receptors.

Generally speaking, each fungiform papilla contains two kinds of mechanoreceptors: one is fast-adapting and the other is slow-adapting. The nerve action potentials of the fast-adapting unit are usually larger than those of the slow-adapting unit. Only eight fungiform papillae, among 49 observed, contained a single mechanoreceptor.

Figure 2 shows a record obtained when a fungiform papilla was stimulated with a long mechanical pulse of about 1 sec. The fast-adapting mechanoreceptor showed on-and-off response to a long tactile stimulus, while the slow-adapting one fired repetitively throughout the stimulation. Slow-potential changes in the fast-adapting receptor are not clear in this case, but in the slow-adapting receptor they are clearly evident.

The impulses of medium size seen in the fourth, sixth, and seventh columns were from a water fiber, but in the second and seventh columns the nature of the impulses, which appear to be antidromic, is not certain.

The response of the fast-adapting mechanoreceptor to repetitive tactile stimulation is shown in Figure 3, in which A is a record from a normal preparation and B is a record obtained when the propagated impulses had been blocked by tetrodotoxin. Slow-potential changes remained after the propagated action potential had been completely blocked. These slow-potential changes are considered to be the generator potentials that occur at the nonmyelinated free nerve endings of the mechanoreceptor.

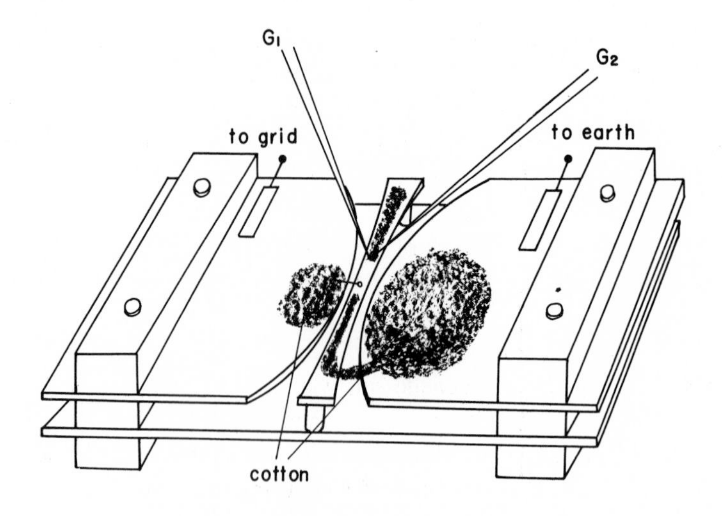

Figure 1 Schematic diagram of the set-up.

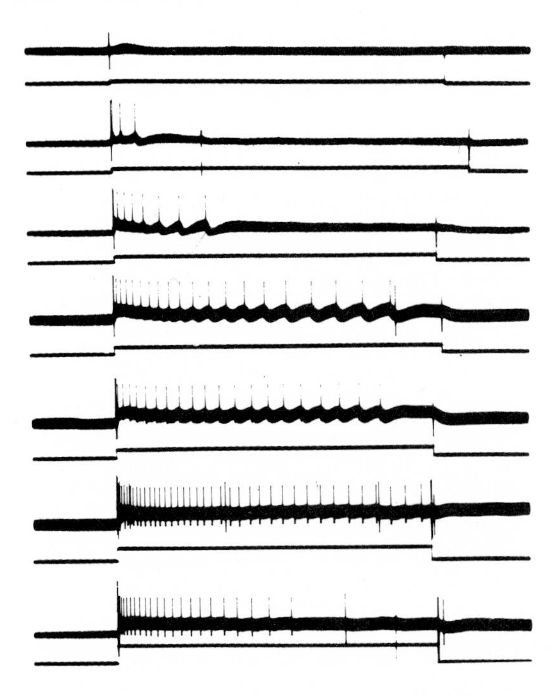

Figure 2 Responses of mechano- and water receptors to long mechanical pulses. The upper trace of each pair is the potential change, and the lower trace is the stimulus pulse of the pulse generator.

Local Potential Change of the Water Receptor

Figure 4 shows slow-potential changes and afferent nerve impulses from the water receptors. The top trace shows the spontaneous activity of the receptor in the Ringer's solution, in which one afferent spike and four slow potentials are seen.

The middle two traces show the response of the same receptor to a high calcium solution, and the bottom traces show the similar response from another preparation, which contained two water receptors. A great number of slow-potential changes were evoked after the application of the taste solution, and the direction of the slow-potential change is the same as that of the generator potential of the mechanoreceptors. These facts indicate that this slow-potential change is local, and is probably caused by a local inward electrical current of the water fiber terminals.

In the top and middle traces, the afferent impulses appear to correspond to the slow-potential change, but in the bottom trace some of them appear to be evoked without the corresponding slow-potential change. (The second to fourth spikes were probably of the slow-adapting mechanoreceptor, which sometimes initiated impulses in response to a rapid flow.)

The effect of tetrodotoxin upon the slow-potential changes of the water receptor is shown in Figure 5. A is the control record, B is the record after the application of

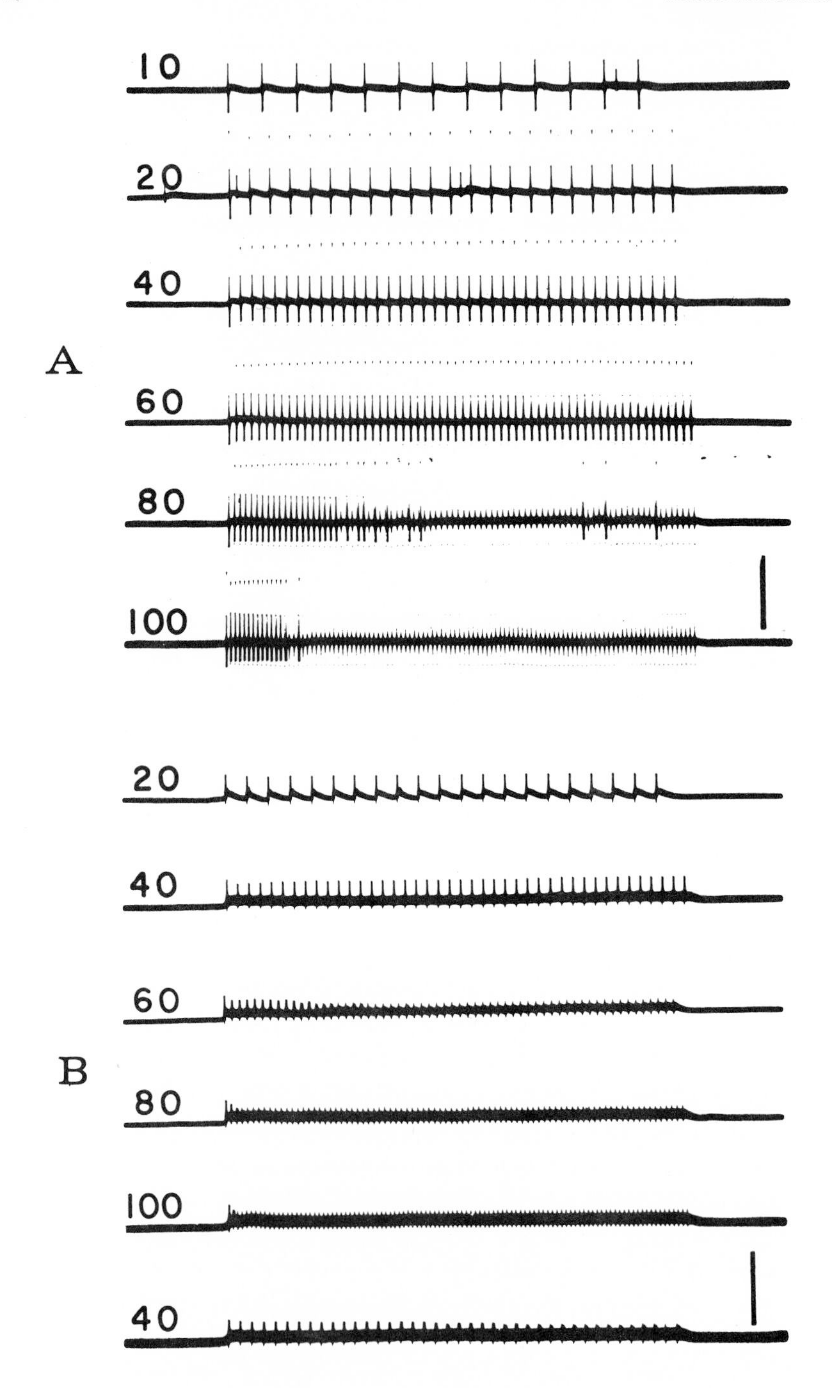

Figure 3 Responses of the fast-adapting mechanoreceptor to repetitive tactile stimulation. A: Control. B: After the application of 5×10^{-8} g/ml tetrodotoxin. Vertical bars indicate 5 mV, and the numbers at the left indicate the stimulus frequency (cycles per second).

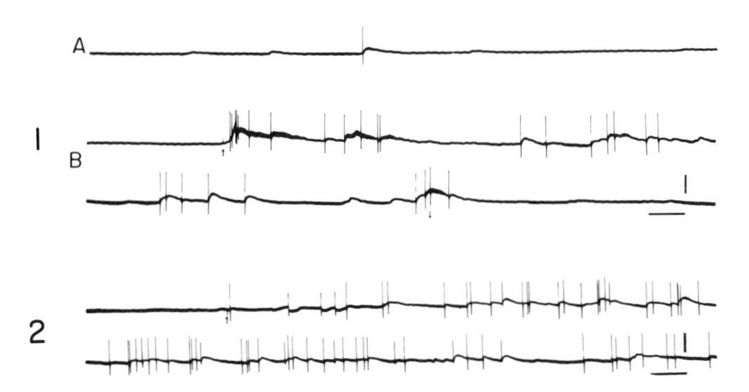

Figure 4 Slow-potential changes and afferent nerve impulses from the water receptors. Upward and downward arrows indicate the respective times of applying and washing out the taste solution. 1A: Record in the Ringer's solution. 1B and 2: Records by taste stimulation. Vertical bars and horizontal bars indicate 2 mV and 100 msec, respectively.

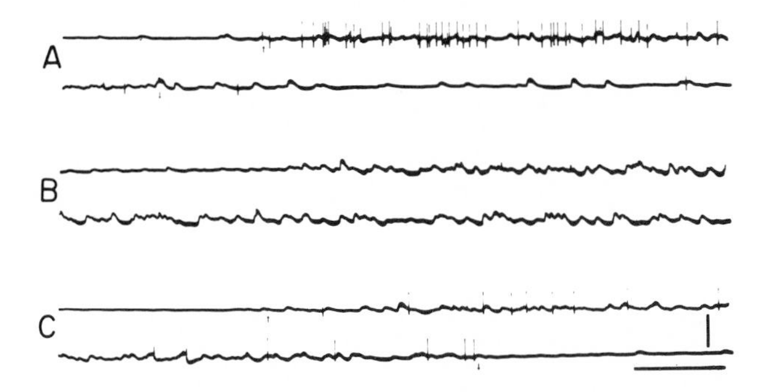

Figure 5 Effect of tetrodotoxin on the slow-potential change of the water receptor. A: Control. B: After application of 5×10^{-8} g/ml tetrodotoxin. C: Recovery. Horizontal bar indicates 100 msec, and vertical bar indicates 10 mV in A and C, and 5 mV in B.

5×10^{-8} g/ml tetrodotoxin, and C is the record of the recovery. The slow-potential changes remained after the propagated impulses had been blocked by tetrodotoxin.

Nomura et al.[1] have shown that the water receptors of the frog tongue respond well to tactile stimulation. Therefore, we decided to record the slow-potential changes of the water receptor by applying tactile stimuli to single fungiform papilla preparations. An example of such a record is shown in the left column in Figure 6. When increasing tactile stimuli, made by short pulses of one per second, were applied to the preparation, it was possible in only a few cases to record the slow-potential changes of the water receptor without a response of the mechanoreceptor. When the afferent impulses did appear they seemed to be associated with the slow-potential changes.

It has been shown that the latency of the mechanoreceptor response of the frog tongue is constant if the stimulus intensity is strong enough, while the latency of the water receptor to tactile stimulation always varies, and the response is sometimes abortive[2] as shown in Figure 6. This suggests that in the frog the afferent system of the water receptor is more complicated than that of the mechanoreceptor.

The record in the right column shows the water response from the same receptor that is shown in the left column. This record is a continuous trace, reading from bottom to top. The records show that the time course of the slow-potential change is rather long: the rise time often exceeds 15 msec.

DISCUSSION

It is generally accepted that there are receptor synapses between the taste cells and the taste nerve fibers in the taste organs of vertebrates. Uga[3] has shown in his electron-microscopic studies that there are chemically transmitting synapses between the taste cells and the connected nerve fiber in the fungiform papilla of the frog. Based on this observation, it is likely that the slow-potential changes occurring with the water response are a kind of synaptic potential evoked at the terminals of the taste-nerve fibers of the frog tongue.

This possibility, however, remains in question, because the correlation between the slow-potential changes and the propagated impulses in the water receptor is not as

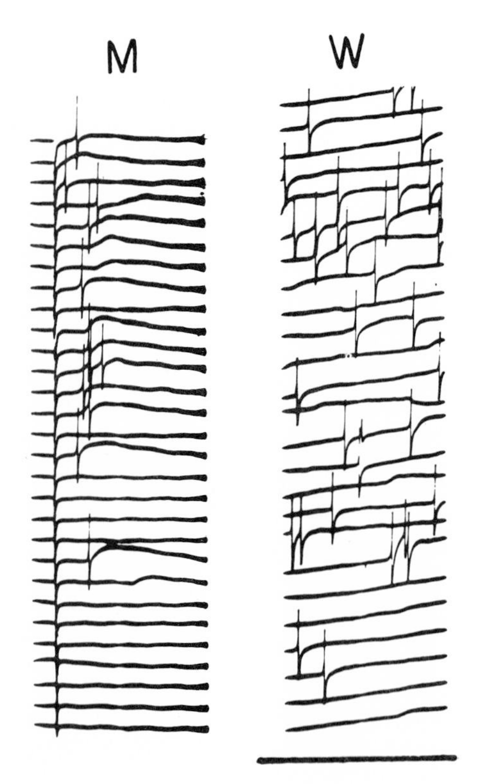

Figure 6 Responses of the water receptor to tactile and taste stimulation. M: Response to tactile stimulation. W: Response to taste stimulation. Horizontal bar indicates 100 msec. The stimulus intensity in the left record was increased from the bottom to the top.

close as it is in the mechanoreceptor. In the mechanoreceptor, the propagated impulses were always preceded by the generator potential and were initiated at a particular amplitude. On the contrary, the impulses of the water receptor were initiated independently of the amplitude of the slow potential, and some impulses appeared to be initiated without a corresponding slow-potential change. If the slow-potential change of the water receptor were a unitary synaptic potential, the propagated impulse should be initiated at a certain level of the slow-potential change. Such a close correlation between the slow-potential change and the impulse initiation has not been observed in the present work.

A tentative explanation for the nature of this slow-potential change is that it is a superposition of a great number of unitary synaptic potentials. Because a fungiform papilla of the frog tongue contains a great number of taste cells, there should be numerous synaptic connections between these cells and the nerve fiber terminals. In addition, the individual unitary synaptic potentials that occur at the terminals are probably too small to be recorded. But if there is a mechanism by which a transmitter is released from the taste cells synchronously, then many unitary synaptic potentials might be superimposed and recorded as a slow-potential change. In this case, the initiation of the impulse might not be closely correlated with the slow-potential change. To date, there is no direct evidence suggesting such a mechanism. Therefore, further investigation is necessary.

REFERENCES

1. Nomura, H., H. Aida, and H. Fukuda. 1966. Branching of the sensory nerve axons in the frog tongue. *Med. Biol.* **73**, pp. 309–312.
2. Nomura, H., H. Aida, and H. Fukuda. 1966. Tactile and electrical stimulation for the single fungiform papilla of the frog tongue. *Med. Biol.* **73**, pp. 313–316.
3. Uga, S. 1966. The fine structure of gustatory receptor and their synapses in frog's tongue. *Symp. Soc. Cell Biol.* **16**, pp. 75–85.

INNERVATION OF RAT FUNGIFORM PAPILLA

DR. LLOYD M. BEIDLER

Department of Biological Science, Florida State University, Tallahassee, Florida

INTRODUCTION

The response of individual taste receptors is of primary concern when studying taste function. It has long been known, however, that single nerve fibers which are used for electrophysiological studies may actually innervate several taste receptors within a taste bud. Systematic study of such innervation has not been previously attempted, primarily because only the electron microscope can reveal the course of the fine nerve endings. In addition, Rapuzzi[16] recently proved that adjacent fungiform papillae of the frog tongue may be innervated by the same nerve fiber. Whether their respective taste buds were also so innervated was not proved. It is the purpose of this paper to present new evidence concerning the anatomy and function of the taste bud complex.

GROSS MORPHOLOGY

The great depth of focus, the wide field, and the large range of magnification of the scanning electron microscope allows one to examine the papillae of the surface of the tongue and obtain a better appreciation of the environment in which the taste buds operate. Segments (about 1 cm^2) of the tongue surface are excised and dropped immediately into a 3:1 propane-isopentane mixture that has been cooled by liquid nitrogen to −175°C. The tissue is fixed over a period of several days by freeze substitution and then either freeze dried or taken through a series of acetone-water mixtures. An alternative method of formalin fixation followed by acetone drying was sometimes used. The base of the dried tissue was mounted onto metal blocks with Duco cement or silver paint, and the surface of the tissue coated with about a 200 Å thickness of gold, using a vacuum evaporator.

The surface structures of the tongue vary greatly from one species to another, particularly the form of the filiform papillae. At low magnifications the fungiform papillae of the rabbit or rat appear like the heads of mushrooms. (See Figures 1 and 2.) At high magnifications, the single cells can be seen to cover the papilla much like the leaves of a cabbage. The taste bud with its inner pore projects into an outer pore formed by the cells covering the fungiform papilla.

In many instances the tip of the taste bud rises slightly above the surface of the papilla. Whether the height of projection of the taste bud can vary with physiological state is not known. The large blood supply beneath the bud suggests a possible

Figure 1 Surface of rabbit tongue shows numerous filiform papillae that look like leaves, and a single fungiform papilla that contains the taste buds. Magnification: 200 ×.

Figure 2 Rat fungiform papilla surrounded by filiform papillae. Note rise of taste bud and numerous leaflike cells covering the fungiform papilla. Magnification: 2080×.

mechanism. It is interesting to note that Graziadei[10] has found muscle cells in the frog fungiform papillae that extend upward toward the base of the taste bud. However, the architecture of the frog fungiform papilla is quite different from that of mammals. The taste buds of all the fungiform papillae studied histologically were located on the dorsal surface and none were found on the papillae walls, in contradiction to that implied by von Békésy[3] for humans.

The use of conventional electron microscopy reveals that the microvilli of the cells of the fungiform taste bud terminate in miniature inner pores which join at different levels with others adjacent until a full-fledged inner pore is formed. The width and length of microvilli differ not only with species but may also differ within any single taste bud of a given species. Most microvilli never reach the top of the outer pore. Solutions placed on the surface of the immobile tongue reach the fungiform microvilli by diffusion or, in some cases, by convection. Movement of the tongue contracts and expands the pores and facilitates movement of the stimulus to and from the microvilli. Korovina[13] showed recently that gustatory stimulation may also initiate a reflex to elicit tongue movements. This feedback may enhance taste-receptor responses.

Taste buds lining the trenches of the circumvallate and foliate papillae are not as easily reached by stimuli. (See Figures 3 and 4.) The latency of stimulation may be slightly greater than that found for fungiform papillae, and the rise takes much longer in an immobile tongue. Glandular secretion at the base of the trenches helps the removal of the stimulus. Tongue movement or stretching greatly increases the transport of taste stimuli within the trenches and facilitates both taste receptor stimulation and the recovery after a water rinse.[1,11,12]

INNERVATION OF RAT FUNGIFORM PAPILLA

How many nerve fibers innervate the fungiform papilla and what are their pathways? A single fungiform papilla from each of five albino rats was prepared for electron microscopy by using glutaraldehyde fixation and osmium postfixation, followed by araldite embedding. After examination of thin sections with the electron microscope, a large (1.4 m^2) montage of the papilla cross section was assembled at every micron from the base to the apex of the papilla. The total number of myelinated and unmyelinated nerve fibers in cross sections was counted. The average papilla contained 224 nerve profiles at its base; 196 were unmyelinated and 28 myelinated. Many of the fibers were grouped together as they rose in the papilla, and smaller groups of these nerves branched toward the top to terminate in the epithelium immediately surrounding the taste bud. Approximately 25 per cent of the total number of nerve fibers within the papilla entered the taste bud. Note that seven times as many unmyelinated fibers enter the papilla as myelinated. Whether all these fibers originated from the unmyelinated fibers seen in the chorda tympani nerve bundle or whether some myelinated chorda tympani fibers lost their myelin before reaching the papilla is not now known, but the latter appears most probable.

It should be remembered that only the total number of nerve profiles were counted. Some fibers may terminate early as they rise in the papilla and others may divide. In

Figure 3 Circumvallate papilla of three-week-old puppy. Taste buds are found in the trenches surrounding the surface papillae. Magnification: 108 ×.

Figure 4 Foliate papillae of rabbit. Taste buds are in grooves. Magnification: 212×.

order to study the papilla innervation more carefully, three additional rat fungiform papillae were sectioned serially, one longitudinally and two tangentially. One of the latter is now described in detail. One hundred seventy-one of the sections were selected for study and large montages were made of each, using up to 35 photographs for each papilla cross section. The nerve profiles within several montages of the papilla were outlined and photographed. (See Figure 5, pages 360, 361.) The change in distribution as the nerves rise up the papilla is shown in Figure 6.

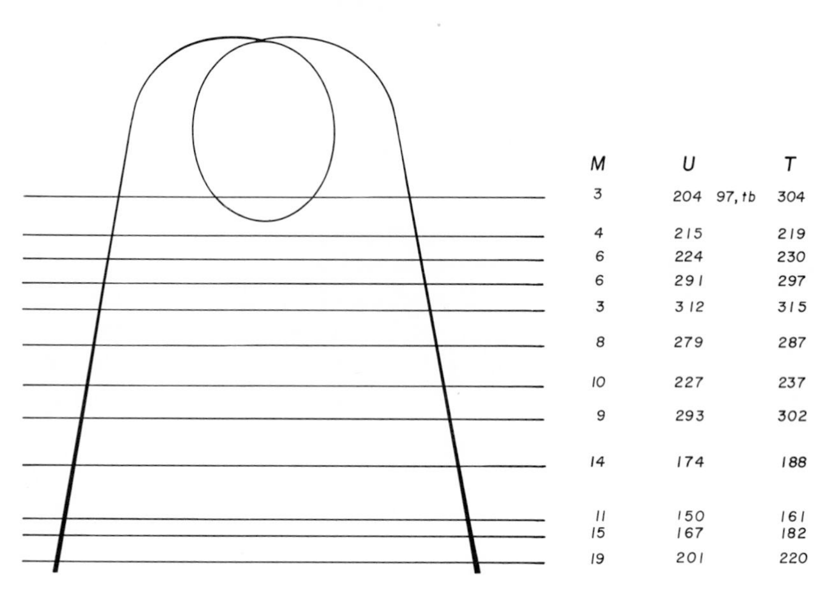

Figure 6 The number of myelinated (M), unmyelinated (U), and total (T) axon profiles counted in rat fungiform papilla cross sections at each level shown.

There were 19 myelinated and 201 unmyelinated nerve fibers entering the base of this particular papilla. The number of branches varied as they progressed up the papilla, some terminating and others branching farther. All but two of the myelinated fibers definitely innervated areas of the papilla other than the taste bud. Trace of the branches of the two remaining was lost immediately below the taste bud and no conclusion as to the area of innervation can be made. Only 54 fibers of the 315 counted about 25 μ below the base of the taste bud actually entered the taste bud. These 54 fibers continued to approach the taste bud in a firm bundle without branching until they reached almost to its base. It was definitely ascertained that 42 of these fibers did not branch after first entering the papilla and did not send axons to other areas of the papilla. Complete traces of the other 12 were not possible. Ninety-seven fibers were counted a few microns above the taste bud base, and these fibers branched farther to provide a maximum of 206 separate and distinct branches within the taste bud.

NERVE FIBERS IN THE CHORDA TYMPANI NERVE OF THE RAT

We have shown that several hundred axons enter a single fungiform papilla, but are there enough fibers in the chorda tympani nerve to innervate all the papillae without duplicate innervation?

Fish, Malone, and Richter[6] made a careful study of the number of fungiform papillae on tongues of individual hooded rats. Each half of 103 tongues studied contained, on the average, 89 fungiform papillae; each of these contained a single taste bud. Lindsley[14] used silver and osmium staining to count the number of myelinated and unmyelinated fibers in the rat chorda tympani. Since electron microscopy allows a better evaluation of unmyelinated fibers, we restudied the chorda tympani. A chorda tympani nerve bundle was removed from each of four albino rats, the connective tissue removed, and the nerve prepared for electron microscopy study by fixing with glutaraldehyde and postfixing with osmium. A large montage, approximately 1.5 m^2, was assembled for the cross section of each of the four nerve bundles, showing profiles of both myelinated and nonmyelinated nerve fibers. The area of each individual nerve fiber, excluding the myelin, was measured with a planimeter. The nerve fiber profiles were seldom circular in shape and varied considerably, so the area values were converted into diameters of circles of equivalent areas.

The summary of the measurements is presented in Table I. Note that the average

TABLE I
NUMBER OF NERVE FIBERS IN RAT CHORDA TYMPANI

Rat	Myelinated	Unmyelinated	Total	Chorda Diameter
#1	626	302	928	115μ
2	530	472	1002	107
3	590	485	1075	126
4	572	259	831	103
Average	580	380	959	113

bundle contains 580 (60 per cent) myelinated and 380 (40 per cent) unmyelinated nerve fibers. Lindsley[14] counted 1139 fibers in the rat chorda tympani, 45 per cent of which were unmyelinated. Foley[7] found 17 per cent unmyelinated fibers in the dog and 19 per cent in the cat. He showed that many unmyelinated fibers innervate the submaxillary gland, although 23 per cent of the sensory chorda tympani axons of the dog were also unmyelinated. The myelinated fibers are known to serve not only taste but also other sensory functions of the tongue. Our studies indicate that almost half

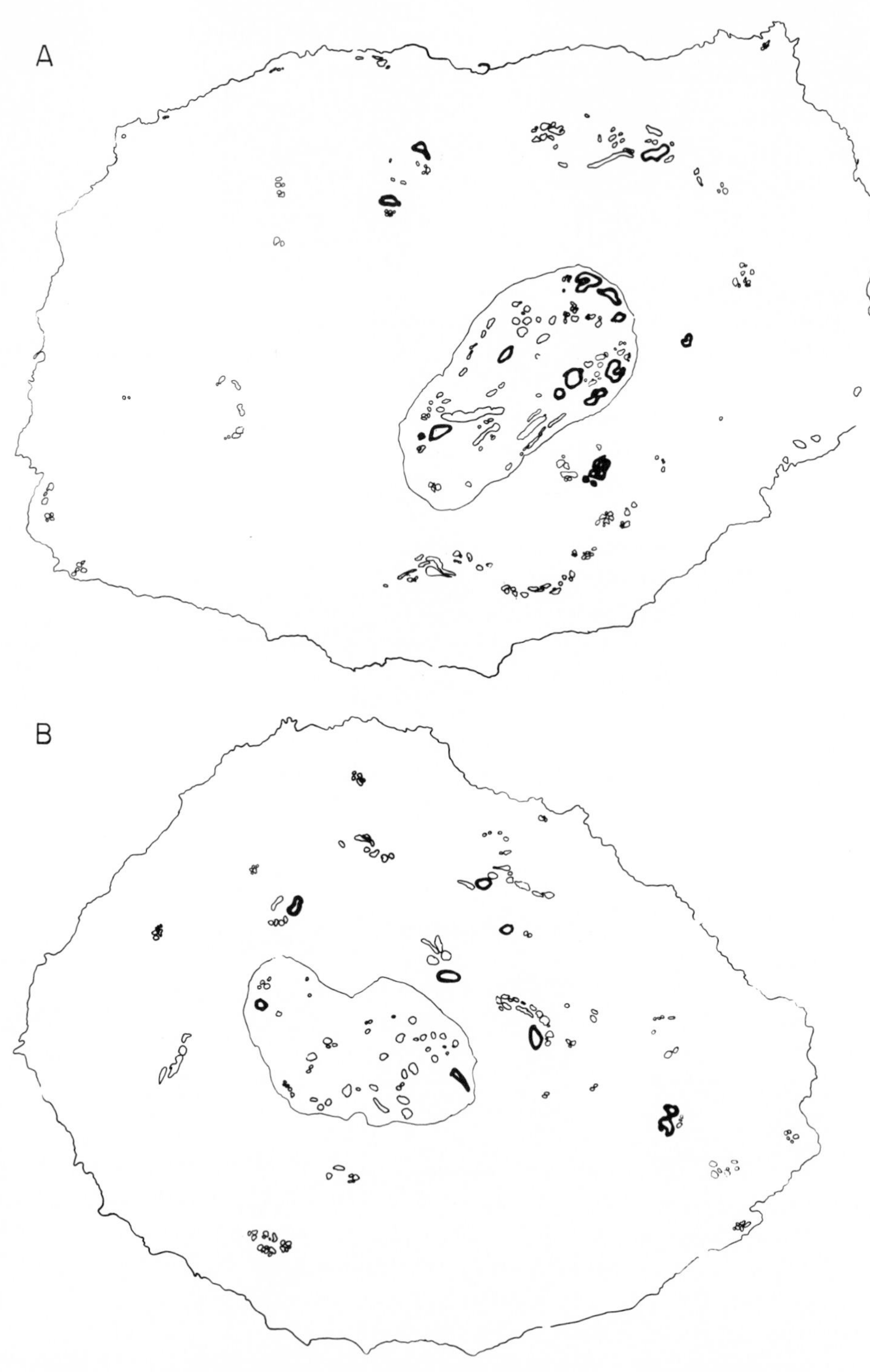

Figure 5 Tracings of inside wall of rat fungiform papilla with nerve profiles inserted. A: base of papilla, 121 μ from taste pore. B: 92 μ from pore. C: 64 μ from pore. D: close to base of taste bud, 39 μ from pore.

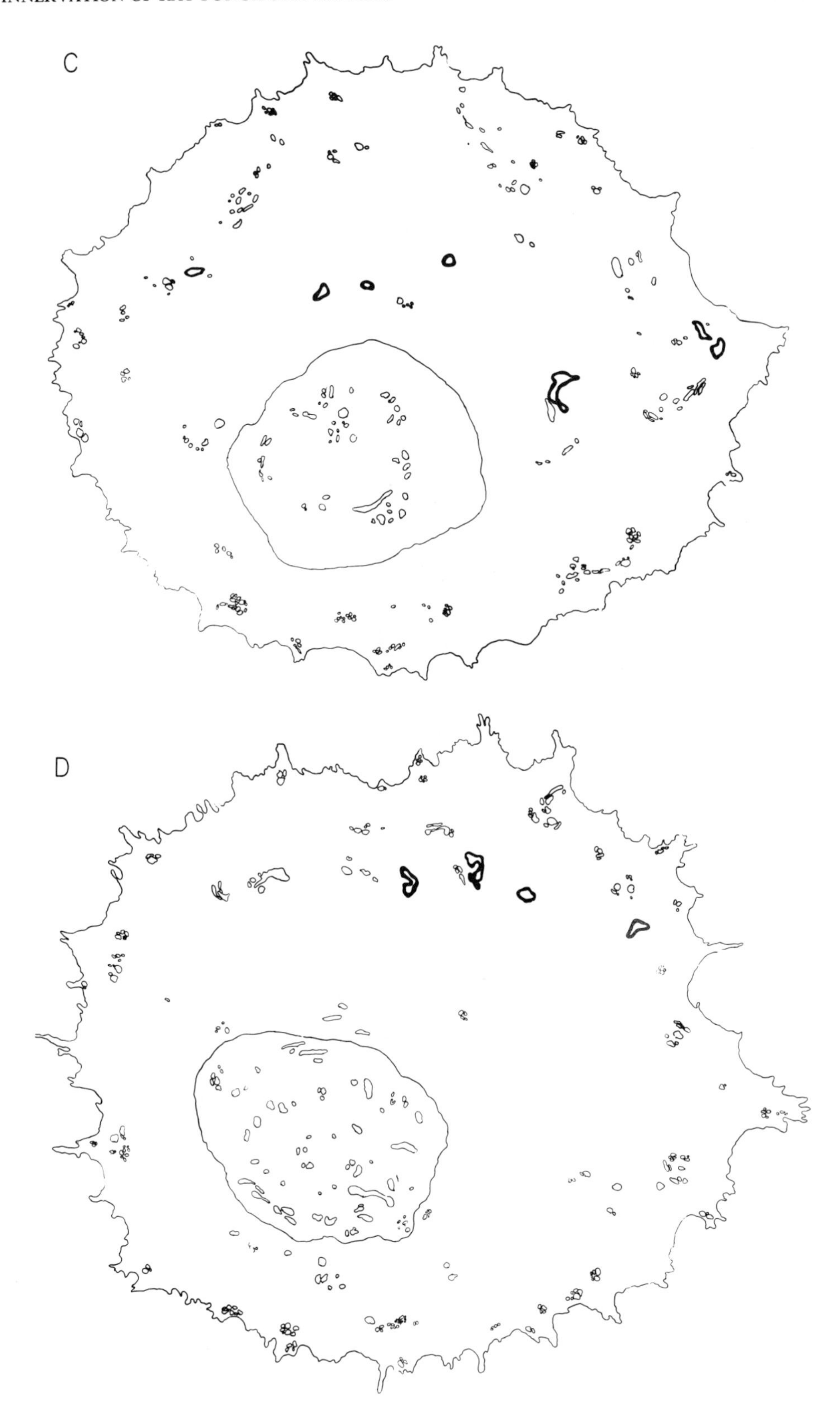
C
D

the myelinated fibers have a diameter of 2.0 to 4.0 μ, excluding the myelin. Figure 7 shows data from one nerve bundle.

If one makes the maximum assumption that all the myelinated fibers of the chorda tympani nerve bundle plus 25 per cent of the unmyelinated are taste fibers, then each fungiform taste bud should be innervated by a maximum of 7 to 8 fibers if no branching occurs before their entrance to the base of the papilla. However, several hundred nerve fibers enter the base of each fungiform papilla of the rat, and many of these

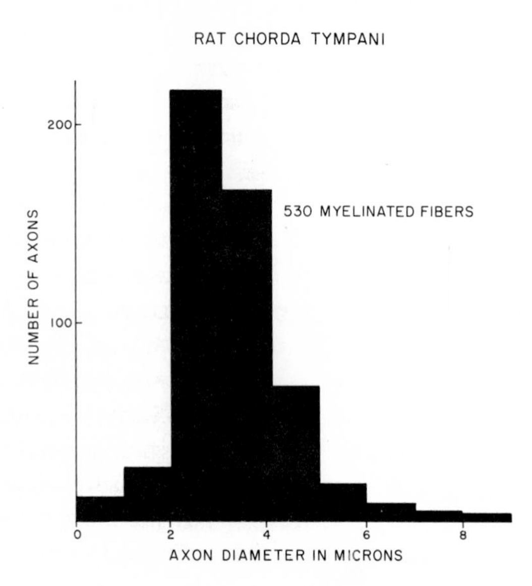

Figure 7 Size distribution of myelinated axons of rat chorda tympani. Myelin of each axon was excluded in measurement.

are chorda tympani fibers innervating the taste bud. Thus, multiple branching of the taste nerve fibers takes place before they enter the papilla, and each taste bud is innervated by a relatively large number of nerve branches.

ORIGIN OF NERVE FIBERS IN FUNGIFORM PAPILLA

It is well known that a component of both the lingual and the chorda tympani nerve innervates receptors of the tongue. The latter, however, is necessary for maintenance of taste buds. What percentage of the fibers entering the base of the fungiform papilla originate in the chorda tympani nerve? To study this, rats with one of their chorda tympani nerves severed were compared with others whose lingual nerves (before chorda tympani joins its bundle) were severed. Both groups were killed six to seven days after nerve section and single fungiform papillae were dissected free and prepared for electron microscopy.

Most of the myelinated nerves that enter the papilla come from the lingual nerve.

For example, an average of two myelinated nerve fibers were observed at the base of the papilla after sectioning the lingual nerve. Furthermore, the total nerve fiber count in these papillae was no more than 25 per cent of that observed in normal papillae. Thus, the vast majority of myelinated fibers and the majority of unmyelinated fibers seen in normal fungiform papillae come from the lingual and not the chorda tympani nerve. The majority of fibers innervating the taste bud definitely come from the chorda tympani nerve, although it has not been established that no lingual fibers enter the bud. The calculated maximum value of 7 to 8 fibers in the chorda tympani nerve bundle allotted to each fungiform papilla must be increased by at least sevenfold by branching before entering the base of the papilla and eventually innervating the taste bud.

Sectioning the chorda tympani results in loss of taste buds. Similar treatment to the lingual nerve, central to the point at which the chorda tympani leaves it, results in intact fungiform papillae and taste buds, but both appear to be reduced in size.

INTERACTIONS BETWEEN FUNGIFORM PAPILLAE

If nerve fibers branch before the chorda tympani nerve fibers enter the base of the fungiform papilla, there is a good possibility that two adjacent papillae are innervated by the same fiber. For this reason, Mr. Inglis Miller in our laboratory investigated papillae interactions through the use of single nerve-fiber responses. He applied chemical solutions to a small area of the tongue to locate the most responsive papilla innervated by the previously dissected single chorda tympani nerve fiber. After the papilla was located, he applied a flow chamber designed to stimulate but a single papilla. I will describe two of the experiments.

A stimulus of 0.1 M NaCl was applied to a single papilla, and various NaCl concentrations were flowed over the area of the tongue surrounding the chamber. Enhancement of the response from the single papilla was obtained as the surround concentration was increased (Figure 8). Because adequate controls were undertaken to prevent the surround solution from leaking toward the chosen single papilla, the results of the above experiment are evidences that papilla-papilla interactions do occur.

Are there suppressions as well as enhancements of taste responses? There are few known substances that produce a suppression of taste neural activity. It has been demonstrated[2] with whole nerve integrated responses that moderate concentrations of potassium benzoate will depress the small neural activity observed during a water rinse of the rat tongue. For this reason, Mr. Miller tried increasing concentrations of potassium benzoate applied to the surround while stimulating the chosen single papilla with 0.1 M NaCl. The single-fiber response declined with moderate potassium benzoate concentrations and increased with high concentrations (Figure 8). This most unexpected finding is difficult to understand, although all controls were thoroughly investigated. From these preliminary but provocative experiments by Mr. Miller, we may conclude that both suppressions and enhancements occur as a result of papilla-papilla interactions.

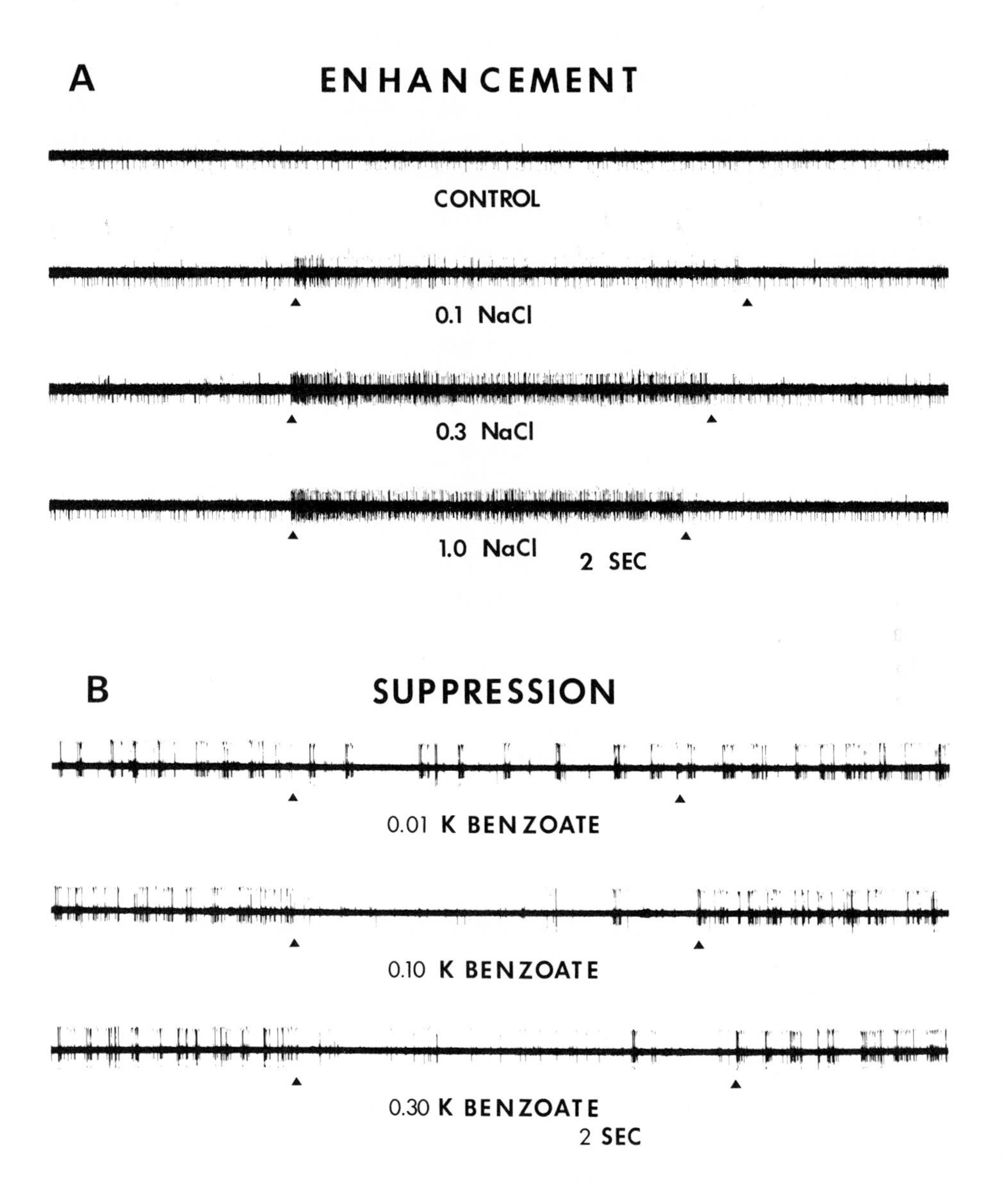

Figure 8 A: Enhancement of single fungiform papilla response to 0.1 M NaCl when various concentrations of NaCl are applied to surround. B: Suppression of single fungiform papilla response to 0.1 M NaCl when various concentrations of K benzoate are applied to surround. Arrows mark application and rinse of surround solutions.

INNERVATION OF SINGLE TASTE BUD

We have shown that 206 unmyelinated nerve fiber branches existed in the taste bud we sectioned serially for careful study. These fibers may turn abruptly as they twist up the taste bud, so that near-serial sections are necessary in order to trace their pathways. The nerve profiles were counted in a number of montages and plotted as a function of distance from the papilla base. Almost a linear decrease was observed, as shown in Figure 9.

To find the anatomical relationship between these fibers and the 59 cells within the taste bud, we numbered each cell (starting with those whose nuclei were nearest the apex of the taste bud and progressing downward) and studied them carefully to determine how many fibers made contact with each one. These contacts were divided into two groups—those fibers that were totally invaginated by the cell and those that made contact by being only partially invaginated. Of all the sections studied, only two showed a classical synaptic structure with vesicles and thickening of membrane. This presents a problem because the usual anatomical evidence for a functional contact between taste cell and nerve in the rat fungiform papilla is apparently ill-defined, except in a very few cases. Murray and Murray[15] found a similar lack of "consistent pattern of specialization" in the taste buds of the rabbit foliate papillae. Gray and Watkins[8] saw no postsynaptic membrane thickening and no consistent evidence of presynaptic vesicles in rat circumvallate taste buds. Farbman,[5] who studied the rat fungiform taste buds, observed a "membranous sac" within the cell near the nerve process. Functional meaning was lacking, however. On the other hand, Graziadei[9] observed typical differentiation of synapses in taste buds of fish and humans. From all of the above literature and from our own observations, we concluded that we

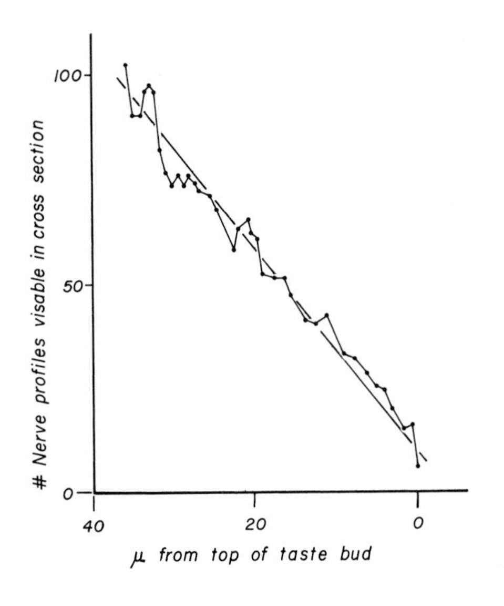

Figure 9 The number of nerve profiles in taste-bud cross section decreases linearly as the nerves progress up toward the pore.

could not determine which of the fibers that made an anatomical contact with a cell also made a functional contact.

Table II lists all contacts made, no single fiber being counted twice in the same cell. Some fibers that contact one cell may also contact another cell. Thus, 550 total contacts were seen, but these were made with only 206 nerve branches within the taste bud; these, in turn, came from only 54 fibers seen below the base of the taste bud.

Note that only one cell of the taste bud had no nerve contacts. Cell number 34 had contacts with 22 different nerve branches. We noted no special relationship between

TABLE II

LISTS OF ALL NEURAL CONTACTS WITH EACH OF 59 CELLS OF RAT TASTE BUDS

Cell #	Partial contact	Envelope	Cell #	Partial contact	Envelope
1	3	2	31	4	3
2	1	0	32	6	5
3	1	1	33	3	4
4	0	0	34	14	8
5	1	0	35	6	5
6	3	3	36	3	12
7	3	0	37	10	4
8	0	6	38	3	1
9	1	4	39	3	1
10	6	3	40	9	10
11	10	5	41	4	0
12	9	6	42	11	1
13	5	9	43	5	0
14	2	10	44	10	0
15	1	0	45	4	0
16	3	0	46	10	6
17	6	6	47	8	5
18	6	9	48	7	1
19	1	4	49	3	10
20	13	0	50	10	0
21	15	6	51	5	0
22	6	1	52	1	0
23	6	0	53	5	5
24	9	2	54	8	0
25	8	5	55	8	1
26	4	2	56	1	0
27	12	7	57	5	6
28	5	1	58	9	8
29	8	10	59	11	4
30	5	9	Totals	339	211

the character or location of a taste cell and the number of contacts. It is interesting that 17 of the cells did not completely envelop any nerve. Since the cells were numbered according to the heights of their nuclei within the bud, it is apparent that no relationship exists with the number or kind of contacts, as shown in Table II. Since each nerve branch was assigned a definite number, we could easily determine that one nerve branch may contact a number of different cells. The courses of three of the 97 nerves counted above the base of the taste bud were examined. The first nerve branched three times, made partial contact with cells numbered 34, 24, 25, 27, 46, 16, 39, 6, 15, 33, and was enveloped by cells numbered 6, 46, 25, and 34. The second nerve branched three times, made partial contact with cells numbered 49, 42, 47, 38, 27, 5, 22, 23, 9, 10, and was enveloped by cells numbered 49, 27, and 22. The third nerve branched three times, made partial contact with cells numbered 13, 34, 44, 29, 30, and was enveloped by cells numbered 29 and 30.

It is interesting to note that not all cells extended from the base to the apex. The nuclei were also of different shapes. Some were "C" shaped so that on one section of the taste bud a cell could appear to have two nuclei, but on observing thin serial sections of the same cell the complete form could be determined. No binuclear cells were observed in the taste bud, in contradiction to such cells observed by Shimamura et al.[17] in rabbit taste buds using light microscopy.

The diameters of nerve fibers found within the taste bud range from 0.3 to 4 μ. The diameter of a given nerve may vary widely as it courses up the taste bud. Thus, one cannot assume that a section showing a nerve of large diameter is necessarily a terminal. Nor is it easy to distinguish some large nerves from taste cells on the basis of size. A study of the cell cytology or a comparison of serial sections are the best means to determine the character of a cell. Gray and Watkins[8] were particularly concerned about these distinctions, and even thought that some of the structures described by De Lorenzo in 1963 as axonal[4] could actually be receptor cell profiles. In addition, it is impossible to classify taste-bud nerves according to size, as dilations followed by contractions are common with many of the nerves. This point has already been stressed by Farbman.[5]

Since the number of nerve profiles in a cross section decreases toward the apex, very few axons are present near the pore region. Six such nerves were counted in the particular papilla described in detail.

All the calculations in this paper were based upon the assumption that all the chorda tympani fibers in the fungiform papilla were afferent and none efferent. No clearcut evidence has been published to show efferents in rat fungiform taste buds.

SUMMARY

The external morphology of the fungiform (rat), the foliate (rabbit), and circumvallate (dog) papillae were examined with the scanning electron microscope and related to stimulus transport.

The innervation of the rat fungiform papilla was studied in detail. Several hundred unmyelinated fibers enter each rat fungiform papilla. About 1/4 of these are chorda tympani fibers and innervate the taste bud. The majority of the remainder are lingual fibers that innervate other areas of the papilla. Twenty to 40 myelinated fibers also enter the papilla and most of them are lingual fibers that do not innervate the taste bud.

The chorda tympani fibers double in number by branching immediately before entering the taste bud. They double again by branching within the taste bud. The majority of taste bud cells make anatomical contact with a number of these branches.

Chorda tympani nerves branch (about sevenfold) before entering the fungiform papilla, and one fiber innervates several papillae. Both suppression and enhancement of the response of a single stimulated papilla can occur if the surrounding papillae are also stimulated.

This study would indicate that when one records electrophysiologically from a single fiber of the chorda tympani nerve bundle, the response is gathered from its 25 or so branches in the taste bud that contact at least an equal number of taste cells. In addition, the response may be enhanced or suppressed by simultaneous stimulation of neighboring papillae that are innervated by other branches of the same single nerve fiber. Thus, it is not surprising that highly specific single taste fibers are seldom seen in the rat.

ACKNOWLEDGMENTS

This paper was supported by National Science Foundation Grant No. GB-4068X and U.S. Public Health Service Grant No. NB07468-02.

I appreciate the technical help of Mrs. Judith Hunter, electron microscopist, Mr. Joe Metcalf, senior histologist, and Mr. David Fleming, photographer. I thank Mr. Inglis Miller for the summary of his preliminary research as described in this manuscript.

REFERENCES

1. Appelberg, B. 1958. Species differences in the taste qualities mediated through the glossopharyngeal nerve. *Acta Physiol. Scand.* **44**, pp. 129–137.
2. Beidler, Lloyd M. 1962. Taste receptor stimulation. *Prog. Biophys. Biophys. Chem.* **12**, pp. 109–151.
3. Békésy, G. von. 1966. Taste theories and the chemical stimulation of single papillae. *J. Appl. Physiol.* **21**, pp. 1–9.
4. DeLorenzo, A. J. D. 1963. Studies on the ultrastructure and histophysiology of cell membranes, nerve fibres and synaptic junctions in chemoreceptors. *In* Olfaction and Taste I (Y. Zotterman, editor). Pergamon Press, Oxford, pp. 5–17.
5. Farbman, A. I. 1965. Fine structure of the taste bud. *J. Ultrastruct. Res.* **12**, pp. 328–350.
6. Fish, H. S., P. D. Malone, and C. P. Richter. 1944. The anatomy of the tongue of the domestic Norway rat. *Anat. Record.* **89**, pp. 429–440.
7. Foley, J. O. 1945. The sensory and motor axons of the chorda tympani. *Proc. Soc. Exp. Biol. Med.* **60**, pp. 262–267.
8. Gray, E. G., and K. C. Watkins. 1965. Electron microscopy of taste buds of the rat. *Histochemie* **66**, pp. 583–595.
9. Graziadei, P. 1968. (Personal communication)
10. Graziadei, P. 1968. Synaptic organization in vertebrate taste buds. *Proc. Int. Union Physiol. Sci.* **3**, Abstract.

11. Ishiko, N., and M. Amatsu. 1964. Effects of stretch of the tongue on taste responses in glossopharyngeal and chorda tympani nerves of the cat. *Kumamoto Med. J.* **17**, pp. 5–17.
12. Kitchell, R. L. 1963. Comparative anatomical and physiological studies of gustatory mechanisms. *In* Olfaction and Taste I (Y. Zotterman, editor). Pergamon Press, Oxford, etc., pp. 235–255.
13. Korovina, M. V. 1967. Elicitation of tongue movements by gustatory stimuli in animals. *Fiziologischeskii Zhurnal SSR imeni I.M. Sechenova* **54**, pp. 1432–1438.
14. Lindsley, O. R. 1950. Neural components of the chorda tympani of the rat. Unpublished M. A. Thesis, Brown University, Providence, Rhode Island.
15. Murray, R. G., and A. Murray. 1967. Fine structure of taste buds of rabbit foliate papillae. *J. Ultrastruct. Res.* **19**, pp. 327–353.
16. Rapuzzi, G., and C. Casella. 1965. Innervation of the fungiform papillae in the frog tongue. *J. Neurophysiol.* **28**, pp. 154–165.
17. Shimamura, A., T. Kawano, and T. Kimura. 1967. Three-dimensional observations on the taste buds in rabbits. *Arch. Histol. Jap.* **28**, pp. 471–481.

ELECTRICAL SIGNS OF TASTE RECEPTOR ACTIVITY

HIROMICHI MORITA

Department of Biology, Faculty of Science, Kyushu University, Fukuoka, Japan

INTRODUCTION

Sensory physiology has established that different qualities of stimuli are received by an animal through different receptors, and that different strengths of the same quality of stimuli are coded by impulse frequency in the same sensory neuron. In the present article, I will discuss the latter problem, i.e., quantitative aspect of receptor activity in terms of electrical signs.

INTEGRATED RESPONSES IN TASTE NERVES

Electrophysiological studies on the taste receptor have supplied us with the quantitative results on its activity. Among them the work by Beidler[1] has clearly shown that the integrated response in the chorda tympani nerve of the rat is good for quantitative analysis. The next year (1954), Beidler[2] published his theory of taste stimulation which was supported by his own quantitative data mentioned above. The integrated response is regarded as a summation of impulses per unit time in the taste neurons that innervate the stimulated receptors. According to Beidler's theory, the relative response r' ($\equiv r/r_m$, where r and r_m is the response magnitude and its maximum, respectively) is expressed as

$$r' = \frac{1}{1 + K_b/a}, \tag{1}$$

where K_b represents a constant corresponding to a dissociation constant in Beidler's taste equation and a is the concentration of stimulus substance.

Let us suppose that there are m active neurons in the integrated response. Then, we can define the relative value, R', in the integrated response as

$$R' = \frac{1}{m} \sum_{i=1}^{m} r_i' = \frac{1}{m} \sum \frac{1}{1 + K_{bi}/a}, \quad 0 \leqq R' \leqq 1.0. \tag{2}$$

This equation indicates that the integrated response is described by Beidler's theory only when all the active neurons have the same K_b value for the stimulus. Therefore, as far as we accept that the integrated response is described by Beidler's theory, we must assume that almost all neurons contributing the integrated response have the same K_b value, and that the neurons of different K_b values are negligible in number. In fact, Fishman[3] studied the single-fiber activity in the chorda tympani nerve of the rat, and reported that the analysis based on Beidler's theory resulted in the same order of K_b value (1/10.2 to 1/19.2 for NaCl) as that Beidler obtained (1/9.8).

RECEPTOR POTENTIAL RECORDED INTRACELLULARLY

Several years later, Kimura and Beidler[7] intracellularly recorded the receptor potential from the rat taste cell. They explained their results with Beidler's theory, and compared their own results with those of the integrated response obtained by Beidler.[2] In Table I, the values of K_b for NaCl obtained with different methods are compared.

TABLE I

VALUES OF K_b FOR NaCl, OBTAINED WITH DIFFERENT METHODS IN RAT

Method	K_b		Reference
	Average	Range	
Integration	1/9.8		Beidler[2]
Single axons		1/10.2–1/19.2	Fishman [3]
Receptor potential	1/16.7	1/7.1–1/40.5	Kimura and Beidler[7]

It is shown that the values of K_b obtained from the receptor potential ranged from 1/7.1 to 1/40.5 with a mean value of 1/16.7 in contrast with the value of 1/9.8 previously found by Beidler[2] in the integrated responses. Considering the variation in K_b in the former case, the difference between the results obtained with different methods may be insignificant. If such is the case, the average of the amplitude of the receptor potential is to be proportional to the average of the frequency of impulses in the taste neurons in the rat. We cannot, however, directly investigate the relation between the receptor potential and impulse frequency in vertebrate preparations. This relation has been studied in insect contact chemoreceptors.

RECEPTOR POTENTIAL AND IMPULSE FREQUENCY

Impulses of insect contact chemoreceptors were first recorded by Hodgson, Lettvin, and Roeder,[4] and its receptor potential was described by Morita.[9] Recently, Morita and Hori[10] improved the method for recording the electrical activities of the labellar chemoreceptors of the blowfly. They have found, in the recorded spike height, evidence that there are two distinct compartments within a single labellar chemosensory hair of the blowfly, *Phormia regina* (Figure 1). When a microelectrode (R_o in Figure 1A) filled with the blowfly's saline was inserted into the chemosensory hair from the convex side of the curved hair, the height of recorded spikes was less than 1 mV, positive with reference to the hair base ($R_o - E$ in Figure 1B). However, when a microelectrode (R_i in Figure 1A) was inserted from the opposite side—an insertion much more difficult than that from the convex side—the recorded spike height was

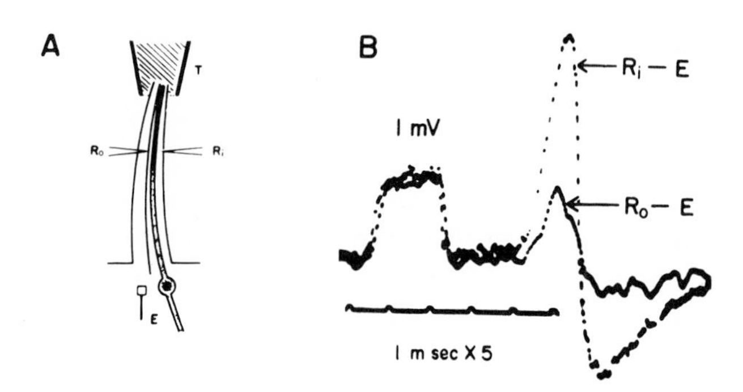

Figure 1 Simultaneous records of a spike potential in a labellar chemoreceptor of the blowfly (*Phormia regina*) with two microelectrodes (R_i and R_o). (A) Electrode R_i was inserted from the side. Indifferent electrode, E, was located on the inside of the hair base. (B) The records obtained with the electrodes R_i and R_o are indicated as $R_i - E$ and $R_o - E$, respectively.[10]

3–4 mV positive to the hair base ($R_i - E$ in Figure 1B). Figure 1B shows the records of receptor cell action potential; these records were simultaneously obtained, at the same level from the hair base, with the two recording electrodes, as mentioned above. Two such distinct compartments in the chemosensory hair apparently correspond to two cavities which have been demonstrated morphologically. I will denote the cavity containing the distal processes of the chemoreceptors as "filled cavity," and the one containing nothing but fluid as "empty cavity." This correspondence between morphological and physiological results leads to a conclusion: large and small spikes are recorded from the filled and the empty cavities, respectively. It follows that the wall of septum between the cavities has high electrical resistance. In Figure 2, the membrane current density (i_m) calculated from the cable theory is shown against the

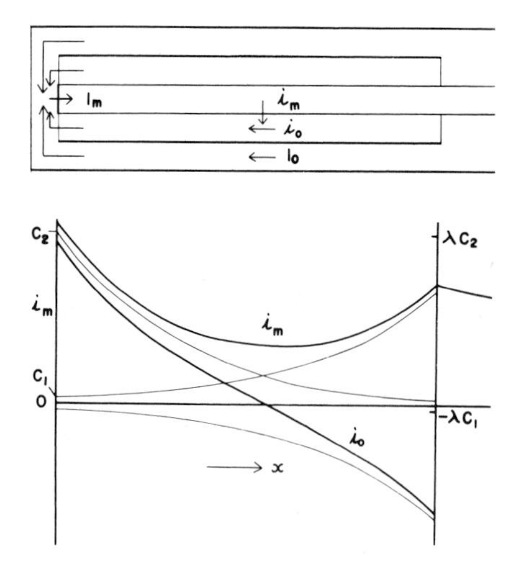

Figure 2 Top. Schematic representation of the structure of the chemosensory hair of the fly. The central cylinder represents the distal process of the chemoreceptor cell and the receptor membrane is represented by thick lines at the left end. The inner cylinder covering the central one represents the septum between the "filled" and "empty" cavities of the chemosensory hair. The outermost cylinder represents the hair wall. I_m, the receptor membrane current ; i_m, membrane current density; i_o, longitudinal current flowing through the extracellular space within the "filled" cavity; I_o, longitudinal current flowing in the "empty" cavity. Bottom. The values of i_m and i_o against the distance from the electromotive force located at the receptor membrane. The "septum" cylinder is assumed to be a perfect insulator except for at both ends.

distance from the locus of the electromotive force at the tip end. The septum is assumed to be a complete insulator, and both the cavities are to be opened at the tip. In fact, when $AgNO_3$ solution was applied to the hair tip, two traces of black precipitation could be identified in some cases. These traces were in the filled and the empty cavities, respectively. From these findings and theoretical considerations, it has been concluded that the displacement of the membrane potential at the receptor locus at the hair tip makes current flow along the distal process of the receptor, and that the outflow of the current is focused at the proximal end of the septum. The spike is initiated in this region. The longitudinal current (i_o) flowing through the filled cavity is also shown in Figure 2. The direction of the current flow is reversed on both sides of a point determined by the membrane constants. This explains why Wolbarsht and Hanson[14] sometimes recorded positive-going receptor potentials in the labellar chemosensory hair of the blowfly.

The receptor potential can be recorded with two microelectrodes, both of which are inserted into the empty cavity. The potential recorded in this way is free from irregular fluctuation, whose origin is supposed to be located somewhere near the hair base. Figure 3 shows several records obtained from a single chemosensory hair of the fleshfly, *Boettcherisca peregrina*. We have concluded that the negative-going potential that lasts during the stimulation is a complex of the receptor potentials of the stimulated receptors (the water and the sugar receptors, in this case). It can be seen in this figure that the receptor potential attains stationary level in 0.1 sec after the beginning of stimulus, while impulse frequency does so in 0.15 sec. It will be shown later that the impulse frequency is proportional to the amplitude of the receptor potential, but impulse initiation is sensitive to the slope of the receptor potential, so that the proportionality does not hold for the rising and falling phases of the receptor potential. From records such as those in Figure 3, we can examine the above-mentioned assumption of the proportionality in the stationary phase of the impulse frequency. The

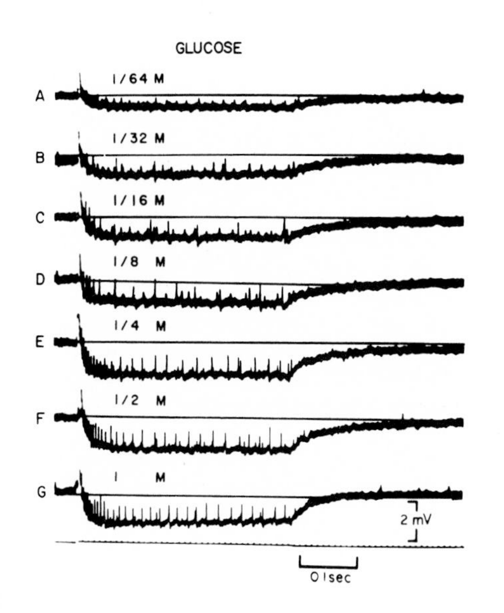

Figure 3 The receptor responses (water and sugar receptors) to glucose solutions. Two microelectrodes were inserted into the "empty" cavity, and the potential difference between the two electrodes was d-c amplified.[10]

number of impulses of the water receptor is plotted against the amplitude of the receptor potential in the cases where only the water receptor discharged impulses when stimulated by water or by dilute solutions of sugars (open circles in Figure 4A). Then a straight line was drawn through groups of the open circles and the origin of the co-ordinates. We assume that this straight line fulfills the relation between the impulse frequency and the amplitude of the receptor potential in the water receptor, even when both the water and the sugar receptors are stimulated by aqueous solutions of high concentrations of sugars. Then the amplitude of the receptor potential of the sugar receptor can be estimated, because the summated amplitude of the receptor potentials of the two receptors can be measured on the record, and because the receptor potential of the water receptor is estimated from the straight line in Figure 4A, counting its impulse frequency in the record. The impulse frequency was plotted against these estimated values of the receptor potential in the sugar receptor, as shown in Figure 4B. The relation between the impulse frequency and the amplitude of the receptor potential in the sugar receptor can thus be regarded as the proportional one. Based on this proportional relationship in the sugar receptor (i.e., the straight line in

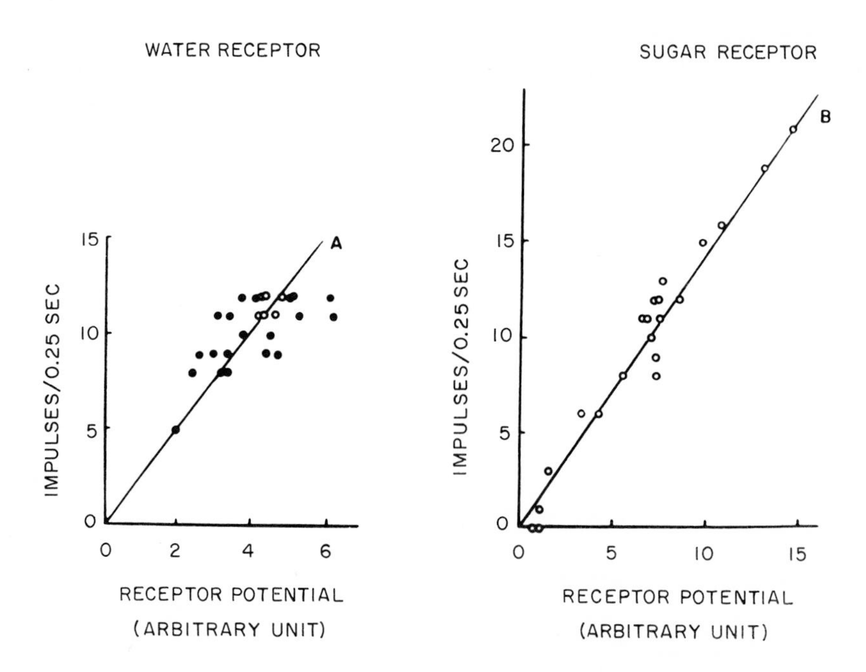

Figure 4 Relation between receptor potential and impulse frequency in stimulations by sugar solutions. The open circles in A represent the values measured directly on the records, in which only the water receptor discharged impulses on stimulation by distilled water or by dilute solutions of sugar. The values for closed circles in A and those for the open circles in B were calculated, based on the assumption that the impulse frequency and the amplitude of the receptor potential is proportional in both the receptors even when the two receptors discharged impulses on stimulation by high concentrations of sugar (see text).[10]

Figure 4B), the amplitude of the receptor potential of the water receptor can also be estimated and is shown by closed circles in Figure 4A. Thus, we can safely assume that the receptor potentials of the water and the sugar receptors are, respectively, proportional to their impulse frequencies in a stationary state of responses to aqueous solutions of sugars. This assumption is proved if the proportional relationship is ascertained in one of the receptors. In fact, Morita and Yamashita[12] showed this in the sugar receptor of the green bottlefly, *Lucilia*, where only the sugar receptor happened to be functional in the same chemosensory hair.

In the salt receptor, the recorded potential is influenced by a certain factor, which can be regarded as diffusion potential caused by high concentration of an applied salt solution. It is possible, however, to obtain a relation such as shown in Figure 4, based on some approximations.

In Figure 5, also, the open circles represent the values obtained when only one receptor (the water receptor or the salt receptor) discharged impulses, whereas the closed circles represent those when both the receptors discharged impulses. Here, too, a straight-line relationship exists in the salt receptor, but the line does not pass through the origin of the co-ordinates. This apparently corresponds to the fact that the salt receptor discharged impulses "spontaneously" under the experimental conditions.

From these results, the impulse frequency is considered to be equivalent to the

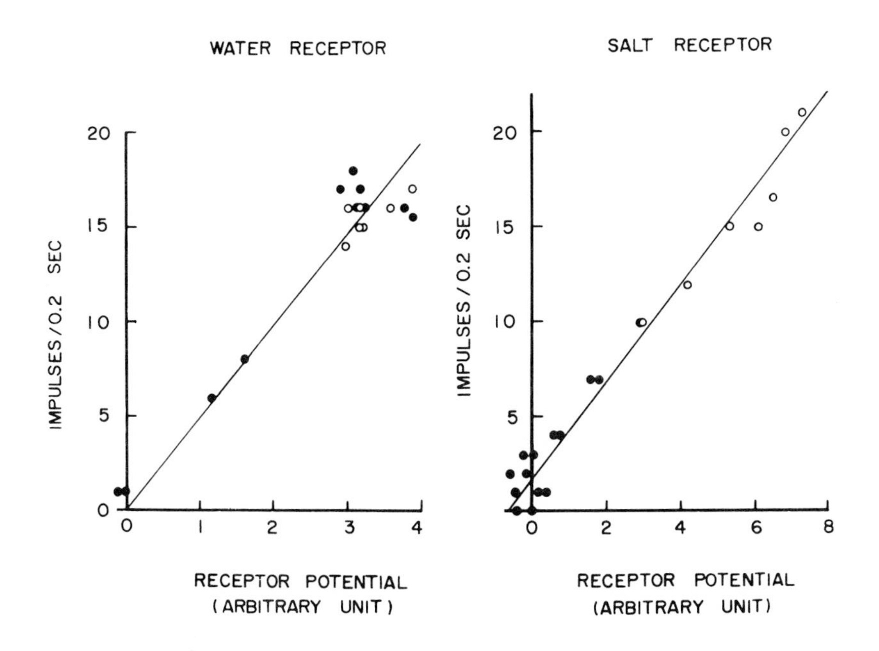

Figure 5 Relation between receptor potential and impulse frequency in stimulations by salt solutions. Open circles represent the values measured directly when only one of the receptors discharged impulses, and closed circles represent those calculated on the assumption as mentioned in explanation of Figure 4.[10]

amplitude of the recorded receptor potential in the stationary state of response, as a measure of the response magnitude in the relative value. The amplitude of the recorded receptor potential in the stationary state of response is, in turn, proportional to the displacement, from the resting value, of the potential across the receptor membrane (membrane receptor potential). Here, it is assumed that the characteristics of the membrane, except for the receptor membrane, are constant whether the states are resting or stimulated. Then we can quantitatively discuss, although not in absolute values, the stationary state of the membrane receptor potential by measuring the impulse frequency.

STIMULUS STRENGTH AND RESPONSE MAGNITUDE

In Figure 6, the responses of a single labellar sugar receptor of the fleshfly to sucrose, fructose, and glucose are shown in their stationary phases.[11] As discussed above, the maximum membrane receptor potentials are concluded to be different with different sugars, although the response magnitude is expressed as the impulse frequency. Morita and Shiraishi[11] have found that the magnitudes of responses to disaccharides can be described by the original form of Beidler's theory, but the ones to monosaccharides cannot.

RECEPTOR POTENTIAL AND NUMBER OF ACTIVATED RECEPTOR SITES

Many investigators of sensory cells, muscle end-plate, etc. have adopted the electric circuit diagram as shown in Figure 7B as a model illustrating the potential changes in the cells they used.

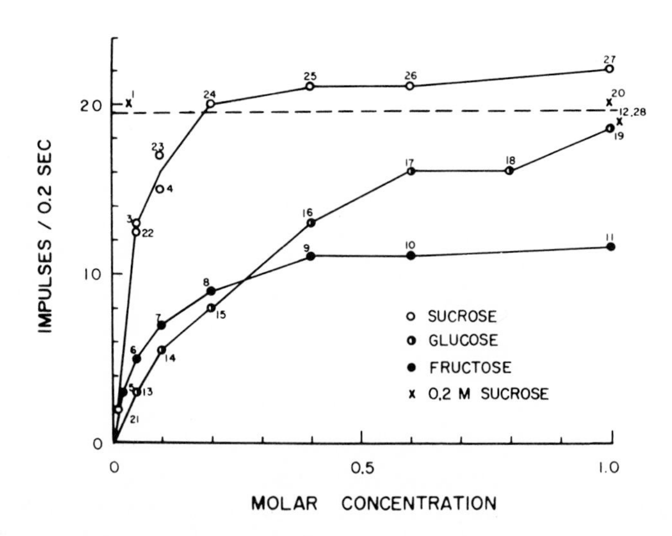

Figure 6 Response magnitude at the stationary level vs. stimulus strength. Stimulation of a single receptor by sucrose, glucose, and fructose. The number attached to each symbol represents the order of stimulation.[11]

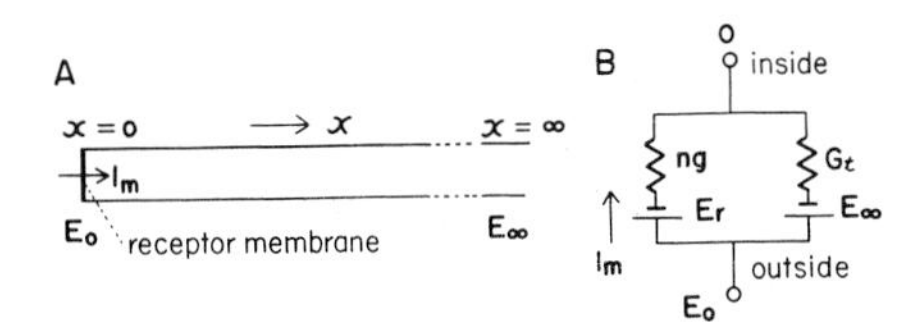

Figure 7 Distal process of the receptor (A) and its equivalent circuit diagram (B). n, the number of activated sites; g, the conductance per activated site; G_t, the conductance across the receptor membrane when $n = 0$; E_o, the receptor membrane potential; E_r, the value of E_o when $n = \infty$; E_∞, the value of E_o when $n = 0$.

In this diagram,

g = conductance per activated site,
n = number of activated sites,
G_t = conductance measured across the receptor membrane in the resting state when $n = 0$,
E_o = potential across the receptor membrane,
E_r = the value of E_o when $n = \infty$ (the maximum value of E_o),
E_∞ = the value of E_o when $n = 0$ (the resting membrane potential).

Defining that $V \equiv E_o - E_\infty$ and $V_m \equiv E_r - E_\infty$, we obtain the following relation from the diagram in Figure 7B:

$$V/V_m = (E_o - E_\infty)/(E_r - E_\infty) = 1/(1 + G_t/ng). \tag{3}$$

According to Kimizuka,[5] the receptor membrane current in a stationary state, I_m, is expressed as

$$I_m = 4RTG_tF^{-1} \sinh [F(E_o - E_\infty)/4RT]. \tag{4}$$

Applying the equation of the stationary membrane current by Kimizuka and Koketsu[6] to the receptor membrane, we can write

$$I_m = -2RTF^{-1}\, ng \sinh [F(E_o - E_r)/2RT]. \tag{5}$$

From Equations 4 and 5, we obtain

$$2 \sinh [F(E_o - E_\infty)/4RT]/\sinh [F(E_r - E_o)/2RT] = ng/G_t. \tag{6}$$

Equation 6 can be reduced to

$$(E_o - E_\infty)/(E_r - E_o) = ng/G_t \tag{7}$$

when $F(E_o - E_\infty)$ and $F(E_r - E_o)$ are small compared with $4RT$ and $2RT$, respectively. Equations 4 and 5 can be reduced, respectively, in such a case, to

$$I_m \doteqdot G_t(E_o - E_\infty)$$

and

$$I_m \doteqdot ng(E_r - E_o).$$

From Equation 7, we can derive completely the same equation as Equation 3. Thus, the diagram in Figure 7B represents the situation deduced from the approximated equations which Kimizuka and Koketsu[6] derived without the assumption of constant field within the membrane. Figure 8 shows comparison between the results calculated from Equation 6 (solid line) and Equation 3 (broken line) for $E_\infty = 60$ mV and $E_r = 0$. This indicates that Equation 3 is good enough for our purpose, since the

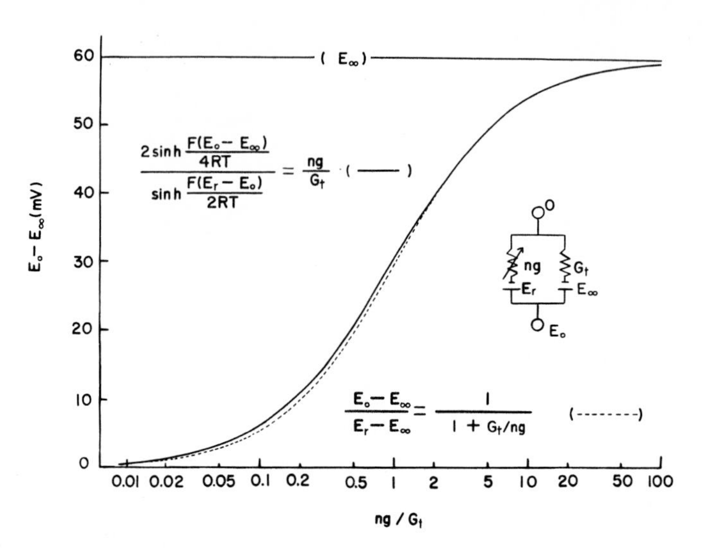

Figure 8 Comparison between theoretical curve calculated from the equations by Kimizuka and Koketzu[6] and Kimizuka[5] (continuous line) and that calculated from the equivalent circuit diagram (broken line).

difference shown by Figure 8 is within the experimental error in the study on the chemoreceptor of insects.

Now, let us assume that the receptor site is activated (having conductance, g) when the stimulant molecule combines with it. The reaction between the stimulant molecule, A, and the receptor site, S, will be expressed as

$$\underset{a}{A} + \underset{s\ \ n}{S} \overset{K}{\rightleftharpoons} \underset{n}{AS}, \tag{8*}$$

$$K = \frac{a(s - n)}{n}, \tag{9}$$

where a is the concentration of A; n, the number of activated sites; s, the total number of receptor sites; K, the dissociation constant between A and S.
Then, n is expressed as

$$n = s/(1 + K/a). \tag{10}$$

Introduction of Equation 10 into Equation 3 gives

$$V(=r) = V_m/(1 + G_t/sg + KG_t/asg). \tag{11}$$

* It should be noticed that the symbols A and S are used also in the article by Dr. Kaissling in this book, but his denotation is quite different from the present one.

The maximum response, r_m when $a = \infty$, is expressed as

$$r_m = V_m/(1 + G_t/sg). \tag{12}$$

Accordingly, the relative response, r', is

$$r' = r/r_m = 1/[1 + KG_t/(sg + G_t)a]. \tag{13}$$

Comparing Equations 1 and 13, we obtain the relation

$$K_b = KG_t/(sg + G_t). \tag{14}$$

Equation 13 describes completely the same results explained by Beidler's theory. The constant K_b, which is the concentration inducing the response of a half of the maximum, is then expressed by Equation 14.

According to Equation 14, the value of the dissociation constant, K, is different from the one of K_b by a factor of $(sg + G_t)/G_t$ (>1). The value of s may be different from receptor to receptor, so that the K_b value may vary with different receptors. Many investigators have pointed out that the free energy of formation of the complex, AS, in Equation 8, is relatively low (about -1 kcal/mole), calculating from the value of $1/K_b$. If the present hypothesis of chemical stimulation is correct, the free energy of formation of the complex should be still lower than the previously proposed values. However, one more plausible model for chemical stimulation is possible, and then the free energy change of the complex formation would be enormous compared with the conventional estimation (see reference 11 and the following Appendix).

APPENDIX

The change of state in the receptor site may be assumed to be induced by combination with a stimulant molecule, as proposed in Equation 8. We may call this assumption the "complex" model. The state change, however, is possible in the following model, which may be called the "regulator" model (cf. Monod, Wyman, and Changeux;[8] Oosawa and Higashi[13]):

$$S^* \overset{L}{\rightleftharpoons} S,$$

$$AS^* \overset{K^*}{\rightleftharpoons} A + S^*, \quad S + A \overset{K}{\rightleftharpoons} AS,$$

where $L = [S]/[S^*]$, $K^* = [A][S^*]/[AS^*]$ and $K = [A][S]/[AS]$. With these constants we can calculate the number of the sites in S^* state (activated state, i.e., $[S^*] + [AS^*] = n$) as

$$n = s/\{1 + L(1 + c\alpha)/(1 + \alpha)\}, \tag{A-1}$$

where s is the total number of the sites in both states, $\alpha = [A]/K^*$ and $c = K^*/K$. The value of L should be very large compared with unity since there is practically no spontaneous activities in our sugar receptor. Therefore, an appreciable value of n in Equation A-1 is obtained only when the value of α is large enough compared with unity. Thus, Equation A-1 can be reduced, as a closest approximation, to

$$n \doteqdot s/\{1 + L(1 + c\alpha)/\alpha\}. \tag{A-2}$$

Introducing Equation A-2 into Equation 3, we obtain

$$r(=V) = V_m/\{1 + G_t(1 + Lc + L/\alpha)/sg\}, \tag{A-3}$$

and

$$r_m = V_m/\{1 + G_t(1 + Lc)/sg\}. \tag{A-4}$$

The relative response, r', is, accordingly,

$$r' = r/r_m = 1/\{1 + G_tLK^*/(sg + G_t + G_tLc)a\}. \tag{A-5}$$

Now, comparing Equation 1 with Equation A-5, it can be written as

$$K_b = K^*LG_t/(sg + G_t + G_tLc). \tag{A-6}$$

It is evident that Equation A-5 also perfectly describes the same results which are explained by Beidler's theory. We can estimate the value of K^* in the sugar receptor of the fleshfly. The maximum response to sucrose is about 100 impulses/sec, and, therefore, it follows from Equation A-4 that

$$r_m = 100 \text{ impulses/sec} = kV_m sg/(1 + G_t + G_tLc). \tag{A-7}$$

Since there is practically no spontaneous discharge in the sugar receptor, it is assumed that the spontaneous discharging rate, r_o, is less than 1×10^{-2} impulses/sec, so that (from Equation 3)

$$r_o \leqq 1 \times 10^{-2} \text{ impulses/sec} = kV_m/(1 + G_t/n_og), \tag{A-8}$$

where n_o is the number of the activated state (S^*) under no stimulus condition. From Equations A-7 and A-8, we obtain that

$$[G_t/(sg + G_t + G_tLc)] \times 10^{-4} \geqq 1/L, \tag{A-9}$$

where approximations are made as

$$n_og + G_t \doteqdot G_t \text{ and } n_o/s \doteqdot n_o/(s - n_o) = 1/L.$$

From Equations A-6 and A-9, we obtain that

$$K^* \leqq K_b \times 10^{-4}. \tag{A-10}$$

The value of K_b was estimated to be 0.06 for sucrose,[11] and hence

$$K^* \leqq 6 \times 10^{-6}.$$

The free energy (ΔG^o) of formation of the complex, AS^*, is therefore

$$\Delta G^o \leqq -7.1 \text{ kcal/mole, at } 25^\circ\ C.$$

This large amount of free energy release is evidently to be utilized for transition from S to S^*, i.e., for activation of the receptor site.

REFERENCES

1. Beidler, L. M. 1953. Properties of chemoreceptors of tongue of rat. *J. Neurophysiol.* **16**, pp. 595–607.
2. Beidler, L. M. 1954. A theory of taste stimulation. *J. Gen. Physiol.* **38**, pp. 133–139.
3. Fishman, I. Y. 1957. Single fiber gustatory impulses in rat and hamster. *J. Cell. Comp. Physiol.* **49**, pp. 319–334.
4. Hodgson, E. S., J. Y. Lettvin, and K. D. Roeder. 1955. Physiology of a primary chemoreceptor unit. *Science* **122**, pp. 417–418.
5. Kimizuka, H. 1966. Ion current and potential across membranes. *J. Theor. Biol.* **13**, pp. 145–163.
6. Kimizuka, H., and K. Koketsu. 1964. Ion transport through cell membrane. *J. Theor. Biol.* **6**, pp. 290–305.
7. Kimura, K., and L. M. Beidler. 1961. Microelectrode study of taste receptors of rat and hamster. *J. Cell. Comp. Physiol.* **58**, pp. 131–139.
8. Monod, J., J. Wyman, and J.-P. Changeux. 1965. On the nature of allosteric transitions: a plausible model. *J. Mol. Biol.* **12**, pp. 88–118.
9. Morita, H. 1959. Initiation of spike potentials in contact chemosensory hairs of insects. III. D.C. stimulation and generator potential of labellar chemoreceptor of *Calliphora*. *J. Cell. Comp. Physiol.* **54**, pp. 189–204.

10. Morita, H., and N. Hori. 1969. (In preparation)
11. Morita, H., and A. Shiraishi. 1968. Stimulation of the labellar sugar receptor of the fleshfly by mono- and disaccharides. *J. Gen. Physiol.* **52**, pp. 559–583.
12. Morita, H., and S. Yamashita. 1966. Further studies on the receptor potential of chemoreceptor of the fleshfly. *Mem. Fac. Sci., Kyushu Univ., Ser. E (Biol.)* **4**, pp. 83–93.
13. Oosawa, F., and S. Higashi. 1967. Statistical thermodynamics of polymerization and polymorphism of protein. *In* Progress in Theoretical Biology, Vol. 1. (F. M. Snell, editor). Academic Press, Inc., New York, pp. 79–164.
14. Wolbarsht, M. L., and F. E. Hanson. 1967. Electrical and behavioral responses to amino acid stimulation in the blowfly. *In* Olfaction and Taste II (T. Hayashi, editor). Pergamon Press, Oxford, etc., pp. 749–760.

THE MECHANISM OF INSECT SUGAR RECEPTION, A BIOCHEMICAL INVESTIGATION

KAI HANSEN
Zoologisches Institut der Universität Heidelberg, Germany

INTRODUCTION

Many insects possess contact chemoreceptors on their mouthparts, legs, and antennae. The sense organs are hair-like sensilla of the sensilla basiconica type. They inform the insect about the presence and the quality of a food while contacting it.[10,13,22,34] The sense hairs on the tarsi of the blowfly *Phormia regina* seem to be anatomically similar to those located on the labellae. They are supplied with five sense cells, which lie near the hair base below the cuticle and send dendrites through the hollow hair up to the tip. There the dendrites terminate in swollen lobes. These lobes contact the outside through a pore in the hair tip and seem to be the site of sugar reception.[1,23,33,40]

As revealed by electrophysiological investigations, there exist one sugar, two salt, and one water receptor within the same taste hair.[11,19,32] In contrast to the multifunctional vertebrate taste cells (cf. Oakley and Benjamin[39]), these insect cells are specialized to the reception of certain chemically related substances. For instance, the sugar receptor of the blowfly responds only to certain pentoses, hexoses, disaccharides, and sugar alcohols.[7,9,18,31,35,36] For vertebrates, however, many compounds are sweet although structurally they are not related to carbohydrates, as, for example, saccharine and L-amino acids.[44] Considering these facts, the sugar receptor of the blowfly seems to be particularly well-suited to studies concerning the unknown receptor mechanism. The elucidation of this primary process is the main topic of this paper.

My biochemical investigations on the receptor mechanism were initiated by Dethier's observation[8] that the legs of *Phormia* contain a sugar-splitting enzyme. Wiesmann[47] confirmed this result for the fly *Musca*. Dethier did not discuss the functional meaning of this enzyme. I measured the morphological distribution of the activities in the legs and the kinetic properties of this α-glucosidase (Enzyme Commission No. 3.2.1.20), splitting disaccharides with α-glucosidic bonds into monosaccharides. The distribution and the properties of this enzyme correspond to those of the sugar receptors on the tarsi.[25,28–30] These results led to the working hypothesis that the primary process of sugar reception is identical with the formation of a sugar-glucosidase complex similar to the enzyme-substrate complex of Michaelis and Menton (cf. Dixon and Webb[16]).

MATERIAL AND METHODS

Laboratory-reared, three- to seven-day-old imagoes of the blowfly *Phormia regina* Meig. were used. Twenty tarsi were cut from the flies at 4°C and homogenized with 100μl of a medium containing 20mM phosphate buffer pH 7.0 and 0.1 per cent bovine serum albumin. Thereafter the homogenate was centrifuged for 2 min at 15,000 *g*. The supernatant contained about 90 per cent of the total glucosidase activity, and was sufficient for 30 single activity determinations. It was filtered on Sephadex G25 to determine kinetic data (section B) or used directly to estimate absolute activities (section A). In brief, the glucosidase activity was measured at pH 5.6 in the following way: (*a*) with disaccharides—mainly sucrose—as substrates, and enzymatic micro-estimation of the released glucose with glucose oxidase, peroxidase, and *o*-dianisidine,[4,26] or (*b*) with *p*-nitrophenyl-α-glucoside as substrate and photometric determination of released *p*-nitrophenyl at 402 mμ.[24] The protein content of the legs was determined in separate assays by the biuret reaction. The enzyme activity is expressed in milliunits (mU, 1 mU splits 10^{-9} moles per min) per mg protein (spec. activity) or in mU per mg dry weight (chitin + protein). For further methodological details see Hansen.[29]

RESULTS

Localization of the Enzyme Activity and the Sugar Receptors

Figure 1 shows a correlation between the morphological distribution of the glucosidase activities and the number of sense hairs on the legs of *Phormia*.[23] The tarsi of the first leg have highest activities and also most of the sense hairs. The tarsi of the second and third leg and the tibiae have lower activities and possess proportionally fewer sense hairs. The femora of the three legs have no sense hairs at all, but a low enzyme activity is found, which is perhaps of muscular origin, as has been described several times.[43,46,48]

A more detailed distribution pattern of the activities and the number of hairs of the first leg is shown in Figures 2 and 3. As morphologically revealed by Grabowski and Dethier,[23] four sensory hair types (A–D) exist on the tarsi of *Phormia*. The C- and D-type hairs are not further mentioned because of their small number (only 5 per cent of the total). The short (36 μ) hairs of the A-type are on the lower (ventral) side of tarsomeres 2–5. Most of the long (60 μ) hairs of the B-type are localized on the ventrolateral parts of tarsomeres 1–5, while a few of them are distributed over the whole tarsi and the distal part of the tibiae. Figure 2 shows the sum of A- and B-hairs of single tarsomeres compared with the enzyme activities. Both have the same distribution pattern. Figure 3 gives the activities in the upper and lower halves of tarsomere 1 and tarsomeres 2–5. The lower halves of tarsomeres 2–4 that possess all receptors of the A-type have, correspondingly, four times higher activities than the upper halves.

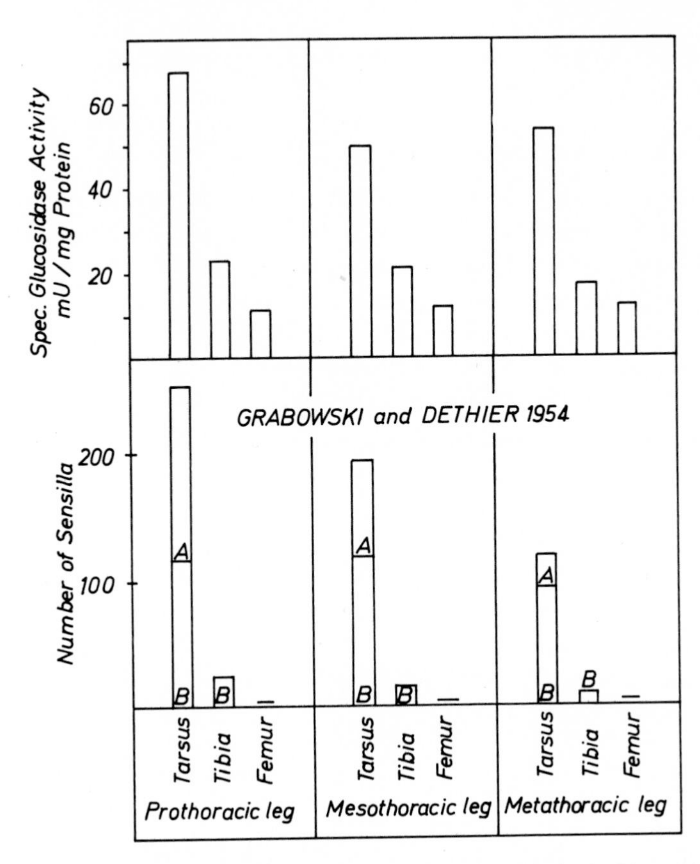

Figure 1 The morphological distribution of the glucosidase activity in the *Phormia* legs (upper ordinate, expressed as activity per mg protein) is compared with the number of sensilla (lower ordinate) counted by Grabowski and Dethier.[23]

The same correlation between enzyme activity and sense hairs as revealed for the legs was also found on the proboscis of the blowfly (Figure 4). The distal part of the proboscis with the marginal chemoreceptor hairs gives the highest enzyme activities. Receptor-free parts—such as small pieces of scutellar or abdominal cuticle—contain only small activities. The receptor-free first leg of the cockroach *Blatta*[21] contains some enzyme activity, but lacks the special distribution pattern found in *Phormia*.

Control experiments showed the following results: (1) The enzyme activity is not localized on the outside of the tarsi, but exists inside. (2) The activity of the hemolymph amounts to only 5 per cent of the total activity. (3) The pulvilli on the leg's tip have no activity. An exact histochemical localization failed because of the enzyme's solubility (cf. Hansen[27]), but the tarsi contain no other tissues except the sense organs and the epidermal cells.

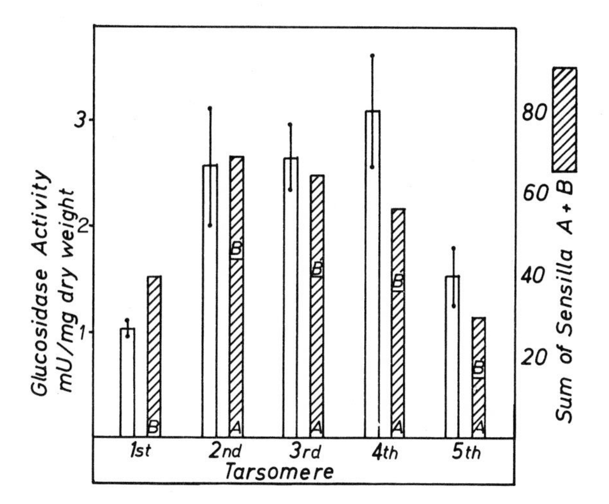

Figure 2 Comparison of glucosidase activity (left ordinate) and number of sensilla[23] (right ordinate) of single tarsomeres.

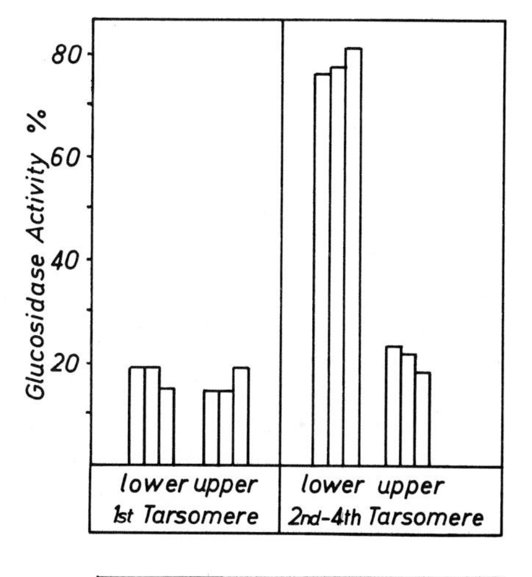

Figure 3 The activity of lower and upper halves of tarsomeres 1 and 2–4. Only the lower halves of tarsomeres 2–4 have sensilla of the A-type. The activity is given in percentages from activity (mU/mg dry weight) of whole tarsomeres.

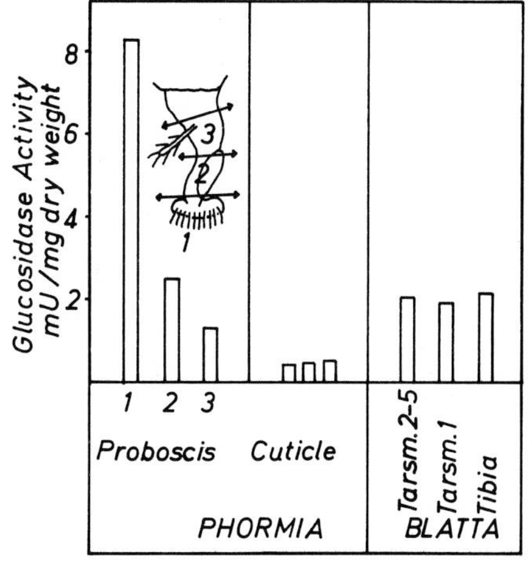

Figure 4 *Left:* The activity pattern of *Phormia* proboscis divided into three parts according to the figure. The distal part 1 bears the chemosensory marginal hairs. *Middle:* Abdominal and scutellar parts of blowfly cuticle contain only small activities. *Right:* The glucosidase activity of receptor-free legs of the cockroach *Blatta orientalis* does not show the characteristic pattern of *Phormia*.

TABLE I

Substrate		Enzyme	Hydrolysis	Sensitivity
Sucrose	Fruβ2-1αGlc	α-Glucosidase 3.2.1.20	+	+++
Turanose	Glcα1-3Fru	α-Glucosidase 3.2.1.20	+	+++
Maltose	Glcα1-4Glc	α-Glucosidase 3.2.1.20	+	+++
Melezitose	Glcα1-3Fruβ2-1αGlc	α-Glucosidase 3.2.1.20	+	++
p-Nitrophenyl-α-glucoside		α-Glucosidase 3.2.1.20	+	+++
Trehalose	Glcα1-1Glc	Trehalase 3.2.1.28	+	+
Cellobiose	Glc 1-4Glc	β-Glucosidase 3.2.1.21	—	(+)
Gentiobiose	Glcβ1-6Glc	β-Glucosidase 3.2.1.21	—	(—)
Raffinose	Galα1-6Glc 1-2 Fru	α-Galactosidase 3.2.1.22	—	+
Melibiose	Galα1-6Glc	α-Galactosidase 3.2.1.22	—	—
Lactose	Galβ1-4Glc	β-Galactosidase 3.2.1.23	—	—

Substrate specificity of the tarsal disaccharidase activity. In the first column the tested substrates are listed, in the second the adequate enzyme, and in the third column is noted whether the substrate is split (+) or not (—). The last column gives the sensitivity of *Phormia* to the substrates measured by Dethier[7] (+++ Threshold 1–20 mM, ++ 20–100 mM, +> 100 mM).

Identical Properties of the Glucosidase and the Sugar Receptors

Table I shows that the investigated enzyme is an α-glucosidase splitting only di- and trisaccharides with an α-glucosidic bond. Such glucosidases generally have a broad substrate specificity and are well known from digestive tracts. To the same di- and trisaccharides—which are split by the tarsal glucosidase—the tarsal sugar receptors respond with high sensitivity. Sugars with other bond types are perceived badly (raffinose, cellobiose) or not at all (lactose, melibiose, gentiobiose). In conformity, other disaccharidase activities are lacking in the tarsal extracts. The trehalose represents an exception that is discussed below.

It is important that the Michaelis constants of five tested sugars (Table II) are correlated with the behavioral threshold values determined by Dethier.[7] The Michaelis constant is a reciprocal measure for the affinity between enzyme and substrate, and is identical with the dissociation constant of the enzyme-substrate complex. The behavioral threshold is defined as the millimolar concentration of a sugar solution that causes 50 per cent of a fly population to react positively. The absolute threshold values are variable; they depend upon central nervous system influences as well as on the number of stimulated receptors. But the relations of the absolute values for different sugars to each other are constant and represent real properties of the sugar receptor cell. Therefore, we can state that the disaccharide pattern of glucosidase

TABLE II

Substrate	Maximal Velocity [%]	Michaelis Constant [mMoles/liter]	Behavioral Threshold [mMoles/liter]
Sucrose N = 9	= 100%	9	10
Maltose N = 5	42	< 2 K_i = 1.5*	4
Turanose N = 2	30	7	11
Melezitose N = 2	44	32	64
p-Nitrophenyl-α-glucoside N = 5	110	0.7	(4)**
Trehalose	---	1***	130

Comparison of the Michaelis constants and maximal velocities for different substrates of the tarsal glucosidase with the behavioral thresholds.[8] The Michaelis constants are determined graphically using Lineweawer-Burk plots (the values are averaged, N = number of experiments). The maximal velocities of the glucosidase are relative to sucrose, whose velocity is called 100%. (N = 6)

* Only maltose gives no linear Lineweawer-Burk plots, because it seems to interfere with the enzymatic glucose determination. Therefore, the inhibitor constant of maltose in a *p*-nitrophenyl-α-glucoside assay was determined, which theoretically should be nearly identical with the Michaelis constant.

** *p*-Nitrophenyl-α-glucoside has not yet been tested behaviorally; in electrophysiological tests the spike frequencies caused by submaximal concentrations of this sugar are in the same range as those of maltose.

*** The Michaelis constant for trehalose is also listed for comparison. The given value is derived from *Phormia* flight muscle,[26] agreeing with that of purified *Phormia* enzyme.[20]

and sugar receptors is identical. Electrophysiological investigations show that the behavioral threshold values measured by Dethier lie in the upper part of the working range of single sugar receptors (A-type hairs). Table II further shows that no correlation exists between maximal velocity of the enzyme reaction and the threshold values.

Monosaccharides perceived by the receptors inhibit the glucosidase activity competitively. But there is no correspondence between the inhibition constants of simple monosaccharides and their behavioral threshold values. The presence of a substituent in the α-position on the Carbon 1 (C-1) seems to be a necessary assumption for a correlation, as demonstrated for *p*-nitrophenyl-α-glucoside. For this compound the receptor, as well as the enzyme, shows highest sensitivity and affinity.

Fructose and trehalose are exceptions favoring the hypothesis. The trehalose—being the main blood sugar of insects—is split by a high specific trehalase (Enzyme Commission No. 3.2.1.28, Tables I and II) found in all insect tissues. But it is not split by glucosidases in general,[3,24,41] and shows a high sensory threshold (120 mM). The fructose does not inhibit the enzyme, but is well perceived by the receptor (threshold = 6 mM). This discrepancy finds its explanation in the results of Evans.[17] He identified a specific fructose site existing in the sugar receptor of *Phormia* beside the glucose site.

Dethier and Chadwick,[14,15] as well as Steinhardt, Morita and Hodgson[45] demonstrated that the sugar receptor of *Phormia* is specifically inhibited by higher aliphatic alcohols and amines; best by octylamine. Formally, the inhibition corresponds to a competitive one. In the same way, the enzymatic splitting of sucrose and *p*-nitrophenyl-α-glucoside is reversibly inhibited by octylamine (Figure 5).[28] Because the configuration of octylamine is very different from that of carbohydrates, it necessarily implies the postulation of two enzyme sites, one sugar site, and one lipophilic inhibitor site. Therefore, the inhibition of the glucosidase—and possibly that of one of the sugar receptors, too—belongs to the biochemically well-known allosteric type. As pointed out for the glucosidase and for the sugar receptor in Figure 6, the inhibition by aliphatic alcohols depends upon the chain length. The rejection threshold and the inhibition constant decrease logarithmically with the chain length from one to eight carbon atoms.

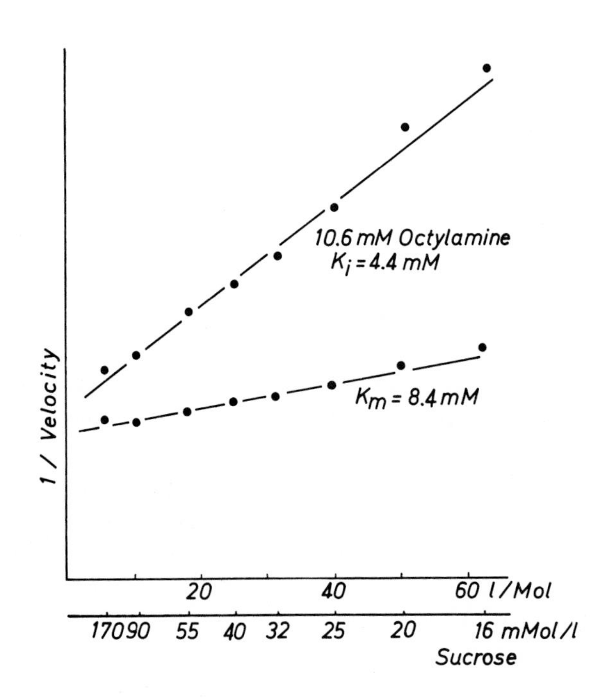

Figure 5 Lineweawer-Burk plot, showing the allosteric (formal competitive) inhibition of the glucosidase (substrate sucrose) by octylamine-HCl at pH 5.6.

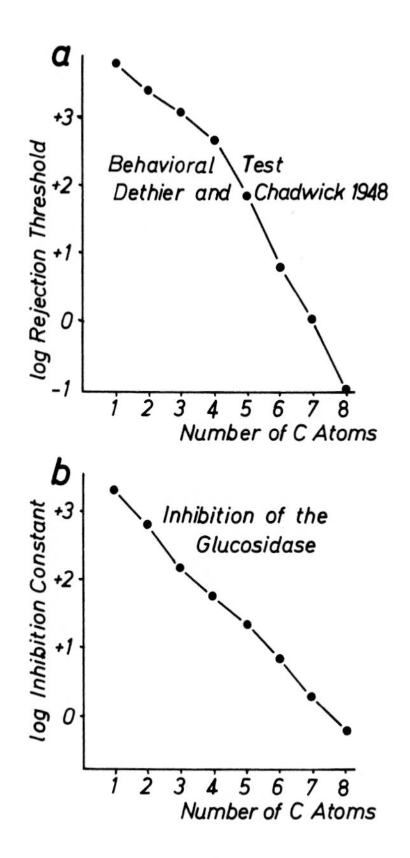

Figure 6 (*a*) The dependence of the rejection thresholds (ordinate, logarithmic scale, mMoles/liter) of *Phormia* tarsi on the number of C atoms of a series of aliphatic *n*-alcohols. (*b*) The dependence of the inhibition constant (logarithmic scale, mMoles/liter) of the glucosidase on the number of C atoms of the inhibiting alcohols.

CONCLUSION

All these results confirm the working hypothesis that the first reaction between the sugar and the receptor is identical with the formation of a sugar-glucosidase complex.

This hypothesis is not inconsistent with earlier interpretations. (1) The formation of an enzyme-substrate complex can be discussed as a special absorption process,[38] analogous to the postulate of Dethier[9] and Evans[18] for the sugar receptor. (2) Dethier and Arab[12] found that the threshold of the sugar receptor is not altered by changing the temperature of the stimulus within the range of 4°C to 35°C. In agreement with this observation, the temperature coefficient of the Michaelis constant of the tarsal glucosidase is slightly negative. At 6°C, it is twice as high as that at 30°C if sucrose and *p*-nitrophenylglucoside and substrates.[29]

Similar biochemical investigations on the receptor mechanism have been done by Dastoli and Price.[5,6,42] They isolated a "sweet-sensitive protein" from bovine tongues. This protein can combine both with saccharine and with various sugars, according to their degree of sweetness. Therefore, it is assumed that this protein represents the sugar receptor protein.

How the sugar-glucosidase complex controls the permeability of the receptor membrane and, thus, the height of the receptor potential remains unresolved.[37] The recent results of Del Castillo and collaborators,[2] however, indicate a solution. They loaded artificial black lipid membranes with purified enzymes, e.g., lactate dehydrogenase, and demonstrated that by adding substrate the molecular texture of the lipid membrane is changed within milliseconds by the formation of an attached enzyme-substrate complex, measured as a sudden decrease of membrane impedance.

ACKNOWLEDGMENTS

I thank Dr. Karl Ernst Kaissling for many helpful discussions and Dr. Elisabeth Hansen-Delkeskamp as well as Stan Skordilis for preparing the English manuscript. These investigations were supported by a grant from the Deutsche Forschungsgemeinschaft.

REFERENCES

1. Adams, J. R., and P. Holbert. 1963. The fine structure of insect contact chemoreceptors. *Proc. 16th Int. Congr. Zool.* **3**, pp. 93–95.
2. Del Castillo, J., A. Rodriguez, C. A. Romero, and V. Sanchez. 1966. Lipid films as transducers for detection of antigen-antibody and enzyme-substrate reactions. *Science* **153**, pp. 185–188.
3. Dahlqvist, A. 1959. The separation of intestinal invertase and three different intestinal maltases on TEAE-cellulose by gradient elution, frontal analysis and mutual displacement chromatography. *Acta Chem. Scand.* **13**, pp. 1817–1827.
4. Dahlqvist, A. 1964. Method for assay of intestinal disaccharidases. *Anal. Biochem.* **7**, pp. 18–25.
5. Dastoli, F. R., D. V. Lopiekes, and S. Price. 1968. A sweet-sensitive protein from bovine taste buds. Purification and partial characterization. *Biochemistry* **7**, pp. 1160–1164.
6. Dastoli, F. R., and S. Price. 1966. A sweet-sensitive protein from bovine taste buds: isolation and assay. *Science* **154**, pp. 905–907.
7. Dethier, V. G. 1955. The physiology and histology of contact chemoreceptors of the blowfly. *Quart. Rev. Biol.* **30**, pp. 348–371.
8. Dethier, V. G. 1955. Mode of action of sugar-baited fly traps. *J. Econ. Entomol.* **48**, pp. 235–239.
9. Dethier, V. G. 1956. Chemoreceptor mechanisms. *In* Molecular Structure and Functional Activity of Nerve Cells (R. G. Grenell and L. V. Mullins, editors). American Institute Biological Sciences, Washington, D.C. Publication No. 1, pp. 1–34.
10. Dethier, V. G. 1963. The Physiology of Insect Senses. Methuen and Co. Ltd., London, England.
11. Dethier, V. G. 1968. Chemosensory input and taste discrimination in the blowfly. *Science* **161**, pp. 389–391.
12. Dethier, V. G., and Y. M. Arab. 1958. Effect of temperature on the contact chemoreceptors of the blowfly. *J. Insect Physiol.* **2**, pp. 153–161.
13. Dethier, V. G., and L. E. Chadwick. 1948. Chemoreception in insects. *Physiol. Rev.* **28**, pp. 220–254.
14. Dethier, V. G., and L. E. Chadwick. 1948. The stimulating effect of glycols and their polymers on the tarsal receptors of blowflies. *J. Gen. Physiol.* **32**, pp. 139–151.
15. Dethier, V. G., and L. E. Chadwick. 1950. An analysis of the relationship between solubility and stimulating effect in tarsal chemoreception, *J. Gen. Physiol.* **33**, pp. 589–599.
16. Dixon, M., and E. C. Webb, 1964. Enzymes. Longmans Green and Co. Ltd., London. 2nd edition.
17. Evans, D. R. 1961. Depression of taste sensitivity to specific sugars by their presence during development. *Science* **133**, pp. 327–328.
18. Evans, D. R. 1963. Chemical structure and stimulation by carbohydrates. *In* Olfaction and Taste I (Y. Zotterman, editor). Pergamon Press, Oxford, etc. pp. 165–176.
19. Evans, D. R., and D. Mellon, Jr. 1962. Electrophysiological studies of a water receptor associated with the taste sensilla of the blowfly. *J. Gen. Physiol.* **45**, pp. 487–500.
20. Friedman, S. 1960. The purification and properties of trehalase isolated from *Phormia regina*. *Arch. Biochem. Biophys.* **87**, pp. 252–258.

21. Frings, H., and M. Frings. 1949. The loci of contact chemoreceptors in insects. *Amer. Midland Natur.* **41**, pp. 602–658.
22. Frisch, K. von. 1935. Über den Geschmackssinn der Biene. Ein Beitrag zur vergleichenden Physiologie des Geschmacks. *Z. Vergleich. Physiol.* **21**, pp. 1–155.
23. Grabowski, C. T., and V. G. Dethier. 1954. The structure of the tarsal chemoreceptors of the blowfly, *Phormia regina* Meigen. *J. Morphol.* **94**, pp. 1–20.
24. Halvorson, H., and L. Ellias. 1958. The purification and properties of an α-glucosidase of *Saccharomyces italicus*. *Biochim. Biophys. Acta* **30**, pp. 28–40.
25. Hansen, K. 1963. Über Carbohydrasen in den Tarsen von *Calliphora* und *Phormia*. *Verh. Deut. Zool. Ges. München*, Zool. Anz. Suppl. **27**, pp. 628–634.
26. Hansen, K. 1966. Zur cytologischen Lokalisation der Trehalase in der indirekten Flugmuskulatur der Insekten. *Biochem. Z.* **344**, pp. 15–25.
27. Hansen, K. 1966. Zum histochemischen Nachweis der Trehalase in den Mitochondrien von Insekten-Flugmuskeln. *Histochemie* **6**, pp. 290–300.
28. Hansen, K. 1967. Zur kompetitiven Hemmung der tarsalen Glucosidase und der Zucker-Rezeptoren bei *Phormia regina* Meig. durch Amine. *Verh. Deut. Zool. Ges. Heidelberg*, Zool. Anz. Suppl. **31**, pp. 558–564.
29. Hansen, K. 1968. Untersuchungen über den Mechanismus der Zucker-Perzeption bei Fliegen. Habilitationsschrift der Universität Heidelberg. (In preparation)
30. Hansen, K. 1968. Wie nehmen Fliegen Zucker wahr? *Umsch. Wiss. Tech.* **68**, pp. 499–500.
31. Hodgson, E. S. 1957. Electrophysiological studies of arthropod chemoreception II. Responses of labellar chemoreceptors of the blowfly to stimulation by carbohydrates. *J. Insect Physiol.* **1**, pp. 240–247.
32. Hodgson, E. S., and K. D. Roeder. 1956. Electrophysiological studies of arthropod chemoreception I. General properties of the labellar chemoreceptors of Diptera. *J. Cell. Comp. Physiol.* **48**, pp. 51–75.
33. Larsen, J. R. 1962. The fine structure of the labellar chemosensory hairs of the blowfly *Phormia regina* Meig. *J. Insect Physiol.* **8**, pp. 683–691.
34. Minnich, D. E. 1921. An experimental study of the tarsal chemoreceptors of two nymphylid butterflies. *J. Exp. Zool.* **33**, pp. 173–203.
35. Morita, H. 1969. Electrical signs of taste receptor activity. Olfaction and Taste III (C. Pfaffmann, editor). Rockefeller Univ. Press, New York, pp. 370–381.
36. Morita, H., and A. Shiraishi. 1968. Stimulation of the labellar sugar receptors of the fleshfly by mono- and disaccharides. *J. Gen. Physiol.* **52**, pp. 559–583.
37. Morita, H., and S. Yamashita. 1959. Generator potential of insect chemoreceptor. *Science* **130**, p. 922.
38. Netter, H. 1959. Theoretische Biochemie. Springer-Verlag, Berlin, Göttingen, Heidelberg, Germany.
39. Oakley, B., and R. M. Benjamin. 1966. Neural mechanisms of taste. *Physiol. Rev.* **46**, pp. 173–211.
40. Peters, W. 1965. Die Sinnesorgane an den Labellen von *Calliphora erythrocephala*. *Z. Morphol. Olkol. Tiere* **55**, pp. 259–320.
41. Phillips, A. W. 1959. The purification of a yeast maltase. *Arch. Biochem. Biophys.* **80**, pp. 346–352.
42. Price, S., and R. M. Hogan. 1969. Glucose dehydrogenase activity of a "sweet-sensitive" protein from bovine tongues. Olfaction and Taste III (C. Pfaffmann, editor). Rockefeller Univ. Press, New York, pp. 397–403.
43. Sacktor, B. 1955. Cell structure and the metabolism of insect flight muscle. *J. Biophys. Biochem. Cytol.* **1**, pp. 29–46.
44. Schutz, H. G., and F. J. Pilgrim. 1957. Sweetness of various compounds and its measurement. *Food Res.* **22**, pp. 206–213.
45. Steinhardt, R. A., H. Morita, and E. S. Hodgson. 1966. Mode of action of straight chain hydrocarbons on primary chemoreceptors of the blowfly, *Phormia regina*. *J. Cell. Comp. Physiol.* **67**, pp. 53–62.
46. Van Handel, E. 1968. Utilization of injected maltose and sucrose by insects: evidence for non-intestinal oligosaccharidases. *Comp. Biochem. Physiol.* **24**, pp. 537–541.
47. Wiesmann, R. 1960. Zum Nahrungsproblem der freilebenden Stubenfliegen. *Musca domestica* L. *Z. Angew. Zool.* **47**, pp. 159–182.
48. Zebe, E., and W. H. McShan. 1959. Trehalase in the thoracic muscles of the woodroach, *Leucophaea maderae*. *J. Cell. Comp. Physiol.* **53**, pp. 21–29.

RELATION BETWEEN ATP AND CHEMICAL TRANSMITTERS IN THE FROG'S RECEPTORS

GIOVANNI RAPUZZI, ANNA VIOLANTE *and* CESARE CASELLA
Istituto di Fisiologia Generale, Università di Pavia, Italy

INTRODUCTION

The importance of adenosine triphosphate (ATP) for receptor activation is well established. Electrophysiological researches carried out on different receptors in several animal species have shown that administration of ATP produces an intense activating effect.[6–8] Endogenous ATP also has seemed to be necessary for receptor discharge, because the discharge is clearly modified by influencing the metabolic activity of the receptors.[5,9,10]

So evident a relationship between the receptor discharge and ATP is described as consistent with the relevant importance of the metabolic activity for receptor function or enzymic activity at the excitable membranes.[3,13] At the present time, however, the mechanism by which ATP activates the receptors is little understood, and helpful experimental data are lacking. Therefore, it is interesting to report some observations that seem to elucidate, in some degree, the mechanism by which ATP activates the frog's lingual receptors.

The massive afferent discharge of impulses that results from perfusing the frog's tongue vessels with physiological fluids containing ATP has been observed concomitantly, but independently, by Duncan[2] and Rapuzzi et al.[12] This effect of ATP is often defined as prompt, specific, and direct: it must be noted, however, that it occurs with rather complex features.

In this respect, the frequent ineffectiveness of ATP administered to these receptors, as observed by Rapuzzi et al.,[12] is meaningful. For example, the intensity with which the lingual receptor responds to ATP is reduced or nonexistent either after repeated injections or after an injection of Tyrode preliminary to ATP administration. Moreover, the ATP is ineffective when the phosphate content is kept normal in the Tyrode. Ultimately, the ATP stimulates the lingual receptors only through the tongue vessels, but it is fully inactive when applied on the lingual surface.

These limitations in ATP effectiveness suggest that its activating effect on the frog's lingual receptors cannot be obtained directly, but in correlation with other activating factors. The most suitable substance could be acetylcholine, since our laboratory work has recently ascertained that chemical stimulation of the frog's lingual receptors with their specific stimulus (i.e., $CaCl_2$ solutions) is associated with the release of acetylcholine in the medium surrounding receptor apparatus.[14] Thus it might be supposed that ATP can produce the afferent discharge of these receptors by interfering with the activating action of the released chemical transmitters. If such interference does take place, the receptor response to ATP may, first, be influenced by substances like eserine or atropine, which are agonistic or antagonistic to the receptor chemical mediators. Second, the response may become more regular when ATP is administered with acetylcholine. If one could find these expected effects, and so obtain clear evidence for the coupling of ATP activating action and receptor chemical transmitters, it would help elucidate the mechanism of ATP activating action in the frog's lingual receptors.

This paper describes the behavior of the frog's lingual receptor response that we observed after one or repeated injections of ATP, either alone or combined with acetylcholine, eserine, or atropine.

MATERIALS AND METHODS

Experiments carried out on frog's isolated lingual preparations have been described in previous papers.[14] The substances under examination were injected manually, under microscopic control, by means of a thin glass pipette connected to an insulin syringe and inserted into the external carotid. At each single injection, 0.2 ml of one of the following mixtures was introduced into the tongue vessels:

Tyrode + ATP 3.10^{-4} M
Tyrode + ATP + acetylcholine 1.10^{-5} M
Tyrode + ATP + eserine 7.10^{-6} M
Tyrode + ATP + acetylcholine + eserine
Tyrode + ATP + atropine 1.10^{-4} M.

The phosphates in the Tyrode were reduced to 1/5 of their normal content.

The receptor response was evaluated by counting the discharge impulses electronically. They were driven from the glossopharyngeal nerve of the injected side of the tongue and sucked into the micropipette as described in previous papers.[11] The mean values of the responses to each mixture were calculated from at least five experiments on different preparations so their differences could be compared by the *t* test. Each mixture was injected in different groups of experiments, as follows: (1) a single mixture injected once; (2) a single mixture injected 10 min after an ATP injection; (3) mixtures injected repeatedly. The mean values of the responses obtained in these groups of experiments are shown as histograms in Figure 1, and expressed in percentages of the response to ATP.

RESULTS

The results of group 1 (Figure 1A) show that the responses to ATP and to ATP plus the mixtures have different intensities. These are statistically significant at $p \leq 5$ per cent. Acetylcholine, eserine, or both, associated with ATP, increases the receptor response, while the atropine decreases it. All the substances, when injected singly as controls, were ineffective. Figure 1B shows that the response to ATP is high with the first injection but of small intensity with the next. This weak response, however,

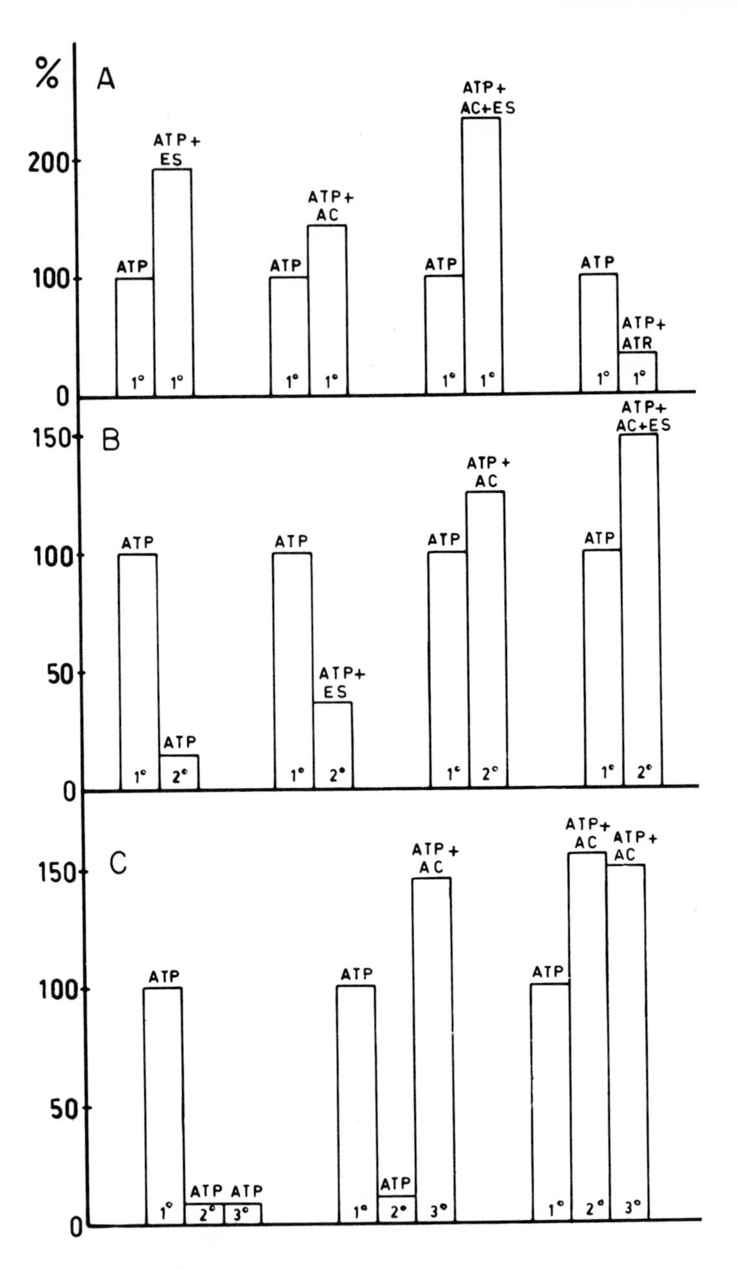

Figure 1 The histograms are the mean values of the responses of the frog's lingual receptors to injections into the frog's vessels of the Tyrode containing respectively: ATP $3 \cdot 10^{-4}$ M; acetylcholine $1 \cdot 10^{-5}$ M; eserine $7 \cdot 10^{-6}$ M; atropine $1 \cdot 10^{-4}$ M. The values are expressed in percentages of the response to ATP. At the base of the histograms is indicated the succession of the injections into the same preparation. A) mixtures injected only once; B) mixtures injected after one injection of ATP; C) mixtures injected repeatedly. 1° = single injection; 1° 2° = two successively; 1° 2° 3° = three successively.

increases feebly when eserine is added to ATP, and is fully restored when acetylcholine or acetylcholine plus eserine is added. Repeated injections of ATP (Figure 1C) are ineffective; the receptor discharge, however, recovers or maintains its normal intensity when the ATP and acetylcholine mixture is injected.

These results show that the lingual receptor response to ATP is influenced positively by acetylcholine and eserine and negatively by atropine, and that ATP plus acetylcholine is effective at every injection. The ability of both agonistic and antagonistic substances of acetylcholine to influence the lingual receptor response to ATP suggests that its activating action is correlated with that of the chemical transmitters that are present in the receptor apparatus. Such an interaction is explicitly disclosed by the persistent receptor response to repeated injections of the ATP-acetylcholine mixture. In fact it demonstrates the essential role of acetylcholine in the effect of ATP.

It would be interesting to know the mechanism by which acetylcholine participates in the indirect activating action of ATP to cause a specific chemical response.

DISCUSSION

With ATP a facilitating agent, it might be said that this energy donor facilitates the specific activating action of the chemical mediators on the afferent nerve endings of the frog's lingual receptors. It is difficult, therefore, to accept the suggestion of Duncan[3,4] that the energy released by the ATP activates the lingual receptors.

ATP seems able to act closely on receptor activation, but in a very complex way. The action of ATP could be considered as preparatory to the depolarizing effect of acetylcholine, which seems able to trigger the impulse discharge at very ineffective concentrations. The importance of the ATP-acetylcholine interaction for the activating mechanism of the frog's lingual receptors can be demonstrated by the observed release of acetylcholine during receptor stimulation[14,15] and by the histochemical observation—by electron microscopy—that the enzyme adenosine triphosphatase is present on the outside of the lingual receptor nerve endings.[16] It is now essential, however, to establish whether the ATP-acetylcholine interaction is only an experimental effect or is a physiological event as well, and, therefore, a physiological device for the normal activating mechanism of the frog's lingual receptors. At the present time, a consideration favoring the latter hypothesis is the observation that such interaction seems to take place elsewhere in the organism, as, for instance, at the neuromuscular junctions.[1] A further element favoring this hypothesis is the result of recent research conducted in our laboratory: they show that ATP becomes detectable at the tongue because of receptor stimulation, and that a complex enzymic system facilitates ATP-acetylcholine interaction in the receptor. This we have ascertained by injecting SH^- donor groups associated with acetylcholine.

Therefore, in the activation of the afferent nerve endings of the frog's lingual receptors, the action of ATP appears to be intimately coupled to that of acetylcholine, but in a way that still must be elucidated.

REFERENCES

1. Buchtal, F., and B. Folkow. 1948. Interaction between acetylcholine and adenosine triphosphate in normal, curarised and denervated muscle. *Acta Physiol. Scand.* **15**, pp. 150–160.
2. Duncan, C. J. 1964. The transducer mechanism of sense organs. *Naturwissenschaften* **51**, pp. 172–173.
3. Duncan, C. J. 1965. Cation-permeability control and depolarization in excitable cells. *J. Theor. Biol.* **8**, pp. 403–418.
4. Duncan, C. J. 1967. The molecular properties and evolution of excitable cells. Pergamon Press, Oxford, New York, etc.
5. Emmelin, N., and W. Feldberg. 1948. Systemic effects of adenosine triphosphate. *Brit. J. Pharmacol. Chemother.* **3**, pp. 273–284.
6. Heymans, C. 1955. Action of drugs on carotid body and sinus. *Pharmacol. Rev.* **7**, pp. 119–142.
7. Jarisch, A., S. Landgren, E. Neil, and Y. Zotterman. 1952. Impulse activity in the carotid sinus nerve following intra-carotid injection of potassium chloride, veratrine, sodium citrate, adenosine triphosphate and α-dinitrophenol. *Acta Physiol. Scand.* **25**, pp. 195–211.
8. Jarisch, A., and Y. Zotterman. 1948. Depressor reflex from the heart. *Acta Physiol. Scand.* **16**, pp. 31–51.
9. Joels, N., and E. Neil. 1963. The excitation mechanism of the carotid body. *Brit. Med. Bull.* **19**, pp. 21–24.
10. Landgren, S., G. Liljestrand, and Y. Zotterman. 1953. Impulse activity in the carotid sinus nerve following intracarotid injection of sodium-iodoacetate, histamine hydrochloride, lergitin and some purine and barbituric acid derivatives. *Acta Physiol. Scand.* **30**, pp. 149–160.
11. Rapuzzi, G. 1964. Metodo di perfusione della lingua di rana per lo studio della funzione recettrice. *Boll. Soc. Ital. Biol. Sper.* **40**, pp. 1051–1053.
12. Rapuzzi, G., and J. Bernini. 1964. L'azione dell'ATP sui ricettori linguali di rana. *Boll. Soc. Ital. Biol. Sper.* **40**, pp. 2021–2024.
13. Rapuzzi, G., and C. Casella. 1965. L'action du phosphate créatine au niveau des papilles fungiformes de la langue de Grenouille. *J. Physiol.* (*Paris*) **57**, pp. 684–685.
14. Rapuzzi, G., and G. Ricagno. 1965. La mediazione chimica nell'eccitamento dei ricettori gustativi. *Boll. Soc. Ital. Biol. Sper.* **41**, number 20, abstract 163.
15. Rapuzzi, G., and A. Violante. 1967. Influenza del Ca^{++} e Mg^{++} sulla risposta dei recettori linguali allo stimolo chimico. *Boll. Soc. Ital. Biol. Sper.* **43**, pp. 559–562.
16. Scalzi, H. A. 1967. The cytoarchitecture of gustatory receptors from the rabbit foliate papillae. *Histochemie* **80**, pp. 413–435.

GLUCOSE DEHYDROGENASE ACTIVITY OF A "SWEET-SENSITIVE PROTEIN" FROM BOVINE TONGUES

STEVEN PRICE *and* RICHARD M. HOGAN
Department of Physiology, Medical College of Virginia, Health Sciences Division, Virginia Commonwealth University, Richmond, Virginia

INTRODUCTION

Dastoli and Price[4] reported the isolation, from homogenates of bovine tongue epithelium, of a protein fraction which formed complexes with sugars. The dissociation constants were determined for the interaction of this material with each of five sugars, and were found to parallel the taste thresholds. It was postulated that this protein represented the taste receptor molecule for sweet compounds, and it was referred to as the "sweet-sensitive protein." Dastoli, Lopiekes, and Price[3] subsequently purified this material to what appeared to be a state of near homogeneity, and characterized it as being a protein with a molecular weight of about 150,000, and an isoionic point at pH 9.1.

In considering the possible evolutionary origin of a mammalian sensory glucoreceptor, it occurred to us that the receptor protein might have evolved from some protein already present in cells, which had a structure appropriate to the formation of complexes with sugars. Any enzyme for which a sugar is a substrate would, of course, form such complexes. Modification of such an enzyme into a glucoreceptor would, in effect, represent the evolutionary exploitation of the receptor's tendency to form complexes with sugars by adapting this property to a new function. This rather speculative line of reasoning led us to attempt to determine if the sweet-sensitive protein had any catalytic activity toward sugars. We have found that the sweet-sensitive protein does, indeed, possess catalytic activity, catalyzing the dehydrogenation of reducing sugars. It should be clear that our finding of such catalytic activity does not prove that our evolutionary speculation is correct. However, it does make it possible to employ the arsenal of enzymological techniques to study the nature of the interaction of the protein with sugars.

MATERIALS AND METHODS

Preparation of the Sweet-Sensitive Protein

The procedure used for preparing the sweet-sensitive protein is essentially that described previously.[4] The epithelium from the dorsal surface of the anterior one-third of the tongue is separated from the underlying muscle, and sliced into small pieces (approximately 1 cm × 1 cm). The exact dimensions of the pieces are not critical, although very long pieces should be avoided as they tend to wrap around the blades of the blender in the next step. The pieces of tongue epithelium are homogenized with four volumes of 0.1 M sodium phosphate buffer, pH 7.0, in a Waring Blendor. This operation is carried out in a cold room, and the blender is turned off and allowed to cool after each 5 min of operation. After a total of 20 min of operating time in the blender, the homogenate is passed through a suction filter, using a coarse grade of filter paper, and the particles remaining on the filter are washed with additional phosphate buffer. The filtrate is centrifuged at 25,000 *g* for 20 min at 2–4°C. Longer centrifugation times and centrifugal forces of 50,000 *g* were used in some preparations, but yielded final products indistinguishable from those prepared by the procedure described above. 20 g of solid $(NH_4)_2SO_4$ were added to each 100 ml of the supernatant solution, and stirred in the cold overnight. The precipitate which formed was removed by centrifugation at 45,000 *g* for 1 hr. Another portion of $(NH_4)_2SO_4$, equal to the first, was added to the supernatant solution, and stirred in the cold overnight. The resulting precipitate was collected by centrifugation at 45,000 *g* for 1 hr. This material is the "40% $(NH_4)_2SO_4$ fraction," the sweet-sensitive protein fraction described by Dastoli and Price.[4] Typically, the yield is about 400 mg of protein from 200 g of tongue epithelium.

Gel Filtration

A column (1.5 cm × 22 cm) was prepared with Bio-Gel P-300 (50–100 mesh) as the packing material, and equilibrated with 0.1 M sodium phosphate buffer, pH 7.0. The column was operated at room temperature (approximately 22°C) with a flow rate of 0.2 ml per min, and the effluent was collected in fractions of 2 ml each.

Glucose Dehydrogenase Assay

The assay procedure was based upon the increase in absorbance at 340 mμ accompanying the reduction of nicotinamide adenine dinucleotide (NAD^+), the electron acceptor used in these studies. Assays were carried out in cuvettes of 1 cm path length in a Beckman DB spectrophotometer equipped with a scale expander and recorder, the cell compartments being thermostatted by the flow of water from a bath held at 37.0± 0.1°C. Each cuvette contained protein (0.44 mg/ml), NAD^+ (3×10^{-4} M), sugar, and either 0.2 M sodium phosphate buffer or 0.2 M Tris-HCl buffer. In no instance did we observe changes in absorbance at 340 mμ if either the protein or the NAD^+ was omitted from the cuvette. In the preparations on which the data reported here were obtained, no changes in absorbance at 340 mμ occurred in the absence of added sugar. However, we have encountered some preparations in which small apparent rates of NAD^+ reduction took place without adding sugar, presumably due to contamination of the protein with some substrate from the tongue tissue.

RESULTS AND DISCUSSION

Dastoli, Lopiekes, and Price[3] reported that gel filtration of the "40% $(NH_4)_2SO_4$ fraction" on Bio-Gel P-150 resulted in two peaks in the effluent. One, amounting to 80 per cent of the total material, contained all of the sugar-binding activity. This peak had the same retention volume as yeast alcohol dehydrogenase (referred to in that paper simply as the standard protein of molecular weight 150,000). In the gel filtration study reported here, we used Bio-Gel P-300, which gives finer resolution than does Bio-Gel P-150 in the molecular weight range above 100,000. Our results are shown in Figure 1. The major peak occurs within one fraction of the elution peak

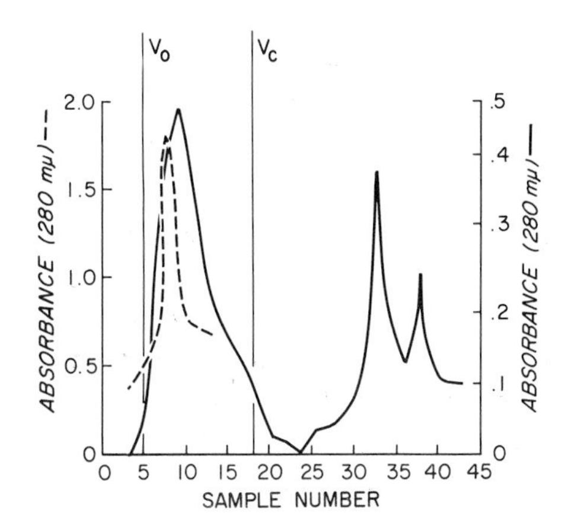

Figure 1 Elution pattern of 1 mg of the "40% $(NH_4)_2SO_4$ fraction" on Bio-Gel P-300, represented by the solid line and the ordinate on the right. The dotted line and the left ordinate represent the elution pattern of 5 mg of yeast alcohol dehydrogenase. V_0 and V_c indicate the void volume and total column volume, respectively.

for yeast alcohol dehydrogenase, and amounts to 70 per cent of the total material. It contains all of the catalytic activity to be described below. Indeed, the only apparent difference between these results and those of Dastoli, Lopiekes, and Price[3] is that the inactive material emerged as two peaks on our Bio-Gel P-300 column, while they reported only a single peak of inactive material on Bio-Gel P-150.

We surveyed the substrate specificity of the protein-catalyzed reaction by measuring the rates of reduction of NAD^+ using various reducing sugars, all at 0.5 M concentrations, as substrates. The results are shown in Table I. Although the reaction is not absolutely specific for D-glucose, none of the other sugars tested was oxidized at a rate as high as the rate of oxidation of D-glucose. This provides the justification for our referring to the catalytic property as a glucose dehydrogenase activity. We have not studied the specificity of the reaction with regard to different electron acceptors, although we have found that 2,6-dichloro-phenol indophenol can be substituted for NAD^+ in the reaction.

When the rates of the reaction were measured with varying substrate concentrations, simple Michaelis-Menton kinetics appeared to describe the results adequately.

TABLE I

RELATIVE RATES OF DEHYDROGENATION OF SUGARS BY THE "SWEET-SENSITIVE PROTEIN FRACTION"

Sugar	Relative Rate
D-glucose	100
D-xylose	43
D-fructose	19
L-xylose	7
D-arabinose	7
D-mannose	4
L-glucose	2
L-mannose	0
L-arabinose	0

Reaction mixtures contained protein at a concentration of 0.44 mg/ml, NAD^+ at 3×10^{-4} M, Tris at 0.2 M (pH 9.1), and the sugar at 0.5 M.

Using D-glucose at 0.5 M concentration as substrate and varying the concentration of NAD^+, we determined K_m for NAD^+ at pH 8.5 to be 6×10^{-6} M. Thus, the 3×10^{-4} M concentration of NAD^+ used in our standard assay mixture represents an excess. With D-glucose as the substrate, cuprous ions were found to accelerate the reaction rates, and 1×10^{-4} M CuCl was included in the reaction mixtures for the study of the kinetics of the reaction with this substrate. Figure 2 shows the kinetics of glucose oxidation at pH 9.1, the data plotted according to the following linear rearrangement of the Michaelis-Menton equation:

$$(S)/v = (S)/V_m + K_m/V_m.$$

This form of the equation is homeomorphic to the basic equation for taste proposed

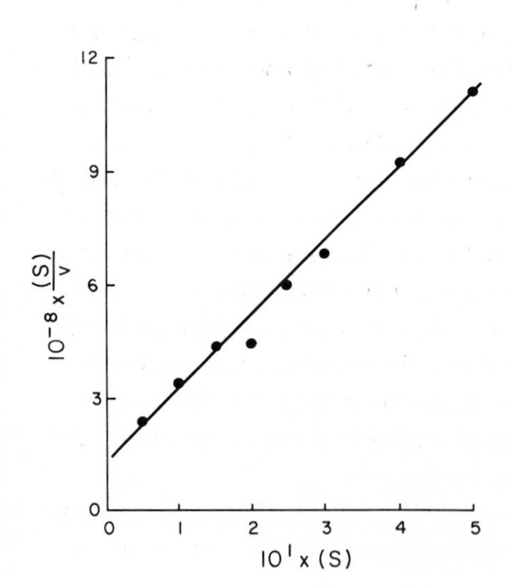

Figure 2 Kinetics of dehydrogenation of D-glucose by a preparation of "sweet-sensitive protein." The units of (S) are moles of glucose per liter. The units of v are moles of NAD^+ reduced per minute per ml.

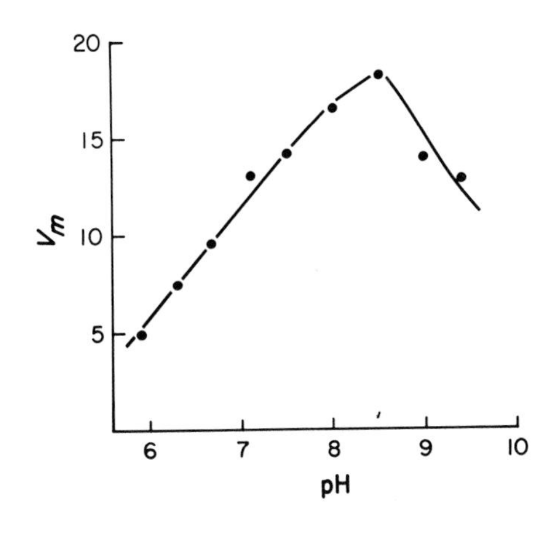

Figure 3 Variation of V_m with varying pH, with D-glucose as substrate. The units of V_m are $10^{10} \times$ moles of NAD^+ reduced per minute per ml.

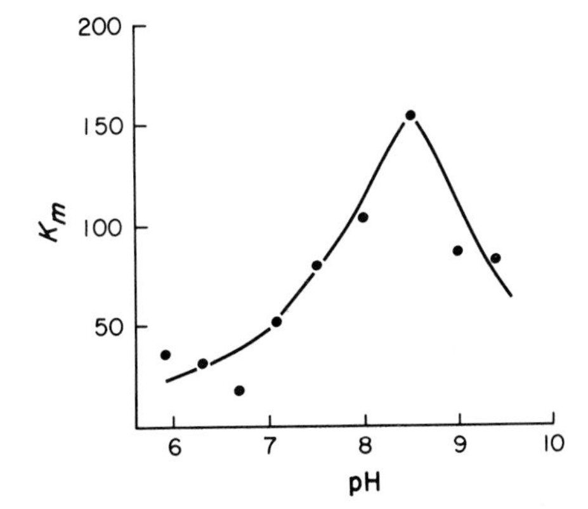

Figure 4 Variation of K_m with varying pH, with D-glucose as substrate. The units of K_m are millimoles per liter.

by Beidler,[1,2] except that K in Beidler's equation is an association constant, while K_m is a dissociation constant (assuming equilibrium conditions in both cases). Values of the kinetic constants, K_m and V_m, were obtained over the pH range 5.9–9.4. The variation of V_m with varying pH is shown in Figure 3. The optimal pH with respect to V_m is at pH 8.5, but there is less than a fourfold variation over the entire range studied. The values of K_m are plotted vs. pH in Figure 4. The curve shows a well-defined maximum at pH 8.5, decreasing more rapidly than the curve in Figure 3 at higher and lower pH. It should be noted that a maximum in K_m is not an optimum in any sense, the affinity of the protein for the substrate being in inverse proportion to K_m. The value of K_m at pH 7.1 is in good agreement with the value of K reported by Dastoli and Price[4] for complexing of glucose by the sweet-sensitive protein at pH 7.0. From our data, K_m is 51 mM, while $1/K$ from the data of Dastoli and Price[4] is 37 mM. The kinetics of fructose oxidation at pH 9.1 are shown in Figure 5. The scatter in the data is due to the fact, noted above, that fructose is oxidized at a far lower rate than is D-glucose (cf. Table I). Despite the scatter, the data appear to be adequately described by the Michaelis-Menton equation, although there is a considerable uncertainty in the value obtained for K_m, which we can only conclude to be within the range 1–3 mM. The value for $1/K$ for fructose, from the data reported by Dastoli and Price,[4] was 1 mM at pH 7.0. Again, the agreement is satisfactory.

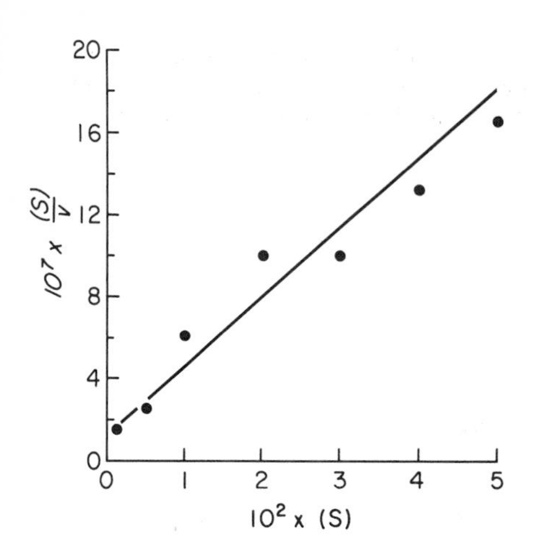

Figure 5 Kinetics of dehydrogenation of D-fructose by a preparation of "sweet-sensitive protein." The units of (S) are moles of fructose per liter. The units of v are moles of NAD$^+$ reduced per minute per ml.

Our finding of this catalytic activity of the sweet-sensitive protein raises the disturbing possibility that the sugar-binding property reported by Dastoli and Price[4] was nothing more than the formation of complexes between an enzyme and substrates or substrate analogues. Since those reaction mixtures contained no electron acceptor, the reported stability of the complexes is not, in any way, in disagreement with our present findings. Although our observation that the values of K_m vary with pH appears to be at variance with the reported independence of pH for sugar binding by the sweet-sensitive protein,[4] closer examination of the data in that paper shows this discrepancy to be unreal. In that report, the amount of fructose-protein complex formed in the presence of 100 mM fructose was virtually independent of pH from pH 6 to pH 10. However, $1/K$ for fructose at pH 7 was only 1 mM. Thus, the data show only that $1/K$ does not increase as much as fiftyfold over that range. The dependence of K_m on pH reported in this paper (Figure 4) is entirely consistent with that statement.

An alternative to the possibility that the sweet-sensitive protein is a tissue enzyme, is that the dehydrogenase activity reported here is of no physiological significance, and that its physiologically significant property is its affinity for sugars. A similar view has been advanced for insect glucoreceptors, in which it has been proposed that a glucosidase is the receptor protein for disaccharides.[5,6] As a working hypothesis, we have adopted the view that this protein is related to taste. According to this hypothesis, the significant property of the protein is that it forms complexes with sugars, and the catalytic activity is simply a property which simplifies measurement of its interactions with sugars. The catalytic activity is, indeed, very low, which is entirely in accord with this hypothesis. The low activity could also mean that the sugar-binding activity reported by Dastoli and Price[4] is not due to the same protein as the catalytic activity reported here; that is, the dehydrogenase may simply be a contaminant. This is not an easy point to test experimentally, but the similar binding constants for sugars and the similar behavior on gel filtration argue against it.

SUMMARY

Preparations of sweet-sensitive protein from bovine tongue epithelium, previously reported to form complexes with sugars, are found to catalyze the dehydrogenation of reducing sugars. The sugar oxidized most rapidly is D-glucose. The previously reported dissociation constants for complexing of glucose and of fructose by such preparations are in good accord with the values of K_m determined from the kinetics of oxidation of these sugars. We suggest that the sugar binding reported previously and the catalytic activity reported here are properties of the same protein, and have speculated that this protein is a taste receptor protein which may have evolved from a glucose-metabolizing enzyme.

REFERENCES

1. Beidler, L. M. 1954. A theory of taste stimulation. *J. Gen. Physiol.* **38**, pp. 133–139.
2. Beidler, L. M. 1961. Taste receptor stimulation. *Progr. Biophys. Biophys. Chem.* **12**, pp. 107–151.
3. Dastoli, F. R. D., D. V. Lopiekes, and S. Price. 1968. A sweet-sensitive protein from bovine taste buds. Purification and partial characterization. *Biochemistry* **7**, pp. 1160–1164.
4. Dastoli, F. R., and S. Price. 1966. Sweet-sensitive protein from bovine taste buds: isolation and assay. *Science* **154**, pp. 905–907.
5. Hansen, K. 1968. Zur kompetitiven Hemmung der tarsalen Glucosidase und der Zucker-Rezeption bei *Phormia regina* Meig. *Zool. Anz.* (Suppl.). pp. 558–564.
6. Hansen, K. 1969. The mechanism of insect sugar reception, a biochemical investigation. *In* Olfaction and Taste III (C. Pfaffmann, editor). Rockefeller Univ. Press, New York, pp. 382–391.

V PSYCHOPHYSICS AND SENSORY CODING

V · PSYCHOPHYSICS AND SENSORY CODING

EFFECTS OF ADAPTATION ON HUMAN TASTE FUNCTION

DONALD H. McBURNEY
University of Pittsburgh, Pittsburgh, Pennsylvania

INTRODUCTION

Although gustatory adaptation has been recognized for many years, adaptation state has largely been ignored in taste research since the pioneering work of Hahn during the thirties and forties. Often adaptation is considered to be an abnormal condition of the receptor, to be avoided. This is in spite of the fact that adaptation state has been recognized as an important parameter in the function of other sensory systems, most notably vision.

The steady application of a stimulus to the tongue has a number of outcomes, of which (1) *decreased responsivity* to the same stimulus as measured by response decrement and increase in threshold is only one. The others include (2) an increase in threshold for a chemically different stimulus and a decrease in the response to the different stimulus. This quality generalization of adaptation, or *cross-adaptation*, is often taken to indicate a common receptor mechanism for the two stimuli.[10] On the other hand, (3) the threshold for another stimulus may be decreased and the response to the different stimulus increased. This *adaptation-produced potentiation* is taken as evidence for opponent processes underlying the responses to these stimuli.[2] The slope of the subjective magnitude function varies with adaptation (4); it is steepest when the adapting level is high.[7] This *steepening* means that the responses to stimuli near the adapting level are affected the most and more intense stimuli are affected the least. The just noticeable difference threshold (jnd) is smaller for a given standard stimulus when the receptor is adapted to that stimulus (5).[8] This has the effect of making the point of maximum differential sensitivity vary with adapting level[6] and, perhaps more than any of the other outcomes, the *jnd change* illustrates that adaptation does not mean a loss of sensory information capacity. All of these results are known for vision, and several of them have been demonstrated for the other well-studied senses. Together they may be considered to be the detailed operational definition of adaptation, or to constitute "laws of adaptation."

The various outcomes of adaptation have been useful tools in the study of sensory coding problems in the other senses, particularly in vision. The purpose of this paper is to review some recent experiments that manipulate adaptation state in an effort to understand the gustatory system.

APPARATUS

The experiments to be reported all employed a flow system that delivered a 34°C stream of solution to the anterior dorsal tongue, extended between the lips.[7] This apparatus, although simple, has a number of advantages over other methods, including the ability to switch rapidly from one stimulus to another; giving a relatively large area to be stimulated; and eliminating dilution by saliva.

PROCEDURE

The adapting stimulus flowed, generally for 20 sec, between each stimulus presentation. Then the stimulus was presented for about 2 sec after a break of about 1 sec. The responses measured were direct estimates of numerical magnitude. The subjects were primarily naive volunteers, but in most experiments I have been one of the subjects. If an experiment involved comparison of two or more adapting conditions, each subject was his own control.

CROSS-ADAPTATION

Hahn[5] found generalization of adaptation between compounds within the sour, sweet, or bitter taste qualities, but he found no evidence of cross-adaptation between salts. Because generalization of adaptation implies a common mechanism for those compounds that cross-adapt, this result has been interpreted to mean that there must be a separate mechanism for every salty compound. However, when subjects were instructed to make magnitude estimates of the over-all intensity of salts matched in subjective intensity to 0.1 M NaCl, adaptation was found to reduce the intensity of each of the eight salts tested compared with HOH adaptation. Twenty subjects rated the over-all intensity of eight salts, once under HOH adaptation and once under 0.1 M NaCl adaptation. These results are shown in Figure 1. Similar results were found for adaptation to 0.11 M KCl, 0.051 M NH_4Cl, and 0.026 M $CaCl_2$ on these salts, except that as NaCl reduced the intensity of NaBr more than it reduced the others, so KCl affected KBr, NH_4Cl affected NH_4Br, and $CaCl_2$ affected $CaBr_2$ more than the others.[9] Smith and McBurney[13] suspected that over-all intensity might not be the appropriate measure. We asked subjects to estimate the over-all subjective magnitudes of 12 salts and then to divide their rating among the four "basic taste" qualities. In this manner we obtained a taste profile for each of the salts. Five estimates by each of 10 subjects under HOH and NaCl adaptation are shown in Figure 2 for four of the 12 salts tested. If we look at the total intensity we come to the same conclusion as in the previous experiment. When we look at the change in saltiness, however, we find a rather striking result: the saltiness is changed more than the over-all intensity. Figure 3 shows the results for the salty quality only. The saltiness of all salts tested fell essentially to zero and below threshold in all cases, as shown by the unfilled triangles.

This experiment suggests a more congenial conclusion—there appears to be a single mechanism responsible for coding saltiness. The fact that most of these salts have other qualities as well as saltiness and the use of threshold as a measure may be the reason Hahn found no cross-adaptation.

We next studied the effect of adaptation to citric acid on the taste of a series of other compounds. Fifteen subjects judged the solutions four times each under HOH and 0.004 M citric acid adaptation by the taste profile method. Figure 4 shows the results for the sourness of those compounds that were predominantly sour. The sourness of all the acids was reduced significantly. This confirms Hahn's findings[5] for acids. It should be mentioned that the different intensity of sourness under HOH adaptation for the various acids is largely a matter of different over-all intensity due to a failure of our matching procedure. All of the acids are predominantly sour. This probably does not affect the conclusion.

The picture is not as clear, however, when we look at bitter cross-adaptation. Fifteen subjects judged the intensity of each compound four times under HOH and under 0.0001 M quinine hydrochloride adaptation, using the taste-profile method. Figure 5 shows the results for bitterness of these compounds. It may be seen that QHCl adaptation reduced the bitterness of QHCl, quinine sulfate, caffeine, and sucrose octa acetate, but did not affect urea, KNO_3, PTC, or $MgSO_4$. This seems to

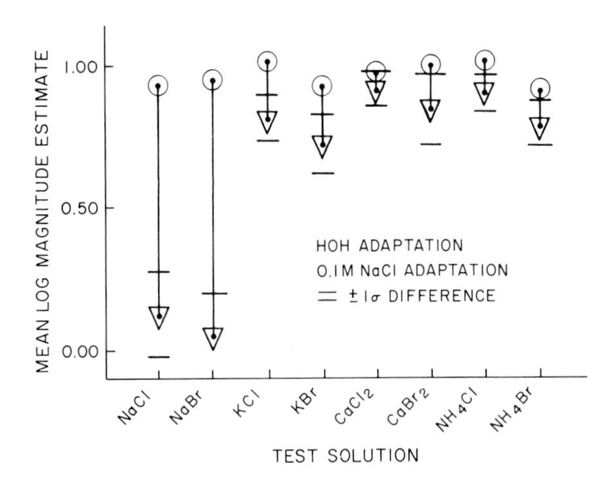

Figure 1 Over-all subjective intensity of the taste of eight salts after adaptation to HOH (circles) and after adaptation to 0.1 M NaCl (triangles). The length of the lines indicates amount of change. The horizontal lines indicate ±1 standard error of the difference.[9]

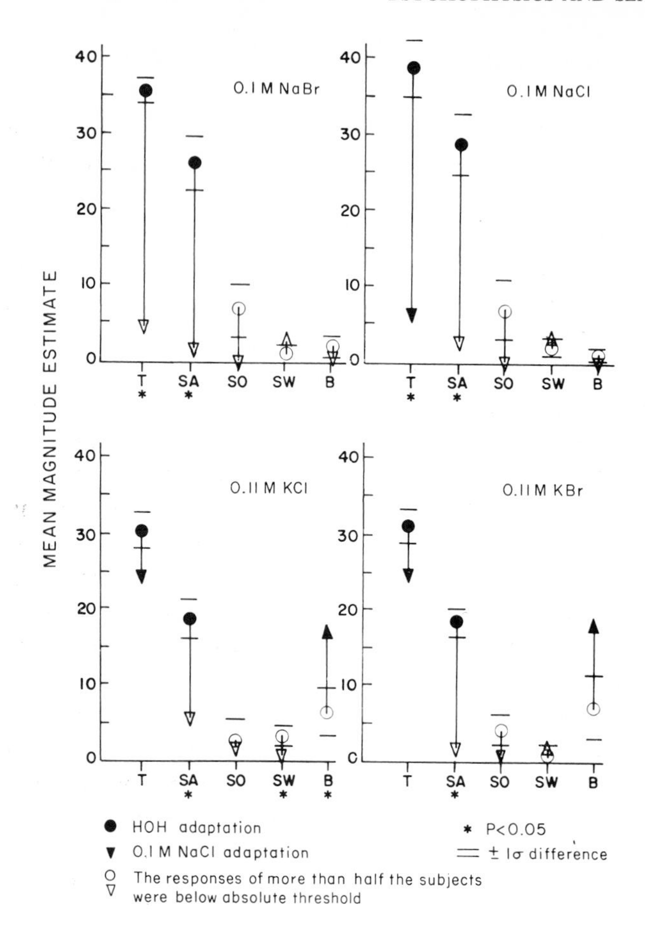

Figure 2 Total subjective intensity and taste profile for four of the salts tested, after adaptation to HOH (circles) and after adaptation to 0.1 M NaCl (triangles).[13]

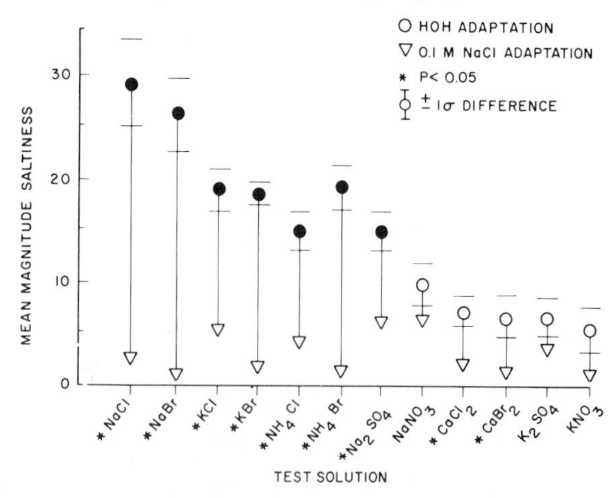

Figure 3 The saltiness of twelve salts after adaptation to water and after adaptation to 0.1 M NaCl.

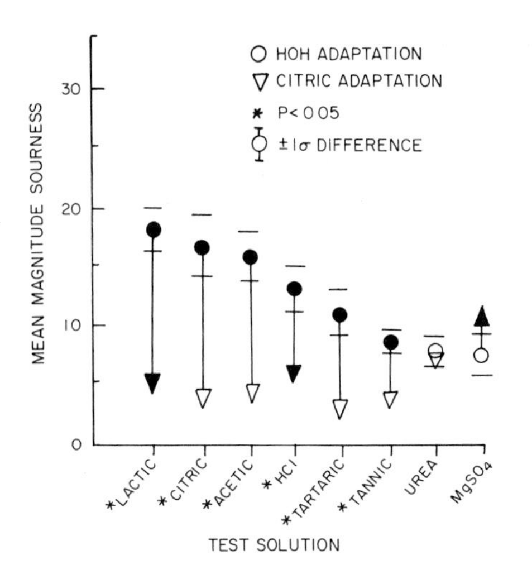

Figure 4 The sourness of eight compounds after adaptation to water and after 0.004 M citric acid adaptation. (McBurney and Smith, unpublished)

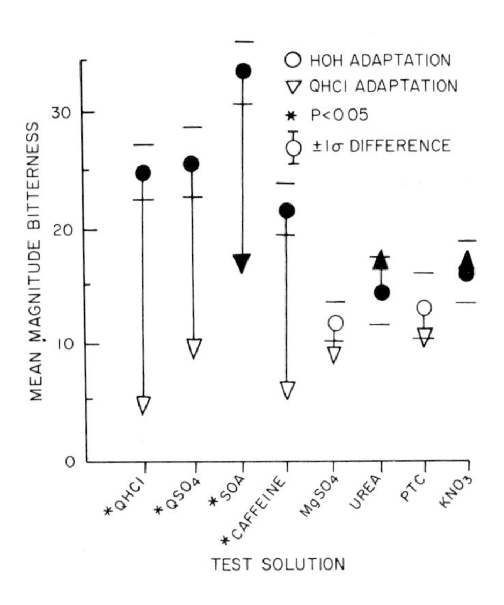

Figure 5 The bitterness of eight compounds after adaptation to water and after adaptation to 0.0001 M quinine hydrochloride. (McBurney and Smith, unpublished)

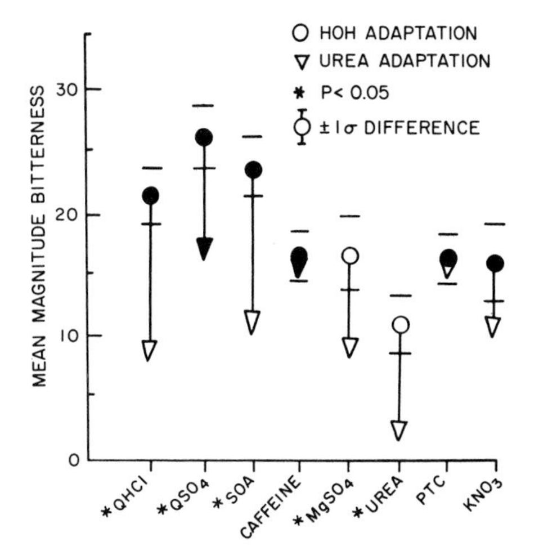

Figure 6 Same as Figure 5, after HOH or 1.0 M urea adaptation. (McBurney and Smith, unpublished)

suggest the existence of at least two bitter mechanisms. The obvious experiment would be to adapt to one of the compounds that was not affected by QHCl adaptation. Fifteen subjects judged the bitterness of these compounds four times each under HOH and 1.0 M urea adaptation by the same method. Figure 6 shows the results.

To our surprise, the bitterness of all but one (caffeine) of the substances tested was reduced, even though urea was not as bitter as some of the other substances. This is rather puzzling, but I have two thoughts on the matter. The first and most satisfying is that there may be two or more bitter mechanisms, and urea stimulates two of them while quinine stimulates only one. Another possible reason is the well-known psychophysical confusion between sour and bitter. Urea is actually more sour than bitter, as measured in this experiment and it may be that the subjects are confusing the two qualities. In relation to this, let us look at the effect of citric acid adaptation on bitterness.

Figure 7 shows the change in bitterness for those compounds that were predominantly bitter. We can see that citric acid adaptation reduced the bitterness of the same compounds that were affected by urea (except for urea) even though citric acid, like the other acids tested, does not have an appreciable amount of bitterness. Unfortunately, I cannot make a similar comparison for the effect of bitter adaptation on sourness, because there were not enough sour stimuli included in the quinine or urea adaptation experiments. It is perhaps worth noting that although urea tastes sour, the pH of a 1.0 M solution is about 8.0 and its sourness was not reduced by citric acid adaptation. Hahn also found that some bitter substances cross-adapted and some did not.

ADAPTATION-PRODUCED POTENTIATION

It is known that adaptation to sour compounds will cause water to have a sweet taste and, conversely, adaptation to sweet compounds will cause water to taste sour.

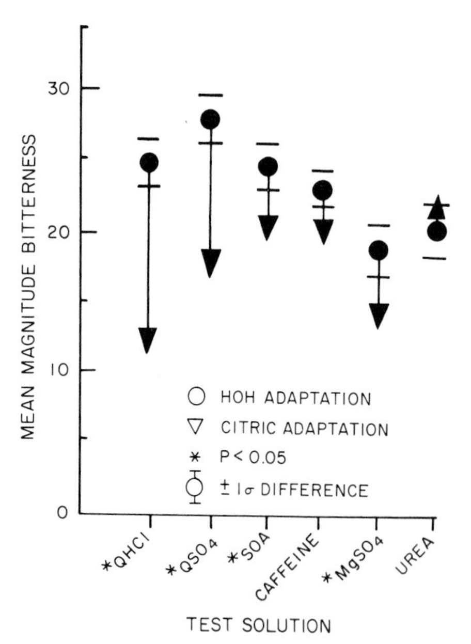

Figure 7 The bitterness of six of the compounds shown in Figure 6, after adaptation to HOH or citric acid adaptation. (McBurney and Smith, unpublished)

Bartoshuk et al.[2] showed that sodium chloride adaptation makes water taste sour-bitter.

Bartoshuk, in a recent experiment,[1] found that adaptation to sucrose made water and weaker concentrations of sucrose taste sour-bitter. Adaptation to quinine hydrochloride made water and weaker concentrations of quinine hydrochloride taste sweet. Sodium chloride adaptation made water and weaker NaCl solutions taste sour-bitter. HCl adaptation produced sweet and sour responses in water and weaker concentrations of HCl. These results make an opponent-process theory of taste attractive, but a problem has been to find a compound adaptation that will produce a salty taste in water. V. Skramlik[12] states that adaptation to $CoCl_2$ and $NiCl_2$ will make water taste salty, but we have not been able to confirm this. McBurney and Bogart asked subjects to rate the taste of water using the taste profile method. The 10 subjects received HOH on two trials after 30-sec adaptation to each of 12 compounds matched in intensity to 0.1 M NaCl. Figure 8 shows the results. Water had a more intense taste as a secondary stimulus after most of these compounds than it did after HOH adaptation. However, they did not all give the same magnitude of secondary taste in HOH, although they had been matched for primary taste. The other rows show the intensity for each quality. In general, the results are similar to those of the earlier studies. Sweet is the most common taste of water after adaptation to these compounds; it is produced by caffeine, QHCl, $NaNO_3$, Na_2SO_4, citric acid, lactic acid, NaCl, and KCl. Sucrose makes water taste sour; Na_2SO_4 and NaCl make it taste bitter. The most interesting result is that urea makes water taste salty. This aftertaste is quite reliable, and has been replicated. An opponent-process theory of taste seems to be confirmed here, since sodium chloride, which is salty, makes water taste sour-bitter and urea, which is sour-bitter, makes water taste salty. (In this experiment, NaCl produced a bitter-sweet taste but a sour-bitter taste is more typical.) However, there is a good deal more that is rather confusing. For example, sucrose makes water sour-bitter but, of course, it tastes sweet rather than salty. Not all compounds with the same primary taste produce the same secondary taste in water. We find, for example, that HCl does not produce a significant sweet taste in water, although citric acid and a number of other compounds do. Sucrose makes water taste sour but fructose does not.

In short, these results seem to defy any attempt at organization into a two-dimensional opponent-process scheme analogous to the color circle. In the same experiment we also presented weak concentrations of 0.01 M NaCl, 0.0018 M caffeine, 0.0057 M sucrose, or 0.000003 M tartaric acid as secondary stimuli after adaptation to stimuli that, we found, produced the same taste in water. Our purpose was to see whether adaptation would potentiate the typical taste of the weak secondary stimulus; for example, would 0.0057 M sucrose given after adaptation to 0.1 M caffeine taste sweeter than when the tongue was adapted to water. Figure 9 shows the taste profile for the four weak solutions after adaptation to water and the appropriate adapting compound. In each case the baseline for the profiles is the intensity of HOH after HOH

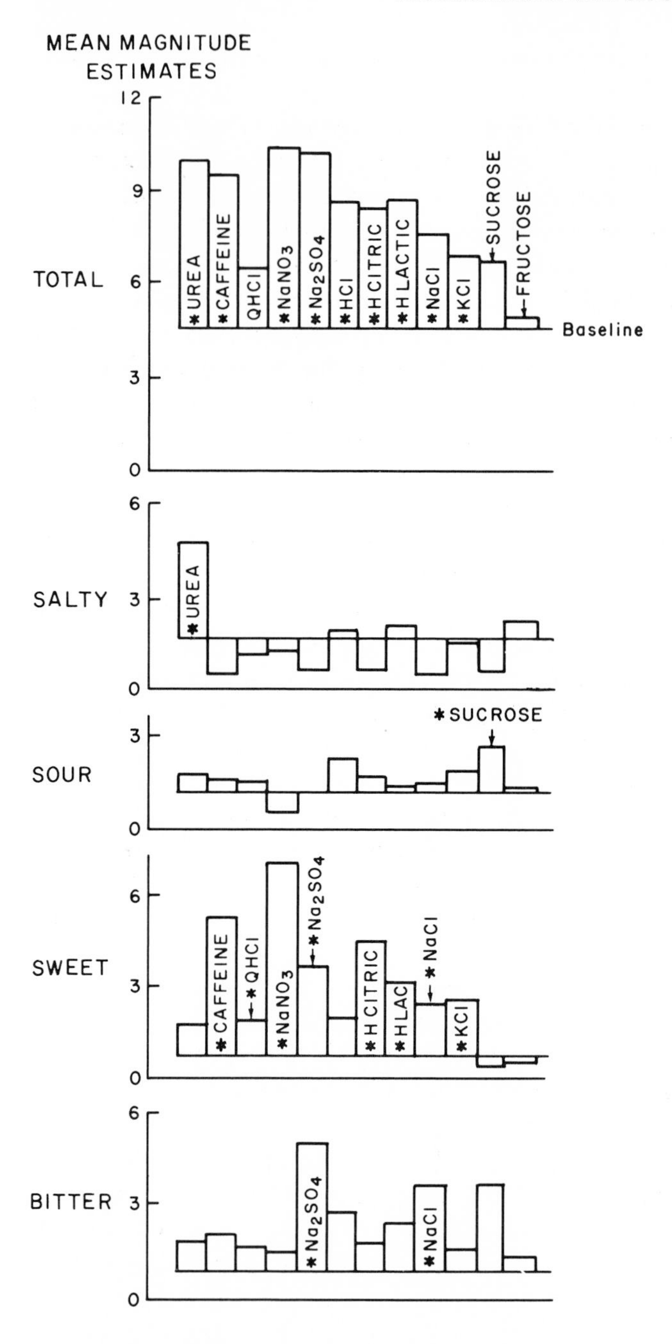

Figure 8 Total intensity and intensity of sourness, saltiness, sweetness, and bitterness of water after adaptation to 12 compounds matched in subjective intensity to 0.1 M NaCl. The baseline in each section is the taste of HOH following HOH adaptation. The asterisks indicate that the subjective magnitude of HOH was significantly ($p < 0.05$) more intense for that quality and condition than HOH after HOH adaptation. (McBurney and Bogart, unpublished)

adaptation. Comparing the filled with the unfilled bars shows that, in each case, the appropriate potentiation occurs. For example, 0.0057 M sucrose is sweeter after adaptation to 0.1 M caffeine than after adaptation to HOH. The fact that KCl adaptation potentiated the bitterness as well as the sourness of tartaric acid and sucrose adaptation increased the sourness as well as the bitterness of caffeine is probably a reflection of the mixed taste induced in water by these substances.

In this context it is relevant to return for a moment to the NaCl cross-adaptation study. We found that the decrease in saltiness was greater than the decrease in over-all intensity. This requires that some other quality had to increase in intensity. Figure 10 shows the changes in saltiness plotted against the changes in bitterness. For most of the stimuli a decrease in saltiness is accompanied by an increase in bitterness. Figure 11 shows the same relationship for sourness. The bitterness or sourness of these salts was potentiated by adaptation to NaCl.

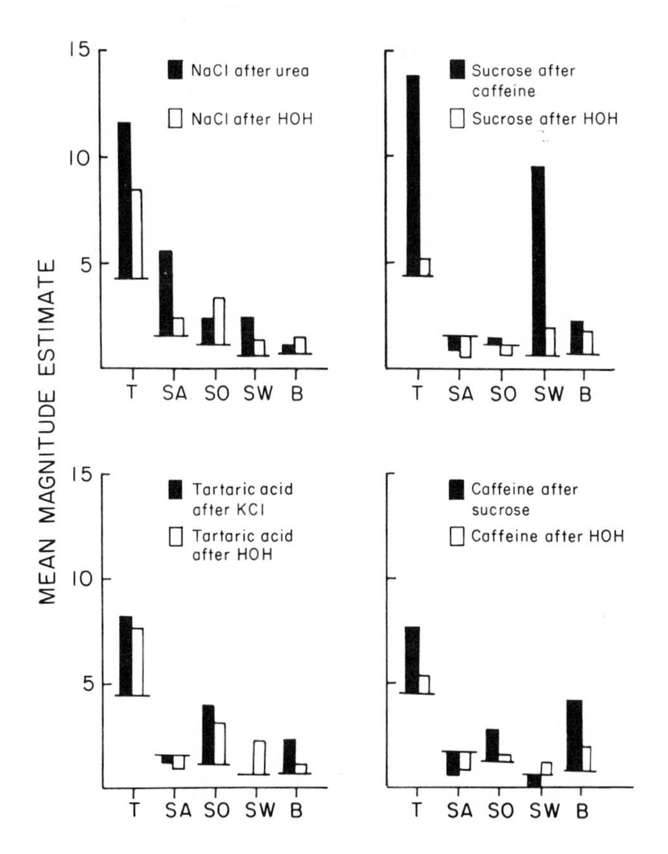

Figure 9 Total intensity (T) and intensity of saltiness (SA), sourness (SO), sweetness (SW), and bitterness (B) of four compounds after adaptation to water (open bars) or another compound (filled bars). Each baseline is the taste of HOH after HOH adaptation for that quality. When the filled bar exceeds *R* height of the unfilled bar, adaptation has potentiated the taste of the test solution. (McBurney and Bogart, unpublished)

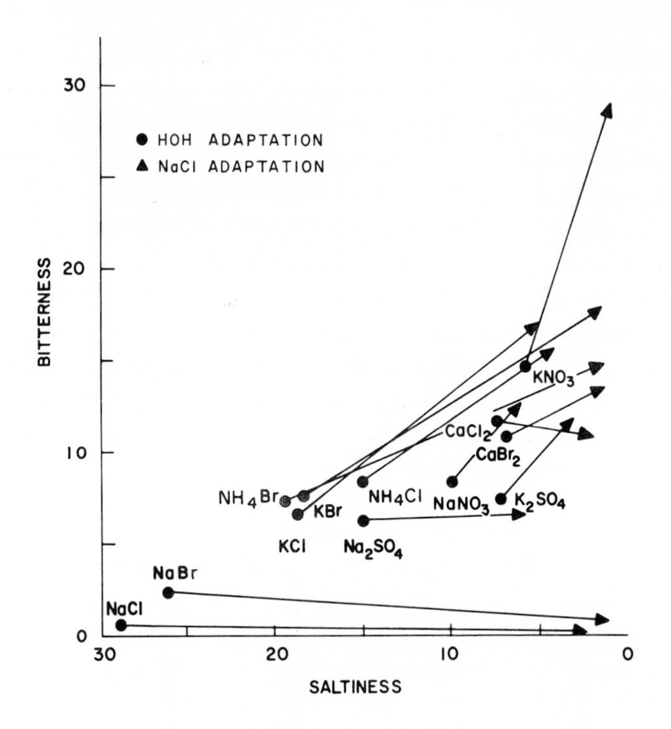

Figure 10 Saltiness vs. bitterness of 12 salts after adaptation to HOH or 0.1 M NaCl.[13]

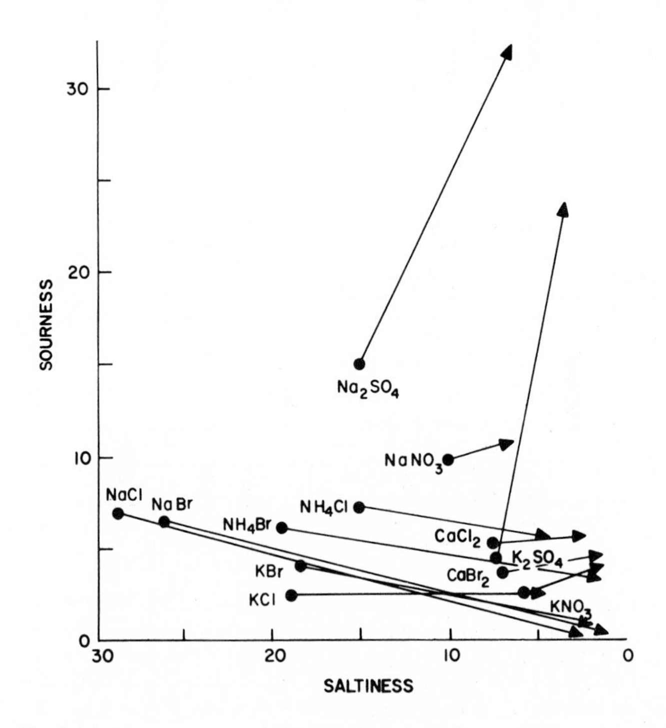

Figure 11 Saltiness vs. sourness of 12 salts after adaptation to HOH or 0.1 M NaCl.[13]

An obvious question here would be what the effect is of *Gymnema sylvestre* extract on the taste of water after adaptation to these compounds. Bartoshuk and I had made some casual (unpublished) observations and found that *Gymnema* extract reduced the sweet secondary taste in water after a sour-adapting stimulus. Bogart and I presented to four subjects for 2 min either a 130 g/liter extract of *Gymnema sylvestre* leaves (generously supplied by Bartoshuk) or a brew of tea and food coloring that we had made as a control for color and taste of the *Gymnema* extract. This was repeated before every sixth stimulus. The subjects—all members of our laboratory, because the taste of the extract was rather unpleasant—received each stimulus twice under each condition. The tea was a fairly good control as far as taste was concerned, but the subjects did notice that on some days, after they spit out the bitter mixture, the water with which they rinsed their mouth was sweet and on some days it was not, corresponding to the tea or *Gymnema* conditions, respectively. This pretty well summarizes the data for the whole experiment, for the *Gymnema* reduced the sweet taste of water after adaptation but did not significantly affect the other qualities. You will note that the concentration of *Gymnema* was not very strong. This was done because of the unpleasant quality of the extract and the possibility that if the extract were too strong its residual bitterness would mask the qualities of the secondary water stimuli.

JUST NOTICEABLE DIFFERENCE (jnd) CHANGE

McBurney, Kasschau, and Bogart[8] found that the jnd for the 0.1 M NaCl was 0.018 M when the tongue was adapted to HOH and 0.009 M when adapted to 0.1 M NaCl, a twofold difference. This shows that the tongue is more sensitive to changes in intensity around the concentration value of 0.1 M when the tongue is adapted to 0.1 M than when adapted to water. It does not say anything about relative sensitivity at other concentrations. Bogart[3] extended this experiment by adapting at 0.001 M and 0.098 M and testing at 0.1 M and 0.3 M. (0.098 M is almost certainly a trivial difference from 0.1 M, but was chosen as an adapting solution in pilot work to be as close to 0.1 M as possible without making the standard 0.1 M solution tasteless.) Since this is a difference of about $\frac{1}{5}$ of a jnd from 0.1 M, it probably proves only that adaptation to 0.1 M NaCl is not complete. The same apparatus and procedure were used as in the previous study. A dam divided the extended tongue along the midline. Both sides of the tongue were adapted to either 0.001 M or 0.098 M NaCl. Then the standard (0.1 M or 0.3 M) was presented to one side and the comparison to the other. The subject indicated which stimulus was the more intense. The method of constant stimuli was used. There were 20 trials for each of three subjects with each comparison stimulus. Figure 12 shows relative sensitivity as measured by $I/\Delta I$ for the two standard stimuli as a function of adaptation. The two points at 0.098 M nicely replicate the 0.1 M-adapted data from the previous experiment. We can see that the effect of 0.098 M adaptation on the sensitivity was relatively greater at 0.1 M than at 0.3 M.

The data of Schutz and Pilgrim[11] are plotted with our data for comparison. One conclusion that may be drawn is that any theory that would measure intensity of taste

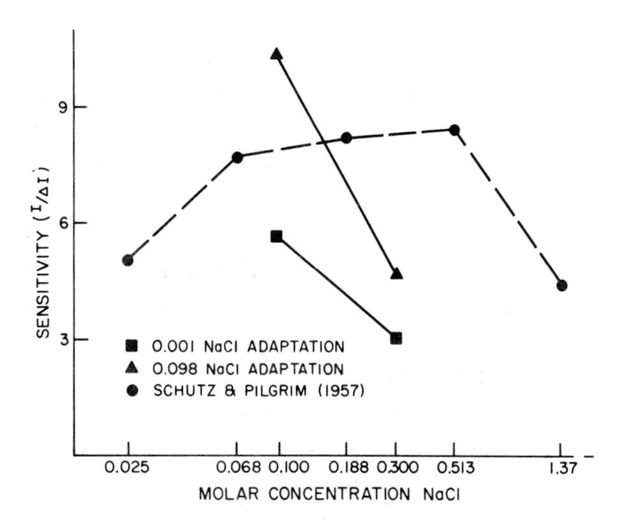

Figure 12 Relative sensitivity (reciprocal of the Weber fraction) to NaCl as a function of adapting concentrations and standard. The data of Schutz and Pilgrim[11] are shown for comparison. (Bogart and McBurney, unpublished)

based on cumulated jnds needs a parameter to account for adaptation state. Green[4] has suggested that this is the major reason for conflicting results between indirect and direct methods. Beidler's taste equation may be modified to account for the effect of adaptation by subtracting a constant from the stimulus intensity. It is interesting that subtractive constants on the stimulus intensity are used to account for the effects of adaptation in the power function that result from direct magnitude estimation. Also, in the complete form of Fechner's equation a constant is subtracted from the logarithm of the stimulus measure to account for threshold. In none of these cases does the subtractive constant seem to relate to an identifiable physiological process.

SUMMARY

The gustatory system is in a dynamic state with respect to adaptation. The experiments reviewed here indicate that we may capitalize on this fact by manipulating adaptation state and determining its effect on taste function. They also demonstrate the usefulness of direct psychophysical methods for measurement of sensory magnitude. Most of these experiments could not have been done, or could only have been done very tediously, by the indirect methods.

ACKNOWLEDGMENTS

Some of the experiments were supported by USPHS Grant No. NB-07873-01. The technical assistance of Lawrence Glanz and Thomas Shick in certain of these experiments is gratefully acknowledged.

REFERENCES

1. Bartoshuk, L. M. 1968. Water taste in man. *Percept. Psychophys.* **3**, pp. 69–72.
2. Bartoshuk, L. M., D. H. McBurney, and C. Pfaffmann. 1964. Taste of sodium chloride solutions after adaptation to sodium chloride: implications for the "water taste." *Science* **143**, pp. 967–968.
3. Bogart, L. M. 1967. Differential effects of adaptation on taste jnd's. Unpublished Master's thesis. University of Tennessee, Knoxville, Tenn.
4. Green, E. E. 1962. Correspondence between Stevens' terminal brightness function and the discriminability law. *Science* **138**, pp. 1274–1275.
5. Hahn, H. 1949. Beiträge zur Reizphysiologie. Scherer, Heidelberg, Germany.
6. Keidel, W. D., U. O. Keidel, and M. E. Wigand. 1961. Adaptation: Loss or gain of sensory information? *In* Sensory Communications (W. A. Rosenblith, editor). MIT Press and John Wiley and Sons, New York, pp. 319–338.
7. McBurney, D. H. 1966. Magnitude estimation of the taste of sodium chloride after adaptation to sodium chloride. *J. Exp. Psychol.* **72**, pp. 869–873.
8. McBurney, D. H., R. A. Kasschau, and L. M. Bogart. 1967. The effect of adaptation on taste jnds. *Percept. Psychophys.* **2**, pp. 175–178.
9. McBurney, D. H., and J. A. Lucas. 1966. Gustatory cross adaptation between salts. *Psychonomic Sci.* **4**, pp. 301–302.
10. Pfaffmann, C. The sense of taste. *In* Handbook of Physiology (J. Field, H. W. Magoun, and V. E. Hall, editors), Section 1: Neurophysiology, Vol. I. American Physiological Society, Washington, D.C., pp. 507–533.
11. Schutz, H. G., and F. J. Pilgrim. 1957. Differential sensitivity in gustation. *J. Exp. Psychol.* **54**, pp. 41–48.
12. Skramlik, E. v. 1926. Handbuch der Physiologie der niederen Sinne: Bd. 1, Die Physiologie des Geruchs-und Geschmackssinnes. Thieme, Leipzig, East Germany.
13. Smith, D. V., and D. H. McBurney. 1969. Gustatory cross adaptation: does a single mechanism code the salty taste? *J. Exp. Psychol.* (In press)

BASIS FOR TASTE QUALITY IN MAN

ERNEST DZENDOLET
Department of Psychology, University of Massachusetts, Amherst, Massachusetts

In any discussion, it is well to have important terms defined carefully. This is certainly true of gustatory quality. The general definition of a quality as opposed to a modality was first given by Helmholtz (Fick,[9] pp. 165–166).

Between sensations of different kinds, there occur two different degrees of distinction. The more encompassing is the distinction between sensations which belong to different senses, as between blue, sweet, high-pitched. I have designated these as a distinction in the "modality" of the sensation. It is so encompassing that it excludes every transition from one to the other, every relation of greater or lesser similarity. Whether, for example, sweet be more similar to blue or red, can not be asked at all. The second kind of distinction, the less encompassing, is that between different sensations of the same sense. I designate this by a difference in "quality." . . . Within every such range, transition and equating is possible. From blue we can go over to scarlet-red by way of violet and carmine-red, and, for example, state that yellow is more similar to orange-red than to blue.

Such a definition is not at all quantitative, and could easily lead to arbitrary decisions concerning the number of qualities within a modality, and which ones they are. There has been, however, a recent direct attempt at a quantitative evaluation of the existence of a quality. This was carried out by Sternheim and Boynton[22] in relation to visual hue. They used a procedure in which the subject had available to him only a given number of specific quality names. The subject was presented with a particular wavelength of light, and was required to assign both a quality name and a number to the stimulus. The number was to specify the relative amount of that particular quality as a fraction of the over-all stimulus that was present on that trial. The investigators used wavelengths covering most of the visible spectrum, and employed three criteria for evaluating their results. The first criterion was that a quality name had to be used with high reliability; the second, that the function defined by the numbers assigned to a given quality name be at a maximum when those of the other names were at a minimum. The third criterion had to be evaluated in two stages. In one of these, the use of the quality in question was allowed, and resulted in a function over a particular stimulus range. In the other stage, the quality was not allowed, and there should now have been no function in that same stimulus range. Sternheim and Boynton's task in their experiment was to determine whether the quality "orange" is a unique hue. Their conclusion, using the criteria given above, was that it is not.

Over the years, since the time of Aristotle, various and numerous gustatory qualities have been postulated. Aristotle himself said that there were two primary ones, sweet and bitter. Other taste qualities also existed, and were believed to occur between these two poles. In the sixteenth, seventeenth, and eighteenth centuries, the number of primary qualities rose to nine or ten. The current view of bitter, salty, sour, and sweet as the primary qualities was advanced by Fick in 1864, and supported by Vintschgau in 1879.[17] However, Öhrwall[17] did not agree that these four names referred to qualities. For two reasons, he was of the opinion that they were modalities, i.e., separate sensory systems. The first reason was that taste mixtures were usually reported as such, and not as new qualities; the second was that there was no way to change one quality into another in a continuous manner, as can be done with spectral hues.

More recently, Frings[10] has discussed this problem. It is difficult to be certain of his argument, however, for he speaks of the four primary modalities of bitter, salty, sour, and sweet. It appears that he had accepted Öhrwall's conclusion as an initial premise. Then, on the basis of his work with insects, he suggested that these four tastes are, instead, merely "points of familiarity in a continuous taste-spectrum." Such a conclusion is, of course, essentially the definition of quality as originally proposed by Helmholtz.

Although there do not seem to have been any direct experiments of the Sternheim and Boynton type carried out in gustation, there are a number of experiments which do have a bearing on the problem of quality. Before I discuss them, I should like to mention briefly some theories and controversies that have resulted from the consideration of such issues as I have been describing.

The existence of qualities, no matter how many or which ones, implies two things: the existence of specific structures or properties in gustatory stimuli; and the reflection of these structures in gustatory qualities. This point will not be discussed here, except to note that two new theories concerning the structure of sweet compounds have recently been proposed independently by Dzendolet[7] and by Shallenberger and Acree.[21] The second implication concerns the type of mechanism used to signal the presence of these properties to the central nervous system (CNS), when they appear at the receptors.

As we all know, one long-lived theory concerning the signaling mechanism is currently known to us as the "specificity" theory. An older term, which I prefer for the same theory, is the "neoMüllerian" view. (The Müllerian part of this term, of course, refers to the earlier doctrine of specific nerve energies.) This neoMüllerian view states that a specific class of stimuli reacts with a specific receptor, and that there are a number of these specific receptor types. Each receptor has a nerve fiber connected to it, and the fiber travels to a specific portion of the CNS—that portion being, presumably, different for the various receptor types. The only information that the fiber carries to the CNS is that the specific fiber has fired, and at some intensity. The CNS, according to this view, acts only as a relay point.

More recently, as we also all know, another theory of signaling has come to compete with the neoMüllerian. This is the "patterning" theory. In brief, this theory states that all receptors within a modality are essentially alike in that they are responsive to all stimuli, although not equally so. If a specific stimulus is presented, all the receptors react, and information is carried to the CNS by the whole nerve, or by a large portion of it. The pattern of firing in the nerve, presumably as it passes a given point in time, is analyzed by the CNS, and identified as indicating the presence of a specific stimulus class. Because of the role of the CNS in this mechanism, a clearer term for it, I feel, is the "neural interpretation" theory or view, which would substitute for the phrase "interpretation by the CNS of the firing pattern of a nerve."

The neural interpretation view in gustation began with the electrophysiological investigations of Pfaffmann,[19] and, as a general sensory mechanism, has received support from investigators in other sensory systems. Chief among these have been Weddell and his colleagues in their work on the cutaneous system. I shall not review any of the electrophysiological work in gustation here, because that will be done ably in the following paper by Diamant and Zotterman. My primary concern, as indicated in the title, will be the gustatory qualities in man.

The first experiment I wish to discuss is by Békésy,[4] in which he used electrical stimulation. The electrode was gold, about 0.3 mm in diameter. The stimulating current was a train of rectangular pulses of 0.5 msec duration. This duration was short enough to prevent electrolytic decomposition of the saliva. The electrode was small enough to rest on a single fungiform papilla at the tip of the tongue. Békésy reported that the only tastes mentioned by his subjects were the classical four, or no taste at all. In addition, only one taste was reported from each papilla, except from some large papillae near the edge of the tongue. From these, two tastes, invariably salty and sour, were reported; one taste was evoked by stimulating one side of the papilla, and the other by stimulating the opposite side. Curves of absolute threshold for the different papillae types as a function of stimulus frequency were obtained. These are shown in Figure 1. It can be seen that the curves for salty and sour resemble one another, as do those for bitter and sweet, but the two pairs have different frequency minima. Békésy also reported in this paper that the gross shapes of the fungiform papillae responding to the different qualities appear to be distinguishable from one another, primarily within a given subject.

In a second series of experiments, Békésy[5] was able to stimulate single papillae with a droplet of various solutions representing the four basic qualities. He reported that the papillae were most sensitive when the droplet was placed on a side, rather than on the top. As with electrical stimulation, most papillae were found to be sensitive to only one of the stimulating solutions, i.e., to one quality. As before, some large papillae near the edge of the tongue were sensitive to two solutions, and a few to more.

The test solutions that were used in this experiment have been thought to be somewhat low. However, the test solutions used to screen subjects for an experiment by Meiselman and Dzendolet[15] were similar to the upper values used by Békésy, except that his quinine sulfate was higher by about a factor of ten, but his sucrose was lower by about the same factor.

There has also been some question concerning his location of the taste buds on the fungiform papillae. Most reviews have pointed out that the buds are found only on the dorsal surface. However, Hoffmann[14] and Henkin[12] have reported that buds do occur at the upper sides and the circumferential edges of these papillae. It is unfortunate that most microscopists have not used the fungiform papillae for their taste bud preparations, but have instead used the foliate papillae that generally have a much higher density of buds.

An attempt to confirm Békésy's chemical observations was carried out by Harper, Jay, and Erickson.[11] They used a flow-chamber arrangement small enough to cover a single large papilla from the dorsum, or edge, of the tongue. The chamber was firmly affixed to the required tongue area by a slight vacuum. Of the four subjects used, only two were able to identify the various stimuli presented. The remaining two subjects had 10 sensitive papillae between them. Of the 10, four responded to only one stimulus, whereas the other six responded to either two or four stimuli. The authors concluded that these results argued against Békésy's findings, and against the neo-Müllerian mechanism.

In my own opinion, I do not think that there is that much disagreement between the results of the two investigations. Harper et al. did find papillae with a single sensitivity, and Békésy did point out that large papillae near the edge of the tongue are sensitive to more than one stimulus class. The reason for the general decrease in sensitivity of the subjects in the Harper investigation might have been due to two possible sources. The first is a choice of subjects. Meiselman and Dzendolet[15] concluded, on the basis of sampling the gustatory discriminating power of 120 college students, that only 16 per cent of the males, and 34 per cent of the females could be considered to be discriminating tasters. In addition, 28 per cent of the males and 10 per cent of the females consistently confused sour and bitter. Two of the subjects used by Harper et al. could have belonged to the large insensitive group. Békésy, too, did not

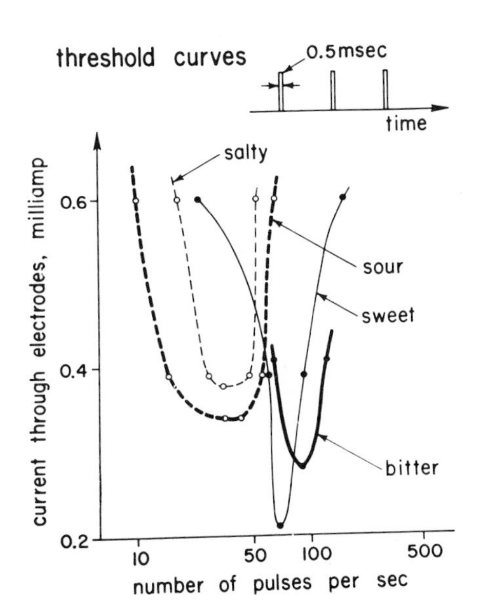

Figure 1 Absolute thresholds of single fungiform papillae to electrical stimulation by short rectangular pulses. Note that the salty and sour curves appear to form a family, separate from the bitter and sweet, which form their own. (From Békésy[4])

report any preliminary screening of subjects in his work. However, in a personal discussion, it was clear that a criterion of internal consistency of report was applied to the early subjects before they were allowed to continue and become the major ones in his experiments. The second possibility is that the slight vacuum of the stimulator decreased the blood supply to the papilla, and decreased its sensitivity in that manner. Such an effect is well known in the visual receptors.

The final experiment I wish to discuss is one from my own laboratory, and is a reinvestigation of the change in quality with concentration of some inorganic salts.[8] The four subjects had been screened, and were allowed only one of five possible responses to each stimulus presentation; these were the four qualities and that of "no taste." The results for the compound LiCl are shown in Figure 2. The ordinate is the mean percentage of times that a particular quality name was given at the indicated concentration. The sweet quality is pronounced at low concentrations; it is replaced by sour, and then by salty, as concentrations increase. Bitter appears rarely. The results are similar for KCl, except that the central sour response is replaced by one of bitter, as indicated in Figure 3.

These results of one quality being replaced by another are consistent within the neoMüllerian view if one assumption is made—that localization or local sign is a quality, as is hue or pitch. Such an assumption was implicit in the arguments used by Weddell, Taylor, and Williams[23] for their patterning theory of cutaneous localization, and Békésy[2] made the assumption explicit in some of his own work on the skin. This discussion is necessary because it seeks to apply the results of an experiment by Békésy[1] to the gustatory one under discussion. In brief, Békésy found that a vibra-

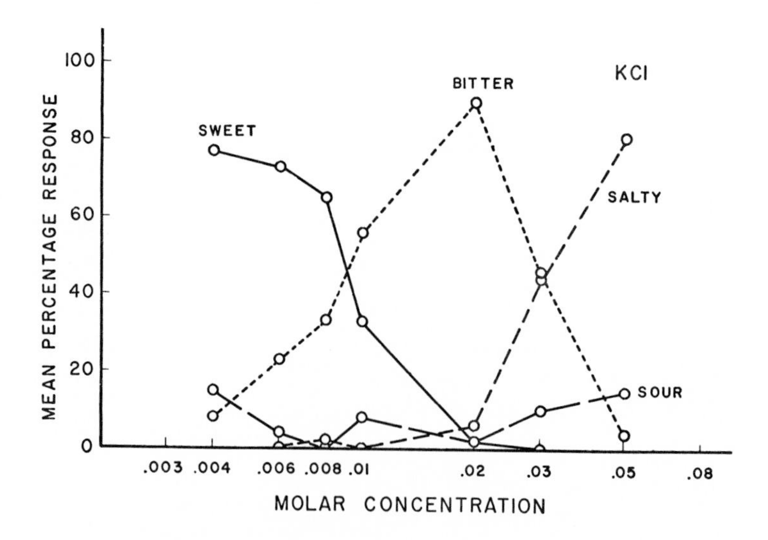

Figure 2 A quantitative indication of the change in quality with increase in concentration. Only three qualities are represented in this range with this salt (LiCl). (From Dzendolet and Meiselman[8])

tion at one point on the skin was inhibited by a vibration at a second point on the skin, when the absolute threshold of the second point was greater than that of the first. It is my conclusion that in this taste experiment a similar shift occurs from one gustatory quality to another, and for the same general reason of difference in thresholds, particularly for the two higher qualities.

By now, I hope I have shown that there is a body of evidence supporting the neoMüllerian view in gustation. It is equally obvious that there is a body of evidence supporting a neural interpretation mechanism. A controversy over these two mechanisms is not unique to gustation, so I would like to discuss some other senses briefly, and argue by analogy.

The situation exists, for example, with the skin senses. In this system, Hensel and Boman[13] have attempted to clarify the issue by pointing out that the idea of specificity has two aspects. The first of these is the neoMüllerian, in which the concern is over the particular sensation or quality response to a given stimulus. This can be investigated only with human subjects. The second concerns the electrophysiological response of a receptor to a stimulus, and here a receptor may be sensitive to different stimulus classes. To relate the two aspects, these investigators suggested that a particular receptor may be specific in the first sense, in that a stimulation may lead to only one quality response, e.g., a mechanical one, but be nonspecific in the second sense, in that the receptor may be responsive electrophysiologically to both mechanical and thermal stimulation. Hensel and Boman have suggested a possible method of evaluating which quality response may be associated with a receptor by using the concept of the "dynamic sensitivity" of the unit. This concept is defined as the change

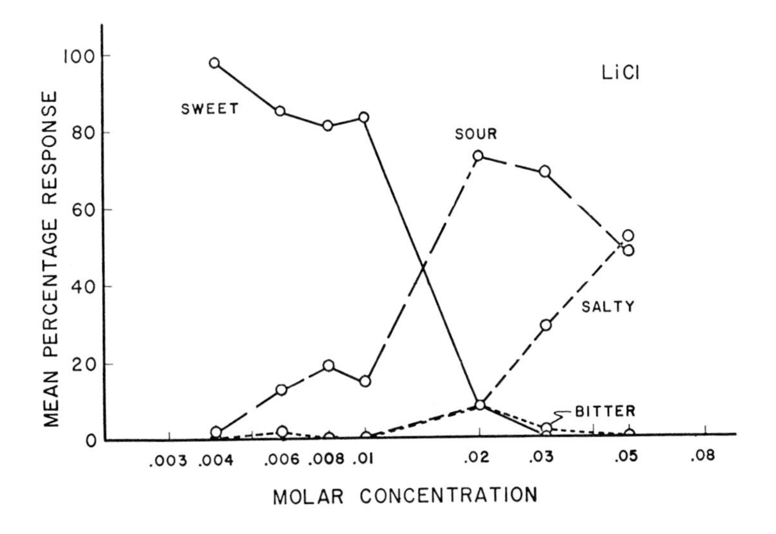

Figure 3 Similar to Figure 2, but for a different salt (KCl). The central function of this salt is bitter, and not sour as it was for the salt in the earlier figure. (From Dzendolet and Meiselman[8])

in neural impulse frequency as a function of change in stimulus magnitude. A unit with a high rate of change for a particular stimulus class would most probably signal that quality, even though it would also be responsive to other stimulus classes. This interpretation is not accepted, however, by all workers in the area.

There is another system in which the controversy does appear to have been settled. I refer to the mechanism for auditory pitch. Over the years, two mechanisms have been suggested as being active. One is the "telephone" theory, which is clearly a form of the neural interpretation mechanism. The other is "place" theory, which is equally clearly the neoMüllerian. I do not have to detail the work or the arguments both sides went through in support of their theories. I think all that is necessary is to report the current status of the controversy according to Békésy.[3] He concluded that, in humans, the telephone mechanism is active solely below 50 Hz. Between 50 and about 3,000 Hz, both mechanisms appear to play a role. Above 3,000 Hz, only the place mechanism signals pitch.

At this point I should like to go somewhat further afield, since I am dealing with generalities, and consider a principle enunciated by the physicist Niels Bohr. It is called the principle of complementarity, and was developed to help answer the problems raised by the apparent dualistic nature of light and of atomic particles. A simple version of this principle (reference 6, p. 92) states that "Evidence obtained under different conditions and rejecting comprehension in a single picture must, notwithstanding any apparent contrast, be regarded as complementary in the sense that together they exhaust all well-defined information about the atomic object." Bohr did not wish this principle restricted to atomic physics, so that we can ignore the reference to atomic object, and apply the principle to our current discussion. Part of the above statement may be paraphrased by saying that conflicting evidence obtained under differing experimental conditions should be considered as being equally valid in explaining the phenomenon in question. Such an application of complementarity has obviously been carried out in the example of pitch perception cited earlier. It is also inherent in the dichotomy concerning specificity proposed by Hensel and Boman.

My own position follows the lines just indicated. We are obviously dealing with separate experimental conditions, and it is inefficient use of our efforts to ask if one view is correct and the other is incorrect. There is enough evidence to show that they are both valid. It would be more fruitful to work towards an eventual synthesis of these views, however that may ultimately be brought about.

In passing, it is interesting to note that the Japanese physicists have had no trouble assimilating Bohr's idea of complementarity. Yukawa, for example, has been quoted as saying: "No, Bohr's argumentation has always appeared quite evident to us. You see, we in Japan have not been corrupted by Aristotle."[20]

As a final point, there is an amusing aspect, at least for me, in the controversy between the two views. A clear early statement of the neoMüllerian view was given as: "Nor will it be sufficient to consider each organ in its entire state, or taken as a whole, but we must look to the primitive textures that enter into its composition, each of these having a different way of modifying and transmitting the impressions it receives.

While there is one mode of sensation belonging to the entire organ, there are others belonging to its component parts . . ." An early statement of the neural interpretation view, first given with regard to the thermal sense, was as follows: "Whatever causes these vessels to contract or shrink, excites the sense of cold, and whatever distends or fills them produces the sensation of warmth . . ." The latter quotation is very close to the statement given much later by Nafe (reference 16, p. 1055). The amusement that I referred to earlier comes from the fact that the above two quotations come from a single journal article, entitled, "An inquiry into the varieties of sensation, resulting from a difference of texture in the sentient organ," which was written by an English surgeon named Park in 1817.[18]

REFERENCES

1. Békésy, G. von. 1959. Neural funneling along the skin and between the inner and outer hair cells of the cochlea. *J. Acoust. Soc. Amer.* **31**, pp. 1236–1249.
2. Békésy, G. von. 1962. Synchrony between nervous discharges and periodic stimuli in hearing and on the skin. *Ann. Otol. Rhinol. Laryngol.* **71**, pp. 678–692.
3. Békésy, G. von. 1963. Hearing theories and complex sounds. *J. Acoust. Soc. Amer.* **35**, pp. 588–601.
4. Békésy, G. von. 1964. Sweetness produced electrically on the tongue and its relation to taste theories. *J. Appl. Physiol.* **19**, pp. 1105–1113.
5. Békésy, G. von. 1966. Taste theories and the chemical stimulation of single papillae. *J. Appl. Physiol.* **21**, pp. 1–9.
6. Bohr, N. 1963. Essays 1958–1962 on Atomic Physics and Human Knowledge. Interscience Publishers, New York.
7. Dzendolet, E. 1968. A structure common to sweet-evoking compounds. *Percept. Psychophys.* **3**, pp. 65–68.
8. Dzendolet, E., and H. L. Meiselman. 1967. Gustatory quality changes as a function of solution concentration. *Percept. Psychophys.* **2**, pp. 29–33.
9. Fick, A. 1879. Die Lehre von der Lichtempfindung. *In* Handbuch der Physiologie III/1 (L. Hermann, editor). Vogel, Leipzig.
10. Frings, H. 1946. Gustatory thresholds for sucrose and electrolytes for the cockroach, *Periplaneta americana* (Linn.). *J. exp. Zool.* **102**, pp. 23–50.
11. Harper, H. W., J. R. Jay, and R. P. Erickson. 1966. Chemically evoked sensations from single human taste papillae. *Physiol. Behav.* **1**, pp. 319–325.
12. Henkin, R. I. 1967. On the mechanisms of the taste defect in familial dysautonomia. *In* Olfaction and Taste II (T. Hayashi, editor). Pergamon Press, Oxford, etc., pp. 331–335.
13. Hensel, H., and K. K. A. Boman. 1960. Afferent impulses in cutaneous sensory nerves in human subjects. *J. Neurophysiol.* **23**, pp. 564–578.
14. Hoffmann, A. 1875. Ueber die Verbreitung der Geschmacksknospen beim Menschen. *Arch. Pathol. Anat. Physiol. Klin. Med.* (*Virchows*). **62**, pp. 516–530.
15. Meiselman, H. L., and E. Dzendolet. 1967. Variability in gustatory quality identification. *Percept. Psychophys.* **2**, pp. 496–498.
16. Nafe, J. P. 1934. The pressure, pain and temperature senses. *In* A Handbook of General Experimental Psychology (C. Murchison, editor). Clark Univ. Press, Worcester, Mass., pp. 1037–1078.
17. Öhrwall, H. 1891. Untersuchungen über den Geschmackssinn. *Skand. Archiv. Physiol.* **2**, pp. 1–69.
18. Park, J. R. 1817. An inquiry into the varieties of sensation, resulting from a difference of texture in the sentient organ. *Quart. J. Sci. Arts* **2**, pp. 1–25.
19. Pfaffmann, C. 1941. Gustatory afferent impulses. *J. Cell. Comp. Physiol.* **17**, pp. 243–258.
20. Rosenfeld, L. 1963. Niels Bohr's contribution to epistemology. *Phys. Today* **16**, #10, pp. 47–54.
21. Shallenberger, R. S., and T. E. Acree. 1967. Molecular theory of sweet taste. *Nature* **216**, pp. 480–482.
22. Sternheim, C. E., and R. M. Boynton. 1966. Uniqueness of perceived hues investigated with a continuous judgmental technique. *J. Exp. Psychol.* **72**, pp. 770–776.
23. Weddell, G., D. A. Taylor, and C. M. Williams. 1955. Studies on the innervation of skin. III. The patterned arrangement of the spinal sensory nerves to the rabbit ear. *J. Anat.* **89**, pp. 317–342.

A COMPARATIVE STUDY ON THE NEURAL AND PSYCHOPHYSICAL RESPONSE TO TASTE STIMULI

HERMAN DIAMANT *and* YNGVE ZOTTERMAN

Department of Otorhinolaryngology, Umeå University, and the Department of Physiology, Veterinärhögskolan, Stockholm, Sweden

INTRODUCTION

Electrophysiological records from the human chorda tympani nerve can provide information about the relationship between taste stimulus and neural response. The choice of the electrophysiological recording technique is particularly important. Is the rectified, summated, and filtered transformation of the activity of the whole chorda tympani nerve related to the intensities and qualities of the taste sensations we experience? We do not know if the central nervous system "sees" the responses from the chorda tympani in the same way as does electronic apparatus. What we record might be just an epiphenomenon that parallels in time the true and unrecorded information-bearing responses of the peripheral taste nerve. However, if there is a good correspondence between the summated chorda tympani discharge and the psychophysical responses, we can infer not only that we have chosen a meaningful recording technique to "tap into the peripheral input live," but also that the central mechanisms do not seriously alter the input. If such a correspondence is lacking, we must conclude that an inappropriate recording technique was used and/or the sensory input was modified in the central nervous system.

PSYCHOPHYSICAL EXPERIMENTS

The human chorda tympani nerve was exposed for recording in the course of otological surgery to free the otosclerosed ossicle bones of the middle ear. This operation is performed under general anesthetic. Two days before the operation, psychophysical taste experiments with citric acid, NaCl, and sucrose were carried out on the patients. The method of magnitude estimation was used. This method, which was introduced by Stevens,[15] requires the subject to handle numbers and make quantitative estimations on the ratio of sensory intensities. In a trial experiment, all subjects were asked to make quantitative estimations of surfaces of different sizes so that we could screen out those who obviously could not estimate magnitude. The same stimuli and the same random order of presentation were used in the electrophysiological experiments. The stimuli were presented in pairs—the standard and one comparison stimulus.

The results of psychophysical experiments with a group of fourteen students are seen in Figure 1. A straight line may be fitted nicely to the psychophysical response to citric acid when plotted against molarity in log-log co-ordinates. A simple power function $R = cM^n$, with $n = 0.67$, describes the relation. Although the fit of a straight line to the salt values is not so good, the relation may also be described with a power function of the form $R = d + cM^n$, with rather high d value and $n = 1.0$. The result for salt is in accordance with previous results by Ekman, who introduced the additive constant d and found a power function of this form applicable to salt and sucrose,[9,10] but with n slightly above 1.0.

The results of the first successful experiments in which we obtained both relative psychophysical and neural responses to citric acid from the same patient are presented in a log-log diagram (patient C.L. in Figure 2). Straight lines may be adjusted to the values, i.e., a power function may describe the relations, although a Fechnerian log-function may give a better fit to the variation in neural activity.[4] If, as a first rough estimation, we describe both the neural and the psychophysical responses with power functions of the simple form $R = cM^n$, we find an astonishingly good agreement between them. The exponent of the psychophysical function $n_R = 0.5$ is the same as that of the neurophysical function $n_N = 0.5$.

In November 1965 only one (V.R.) out of three patients was able to perform the psychophysical tests (Figure 3). Unfortunately the chorda tympani response was very poor. We obtained, however, good response to citric acid and NaCl solutions from

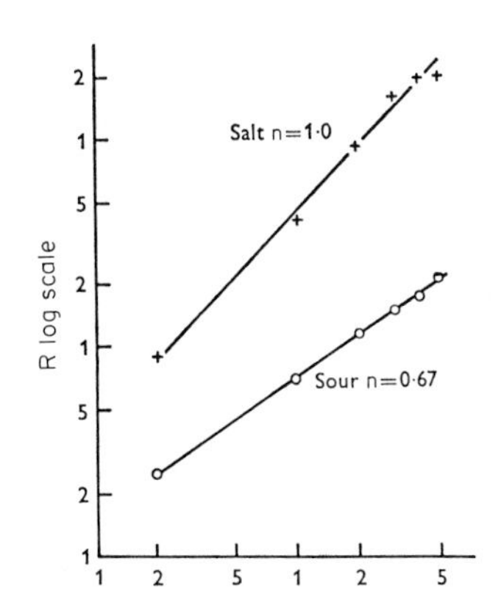

Figure 1 Result of a psychophysical experiment on fourteen students. The perceptual intensity plotted against molarity of citric acid in log-log scale. (Borg et al.[4])

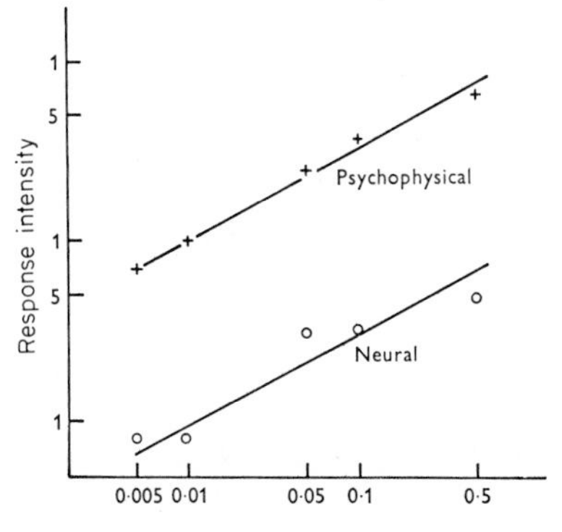

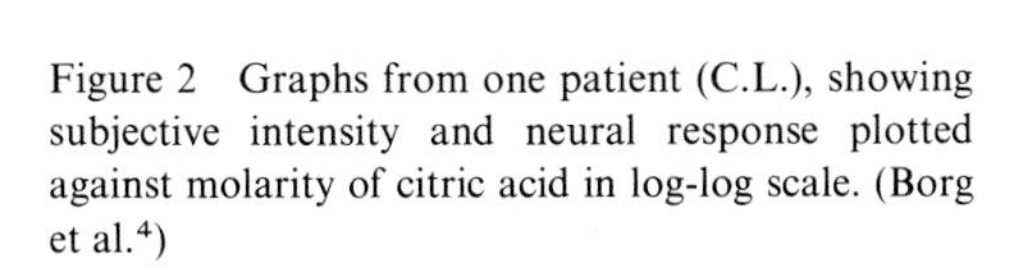

Figure 2 Graphs from one patient (C.L.), showing subjective intensity and neural response plotted against molarity of citric acid in log-log scale. (Borg et al.[4])

another patient (H.N. Figure 4), although she failed in the subjective tests. When plotted in a log-log diagram, her neural responses gave a good fit to straight lines of about the same slopes as those of the subjective data from the patient in Figure 3. It is apparent that in all these cases the slope of the salt line is definitely steeper than that of the acid line.

The next year we obtained magnitude estimations as well as electrical responses to sucrose and citric acid. In Figure 5, both the relative psychophysical and the electrical responses to citric acid are plotted in a log-log diagram (patient S.P.). The diagram in Figure 6 gives the relations for sucrose for patient I.J. Each point is the median of three observations in each series of tests on the same individual. How well the psychophysical functions follow the neural functions will be seen in Figure 7, where the diagram gives the functions of the mean values obtained for these two patients.

Quite aside from the question of whether the function describing the relation between the strength of the sapid solution and the subjective estimation satisfies a Stevens' power function or a Fechnerian log function, it is apparent from the diagrams presented above that there is a remarkably close correlation between the subjective and neural data. When describing the relation as a power function, it is clearly seen that for each of the three sapid substances, acid, sucrose and salt, the exponent n of a simple power function, $R = cM^n$, will obtain a different value in such a way that the exponent for citric acid always is lower than 1, while that for sucrose and NaCl is equal to or higher than 1.

Previously, Diamant et al.[7] demonstrated that, in spite of individual variations in the responses of the chorda tympani, there was a good correspondence between the psychophysical and neural data on the sweetness of a series of biological sugars (Table I). The values in each column were rounded off to the nearest 5 per cent and

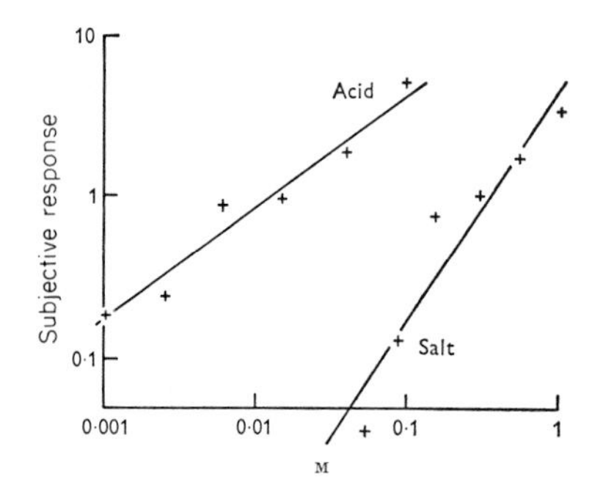

Figure 3 Log-log diagram of relation between the subjective response and the molarity of salt and citric acid (patient V.R.). (Borg et al.[5])

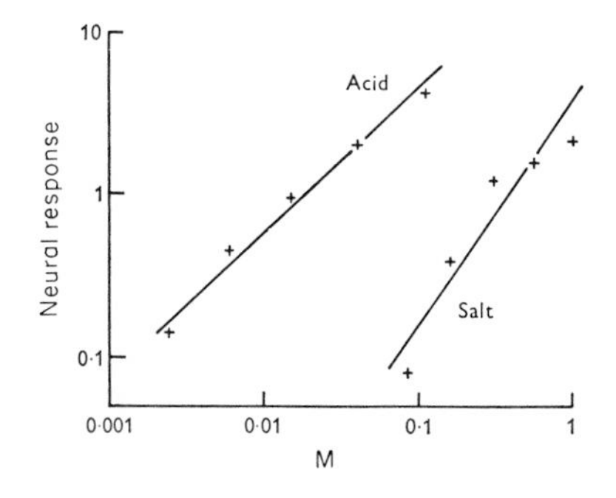

Figure 4 Nerve response to salt and citric acid plotted against molarity (patient H.N.). (Borg et al.[5])

are relative to the response to 0.5 M sucrose, which has been set at 100. The subjective reports are means of two determinations; the neural values are based upon a single determination. The correspondence seems good for the sugars. The artificial sweeteners saccharin and cyclamate have quite different tastes from the sugars and this may have affected the judgments of sweetness. The better agreement for patient 4 may depend upon the fact that this patient received the standard before each of the comparison stimuli, while the other patient received the standard only at the beginning and end of the two runs.

We have, however, another tentative explanation for the subjective overestimation of the sweeteners in comparison to the neural responses. Figure 8 shows records from a fine strand of the peripheral part of the chorda tympani of the rhesus monkey.[11] In this preparation, quinine elicits an electrical response that consists of small spikes; sweet-tasting substances like sucrose, saccharin, glycerol, and ethyleneglycol produce large spikes. If the records are carefully scrutinized it will be found that, in addition

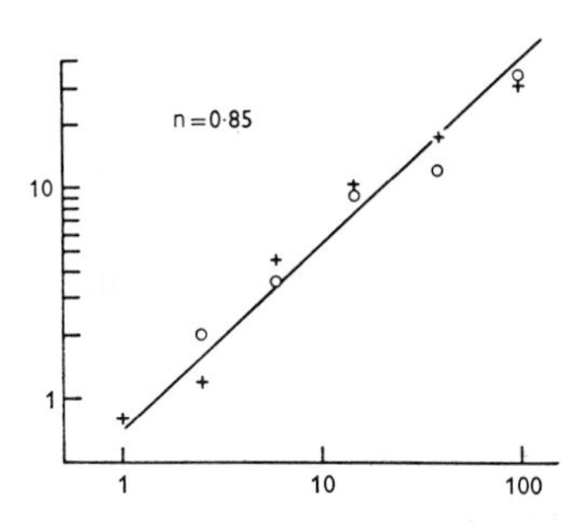

Figure 5 Nerve response to citric acid solutions (open circles) and psychophysical estimations (crosses) plotted against the molarity in log-log scale (patient S.P.). (Borg et al.[5])

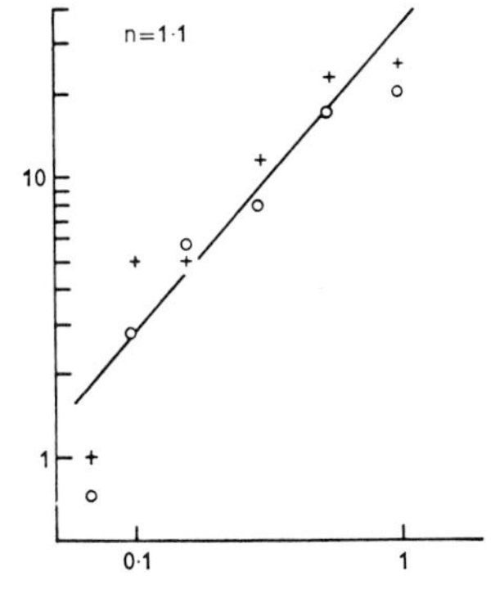

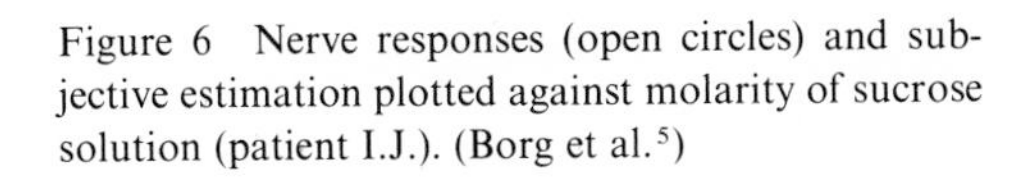
Figure 6 Nerve responses (open circles) and subjective estimation plotted against molarity of sucrose solution (patient I.J.). (Borg et al.[5])

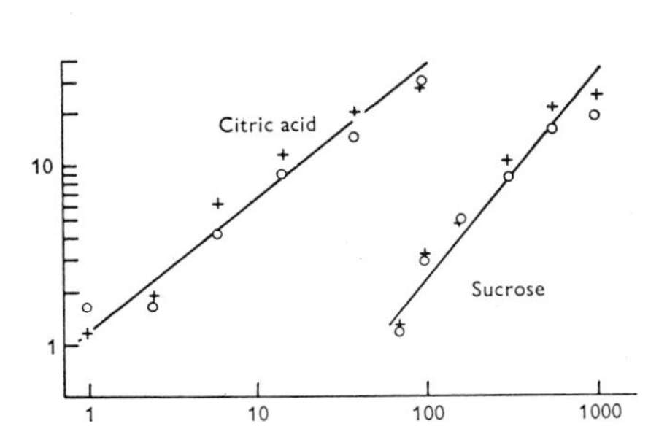

Figure 7 Mean values of neural response (open circles) and of subjective response from two patients (I.J. and S.P.) plotted against molarity of citric acid and sucrose solution. (Borg et al.[4])

TABLE I

COMPARISON OF PSYCHOPHYSICAL AND NEURAL RESPONSE TO SWEET-TASTING SUBSTANCES

The values in each column are relative to 0.5 M sucrose set at 100. The maximum height of the summator record was measured. (Diamant et al.[7])

Stimulus	Patient 3		Patient 4	
	Psy.	Neur.	Psy.	Neur.
0.5 M sucrose	100	100	100	100
0.5 M fructose	100	100	80	80
0.5 M maltose	—	40	75	60
0.5 M galactose	40	45	45	40
0.5 M lactose	45	45	30	30
0.5 M glucose	25	45	35	40
0.004 M Na saccharin	100	65	125	105
0.03 M Na cyclamate	55	80	115	100

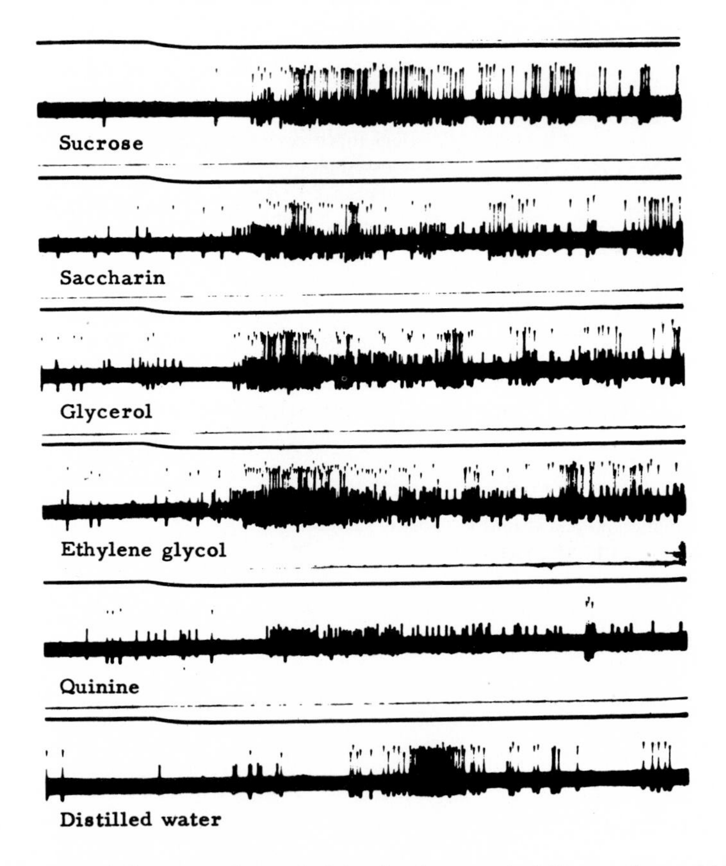

Figure 8 Records from a small strand of the chorda tympani of *Macaca rhesus* which contained few active fibers. Note particularly the large spikes, and also those of intermediate size which project both above and below the baseline. All solutions made up in Ringer's solution. Time: 10 per sec. (Gordon et al.[11])

to the large spikes, sucrose produces a few small spikes; saccharin, and, even more noticeably, glycerol and glycol, elicit a large number of small spikes. It is interesting to compare the animal records with taste sensations a human experiences. Of these substances, sucrose is considered to give a pure sweet taste; saccharin, glycerol, etc. give a somewhat bitter taste in addition to the sweetness. Assuming that man has the same kinds of neural mechanisms as the monkey, it is conceivable that the different sensations experienced may depend upon the signaling of two specific kinds of nerve fibers, one of which carries the information "sweet" and the other, smaller fibers, the information "bitter." On this assumption we may also explain a phenomenon described by von Békésy. A special technique of electrically stimulating a single taste papilla in the human tongue is used, and the patient is asked to give the quality of the gustatory sensation. When the report is sweet, the subject says that it is "heavenly sweet"—more pure sweet and pleasant than the taste of sugar. As Békésy applies weak electrical pulses he may stimulate the larger nerve fibers only, thus eliciting a sweet sensation of hitherto unknown quality—one more sweet than the taste of saccharose, which stimulates some "bitter fibers" as well. This assumption is strengthened by the observations reported by Andersson et al.[1] that quinine in the dog's tongue stimulates only small gustatory nerve fibers, and by Iriuchijima and Zotterman[12] that "bitter fibers" in the dog conducted at velocities of only a few meters per second. If we assume that this is also true in man, it is conceivable that the summated electrical responses to saccharin that contain a greater number of small spikes will attain relatively lower values than those of saccharose, which contains more large spikes.

It is important to keep in mind that the largest gustatory nerve fibers dominate the summated electrical response. So when we speak of a close relationship between the summated electrical response and the subjective response we mean the relation between the large fiber activity and the perceptual estimation. Assuming that the activity of the tiny fibers has as strong an influence as the large fibers on the perceptual response, we should predict a relatively stronger subjective response to saccharine. The estimation of the subjective strength of saccharine is most likely built up by a summation of information carried in large "sweet" fibers and small "bitter" fibers. Thus, the lack of correspondence between the neural and the perceptual responses to sweeteners (as seen in Table I) may really be due to the failure of our electronic recording system, and not to any misjudgments by the subjects.

ADAPTATION

The summated chorda tympani response to a 0.2 M NaCl solution is shown in Figure 9. The decline in neural activity in response to a continuous flow of salt solution over the tongue for three minutes may be seen in the records A, B, and C taken from three human subjects. The initial peak response to the application of the salt solution is indicated by an arrow, and the application of distilled water is indicated by a dot above the curve. The responses to water were due primarily to the low temperature of the water and also, to some degree, to mechanical stimulation of the

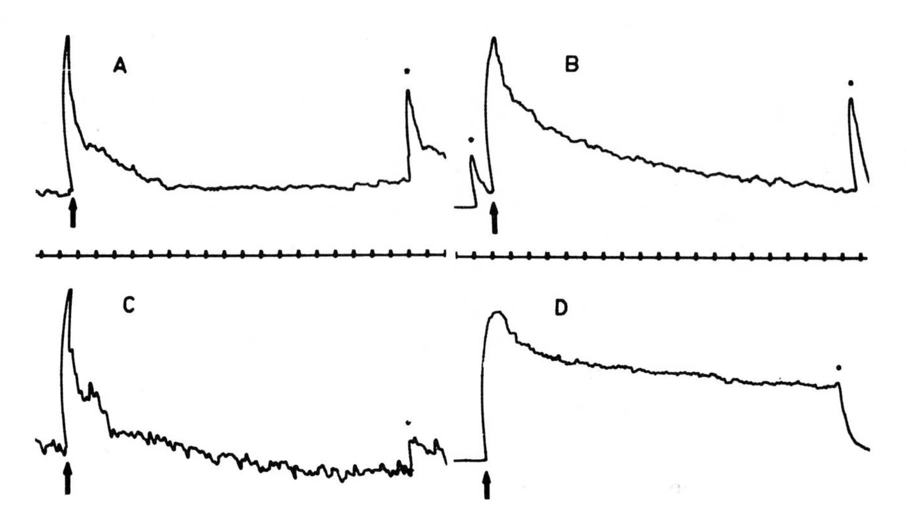

Figure 9 The summated chorda tympani response to a continuous 3 min flow of 0.2 M NaCl. A, B, C are human responses for patients no. 1, 2, 3, respectively, and D is a rat response. Dots indicate response during application of distilled water: arrows onset of salt. Tape-recorded data processed under identical conditions with rise and fall time constants of 1.5 sec. The tape recorder was off at beginning of B. Time base in 10 sec intervals. (Diamant et al.[7])

tongue. Record A is the most satisfactory from the technical point of view, for the signal-to-voice ratio was good and the baseline was relatively stable. The response was 95 per cent adaptation within 50 sec. In addition to the three records shown here, the adaptation in two more patients was very similar to that shown in A and C. Our records of human neural adaptation to 0.2 M NaCl contrast distinctly with record D, which is taken from rat chorda tympani. Here the response declines slowly over a 3 min period and, when the record is continued for some minutes more, little or no further decline in amplitude is seen.[2,16] The patient whose neural response is shown in record C indicated in the psychophysical test that the salt taste disappeared after 90 sec. Patient no. 4, whose record is not shown, indicated that he could no longer taste salt after 79 sec, which corresponded to a 95 per cent reduction in the magnitude of his neural response.

Bujas[6] studied psychophysical salt adaptation in two subjects. Using 0.15 M NaCl to stimulate a tongue area 1 cm in diameter, he found complete adaptation in 50 and 54 sec! These values are of the same order of magnitude as those given by our patients (79, 90, and 122 sec). Our records of neural adaptation to 0.2 M NaCl solution suggest that adaptation is complete, i.e., the activity decreases until it reaches the resting level of activity. In addition, there is a reasonable correspondence between the neural and psychophysical records for the time necessary for complete adaptation. Thus we may conclude that the human psychophysical observation of rapid and complete salt adaptation can be accounted for by diminished activity in the chorda tympani nerve. There is no need to postulate the existence of central adaptation mechanisms.

The close agreement between the subjective and neural function is not surprising, considering that our subjective estimation, carried out by the neural analyses in the central nervous system, must work on the information it receives from the peripheral receptors. Katz[13] found that there is a linear relation between the height of the receptor potential of the muscle spindle and the peak frequency of the nerve discharge, and Döving[8] found such a relation in the frog between the peak amplitude of the electro-olfactogram (Ottoson's EOG) and the discharge of impulses from secondary neurons of the olfactory bulb. Such a linear relation between receptor potential and impulse frequency in the labellar sugar receptor of the blowfly was recently reported by Morita and Yamashita.[14] Thus, the receptor potential evoked by the rapid solution is transformed into a volley of spikes propagated to the next neuron, where this volley sets up a postsynaptic potential that is transmitted to the next relay station, in the characteristic method of impulse-frequency modulation. Therefore, it should not be surprising that the summated electrical response which we obtained from the chorda tympani varied linearly with the amplitude of the postsynaptic potential evoked in the cerebral cortex.

REFERENCES

1. Andersson, B., S. Landgren, L. Olsson, and Y. Zotterman. 1950. The sweet taste fibres of the dog. *Acta Physiol. Scand.* **21**, pp. 105–119.
2. Beidler, L. 1953. Properties of chemoreceptors of tongue of rat. *J. Neurophysiol.* **16**, pp. 595–607.
3. Borg, G., H. Diamant, L. Ström, and Y. Zotterman. 1967. A comparative study of neural and psychophysical responses to gustatory stimuli. *In* Olfaction and Taste II (T. Hayashi, editor). Pergamon Press, Oxford, etc., pp. 253–264.
4. Borg, G., H. Diamant, L. Ström, and Y. Zotterman. 1967. Neural and psychophysical responses to gustatory stimuli. *In* The Skin Senses (Dan R. Kenshalo, editor). Charles C Thomas, Springfield, Ill., pp. 368–383.
5. Borg, G., H. Diamant, L. Ström, and Y. Zotterman. 1967. The relation between neural and perceptual intensity: a comparative study on the neural and psychophysical response to taste stimuli. *J. Physiol. (London)* **192**, pp. 13–20.
6. Bujas, Z. 1953. L'adaptation gustative et son méchanisme. *Acta Inst. Psychol. Univ. Zagreb* **17**, pp. 1–10.
7. Diamant, H., B. Oakley, L. Ström, C. Wells, and Y. Zotterman. 1965. A comparison of neural and psychophysical responses to taste stimuli in man. *Acta Physiol. Scand.* **64**, pp. 67–74.
8. Döving, K. 1964. Studies of the relation between the frog's electro-olfactogram (EOG) and single unit activity in the olfactory bulb. *Acta Physiol. Scand.* **60**, pp. 150–163.
9. Ekman, G. 1961. Methodological note on scales of gustatory intensity. *Scand J. Psychol.* **2**, pp. 185–190.
10. Ekman, G., and C. Åkesson. 1965. Saltness, sweetness and preference; a study of quantitative relations in individual subjects. *Scand. J. Psychol.* **6**, pp. 241–253.
11. Gordon, G., R. Kitchell, L. Ström, and Y. Zotterman. 1959. The response pattern of taste fibres in the chorda tympani of the monkey. *Acta Physiol. Scand.* **46**, pp. 119–132.
12. Iriuchijima, J., and Y. Zotterman. 1961. Conduction rates of afferent fibres to the anterior tongue of the dog. *Acta Physiol. Scand.* **51**, pp. 283–289.
13. Katz, G. 1950. Depolarization of sensory terminals and the initiation of impulses in the muscle spindle. *J. Physiol. (London)* **111**, pp. 261–282.
14. Morita, H., and S. Yamashita. 1966. Further studies on the receptor potential of chemoreceptors of the blowfly. *Mem. Fac. Sci. Kyushu Univ. Ser. E* **4**, pp. 83–93.
15. Stevens, S. S. 1957. On the psychophysical law. *Psychol. Rev.* **64**, pp. 153–181.
16. Zotterman, Y. 1956. Species differences in the water taste. *Acta Physiol. Scand.* **37**, pp. 60–70.

EFFECTS OF *GYMNEMA SYLVESTRE* AND *SYNSEPALUM DULCIFICUM* ON TASTE IN MAN

L. M. BARTOSHUK, G. P. DATEO, D. J. VANDENBELT,
R. L. BUTTRICK *and* L. LONG, JR.
Pioneering Research Laboratory, U.S. Army Natick Laboratories, Natick, Mass.

INTRODUCTION

Leaves from the plant *Gymnema sylvestre* have been known to the Western world for their taste-altering properties since they were introduced to a British officer, Captain Edgeworth, and his wife more than a century ago by the inhabitants of an Indian village. Upon chewing the leaves, the couple found that the sweetness of their tea vanished.[20] In 1887 Hooper verified the suppression of sweet produced by the leaves and in addition noted the suppression of bitter. After chewing the leaves, quinine sulphate tasted like chalk. However, Hooper found no effects on salts or acids. Hooper also investigated the leaves chemically. He attributed the suppression of sweet and bitter to "gymnemic acid" (HG) which was insoluble in water but soluble in alcohol. The HG was stated to be in the form of potassium salts (potassium gymnemate, KG) in the leaves, the KG being soluble in water and thus easily obtained in an aqueous decoction.[12] Shore[15] used both an aqueous decoction of the leaves and the sodium salt of gymnemic acid to examine the effects on glycerine, quinine, H_2SO_4, and NaCl. He first determined the minimal concentration necessary to produce a taste on specific areas of the tongue; then he applied the decoction for 20 sec, washed out the mouth, and tested with various concentrations of the taste stimuli. He concluded that the sweet taste of glycerine could be prevented entirely. The bitter taste of a 1 per cent (0.013 M) solution of quinine sulphate (which is normally extremely bitter) could also be prevented, but sour was not affected and salt was only slightly depressed. Kiesow[8] investigated the effects of gymnemic acid dissolved in alcohol on sucrose, NaCl, HCl, and quinine sulphate. He painted the solution on the tongue with a brush and observed the concentrations that could be tasted at various intervals after the application. Like Shore's results, he found that the sweetness of sucrose was completely abolished, the bitterness of quinine sulphate was markedly depressed (0.1 per cent or 0.0013 M, which is normally very bitter, was barely recognizable), the saltiness of NaCl was slightly depressed, and the sourness of HCl was essentially unaffected. Warren and Pfaffmann[19] examined the effects on sucrose more extensively, as well as observing the effects on the artificial sweetener, sodium saccharin. They observed the increases in the thresholds after exposure to different concentrations of both an aqueous extract of the leaves and purified KG, and also obtained sweetness matches between suprathreshold concentrations of sucrose and sodium saccharin. The sweetness of both substances decreased by the same amount after exposure to either the aqueous extract or KG. Diamant et al.[3] obtained electrophysiological responses in the human chorda tympani to NaCl, sucrose, citric acid, quinine hydrochloride (QHCl), and saccharin both before and after treatment with an aqueous extract of *Gymnema sylvestre*. The responses to sucrose and saccharin were completely abolished, confirming the early observations and the systematic study of Warren and Pfaffmann. However, the response to 0.002 M QHCl (normally a very strong bitter) was almost completely unaffected. This result does not confirm the early observations and generally cited conclusion that *Gymnema sylvestre* leaves strongly suppress suprathreshold bitter tastes.

The existence of a substance that appears to differentially suppress some taste qualities and not others is of considerable importance for theories of receptor mechanisms. Interest in *Gymnema sylvestre* has been enhanced by the efforts of several investigators to purify and determine the exact structure of the active components.[9,17–19] The purpose of part of the present study was to reexamine the specificity of the effects of the *Gymnema sylvestre* leaves, particularly with regard to the effect on bitter.

The studies conducted with the effects of *Gymnema sylvestre* leaves on sweet have so far dealt with substances that normally produce a sweet taste: sucrose, glycerine, and saccharin. However, sweet sensations can also be produced under certain circumstances by substances not normally associated with sweet. Salt solutions have been shown to produce sweet sensations at weak suprathreshold concentrations.[2,4,10,13,14] Distilled water can produce sweet sensations when preceded by adaptation to citric acid, HCl, or QHCl.[1,2] These sweet tastes resulting from NaCl or water are abolished after exposure to an aqueous decoction of *Gymnema sylvestre* leaves (L. M. Bartoshuk and D. H. McBurney, unpublished). The present study examines the effects of *Gymnema sylvestre* on an additional source of sweet tastes—those produced when tasting acids after the tongue has been exposed to miracle fruit.

The miracle fruit plant (*Synsepalum dulcificum*) is a shrub that is indigenous to tropical west Africa. It produces berries that turn red when ripe, are about the size and shape of Spanish peanuts, and consist of a thin layer of pulp over a large seed. In 1919 Fairchild sampled the berries in Africa, paying little attention to the "miraculous power" they were supposed to have; later he noticed that the beer he was drinking was too sweet. After also finding a lemon excessively sweet he collected miracle fruit seeds and thus introduced the plants into the United States.[11] Inglett et al.[7] observed the effects of miracle fruit on the tastes of several different fruits as well as several acids. They concluded that generally any sour material tasted for several hours after exposure would taste sweet and a potent berry could even replace all the sourness in a lemon slice with sweetness. In the present study, the effects of miracle fruit on NaCl, sucrose, QHCl, HCl, and citric acid were observed first. Then the effects of *Gymnema sylvestre* leaves on the taste alterations produced by the miracle fruit were observed.

METHOD

Five members of the Behavioral Sciences Division of Natick Laboratories served as subjects. They were not informed of the purpose of these studies but were experienced with direct-magnitude estimation procedures. Each S served in nine sessions, one for each of the four stimuli tested with *Gymnema sylvestre* and one for each of the five stimuli tested with miracle fruit.

Apparatus

All solutions were maintained in a water bath at 34°C and were delivered to the anterior third of the extended tongue through a gravity flow system. Distilled water and test solutions were delivered into the flow system through separate funnels but emptied into the same glass tubing near the tongue. The experimenter said "stimulus" when the stopcock at the junction of the funnels was turned to switch from the adapting solution to the test solution. The test solution forced the adapting solution that remained in the delivery tube onto the tongue. Approximately 2 sec after switching, the stimulus reached the tongue, so there were no cues for the exact onset of the test stimulus except a change from the adapting solution to the test stimulus.

Solutions

NaCl, QHCl, HCl, and citric acid solutions were made with distilled water and reagent grade chemicals. Sucrose solutions were made with distilled water and commercial grade sucrose. Five concentrations of each substance were selected to be equally spaced on a log molar scale and to vary in intensity from very weak to moderately strong. The extreme values were as follows: NaCl, 0.001 to 0.300 M; QHCl, 3.0×10^{-6} to 1.8×10^{-3} M; HCl and citric acid, 0.001 to 0.013 M; sucrose, 0.03 to 0.30 M.

Preparation of a Gymnemic Acid Fraction

Leaves of *Gymnema sylvestre* (1558 g obtained from the Himalaya Drug Company, Bombay, India) were powdered in a Wiley mill. The powder was extracted first with ethanol-water (2:1), then three times with 60 per cent ethanol. The combined extracts (24 liters) were concentrated in vacuum to a volume of 1.5 liters. The aqueous concentrate was extracted four times with an equal volume of hexane, then four times with ether. Chloroform-ethanol (2:1) extraction was performed four times with an equal volume of chloroform-ethanol. The combined chloroform-ethanol extract was treated with 10 per cent Na_2SO_4 and concentrated to dryness. The residual syrup was dissolved in absolute ethanol (400 ml) and precipitated by adding 3 to 4 volumes of ether. After it was dried in vacuum the material weighed 66.3 g. The major HG component appeared to be A_1 (Stöcklin et al.[18]) by chromatographic evidence. The HG preparation in the form of water-soluble salts contains at least 40 per cent inactive material (G. P. Dateo and L. M. Bartoshuk, unpublished data). The solution used in all tests consisted of 0.5 g of HG powder per liter of cooled, boiled distilled water. The solution was easily effected at room temperature. The above procedure produces the active material in a water soluble salt form essentially free of carbohydrate, fats, inorganic salts, and proteinaceous material, avoids thermal degradation, and removes a large portion of the inactive water soluble constituents.

Source and Maintenance of Miracle Fruit

The berries used in initial observations were obtained from Florida. The berries used in the following experiments were obtained from plants originally grown from seed in Florida and maintained in the Wellesley College greenhouse.* The berries were stored at −20°F for approximately three months before use.

Procedure

Each trial consisted of administering 40 sec of distilled water followed by the test solution, both flowing at the rate of 4 ml per sec. Subjects were instructed to report both the magnitude and quality of the test stimulus. A magnitude standard of 0.13 M NaCl after 40 sec of distilled water was provided at the beginning of each session and arbitrarily given a magnitude of 100. Subjects wrote magnitude estimates on response sheets under columns headed "tasteless," "salty," "sweet," "sour," or "bitter." An additional column was provided for tastes not described by one of the headings.

Experiment 1: Effects of *Gymnema sylvestre*. In the first half of the session (12 trials) each stimulus was presented twice in random order, as described above. In the second half, for each trial the subject held 8 ml of the HG fraction in his mouth for 30 sec; then the water rinse began and the trial continued as above.

Experiment 2: Effects of miracle fruit and *Gymnema sylvestre*. In addition to the instructions above, subjects were instructed to report all qualities present in the stimulus and break the total magnitude into appropriate amounts for the different qualities. The session was divided into three parts. In the first part (six trials) each stimulus was presented once in random order. Before the second part the subject chewed a berry for one minute then the six trials followed as above. In the third part each trial was preceded by the HG fraction.

RESULTS AND DISCUSSION

Experiment 1: Effects of Gymnema sylvestre

The results are shown in Figures 1 and 2 and Table I. The sweetness of sucrose was almost completely depressed by the HG fraction, but none of the other stimuli was significantly influenced. The QHCl results are particularly interesting in light of the

* We thank Professor T. Reichstein and Dr. W. Stöcklin for a sample of gymnemic acid A that contained 90 per cent A_1; Mrs. O. Churney for providing a sample of miracle fruit; and Dr. H. Creighton, Mr. W. J. Jennings, and Mr. A. E. Knickerson for providing greenhouse facilities at Wellesley College and for the maintenance of the miracle fruit plants.

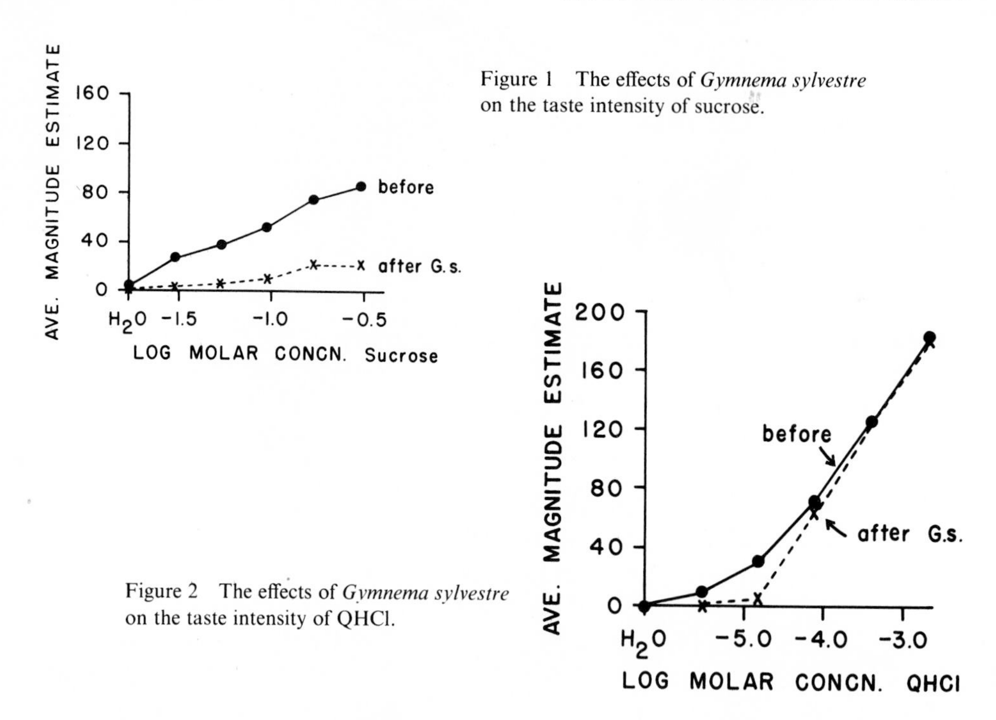

Figure 1 The effects of *Gymnema sylvestre* on the taste intensity of sucrose.

Figure 2 The effects of *Gymnema sylvestre* on the taste intensity of QHCl.

early reports of bitter suppression. The concentrations of quinine sulphate, which showed complete suppression in the work of Shore and Kiesow, are normally of equal or of greater bitterness than the high concentrations shown in Figure 2. The concentration of QHCl used by Diamant et al. that did not show suppression is slightly higher than the highest value in Figure 2.

An examination of the procedure in the different experiments suggests an explanation for the discrepancies. Hooper, Shore, and Kiesow all observed the effects of *Gymnema sylvestre* on quinine sulphate, but only Shore mentions using a rinse between the extract and the quinine and this rinse was apparently very brief. Extracts from *Gymnema sylvestre* taste very bitter in themselves and apparently cross-adapt with quinine sulphate. This cross-adaptation can be easily and dramatically observed by following Hooper's original procedure. After chewing the *Gymnema sylvestre* leaves, quinine sulphate powder is tasteless. However, if the mouth is rinsed thoroughly between chewing the leaves and tasting the powder, cross-adaptation is prevented and the powder will taste very bitter. The present study and the study by Diamant et al. used QHCl rather than quinine sulphate, but both quinines can apparently cross-adapt with *Gymnema sylvestre*. Without a rinse, strong QHCl applied after an aqueous decoction of *Gymnema sylvestre* is nearly tasteless. Diamant et al. applied other stimuli between the extract and quinine and effectively removed the extract, preventing cross-adaptation. In the present study, the 40 sec distilled water rinse removed the HG fraction and prevented cross-adaptation.

TABLE I
AVERAGE MAGNITUDE ESTIMATES OF TOTAL INTENSITY

Solution	Treatment	Concentration						p*
		H_2O	1	2	3	4	5	
NaCl	before G.s.	14	18	16	25	66	123	n.s.
	after G.s.	2	29	31	47	89	142	
QHCl	before G.s.	2	9	30	72	126	183	n.s.
	after G.s.	0	3	7	65	124	181	
HCl	before G.s.	6	36	55	63	102	145	n.s.
	after G.s.	2	49	80	91	149	174	
Sucrose	before G.s.	4	27	38	52	75	83	<0.01
	after G.s.	2	4	6	11	22	22	

* Friedman two-way analysis of variance (Siegel[16]).
Testing difference between "before *Gymnema sylvestre* (G.s.)" and "after G.s.".
n.s. = no suppression.

The slight suppression of saltiness in the work of Shore and Kiesow also may have resulted from cross-adaptation. Shore used HG in the form of salts. Although the taste of even purified HG or HG salts is predominantly bitter, the salt present would be adapting the sites responsive to these salts. Kiesow does not report the additional components of his HG solution, but he reports that the barely recognizable concentration of NaCl (0.6 per cent or 0.10 M) is very close to the value given by Shore (0.5 per cent or 0.08 M), and these concentrations were clearly not suppressed in the present study.

The relatively long term effects of *Gymnema sylvestre* are specific to the sweet quality in the studies so far conducted.

Experiment 2: Effects of miracle fruit and Gymnema sylvestre

The results are shown in Figure 3 and Tables II and III. Miracle fruit altered the sour taste of citric acid to part sweet and part sour; however, the total taste intensity did not change. The solid line in Figure 3 shows the intensity of the sour taste of the citric acid before exposure to miracle fruit. The broken line connected by open circles shows the intensity of the sweet and sour tastes after miracle fruit. The shading shows the relative contributions of sweet and sour to the total. When the HG fraction was applied, the sweet taste was abolished and the intensity of the sour taste returned to its original value. This intensity is shown by the dashed line connected by crosses in Figure 3. Two of the five subjects showed similar responses to HCl but three did not report a sweet taste to HCl. This could be because the berries were not of equivalent potency. An exact comparison of several acids would be of considerable value. Miracle fruit did not significantly affect NaCl, QHCl, or sucrose in this study.

Gymnema sylvestre appears to act peripherally[3] and to be specific to sweet-tasting substances. Its effectiveness in also suppressing the sweetness produced by acids after exposure to miracle fruit supports the suggestion that this sweetness is mediated by at least some of the same receptor sites that mediate normal sweetness. The return of the intensity of the sour of citric acid to its original value after *Gymnema sylvestre* is particularly noteworthy. *Gymnema sylvestre* does not usually enhance the sourness of acid. Therefore the increase in sourness after *Gymnema sylvestre* would seem to be specifically associated with the removal of sweetness. The decrement in sourness accompanying the addition of sweetness following miracle fruit is similar to decrements in the components of a mixture. Halpern[6] pointed out that the components in a mixture are usually judged less intense than if they are judged separately. In addition, he has recorded responses in the anterior tongue zone of the solitary nucleus in the medulla of the rat to mixtures of sucrose and acetic acid.[5] The responses to the mixture did not show algebraic summation, but rather were less than the sum of the components. The addition of sweetness, through miracle fruit, to the normal sourness of citric acid might then be expected to decrease the perceived intensity of sourness. Miracle fruit could be influencing the taste of acids primarily by causing the excitation of sites that usually mediate sweetness and not by causing any peripheral suppression of responses to acid. Electrophysiological investigations are planned to examine further the peripheral involvement of nerve fibers responsive to those substances that taste sweet or sour to man.

Studies of substances that have differential effects on taste qualities are clearly important for taste theories; they also have interesting implications for the controlled alteration of tastes in man. In addition, a cross-species approach may hopefully provide tools for the control of taste variables in animal research.

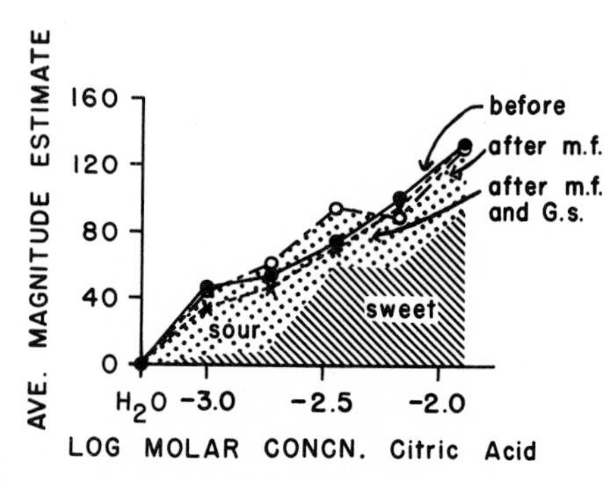

Figure 3 The effects of miracle fruit and *Gymnema sylvestre* on the taste intensity and quality of citric acid. The three lines show the total intensity of the taste of citric acid under the three conditions. The shading shows the relative contributions of sour and sweet to the total intensity of the taste of citric acid after miracle fruit. Nearly all of the taste intensity shown by the other two curves is contributed by sour.

TABLE II
AVERAGE MAGNITUDE ESTIMATES OF TOTAL INTENSITY

Solution	Treatment	Concentration						p*
		H_2O	1	2	3	4	5	
NaCl	before	2	6	11	25	61	115	
	after m.f.	2	4	12	18	60	146	n.s.
	after m.f. and G.s.	0	4	8	13	61	160	n.s.
QHCl	before	0	0	11	71	125	199	
	after m.f.	2	1	4	47	111	146	n.s.
	after m.f. and G.s.	0	10	6	40	118	178	n.s.
HCl	before	14	28	48	66	89	93	
	after m.f.	2	23	48	54	65	75	n.s.
	after m.f. and G.s.	0	18	34	51	70	100	n.s.
Citric acid	before	2	46	54	75	91	133	
	after m.f.	0	44	62	95	90	132	n.s.
	after m.f. and G.s.	0	33	46	71	95	135	n.s.
Sucrose	before	0	10	20	23	41	60	
	after m.f.	2	9	20	51	65	84	n.s.
	after m.f. and G.s.	0	2	4	5	19	12	<0.01

* Friedman two-way analysis of variance (Siegel[16]).
Testing difference between "before" and "after miracle fruit (m.f.)" and "before" and "after m.f. and G.s.".

TABLE III
AVERAGE MAGNITUDE ESTIMATES OF SWEET AND SOUR TASTES OF CITRIC ACID*

Taste	Treatment	Concentration						p**
		H_2O	1	2	3	4	5	
Sour	before	1	44	52	74	81	133	
	after m.f.	0	35	49	33	24	37	<0.05
	after m.f. and G.s.	0	19	46	71	93	135	n.s.
Sweet	before	0	0	1	1	0	0	
	after m.f.	0	7	12	61	62	95	<0.05
	after m.f. and G.s.	0	14	0	0	0	0	n.s.

* The total of sour plus sweet does not always equal the value for total intensity in Table II. The total intensity includes some small responses of other qualities.
** Friedman two-way analysis of variance (Siegel[16]).
Testing difference between "before" and "after m.f." and "before" and "after m.f. and G.s." for sour and sweet.

CONCLUSIONS

The results of this study suggest that *Gymnema sylvestre* suppresses the sweetness of sucrose but does not suppress the bitterness of QHCl, the saltiness of NaCl, or the sourness of HCl. These results confirm the electrophysiological results of Diamant et al.[3] and contradict the conclusion from the early literature that *Gymnema sylvestre* suppresses bitter. The earlier results are explained as resulting from cross-adaptation with the bitter taste of *Gymnema sylvestre* leaves and extracts. In addition, miracle fruit (*Synsepalum dulcificum*) was shown to leave the magnitude of the taste of citric acid unchanged but to change the quality from sour to sweet and sour. *Gymnema sylvestre* abolished this sweetness produced by citric acid and the sour intensity returned to normal. This decrement in sourness after tasting miracle fruit was compared to the decrement in sourness observed when sucrose is mixed with acid.

REFERENCES

1. Bartoshuk, L. M. 1968. Water taste in man. *Percept. Psychophys.* **3**, pp. 69–72.
2. Bartoshuk, L. M., D. H. McBurney, and C. Pfaffmann. 1964. Taste of sodium chloride solutions after adaptation to sodium chloride: implications for the "water taste." *Science* **143**, pp. 967–968.
3. Diamant, H., B. Oakley, L. Ström, C. Wells, and Y. Zotterman. 1965. A comparison of neural and psychophysical responses to taste stimuli in man. *Acta Physiol. Scand.* **64**, pp. 67–74.
4. Dzendolet, E., and H. L. Meiselman. 1967. Gustatory quality changes as a function of solution concentration. *Percept. Psychophys.* **2**, pp. 29–33.
5. Halpern, B. P. 1959. Gustatory responses in the medulla oblongata of the rat. *Diss. Abstr.* **20**, p. 2397.
6. Halpern, B. P. 1967. Some relationships between electrophysiology and behavior in taste. *In* The Chemical Senses and Nutrition (M. R. Kare and O. Maller, editors). The Johns Hopkins Press, Baltimore, Md., pp. 213–241.
7. Inglett, G. E., B. Dowling, J. J. Albrecht, and F. A. Hoglan. 1965. Taste-modifying properties of miracle fruit (*Synsepalum dulcificum*). *J. Agr. Food Chem.* **13**, pp. 284–287.
8. Kiesow, F. 1894. Über die Wirkung des Cocain und der Gymnemasäure auf die Schleimhaut der Zunge und des Mundraums. *Phil. Stud.* **9**, pp. 510–527.
9. Manni, P. E., and J. E. Sinsheimer. 1965. Constituents from *Gymnema sylvestre* leaves. *J. Pharm. Sci.* **54**, pp. 1541–1544.
10. McBurney, D. H. 1964. A psychophysical study of gustatory adaptation. No. 65–2223. University Microfilms, Ann Arbor, Mich.
11. Menninger, E. A. 1967. Fantastic Trees. The Viking Press, Inc., New York.
12. Mhaskar, K. S., and J. F. Caius. 1930. A study of Indian medicinal plants. II: *Gymnema sylvestre*, Brown. *Indian J. Med. Res.* **17**, pp. 1–49.
13. Renqvist, Y. 1919. Über den Geschmack. *Skand. Arch. Physiol.* **38**, pp. 97–201.
14. Richter, C. P., and A. MacLean. 1939. Salt taste threshold of humans. *Amer. J. Physiol.* **126**, pp. 1–6.
15. Shore. L. E. 1892. A contribution to our knowledge of taste sensations. *J. Physiol.* (*London*) **13**, pp. 191–217.
16. Siegel, S. 1956. Nonparametric Statistics for the Behavioral Sciences. McGraw-Hill Book Co., New York.
17. Stöcklin, W. 1967. Gymnemagenin, vermutliche Struktur. *Helv. Chim. Acta.* **50**, pp. 491–503.
18. Stöcklin, W., E. Weiss, and T. Reichstein. 1967. Gymnemsäure, das antisaccharine Prinzip von *Gymnema sylvestre* R. Br. *Helv. Chim. Acta* **50**, pp. 474–490.
19. Warren, R. M., and C. Pfaffmann. 1959. Suppression of sweet sensitivity by potassium gymnemate. *J. Appl. Physiol.* **14**, pp. 40–42.
20. Yackzan, K. S. 1966. Biological effects of *Gymnema sylvestre* fractions. *Ala. J. Med. Sci.* **3**, pp. 1–9.

MIRACULIN, THE SWEETNESS-INDUCING PRINCIPLE FROM MIRACLE FRUIT

G. J. HENNING, J. N. BROUWER, H. VAN DER WEL, *and* A. FRANCKE

Unilever Research Laboratory, Vlaardingen, The Netherlands

Although a number of flavor enhancers have been discovered in recent years, the number of naturally occurring substances that show a more specific action on one or more of the basic tastes is still relatively small. One of the oldest examples is gymnemic acid, occurring in the plant *Gymnema sylvestre*, which in man entirely suppresses the response to sweet substances without affecting the responses to salty, acid, or bitter materials. Electrophysiological investigations in human beings by Borg et al.[4] showed that gymnemic acid blocks the chorda tympani response to sugars and saccharin. The main structure of the aglycone of gymnemic acid has recently been elucidated by Stöcklin[10] and Stöcklin et al.[11] (see also Sinsheimer et al.[8]).

We have been carrying out research on another, perhaps more interesting example of a taste modifier, which is found in the berries of a shrub native to tropical West Africa. In the course of the years, this shrub has been called by various botanical names (*Bumelia dulcifica, Sideroxylon dulcificum, Bakeriella dulcifica, Synsepalum dulcificum*), but recently it has been shown that it belongs to the genus *Richardella* and, consequently, is now called *Richardella dulcifica* (Schum. and Thonn.) Baehni.[1] According to a more recent taxonomic survey, the name *Synsepalum dulcificum* is preferred (N. W. J. Borsboom and H. C. D. de Wit, personal communication). The berries of this plant are red, 1–2 cm long, olive-shaped, and consist of a relatively large seed surrounded by a thin layer of fruit pulp. This fruit pulp has the interesting property of changing the taste of sour foods and dilute mineral and organic acids into a sweet taste after it has been chewed for some time. Because of this taste-modifying effect, which generally lasts from 1 to 2 hr, the berries are known as "miraculous berries" or "miracle fruit."

The first adequate description of the plant and its peculiar properties was given in 1852 by Daniell, a British surgeon stationed in West Africa.[5] It was, however, not until 1965 that the first trials were made by Inglett, Dowling, Albrecht, and Hoglan[6] to isolate the active principle from the berries. Their attempts were unsuccessful; they found the active principle to be rather labile and insoluble in the ordinary aqueous and organic solvents. They did succeed, however, in obtaining a fivefold concentration of the active material by removal of inactive matter.

Our own efforts to extract the sweetness-inducing principle by homogenizing the fruit pulp in aqueous buffer solutions of varying pH and ionic strength confirmed the insolubility in these media. After centrifugation at 30,000 *g*, the physiological activity was entirely confined to the pellet; none of it was present in the supernatant. In analogy with the physiological conditions in the mouth, the berries were extracted with saliva at pH $\geq$ 7, which yielded a solution with strong sweetness-inducing activity, even after centrifugation at 100,000 *g*. Later it was found that the same result could be obtained by incorporating small amounts of extraction aids, e.g., polyethylene glycols, gelatin, or casein, into the aqueous extraction medium.

On further purification of the fairly stable and clear solutions thus obtained, notably by gel filtration, it appeared that the molecular weights of the active compounds in these extracts were of the order of $2 \cdot 10^5$–10^6, which makes characterization of structure less promising. But although these substances behaved as seemingly homogeneous materials, even on electrophoresis on starch gel and paper, we had reason to believe that they still represented bound forms of the active principle. By rather drastically changing the extraction conditions, viz., by extracting the berries with an aqueous solution of highly basic compounds like salmine (a protamine), one of us (J.N.B.) succeeded in obtaining a solution of the presumably free form of the active principle. The substance obtained in this way had a much lower molecular weight and behaved quite differently on electrophoresis, being much more basic in character. Further purification of these solutions, according to the scheme outlined in Table I, yielded a nearly pure preparation of the active principle, which we propose to call *miraculin*. Instead of salmine, the naturally occurring polyamine spermine (*N,N′-bis*-[3-aminopropyl]-1,4-diaminobutane) was used as an extraction aid. The purification achieved in the various steps was checked by electrophoresis on starch and polyacrylamide gels. After ammonium sulfate fractionation and gel filtration on Sephadex G-50, the active principle was still contaminated with material of a lower

TABLE I
PROCEDURE FOLLOWED IN ISOLATING MIRACULIN

Miracle Fruit (1 *kg*)

1. Deseeding
2. Homogenization of fruit pulp in aqueous solution of spermine
3. Centrifugation

Solution of Active Principle

1. Ammonium sulfate fractionation; fraction of 40–60% saturation retained
2. Gel filtration on Sephadex G-50
3. Gel filtration on Sephadex G-25

Miraculin (50–100 *mg*)

molecular weight, including some spermine. Further purification could be achieved by gel filtration on Sephadex G-25, as illustrated in Figure 1. As can be seen from the figure, the purification attained is quite satisfactory. In our purest preparations, with which structural investigations were carried out, a small amount (less than 5 per cent) of a closely related material was still present.

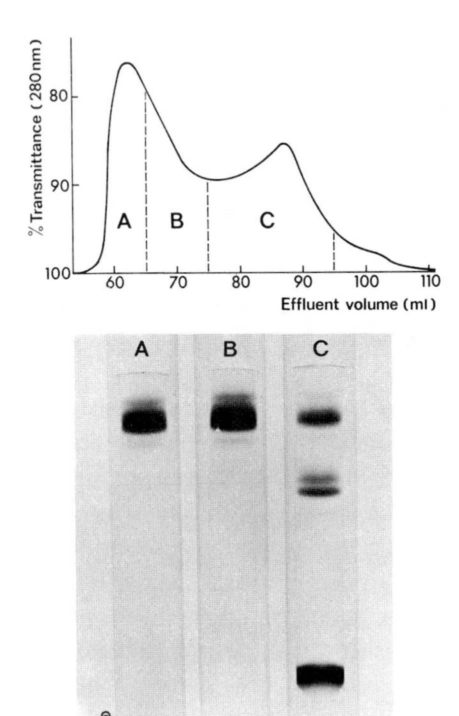

Figure 1 Upper part: Gel filtration of 35 mg of partly purified miraculin on a 2 × 42 cm column of Sephadex G-25 with water as eluent. Fractions were pooled as indicated by A, B, and C.
Lower part: Polyacrylamide gel electrophoresis of the pooled fractions on a 15% gel at pH 4.3, in the buffer system according to Reisfeld et al.[7]

Miraculin is a glycoprotein, as follows from the loss of activity on treatment with proteolytic enzymes, from the positive test for sugars before and after electrophoresis on polyacrylamide gel, and from the formation of amino acids and sugars on hydrolysis with acids. Thus we could substantiate the suggestion of Inglett et al.[6] regarding its structure. Depending on the method of determination used, the sugar content of miraculin relative to glucose varies from 7.5 to 21 per cent. On ultracentrifugal analysis, miraculin appears to be nearly homogeneous in sedimentation experiments; its molecular weight is 42,000 ± 3000. It is nearly homogeneous on starch-gel electrophoresis on either side of its isoelectric point, which is about 9.

Miraculin is soluble in dilute buffer solutions; it is thermolabile, stable at room temperature between pH 3 and 12, but rapidly inactivated below pH 2. It has no taste of its own, but 100 μg is sufficient to change the sensation of sourness into sweetness for 1–2 hr.

Some speculations on the mechanism of action of the active principle have been made by Beidler.[2] That it acts as an anesthetic does not seem to be very likely, either from its structure, or from the fact that anesthetics usually affect all basic tastes, although to different degrees.[9] Miraculin converts the acid taste into a sweet one, without, however, impairing the bitter, salty, or sweet response. There is at least one example of a similar taste conversion, viz., the sweet aftertaste we experience on rinsing the mouth with water after previously tasting a dilute solution of, for instance, citric acid. Although the phenomena may have different explanations, one is apt to conclude that the sour taste response, ionic in character, and the sweet taste response, predominantly nonionic, are somehow interconnected.

If we assume that the discrimination of taste qualities depends upon the recognition of stimulation patterns across many types of taste fibers or receptors, the simplest explanation for the action of miraculin would be that it influences the response of one or more of these fibers to acids in such a way as to change the pattern "sour" into "sweet." A decrease in response to acids seems more probable than an increase, because on tasting dilute solutions of acids (e.g., 0.002 M citric acid) after the application of small amounts of miraculin to the tongue, the acid is not perceptible, although sweetness is not yet tasted.

Can we explain this inhibitory effect in terms of the site theory, discussed by Beidler[3] at the second Olfaction and Taste Symposium? In the response to acids, both the protons and the accompanying anions are important. A decrease in the response to acids could be brought about either by an increase of the interaction of anions with the cationic membrane sites or a decrease in the interaction of the protons with the anionic sites. Although miraculin carries a positive charge and therefore could very well shield the anionic sites from their interaction with protons, the positive charge alone is not sufficient, since with other basic proteins or glycoproteins no similar taste conversion is observed. Miraculin, therefore, seems to have a much more specific action. Moreover, any influence of miraculin on the interaction with the ionic sites in the taste cell membrane probably would also reflect itself in an influence on the salty taste, which, up to the present, we have not observed.

We need more factual information regarding the interaction of this peculiar substance with the taste cells. As miraculin so profoundly changes our taste perception, it is to be expected that this information will also provide us with new clues regarding the mechanism of taste perception.

ACKNOWLEDGMENT

We gratefully acknowledge the help and stimulating interest of Dr. J. G. Collingwood, Prof. Dr. J. Boldingh, Mr. A. N. R. Grant, and Dr. J. H. van Roon.

REFERENCES

1. Baehni, C. 1965. Mémoire sur les Sapotacées III. Inventaire des genres. *Boissiera.* **11**, pp. 1–262.
2. Beidler, L. M. 1966. A physiological basis of taste sensation. *J. Food Sci.* **31**, pp. 275–281; cf. Anonymous. 1965. "Miraculous Fruit" makes sour foods taste sweet. *Sci. News Letter* **88**, p. 329.
3. Beidler, L. M. 1967. Anion influences on taste receptor response. *In* Olfaction and Taste II (T. Hayashi, editor). Pergamon Press, Oxford, etc., pp. 509–534.

4. Borg, G., H. Diamant, B. Oakley, L. Ström, and Y. Zotterman. 1967. A comparative study of neural and psychophysical responses to gustatory stimuli. *In* Olfaction and Taste II (T. Hayashi, editor). Pergamon Press, Oxford, etc., pp. 253–264.
5. Daniell, W. F. 1852. On the *Synsepalum dulcificum*, De Cand.; or, miraculous berry of Western Africa. *Pharm. J.* **11**, pp. 445–448.
6. Inglett, G. E., B. Dowling, J. J. Albrecht, and F. A. Hoglan. 1965. Taste-modifying properties of miracle fruit (*Synsepalum dulcificum*). *J. Agr. Food Chem.* **13**, pp. 284–287.
7. Reisfeld, R. A., U. J. Lewis, and D. E. Williams. 1962. Disk electrophoresis of basic proteins and peptides on polyacrylamide gels. *Nature* **195**, pp. 281–283.
8. Sinsheimer, J. E., G. Subba Rao, H. M. McIlhenny, R. V. Smith, H. F. Maassab, and K. W. Cochran. 1968. Isolation and antiviral activity of the gymnemic acids. *Experientia* **24**, pp. 302–303.
9. Skramlik, E. von. 1962/63. Die Beeinflussung der Substrate für die Geschmacksgrundempfindungen. *Z. Biol.* **113**, pp. 293–322.
10. Stöcklin, W. 1967. Gymnemagenin, vermutliche Struktur. *Helv. Chim. Acta* **50**, pp. 491–503.
11. Stöcklin, W., E. Weiss, and T. Reichstein. 1967. Gymnemasäure, das antisaccharine Prinzip von *Gymnema sylvestre* R.Br. Isolierungen und Identifizierungen. *Helv. Chim. Acta* **50**, pp. 474–490.

ISOLATION AND MECHANISM OF TASTE MODIFIERS; TASTE-MODIFYING PROTEIN AND GYMNEMIC ACIDS

KENZO KURIHARA, YOSHIE KURIHARA, *and* LLOYD M. BEIDLER
Department of Biological Science, The Florida State University, Tallahassee, Florida

INTRODUCTION

The leaves of *Gymnema sylvestre* suppress the sensitivity to sweet substances and miracle fruit gives sweet taste to sour substances. The studies on the mechanisms of these amazing actions on taste sense may furnish a key for elucidating the mechanism of taste stimulation. The present study was designed to elucidate the mechanism of the actions of the taste modifiers on the basis of molecular structure of the modifiers.

TASTE-MODIFYING PROTEIN FROM MIRACLE FRUIT

A shrub (*Synsepalum dulcificum*) native to Nigeria and Ghana yields a red berry the size of an olive. After it is chewed, acidic foods taste sweet. The natives chew the berry before eating their sour maize bread or drinking sour palm wine and beer. In this way the sour food tastes sweet. Daniell[5] published an account of the amazing properties of the red berry and described the properties of the shrub. The berry was named miracle fruit. It makes lemons taste much like oranges, tomatoes taste as if saccharin was sprinkled over them, and causes grapefruit to taste sweet without the addition of sugar. An attempt to isolate the active principle was made by Inglett et al.[9] but they never found a way to dissolve it.

In this paper we describe the isolation and characterization of the active principle from miracle fruit and suggest a mechanism by which the active principle may act to elicit a sweet sensation.

METHOD

Assay of the Activity of the Active Principle

Six or seven male and female subjects were subjected to the taste testing and every testing was repeated twice. The effort was made to give the subjects no preconception of a result either through the experimenter or through communication between subjects.

Five milliliters of the solution containing the active principle, which was enough volume to cover the tongue of the subjects, was held in the mouth for 5 minutes and then spit out. The mouth was rinsed with distilled water and the sweetness to 0.02 M citric acid was compared with that of a series of sucrose solution. The subjects were asked to choose the one of the 10 sucrose solutions (0.1–1.0 M) that best approximated the intensity of sweetness induced by the acid.

RESULTS AND DISCUSSION

Isolation and Characterization of the Active Principle[10]

Plants were cultivated in the greenhouse at Florida State University for two to three years until berry production was prolific (Figure 1). Berries were harvested twice a

Figure 1 Miracle fruit grown in a greenhouse at The Florida State University. Berries are about the size of small grapes.

week during the height of the season. Since the active principle is labile, the berries were stored in a deep freezer (−70°C) until needed.

The isolation procedures are summarized in Figure 2. The skin and seed of the berries were removed by hand. The pulp was homogenized in a Waring blendor with water and the homogenate was centrifuged. The active principle did not appear in the supernatant. The insoluble slurry was extracted with carbonate buffer (pH 10.5). The extracts with carbonate buffer were applied to the diethylaminoethyl (DEAE)-Sephadex column and the column was eluted with 0.1 M carbonate buffer. The effluent

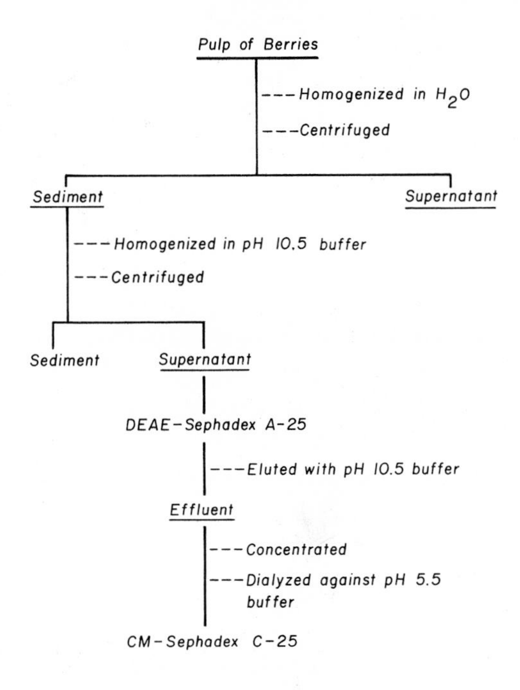

Figure 2 Procedure by which the active principle of miracle fruit is isolated.

was colorless and had a strong activity. The effluent, concentrated with Aquacide, was used to obtain preliminary information on the properties of the active principle.

Table I shows the stability of the active principle under various conditions. Activity was destroyed when the solution was boiled or exposed to a high concentration of organic solvents such as ethanol or acetone at room temperature. Activity was greatly decreased by exposure to pH above 12.0 or below 2.6 at room temperature, whereas activity was stable at pH 3.7 and 4°C for at least 1 month. Addition of trypsin or pronase destroyed the activity. When the active solution was dialyzed against distilled water for 48 hours, no activity was observed in the dialyzate. These characteristics suggest that the active principle is protein.

The molecular weight of the active principle was estimated with Sephadex G-75

TABLE I
STABILITY OF ACTIVE PRINCIPLE
Stable at pH 3.7 and 4°C for at least one month

	Relative Sweetness
pH 1.2*	0
pH 2.6*	3
pH 3.5*	4
pH 7.0*	4
pH 10.5*	4
pH 11.0*	2
pH 12.0*	0
Boiled	0
Acetone Treatment*	0
Trypsin Treatment	1
Pronase Treatment	0

* Stability at 25°C for one hour.

column. As shown in Figure 3, the elution pattern of the active principle was typical of that of a single homogeneous polymer. Elution volumes of proteins with known molecular weights were determined on the same column, and the observed elution volumes were plotted against the logarithm of the molecular weights according to the method of Andrews[1] (Figure 4). The molecular weight of the active principle was estimated as 44,000.

Further purification of the active principle was carried out on a cation-exchange column. The concentrated effluent from the DEAE-Sephadex column was dialyzed

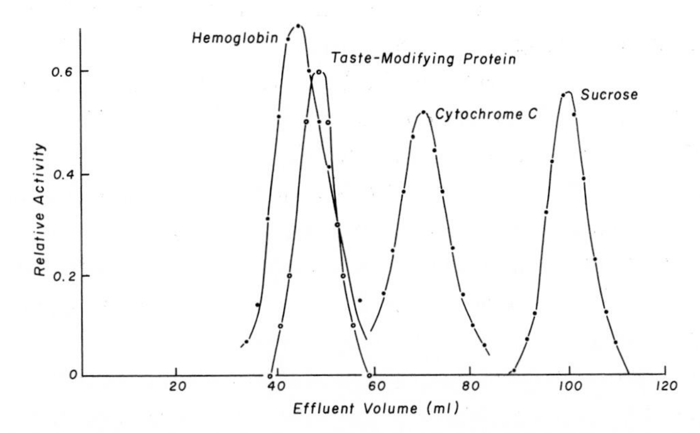

Figure 3 An elution pattern of the active principle on a Sephadex G-75 column. One ml of a solution of the active principle (taste-modifying protein) and 0.5 per cent of protein solution were applied to a column (1.5 × 61.5 cm) of Sephadex G-75 (0.1 M phosphate buffer, pH 7.0: flow rate 20 ml/hour). Activity was measured with 1 ml of each fraction after dilution with water to 5 ml.

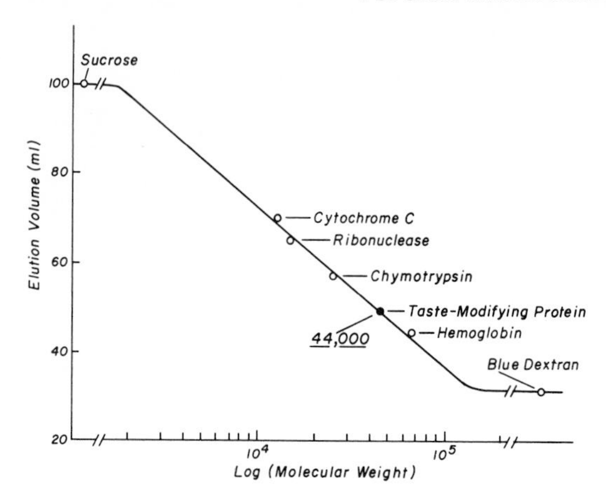

Figure 4 Plots of elution volumes against log (molecular weight) for proteins on the Sephadex G-75 column. The molecular weight of the active principle (taste-modifying protein) was estimated as 44,000.

against 0.03 M acetate buffer and applied to a column of carboxymethyl (CM)-Sephadex C-25. The active principle was adsorbed on the column and eluted with a gradient of phosphate buffer. The active fraction from the column had characteristics typical of proteins; its reaction to biuret reagent was positive, and it had absorption maximum at 278 mμ. The elution pattern of the active protein, measured by ultraviolet absorption, correlates well with the activity curve (Figure 5). This purified protein seems to be an aggregate of the protein monomer, because application of the protein to the Sephadex G-75 column indicated that the molecular weight of the purified protein, measured by activity and ultraviolet absorption, was more than 80,000. Aggregation proceeded during the dialyzing process against acetate buffer of low ionic strength, although this process did not change the activity of the protein.

Purity of the isolated protein was checked with disc electrophoresis on polyacrylamide gel. The protein migrated to the cathode at pH 4.5 (Figure 6) but did not migrate at pH 8.3. This electrophoretic behavior was in agreement with that expected from the result obtained with the cation-exchange column, and indicated that the protein is basic. A single sharp band accompanied a light-diffused band. These bands changed into a single band (Figure 6) when electrophoresis was carried out in 8 M urea. This indicated that the isolated protein does not contain any other protein than the active component.

Sugar components comprise 6.7 per cent of the purified protein. The sugar content of each fraction from a CM-Sephadez column was determined by the tryptophan-sulfuric acid reaction[2] (Figure 5C). The possibility that the sugars are an impurity was eliminated by the facts that the curve of sugar content correlated well with the chromatographic pattern of the active protein (Figure 5) and that the sugar content was not changed by dialysis or rechromatography of the fractions on a CM-Sephadex

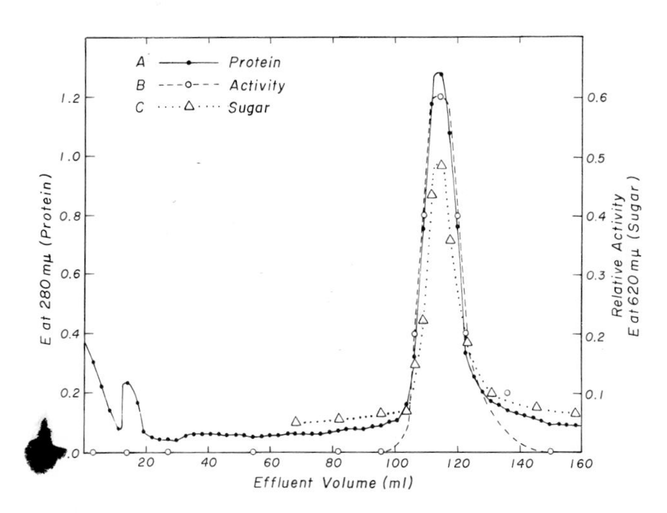

Figure 5 Chromatography of taste-modifying protein on CM-Sephadex C-25. A sample solution (100 ml) dialyzed against 0.03 M acetate buffer (pH 5.5) was applied to a column (1.1 × 42 cm) of CM-Sephadex C-25 (100 to 270 mesh) buffered with 0.03 M acetate buffer, pH 5.5. The column was eluted with a linear gradient between 0.03 M phosphate, pH 6.0 (80 ml), and 0.02 M phosphate, pH 8.8 (80 ml). Flow rate 20 ml/hour. The volume of each fraction was 2.7 ml. Protein was detected by absorption at 280 mμ. Sugar content was determined at 620 mμ after 0.25 ml of each fraction was colored by reaction with the tryptophan-sulfuric acid. Activity of the protein was assayed with 0.1 ml of each fraction after dilution with water to 5 ml. E = extinction coefficient

Figure 6 Polyacrylamide disc electrophoresis of taste-modifying protein. Left: without urea, Right: in 8 M urea. A concentrated active fraction from CM-Sephadex column was applied to gels. Electrophoresis was conducted at 4°C for 3 mA for 2 hours at pH 4.5. The gels were stained with 1 per cent amido black in 7.5 per cent acetic acid solution.

column. Paper chromatography of the protein hydrolyzate was carried out by the method of Masamune and Yoshizawa[13] to identify the sugar component. The sugars were detected by the aniline hydrogen phthalate-spray method.[16] The red spots were identified as L-arabinose and D-xylose, respectively, by comparison with known spots of sugars (Figure 7). An amino acid analysis of the protein was carried out after the lyophilized protein was hydrolyzed in 6 N HCl, and the result is shown in Table II. The above results indicated that the active principle of miracle fruit is a glycoprotein. We call the protein "taste-modifying protein".

The Mechanism of the Action of Taste-Modifying Protein[11]

The purified taste-modifying protein itself has no inherent taste. A mixture of the protein solution with sour substances initially tasted sour, even after the protein was contacted with acid for certain periods of time, and slowly changed to sweet if the mixture was held in the mouth. This suggested that the modification of the taste function by the protein proceeds slowly. In order to know how fast the taste modification by the protein proceeds, the protein solution in 0.01 M $NaHCO_3$ was held in the mouth for 1/3, 2/3, 1, 2, 3, 4, and 5 min, respectively. The solution in the mouth was spit out and the sweetness of 0.02 M citric acid was compared with that of a series of sucrose solutions. As shown in Figure 8, the sweetening effect reached the maximum level after being held in the mouth for about 3 min.

Figure 9 shows the sweetening effect with different concentrations of the protein solution. The effect reached a maximum level at 4×10^{-7} M of the protein solution. The maximum sweetness to the acid solution was equivalent to the sweetness of about

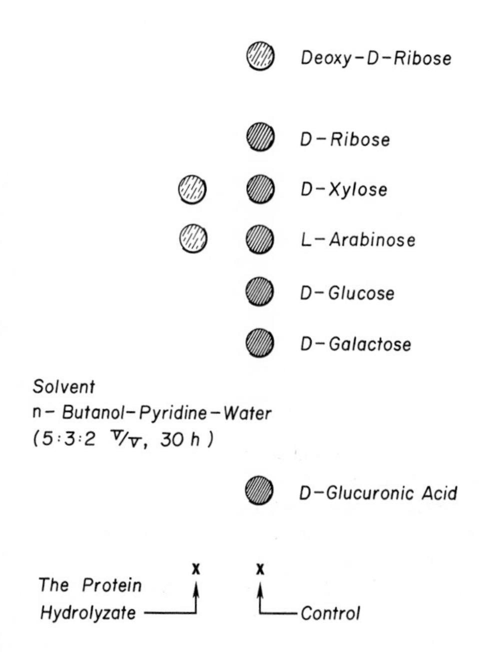

Figure 7 Paper chromatography of sugar components in taste-modifying protein. Ten milligrams of the lyophilized taste-modifying protein was hydrolyzed in 2 ml of 2 N HCl at 100°C for 5 hours. The solution was then brought to pH 5.8 by addition of saturated $Ba(OH)_2$, and the $BaSO_4$ was removed by centrifugation. The supernatant was evaporated to dryness and the sugars were redissolved in a small volume of pyridine. This sugar solution was applied to Whatman No. 1 filter paper along with the proper standard mixtures of monosaccharides. Ascending chromatography with multiple development was carried out in n-butanol-pyridine-water (5:3:2, v/v) for 30 hours. The sugars were detected by the aniline hydrogen phthalate-spray method.

TABLE II

AMINO ACID COMPOSITION OF TASTE-MODIFYING PROTEIN

The data are given as residues per 100 total residues. Three mg of the lyophilized taste-modifying protein were hydrolyzed in 6 N HCl at 100°C for 22 hours. The amino acid analysis was performed on the Beckman Model 120 amino acid analyzer.

	Amino Acid Residue/100 Total Residues		Amino Acid Residue/ Total Residues
Lysine	7.9	Half Cystine	2.3
Histidine	1.8	Valine	8.0
(Ammonia)	17.4	Methionine	1.0
Arginine	4.7	Isoleucine	4.7
Aspartic Acid	11.3	Leucine	6.5
Threonine	6.1	Tryptophan	—
Serine	6.1	Tyrosine	3.6
Glutamic Acid	9.2	Phenylalanine	5.0
Proline	6.0		
Glycine	9.8		
Alanine	6.3	Arabinose + Xylose	5.4

0.4 M sucrose solution, and the sweetening effect did not increase, even when a higher concentration of the protein solution was applied to the mouth. We could not perform quantitative studies on the sweetening effect at concentrations higher than 0.02 M citric acid, because at high concentrations the acid was painful to the mouth. Preliminary studies, however, indicate that an increase of the acid concentration above 0.02 M did not increase sweetness. Sour taste, however, became noticeable again.

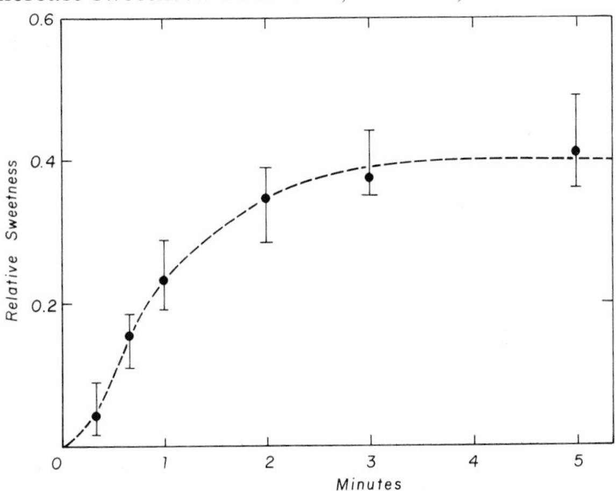

Figure 8 The time course of the taste modification. The sweetening effect was assayed after 5 ml of 7.0×10^{-7} M taste-modifying protein in 0.01 M $NaHCO_3$ was held in the mouth for certain periods of time. Abscissa is the period during which the protein solution was held in the mouth. Each point in the Figure is the average value of the data obtained from seven subjects and the bars indicate the range of average variation.

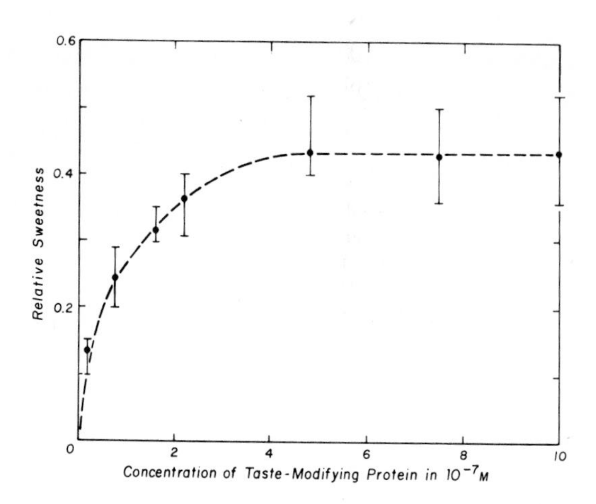

Figure 9 The sweetening effect as a function of the concentration of the taste-modifying protein. The sweetening effect to 0.02 M citric acid was assayed after the different concentrations of the protein were held in the mouth for 5 minutes.

The sweetening effect induced from one application of the protein persists for a certain period of time. The persistency of the effect was measured at 10- or 20-min intervals after the protein solution was applied to the mouth. The effect from application of 2.0×10^{-8} M protein solution disappeared in 20 min, while the effect from a solution of 2.3×10^{-6} M persisted for over 3 hours (Figure 10). Provoking secretion of saliva by chewing gum or eating foods after application of the protein caused the sweetening effect to decline more rapidly.

The sweetness, which resulted after application of the protein to the mouth, was induced by a sour substance tested. Sourness is mainly related to the proton concentration, but different acid solutions at the same pH give different intensities of sour-

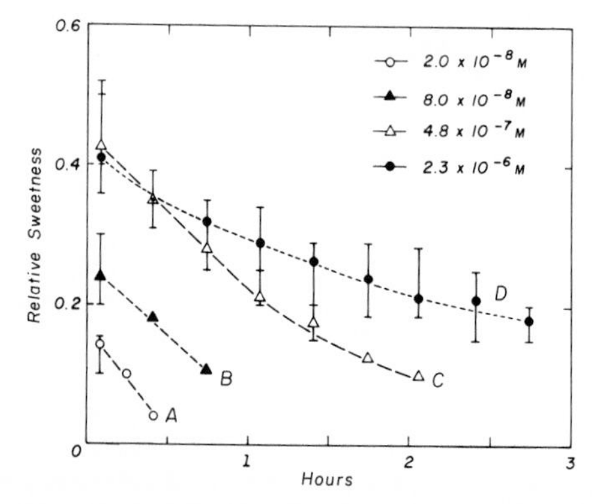

Figure 10 The time dependency of persistency of the sweetening effect. The effect to 0.02 M citric acid was assayed at 10- or 20-minute intervals after the taste-modifying protein was held in the mouth for 5 minutes. Abscissa is the period after the protein was applied to the mouth.

ness.[4] The sweet-inducing potency of acids was compared with the intensity of their sourness. Figure 11 shows the intensity of sourness for acids at different pHs, compared to the sourness of a series of citric-acid solutions. The figure indicated that the sourness of the acid solutions at the same pH decreases in order of acetic, formic, lactic, oxalic, and hydrochloric acids. This order of sourness corresponds to that obtained by electrophysiological studies with rats.[4] Beidler explained the result by the theory that neutralization of the positive charge on the taste receptor membrane by simultaneous anion absorption allows for easier proton binding to the receptor membrane. Figure 12 shows the sweet-inducing potency of the acids. The sweetness

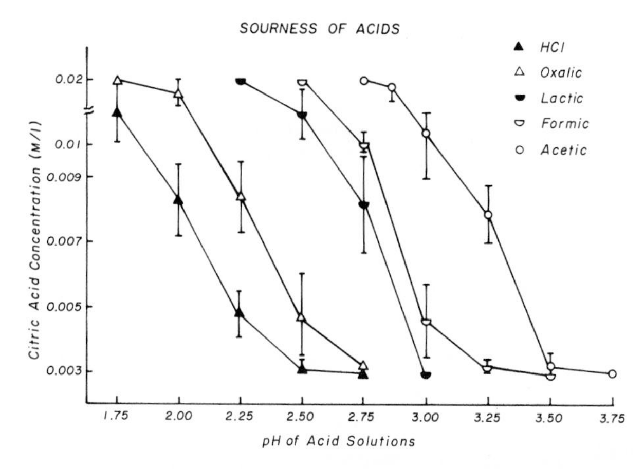

Figure 11 Sourness of acids. The sourness of acids at the different pH was compared with that of citric acid. The subject was asked to choose one of 0.003, 0.005, 0.008, 0.009, 0.01 and 0.02 M citric acid solutions which best approximated the intensity of sourness of the given acid solution.

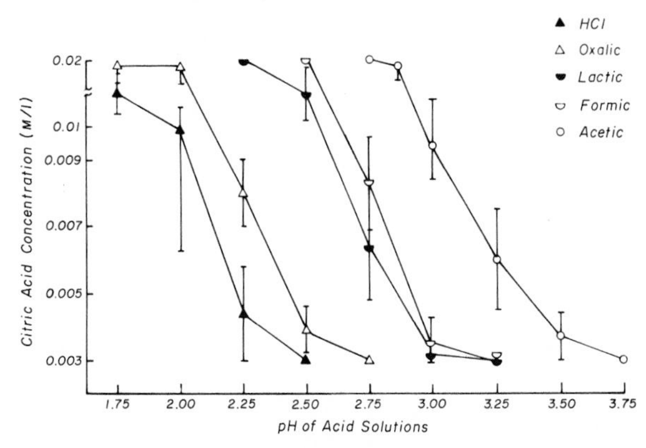

Figure 12 Sweetening effect of acids. The sweetness to acids at the different pH was compared with the sweetness to citric acid after the subject held 2.0×10^{-6} M protein solution in the mouth for 5 minutes. The subject was asked to choose, out of a series of citric acid solution, one which best approximated the intensity of sweetness of the given acid solution.

of the acids at different pHs, after application of the taste-modifying protein, was compared to a series of citric acid solutions. Each of the sweetness curves is very similar to the sourness curves in Figure 11. This result suggests that the mechanism of sweet induction by acid is closely associated with the mechanism of sourness. This close relation between the intensity of sourness of an acid and its sweet-inducing potency will be explained in the following two ways. One explanation is that the taste-modifying protein changes the coding of responses from taste-receptor cells and sour taste is converted into sweet taste. Another explanation is as follows: the taste-receptor membrane changes conformation when sour taste is induced at low pH. This conformational change of the membrane modified by the taste-modifying protein allows the induction of a sweet taste at the sweet receptor site. The former explanation is not likely, on the basis of the following experiments.

How does the taste-modifying protein affect the taste thresholds? The thresholds for salty, bitter, and sweet tastes were determined before and after the protein was applied to the mouth according to the method of Harris and Kalmus[7] as modified by Fischer et al.[6] As seen from Table III, no change in these thresholds was observed after the application of the protein. Studies on thresholds for sour taste were considered to be complex, because acids induce sweetness after application of the protein. We found that one tastes only sour at very low concentrations of acid even after

TABLE III

TASTE THRESHOLDS BEFORE AND AFTER APPLICATION OF TASTE-MODIFYING PROTEIN

Taste thresholds were determined before and after application of 2.0×10^{-6} M taste-modifying protein. Taste thresholds are expressed by solution number, which represents concentration of solution as shown at the lower part of the table.

Subject No.	(Age)	Sucrose		NaCl		Quinine		Citric Acid	
		Before	After	Before	After	Before	After	Before	After
1	(15)	15	16	12	11	2	3	6	6
2	(13)	15	16	13	12	3	1	6	7
3	(10)	15	16	9	9	4	5	7	7
4	(9)	15	16	11	13	0	2	6	6
5	(16)	14	15	10	12	2	1	5	5
6	(11)	16	17	12	13	5	5	8	9
7	(10)	13	14	13	14	8	8	8	7

Solution No.	Molarity	Solution No.	Molarity	Solution No.	Molarity
17	4.80×10^{-2}	11	7.50×10^{-4}	5	1.17×10^{-5}
16	2.40×10^{-2}	10	3.75×10^{-4}	4	5.86×10^{-6}
15	1.20×10^{-2}	9	1.88×10^{-4}	3	2.93×10^{-6}
14	6.00×10^{-3}	8	9.38×10^{-5}	2	1.46×10^{-6}
13	3.00×10^{-3}	7	4.69×10^{-5}	1	7.32×10^{-7}
12	1.50×10^{-3}	6	2.34×10^{-5}	0	3.66×10^{-7}

application of the protein. As shown in Table IV, one tastes only the sourness of the citric acid solution below 1.50×10^{-3} M. The threshold to citric acid also did not change after application of the protein. From the above results, it is clear that the taste-modifying protein does not affect thresholds for any quality of taste.

The taste of 0.02 M citric acid was very sour, but became pleasantly sweet after application of the protein. Is the sourness of the acid solution depressed when one tastes sweet to the acid solution after taste modification? The use of gymnemic acid, which is known as a depressor of sweetness[21] (Table IV) solved this problem. Five milliliters of a gymnemic acid A_1 solution of 1.0×10^{-3} M was held in the mouth for 2 min after a taste-modifying protein solution of 2.0×10^{-6} M was applied.[12,20] The solution in the mouth was spit out, the mouth was rinsed with distilled water and 0.02 M citric acid solution was tasted. No sweetness was detected upon sampling the acid solution, which now tasted very sour. The sweetening effect reappeared about 10 min after the application of gymnemic acid. This occurred because the gymnemic acid effect decreases much faster than the effect of the taste-modifying protein. The fact that one can still taste the sourness of the acid solution after application of the protein indicates that the protein does not depress the sourness. This suggests that the pleasant sweet taste of the acid after taste modification was brought about by addition of a sweet taste to the sour taste.

The taste-modifying protein contained 6.7 per cent of arabinose and xylose, which have a sweet taste.[10] The sweetness of 1 M arabinose and 1 M xylose was compared with a series of sucrose solutions, and was equivalent to the sweetness of 0.35 ± 0.07 M and 0.34 ± 0.06 M sucrose solution, respectively. The taste of 1 M arabinose or xylose dissolved in 0.02 M citric acid solution was very similar to the taste of the acid solution after taste modification. Since the linkage between protein and sugar component in

TABLE IV

THE SWEET-INDUCING POTENCY OF DIFFERENT CONCENTRATIONS OF CITRIC ACID AND THE EFFECT OF GYMNEMIC ACID ON THE SWEET TASTE OF THE ACID BROUGHT ABOUT BY THE TASTE-MODIFYING PROTEIN

The solutions from No. 5 to No. 16 (see Table III) were tasted after 2.0×10^{-6} M taste-modifying protein was applied to the mouth. The effect of gymnemic acid was studied by applying 1×10^{-3} M gymnemic acid A_1 solution to the mouth after application of the protein solution.

	Concentration of Citric Acid		
	7.5×10^{-4} M	1.5×10^{-3} M	2.0×10^{-2} M
Before Taste-Modifying Protein	Sour	Sour	Very Sour
After Taste-Modifying Protein	Sour	Sweet	Very Sweet
After Taste-Modifying Protein and Gymnemic Acid A_1	Sour	Sour	Very Sour

glycoprotein is stable in a mild acidic condition such as 0.02 M citric acid, it is not likely that the sugar component in the taste-modifying protein is liberated by the action of the acid solution for the taste testing. Probably, sugar residues themselves in the protein molecule have a binding ability to the sweet receptor site on the taste buds. As seen in Figure 9, 2×10^{-8} M taste-modifying protein showed an appreciable sweetening effect. Since one molecule of the taste-modifying protein contains about 20 sugar residues, 4×10^{-7} M sugar residue is contained in a 2×10^{-8} M protein solution. This concentration of sugar is much lower than the threshold concentrations of arabinose and xylose.

The results obtained in this paper are explained by the scheme in Figure 13. Since the taste-modifying protein has a molecular weight of 44,000, it is unlikely that the protein penetrates so easily into taste cells. Probably, the protein binds to the membrane surface of taste cells. It has become increasingly apparent that proteins called *allosteric proteins* have other binding sites for substrates or metabolites than the catalytic (substrate) binding sites.[15] The idea of the *allosteric sites* was introduced to the taste-modifying protein. In the scheme, the protein has two binding sites, one to bind the receptor membrane and the other to bind the sweet-receptor site. The site of the protein which binds to the sweet-receptor site is postulated to be arabinose and (or) xylose. Unless acid is applied, the protein should not occupy any of the receptor sites of the four taste qualities, since the binding of the protein did not affect their thresholds. Probably the protein binds to the receptor membrane near the sweet-receptor site. The receptor membrane will change its conformation at the low pH that induces sourness, and this conformational change of the membrane results in "fitting" the sugar part of the protein into the sweet-receptor site. The conformational change in most proteins occurs near the pK value (3.4–4.5) of carboxyl groups in proteins.[17] Since the sweetness of acids, especially of HCl is induced at much lower pH

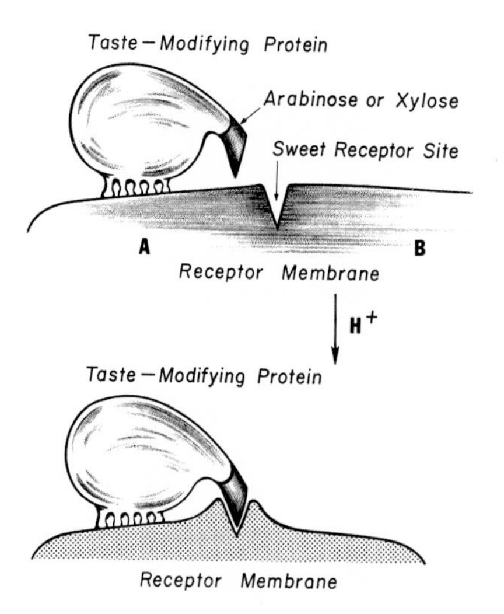

Figure 13 A scheme for the sweet-inducing mechanism of the taste-modifying protein.

than the pK values of the carboxyl groups in proteins, the idea that the conformational change of the taste-modifying protein at low pH is responsible for the induced sweetness is not likely. The idea of conformational change of the receptor membrane in the scheme can explain much better the close relation between the sweet-inducing potency and the sourness of acids.

The scheme does not mean that the protein binds only to a specific site of the membrane. The protein may bind to many places on the membrane, but the sugar part of the protein will not "fit" the sweet-receptor site except for that protein bound to a specific site. The sweetening effect after application of 4.8×10^{-7} M protein solution (curve C in Figure 10) decreased much faster than curve D in the Figure for 2.3×10^{-6} M solution, even though the sweetness of the acid was the same just after application of the protein. This fact was explained by postulating that the protein also binds to site B in the scheme. The receptor site for sweetness can be occupied with the sugar part of either protein, but the maximum sweetness of the acid does not change, even if both sites A and B are occupied. The sweetening effect will not change even if the protein bound to one of the sites comes off. The protein molecules bound at one site may come off the membrane more easily than those at another site, and this postulation explains the difference between the slope of curve B and that of curve D at the same level of sweetness, for example, the level of 0.24 sucrose sweetness.

SUMMARY

The "taste-modifying protein" was isolated from miracle fruit (*Synsepalum dulcificum*). The protein was a basic glycoprotein with a probable molecular weight of 44,000. L-arabinose and D-xylose were identified as sugar components of the protein. Application of the protein to the tongue modifies the taste so that one tastes sour substances as sweet. The mechanism of the action of the protein was discussed on the basis of the following results.

1. The taste-modifying protein itself has no inherent taste, and a premixture of the protein with acid has no sweet taste. In order to obtain the maximum sweetening effect, it was necessary to hold the protein solution in the mouth for at least 3 min.
2. The maximum sweetening effect was brought about by application of the protein solution above 4×10^{-7} M concentration.
3. The sweetening effect brought about by one application of high concentration of protein solution persisted for over 3 hours.
4. All inorganic and organic acids tested were found to have sweet-inducing potency. The sweet-inducing potency of an acid was closely associated to the intensity of sourness of that acid.
5. The application of the taste-modifying protein to the tongue did not affect thresholds of any taste quality.
6. The sour taste of acids was not depressed by application of the protein. The pleasant sweet taste of the acid solution after application of the protein was brought about by addition of a sweet taste to the sour taste of the acid.

The *allosteric* model of the taste-modifying protein was proposed on the basis of the above results. The protein was postulated to bind closely to the vicinity of the sweet receptor site on the taste receptor membrane. The conformational change of the receptor membrane by acid will result in "fitting" the sugar residues of the taste-modifying protein to the sweet receptor site.

ANTISWEET ACTIVITY OF GYMNEMIC ACID A_1 AND ITS DERIVATIVES

The leaves of *Gymnema sylvestre* temporarily suppress the taste sensitivity to sweet substances. Since 1887, several investigators[8,21,22] have tried to isolate the active principle "gymnemic acid" from the leaves. Stöcklin et al.[20] reported the isolation method of gymnemic acid, and proposed that its structure was D-glucuronide of hexahydroxytriterpene which is esterified with acids. Using thin-layer chromatography, they found that "gymnemic acid" (gymnemic acid A) consists of gymnemic acid A_1 (a main component), A_2, A_3, and A_4 (Figure 14).

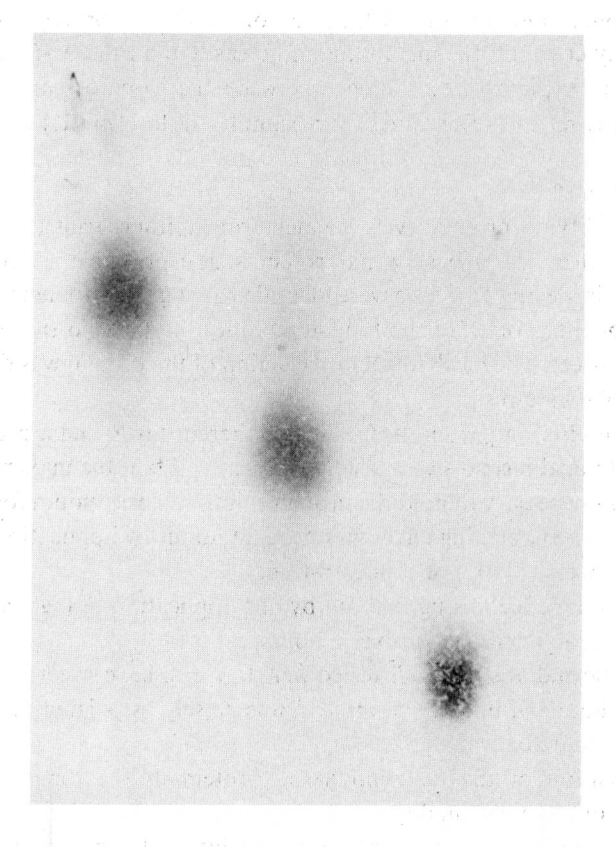

Figure 14 Thin-layer chromatography of gymnemic acids. Left: A_1, Middle: A_2, Right: A_3. The chromatography was carried out in the system of butyl formate-methyl ethyl ketone-formic acid-H_2O (5:3:1:1) on the silica gel G, and 20% H_2SO_4 was used as spray reagent.

In order to elucidate the relation between the structure and the ability to suppress sweetness of gymnemic acids, the antisweet activities of these different gymnemic acid components were compared, based on studies of their structural differences in the present paper. Since the reported isolation methods[18,20] are not satisfactory to obtain sufficient quantities of gymnemic acid A_1 for preparative purpose, we devised a new isolation method to obtain high yields of the substance. The effect of gymnemic acid A_1 on the sweet taste of the substances of quite diverse chemical structure was also studied in this paper.

METHOD

Assay of Antisweet Activity

The antisweet of gymnemic acid was assayed as follows. Four subjects undertook the following taste test. Five ml gymnemic acid solution in 0.01 M $NaHCO_3$ was held in the mouth for 2 min. The solution was spit out and the mouth was rinsed with distilled water. The subjects were directed to taste 10 sucrose solution from 0.1 to 1.0 M. The activity of a gymnemic acid solution was expressed as the maximum concentration of a sucrose solution whose sweetness was depressed completely. The persistency of the depressive effect on sweet taste was assayed at 5-min intervals after application of the gymnemic acid solution.

The antisweet activity of gymnemic acid on other sweet compounds was assayed as follows: A 1×10^{-3} M gymnemic acid A_1 solution was applied to the mouth and the following solutions, which give proper sweet taste, were tasted: 0.025 M cyclamate (sodium cyclohexanesulfamide), 0.025 M D-tryptophan, 0.075 M D-leucine, 0.038 M beryllium chloride, 0.066 M lead acetate, and chloroform.

RESULTS AND DISCUSSION

For the isolation of gymnemic acid A_1, Stöcklin et al.[20] used a silica-gel column eluting with alkaline or acidic solution. On repeating their isolation method, we noticed that the presence of acid brought about a modification of gymnemic acid A_1 into a substance which gave a larger R_f value than that of A_1 on thin-layer chromatography, and that the presence of alkali brought about a modification of A_1 into a substance which gave an R_f value *less* than that of A_1. The poor yield of gymnemic acid A_1 also arose from difficulty in detaching gymnemic acids from a silica-gel column. The isolation method without the use of acid, alkali, and silica-gel column was devised.

The leaves of *Gymnema sylvestre* were extracted with water and gymnemic acids were precipitated by acidification of the extracts. The precipitates were dissolved in ethanol and acetone, and the insoluble materials in these solvents were eliminated. The solvents were evaporated and the residue was extracted by diethyl carbonate. Gymnemic acid A was crystallized from the solvent. Thin-layer chromatography of the obtained gymnemic acid A indicated that it consisted predominantly of gymnemic acid A_1, which is accompanied by A_2. Gymnemic acid A was applied to a DEAE-Sephadex column. Though elution with absolute ethanol gave poor yield of gymnemic acid A_1, elution with 95 per cent ethanol brought good recovery.

The fractions containing only gymnemic acid A_1 were collected, and the combined fractions were evaporated to dryness. The residue was crystallized from diethyl carbonate. The genin part of gymnemic acid is esterified with acids.[20] The acids in gymnemic acid A_1 were liberated by treatment with 3 per cent KOH methanol solution under refluxing for 15 min. The paper chromatography of the hydrolyzate was carried out according to the method of Bayer and Reuther[3] and gave three spots of acetic, isovaleric, and tiglic acid. By using gas chromatography of the hydrolyzate according to the method of Metcalfe,[14] 1 M of acetic, 2 M of isovaleric and 1 M of tiglic acid were identified. Stöcklin et al. identified five acids; formic, acetic, n-butyric, isovaleric, and tiglic acid, from alkaline hydrolysis of gymnemic acid A. Apparently, the difference between the results arose from the fact that we used purified A_1, while they used a mixture of gymnemic acids. The chemical analysis of the obtained gymnemic acid A_1* corresponded to the structure proposed by Stöcklin, including our results on acid components in the genin.

The presence of potassium bicarbonate modified gymnemic acid A_1 into a substance which gave a smaller R_f than that of A_1 on the thin-layer chromatography. After gymnemic acid A_1 was treated with a potassium bicarbonate solution (0.5 g $KHCO_3$ in 18 ml of water and 3 ml of ethanol) at room temperature, the solution was evaporated and the residue was dissolved in water. The solution was acidified and the precipitate centrifuged. The obtained precipitate was applied to DEAE-Sephadex column and eluted with 95 per cent ethanol. The purified substance on the DEAE-Sephadex column was identified as gymnemic acid A_2† by co-chromatography on thin layer with gymnemic acid A_2 isolated by the Stöcklin et al. method. Refluxing gymnemic acid A_1 in 3 per cent KOH methanol solution yielded a substance whose behavior on the thin-layer chromatography was identical to gymnemic acid A_3‡. These modifications of gymnemic acid A_1 liberated acids which are esterified to the genin. Acids liberated by the modification were identified by paper- and gas-chromatography. The results indicated that one mole of acetic acid was liberated by the conversion of A_1 to A_2 and 2 moles of isovaleric acid and 1 mole of tiglic acid by the conversion of A_2 to A_3. In the chemical analysis of the

$$A_1 \xrightarrow{-1 \text{ mole of acetic acid}} A_2 \xrightarrow[-2 \text{ mole of isovaleric acids}]{-1 \text{ mole of tiglic acid}} A_3,$$

isolated A_2 and A_3 corresponded to the structures expected from the postulation that only ester groups of genin were modified by the alkaline treatment of A_1.

Antisweet activities of gymnemic acid A_1, A_2, and A_3 were assayed. The antisweet

* Gymnemic acid A_1, m.p. 215°C (decomp.). Anal. Calcd. for $C_{53}H_{82}O_{16}2H_2O$; C, 62.97; H, 8.51; O, 28.51, Found: C, 62.55; H, 8.83; O, 28.80.

† Gymnemic acid A_2, m.p. 205–7°C (decomp.). Anal. Calcd. for $C_{51}H_{80}O_{15}H_2O$; C, 59.88; H, 8.80; O, 31.31. Found: C, 60.13; H, 8.62; O, 30.80.

‡ Gymnemic acid A_3, m.p. 250°C (decomp.). Anal. Calcd. for $C_{36}H_{58}O_{12}H_2O$. C, 61.71; H, 8.56; O, 29.71. Found: C, 62.36; H, 9.05; O, 29.20.

activity of gymnemic acids reached the maximum level after being held in the mouth in solution form for about 30 sec. The activity was assayed by tasting a series of sucrose solution after a gymnemic acid solution was held in the mouth for 2 min and the activity was expressed by the maximum concentration of sucrose solution whose sweetness was depressed completely. Figure 15 shows the activity of gymnemic acid A_1 and A_2. Gymnemic acid A_3 did not show any antisweet activity. As seen from the

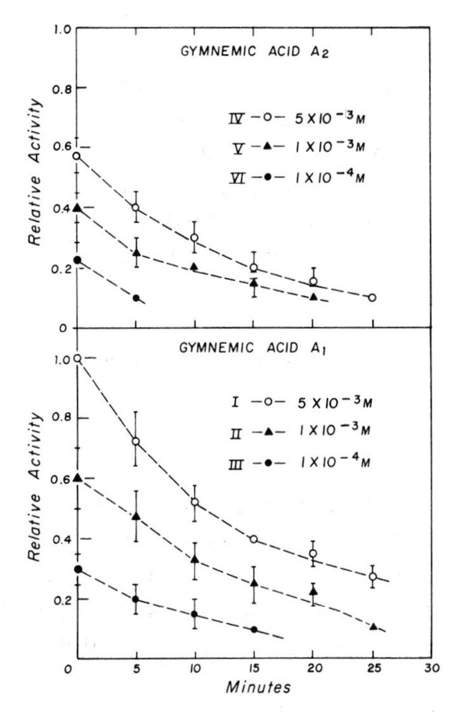

Figure 15 Antisweet activity of gymnemic acids. The activity was assayed at 5-minute intervals after gymnemic acid solution was held in the mouth for 2 minutes. The activity is expressed by the maximum concentration of sucrose solution whose sweetness was depressed completely. The molecular weights of gymnemic acid A_1 and A_2 were calculated from their structures as 974 and 932, respectively.

curve III, 1×10^{-4} A_1 depressed the sweetness of 0.3 M sucrose solution and this effect decreased in 15 minutes. Application of 5×10^{-3} M A_1 brought the complete depression of sweetness of solid sucrose as well as that of 1.0 M sucrose solution and the effect decreased as shown by curve I. The activity of gymnemic acid A_2 was much lower than that of A_1. The activity of 5×10^{-3} M A_2 (curve IV) was a little lower than that of 1×10^{-3} M A_1 (curve II). This indicated that the activity of A_2 is less than 1/5 of the activity of an equal concentration of A_1. It is interesting that the activity decreased greatly by the conversion of A_1 into A_2, even though the conversion resulted in the loss of only one mole of acetic acid. Apparently, the ester group in the genin has an important role in manifestation of antisweet activity of gymnemic acid A_1. Since gymnemic acid A_1 has a rather large molecular weight of about 1000, the elimination of acetic acid from A_1 will not have a large effect on the whole molecule of A_1. The moiety of the ester group of gymnemic acid A_1 may be related directly to the binding of A_1 to the receptor site in taste buds.

It is known that a great variety of other compounds besides sugars, such as synthetic sweetening agents, D-amino acids, and some inorganic compounds, taste sweet. The antisweet activity of gymnemic acid A_1 on cyclamate, D-tryptophan, D-leucine, beryllium chloride, lead acetate, and chloroform was examined. The sweet taste of all these compounds except chloroform was suppressed by the action of A_1. Although it is not known whether these compounds of different chemical structure induce sweet taste by the same mechanism, it can be concluded that at least a part of the sweet mechanism is common to substances of quite different structures. The fact that chloroform is the only sweet substance which has a high vapor pressure and is a good solvent for lipids may be related to its different mechanism of sweet induction.

Sodium glutamate has a unique flavor. Application of 5×10^{-3} M gymnemic acid A_1 changed the unique taste of crystalline sodium glutamate to a taste much like that of sodium chloride.

SUMMARY

A method was devised to isolate gymnemic acid A_1 (a main component of gymnemic acid A) for preparative purpose. The acids esterified in the genin of gymnemic acid A_1 were determined to be 1 mole of acetic, 2 moles of isovaleric and 1 mole of tiglic acids. Gymnemic acid A_1 was converted into A_2 by hydrolysis of 1 mole of acetic acid in the genin and into A_3 by hydrolysis of all acids in the genin. The antisweet activity of A_1 was greatly decreased by conversion into A_2. A_3 had no activity. It was suggested that ester groups in the genin have an important role in manifesting the antisweet activity of gymnemic acid A_1.

The studies of antisweet activity of gymnemic acid on other sweet compounds besides sugars indicated that gymnemic acid A_1 suppresses the sweet taste of cyclamate, D-amino acids, beryllium chloride, and lead acetate, and does not suppress that of chloroform.

ACKNOWLEDGMENTS

This work was supported by National Science Foundation Grant NB GB-4068X and Atomic Energy Commission Grant No. AT-(40-1)-2690. The authors wish to thank Mr. Cuyler V. Smith for valuable technical assistance.

REFERENCES

1. Andrews, P. 1964. Estimation of the molecular weights of proteins by Sephadex gel-filtration. *Biochem. J.* **91**, pp. 222–233.
2. Badin, J., C. Jackson, and M. Shubert. 1953. Improved method for determination of plasma polysaccharides with trypotophan. *Proc. Soc. Exp. Biol. Med.* **84**, pp. 288–291.

3. Bayer, E., and K. H. Reuther. 1956. Papierchromatographische Analyse von Carbonsäureester-Gemischen sowie deren Anwendung zur Untersuchung von Aromastoffen. *Angew. Chem.* **68**, pp. 698–701.
4. Beidler, L. M. 1967. Anion influences on taste receptor response. *In* Olfaction and Taste II (T. Hayashi, editor). Pergamon Press, Oxford, etc., pp. 509–534.
5. Daniell, W. F. 1852. On the *Synsepalum dulcificum* De Cand.; or, miraculous berry of Western Africa. *Pharm. J.* **11**, pp. 445–448.
6. Fischer, R., F. Griffin, and A. R. Kaplan. 1963. Taste thresholds, cigarette smoking, and food dislikes. *Med. Exp.* **9**, pp. 151–167.
7. Harris, H., and H. Kalmus. 1949. The measurement of taste sensitivity to phenylthiourea. *Ann. Eugen. (Cambridge)* **15**, pp. 24–31.
8. Hooper, D. 1887. An examination of the leaves of *Gymnema sylvestre*. *Nature* **35**, pp. 565–567.
9. Inglett, G. E., B. Dowling, J. J. Albrecht, and F. A. Hoglan. 1965. Taste-modifying properties of miracle fruit (*Synsepalum dulcificum*). *J. Agr. Food Chem.* **13**, pp. 248–287.
10. Kurihara, K., and L. M. Beidler. 1968. Taste-modifying protein from miracle fruit. *Science* **161**, pp. 1241–1243.
11. Kurihara, K., and L. M. Beidler. 1969. The mechanism of the action of taste-modifying protein. *Nature* **222**, pp. 1176–1179.
12. Kurihara, Y. The antisweet activity of gymnemic acid A_1 and its derivatives. *Life Sciences* **8**, pp. 537–543.
13. Masamune, H., and Z. Yosizawa. 1953. Biochemical studies on carbohydrates CLXI. Paper partition chromatography of sugars, in particular of hexosamins and hexuronic acids. *Tohoku J. Exp. Med.* **59**, pp. 1–9.
14. Metcalfe, L. D. 1960. Gas chromatography of unesterified fatty acids using polyester columns treated with phosphoric acid. *Nature* **188**, pp. 142–143.
15. Monod, J., J. P. Changeux, and F. Jacob. 1963. Allosteric proteins and cellular control systems. *J. Mol. Biol.* **6**, pp. 306–329.
16. Partridge, S. M. 1949. Aniline hydrogen phthalate as a spraying reagent for chromatography of sugars. *Nature* **164**, p. 443.
17. Scheraga, H. A. 1961. Protein Structure. Academic Press, New York.
18. Sinsheimer, J. E., G. Subba Rao, H. M. McIlhenny, R. V. Smith, H. F. Maassab, and K. W. Cochran. 1968. Isolation and antiviral activity of the gymnemic acids. *Experientia* **15**, p. 302.
19. Stöcklin, W. 1967. Gymnemagenin, vermutliche Struktur. *Helv. Chim. Acta* **50**, pp. 491–503.
20. Stöcklin, W., E. Weiss, and T. Reichstein. 1967. Gymnemasäure, das antisaccharine Prinzip von *Gymnema sylvestre* R. Br. Isolierungen und Identifizierungen. *Helv. Chim. Acta* **50**, pp. 474–490.
21. Warren, R. M., and C. Pfaffmann. 1959. Supression of sweet sensitivity by potassium gymnemate. *J. Appl. Physiol.* **14**, pp. 40–42.
22. Yackzan, K. S. 1966. Biological effects of *Gymnema sylvestre* fractions. *Ala. J. Med. Sci.* **3**, pp. 1–9.

AFFERENT SPECIFICITY IN TASTE

MASAYASU SATO, SATORU YAMASHITA, *and* HISASHI OGAWA
Department of Physiology, Kumamoto University Medical School, Kumamoto, Japan

INTRODUCTION

In the gustatory nervous system, as in other sensory systems, the concept of stimulus-specificity of endorgans and of nerve fibers prevailed for a long time. Each of the four kinds of stimuli, representing the four basic taste qualities, stimulates a specific type of receptor and sends a message along a nerve fiber specific for this particular stimulus, producing a specific quality of taste. However, the hypothesis that there were just four kinds of fibers with specificities directly corresponding to the four qualities of sweet, salt, sour, and bitter was rejected in 1941 by Pfaffmann,[9] who found that many cat chorda tympani single fibers responded to more than one kind of chemical, e.g., acid and salt or acid and quinine. In the later studies,[5,10] a broader multiple sensitivity of chorda tympani fibers for various chemicals was found in cat, rat, rabbit, and hamster, indicating that chemical specificity of the receptor cells is not absolute but relative. In addition, thermal sensitivity of taste fibers was found in cats by Nagaki, Yamashita, and Sato.[7] On the other hand, a few investigators still keep the concept of stimulus specificity for taste fibers in the cat[1] and in the monkey,[6] although by a terminology such as "salt fiber," "water fiber," "acid fiber," and "quinine fiber" they do not mean an absolute specific sensitivity of single gustatory nerve fibers to only one kind of chemical.

In this paper, the results of experiments on impulse discharges in 50 chorda tympani fibers of rats and 28 chorda tympani fibers of hamsters[8,12,13] are described, and an attempt is made to determine whether chorda tympani fibers of rats and hamsters can be classified into certain classes, which differ from one another in their sensitivities to various chemicals, to thermal changes, and in their rate of spontaneous discharges.

As stimuli to test the responsiveness of chorda tympani fibers, four chemical stimuli—0.1 M NaCl, 0.5 M sucrose, 0.01 N HCl, and 0.02 M quinine hydrochloride—as well as thermal changes of the tongue by cold water (20°C below tongue temperature) and warm water (20°C above tongue temperature) were employed. In addition, sodium saccharin of varying concentrations was used as a test stimulus.

MULTIPLE SENSITIVITY OF CHORDA TYMPANI FIBERS

A majority of chorda tympani fibers of rats and hamsters responded to a variety of gustatory stimuli as well as to thermal changes of the tongue. Impulse discharges in a single chorda tympani fiber of a rat are presented in Figure 1. This particular unit responded to 0.1 M NaCl and 0.02 M sodium saccharin massively, but, in addition, to

0.5 M sucrose, 0.01 N HCl, 0.02 M quinine, and to cooling, though its sensitivities to the latter four stimuli were small.

Response profiles of 50 chorda tympani fibers of rats and of 28 fibers of hamsters for four basic gustatory stimuli and thermal changes of the tongue are shown in Figures 2 and 3, which reveal that a certain number of fibers responded to only one kind of stimulus, but that a majority of fibers gave responses to more than one. Careful examination of these figures indicates further that some fibers responded predominantly to NaCl or to sucrose, while some other fibers also gave responses to HCl, quinine, and cooling.

Figure 4 represents the concentration-response curves of three major types of chorda tympani fibers of rats. The unit predominantly sensitive to NaCl (top) re-

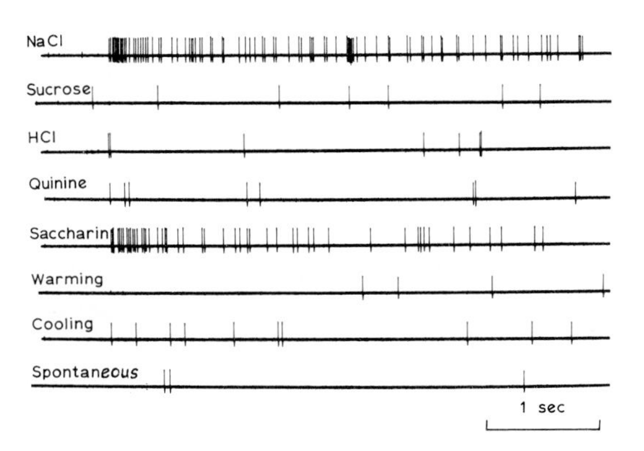

Figure 1 Impulse discharges in a single chorda tympani nerve fiber of rat elicited by application to the tongue of 0.1 M NaCl, 0.5 M sucrose, 0.01 N HCl, 0.02 M quinine hydrochloride, 0.02 M sodium saccharin, 40°C water, and 20°C water (from the top). Bottom trace, spontaneous discharge. Temperature of the tongue was kept at 25°C throughout the experiment by rinsing it with 25°C water, but for warming, 40°C water was applied to the tongue at 20°C and for cooling, 20°C water was applied at 40°C. (From Ogawa et al.[8])

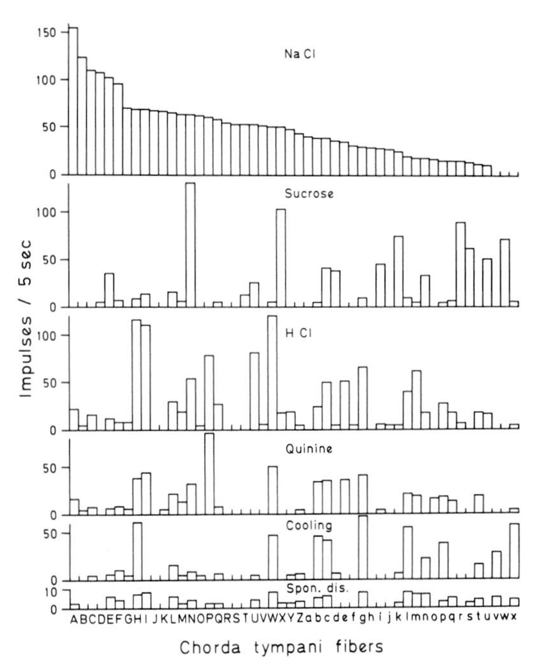

Figure 2 Response profile of 50 chorda tympani fibers of rat (A, B, C, etc., to x), in which fibers were arranged in the order of responsiveness to 0.1 M NaCl. Stimuli were, from the top, 0.1 M NaCl, 0.5 M sucrose, 0.01 N HCl, 0.02 M quinine hydrochloride, and cooling (20°C water to 40°C tongue). Bottom: spontaneous discharge. (From Ogawa et al.[8])

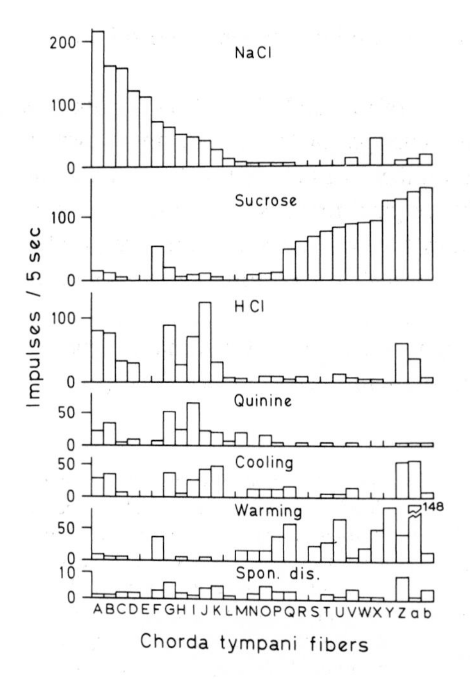

Figure 3 Response profile of 28 hamster chorda tympani fibers. Half are arranged in the order of responsiveness to 0.1 M NaCl; the remaining half in the order of responsiveness to 0.5 M sucrose. Stimuli were, from the top, 0.1 M NaCl, 0.5 M sucrose, 0.01 N HCl, 0.02 M quinine hydrochloride, cooling (20°C water to 40°C tongue), and warming (40°C water to 20°C tongue). Bottom: spontaneous discharge. (From Ogawa et al.[8])

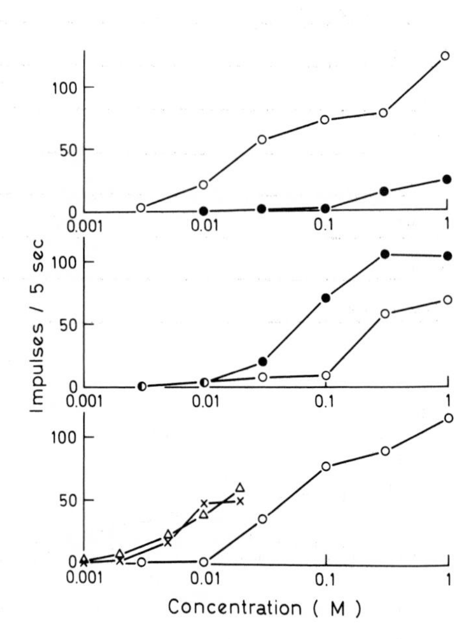

Figure 4 Concentration-response magnitude relationships in three typical chorda tympani fibers of rats. Top, unit predominantly sensitive to NaCl; middle, unit more sensitive to sucrose than to NaCl; bottom, unit sensitive to NaCl, quinine, and HCl. Circle, responses to NaCl; dot, responses to sucrose; triangle, responses to quinine hydrochloride; cross, responses to HCl. Ordinate indicates number of impulses elicited in the first 5 sec after stimulation. Number of spontaneously discharged impulses was subtracted from the magnitude of response to stimuli. (From Ogawa et al.[8])

sponded to 0.3 M sucrose but not to 0.1 M sucrose, while the unit more sensitive to sucrose than to NaCl (middle) gave responses to 0.3 M NaCl but scarcely responded to 0.1 M NaCl. The unit shown at the bottom was sensitive to NaCl, HCl, and quinine, but not to sucrose. These facts indicate that there exist few fibers that possess absolute sensitivity to only one kind of chemical. The results presented in Figures 2 and 3 suggest a possibility of categorizing chorda tympani fibers into certain classes based on the relative sensitivity to one or to a combination of gustatory and thermal stimuli, but those in Figure 4 indicate that categories of units classified according to sensitivity to chemicals may depend greatly on the concentration of chemicals.

THERMAL SENSITIVITY AND SPONTANEOUS DISCHARGES

Some fibers showed spontaneous discharges of fairly high rate, as seen in Figures 2 and 3, and because of this it was difficult to determine whether such fibers actually responded to a particular stimulus when the magnitude of response was small. In rats and hamsters, cold-sensitive units usually showed marked spontaneous discharges, compared with those insensitive to thermal change. As shown in Table I,

TABLE I
RATE OF SPONTANEOUS DISCHARGES IN CHORDA TYMPANI FIBERS OF RATS AND HAMSTERS
(mean number of impulses/5 sec ± SD)

	Rate of spontaneous discharge (imp/5 sec)		Thermal sensitivity (imp/sec °C)	
	Rat	Hamster	Rat	Hamster
Cold-sensitive units	4.95 ± 2.50(19)	3.38 ± 2.21(16)	−0.42(19)	−0.44(4)*
Units insensitive to cooling	1.03 ± 1.22(29)	0.83 ± 0.97(12)		
t test	$P < 0.001$	$P < 0.001$		
Units sensitive to warming		2.60 ± 2.31(20)		0.67(8)*
Units insensitive to warming		1.50 ± 1.41(8)		
t test		$0.1 < P < 0.5$		

* Remaining 20 units responded to both cooling and warming, their average sensitivities being −0.26 and 0.31 imp/sec °C, respectively.

the mean rate of impulses in spontaneous discharges in both rats and hamsters was significantly greater in units sensitive to cooling than in those insensitive. In agreement with this, there exists a highly significant positive correlation between the number of impulses discharged in 5 sec after stimulation by cold water 20° C below the tongue temperature and the rate of spontaneous discharges: 0.781 in the rat and 0.503 in the hamster.

SPECIFIC SENSITIVITY OF CHORDA TYMPANI UNITS TO A PARTICULAR STIMULUS OR A PARTICULAR COMBINATION OF STIMULI

In order to determine whether any units actually responded to a particular stimulus, the following criterion was adopted: when the number of impulses discharged in the initial 5 sec after stimulation in a particular unit was greater than the mean number, plus SD, of spontaneously discharged impulses during 5 sec, this unit was considered to have responded to the stimulus. However, because of the difference in the rate of spontaneous discharges between cold-sensitive and cold-insensitive units, two different mean rates of spontaneous discharges, as shown in Table I, were applied to each class of units.

The results of determinations of sensitivities of 48 rat units and 28 hamster units are presented in Table II. In the rat nearly all of the units responded to 0.1 M NaCl, 70 per cent of the units to 0.01 N HCl, 50 per cent of the units to 0.5 M sucrose and to 0.02 M quinine, while in the hamster a majority of the units responded to 0.5 M sucrose, 0.01 N HCl, and 0.1 M NaCl and half of the units to 0.02 M quinine. In addition, about 40 per cent of the units in the rat and 60 per cent of the units in the hamster gave responses to cooling and about 70 per cent of the hamster units showed sensitivity to warming. In addition to 48 units of the rat, two units responded to cooling, but scarcely to gustatory stimuli.

TABLE II
RATE OF CHORDA TYMPANI NERVE FIBERS OF RATS AND HAMSTERS SENSITIVE TO ONE OF SIX FUNDAMENTAL STIMULI
(From Ogawa et al.[8])

Stimuli	48 Units (Rat)*	28 Units (Hamster)**
0.1 M NaCl	47/48 (0.979)	18/28 (0.643)
0.5 M Sucrose	24/48 (0.500)	22/28 (0.786)
0.01 N HCl	34/48 (0.708)	22/28 (0.786)
0.02 M Quinine	25/48 (0.521)	14/28 (0.500)
20°C Water to 40°C Tongue	19/48 (0.396)	16/28 (0.571)
40°C Water to 20°C Tongue		20/28 (0.714)

* Mean ± SD of spontaneous discharges of 48 units: (2.58 ± 2.67)/5 sec.
Mean ± SD of spontaneous discharges of 19 units, which responded to cooling: (4.95 ± 2.50)/5 sec.
Mean ± SD of spontaneous discharges of 29 units, insensitive to cooling: (1.03 ± 1.22)/5 sec.
** Mean ± SD of spontaneous discharges of 28 units: (2.90 ± 1.32)/5 sec.
Mean ± SD of spontaneous discharges of 16 units, which responded to cooling: (3.38 ± 2.15)/5 sec.
Mean ± SD of spontaneous discharges of 12 units, insensitive to cooling: (0.83 ± 0.99)/5 sec.

The numbers of units which responded to a particular stimulus or a particular combination of stimuli are represented by blocks with hatched lines in Figure 5. In the rat, nine units among 48 responded to only one kind of stimulus (19 per cent), eight units to two kinds (17 per cent), 19 units to three kinds (39 per cent) and 12 units to all the four stimuli (25 per cent); in the hamster, five units gave responses to one kind of stimulus (18 per cent), five units to two kinds (18 per cent), 11 units to three kinds (39 per cent) and seven units to all the four stimuli (25 per cent). As mentioned above, the number of units responding to one or two kinds of stimuli shown in Figure 5 would be reduced if concentrations of test stimuli were increased.

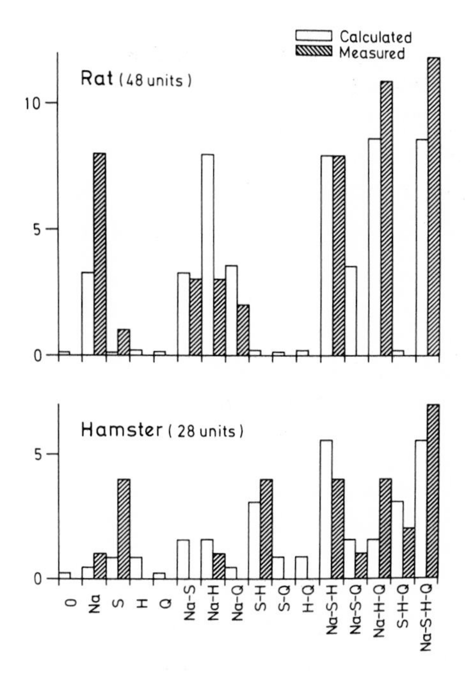

Figure 5 Observed numbers of units in rats and hamsters responding to combinations of the four basic gustatory stimuli (hatched blocks) and expected numbers based on the data shown in Table I (empty blocks). Na, S, H, and Q denote 0.1 M NaCl, 0.5 M sucrose, 0.01 N HCl and 0.02 M quinine hydrochloride.

Assuming that the probability of response to each of the four taste stimuli in chorda tympani units (Table II) is independent of their responsiveness to the other three stimuli, one can calculate the expected frequencies of units responding to combinations of the four stimuli. In Figure 5, the predicted numbers of units are represented by empty blocks. One may easily find discrepancies between the observed and predicted numbers of units. For example, in the rat, observed numbers of units responding to NaCl alone, to NaCl, HCl, and quinine, and to all the four stimuli are greater than the predicted ones. In the hamster, the observed numbers of units responding to sucrose alone, to NaCl, HCl, and quinine, and to all the four stimuli are also greater. This would suggest that sensitivity of chorda tympani units to each of the four fundamental stimuli is not independent of that to other kinds of stimuli. Therefore, probabilities of independent occurrences of responses of chorda tympani units to a pair of stimuli were calculated according to the Fisher exact probability test,[11] assuming that their sensitivity to a stimulus is independent of that to the other of a pair. The results are shown in Table III. As shown in this table, responses to HCl, quinine and cooling tend to occur concomitantly in chorda tympani units of both rats and hamsters, while in hamsters responses to sucrose and warming occur concomitantly but those to NaCl and sucrose reciprocally.

Similar conclusions may be obtained by calculating correlation coefficients between the amounts of responses to a pair of stimuli: as shown in Figure 6, highly significant positive correlations exist among the responses to HCl, quinine, and cooling in both rats and hamsters, and a negative correlation between responses to sucrose and those

TABLE III
PROBABILITY OF INDEPENDENT OCCURRENCE OF RESPONSES IN CHORDA TYMPANI FIBERS TO A PAIR OF STIMULI***
(From Ogawa et al.[8])

Interaction between a pair of stimuli	Rat (48 units)	Hamster (28 units)
NaCl and sucrose	$0.05 < P$	$0.01 < P < 0.05$**
NaCl and HCl	$0.05 < P$	$0.05 < P$
NaCl and quinine	$0.05 < P$	$0.01 < P < 0.05$*
Sucrose and HCl	$0.05 < P$	$0.05 < P$
Sucrose and quinine	$0.05 < P$	$0.05 < P$
HCl and quinine	$P < 0.001$*	$0.05 < P$
Cooling and NaCl	$0.05 < P$	$0.05 < P$
Cooling and sucrose	$0.05 < P$	$0.05 < P$
Cooling and HCl	$P < 0.001$*	$0.01 < P < 0.05$*
Cooling and quinine	$0.01 < P < 0.05$*	$0.05 < P$
Cooling and warming		$0.05 < P$
Warming and NaCl		$0.05 < P$
Warming and sucrose		$0.01 < P < 0.05$*
Warming and HCl		$0.05 < P$
Warming and quinine		$0.05 < P$

* Responses to a pair of stimuli are not independent of each other and occur concomitantly.
** Responses to a pair of stimuli are not independent of each other, but occur reciprocally.
*** Calculations of probabilities were made according to the Fisher exact probability test (Siegel.[11]).

to NaCl, although the latter correlation is not significant in the rat. In addition, a significant positive correlation exists between responses to sucrose and those to warming. From such facts, it appears that chorda tympani units in both rats and hamsters tend to respond more easily or less easily to some combinations of stimuli than to others, as shown in Figure 5.

SPECIFIC SENSITIVITY AND SPONTANEOUS DISCHARGE

A significant difference in the rate of spontaneous discharges and in the magnitude of response to cooling exists between the units sensitive to a variety of gustatory

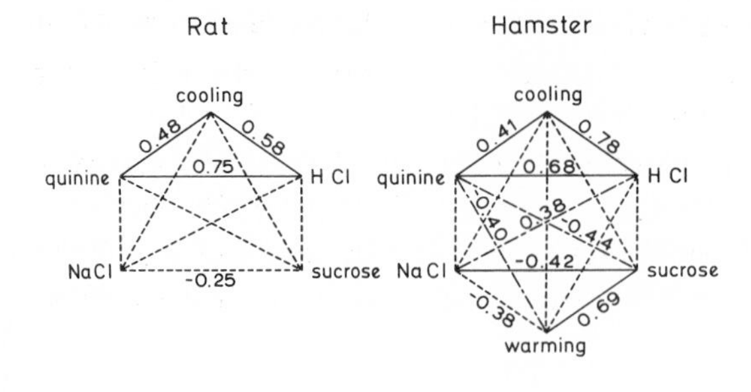

Figure 6 Correlation coefficients among amounts of responses in 45–47 rat chorda tympani fibers and in 26–27 hamster fibers produced by six kinds of stimuli: 0.1 M NaCl, 0.5 M sucrose, 0.01 N HCl, 0.02 M quinine hydrochloride, cooling, and warming. Continuous lines indicate significant or highly significant correlations, broken line probably significant correlations and dotted line no significant correlations.

stimuli (nonspecific type) and those sensitive to one or two kinds of stimuli (specific type). The nonspecific unit showed a significantly greater magnitude of response to cooling than did the specific type unit (Table IV). Therefore, almost all the cold-sensitive units belong to the nonspecific type; in the rat, 18 out of 19 cold-sensitive units were among 31 units of the nonspecific type, and 12 out of 16 cold-sensitive units in the hamster were among 18 such units. In addition, the nonspecific unit

TABLE IV
DIFFERENCE IN THE RATE OF SPONTANEOUS DISCHARGE IN THE SPECIFIC AND NON-SPECIFIC TYPES OF CHORDA TYMPANI FIBERS
(From Ogawa et al.[8])

	Spontaneous discharge (imp/5 sec)	Response to cooling (imp/5 sec)
Rat		
Specific type (17)	0.82 ± 0.99	1.38 ± 2.47
Non-specific type (31)	3.55 ± 2.78	17.00 ± 20.48
t test	$P < 0.001$	$P < 0.001$
Hamster		
Specific type (10)	1.40 ± 1.11	5.30 ± 4.60
Non-specific type (18)	2.78 ± 2.41	18.67 ± 18.49
t test	$P > 0.05$	$0.01 < P < 0.05$

Numerals indicate the mean ± SD of numbers of impulses discharged during 5 sec; those in the parentheses the number of fibers.
Response to cooling: Response to cold water 20°C below tongue temperature.

showed a greater rate of spontaneous discharges than the specific type unit, as shown in Table IV. These are understandable from the fact that sensitivity to HCl and quinine is not independent of that to cooling, and that all the nonspecific units are sensitive to both quinine and HCl in addition to other kinds of stimuli.

CATEGORIES OF CHORDA TYMPANI UNITS

Categories of chorda tympani units of both rats and hamsters, classified from their responsiveness to the four basic gustatory stimuli are presented in Table V, and representative examples of each category are shown in Figures 7 and 8. In the rat, units predominantly sensitive to NaCl (Na type) occupied about 17 per cent of the total population, and those sensitive to NaCl and sucrose (Na-S type), to NaCl and HCl (Na-H type), and to NaCl and quinine (Na-Q type), also totaled 17 per cent. They possess little sensitivity to cooling. Units sensitive to NaCl, sucrose and HCl (Na-S-H type), those sensitive to NaCl, HCl and quinine (Na-H-Q type), and those sensitive to all the four basic stimuli (Na-S-H-Q type) occupied 17 per cent, 23 per cent and 25 per cent of the total population, respectively. A majority of these units possesses a marked sensitivity to cooling.

TABLE V
CLASSES OF CHORDA TYMPANI FIBERS

Sensitivity	Number of units			
	Rat		Hamster	
Na	8(1)*	5(2)**	1(0,0,0)*	1(1,0,0)**
S	1(0)	1(0)	4(0,3,1)	3(0,2,1)
Na-S	3(0)	3(0)	0	1(0,0,1)
Na-H	3(0)	5(0)	1(1,0,0)	1(1,0,0)
Na-Q	2(0)	1(0)	0	0
S-H	0	0	4(0,2,2)	3(0,1,2)
Na-S-H	8(4)	6(2)	4(0,1,3)	2(0,1,1)
Na-S-Q	0	0	1(0,1,0)	0
Na-H-Q	11(6)	12(5)	4(0,1,1)	3(0,0,1)
S-H-Q	0	0	2(0,0,1)	0
Na-S-H-Q	12(8)	15(10)	7(3,0,4)	14(1,2,10)
Total	48(19)	48(19)	28(4,8,12)	28(3,6,16)

Na, S, H and Q denote 0.1 M NaCl, 0.5 M sucrose, 0.01 N HCl, 0.02 M quinine hydrochloride.
Numerals in the parentheses in Rat column indicate the number of units sensitive to cooling, while those in the Hamster column indicate the number of units sensitive to cooling, to warming, and to both cooling and warming, respectively.
* Number of units classified according to the criterion that, when the number of impulses discharged in 5 sec in response to a particular stimulus in any unit is greater than the mean plus SD of spontaneously discharged impulses during 5 sec, the unit was considered to have responded to the stimulus.
** Number of units responding to the four basic stimuli when units showing a number of impulses to a stimulus greater than the mean of numbers of spontaneously discharged impulses were assumed to have responded to the stimulus.

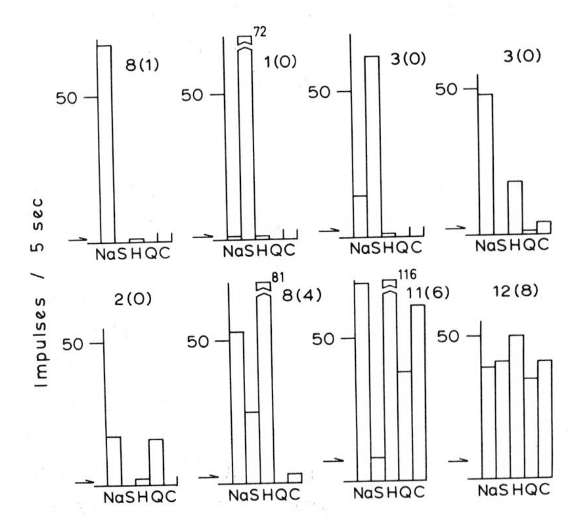

Figure 7 Response profiles of eight rat chorda tympani fibers representative of eight classes found in the total population of 48. Na, S, H, Q, and C denote 0.1 M NaCl, 0.5 M sucrose, 0.01 N HCl, 0.02 M quinine hydrochloride, and cooling, respectively. Arrows indicate rate of spontaneous discharges. Numerals at the top of each profile represent number of units belonging to each class and those inside parentheses indicate number of units responding to cooling.

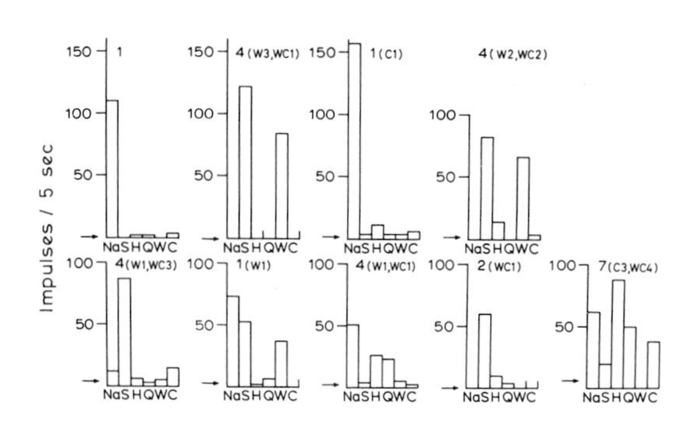

Figure 8 Response profiles of nine hamster chorda tympani fibers representative of nine classes found in the total population of 28. Na, S, H, Q, C, and W denote 0.1 M NaCl, 0.5 M sucrose, 0.01 N HCl, 0.02 M quinine hydrochloride, cooling, and warming, respectively. Arrows indicate rate of spontaneous discharges. Numerals at the top of each profile represent number of units belonging to each class and those inside parentheses indicate number of units responding to warming (W), cooling (C), or warming and cooling (WC).

In the hamster, units predominantly sensitive to sucrose (S type), and to sucrose and HCl (S-H type) each occupied about 14 per cent of the total population. They are more sensitive to warming than to cooling. Units sensitive to NaCl, sucrose and HCl (Na-S-H type), and those sensitive to NaCl, HCl, and quinine (Na-H-Q type) each occupied 14 per cent of the total population. A majority of them seems to have sensitivity to both warming and cooling. Twenty-five per cent of the total population of hamster units contain those sensitive to all the four gustatory stimuli, and all of them possess sensitivity to cooling.

When responsiveness of individual units to each stimulus was determined by employing just the mean number of spontaneously discharged impulses during 5 sec instead of the mean plus SD in cold-sensitive groups and insensitive groups of units, respectively, categories of the chorda tympani units did not vary. However, the number of units sensitive to only one kind of stimulus decreased and those responding to all the four gustatory stimuli increased in number (Table V).

That the categories presented in Table V are fixed is also shown in Table VI. Except for 48 rat chorda tympani fibers, on which statistical analyses had been made, we had 43 rat fibers, the responses of which to the four basic stimuli had been recorded but not their responses to thermal stimuli or their spontaneous discharges. Classifications of both the first and the second groups of units were tentatively made by assuming that when the number of impulses discharged 5 sec after application of any stimulus in any unit is greater than the mean plus SD of the rate of spontaneous discharges in the first group of units, i.e., $(2.70 + 3.46)/5$ sec, the unit responded to the stimulus. As seen in Table VI, the groups gave similar results in the categories of units and in the number of units belonging to each category. Furthermore, when Tables V and VI are compared with each other, categories of the units are the same, although different standards were made for each classification.

DIFFERENCE IN THE RESPONSE TO SODIUM SACCHARIN BETWEEN SUCROSE-SENSITIVE AND NaCl-SENSITIVE UNITS

As shown in Figure 3, hamster chorda tympani units are sharply divided into two distinct classes, one predominantly sensitive to sucrose and one to NaCl. The former

TABLE VI
CLASSES OF CHORDA TYMPANI FIBERS

Sensitivity	Group I	Group II
Na	13	13
S	1	3
Na-S	4	5
Na-H	2	1
Na-Q	1	0
S-H	0	0
Na-S-H	7	8
Na-S-Q	0	2
Na-H-Q	14	6
S-H-Q	0	0
Na-S-H-Q	6	4
Total	48	43

Both groups of units were categorized according to the criterion that when the magnitude of response in any unit to a stimulus was greater than the mean plus SD of the rate of spontaneous discharges in the first group of units, i.e., (2.70 + 3.46)/5 sec, the unit was considered to have responded to this stimulus.

showed sensitivity to warming, as well. Both kinds gave responses to 0.02 M sodium saccharin (Figure 9). The magnitude of response to saccharin is correlated both with that for NaCl and with that for sucrose. However, a difference in the concentration-response curve for sodium saccharin was observed between the two. The sucrose-sensitive units showed a marked increase in the magnitude of response from 0.0003 M sodium saccharin to 0.01 to 0.03 M, at which the magnitude of response was greatest, and with further increase in concentration, the magnitude of response was decreased. On the other hand, the units predominantly sensitive to NaCl sodium saccharin

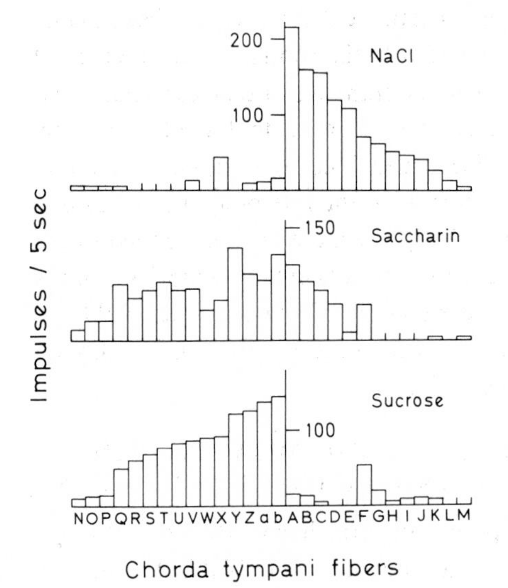

Figure 9 Response profile for 0.02 M sodium saccharin of 28 hamster chorda tympani fibers. On the left side of the vertical axis, units predominantly sensitive to sucrose were arranged in the order of responsiveness to 0.5 M sucrose; on the right side, units primarily sensitive to 0.1 M NaCl were arranged in the order of responsiveness to 0.1 M NaCl.

reacted as though it were an ordinary sodium salt; its concentration-response curve was similar in shape to the NaCl curve. In the rat, the distinction between the two types of units is not as marked as in the hamster (Figure 2), but a similar difference in the response to sodium saccharin was observed between units predominantly sensitive to NaCl and those more sensitive to sucrose than to NaCl. This is illustrated in Figure 10, in which two units of rats are exclusively sensitive to NaCl while the other two are more sensitive to sucrose than to NaCl. This difference in behavior for sodium saccharin may result from a difference in the receptor molecules over the surface of

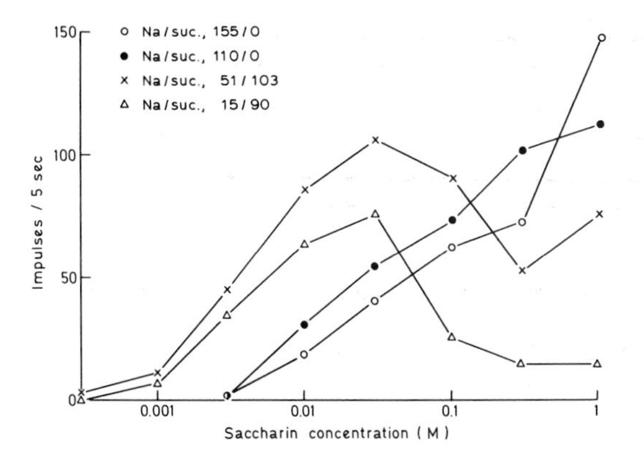

Figure 10 Concentration-response curve for sodium saccharin in four units in rats. Two units are predominantly sensitive to NaCl, while the remaining two are more sensitive to sucrose than to NaCl. Numerals at the top of the figure indicate number of impulses discharged in each unit during 5 sec after stimulation by 0.1 M NaCl and 0.5 M sucrose.

gustatory receptors; probably the receptor molecules in the NaCl-sensitive units react with sodium ions, while those in the sucrose-sensitive units combine with the anionic base of saccharin. This possibility can be supported by the experiment shown in Figure 11, where the unit responding to both 0.1 M NaCl and 0.01 M sodium saccharin did not give any response to 0.01 M saccharin solution buffered with tris because of the absence of Na ions. On the other hand, units predominantly sensitive to sucrose produced impulse discharges in response to tris-buffered 0.01 M saccharin as well as to 0.01 M sodium saccharin (Figure 12).

Impulse discharges elicited by sodium saccharin in the units dominantly sensitive to NaCl consisted of an initial dynamic phase of high rate and a subsequent steady phase (Figure 13 B), while those predominantly sensitive to sucrose showed rhythmic bursts of impulse discharges to both sucrose and saccharin, with intervals between bursts of 250–750 msec (Figure 13 D and E). Analyses of temporal pattern of impulses were made by calculating the mode value of impulse intervals and the autocorrelogram, using a computer. The results of analyses on 11 units are shown in Table VII. Seven units among 11 were more sensitive to sucrose than to NaCl, and five units presented a regular rhythm of 250–750 msec in impulse discharges to both 0.5 M sucrose and 0.02 M sodium saccharin. On the other hand, no regular rhythms were found in impulse discharges produced by saccharin in units which responded to NaCl better than to sucrose, although they showed similar mode values of impulse intervals to those in units dominantly sensitive to sucrose. Such differences in the temporal

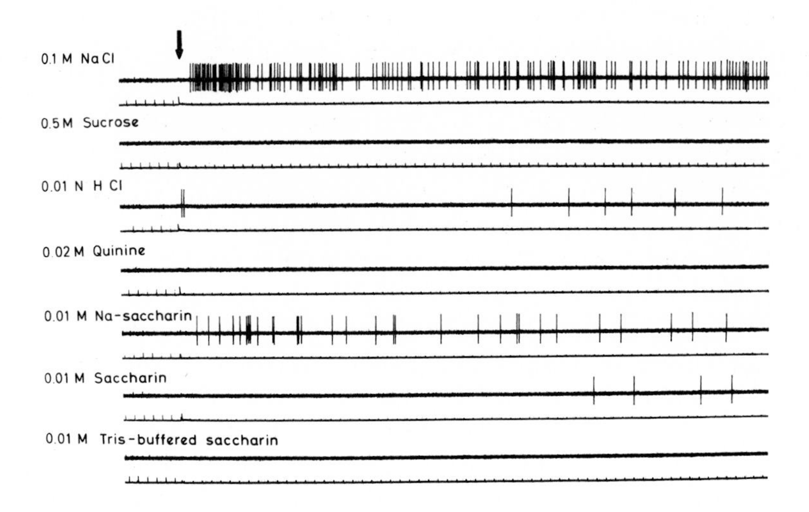

Figure 11 Impulse discharges produced by seven kinds of chemicals in a single rat chorda tympani fiber. Note the absence of response to 0.01 M saccharin solution buffered with tris. Delayed impulse discharges in the response to 0.01 M saccharin are considered to result from its low pH because of their similarity to those in the response to HCl. Time signal: 0.1 sec.

pattern of impulse discharges to sodium saccharin between the unit types indicate a difference in the mechanism of the reaction occurring between receptors and sodium saccharin; in the NaCl-sensitive unit sodium ions react with the receptor-molecules, while in the sucrose-sensitive unit saccharin combines with the receptor-molecules.

A majority of chorda tympani units in the rat responded to 0.1 M NaCl, while only half of the total population showed sensitivity to 0.5 M sucrose. In addition, eight units among 48 responded to 0.1 M NaCl alone, while only one unit showed exclusive sensitivity to 0.5 M sucrose. This indicates that, in the rat, NaCl-receptor molecules or, rather, Na^+-receptor molecules dominate in number over sucrose-receptor molecules. On the other hand, in the hamster the situation is reversed: units sensitive to sucrose are greater in number than those sensitive to NaCl. Four units among 28 responded to 0.5 M sucrose alone and only one unit gave responses to 0.1 M NaCl

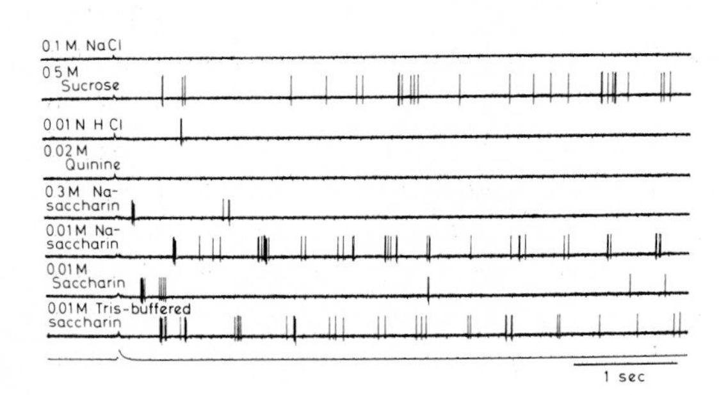

Figure 12 Impulse discharges produced by seven kinds of chemicals in a single rat chorda tympani fiber. Note responses to 0.01 M saccharin solution as well as to sucrose.

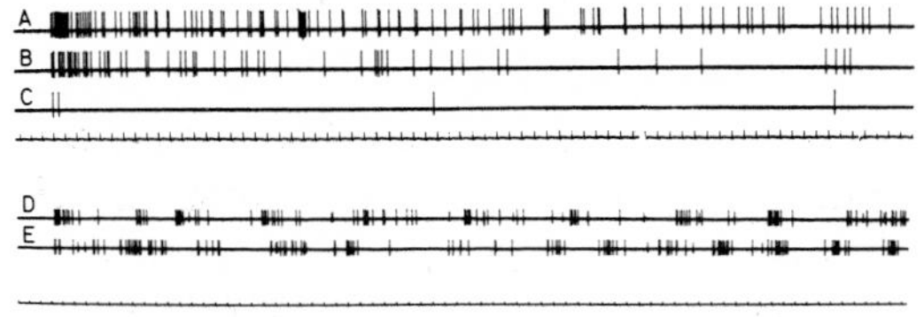

Figure 13 Impulse discharges in two rat chorda tympani fibers, the one predominantly sensitive to NaCl (A-C) and the other more sensitive to sucrose than to NaCl (D and E). A, impulse discharges elicited by 0.1 M NaCl; B, those by 0.02 M sodium saccharin; C, spontaneous discharges; D, impulse discharges by 0.5 M sucrose and E, those by 0.02 M sodium saccharin. Time signal: 0.1 sec.

exclusively. Thus sucrose-receptor molecules dominate over Na^+-receptor molecules in the hamster. Presence of negative correlation between the response to sucrose and that to NaCl suggests reciprocal distribution of these two kinds of receptor molecules in the receptor. If one examines the categories of chorda tympani fibers shown in Table V, one finds a certain number of Na, Na-S, Na-H and Na-Q types in the rat, but in the hamster S and S-H types occupy a similar proportion of the total population.

TABLE VII
PERIODICITY OF IMPULSE DISCHARGES IN CHORDA TYMPANI NERVE FIBERS

No.	Animals	Magnitude of response to four basic stimuli NaCl:Suc.:HCl:QHCl*	Stimulus	Mode of impulse intervals (msec)	Periodicity found from auto-correlogram (msec)
1.	Rat	66: 15:29:11	0.5 M Sucrose	17	(—)
			0.02 M Saccharin	8	(—)
2.	Rat	18: 33:17: 7	0.5 M Sucrose	8	(—)
3.	Rat	1: 72: 1: 0	0.5 M Sucrose	10	750
			0.02 M Saccharin	10	750
4.	Rat	25: 75: 3: 1	0.5 M Sucrose	10	420
			0.02 M Saccharin	10	480
5.	Rat	51:103:17: 0	0.5 M Sucrose	10	(—)
			0.02 M Saccharin	10	450
6.	Rat	96: 6: 7: 8	0.1 M NaCl		(—)
			0.02 M Saccharin		(—)
7.	Rat	15: 90: 6: 0	0.5 M Sucrose	12	560
			0.02 M Saccharin	10	560
8.	Rat	110: 0:15: 7	0.1 M NaCl	20	(—)
			0.3 M NaCl	15	(—)
			0.1 M Saccharin	18	(—)
			0.3 M Saccharin	15	(—)
			1 M Saccharin	20	(—)
9.	Rat	36: 38: 4: 2	0.5 M Sucrose	15	400
			0.02 M Saccharin	10	250
10.	Rat	103: 35:11: 6	0.5 M Sucrose	10	(—)
			0.02 M Saccharin	10	(—)
11.	Hamster	15:145: 6: 3	0.5 M Sucrose	15	630
			0.02 M Saccharin	15	660

* 0.1 M NaCl:0.5 M Sucrose:0.01 N HCl:0.02 M QHCl.

CATEGORIES OF CHORDA TYMPANI FIBERS IN THE CAT AND MONKEY

Nagaki et al.[7] examined the responsiveness of 28 chorda tympani fibers in the cat, and found that 23 units among 28 responded to a variety of gustatory stimuli as well as to thermal changes (either cooling or both cooling and warming). Sensitivity of cat chorda tympani fibers is different from that of taste fibers of rats and hamsters in that the former do not possess sensitivity to sucrose but respond to water. In the cat, the response to NaCl is negatively correlated with that to water, as though sucrose-receptor molecules in the gustatory receptors of the hamster were replaced by water-receptor molecules in the cat. In the cat, responses to cooling seem to be positively correlated with those to HCl and quinine hydrochloride, although definite evidence on this point has not been presented by Nagaki et al.[7] The response to warming in the cat is not correlated with responses to any chemical stimuli. Sodium saccharin produced responses in both fibers dominantly sensitive to NaCl and those to water. These results by Nagaki et al. seem to indicate the presence of distinct categories in cat chorda tympani fibers.

Responses of chorda tympani fibers of monkeys were examined by Gordon et al.[6] They found that, as in other mammals, a multiple responsiveness of single units to a variety of gustatory stimuli such as NaCl, sucrose, quinine, HCl, saccharin, and water. However, they have classified units simply into "sweet fibers," "salt fibers," "acid fibers," "quinine fibers," and "water fibers," according to relatively specific responsiveness to chemicals. Therefore, a quantitative analysis of responsiveness of the chorda tympani fibers of monkeys would be necessary for categorizing them.

NEURAL INFORMATION FOR QUALITIES OF TASTE

The results presented above do not indicate the presence of absolutely specific sensitivity of any chorda tympani fibers to only one kind of chemical, but suggest relatively specific sensitivity to one or a combination of chemical stimuli. Therefore, information for a chemical stimulus would be carried along a number of fibers, which are relatively sensitive to this stimulus, but in addition are responsive to other kinds of chemicals. Consequently, neural information for a quality of taste depends on the pattern of impulse discharges across many neurons, as proposed by Pfaffmann,[10] Erickson,[2,3] and Erickson, Doetsch and Marshall.[4] However, because of the presence of several categories of chorda tympani fibers with relatively specific sensitivity to one, or a combination, of four basic gustatory stimuli, some modification of the "across-fiber pattern theory" by Erickson may be necessary.

As shown in Table VIII, simple correlation coefficients across many neurons between the amount of response to 0.02 M sodium saccharin and that for 0.1 M NaCl are not significant in both rats and hamsters, although there exists a highly significant correlation between the response to 0.02 M sodium saccharin and that for 0.5 M sucrose. It can be considered from the data shown in Figure 9 that neural information for 0.02 M sodium saccharin consists mainly of that for 0.5 M sucrose and that

TABLE VIII
CORRELATION COEFFICIENTS BETWEEN AMOUNTS OF RESPONSES TO A PAIR OF STIMULI

	0.1 M NaCl	0.5 M Sucrose	0.01 N HCl	0.02 M Quinine	
0.02 M Sodium saccharin	0.10	0.74	−0.15	−0.45	Hamster (27)
0.02 M Sodium saccharin	0.13	0.83	0.00	−0.02	Rat (47)
0.1 M KCl	0.00	0.21	0.82	0.88	Rat (25)
0.3 M KCl	0.08	0.12	0.72	0.76	Rat (18)
0.3 M $CaCl_2$	0.13	0.11	0.91	0.52	Rat (17)
0.3 M $MgCl_2$	0.43	0.05	0.78	0.74	Rat (17)
0.01 N Tartaric acid	−0.05	0.01	0.85	0.56	Rat (25)

Correlation coefficients were based on numbers of impulses elicited in the first 5 sec after stimulation. Numerals in parentheses indicate numbers of units employed for calculation.

for 0.1 M NaCl. This would explain why any significant correlation is absent between the amount of response to saccharin and that for NaCl, in spite of the fact that units predominantly sensitive to NaCl respond very well to sodium saccharin. Consequently, to estimate the correlation between the response to saccharin and that to NaCl or sucrose, one should calculate a partial correlation coefficient among the three. The results of calculations of the partial correlation coefficient between the response to saccharin and those to NaCl and sucrose are shown in Table IX, which indicates highly significant correlations between saccharin and NaCl and between the former and sucrose. The regression equations that express the relationship among the responses to the three chemicals, shown in Table IX, indicate how the neural information for sodium saccharin consists of that for NaCl and that for sucrose.

Extending this concept further, one is able to obtain equations that express the neural information for any chemicals by those for four basic gustatory stimuli. Equations for various chemicals, presented in Table X, were obtained by calculating the relationships between the magnitudes of responses to these chemicals and those for the four basic stimuli according to the least-square method. The equations reveal

TABLE IX
CORRELATION BETWEEN THE AMOUNT OF RESPONSE TO SACCHARIN (z) AND THOSE FOR NaCl (x) AND SUCROSE (y) IN 27 HAMSTER UNITS

	0–5 sec		5–10 sec	
	0.1 M NaCl	0.5 M Sucrose	0.1 M NaCl	0.5 M Sucrose
Simple correlation coefficient	0.10	0.74	0.19	0.55
Partial correlation coefficient	0.67	0.865	0.55	0.70
Multiple correlation coefficient	0.87		0.72	
Regression equation	$z = 0.326x + 0.758y - 0.462$		$z = 0.294x + 0.623y + 0.110$	

TABLE X
TASTE EQUATIONS FOR VARIOUS CHEMICALS

Stimuli	NaCl	Sucrose	HCl	QHCl		
0.02 M Sodium saccharin	$z = 0.356x$	$+0.705y$	$-0.090u$	$-0.368v + 7.289$	Hamster	(27)
0.02 M Sodium saccharin	$z = 0.335x$	$+0.905x$	$-0.136u$	$+0.259v - 2.383$	Rat	(47)
0.1 M KCl	$z = 0.016x$	$+0.037y$	$+0.164u$	$+0.672v + 2.486$	Rat	(25)
0.3 M KCl	$z = -0.070x$	$+0.005y$	$+0.179u$	$+0.827v + 10.959$	Rat	(26)
0.3 M $MgCl_2$	$z = 0.329x$	$-0.067y$	$+0.373u$	$+0.938v + 2.673$	Rat	(25)
0.3 M $CaCl_2$	$z = -0.076x$	$-0.579y$	$+0.999u$	$-0.259v + 10.296$	Rat	(25)
0.01 N Tartaric acid	$z = -0.154x$	$-0.117y$	$+0.897u$	$-0.486v + 10.828$	Rat	(29)

x: Number of impulses elicited in the first 5 sec after stimulation by 0.1 M NaCl.
y: Number of impulses elicited in the first 5 sec after stimulation by 0.5 M sucrose.
u: Number of impulses elicited in the first 5 sec after stimulation by 0.01 N HCl.
v: Number of impulses elicited in the first 5 sec after stimulation by 0.02 M quinine hydrochloride.
z: Number of impulses elicited in the first 5 sec after stimulation by the chemical shown in the left-hand column of the Table.

well how neural information for various chemicals consists of those for four fundamental stimuli. In comparing Table X with Table VIII, one may notice some discrepancies. For example, the amount of responses to KCl is significantly correlated with that for HCl and quinine (Table VIII), but equations for KCl in Table X show that information for KCl is predominantly made up by that for quinine. The latter fact indicates a bitter taste for KCl rather than a bitter-acid taste. Similarly, in Table VIII information for 0.02 M sodium saccharin is correlated only with that for sucrose, but according to the equations shown in Table X it consists of information for sucrose (sweet) as well as of those for NaCl (salty) and quinine (bitter). Consequently, one may be able to obtain equations expressing quality of taste for any chemical by those for four basic gustatory stimuli. However, the coefficients in the equation would change with changes in concentration of any one of x, y, u, v and z.

SUMMARY

Responsiveness of chorda tympani fibers of rats and hamsters to the four basic gustatory stimuli, sodium saccharin, and thermal stimuli was examined in detail, and an attempt was made to classify units into several categories according to their responsiveness to the four basic stimuli as well as thermal stimuli.

In the rat, nine units among 48 responded to only one kind of stimulus, and eight units to two kinds. Such specific units barely responded to cooling of the tongue and showed a low rate of spontaneous discharge. Nineteen units in the rat responded to three kinds of stimuli and 12 units to all the four basic stimuli. Such nonspecific units are sensitive to cooling and showed a high rate of spontaneous discharge.

In the hamster, five units among 28 responded to only one kind of stimulus, and five units to two. They did not respond to cooling and showed a low rate of spontaneous discharge. Eleven units responded to three kinds of stimuli and seven units to all the four basic ones. These 11 showed a high rate of spontaneous discharge and responded to cooling.

Eight hamster units responded to warming of the tongue and 12 to both cooling and warming. Responsiveness to warming is correlated with that to sucrose.

Units predominantly sensitive to sucrose as well as those primarily sensitive to NaCl responded to 0.02 M sodium saccharin. However, the two types of units are different in the concentration-response curve for saccharin as well as in the temporal pattern of impulse discharges, indicating differences in the mechanism of reactions between sodium saccharin and gustatory receptors. The present paper has thus revealed that, although most chorda tympani fibers showed multiple responsiveness to the four basic gustatory and thermal stimuli, they can be classified statistically into several distinct categories.

Neural information for various chemicals was expressed by equations relating it to those for four basic gustatory stimuli.

ACKNOWLEDGMENTS

This work was supported by the US Air Force Office of Scientific Research through USA and DGp (Far East) Grants No. DA-CRD-AG-S92-544-62-G30, No. DA-CRD-AFE-S92-544-68-G121 and Contract No. DAJB17-69-C-0073. Authors are grateful to the technical staffs in Computer Division, Japan Electron Optical Laboratory Co., Ltd. for the analysis of spike trains with data-processing computer JRA-5.

REFERENCES

1. Cohen, M. J., S. Hagiwara, and Y. Zotterman. 1955. The response spectrum of taste fibres in the cat: A single fibre analysis. *Acta Physiol. Scand.* **33**, pp. 316–332.
2. Erickson, R. P. 1963. Sensory neural patterns and gustation. *In* Olfaction and Taste I (Y. Zotterman, editor). Pergamon Press, Oxford, etc., pp. 205–213.
3. Erickson, R. P. 1967. Neural coding of taste quality. *In* The Chemical Senses and Nutrition (M. R. Kare and O. Maller, editors). The Johns Hopkins Press, Baltimore. Md., pp. 313–327.
4. Erickson, R. P., G. S. Doetsch, and D. A. Marshall. 1965. The gustatory neural response function. *J. Gen. Physiol.* **49**, pp. 247–263.
5. Fishman, I. Y. 1957. Single fiber gustatory impulses in rat and hamster. *J. Cell. Comp. Physiol.* **49**, pp. 319–334.
6. Gordon, G., R. Kitchell, L. Ström, and Y. Zotterman. 1959. The response pattern of taste fibres in the chorda tympani of the monkey. *Acta Physiol. Scand.* **46**, pp. 119–132.
7. Nagaki, J., S. Yamashita, and M. Sato. 1964. Neural response of cat to taste stimuli of varying temperatures. *Japan. J. Physiol.* **14**, pp. 67–89.
8. Ogawa, H., M. Sato, and S. Yamashita. 1968. Multiple sensitivity of chorda tympani fibres of the rat and hamster to gustatory and thermal stimuli. *J. Physiol.* (*London*) **199**, pp. 223–240.
9. Pfaffmann, C. 1941. Gustatory afferent impulses. *J. Cell. Comp. Physiol.* **17**, pp. 243–258.
10. Pfaffmann, C. 1955. Gustatory nerve impulses in rat, cat, and rabbit. *J. Neurophysiol.* **18**, pp. 429–440.
11. Siegel, S. 1956. Nonparametric Statistics for the Behavioral Sciences. McGraw-Hill Book Co., New York.
12. Yamashita, S., H. Ogawa, and M. Sato. 1967. Multimodal sensitivity of taste units in the rat. *Kumamoto Med. J.* **20**, pp. 67–70.
13. Yamashita, S., H. Ogawa, and M. Sato. 1967. Analysis of responses of hamster taste units to gustatory and thermal stimuli. *Kumamoto Med. J.* **20**, pp. 159–162.

THE DISTRIBUTION OF TASTE SENSITIVITIES AMONG SINGLE TASTE FIBERS

MARION FRANK *and* CARL PFAFFMANN

The Rockefeller University, New York, N.Y.

Most studies of single mammalian taste fibers have been restricted to fibers in the chorda tympani nerve. The rat, in particular, has been studied extensively; but its chorda tympani primarily mediates responses to only two of four common taste stimuli. It responds strongly to sodium chloride and acid, but weakly to quinine and sucrose. Therefore, the distribution of sensitivities to these four substances among single fibers innervating the rat's anterior receptive field would necessarily be biased. The posterior glossopharyngeal receptive field of the rat's tongue, however, is more equally sensitive to all four tastes.

In our study, we isolated 31 taste fibers of the rat glossopharyngeal nerve, and determined receptive fields and adequate stimuli for each fiber. Taste solutions were flowed through a pipette inserted into the relevant circumvallate or foliate papilla. Test stimuli were 0.3 M sucrose, 0.01 M hydrochloric acid, 0.001 M quinine hydrochloride, and 0.3 M sodium chloride. These stimulus intensities produce approximately 50 per cent of the total nerve's maximum response to these substances.

We chose a criterion of a 50 per cent increase in response rate during the first five seconds of stimulation. Nineteen fibers responded to stimulation of the circumvallate and 12 to stimulation of the foliate papillae. Eight fibers responded to one, 12 to two, five to three, and two to all four taste stimuli. Four responded only to water. Responses of one fiber are shown in Figure 1.

It is possible that sensitivities to the four test stimuli, each representing one of the four taste qualities—sweet, sour, bitter, and salty—are independent and randomly distributed among taste fibers. The binomial distribution can describe a random process comprised of a number of mutually independent events (N) if the probabilities of occurrence (p_x) and nonoccurrence (q_x) of these events are known. If there are independent receptor mechanisms for each of the four taste qualities ($N = 4$), the probabilities of fibers responding to one, two, three, or four qualities (P_n, $n = 1, 2, 3, 4$) will be given by successive terms of the binomial expansion $[(p + q)^N]$. Also, the probabilities that fibers are sensitive to any two of the four tastes will be given by the product of the probabilities for the individual sensitivities ($p_x \cdot p_y$). The probability of occurrence of a sensitivity (p_x) can be estimated by the proportion of the sample of fibers that responds to a particular stimulus.

Of 27 rat glossopharyngeal taste fibers, 6/10 responded to sodium chloride, 4/10

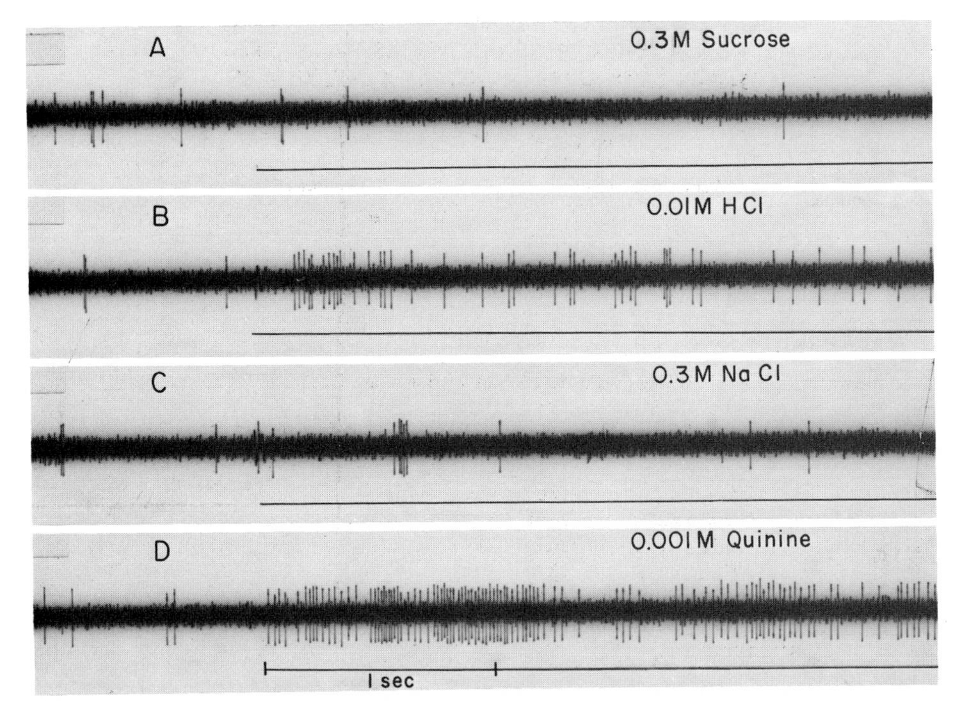

Figure 1 The electrophysiological response of one glossopharyngeal fiber to four stimuli. Four seconds of record is shown. The line beneath each record indicates application of the stimulus.

to quinine, 6/10 to hydrochloric acid, and 4/10 to sucrose. The number of fibers sensitive to one, two, three, or four tastes, predicted by successive terms of the binomial expansion and the number obtained in this set of fibers, are given in Table IA. The predicted probabilities for each of the six combinations of sensitivities to two of the four tastes, the predicted number of fibers, and the obtained number of fibers from the glossopharyngeal nerve with each combination of sensitivities are given in Table IB.

Through the cooperation of R. P. Erickson, responses of 25 rat chorda tympani fibers to the same substances were made available. The intensities of the test stimuli (0.1 M NaCl, 0.03 M HCl, 0.01 M quinine, and 1.0 M sucrose) were different. Eight-tenths of these chorda tympani fibers responded to sodium chloride, 7/10 to hydrochloric acid, 4/10 to quinine, and 2/10 to sucrose. Differences in these proportions for the two taste nerves reflect differences both in the sensitivities of the nerve receptive fields and in the intensities of the test stimuli. Table IIA gives the obtained and predicted numbers of chorda tympani fibers sensitive to one, two, three, or four stimuli; Table IIB gives the probabilities, and both the predicted and the observed number of chorda tympani fibers sensitive to the combination of two stimuli. The assumption that independent sensitivities to four taste qualities are randomly distributed among

TABLE I
DISTRIBUTION OF SENSITIVITIES TO 4 TASTES IN RAT GLOSSOPHARYNGEAL TASTE FIBERS

A. Responses to 1, 2, 3, or 4 tastes

Number of Responses	Number of Fibers	
	Predicted	Obtained
1	7.1	8
2	11.2	12
3	7.1	5
4	1.7	2

B. Responses to each of 6 pairs of the 4 tastes

Pair of Stimuli	(p_x)	(p_y)	Number of Fibers	
			Predicted	Obtained
NaCl, HCl	(0.6)	(0.6)	9.7	9
NaCl, quinine	(0.6)	(0.4)	6.5	6
NaCl, sucrose	(0.6)	(0.4)	6.5	7
HCl, quinine	(0.6)	(0.4)	6.5	8
HCl, sucrose	(0.6)	(0.4)	6.5	6
quinine, sucrose	(0.4)	(0.4)	4.3	3

TABLE II
DISTRIBUTION OF SENSITIVITIES TO 4 TASTES IN RAT CHORDA TYMPANI TASTE FIBERS

A. Responses to 1, 2, 3, or 4 tastes

Number of Responses	Number of Fibers	
	Predicted	Obtained
1	5.4	5
2	11.3	13
3	7.2	6
4	1.0	1

B. Responses to each of 6 pairs of the 4 tastes

Pair of Stimuli	(p_x)	(p_y)	Number of Fibers	
			Predicted	Obtained
NaCl, HCl	(0.8)	(0.7)	14.0	14
NaCl, quinine	(0.8)	(0.4)	8.0	8
NaCl, sucrose	(0.8)	(0.2)	4.0	4
HCl, quinine	(0.7)	(0.4)	7.0	6
HCl, sucrose	(0.7)	(0.2)	3.5	2
quinine, sucrose	(0.4)	(0.2)	2.0	2

innervating fibers allows us to describe distributions of sensitivities in both chorda tympani and glossopharyngeal nerve fibers of the rat.

Mammalian fibers differ from invertebrate taste fibers (e.g., blowfly), which are highly specific. Our descriptions emphasize the characteristic nonspecificity of mammalian taste fibers and provides a quantitative account of the distribution of sensitivities to different taste qualities in such taste fibers.

INFORMATION PROCESSING IN THE TASTE SYSTEM OF THE RAT

GERNOT S. DOETSCH, JUDITH JAY GANCHROW, LINDA M. NELSON, *and* ROBERT P. ERICKSON

Department of Psychology, Duke University, Durham, North Carolina
Dr. Doetsch's present address: Comparative Psychology Division, 6571st Aeromedical Research Laboratory, Holloman A.F.B., New Mexico

INTRODUCTION

In recent years, the processing of sensory information at various synaptic relays has been a particularly rewarding area of research in senses other than olfaction and taste. For example, single-unit studies at progressively higher levels of the visual system have provided data on the coding of information about color or form and the kinds of modifications that may be imposed on neural messages in the process of synaptic transmission.[3,14] Similar findings on the coding and synaptic transfer of sensory input have also been obtained in the somesthetic system[20] and in the auditory system.[15,28]

Analogous data on the synaptic processing of afferent information are not available for the chemical senses. Several reasons for this lack of data are immediately obvious. First, in order to evaluate certain aspects of the synaptic transfer of sensory messages, subtle manipulations of the stimuli with respect to their appropriate dimensions are required. Whereas the stimulus dimensions related to sensory quality in the nonchemical senses are well established, the dimensions underlying taste and olfaction are unknown. The resulting inability to perform the proper stimulus manipulations in the chemical senses has made it rather difficult to determine in detail the effects of synaptic transfer on sensory input.

Second, the central projections of the olfactory and taste systems have not been as accurately defined as those of the nonchemical senses. This is especially true for olfaction, the central projections of which are particularly complex. For taste, the lack of publications on cortical neurons responding to chemical stimulation of the tongue illustrates how little information is available. The only reports on such cortical cells appear to be those of Cohen, Landgren, Ström, and Zotterman[2] and Landgren,[16] who found seven cells in the cortical tongue projection area of the cat that responded to chemical as well as to thermal and mechanical stimulation of the tongue. As long as the locations of the central projections are in doubt, it will be difficult to investigate the processing of sensory input at these higher levels.

This paper reports several efforts to approach the problem of information processing in the taste system in spite of the inherent difficulties. First, the code for taste quality has been described in enough detail[6,7,9] to permit an examination of the manner in which taste information is processed during synaptic transmission. The transfer of taste-quality information was studied at the nucleus tractus solitarius in the medulla—first synaptic relay in the gustatory system of the rat. Second, the transfer of information about taste intensity at this relay was also examined. Third, examination of the central taste projections was pursued by studying neural convergence in the gustatory system and the representation of taste in the cortex.

GENERAL METHODS

The data presented below were obtained from rats deeply anesthetized with sodium pentobarbital. Neural activity was recorded from single fibers of the chorda tympani nerve (CT) by placing small dissected nerve strands over a fine (125 μ) nichrome wire electrode. Glass micropipettes (tip diameter 1.5 to 4.0 μ) filled with 3 M KCl or 3 M NaCl were used to record the responses of single neurons in the nucleus tractus solitarius (NTS), thalamus, and cortex. The response properties of the taste cells were studied by means of chemical stimulation of the tongue. Electrical stimulation of the CT, glossopharyngeal nerve (IX), or cortex was used to investigate the anatomical and functional relationships between various synaptic levels of the taste system.

TRANSFER OF INFORMATION ABOUT TASTE QUALITY

To disclose some of the basic properties of information processing at the first synaptic relay in the NTS, the data obtained from 62 first-order CT fibers were compared with the data obtained from 41 second-order NTS cells. At both levels, the responses of single neurons to chemical stimulation of the tongue were studied by using the following stimuli and concentrations: 1.0 M sucrose; 0.3 M KCl and $CaCl_2$; 0.1 M NaCl, $NaNO_3$, Na_2SO_4, LiCl, Li_2SO_4, NH_4Cl, $MgCl_2$, and NaOH; 0.03 M HCl and HNO_3; and 0.01 M quinine hydrochloride (QHCl). A number of analyses were performed on the data obtained from the two levels in order to examine several aspects of the synaptic transfer of taste-quality information.

The results of this study revealed that certain features of the neural representation of taste quality remained relatively unchanged, whereas others were obviously modified, in the process of synaptic transmission. Those properties which were found to be preserved from the CT to the NTS neurons are summarized below:

1. The cells at both levels were sensitive to a relatively broad range of chemical stimuli, typically representing more than one of the "four basic taste qualities."

2. The form of the neural response function (NRF) of the NTS neurons closely approximated that of the CT fibers. (An NRF is a curve describing the sensitivity or response of a receptor or neuron to stimuli of a particular dimension.[9])

3. The relative degrees of similarity among the taste messages for individual stimuli were essentially the same at both synaptic levels (e.g., the neural inputs for NaCl and LiCl were very similar to each other in the CT and in the NTS).

4. The neural representation of the stimulus dimensions for taste quality was very similar at both levels, i.e., the relative positions of the stimuli in the dimensions derived from the two sets of data were basically the same.

Certain aspects of the sensory input were found to be modified during synaptic transfer in the NTS as follows:

1. The average rate of response to all stimuli was amplified by a factor of about 4.3.
2. The initial evoked burst of activity was attenuated relative to the steady state of response.
3. The neural message for each stimulus showed greater stability over time.
4. The neural messages for different stimuli became less distinct from each other, i.e., the neural inputs for any two stimuli were more highly correlated at the NTS than at the CT level.
5. The neural representation of the stimulus dimensions showed more stability over time, i.e., the relative positions of the stimuli in the dimensions derived from the NTS data were less subject to change over time.

In general, it was found that the afferent input is refined in the process of synaptic transfer so as to produce less "noisy," more stable taste messages in the NTS. A more detailed discussion of these various findings follows.

General Response Properties of NTS Cells

In terms of their over-all sensitivity to chemical stimulation, the second-order NTS cells closely resembled the first-order CT fibers. At both synaptic levels, most neurons were activated by each of the 14 taste stimuli used, but the profile of responsiveness to these stimuli differed from one neuron to another. For example, one cell gave large responses to the sodium and lithium salts, intermediate responses to the other salts and the acids, a relatively small response to QHCl, and little or no response to sucrose. Other neurons showed different relative degrees of sensitivity to these same stimuli. (At both synaptic levels, sucrose and QHCl typically produced relatively weak responses.) Thus, the second-order cells, like the first-order fibers, were characterized by differential sensitivity to a relatively broad range of stimuli and were not specifically sensitive to any particular class of stimuli representing one of the "four basic taste qualities."

The Code for Taste Quality: Across-Neuron Response Patterns

When it became clear that single CT fibers typically respond to stimuli from a variety of taste categories, Pfaffmann[23,24] suggested that the code for taste quality consists of the relative amounts of activity produced in many afferent fibers. Erickson[6,7,8] and Erickson, Doetsch, and Marshall[9] adopted this view and developed an "across-neuron patterning" model as a representation of the code for taste quality. According to this model, any given stimulus may activate a relatively large number of cells, each to a somewhat different degree, producing a pattern of response frequencies across all neurons that is unique to the stimulus and constitutes the neural message for that stimulus. The code for taste quality thus takes the form of the relative amounts or ratios of activity in all neurons, i.e., the form of across-neuron response patterns.

Furthermore, the model predicts that discrimination among taste stimuli depends on differences in these across-neuron response patterns. More specifically, the degree of taste similarity between any two stimuli (and thus the difficulty of discrimination) should be directly related to the similarity between the neural patterns produced by these stimuli. This hypothesis has been supported by a number of behavioral taste experiments performed on rats.[6,19,21]

Figure 1 illustrates the manner in which the neural patterns for different stimuli may be represented. The responses of six CT fibers and six NTS cells to NaCl and

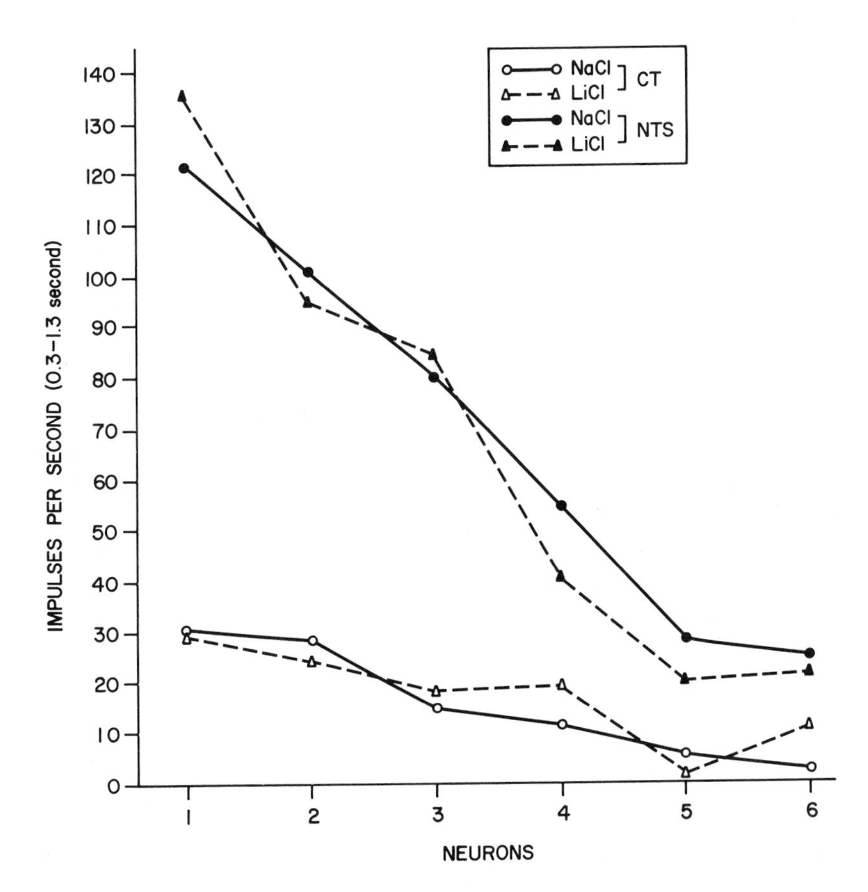

Figure 1 Across-neuron response patterns representing the code for taste quality at the chorda tympani (CT) and nucleus tractus solitarius (NTS) levels. Six CT neurons (open symbols) and six NTS neurons (closed symbols) are shown on the abscissa in order of their decreasing responsiveness to NaCl. The neural patterns produced by NaCl and LiCl were very similar to each other at both levels, which corresponds to behavioral data indicating that NaCl and LiCl taste very much alike. The figure also shows that the general level of neural activity was amplified in the NTS.

LiCl are shown. These units are representative of the 62 CT and 41 NTS neurons tested. The Figure indicates that the across-neuron response patterns associated with NaCl and LiCl were very similar to each other at both levels of the taste system; other stimuli produced quite different patterns of activity across these same neurons. Consequently, both the CT and NTS data presented in Figure 1 lead to the same prediction, namely that the tastes of NaCl and LiCl are very similar.

In general, it may be said that the "across-neuron patterning" model for the representation of taste quality applies to the NTS as well as to the CT, because the results obtained from each level suggest similar predictions concerning taste discrimination. These predictions are supported by behavioral experiments.

Amplification of the Neural Responses in the NTS

One of the most striking differences obtained between the CT and the NTS was in the frequency of discharge produced during the evoked response. This finding is illustrated in Figure 1. The average rate of response of the NTS cells to all stimuli was 4.3 times greater than that of the CT fibers. Thus, the neural activity which carries the taste messages was amplified during the process of synaptic transfer in the NTS.

Increased Similarity of the Neural Messages in the NTS

Further study of the data obtained from the CT and NTS revealed that the across-neuron response patterns for individual taste stimuli became more similar to each other, as measured by product-moment correlations between pairs of stimuli. This increased similarity is shown in Figure 2, in which the correlations obtained from the CT data for a number of stimulus pairs are compared with the correlations from the NTS data. The values indicated are representative of all 90 pairs of interstimulus correlations that were compared. Of these 90 pairs, all except five were higher for the NTS than for the CT, indicating that the neural messages became less distinct from each other during synaptic transmission.

On the other hand, these data also show that the relative degrees of similarity between the neural patterns for the various stimuli were preserved during transfer of the messages from the CT to the NTS. Any two stimuli which were highly correlated at the CT also tended to be highly correlated at the NTS; conversely, any two stimuli which showed a low correlation at the CT also tended to show a low correlation at the NTS. For example, the neural patterns for NaCl and LiCl were among the most similar at both synaptic levels, whereas the neural patterns for Na_2SO_4 and NH_4Cl were among the most dissimilar at both levels. Thus, despite the over-all increase in the interstimulus correlations, the data from the CT and NTS lead to similar predictions concerning the relative degrees of taste similarity among the various stimuli.

The increase in similarity of the taste messages in the NTS presents an interesting problem with respect to taste discrimination. It is very likely that the degree of discriminability is some function of the absolute difference between the neural inputs

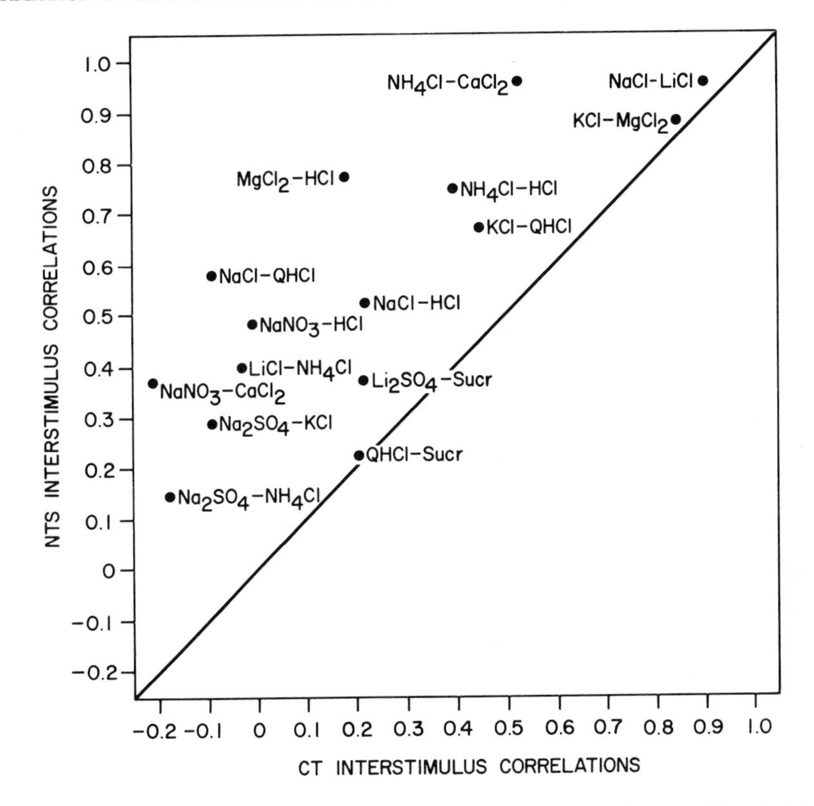

Figure 2 Increase in similarity of the across-neuron response patterns from the CT to the NTS. The similarity between the neural patterns for any two stimuli is expressed by a correlation coefficient, higher correlations indicating greater similarity. The correlations obtained from the CT data for a number of stimulus pairs are given on the abscissa; the correlations from the NTS data for the same pairs of stimuli are shown on the ordinate. The diagonal line represents points of equal correlation for the CT and NTS data; points falling above this line indicate that higher correlations were obtained from the NTS data. Of all 90 pairs of stimuli compared, the correlations for 85 pairs increased from the CT to the NTS, which indicates that the similarity of the neural patterns for these pairs of stimuli increased.

for any two stimuli.[8] The higher correlations obtained from the NTS data indicate that the differences between the forms of the across-neuron response patterns are reduced, whereas the amplification of responses occurring in synaptic transfer tends to counteract this effect by producing an increase in the absolute difference between the two neural inputs. It is possible that these two phenomena reflect a mechanism whereby information about differences in taste quality is preserved during synaptic transfer, but which modifies the form in which the differential input is represented at the two levels. Due to the low response rates in the CT, relatively large differences must exist between the forms of the neural patterns to provide for a given degree of discriminability. On the other hand, due to the high response rates produced in the NTS, the differences between the forms of these patterns need not be as great to provide for the same degree of discriminability.

Temporal Characteristics of the Neural Messages

To evaluate the temporal stability of the neural message for any given stimulus, product-moment correlations were obtained between the neural patterns produced by that stimulus during various time intervals after the beginning of the response. If the taste message, i.e., the across-neuron response pattern, remains relatively unchanged over time, correlations between the neural activity at different time intervals would be high; on the other hand, temporal changes in the message would be indicated by lower correlations. Figure 3 shows such correlations obtained for the CT and NTS

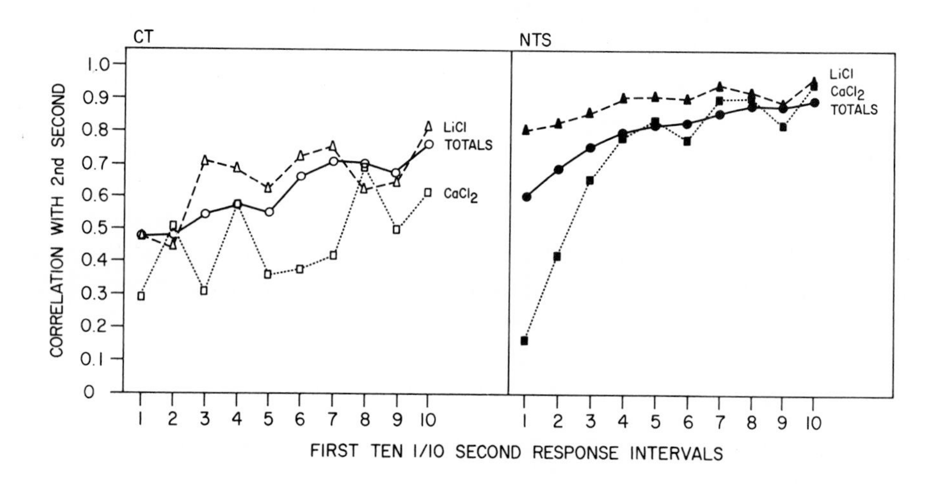

Figure 3 Temporal stability of the taste quality messages. The across-neuron response patterns for a given stimulus during each of the first ten 1/10-second response intervals were correlated with the neural patterns for the same stimulus during the 2nd second of response. Low correlations on the ordinate indicate little similarity with the neural message during the 2nd second; high correlations on the ordinate indicate similar messages. At both synaptic levels, the neural patterns for the sodium and lithium salts (represented by LiCl) were relatively stable over time; the neural pattern for $CaCl_2$ was among the least stable. For most stimuli, the across-neuron response patterns were more stable in the NTS than in the CT (averages for all stimuli given by Totals).

data for LiCl, $CaCl_2$, and for all stimuli averaged together (Totals). Each of the first ten 1/10 seconds of evoked activity were correlated with the 2nd second of response to examine the temporal properties of the initial phase of the neural messages.

Figure 3 illustrates two important characteristics of the neural messages common to both levels of the taste system. First, the temporal stability of the taste messages varied from one stimulus to another. For example, the figure shows that for both the CT and NTS neurons, the neural pattern for LiCl was more stable over time than was the pattern for $CaCl_2$. In general, the neural messages for the lithium and sodium salts were the most stable, whereas the message for $CaCl_2$ was among the least stable.

Second, on the basis of this and other analyses, it was found that the greatest temporal change in the neural patterns for most stimuli occurred during the initial few tenths of a second.

A comparison of the two synaptic levels indicated that the across-neuron response patterns were typically more stable over time in the NTS than in the CT. In the NTS, the initial portions of the neural messages were less "noisy," being generally more predictive of, or similar to, the later portions of the messages. Thus, it appears that taste information is processed in the NTS in such a manner as to produce a temporally more stable message for each stimulus.

The relative instability of the initial portions of the neural messages mentioned above may be related to certain properties of the time course of activity illustrated in Figure 4. This Figure shows the averaged responses produced by KCl during successive 1/10-second intervals in all CT and NTS neurons. Although characteristic differences were found among stimuli in the time course of activity, the averaged response to KCl was typical of the activity evoked by all stimuli in this respect: the response to each solution was characterized by a relatively high transient discharge followed by a rapid decrease to a steady state level of activity. The greatest difference found between the two synaptic levels in the time course of response was that in the NTS the initial transient was attenuated relative to the steady state. In the CT, the

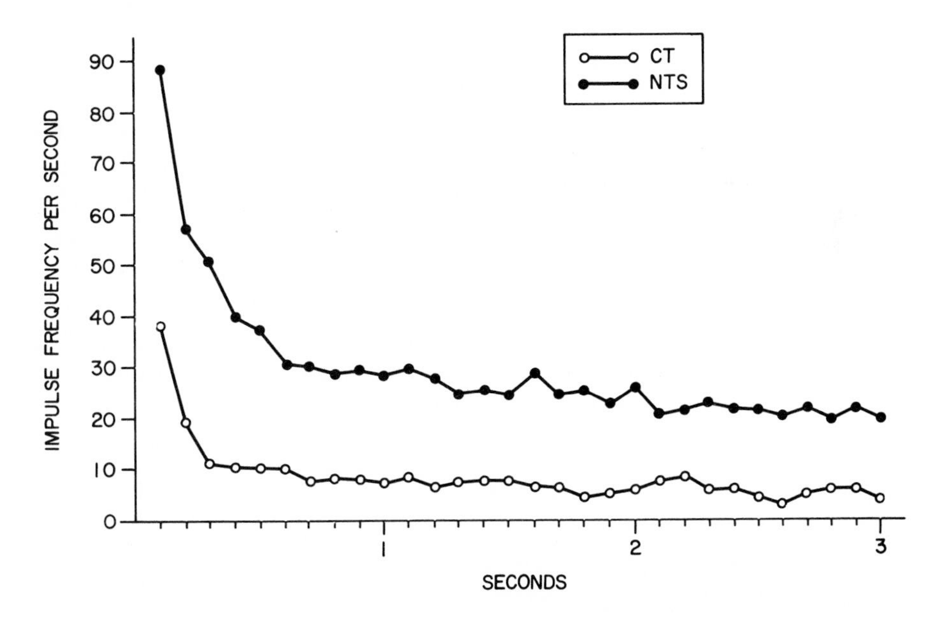

Figure 4 Time course of response in the CT and NTS. The response to KCl shown here was typical of all stimuli in at least one respect: the response to each taste solution consisted of a high initial discharge followed by a rapid decline to a steady state level of activity. In the NTS, the initial transient response was attenuated relative to the steady state. The finding that the over-all level of activity is amplified in the NTS is also evident in this Figure.

initial 1/10-second of activity was, on the average, about five times as great as the steady state, but in the NTS the initial discharge was only about 2.5 times as great as the steady state.

The presence of a high initial transient response (Figure 4) may be related to the finding that the least stable portion of the neural message typically consisted of the first few 1/10-seconds of response (Figure 3). The stabilization of the taste messages for the various stimuli corresponded closely to the decline of the initial discharge. It is possible that the message for a given stimulus is not fully developed during the initial burst of activity and that the significant aspects of the stimulus are represented more clearly in later portions of the response. (The notion that later portions of the neural message may carry more information than the initial portions of the message is supported by Marshall,[18] who found that behavioral taste discrimination in the opossum is much more closely related to the across-neuron response patterns computed for the 2nd second than the neural patterns for the 1st second of activity.) For this reason, the time interval used for the various analyses presented in this paper was the one second of activity beginning after the first 0.3-seconds of response.

The greater stability of the neural patterns in the NTS may be partially due to the attenuation of the initial response relative to the steady state. In any case, it is clear that the temporal course of evoked activity is modified in the process of synaptic transmission and that this transformation is accompanied by an increase in the stability of the taste messages.

Gustatory Neural Response Function

Katsuki[15] has shown in the auditory system that the NRFs of single cells at higher CNS levels tend to be narrower than those of first-order neurons. Since the NRFs of first-order CT fibers are relatively broad, it has often been suggested that an analogous "sharpening" of the NRFs of higher-order taste neurons occurs. In examining this possibility, the methods developed by Erickson et al.[9] and by Schiffman and Falkenberg[26] were used to disclose the approximate form of the NRFs of the first- and second-order taste neurons. It was found that the NRFs of the NTS cells were of the same general form as those of the CT fibers. No evidence of "sharpening" was obtained; in fact, there was a slight indication that the NRFs of the NTS cells were broader than those of the first-order fibers. However, further data are required to describe in more precise quantitative terms the forms of the NRFs of the cells at both synaptic levels.

Erickson has discussed elsewhere[8] why, on theoretical grounds, taste neurons at higher CNS levels, as well as those in the periphery, may be expected to display relatively broad NRFs. Briefly, broad NRFs are required because at each taste-sensitive part of the tongue many taste stimuli must be represented by a rather limited number of neurons.[12] If each cell were specifically sensitive to only a few stimuli, not enough cells would be available to carry the many possible taste messages. On the other hand, a system of relatively few neurons having broad NRFs, such as the color vision system,[3,17] would permit the representation of many stimuli in terms of the relative amounts of activity produced in these neurons, i.e., as across-neuron response patterns.

Gustatory Stimulus Dimensions

Erickson et al.[9] and Schiffman and Falkenberg[26] have developed methods whereby it is possible to approximate the form in which the stimulus dimensions underlying taste quality are represented in the gustatory system. By means of certain analyses of the neural data, these methods permit the taste stimuli to be arranged in dimensions according to the similarity of their neural effects. It was found that, for both the CT and NTS data, all the stimuli, with the exception of sucrose, could be arranged in a two-dimensional space. Furthermore, the positions of the stimuli relative to each other were essentially the same for the CT and NTS data, and were consistent with behavioral findings regarding the taste similarity of these stimuli.[6,19,21] Thus, although certain modifications occurred during synaptic transfer, the neural representation of the stimulus dimensions was well preserved in the NTS.

In order to evaluate the temporal stability of the positions of the stimuli in the dimensions, these positions were compared for several different time intervals. It was found that the relative positions of the stimuli in the dimensions were somewhat more stable over time for the NTS than for the CT data.

The locations of the stimuli with respect to each other were closely related to the similarity between the across-neuron response patterns associated with these stimuli. Consequently, any changes over time in the stimulus placements were related to temporal variations in the neural patterns for these stimuli. As discussed earlier, such temporal fluctuations may simply reflect a certain degree of instability in the taste messages. However, it is also possible that temporal changes in the neural patterns and in the positions of the stimuli in the dimensions reflect some change over time in the taste qualities of the stimuli. For example, the stimulus dimensions derived from the CT data indicate that for the 1st second of response, $CaCl_2$ was located relatively close to KCl ($r = 0.78$) and far from HCl ($r = 0.20$). In the 2nd second, $CaCl_2$ moved away from KCl ($r = 0.72$) and approached HCl ($r = 0.57$). This change in position indicates that the neural pattern for $CaCl_2$ was relatively unstable and suggests that the taste of $CaCl_2$ becomes more similar to that of HCl during the 2nd second of stimulation, perhaps becoming more sour. Thus, any significant temporal change in the taste quality of a stimulus may be revealed by changes in the position of that stimulus relative to the other stimuli in the dimensions and by changes in the correlation of that stimulus with other stimuli.

THE INTENSITY CODE AND THE TRANSFER OF INTENSITY INFORMATION

In order to study the processing of information about taste intensity at the first synaptic relay, the responses of single CT and NTS neurons to each stimulus at different concentrations were recorded. The stimuli used included NaCl, KCl, HCl, QHCl, sucrose, and sodium saccharine. The concentration of each stimulus was varied by four to six intensity steps from near threshold (as determined behaviorally and electrophysiologically) to that concentration which produced approximately a maximum response in the whole CT of the rat.[11] Data were obtained from 40 CT fibers and 10 NTS neurons. This was a separate sample of neurons from that used in the study of taste quality discussed above.

In the gustatory system, as in other sensory systems, stimulus intensity appears to be encoded in terms of the over-all rate of response produced in all neurons. More specifically, at moderate intensities, both CT and NTS neurons were found to respond to increments in stimulus intensity with increases in the rate of discharge. The effect of such changes in stimulus intensity can be represented as shown in Figure 5. This Figure illustrates the responses of five CT neurons and five NTS cells to sodium saccharin at two different concentrations. At each neural level, an in-

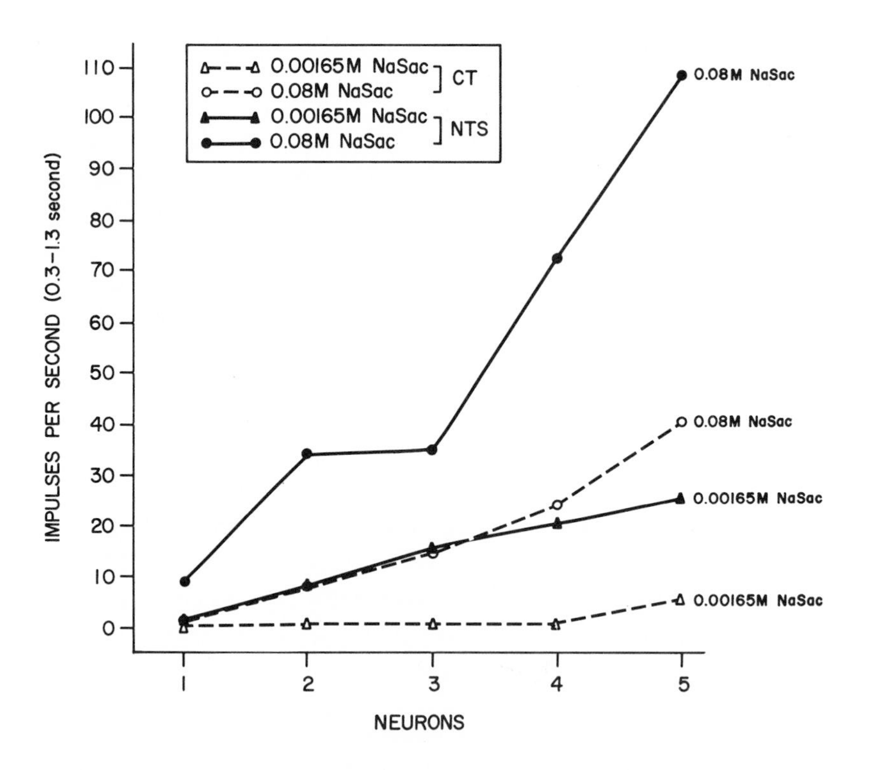

Figure 5 Representation of the code for taste intensity at the CT and NTS levels. Five CT neurons (open symbols) and five NTS neurons (closed symbols) are plotted on the abscissa in order of their increasing responsiveness to sodium saccharin. At each synaptic level, an increase in stimulus intensity produced an increase in the rate of discharge of each neuron. The neural effect of a change in intensity may best be represented as a multiplicative change in the level of the across-neuron response pattern. Since the forms of the neural patterns at each level were relatively unchanged, no change in the taste quality of sodium saccharine is expected. The figure also shows that a given change in intensity produced a greater change in response at the NTS than at the CT level.

crease in stimulus intensity produced an increase in the rate of firing of each neuron. Such changes in response due to an increase in concentration may best be represented as a multiplicative elevation of the entire across-neuron response pattern. This shift in the level of the across-neuron response pattern constitutes the neural representation of a change in stimulus intensity. Since at each level the form of the neural pattern remained essentially the same, no change in the taste quality of the stimulus would be expected.

The data presented in Figure 5 further indicate that for a given change in stimulus intensity, the absolute change in response was greater in the NTS than in the CT. This is consistent with the finding discussed earlier that the average rate of response was amplified at the first synaptic relay. For stimuli of moderate intensity, and for this sample of neurons, the increments in discharge rate produced by changes in concentration were approximately five times as great in the NTS as in the CT. Considering all concentrations used, from near threshold to very strong, the increments in response rate with intensity changes were larger at the NTS level than at the CT level by a factor of 5.9. Thus, the response increments produced by stimulus intensity changes were found to be amplified during synaptic transfer, as was the over-all level of evoked activity.

Intensity-Quality Interactions

Human psychophysical studies have established that a change in the concentration of a given stimulus may be accompanied by a change in the taste quality of that stimulus.[4,13,25] Therefore an attempt was made to relate such intensity-quality interactions to the neural data obtained from the CT and NTS.

In an ascending concentration series, the taste of KCl has been reported to be first sweet, then bitter, then bitter-salty, and then salty. These changes in taste quality should be reflected in corresponding changes in the across-neuron response pattern for KCl. That such changes in the neural pattern occur is shown in Figure 6, which illustrates the patterns produced across six CT and six NTS neurons for two concentrations of KCl and one concentration of QHCl. Table I gives the correlations between the neural patterns for these and other stimuli discussed here.

KCl at a concentration of 0.03 M has been reported to taste bitter, and indeed Figure 6 shows that the neural pattern for KCl at this concentration was very similar to the pattern for 0.01 M QHCl at both CT and NTS levels ($r_{CT} = 0.84$; $r_{NTS} = 0.94$). The correspondence between these neural patterns strongly suggests that the weaker concentration of KCl tastes bitter, which is in agreement with the human psychophysical data.

In order to determine whether the neural pattern, and therefore the taste, of KCl changes at different concentrations, the stronger 0.3 M KCl was compared with the "bitter" 0.03 M KCl. The data presented in both Figure 6 and Table I indicate that some change in the neural pattern did occur. The correlation between the two concentrations of KCl was 0.68 for the CT data and 0.83 for the NTS data. (For no

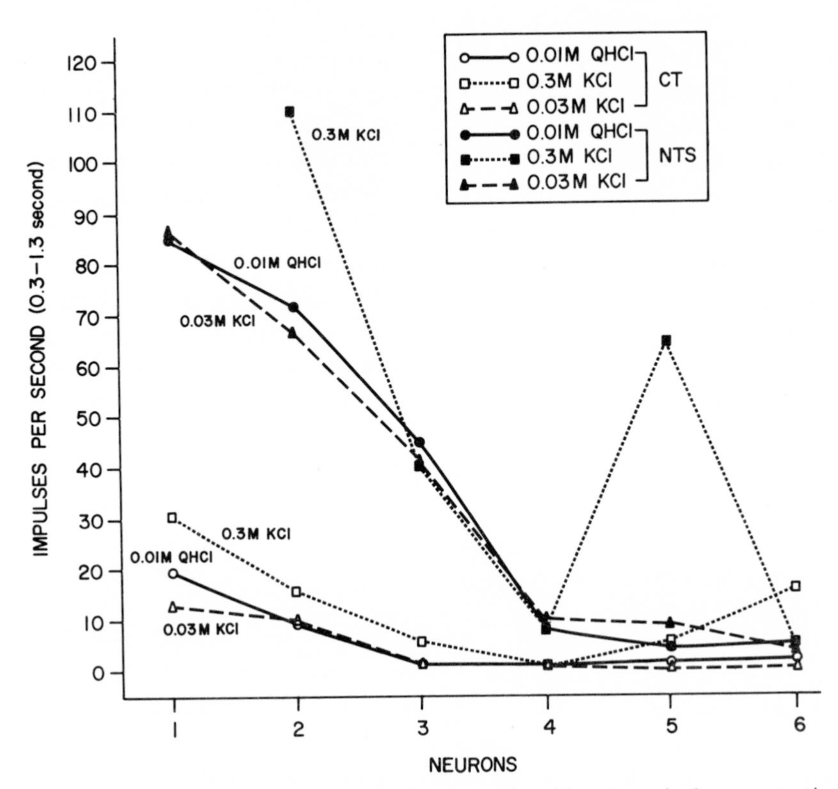

Figure 6 Neural representation of a change in taste quality with a change in the concentration of KCl. Six CT neurons (open symbols) and six NTS neurons (closed symbols) are shown on the abscissa in order of their decreasing responsiveness to 0.03 M KCl. At both synaptic levels, 0.03 M KCl produced across-neuron response patterns similar to those for QHCl, whereas 0.3 M KCl produced neural patterns less similar to those for QHCl, but more similar to NaCl (see text and Table I).

Table I

SIMILARITY AMONG THE ACROSS-NEURON RESPONSE PATTERNS FOR KCl AT TWO CONCENTRATIONS, QHCl, AND NaCl

	"bitter" 0.03 M KCl CT/NTS	"bitter" 0.01 M QHCl CT/NTS	"salty" 0.069 M NaCl CT/NTS
"bitter" 0.03 M KCl		0.84/0.94	0.08/0.77
"salty" 0.3 M KCl	0.68/0.83	0.53/0.80	0.36/0.92

Note: The correlations obtained from both the CT and the NTS data are given for each pair of stimuli. The relatively low correlations between 0.03 M KCl and 0.3 M KCl indicate some difference in taste between them. KCl at a lower concentration of 0.03 M was highly correlated with QHCl, indicating a common bitter taste. KCl at a concentration of 0.3 M was more highly correlated with NaCl, suggesting a common salty taste.

change in the neural pattern the expected correlation would be about 0.91 and 0.98 for the CT and NTS respectively.)

It has further been reported that 0.3 M KCl tastes salty to humans. Therefore, the neural patterns for the two concentrations of KCl were compared with the pattern for NaCl as well as QHCl. As shown in Figure 6 and Table I, the neural pattern for the stronger "salty" 0.3 M KCl became less similar to that of QHCl (r_{NTS} with QHCl decreased from 0.94 for 0.03 M KCl to 0.80 for 0.3 M KCl), and more similar to that for NaCl (r_{NTS} with NaCl increased from 0.77 for 0.03 M KCl to 0.92 for 0.3 M KCl). This change in the neural pattern for KCl thus corresponded well to the human psychophysical data.

As a result of such intensity-quality interactions, the view that changes in stimulus intensity simply produce shifts in the absolute level of the across-neuron response pattern is not altogether true. It appears that changes in the intensity of a stimulus can modify the form of the neural pattern enough to account for differences in taste quality produced by those changes.

NEURAL REPRESENTATION OF TASTE AT HIGHER CNS LEVELS

The research discussed above has provided information concerning the transfer of taste messages at the first synaptic relay in the medulla. We are now conducting similar experiments on information processing at higher levels of the taste system, e.g., the thalamus and cortex. In order to pursue this kind of research we must deal with the questions of convergence of neural input from different taste nerves and the representation of taste in the cortex.

Central Convergence of Taste Messages

Several studies have shown that, in the NTS, fibers of the glossopharyngeal (IX) nerve end in close proximity to the terminations of the CT fibers, with a clear region of overlap.[10,27] Furthermore, in the thalamus and cortex, the IX nerve projection area lies adjacent to the CT projection area with some degree of overlap.[1,5] If the projections of these two nerves actually converge upon a common pool of neurons at any synaptic level, it is likely that the taste messages transmitted by one nerve may be modified by the other.

In order to determine whether such convergence occurs, an experiment was performed to examine whether any interaction could be found between the primary evoked responses of the cortex to electrical stimulation of the CT and IX nerves. The results of this experiment are shown in Figure 7. It was found that, under certain conditions, stimulation of either nerve could block the response to stimulation of the other nerve. Under other conditions, stimulation of both nerves produced a summation of response. These interactions show that the primary evoked response to stimulation of the CT involves some of the same cortical neurons that are responsive to stimulation of the IX nerve. Consequently, these data indicate that the projections of the two nerves converge to some extent at some levels in the CNS. Whether this

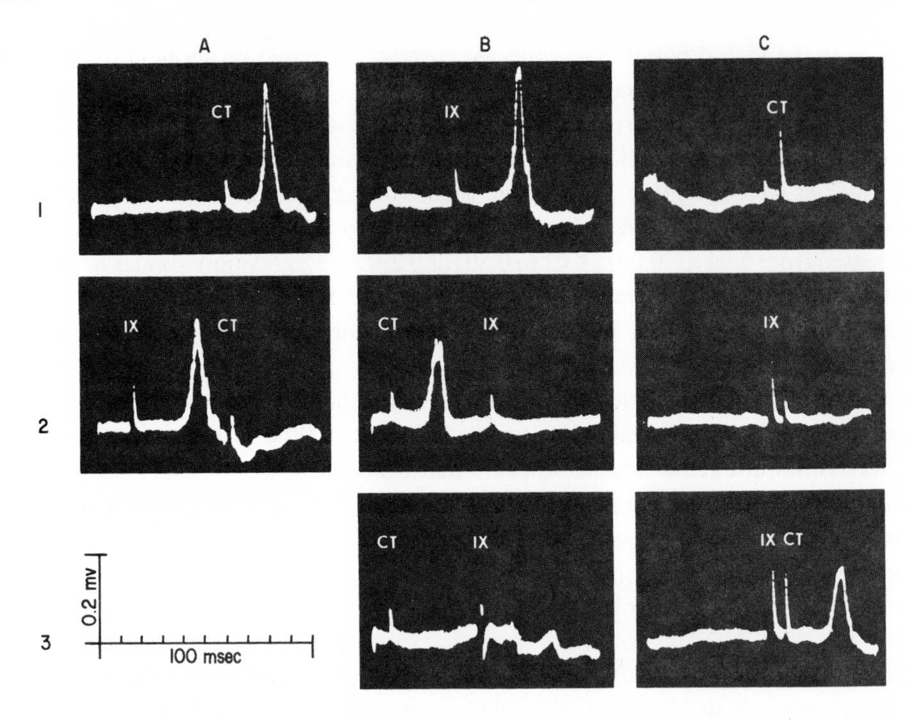

Figure 7 Interactions of the primary evoked responses of the cortex to electrical stimulation of the CT and glossopharyngeal nerves. The responses were recorded ipsilateral to the nerves stimulated. The weakest stimuli required to produce maximal responses were used, except as noted below. Electrical positivity at the recording electrode is indicated by upward deflections. The artifacts showing the time of stimulation are indicated by CT for the chorda tympani nerve and IX for the glossopharyngeal nerve.

Column A: 1. Response to CT stimulation (test stimulus). 2. Conditioning stimulus delivered to the glossopharyngeal nerve followed by the test stimulus to the CT. Interaction between the two nerves is indicated by the absence of a response to stimulation of the CT.

Column B: 1. Response to glossopharyngeal nerve stimulation (test stimulus). 2. Conditioning stimulus delivered to the CT blocked the response to test stimulation of the glossopharyngeal nerve. 3. With the conditioning stimulus to the CT just below the primary response threshold, blocking of the response to test stimulation of the glossopharyngeal nerve still occurred. This suggests that the blocking is not due to the refractory states of cortical cells, but probably depends on some prior subcortical (thalamic ?) interactions.

Column C: 1. Stimulation of the CT just below threshold for the primary response. 2. Stimulation of the glossopharyngeal nerve just below threshold for the primary response. 3. Combining subthreshold stimuli of C_1 and C_2 results in summation, indicating that interactions take place between the two nerves.

convergence involves taste neurons could not be determined by this experiment. However, at this point the evidence strongly suggests that the taste messages transmitted by the CT and IX nerves can converge.

Although convergence of the projections of the two nerves does appear to occur, it is not clear at what synaptic level it begins. Preliminary data obtained from single cell recordings in the NTS using chemical stimulation of the tongue or electrical stimulation of the taste nerves provided no evidence for convergence at that level.

Thus, since the data presented in Figure 7 suggest that convergence occurs prior to the cortical level, the thalamus is implicated as a major site of gustatory convergence.

Taste Representation in the Cortex

Many studies have concurred in defining as "taste cortex" an area subjacent to the projections of the head and tongue regions in the topographically organized somatosensory area.[22] However, as mentioned earlier, only seven cortical neurons have been reported which respond to chemical stimulation of the tongue.[2,16] We therefore performed similar experiments using chemical stimuli, but without success. No cortical cells responding to chemical stimulation of the tongue could be found under the experimental conditions used, although it was not difficult to drive cortical cells in the "taste" area by electrical stimulation of the CT or IX nerves.

Furthermore, there is no evidence that neurons responsive to chemical stimulation of the tongue project directly from the thalamus to the cortex, i.e., it has never been demonstrated that the cells known to project from the taste area of the thalamus to the cortex[29] are the same thalamic neurons that have been shown to be responsive to taste stimulation.[5] In order to approach this problem, we recorded the responses of single thalamic neurons to both chemical stimulation of the tongue and electrical stimulation of the cortex. The results of this experiment are given in Figure 8.

As expected, it was found that thalamic taste neurons do, in fact, project to the cortex of the rat. (The possibility that they may also project elsewhere obviously cannot be ruled out.) The thalamic cells identified as taste neurons by chemical stimulation of the tongue could be antidromically activated by cortical stimulation, which indicates that they project without synapse to the cortex. Antidromic invasion of these thalamic neurons could be produced only by stimulation in the immediate vicinity of the cortical area in which single cells responding to electrical stimulation of the CT were found. This region was very small, about 0.6 mm^2 in the rat, and was centered in the larger cortical "taste" area giving primary evoked responses. (Benjamin and Pfaffmann[1] found that the evoked primary response in the cortex produced by electrical stimulation of the CT in the rat could be recorded over an area of 2 mm^2.)

In addition to antidromic activation, the thalamic taste neurons could also be driven orthodromically by cortical stimulation (Figure 8). This finding indicates the presence of descending cortical influences on the thalamic cells. It also supports the notion that the thalamus may be an important site for gustatory interaction where the projections of different taste nerves may converge along with descending fibers from the cortical "taste" area.

Since taste cells in the thalamus were found to project to the cortex, they were probably responsible, at least in part, for the primary evoked responses illustrated in Figure 7. (Somesthetic neurons could also contribute to these responses.) Furthermore, since strong interactions were found between the primary responses evoked by stimulation of the CT and IX nerves, it is likely that these were actually taste interactions, representing mutual influences between the taste messages transmitted by the two nerves. These neural interactions, as well as the cortico-thalamic influences, may

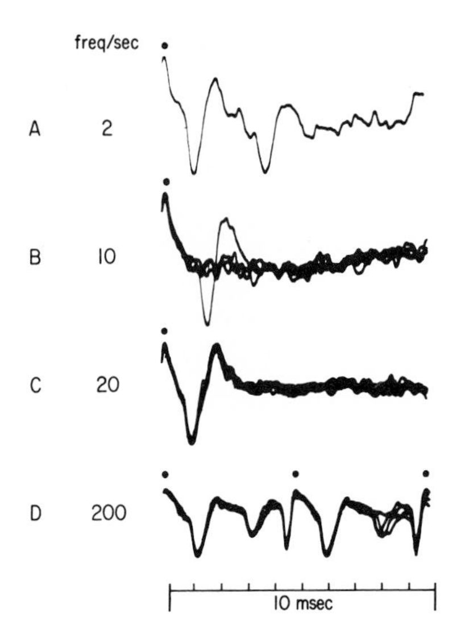

Figure 8 Responses of a single thalamic taste cell to electrical stimulation of the "taste" cortex. This neuron was identified as a taste cell since it responded to chemical stimulation of the tongue. The cortical "taste" area was identified as the region where primary evoked responses to electrical stimulation of the CT could be recorded. The stimulus artifacts are indicated by dots. In records B through D, about four superimposed traces are shown. A. Stimulus intensity = 7V; frequency = 2/sec. Antidromic response followed by an orthodromic response. B. Stimulus intensity = 10V; frequency = 10/sec. Only antidromic responses occurred. The all-or-nothing characteristic of the response is indicated by the difference between the superimposed traces. C. Stimulus intensity = 20V; frequency = 20/sec. Only antidromic responses occurred. D. Stimulus intensity = 20V; frequency = 200/sec. Antidromic responses are shown at the left of the trace and after the artifact in the center of the trace. Additional, possibly orthodromic, responses followed each antidromic response.

thus provide the basis for further modifications or refinements of the taste messages beyond those observed at the first synaptic level.

It should be pointed out, however, that neural convergence in the taste system must be limited to some degree for the following reasons. We have conducted psychophysical experiments that have indicated that the locations of two stimuli (0.59 M NaCl and water, or 0.59 M NaCl and 1.0 M sucrose) administered simultaneously to different areas of the tongue can be identified by human subjects. For example, the subjects could correctly indicate which part of the tongue was stimulated with NaCl and which with sucrose, both the quality and location of each stimulus remaining distinct. Such spatial localization of the taste stimulus is probably mediated by the activation of topographically distinct sets of gustatory neurons, analagous to somesthetic and visual mechanisms. Although spatial discrimination involving taste may not be as fine as in the other spatial senses, any convergence in this system must be limited enough to preserve the topographical relationships of the taste fibers projecting from different regions of the tongue to the CNS.[8]

In conclusion, several predictions concerning the mammalian taste system can be made on the basis of the data presented in this paper as well as on certain theoretical considerations:

1. Neurons responsive to chemical stimulation of the tongue will be found in the cortex.
2. Thalamic and cortical taste cells are topographically arranged according to the location of their receptive fields on the tongue.
3. Individual taste neurons at any synaptic level, and probably in any mammalian species, are sensitive to many diverse chemical stimuli rather than being specifically sensitive to only a few chemical substances.

SUMMARY

Several studies were performed on the transfer of taste quality and intensity information at the first synaptic relay in the gustatory system of the rat. The results of these experiments are briefly summarized in Table II. Based on these findings, and certain theoretical considerations, several predictions about taste information processing in the thalamus and cortex are included.

TABLE II
SUMMARY: INFORMATION PROCESSING IN THE TASTE SYSTEM OF THE RAT

Functions	Chorda Tympani (CT)	Nucleus Tractus Solitarius (NTS)	Predicted Thalamus-cortex
GENERAL			
Initial response relative to steady state		reduced	
Response rates		increased (x5)	
TASTE QUALITY			
Code for taste quality	ANRP*	ANRP	ANRP
Width of neural response function	broad	broad	broad
Similarity of ANRPs between stimuli		increased	
Similarity of ANRPs predicts behavior	yes	yes	yes
Stimulus dimensions consistent with behavioral data	yes	yes	yes
Temporal stability of taste messages		improved	
Temporal stability of stimulus dimensions		improved	
TASTE INTENSITY			
Code for taste intensity	xANRP**	xANRP	xANRP
Change in response rates with change in intensity		increased (x5)	

* ANRP—across-neuron response pattern.
** xANRP—multiplicative change in level of across-neuron response pattern.

ACKNOWLEDGMENTS

We would like to thank Stephanie P. Doetsch for preparing the illustrations and Judith Hughes for her critical comments and preparation of the manuscript. We also wish to express our appreciation to Dr. Marian Frank for her collaboration in certain phases of this research.

The work was supported by U.S.P.H.S. research grant NB-04793.

REFERENCES

1. Benjamin, R. M., and C. Pfaffmann. 1955. Cortical localization of taste in albino rat. *J. Neurophysiol.* **18**, pp. 56–64.
2. Cohen, M. J., S. Landgren, L. Ström, and Y. Zotterman. 1957. Cortical reception of touch and taste in the cat. *Acta Physiol. Scand.* **40**, Suppl. 135, pp. 1–50.
3. De Valois, R. L. 1965. Behavioral and electrophysiological studies of primate vision. *In* Contributions to Sensory Physiology, Vol. 1 (W. D. Neff, editor). Academic Press, New York, pp. 137–178.
4. Dzendolet, E., and H. L. Meiselman. 1967. Gustatory quality changes as a function of solution concentration. *Percept. Psychophys.* **2**, pp. 29–33.
5. Emmers, R., R. M. Benjamin, and A. J. Blomquist. 1962. Thalamic localization of afferents from the tongue in albino rat. *J. Comp. Neurol.* **118**, pp. 43–48.
6. Erickson, R. P. 1963. Sensory neural patterns and gustation. *In* Olfaction and Taste I (Y. Zotterman, editor). Pergamon Press, Oxford, etc., pp. 205–213.
7. Erickson, R. P. Neural coding of taste quality. 1967. *In* The Chemical Senses and Nutrition (M. Kare and O. Maller, editors). The Johns Hopkins Press, Baltimore, Md., pp. 313–327.
8. Erickson, R. P. 1968. Stimulus coding in topographic and nontopographic afferent modalities: On the significance of the activity of individual sensory neurons. *Psychol. Rev.* **75**, pp. 447–465.
9. Erickson, R. P., G. S. Doetsch, and D. A. Marshall, 1965. The gustatory neural response function. *J. Gen. Physiol.* **49**, pp. 247–263.
10. Halpern, B. P., and L. M. Nelson. 1965. Bulbar gustatory responses to anterior and to posterior tongue stimulation in the rat. *Amer. J. Physiol.* **209**, pp. 105–110.
11. Hardiman, C. W. 1964. Rat and hamster chemoreceptor responses to a large number of compounds and the formulation of a generalized chemosensory equation. Unpublished Ph.D. thesis, Florida State University.
12. Harper, H. W., J. R. Jay, and R. P. Erickson. 1966. Chemically evoked sensations from single human taste papillae. *Physiol. Behav.* **1**, pp. 319–325.
13. Höber, R., and F. Kiesow. 1898. Über den Geschmack von Salzen und Laugen. *Z. Physikal. Chem.* **27**, pp. 601–616.
14. Hubel, D. H., and T. N. Wiesel. 1962. Receptive fields, binocular interaction and functional architecture in the cat's visual cortex. *J. Physiol. (London)* **160**, pp. 106–154.
15. Katuski, Y. 1961. Neural mechanism of auditory sensation in cats. *In* Sensory Communication (W. A. Rosenblith, editor). M. I. T. Press, Boston, Mass., pp. 561–583.
16. Landgren, S. 1957. Convergence of tactile, thermal, and gustatory impulses on single cortical cells. *Acta Physiol. Scand.* **40**, pp. 210–221.
17. Marks, W. B. 1965. Visual pigments of single goldfish cones. *J. Physiol. (London)* **178**, pp. 14–32.
18. Marshall, D. A. 1968. A comparative study of neural coding in gustation. *Physiol. Behav.* **3**, pp. 1–15.
19. Morrison, G. R. 1967. Behavioral response patterns to salt stimuli in the rat. *Can. J. Psychol.* **21**, pp. 141–152.
20. Mountcastle, V. B. 1957. Modality and topographic properties of single neurons of cat's somatic sensory cortex. *J. Neurophysiol.* **20**, pp. 408–434.
21. Nachman, M. 1963. Learned aversion to the taste of lithium chloride and generalization to other salts. *J. Comp. Physiol. Psychol.* **56**, pp. 343–349.
22. Oakley, B., and R. M. Benjamin. 1966. Neural mechanisms of taste. *Physiol. Rev.* **46**, pp. 173–211.
23. Pfaffmann, C. 1941. Gustatory afferent impulses. *J. Cell. Comp. Physiol.* **17**, pp. 243–258.
24. Pfaffmann, C. 1955. Gustatory nerve impulses in rat, cat, and rabbit. *J. Neurophysiol.* **18**, pp. 429–440.
25. Renqvist, Y. 1919. Über den Geschmack. *Skand. Arch. Physiol.* **38**, pp. 97–201.

26. Schiffman, H., and P. Falkenberg. 1968. The organization of stimuli and sensory neurons. *Physiol. Behav.* **3**, pp. 197–202.
27. Torvik, A. 1956. Afferent connections to the sensory trigeminal nuclei, the nucleus of the solitary tract, and adjacent structures. *J. Comp. Neurol.* **106**, pp. 51–141.
28. Whitfield, I. C., and E. F. Evans. 1965. Responses of auditory cortical neurons to stimuli of changing frequency. *J. Neurophysiol.* **28**, pp. 655–672.
29. Wolf, G. 1968. Projections of thalamic and cortical gustatory areas in the rat. *J. Comp. Neurol.* **132**, pp. 519–529.

TASTE PSYCHOPHYSICS IN ANIMALS

G. R. MORRISON

Department of Psychology, McMaster University, Hamilton, Ontario, Canada

Most of the existing behavioral data on the rat is for taste preference, which, while interesting in itself, does not indicate much about the more fundamental process of taste discrimination. Consequently, I have been concerned with methods of obtaining psychophysical data for the rat under conditions in which the animal's response is based on the discriminative properties of the taste stimuli and the reinforcing aspects are incidental. The procedure is modified slightly from that which Norrison and I[5] originally described for taste detection, based on the yes-no procedure of signal detection experiments.

The apparatus consists of a two-lever operant conditioning box with the levers and a food-pellet dispenser mounted at one end. A turntable is mounted outside the other end. Ten plastic bottles, each with a drinking tube, are spaced equally around it. The drinking tubes are connected through a drinkometer circuit and arranged so that one tube at a time projects into the box through a slot in the box wall. The turntable is connected to a slow-speed motor. In training, five of the bottles are filled with one particular taste solution, which we can call S_1, and the other five with the second, which we call S_2. These are arranged randomly on the turntable and the order is changed from session to session. The trial starts with the animal licking at the spout. After two seconds of licking, the motor is activated. The turntable revolves a 20th of a revolution, pulling the spout out of the box, and this is the stimulus for bar-pressing. The animal is now forced to select one of the two levers. If the bottle contained S_1, the left lever is correct; if S_2, the right one. The animal is reinforced with a food pellet if it makes the correct response. (These animals are always run while under food deprivation.) If it makes the incorrect response, of course it is not reinforced. Either response terminates the trial. Termination activates the motor to bring up the next bottle for the next round.

The animal, then, must make one of two responses on each trial on the basis of the taste. At the end of training, the typical animal is making between 90 and 100 per cent correct choices. In test sessions, four of the bottles contain the same S_1 solution and another four the same S_2 solution. When any of these eight bottles are present, the reinforcing contingencies are exactly the same as in training. The other two bottles contain two test solutions. When a particular test solution comes up, either response is reinforced, so, in effect, we are asking the animal to taste each test solution and identify it with one or the other of the two training solutions.

In the first experiment,[3] three groups of rats were used, 5 rats per group. All three groups had 0.1 M NaCl as the S_1 solution. Group 1 had 0.5 mM quinine sulfate as the S_2. Group 2 had 0.01 M HCl, and Group 3, 0.2 M sucrose. A variety of salts were then introduced as test stimuli. In this case, the data consist of the percentage of the time that the response conditioned to the S_2 stimulus, i.e., the nonsalt stimulus, was made to each test stimuli for each group. If we take KCL as an example, we find that in Group 1 the animals identify KCL with quinine, rather than NaCl, about 60–70 per cent of the time. In Group 2 they identify KCL with acid about 40 per cent of the time, and in Group 3 they identify it with sucrose about 8 per cent of the time.

Figure 1 shows typical patterns for nine test salts. Patterns for potassium and magnesium salts are similar. Varying the anion makes little difference, although there is a small, but significant, anion effect. The patterns for potassium and magnesium are quite different for those for sodium. In Figure 2, similar patterns are shown for some other salts. The pattern for Na_2CO_3 is again similar to that of NaCl, as is that for LiCl.

When I was carrying out the experiment, I had not seen the Mason and Stafford[2] results on the palatability of sugar of lead, but I had included lead acetate, expecting to get a sucrose component. From our results, shown here, it is not surprising that Mason and Stafford find that this salt is rejected by rats at all suprathreshold concentrations.

Finally, the S_2 solutions were themselves used as test solutions. Figure 3 shows that these solutions are not primary tastes for the rat in the sense that they are equally unlike each other. Quinine is identified about 50 per cent of the time with HCl and 50 per cent of the time with NaCl. It is never identified with sucrose when compared to NaCl. HCl, on the other hand, is consistently identified with quinine rather than with NaCl, but is consistently identified with NaCl rather than sucrose. Sucrose is identified with NaCl, rather than with either quinine or HCl. So these are not the random patterns that you might expect if these four solutions were equally unlike each other.

While this is rather a crude way to go about it, the results are encouraging because most of them are consistent with the neural data of Erickson and Sato. The salts which they find produce similar neural patterns also seem to produce similar behavioral patterns—for example, the sodium-lithium group and the ammonium-potassium-magnesium group. Calcium chloride seems to be quite different from any of these, although more like the magnesium-potassium. So, the results are encouraging to that extent.

At one end of the system, these patterns could be predicted from the neural data. Now I would like to show that they, in turn, can predict preference/aversion data.

Because we found that all sodium salts gave much the same pattern as NaCl, it could be suggested that, if these could be matched in terms of subjective intensity for the rat, we should be able to predict preference behavior for different sodium salts. Rats were first trained to discriminate between distilled water and 1 M sodium chloride and then were tested by probing with various concentrations of different

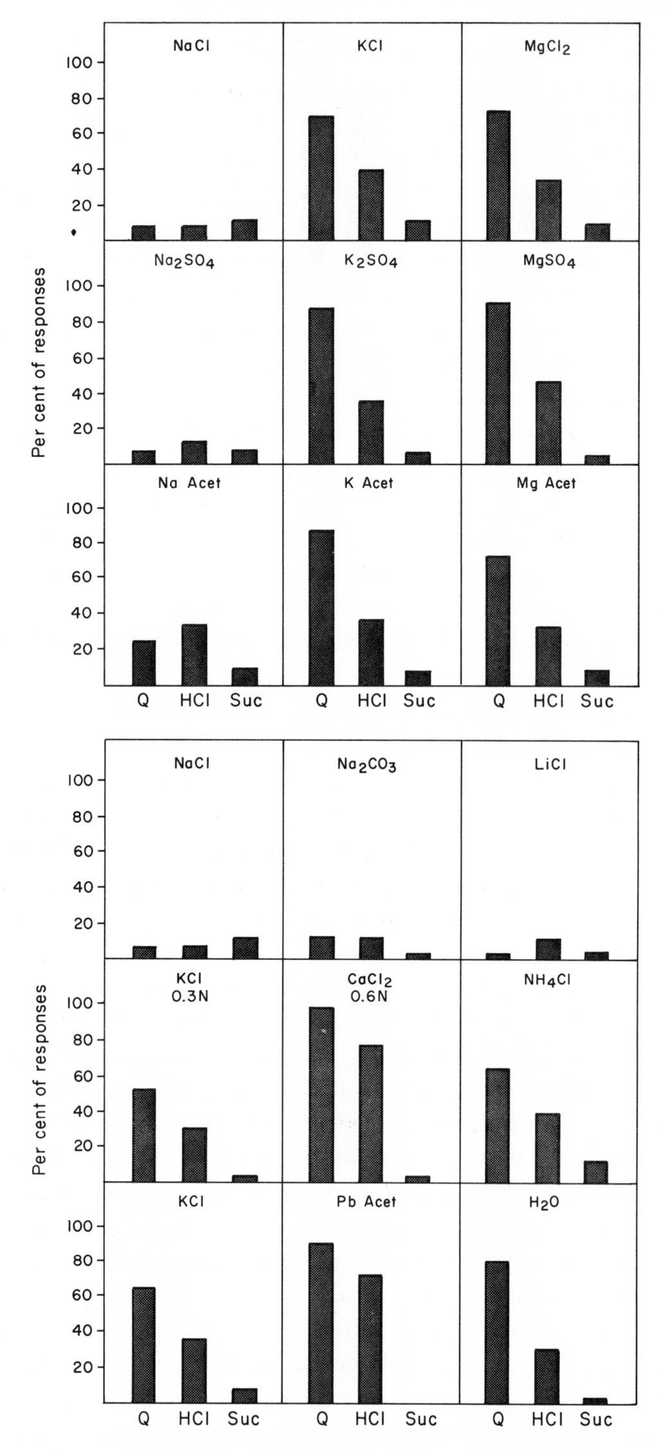

Figures 1 and 2 In these and the following figures, the bars labeled Q, HCl, and Suc show the percentage of responses to each of the test salts that were made on the lever-pressing that had been conditioned previously to quinine hydrochloric acid or sucrose. All test salts are in 0.1 normal concentrations unless otherwise labeled.

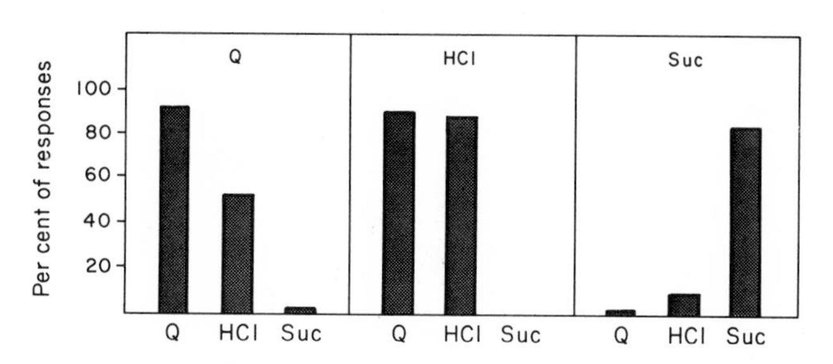

Figure 3 The results of using the three S_2 standards—0.5 mM quinine sulphate 0.01 M hydrochloric acid, and 0.2 M sucrose—as test solutions.

sodium salts. Over the concentration range from 0.01 to 1.0 M, chloride is the least and carbonate the most effective with the sulphate. Bromide is intermediate.

Three concentrations of NaCl—0.1, 0.3, and 0.5 M—were then selected, along with the concentrations of Na_2CO_3 (0.005, 0.03, and 0.06 M) and Na_2SO_4 (0.1, 0.15, and 0.2 M), which had resulted in equal detectability scores. These were then used in a single-bottle preference test to see if, in fact, relative preference could be predicted. The results were surprisingly good. The lowest concentration of all three salts was preferred to water; the mid-concentration was just about the indifference point; the highest concentration was aversive. These results will be of interest in the investigation of factors controlling sodium appetite, because they suggest that it is possible to find two salt solutions (chloride vs carbonate) that may be nearly identical in their taste properties although they differ by a factor of ten in their sodium content.

Guttman[1] found that sucrose and glucose produced two quite different functions when the rate of bar-pressing on an aperiodic schedule was plotted against concentration. If, in this experiment, the reinforcing effects were based on the taste of the sugars rather than their postingestional effects, the two different functions might well represent differences in the effectiveness of the sugars at the receptor level. If this were the case, the two functions should collapse into a single one if bar press rate were plotted against discriminability rather than concentration. That is, reinforcement as a function of concentration might be broken down into two components—discriminability as a function of concentration, which would differ with different sugars, and reinforcement as a function of discriminability, which might be the same for all sugars if postingestional factors are minimized. By first training a discrimination between 1 M sucrose and water and then probing with various concentrations, discriminability functions were obtained for six common sugars.[4] Taking the results for sucrose and glucose and replotting Guttman's data against discriminability confirmed our prediction, as shown in Figure 4. This result suggests a starting point for the investigation of taste vs postingestional factors in reinforcement by sugars.

In summary, we have a method which, I think, is a start in securing some kind of reasonable psychophysical data from the rat. On the one hand, it produces results that are consistent with the neural data; on the other, it can make useful predictions about preference behavior.

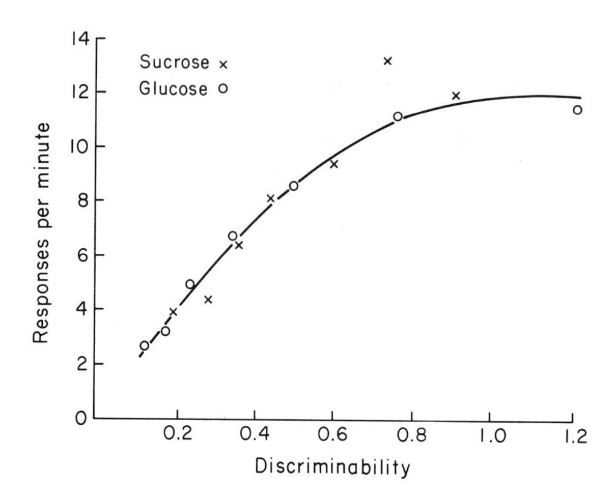

Figure 4 Bar-pressing rates for glucose and sucrose as a function of discriminability. (From Guttman)

REFERENCES

1. Guttman, N. 1954. Equal-reinforcement values for sucrose and glucose solutions compared with equal-sweetness values. *J. Comp. Physiol. Psychol.* **47**, pp. 358–361.
2. Mason, D. J., and H. R. Stafford. 1965. Palatability of sugar of lead. *J. Comp. Physiol. Psychol.* **59**, pp. 94–97.
3. Morrison, G. R. 1967. Behavioural response patterns to salt stimuli in the rat. *Canad. J. Psychol.* **21**, pp. 141–152.
4. Morrison, G. R. 1969. Relative discriminability of sugars for the rat. *J. Comp. Physiol. Psychol.* **68**, pp. 45–49.
5. Morrison, G. R., and W. Norrison. 1966. Taste detection in the rat. *Canad. J. Psychol.* **20**, pp. 208–217.

MODALITY CODING OF LINGUAL AFFERENTS IN THE CAT THALAMUS

RAIMOND EMMERS

Department of Physiology, College of Physicians, and Surgeons, Columbia University, New York City, New York

INTRODUCTION

Knowledge of the cerebral localization of sensory systems has advanced rapidly. Various subcortical and cortical projection areas have been explored and recently[3,4] the cortical area for taste was also localized with certainty. This knowledge, however, has not provided any clues concerning the mechanism responsible for the conversion of neural activity into different sensory experiences. Uncertainty remains as to what neural factors can account for such distinctly different sensory perceptions as touch and taste. A mere topographical separation of neurons into systems, which can do no more than to discharge action potentials of uniform amplitude at various frequencies, does not appear to provide an adequate explanation. Consequently, it has been at least a decade since it was proposed[8] that information about various sensory qualities might be conveyed to the central nervous system in the form of specific sensory codes. The capability of a single neuron for diverse activity seemed to be severely limited, so it was reasonable to assume that a sensory code could consist of multineuron spike discharges. Each neuron, when functioning at its own frequency, could contribute to the formation of a complex spatio-temporal pattern of spike activity.[8,7] Such a pattern could be modified by certain peripheral stimuli and, in this manner, presumably reflect changes in stimulus quality.

The purpose of this paper is twofold: (1) to present evidence indicating that individual neurons relaying lingual modalities at the thalamic level are capable of a more diverse activity than was previously anticipated (they can respond to a single stimulus with multiple, sequentially patterned spike bursts); and (2) to depict as precisely as possible the basic form of sensory codes for touch, pressure, thermal, and gustatory modalities of lingual afferents.

MATERIALS AND METHODS

In order to study the activity of neurons for lingual afferents in the thalamus, the following procedure was adopted. Cats deeply anesthetized with Nembutal were used for the study. They were prepared for stereotactic positioning of microelectrodes in the nuclear regions for lingual afferents, shown diagramatically in Figure 1. According to previous work,[5] thalamic neurons which relay touch from the tongue are situated

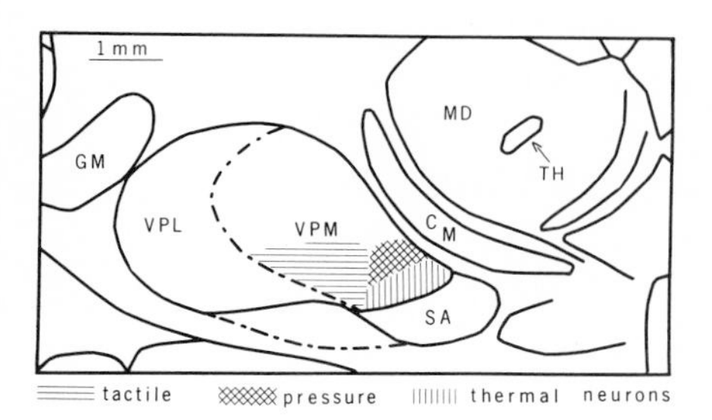

Figure 1 Diagram of a portion of the cat thalamus reconstructed from a coronal section. The location of neurons for three modalities of lingual afferents is indicated by symbols; gustatory neurons occupy the nucleus semilunaris accessorius (*SA*). Abbreviations: *CM*, nucleus centrum medianum; *GM*, nucleus corporis geniculati medialis; *MD*, nucleus medialis dorsalis; *TH*, tractus habenulopeduncularis; *VPL*, nucleus ventralis posterolateralis; *VPM*, nucleus ventralis posteromedialis.

most laterally, whereas those for pressure and thermal modalities are found in a medial position of the nucleus ventralis posteromedialis. Neurons for the gustatory modality are situated in a separate thalamic nucleus known as the nucleus semilunaris accessorius.

Neurons of these nuclear regions were approached by microdriving a tungsten microelectrode with a tip diameter of 1 to 3 μ, and the anatomical position of the electrode tip was verified postexperimentally on histological slides by using a reference microlesion technique.[5] The neurons were electrophysiologically isolated for extracellular recording of spike activity, and their modality specificity was determined by applying physiological stimuli to their peripheral receptive fields on the tongue, as has been described previously.[5] As all sensory receptors can be excited by electrical as well as by intense mechanical, thermal, and chemical stimuli, the concept of modality specificity is usually formulated relative to that physiological stimulus which activates a sensory system most readily.[9]

For the purpose of the present study, four types of thalamic neurons were distinguished: tactile, pressure, thermal, and gustatory. A thalamic neuron was defined as a *touch* neuron if it responded with a phasic burst of 1 to 4 spikes when the tongue was lightly tapped with an electromagnetically activated tactile stimulator and if that response remained essentially the same with other types of intense mechanical, thermal (ice water), or electrical stimulation. A thalamic *pressure* neuron was one that responded with a tonic discharge of spikes for the duration of the time that pressure was applied to the tongue by a thin, but not sharp, glass rod; its response remained essentially the same when ice water was squirted on the tongue's receptive field. Responses to stimulation with ice water were not regarded as indicative of thermal modality, because such a stimulus is one of the intense stimuli that has the capability to excite any sensory receptor. A thalamic *thermal* neuron responded by showing the greatest increase in spike frequency when the tongue was cooled in the range between 35° and 22°C, less within the range between 22° and 3°C; its activity decreased when the tongue was warmed. A thalamic *gustatory* neuron responded by increasing the frequency of its spike discharge when the tongue was squirted with a taste solution or ice water. The taste solution used in this case was a mixture of 0.01 M quinine hydrochloride, 1.0 M sucrose, 1.0 M sodium chloride, and 0.1 N hydrochloric acid.

After the modality of a particular thalamic neuron was thus identified, it was activated by applying single square wave pulses of 0.5 msec duration at a frequency of 1 per sec to the same peripheral field from which the neuron could be driven by a physiological stimulus. Electrical pulses were applied through a double-barreled, concentric, needle electrode with a total diameter of approximately 1 mm and an inner core tip sharpened to approximately 0.1 mm. Such an electrode was chosen to provide for as localized a stimulus as technically feasible, limiting the excitation as far as possible to those receptors which were directly involved in driving a particular thalamic neuron. In fact, with a few gustatory neurons, electrical pulses through such an electrode could not activate a peripheral field large enough for driving the neuron; they were abandoned. The intensity of stimulation was adjusted to maximal in each case. A stimulus of maximal intensity was defined as that stimulus at which the response of the shortest latency consisted of the largest number of spikes, and a further increase in the intensity did not change the response characteristics of a particular neuron. Thalamic touch neurons had the lowest stimulus threshold; thermal and gustatory neurons the highest. Maximal stimulation at 0.5 msec duration was reached at 80 volts. Therefore, in order to keep stimulus intensity the same for all neurons and above any variation in receptor sensitivity to stimulation, the stimulus intensity was always kept at 90 volts. This was not as strong a stimulus as it might appear; when such localized stimuli were applied to the experimenter's tongue they became perceptible at approximately 40 volts. They were not unpleasant unless the voltage was increased to nearly 100 volts.

A Grass *S*8 stimulator was used for delivering the electrical pulses to the tongue and for triggering in synchrony a Tektronix 502 oscilloscope and a computer of average transients (*CAT* model 400 *C* of Technical Instruments, Inc.). Most data were also recorded by an FM tape recorder to reanalyze them on playback. The computer was programed for adding spike potentials as discrete events in 400 separate addresses. A "sweep time" could be selected to scan the addresses. Therefore, 1/400th of the total sweep duration was alloted for counting spikes which fell in a particular address. This was done in sequence, beginning with the moment of triggered zero time. On repeated stimulation the zero times were superimposed, and the computer accumulated spike potentials as dot-like events in separate addresses for the duration of the sweep. An electronic counter was preset to function for 1000 repetitions, after which the computer accumulation was stopped. The accumulated counts per address were typed by a read-out printer and were plotted on a dot-plotting autograph. Should a neuron discharge four spikes at a fixed latency of 7 msecs with each potential being separated from its neighbor by 2.5, 2.5, and 3 msecs, then with 1000 repetitions of stimulation, 1000 events would be accumulated in the 45th, 61st, 77th, and 96th address at a total sweep duration of 62.5 msecs. Since at this sweep duration 0.15625 msecs are alloted per address, values in msecs could be assigned to the four dots on the abscissa, as illustrated in Figure 2A.

The entire procedure of modality testing and recording of spike potentials required

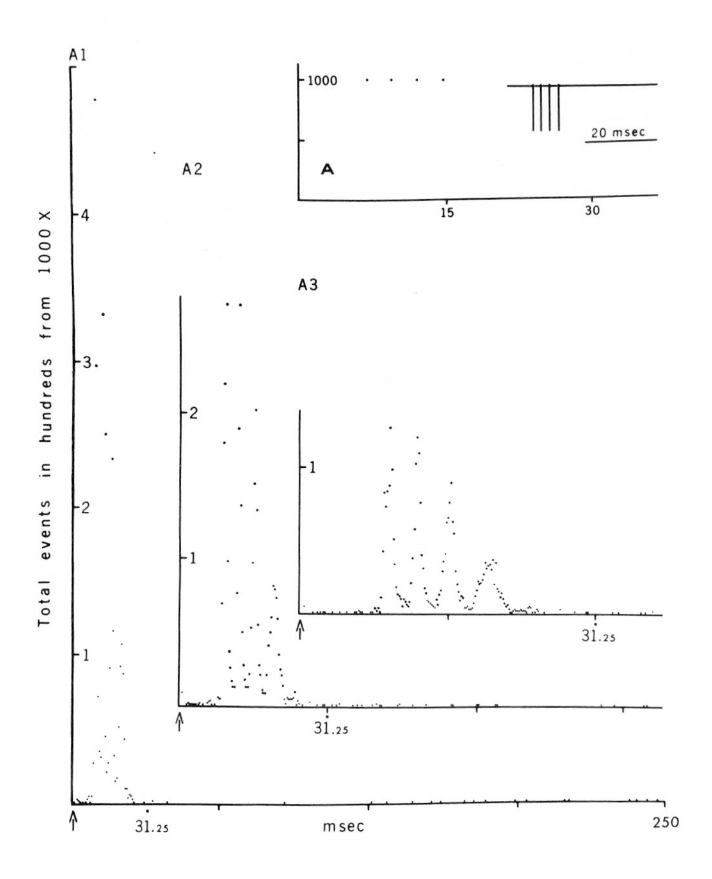

Figure 2 Read-out of a 1000-times accumulated computer count of 4 electronic signals shown in the inset (A), and similar read-outs of spike potentials of a thalamic touch neuron accumulated at the following sweep durations: 250 msecs (A1), 125 msecs (A2), and 62.5 msecs (A3). Except for A1, the abscissas of the other computer plots are cut. Arrow indicates the moment at which the stimulus was applied 1000 times.

a minimum of 1 hr with each thalamic neuron. Although with a few neurons, records could be obtained over a 3-hr period without difficulty, the quality of recording deteriorated rapidly with others, excluding them from the sample. Consequently, the number of neurons on which the results of this study are based is relatively small: 26 touch, 11 pressure, 8 thermal, and 9 gustatory neurons. However, since all neurons of a particular modality studied thus far gave computer dot patterns that were basically alike, it seemed reasonable to assume that this sample represented the population adequately.

RESULTS

A computer read-out of the accumulated evoked activity of a touch neuron is shown in Figure 2 (A1, A2, A3). Although four distinct peaks can be identified readily on the dot plot of A3, this read-out does not conform completely to the hypothetical example of Figure 2A. Variability of the response latency, as well as the interspike intervals, increased progressively with the 2nd, the 3rd, and the 4th spike. Five spikes in a single response occurred only a few times out of 1000 repetitions; there was a tendency for some spikes not to occur with every burst. These factors were represented best in computer accumulations on relatively short sweep durations (Figure 2, A3). When the tape-recorded data were replayed, analysis of the same spike bursts on longer sweep durations (Figure 2, A1 and A2) indicated that the spikes evoked by any single stimulus belonged to a single group, and no other stimulus-related spikes followed this group within 250 msecs. By reanalyzing the same data at 500 msec and 1 sec sweeps it was ascertained that no other stimulus-related spikes were evoked within these intervals. Therefore, it can be stated that the basic form of the sensory code of a thalamic *touch* neuron relaying stimulation is characterized by a single cluster of spike activity.

In contrast, responses of a thalamic *pressure* neuron to electrical stimulation of receptors consisted of two spike clusters (compare Figure 2, A1 with Figure 3B). The composition of the various spike clusters was identified by analyzing the same data on shorter sweep durations, as illustrated for the touch neuron. This analysis revealed

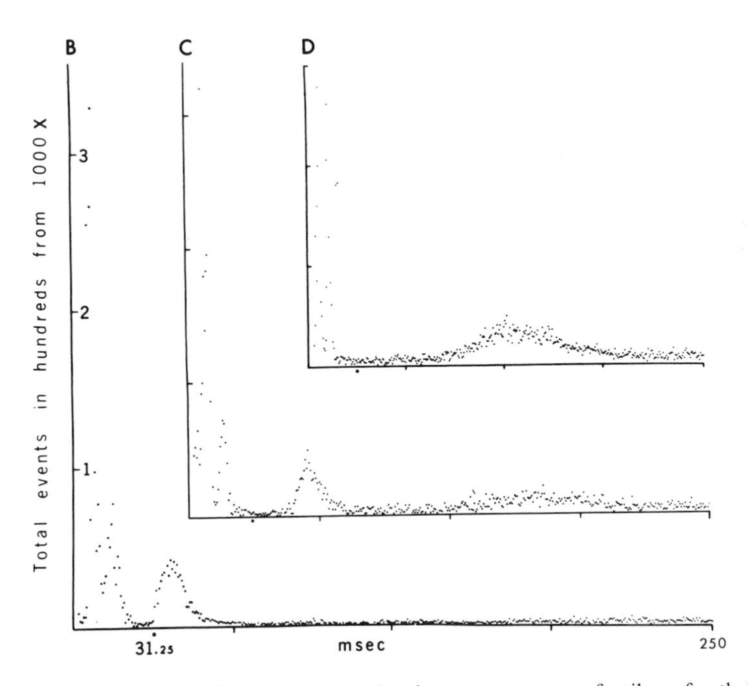

Figure 3 Read-outs of a 1000-times accumulated computer count of spikes of a thalamic pressure neuron (B), thermal neuron (C), and gustatory neuron (D). Ordinate and abscissa divisions indicate the same values for all plots; the absolute size of the plots is a matter of photographic reduction.

that the first cluster consisted of three, or rarely four, spikes; the second was formed mainly by a single spike, which appeared only occasionally in addition to the first cluster of three spikes and which had a latency of between 30 and 60 msecs.

The basic code of a *thermal* neuron was characterized by three clusters of spike activity (Figure 3C). The first cluster consisted of two, or rarely three spikes; the second and the third clusters were formed mainly by a single spike per cluster.

The code of a *gustatory* neuron (Figure 3D) consisted of two-cluster spike activity, but it differed markedly from the two-cluster code of a pressure neuron; the latency of the second cluster of the gustatory code was relatively long, and was usually formed by two spikes. The distribution of latencies for the spikes of the first cluster was very narrow; for the second, it was broad.

As the dot pattern flashed on the screen of the computer with every successive accumulation, it was possible to watch the code emerge. The number of accumulations needed for discerning a definite pattern of a particular sensory code varied with the four sensory systems, but it was considerably below the 1000 accumulations for which the function of the computer was preset. The code of a thalamic touch neuron emerged after approximately 30 accumulations, and its basic form remained unchanged with further accumulations. The code for a thalamic pressure neuron became discernible at approximately 100 accumulations, whereas approximately 150 were needed to see the second and the third clusters of the gustatory and the thermal codes as distinct from random spike acuvity. Consequently, after a few hundred accumulations of the spike responses of any thalamic neuron, it was possible to identify the sensory modality to which it belonged by an inspection of its dot-pattern code. In fact, the dot patterns became more reliable indicators of the modality classification than the test with physiological stimuli. On one occasion, when a neuron, which was mistakenly classified as a touch neuron, gave a two-cluster dot pattern indicative of a pressure neuron, additional testing with pressure stimuli revealed that indeed it was a pressure neuron, although the pressure had to be intense to elicit a tonic type of spike activity; therefore, it was at first overlooked.

The accumulation of the dot pattern on the computer screen could be watched simultaneously with the flashes of the oscilloscope trace revealing the spike activity of a given thalamic neuron. Although the pattern of evoked and spontaneous firings changed from one poststimulus period to another, some information adding to the computer read-out could be obtained. Thus, each of the computer dot-cluster patterns was represented in its entirety by spike-cluster counterparts that flashed on the oscilloscope, during a poststimulus period, on not fewer than 100 occasions out of the 1000 repetitions. Since the oscilloscope trace had to be slow to encompass the entire poststimulus period, spikes of any cluster were not spaced far enough apart for adequate photographic reproduction. However, diagramatic reconstructions (Figure 4) of these spike-cluster counterparts from the computer records depict the forementioned observation clearly. Spikes evoked on most repetitions of stimulation represented only a portion of the entire code. On these occasions, however, spikes

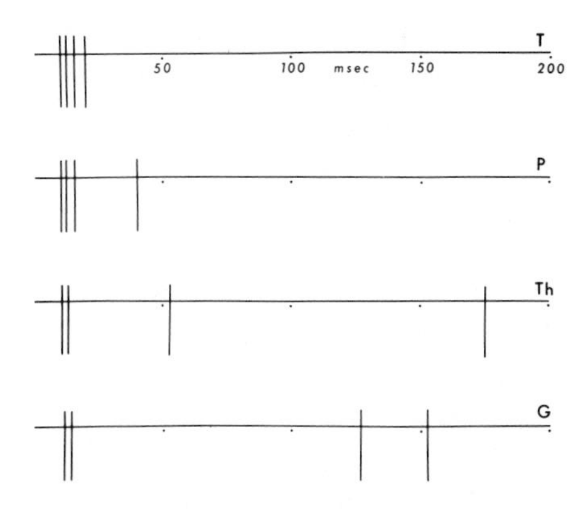

Figure 4 Spike cluster reconstruction of the basic forms of sensory codes from the computer plots for touch (T), pressure (P), thermal (Th), and gustatory (G) modalities. The spikes are placed at the latencies of the various peaks of the dot clusters, and the traces represent the first 200 msecs of a poststimulus period that has a total duration of 1 second. The electrical pulse was applied at the moment the trace began, left.

that formed the second and the third dot cluster (pressure, thermal, and gustatory neurons) never appeared without some spikes of the first cluster. On rare occasions, spikes that formed the first and the third cluster occurred without those forming the second cluster in a three-cluster code.

The form of the modality codes did not change by decreasing the stimulus frequency to less than 1 per sec. However, stimulation at a faster rate encroached upon the time course of the spike distribution, and there was a regrouping of the dot clusters. This regrouping will be described and discussed in a later communication.

Similarly, when the stimulus intensity was varied with a neuron of a particular modality, intricate rearrangements of the dot clusters took place, but they remained modality specific. At low-stimulus intensities, however, the modality codes broke down; latencies of the occasional late spikes became so variable that they formed nothing but a random dot distribution. Details concerning the influence of stimulus intensity on the modality codes are still under study and cannot be included in this presentation.

DISCUSSION

There has been a tendency to study the coding of sensory processes by arranging the experimental situation to become as similar to a real episode in life as is technically feasible. The ultimate situation appears to be one in which spike potentials from thousands of neurons would be picked up by chronically implanted multiple microelectrodes, and telemetrically relayed to a computer that would register the changing pattern of spike activity while a monkey, as the experimental subject, plucks a banana off a tree and consumes it. Although the technical arrangement of such a situation is no longer in the realm of fiction, its usefulness in unraveling definitive principles of sensory function is doubtful.

The present study was designed with a diametrically opposite approach to experimental work—to reduce the situation to one dependent variable totally unlike any episode in life, but abstracted and so related to the factors under the experimenter's control as to provide an answer to one question. *Are there basic differences in the response characteristics of thalamic neurons for lingual afferents that result from some inherent properties of the sensory systems rather than from differences in peripheral stimulation?*

Obviously, the stimulus parameters had to remain constant with neurons of different sensory systems. Because of the ease and the precision of control, electrical pulses were chosen, although any other stimulus that could briefly excite all receptors with nearly equal efficiency could have served. Furthermore, in order to limit the neural activity to the most readily discharged paths of a particular system, deeply anesthetized experimental subjects were preferred to those who were semiconscious or conscious. It is generally known that, under very light barbiturate anesthesia, most thalamic neurons are active in the absence of intentionally applied peripheral stimuli. As the anesthesia deepens, the relation of the spike activity to the stimulus becomes more obvious. Under the conditions of the present experiment, touch and pressure neurons remained silent until peripheral stimuli were applied. The low-level background activity seen on the computer dot plots were also due to fluctuations of excitability brought about by stimulation. Thermal and gustatory neurons were most likely activated by stimuli that were present in the environment during periods when no stimulation was intended; their level of background activity was higher than for the touch and pressure neurons. Injury by the tip of the electrode may give rise to spurious spike discharges. Injured neurons, however, usually die within a few minutes, and therefore they had no part in this study.

With stimulus parameters controlled and the activity in the sensory systems limited, the results clearly indicated that the differences in response characteristics of the four sensory systems represent basic, inherent differences. What we see in the computer dot plots is not the coding of stimulus qualities but the activity of a neuron that belongs to a system which inherently patterns its spike activity in a basically different form than does another system. Since, under natural conditions, one sensory system is preferentially activated by a particular stimulus, the specific activity of this system becomes associated with the particular peripheral stimulus. The specific pattern of spike activity determines the sensory experience, not the stimulus.

In perhaps less cumbersome, although less precise, language Johannes Müller described this phenomenon in his doctrine of "specific nerve energies." Since the "specific energies" of four sensory systems express themselves in the specific patterning of spike activity, one could regard these patterns as *modality codes*. Furthermore, it is possible that these codes are the unit patterns for all the complex spike discharges evoked by physiological stimulation of these sensory systems.

At present, the modality codes can be described only in general terms of spike clusters. It has been impossible to identify any subdivisions of the codes for any particular modality because of the limited number of neurons in the sample. As to the possible existence of separate codes for the four gustatory qualities, it is important to point out that uncertainty exists as to whether the gustatory primaries are conceptual rather than neurophysiological entities.[6] If there were no gustatory primaries, one would expect to find a single gustatory code with as many modulations as there are substances of different gustatory qualities.

Another question awaiting further clarification concerns the level at which the coding takes place. It has been reported that some coding in the olfactory system can take place at the receptor level,[2] and, consequently, the spacing of the spikes might be a matter of the properties of the receptor membrane. However, it is also possible that the spikes which form the second and the third cluster in the dot patterns are discharged by reexcitation of a particular neuron via collateral reverberatory circuits. Such circuits in the thalamus have been described recently.[1]

SUMMARY

Neurons of the ventral posteromedial nucleus and the semilunar accessory nucleus of the cat thalamus were approached stereotactically with tungsten microelectrodes. These neurons were electrophysiologically isolated for extracellular recordings of spike activity, and their modality specificity was determined by applying physiological stimuli to their peripheral receptive fields on the tongue. A thalamic neuron of a particular modality was then driven by single electrical pulses applied at maximal intensity to the same peripheral field from which the neuron could be activated by a specific physiological stimulus. Data were analyzed by a digital computer. This analysis revealed the existence of specific sensory codes for each of the four modalities of lingual afferents. The coding is accomplished by particular groupings of spike discharges. The pattern of spike groupings is constant for neurons of a particular modality but it differs conspicuously from spike groupings of neurons for another modality. Since the electrical pulses used for driving the thalamic neurons were identical, differences in the response characteristics of the four systems represented inherent differences in the properties of the sensory systems.

ACKNOWLEDGMENT

Aided by grant NB-03266 from the National Institute of Neurological Diseases and Blindness, USPHS.

REFERENCES

1. Andersen, P., J. C. McC. Brooks, J. C. Eccles, and T. A. Sears. 1964. The ventro-basal nucleus of the thalamus: potential fields, synaptic transmission and excitability of both presynaptic and postsynaptic components. *J. Physiol.* (*London*) **174**, pp. 348–369.
2. Arvanitaki, A., H. Takeuchi, and N. Chalazonitis. 1967. Specific unitary osmereceptor potentials and spiking patterns from giant nerve cells. *In* Olfaction and Taste II (T. Hayashi, editor). Pergamon Press, Oxford, etc., pp. 573–598.
3. Benjamin, R. M., R. Emmers, and A. J. Blomquist. 1968. Projection of tongue nerve afferents to somatic sensory area *I* in squirrel monkey (*Saimiri sciureus*). *Brain Res.* **7**, pp. 208–220.
4. Benjamin, R. M., and H. Burton. 1968. Projection of taste nerve afferents to anterior opercular-insular cortex in squirrel monkey (*Saimiri sciureus*). *Brain Res.* **7**, pp. 221–231.
5. Emmers, R. Separate relays of tactile, pressure, thermal, and gustatory modalities in the cat thalamus. 1966. *Proc. Soc. Exp. Biol. Med.* **121**, pp. 527–531.

6. Erickson, R. P. 1963. Sensory neural patterns and gustation. *In* Olfaction and Taste I (Y. Zotterman, editor). Pergamon Press, Oxford, etc., pp. 205–213.
7. Moulton, D. G. 1967. Spatio-temporal patterning of response in the olfactory system. *In* Olfaction and Taste II (T. Hayashi, editor). Pergamon Press, Oxford, etc., pp. 109–116.
8. Pfaffmann, C. 1959. The sense of taste. *In* Handbook of Physiology, Vol. I, Section 1 (J. Field, editor). American Physiological Society, Washington, D.C., pp. 507–533.
9. Zotterman, Y. 1959. Thermal sensations. *In* Handbook of Physiology, Vol. I, Section 1 (J. Field, editor). American Physiological Society, Washington, D.C., pp. 431–458.

SUMMARY OF TASTE ROUNDTABLE

Prepared by CARL PFAFFMANN

Issues raised in the preceding papers were discussed further in a roundtable chaired by C. Pfaffmann. Recent electrophysiological recordings add to the earlier evidence, which showed that most individual mammalian taste fibers are multi-quality; that is, they respond to more than one taste quality. The basic taste stimuli that typically represent the four taste qualities are acid, NaCl, sugar, and quinine. The implications of multi-quality response can be examined at several levels: (1) purely peripheral, from receptors to afferent nerve fiber; (2) from afferent fiber to central nervous system; and (3) from afferent nerve discharge pattern to sensation.

The evidence at the first level is by far the most compelling. That individual taste fibers may respond to one, two, three, or all four basic stimuli was well documented in Sato's review of the rat and hamster chorda tympani data in this roundtable. His computations of the frequencies with which all combinations of sensitivities occurred showed, in certain cases, departure from randomness. However, Frank and Pfaffmann found, in the glossopharyngeal nerve, no significant deviation from random combination of sensitivities to the four basic stimuli. Within the taste bud, the multiple branching of afferent nerve fibers that contact many sense cells, described by Beidler earlier in the symposium, might be one contributory factor, but the receptor cells themselves have multiple sensitivities. Kimura and Beidler[4] suggest that this results from the several different chemosensitive sites that are found on each receptor cell.

Further study of the statistical structure of the distribution of sensitivities among receptor cells, taste buds, and innervating fibers is needed to establish the nature and origin of the multiple sensitivity of single taste fibers. Although individual fibers may respond to several basic stimuli, they usually respond "best" to one stimulus. The intensity of test stimuli is an important variable. Wang presented data from cat chorda tympani units which were stimulated over a wide range of intensities. He emphasized the necessity of considering the whole dynamic range, not just one test concentration for each substance.

At the second level, i.e., from afferent nerve through one or more synapses to CNS, other sensory modalities display "sharpening," or a decrease in overlap of unit responses, along such stimulus dimensions as wave length in vision or frequency in audition. The little evidence available indicates this may not be so for taste. Erickson and his colleagues presented new data on the first synaptic relay for taste. They did not find sharpening, but rather a leveling or decrease in fineness of tuning. However, the "across-fiber pattern" appeared more stable and reliable in solitarius cells than in chorda tympani fibers.

The third level, the relationship between the responses of peripheral nerve and sensation, can be studied behaviorally in animals or psychophysically in man. Behavioral methods (see Morrison, this volume) are rapidly being developed to permit "animal psychophysics." In this case, since electrophysiology and behavior are under examination in the same species, a more direct correlation between physiology and behavior can be attempted. Erickson's formulations and behavioral experiments and those of Morrison show that the discrimination can be predicted from across-fiber electrophysiological patterns. But these data are consistent with the view that salt or other stimuli activate several input channels in parallel, so that all contribute to the over-all quality of sensation.

In human psychophysics, the correlation between stated sensation and stimulus is studied directly. McBurney showed that in man adaptation to NaCl eliminated the specific sensation of saltiness from all substances that contained this quality as a component sensation before adaptation. This suggests an inactivation of a single salty receptor system. In the postadaptation test, other taste qualities were unchanged or slightly modified, but they were always present and detectable. Other basic taste stimuli acted in a similar manner for other qualities when they were used as specific adaptants. Gymnemic acid on the human tongue eliminates the sweet sensation of sugar or saccharin by its blocking action, analogous to total adaptation by sugar. At the same time, the sweetness of weak salt solutions and sweetness of water after exposure to acid is eliminated. Again, these results seem to indicate a rather direct stimulus-sensation connection most simply interpreted as inactivation of sweet receptors.

Evidence on the physiological changes produced by gymnemic acid at the level of the receptor to afferent nerve is given in Bartoshuk's data (personal communication) on hamster chorda tympani units responsive to sugar and other stimuli. Figure 1 shows that gymnemic acid blocked only the sugar-sensitive component of the multi-quality fibers' response. Note that fibers D and E respond to only one basic stimulus, but B responds to two, C to three, and A to all four before gymnemate application. Although its response to sugar may be blocked, a fiber still responds well to other stimuli after gymnemate. Hence, the peripheral receptor-site specificity can be dissociated from the less specific fiber response. The receptor site, as such, is probably not multi-quality, but the afferent channels carrying the information are. The response of the acid-sensitive fiber E is of special interest. The drug completely blocks this fiber's small response to sugar and marked response to water after acid. Water after acid tastes sweet to man, and this sweetness, as well as the sweetness of sugar and weak saline, are blocked by gymnemate. The response of fiber E to water after acid may be attributed to a sweet receptor, since it was blocked by gymnemate. Water after acid discharges in fibers A through D may be due to a more general sensitization not related to sweetness.

The meaning of dual or multiple sensitivities in peripheral neural elements for sensation was discussed by Zotterman, who cited relevant experiments in the skin senses. Hensel and Zotterman[3] found nerve fibers specific to cooling, but also found other fibers which responded to both mechanical stimulation and cooling. But the response to cooling was only phasic. Do such fibers give rise to a sensation of cooling

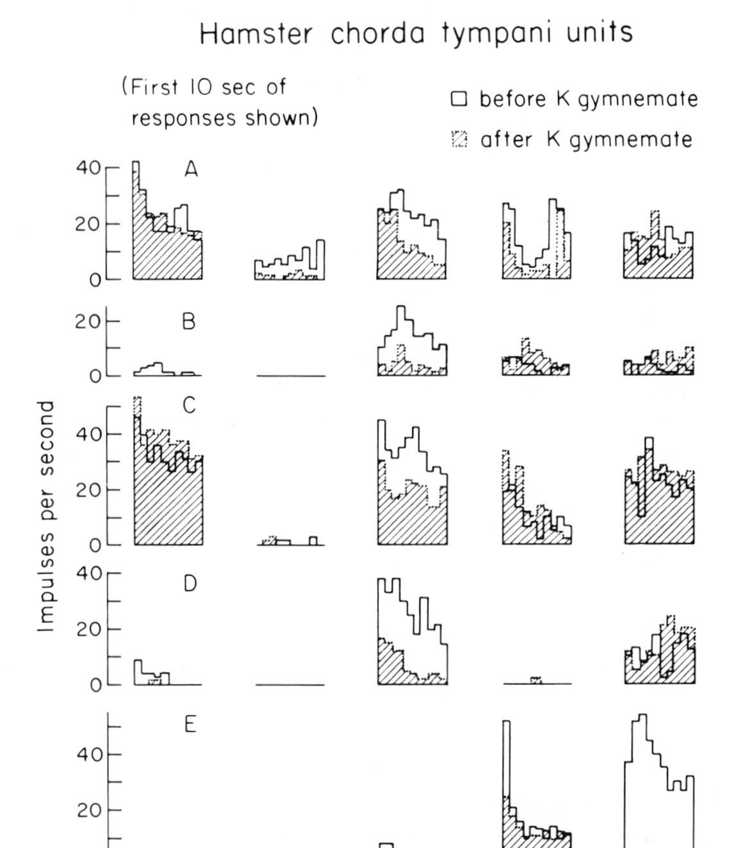

Figure 1 Discharge frequencies of hamster single chorda tympani fibers before and after blockage by K gymnemate. Stimuli were sodium chloride, quinine hydrochloride, sucrose, hydrochloric acid, and (D) distilled water, presented in that order. Each taste stimulus was followed by a distilled-water rinse, but only the post-acid water is shown. The block diagrams represent second-by-second discharges during the first 10 seconds after stimulation.

or only to touch or pressure? In 1884, Blix showed that, in man, stimulation of cutaneous cold spots by a unipolar needle electrode gave rise to a cold sensation; this sensation was not elicited from these points if von Frey stimulating hairs were used, provided the hairs were not cooler than the skin. For this reason, Zotterman[5] concluded that these fibers are often wrongly called dual-modality fibers, for they do not provide information about temperature. Therefore, he differentiated between specificity of the response of the peripheral nerve fiber and specificity of the central response to the activity of that fiber. Hensel and Zotterman[3] believed that the response of these fibers would explain the well-known illusion, first described by Weber in 1846, that if two identical weights were of different temperature, the cooler seemed the heavier.[5]

Dzendolet made the same point in his review of the sensory basis for quality. According to a strict specificity notion, he observed, the influence of cold on the mechanical touch fibers might change the discharge, and hence the apparent intensity of the mechanical sensation, so that a cold mechanical stimulus feels heavier than a warm one.

Zotterman concluded that, similarly, a gustatory fiber responding to sweet and to bitter solutions might produce only one sensation—perhaps that associated with the stimulus which elicits the strongest response. He cited von Békésy's experiment[1] in which electrical stimulation of a single fungiform papilla of a girl's tongue gave rise to a sensation that she described as "mild angelic sweetness," something she had never experienced before. He suggested that electrical stimulation just above threshold selectively excited only the specific sweet fibers, thus causing the sensation of "angelic sweetness." Figure 8 in the Diamant and Zotterman paper in this volume shows single chorda tympani units of the monkey. Sucrose, saccharin, glycerol, and ethylene glycol solutions were seen to elicit fiber discharges (large spikes) of increasing frequency, but these were accompanied by a discharge of small spikes from fibers which responded very strongly to quinine. Zotterman suggested that the same condition holds in man. Thus, even sucrose, the taste of which we describe as sweet, elicits, in addition to its discharge from sweet fibers, a weak but definite response from some bitter fibers as well.

Pfaffmann asked whether fibers which respond to only one basic taste might be mediating the elicitation of a single quality by more intense stimulation of single human papillae. Can a fiber sensitive only to sugar elicit the "angelic sweetness" when stimulated? However, not all fibers responsive to sugar respond only to sugar. It would be interesting to know what taste sensation could be elicited by stimulating fibers with single and multiple sensitivities, if such exist in man.

Zotterman's views, then, seem to reflect a modified specificity concept; that although there are specific "sweet" fibers, the multimodal response or side effects of sugar on other nonsweet fibers contribute to and change taste sensation quality per se. Indeed, von Békésy said something very similar when he theorized that the specific chemical endings are not mono-gustatory to chemicals but only to punctate electrical stimulation.

This is similar to the pattern concept introduced by Pfaffmann, after he found that single fibers were activated by more than one basic taste stimulus. Figure 2 shows discharge frequency as a function of stimulus intensity for two units. Both are stimulated by sugar and salt, but salt is the best stimulus for unit A and sugar the best stimuli for the four basic taste are not mono-gustatory is just what the pattern theory receptor unit, the frequency of discharge increases with concentration.

Pfaffmann suggested that ". . . the relative, rather than the absolute, amount of activity in any one set of afferent fibers may determine the quality of sensation." Low

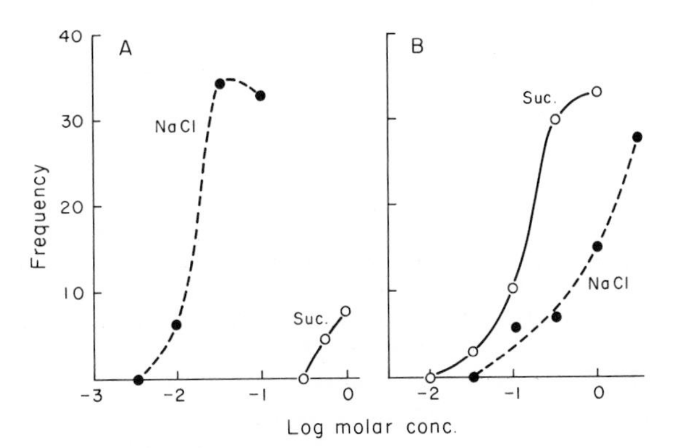

Figure 2 The relation between frequency of discharge and concentration in two fibers, both of which are sensitive to sugar and salt. (Redrawn by permission from the *Journal of Neurophysiology*)

concentrations of salt discharge only A; higher concentrations discharge both A and B, but A will be greater than B. Low concentrations of sugar activate only B; higher concentrations activate both B and A, but B will be greater than A. Thus, the sensory code might read:

Frequency Code

$A > B =$ salty

$B > A =$ sweet

where A or B may go to zero. The dominance of A or B in the afferent input is given by the ratio of such parallel activity.

This emphasis on relative firing rates is a device by which dominance of the primary stimulus for any particular fiber is preserved, even though other stimuli may become effective as concentration is raised. This resembles the situation in single auditory afferent fibers, in which the best-stimulus frequency tends to elicit higher neural firing rates than do the side-band frequencies. The information in any one fiber by itself would not distinguish between a low intensity best frequency or a high intensity neighboring frequency. In addition, the two major stimulus dimensions are frequency and energy. In taste, the relation of chemical class to the number of taste dimensions is not yet settled, but it is unlikely to be a linear scale, as proposed by Frings.[2]

The character of the basilar membrane mechanics make it impossible to stimulate a single auditory fiber by sound stimulation alone. Likewise, the growing morphological evidence of multiple innervation and branching of new fibers within taste buds, as well as interaction among taste buds in neighboring papillae, make it seem unlikely that a stimulus limited to a single papilla would activate only one fiber. Until the relation of single papilla to single fibers and single stimuli is better understood, especially in man, this part of the problem must remain conjecture.

In this form of pattern theory, better called ratio theory, there is little difference operationally from the specificity view expressed above. For instance, that chemical stimuli for the four basic taste are not mono-gustatory is just what the pattern theory is all about. Yet, there really is a subtle theoretical difference which divides direct correspondence between a fiber's "best stimulus" from the taste quality the stimulus elicits. Do all taste fibers elicit only one of the four basic taste qualities? We will know the answer only when the full scale of electrophysiological data on human reactions is available.

REFERENCES

1. Békésy, G. von. 1964. Sweetness produced electrically on the tongue and its relation to taste theories. *J. Appl. Physiol.* **19**, pp. 1105–1113.
2. Frings, H. 1948. A contribution to the comparative physiology of contact chemoreceptors. *J. Comp. Physiol. Psychol.* **41**, pp. 25–34.
3. Hensel, H., and Y. Zotterman. 1951. The response of mechanoreceptors to thermal stimulation. *J. Physiol. (London)* **115**, pp. 16–24.
4. Kimura, K., and L. M. Beidler. 1961. Microelectrode study of taste receptors of rat and hamster. *J. Cell. Comp. Physiol.* **58**, pp. 131–139.
5. Zotterman, Y. 1959. Handbook of Physiology, Section I, Neurophysiology **1**. American Physiological Society, Washington, D.C., pp. 431–458.

VI ROLE OF TASTE IN BEHAVIOR

VI · ROLE OF TASTE IN BEHAVIOR

THE ROLE OF ADRENOCORTICAL HORMONE SECRETION IN SALT APPETITE

D. A. DENTON, J. F. NELSON, ELSPETH ORCHARD, *and* SIGRID WELLER

Howard Florey Laboratories of Experimental Physiology, University of Melbourne, Victoria, Australia

INTRODUCTION

There is evidence that the steroid hormones secreted by the gonads act in the central nervous system to influence the sexually directed behavior of animals—and, presumably, the changes in central excitation and drive result from physiochemical action on the relevant neuronal systems. Michael[11] has shown that radiolabeled female sex hormone which is injected systemically localizes in the preoptic area of the brain—a region which may be important in sexual behavior. On the other hand, there are no data indicating that polypeptide antidiuretic hormone secretion, evoked by water deprivation, acts in the central nervous system and effects thirst.

The question we wish to examine is whether hypersecretion of the steroid hormone, evoked from the adrenal glomerulosa by salt deficiency, is the cause of the salt appetite seen under these circumstances. More specifically, is it the sole cause, a contributory cause, or, alternatively, a necessary or standing condition for the action of other factors which are directly causal?

FIELD STUDIES

First, to consider evidence emerging from a study of wild animals in nature, we have shown that native and introduced animals in the southern Alps of Australia are severely sodium-depleted in spring and early summer.[2] The animals show an avid salt appetite, and rabbits, for example, attack softwood pegs impregnated with Na^+ salts and largely ignore pegs impregnated with Mg^{++}, K^+, Ca^{++}. By use of highly sensitive double-isotope dilution techniques, it has been possible to show that the rabbits exhibiting this behavior have low urine Na^+, high peripheral blood aldosterone concentrations, and high renal renin contents. These metabolic findings, indicative of Na^+ deficiency, contrast with those in desert animals that do not exhibit salt appetite. The specific appetite for Na^+ salts of the mountain rabbits is seasonal. In spring and early summer, when the grass is low in sodium content, the animals eat 100 per cent of the pegs during periods of seven to 14 days, but have little or no appetite for them in late summer or winter. It has been found that blood aldosterone levels fluctuate in a corresponding way.[2]

With man, it is difficult to determine what motivates appetitive behavior. There is an organized salt trade in the highlands of equatorial New Guinea, including tribal ownership of salt sources.[12,17] In the remote areas, people such as the Chimbu live on a vegetarian diet of a very low Na^+ and a high K^+/Na^+ ratio content (Figure 1). In collaboration with Professor MacFarlane, we have measured peripheral blood aldosterone levels of these primitive people and found the mean was fourfold higher than in those New Guineans who had access to a Western diet and in urban Australians. However, as I will discuss later, clinical evidence shows that patients with high blood aldosterone levels as a result of Cohn's tumor do not present a history of salt craving or aberrations of salt intake.

LABORATORY ANALYSIS USING RATS

The laboratory analysis of this problem began with Richter's finding[14] in 1943 that

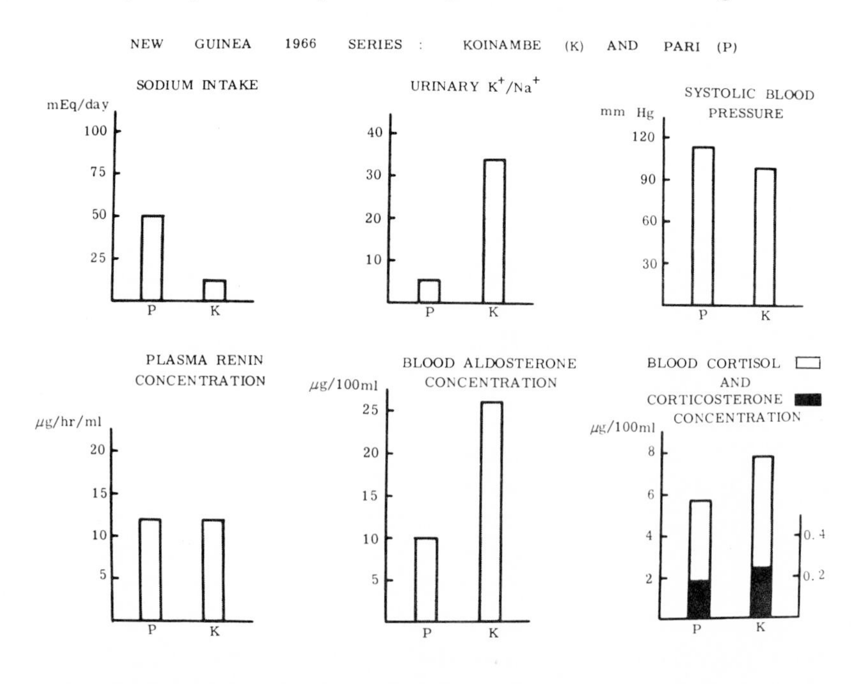

Figure 1 The peripheral blood aldosterone concentration of New Guineans in remote areas of highlands (K) in relation to dietary Na^+ and K^+/Na^+ ratio in the urine as compared with New Guineans with some access to European diet (P). (Collaborative study of Howard Florey Laboratories [Dr. J. P. Coghlan and B. Scoggins], with Professor Victor Macfarlane's group and Dr. S. Skinner of Adelaide.)

injections of desoxycorticosterone acetate (DOCA) increased the need-free intake of saline by caged rats. There appeared to be a threshold with an effect beginning at 1.0 mg per day. Braun Menendez obtained similar results, and also found no direct effects on salt appetite with steroid sex hormones. Subsequently, the important contribution of both Wolf[18] and Fregly[10] and their collaborators have shown that this behavior is elicited by aldosterone administration, and the effect is specific to sodium. This behavior of the sodium-replete animal is, in a sense, paradoxical and maladaptive, because aldosterone increases the Na^+ content of tissues and plasma and, concurrent with K^+ depletion, the Na^+ of intracellular fluid of some tissues.

The natural physiological context, if any, of the appetite-stimulating effect of aldosterone would be in Na^+ deficiency. Thus it could be reasoned that the unphysiologically high doses, reported as necessary to induce appetite in the sodium-replete animal, may overcome the behavior-inhibiting effects of salt accumulation. From this viewpoint, Wolf and Handal[20] examined appetite under a regime of brief access to salt, to prevent salt accumulation from modifying body status. They found no effect on appetite when 30 μg of aldosterone were given daily in divided doses, some effect with 60 μg, and a larger one with 120 μg per day. However, they noted that the effect was small; 60 μg per day gave an increase of 0.7 mEq per day Na^+ intake. Fregly and Waters[9] also noted increased appetite in adrenalectomized rats when the aldosterone dose was increased to about 50 μg per day.

The crucial question is the physiological significance of these doses. Measurements of aldosterone-secretion rate in the rat include that of Fregly et al.,[8] who found a figure of 9.8 μg per 100 g body weight per day—i.e., about 30 μg per day in a 300-g rat. Other values include those cited by Peterson[6] and Singer,[16] who found lower outputs ranging from 10–20 μg per day. Palmore and Mulrow's[13] figure for sodium-deficient rats is about 60 μg per day. All collections were made under anesthesia and, together with trauma and loss of the blood required for analysis, the conditions stimulated aldosterone to levels higher than those found in the conscious animal. Notwithstanding the possible influence of the higher metabolism of the rat, the figure of 10–30 μg per day is similar to the daily output (20–40 μg) of the conscious, trained, sodium-replete sheep—a creature 100 times larger. Man, 200 times larger than the rat, secretes 50–200 μg per day. Unless there is a surprising fiftyfold decrease in target-organ sensitivity to aldosterone in the rat, the figures seem too high.

More recently, Bojeson,[3] using the tosan double-isotope method, has measured the peripheral blood aldosterone concentration of the conscious sodium-replete rat, and finds the same level—about 5 mμg per 100 ml—as found in man and sheep. From the metabolic clearance rate measured concurrently, the secretion rate is about 2 μg per day in a 300-g rat, perhaps a more likely figure. After two days' Na^+ deficiency it rises to 40–60 μg per day—a greater amplification from basal than that seen in man or sheep. At this level, injected aldosterone has small effect on the appetite of a rat, and thus the data may be compatible with those which show that aldosterone contributes to the rat's salt appetite.

Wolf[19] found that dorsolateral hypothalamic lesions in the rat eliminate the appetite-inducing action of aldosterone and the appetitive response to Na^+ deficiency. More recently (personal communication), he has shown that lesions of the gustatory nucleus of the thalamus and large lesions of the ventromedial nucleus of the hypothalamus have the same effect.

RABBITS—THE EFFECT OF DOCA

We have brought young wild rabbits from the mountains, studied their intake of a variety of mineral solutions, and the subsequent effect of DOCA on that intake. The diet had a substantial Na^+ content—5 mEq per day. Figure 2 shows that under control conditions the mean need-free intake of sodium chloride was very small, but was increased tenfold by 2 mg of DOCA per day and thirtyfold by 5 mg daily. There was a larger need-free intake of divalent cations—this was not influenced by DOCA. Figure 3 shows the relations between Na^+ intake and urinary excretion. At the higher DOCA dosage, loss of food appetite occurred. There was little evidence in any of the experiments that DOCA caused urinary retention of Na^+.

With 500 mEq/l solutions (Figure 4), the results were essentially the same, with

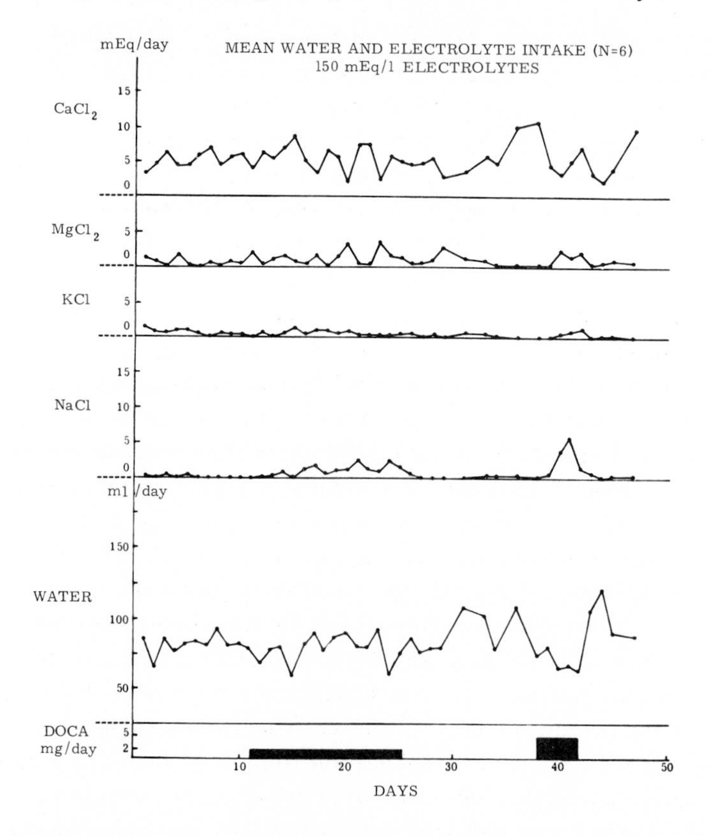

Figure 2 The effect of daily injection of 2 mg and 5 mg of DOCA (bars at bottom) on the mean intake of 150 mEq/l solutions of $CaCl_2$, $MgCl_2$, KCl, NaCl, and water by caged wild rabbits (*Oryctolagus cuniculus*).

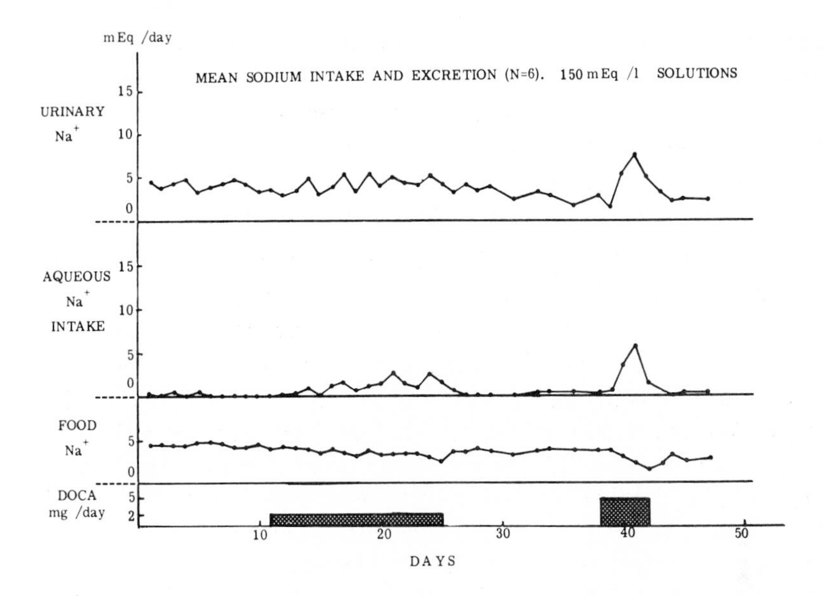

Figure 3 The effect of daily injections of 2 mg and 5 mg of DOCA on the mean voluntary intake of 150 mEq/l Na^+ solution, the urinary Na^+ excretion, and Na^+ intake in food of caged wild rabbits.

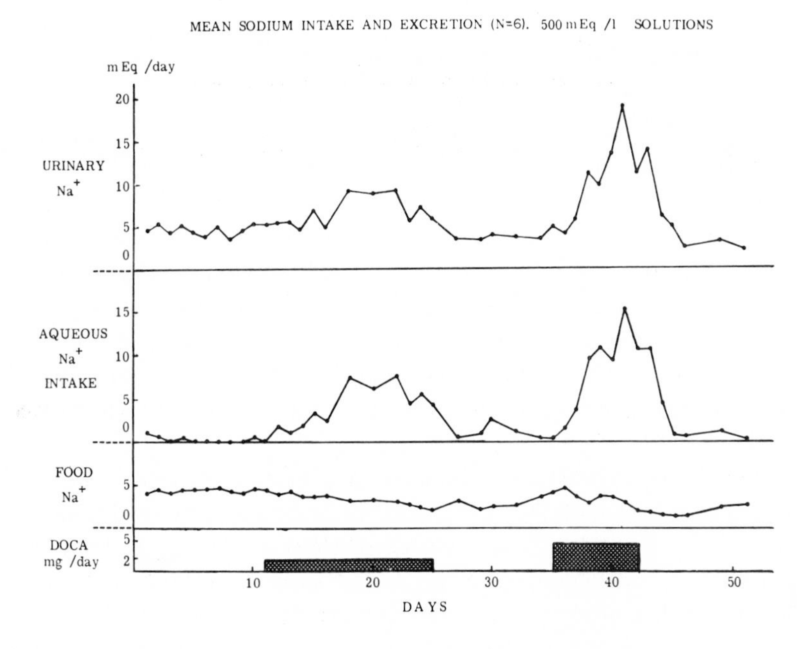

Figure 4 Like Figure 3, except that the solutions of NaCl and other cations offered were 500 mEq/l.

significant increases of Na^+ intake at both 2- and 5-mg dosages. The only difference was a significant inhibition ($p < 1$ per cent) of Mg^{++} intake (and an augmentation of Ca^{++} intake) at the 5-mg DOCA dosage. In terms of Na^+ turnover, the grouped data show a similar effect, as in Figure 3; again there was little evidence of DOCA causing any Na^+ retention. Results with the rabbits differ from the published data on rats as follows. (1) The need-free intake of NaCl by the caged wild rabbits was small (0.1–0.4 mEq per day). The volume of 500 mEq/l NaCl drunk was not different from the 150 mEq/l solution. Also in contrast to the rat,[21] the intake of hypertonic NaCl was increased more with DOCA than was the intake of isotonic NaCl. (2) The effect of DOCA on the salt appetite of the rabbit, although statistically significant, is much less when compared quantitatively with that of the rat. A 250-g rat with about 8 mEq of extracellular Na^+ drinks approximately 10 mEq of Na^+ per day when given 1.0 mg DOCA per day,[15] i.e., equivalent to a daily turnover of their total extracellular sodium. The rabbits (five times heavier), with 40 mEq of intracellular Na^+, seldom take in more than 10 mEq of Na^+ per day when given 5 mg of DOCA daily—an equivalent dose in ratio to body weight. There also appears to be a marked difference in rabbits and rats in their tolerance of the salt-retaining hormone. Richter and Fregly have given 5 mg per day without reporting toxic effects. With this dosage, rabbits, although five times larger than rats, rapidly declined and became anorexic. Four of the 10 in the series died. It is doubtful, in terms of aldosterone equivalent and likely secretion rate in the rabbit, that any physiological significance could attach to effects of dosage much above the 2-mg level, and here the effect on intake was small.

SHEEP—THE EFFECT OF DOCA

Under laboratory conditions, many, but not all, sheep exhibit some need-free intake of hypertonic NaCl solution (300 mEq/l) when the solution is continuously available concurrently with water (control observation blocks on left of Figure 5). With a dosage of 5 mg of DOCA per day (one to two times the maintenance dose required for an adrenalectomized sheep), the mean voluntary Na^+ intake of each animal was little altered. When the same experiment was carried out with the high dose of 20 mg per day (Figure 6), there was clear evidence of increased plasma HCO_3, and also of hypokalemia. After an initial rise, body weight fell because of anorexia, and the general condition deteriorated. One animal (Larry) died at the end of six-days' dosage, and he showed a small increase of Na^+ intake. Two others showed a decline of intake.

Another aspect is illustrated by a study of a renal hypertensive animal (Figure 7). Its mean, need-free, daily NaCl intake of 228 mEq was decreased slightly by DOCA dosage of 20 mg per day. Some increase of HCO_3 occurred, as did an initial hypokalemia. The renal hypertension was exacerbated during the high DOCA dosage, indicating a substantial systemic effect, but no increment of appetite.

A further approach is to study the effect of DOCA on the Na^+ appetite of sheep made salt-deficient by salivary loss from a parotid fistula. $NaHCO_3$ (300 mEq/l) was made available for 15 minutes each day, and the influence of hormone injection on intake determined. The high dose of 20 mg per day did not augment appetite to the level which would be seen, for example, if Na^+ loss had been allowed to continue for two or more days without compensation. There was a slight fall in voluntary intake. With 40 mg per day there was a slight increase of voluntary intake, but the experiment could not be continued because of the decline in the animals' condition.

Over-all, even when the salt-retaining hormone is administered to sheep in doses that are substantially unphysiological, their salt appetite is not augmented in a way similar to that of the laboratory rat.

ADRENALECTOMY AND SALT APPETITE

The main problem in the theory that the blood level of aldosterone determines salt appetite is that we know that sodium-deficient adrenalectomized rats have a sodium appetite. Epstein and Stellar[7] showed that, if naive rats were deprived of salt post-operatively for 10 days until severe deficit developed, intake immediately reflected the

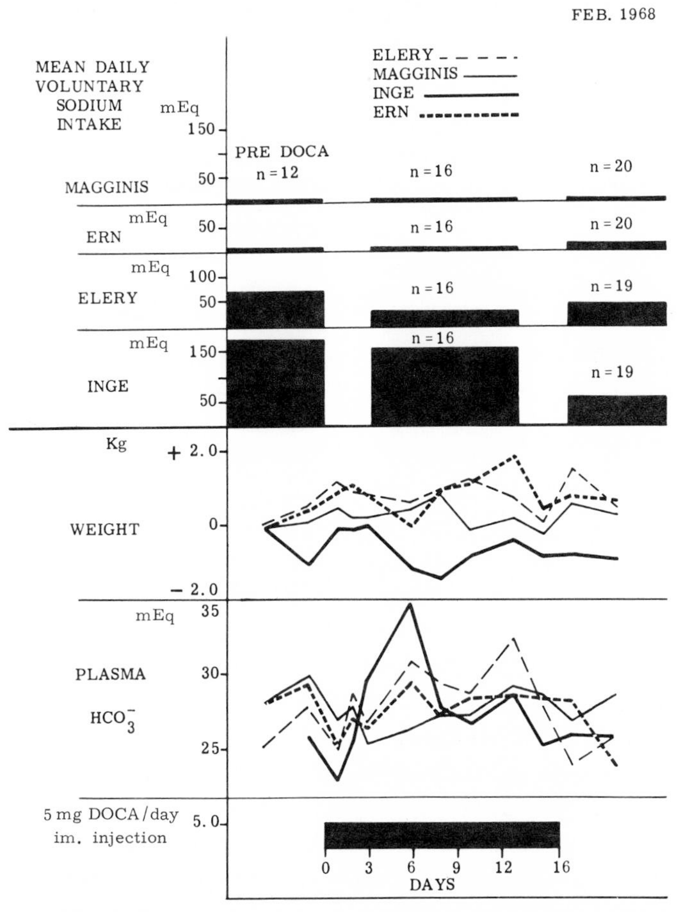

Figure 5 The effect of daily intramuscular injections (i m) of 5 mg DOCA on the mean need-free intake of NaCl by 4 normal sheep (Magginis, Ern, Elery, and Inge). The body weight and plasma HCO_3^- concentration also are shown.

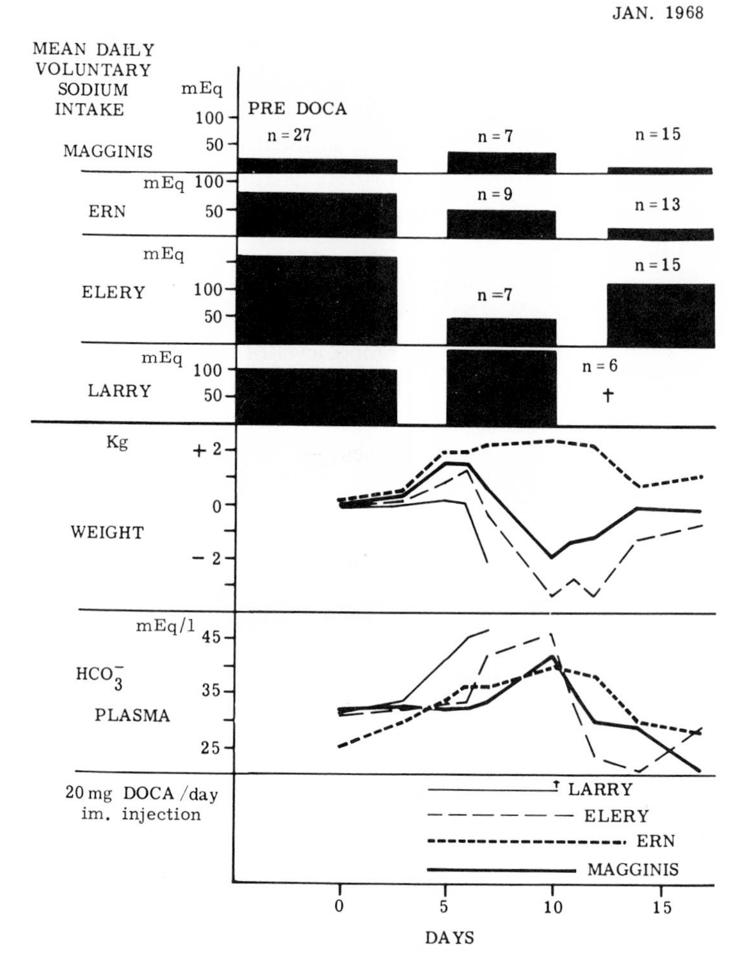

Figure 6 The effect of daily dosage of 20 mg of DOCA over the time interval individually designated on the mean need-free intake of NaCl solution by 4 normal sheep. One sheep, Larry, died after 6 days of injection. The effect on body weight and plasma HCO_3^- also is shown.

extent of the deficit. Wolf and Steinbaum[22] showed that, if adrenalectomized or intact rats were depleted of available Na^+ by acute edema caused by formalin injection, there was no significant difference between the two groups in the amount of Na^+ ingested. Adrenalectomy did not quantitatively modify the response.

In sheep that have lost saliva from a unilateral parotid fistula, the extent of body Na^+ deficit (Figure 8) can be varied according to whether access to a solution of 300 mEq/l $NaHCO_3$ is allowed for 15 minutes each day, every second or third day. Also, saliva can be returned to the animal two or three times over the 24-hour period, so the sheep reaches the test period at near-normal balance. This Figure shows the

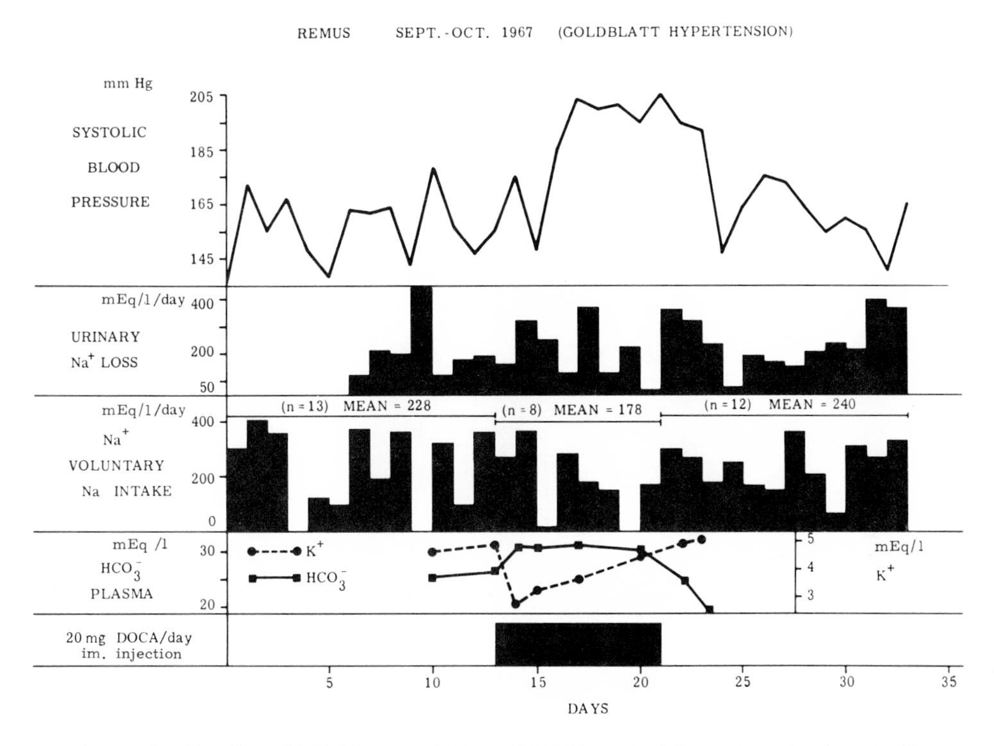

Figure 7 The effect of daily injection of 20 mg of DOCA on the daily need-free intake of NaCl solution and blood pressure of a sheep (Remus) with renal hypertension. The scale for plasma K^+ is at the right; that for HCO_3^- is at left. Urinary Na^+ loss is also shown.

relation between voluntary sodium intake and body sodium deficits ranging from 100–1000 mEq. Each point represents 10–25 observations at each level of deficit made over a study period of three to five months. Voluntary intake and deficit were related and Figure 9 shows the same series of observations conducted over four to five months in four adrenalectomized sheep on constant basal DOCA dosage. The test was carried out 24 hours after the last dose of DOCA when no residual hormone action, as indicated by salivary Na^+/K^+ ration, was apparent. Voluntary intake was again closely related to deficit.

Thus, in the normal animal, blood aldosterone is high, whereas the essentially similar relation of intake and deficit is shown in the adrenalectomized group that lacks the hormone. Another facet of the same phenomenon is evident (Figure 10) when the parotid fistula of an adrenalectomized sheep is surgically closed after one to two years. With cessation of salivary loss, the animal's body Na^+ content increases and Na^+ excretion commences in the urine. With normal body Na^+ status, voluntary intake declines to basal. DOCA dosage was constant throughout, whereas intake was related to body Na^+ status. Figure 10B shows a similar result.

Over-all, it would appear clear that the presence of adrenal salt-active hormone is

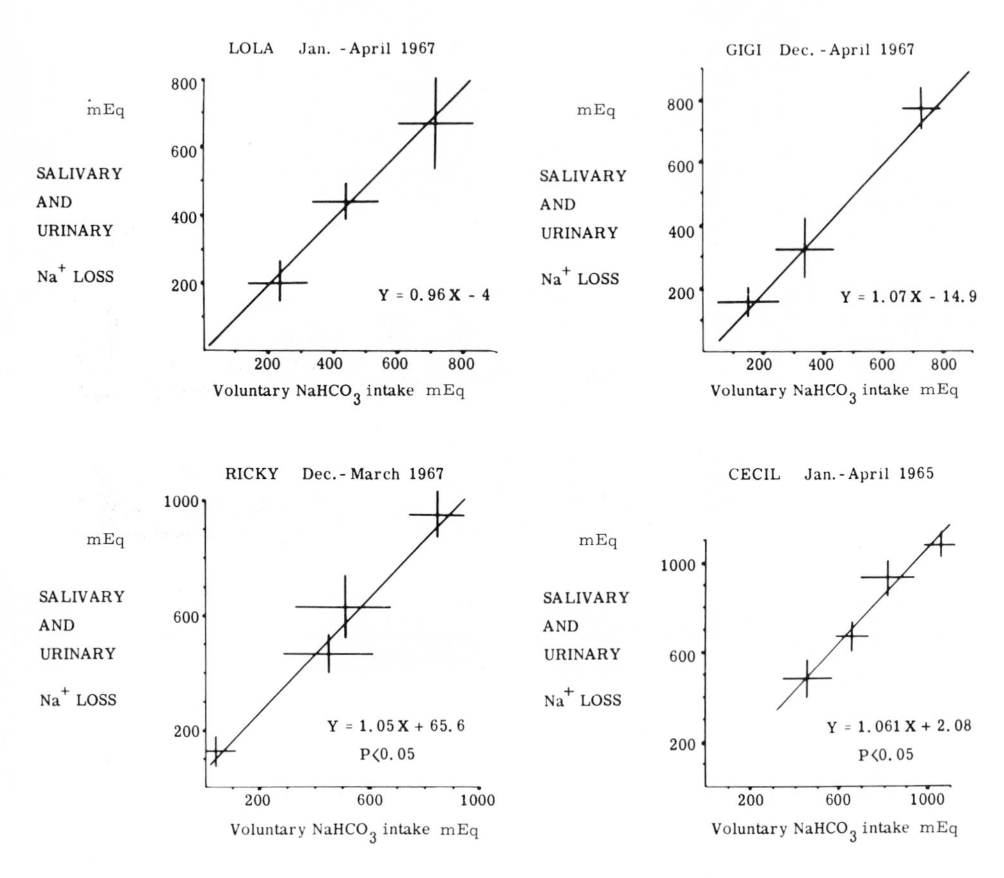

Figure 8 The relation between voluntary intake of 300 mEq/l $NaHCO_3$ solution and body Na^+ loss through saliva and urine in 4 sheep studied over a 3–5 month period. Each point of intersection represents the mean ± 1 SD of 10–25 observations at each level of deficit.

not an essential condition for the development of salt appetite in the species studied. It is possible that it may be a contributory cause of appetite in the laboratory rat with substantial deficit. However, other, unknown causal mechanisms must operate, since adrenalectomized rats can maintain themselves indefinitely. Species appear to differ in the influence of salt-active hormone on appetite. There is little evidence of any effect in sheep, and the clinical data from patients with aldosterone-secreting adenoma does not suggest an effect in man (Conn, personal communication). The doses required to cause appetite in rabbits are high—perhaps beyond the physiological range.

The data on rats are confined to laboratory species, and it would be interesting to know the effects on wild species, of which, for example, there are dozens on the Australian continent.

CONCLUSION

There is still the question of how DOCA or aldosterone have appetitive effects in

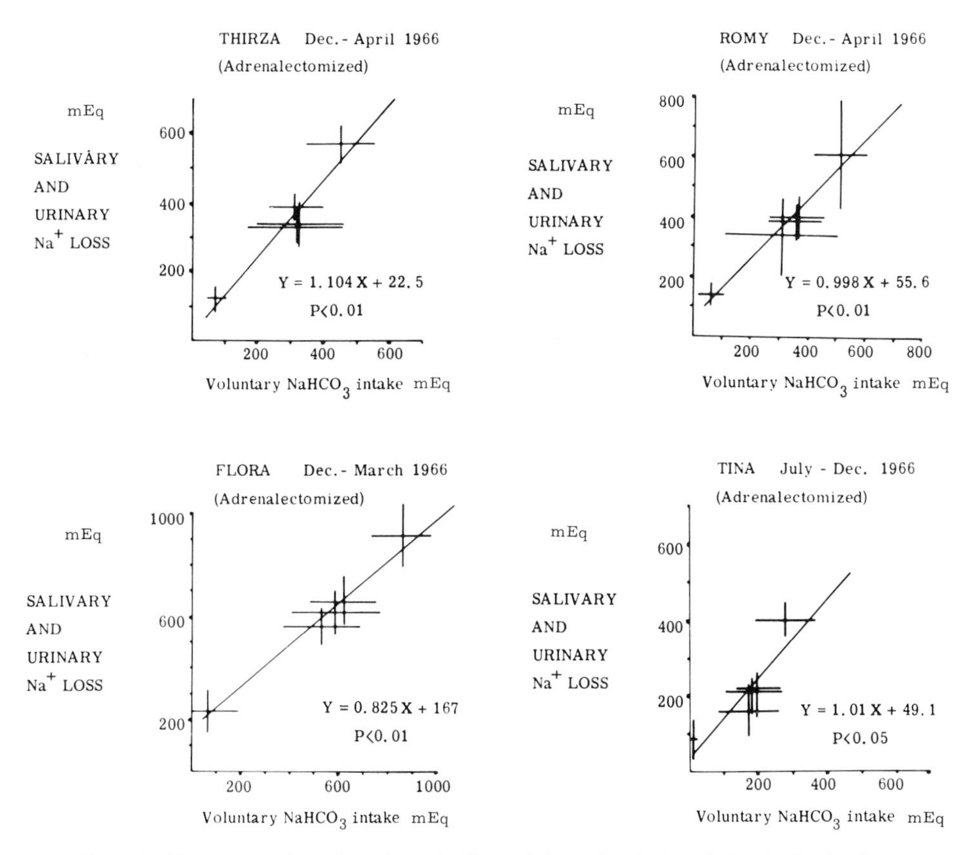

Figure 9 The same study as that shown in Figure 8, but using 4 adrenalectomized animals.

susceptible species. When we make intracarotid or intravenous infusions of saline in sheep deficient in Na^+, there is a delay before change of body balance is translated into changes of appetite and behavior.[1,4] The mode of action of aldosterone on peripheral target tissues is known to be complicated and probably to involve protein synthesis during the latent period between injection and alteration of secretory processes.[5] These two findings may be consistent with the thesis that the significance of the appetite effect of aldosterone is not that it is the main vector of salt appetite but that in high doses its action reproduces the neurochemical events which do, in fact, initiate the appetite under normal conditions.[4] The complex processes of the specific hungers may differ in character from the more direct and rapid action of the osmotic pressure of arterial blood in evoking thirst.

ACKNOWLEDGMENTS

This work was supported by research grants from the U.S.P.H.S.-AM 08701; the National Health and Medical Research Council of Australia; the Wool Industry Research Fund of Australia; the Anti-Cancer Council of Victoria; and the Rural Credits Fund of the Reserve Bank of Australia.

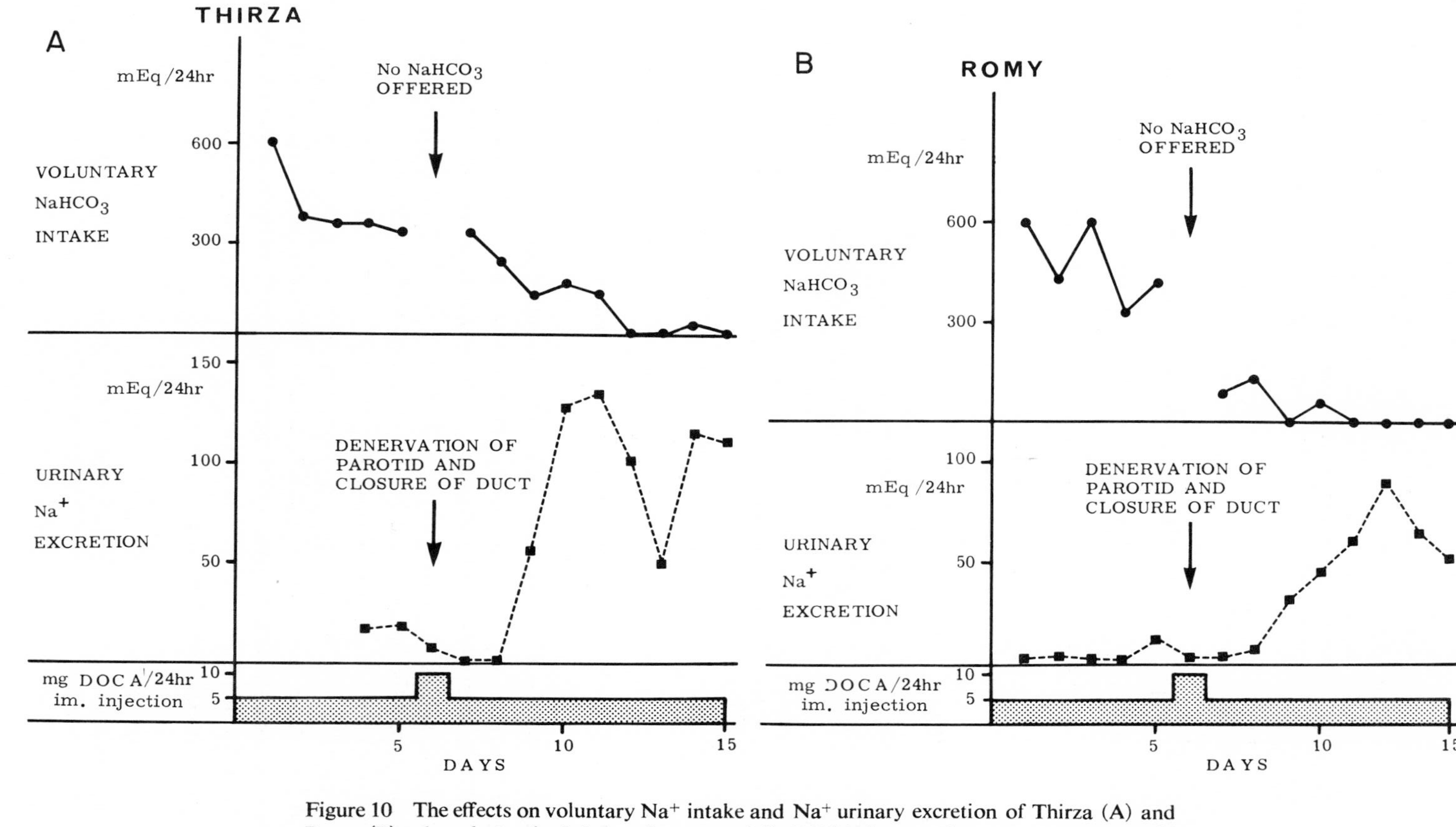

Figure 10 The effects on voluntary Na^+ intake and Na^+ urinary excretion of Thirza (A) and Romy (B), adrenalectomized sheep of constant daily DOCA dosage, when a permanent parotid fistula was closed and salivary loss ceased.

REFERENCES

1. Beilharz, S., E. A. Bott, D. A. Denton, and J. R. Sabine. 1965. The effect of intracarotid infusions of 4M-NaCl on the sodium drinking of sheep with a parotid fistula. *J. Physiol.* (*London*) **178**, pp. 80–91.
2. Blair-West, J. R., J. P. Coghlan, D. A. Denton, J. F. Nelson, E. Orchard, B. A. Scoggins, R. D. Wright, K. Myers, and C. L. Junqueira. 1968. Physiological, morphological and behavioural adaptation to a sodium deficient environment by wild native Australian and introduced species of animals. *Nature* **217**, pp. 922–928.
3. Bojeson, E. 1966. Concentrations of aldosterone and corticosterone in peripheral plasma of rats. The effects of salt depletion, salt repletion, potassium loading and intravenous injections of renin and angiotensin II. *European J. Steroids* I, p. 145.
4. Denton, D. A. 1966. Some theoretical considerations in relation to innate appetite for salt. *Cond. Ref.* **1**, pp. 144–170.
5. Edelman, I. S., R. Bogoroch, and G. A. Porter. 1963. On the mechanism of action of aldosterone on sodium transport: the role of protein synthesis. *Proc. Natl. Acad. Sci. U.S.* **50**, pp. 1169–1177.
6. Eilers, E. A., and R. E. Peterson. 1964. Aldosterone secretion in the rat. *In* Aldosterone (E. E. Baulieu and P. Robel, editors). Blackwell, Oxford, pp. 251–264.
7. Epstein, A. N., and E. Stellar. 1955. The control of salt preference in adrenalectomized rat. *J. Comp. Physiol. Psychol.* **48**, pp. 167–172.
8. Fregly, M. J., J. R. Cade, I. W. Waters, J. A. Straw, and R. E. Taylor, Jr. 1965. Secretion of aldosterone by adrenal glands of propylthiouracil-treated rats. *Endocrinol.* **77**, pp. 777–784.
9. Fregly, M. J., and I. W. Waters. 1965. Hormonal regulation of the spontaneous sodium chloride appetite of rats. *In* Olfaction and Taste II (T. Hayashi, editor). Pergamon Press, Oxford, etc., pp. 439–458.
10. Fregly, M. J., and I. W. Waters. 1966. Effect of mineralocorticoids on spontaneous sodium chloride appetite of adrenalectomized rats. *Physiol. Behav.* **7**, pp. 65–74.
11. Michael, R. P. 1965. Oestrogens in the central nervous system. *Brit. Med. Bull.* **21**, pp. 87–90.
12. Meggit, M. J. 1958. Salt manufacture and trading in the western highlands of New Guinea. *Australian Museum Mag.* **12**, p. 309.
13. Palmore, W. P., and P. J. Mulrow. 1967. Control of aldosterone secretion by the pituitary gland. *Science* **158**, pp. 1482–1484.
14. Rice, K. K., and C. P. Richter. 1943. Increased sodium chloride and water intake of normal rats treated with desoxycorticosterone acetate. *Endocrinol.* **33**, pp. 106–115.
15. Richter, C. P. 1956. Salt appetite of mammals: its dependence on instinct and metabolism. *In* L'instinct dans le Comportement des Animaux et de l'Homme. Masson et Cie, Paris, pp. 577–632.
16. Singer, B., and M. P. Stack-Dunne. 1955. The secretion of aldosterone and corticosterone by the rat adrenal. *J. Endocrinol.* **12**, pp. 130–145.
17. Wills, P. 1958. Salt consumption by natives of the Territory of Papua and New Guinea. *Philippine J. Sci.* **87**, pp. 169–177.
18. Wolf, G. 1964. Sodium appetite elicited by aldosterone. *Psychonomic Sci.* **1**, pp. 211–212.
19. Wolf, G. 1964. Effect of dorsolateral hypothalamic lesions on sodium appetite elicited by desoxycorticosterone and by acute hyponatremia. *J. Comp. Physiol. Psychol.* **58**, pp. 396–402.
20. Wolf, G., and P. J. Handal. 1966. Aldosterone-induced sodium appetite: dose response and specificity. *Endocrinol.* **78**, pp. 1120–1124.
21. Wolf, G., and D. Quartermain. 1966. Sodium chloride intake of desoxycorticosterone-treated and of sodium-deficient rats as a function of saline concentration. *J. Comp. Physiol. Psychol.* **61**, pp. 288–291.
22. Wolf, G., and E. A. Steinbaum. 1965. Sodium appetite elicited by subcutaneous formalin: Mechanism of action. *J. Comp. Physiol. Psychol.* **59**, pp. 335–339.

INNATE MECHANISMS FOR REGULATION OF SODIUM INTAKE

GEORGE WOLF

Department of Anatomy, Mt. Sinai School of Medicine, New York, New York

I am going to review three recent experiments which indicate that the relationship between body sodium deficiency and sodium appetite in the rat is mediated by innate mechanisms. The experiments attempt to show that the emergence of sodium appetite is not dependent upon prior learning that the ingestion of sodium will relieve the symptoms of the deficiency, but that the appetite emerges spontaneously upon the rat's very first experience with sodium deficiency. Before describing the experiments I want to discuss a general assumption underlying their design and interpretation.

It is assumed that the experimental animals never experienced sodium deficiency prior to the experimental depletion of body sodium. The animals used for the experiments were adult male laboratory rats which had been fed standard rat chows ad libitum from the time of weaning. These chows contain a surfeit of sodium in comparison to minimal amounts necessary for normal growth. Thus, while we cannot be certain that the rats never experienced a transient deficiency of body sodium, the assumption that they never did seems quite plausible. It should be noted that rigorous verification of this assumption would require continuous records of body sodium levels from infancy to the time of experimentation. Simply increasing the sodium content of the diet or otherwise administering supplementary sodium would not provide an unequivocal control, for it would still remain possible that a reflexive decrease in the activity of the hormonal system for sodium retention[1] could occasionally result in sodium loss in excess of intake.

There have been a number of experiments, most notably that of Nachman,[5] showing that "inexperienced" sodium-deficient rats will ingest sodium salts but reject nonsodium salts (except lithium) soon after the initial taste and before any significant alteration of body sodium could occur. An experiment by Handal[2] attempted to determine the minimum amount of time necessary for such differential behavior to occur. The rats were depleted of body sodium by subcutaneous formalin injection. (The sodium "depletion" induced by subcutaneous formalin is due to the sequestering of body sodium at the injection site.[10]) They were then washed to remove all salts from the surface of their bodies and put in clean cages with distilled water and salt-free food ad libitum. The rats had no opportunity to taste salt from the time of injection to the time of testing, 24 hours later. The test was a "single stimulus" test, in which each rat was given only one solution, which contained either a sodium or a nonsodium salt. Intake of the solutions was monitored by a drinkometer.

Figure 1 shows the results of the experiment. Sodium-depleted rats given sodium salts drank rapidly and continuously while those given nonsodium salts stopped drinking after the first few licks. The difference between these groups was statistically significant within 5 seconds after the commencement of drinking. Note also that a group of nondepleted rats rejected a strong sodium chloride solution which was accepted by depleted rats.

It seems to me that the most plausible explanation of these results is that sodium-deficient rats immediately recognize sodium by taste and ingest it without previous experience that such behavior will relieve the deficiency. Or, in more operational terms, ingestive behavior is elicited immediately upon conjunction of adequate in-

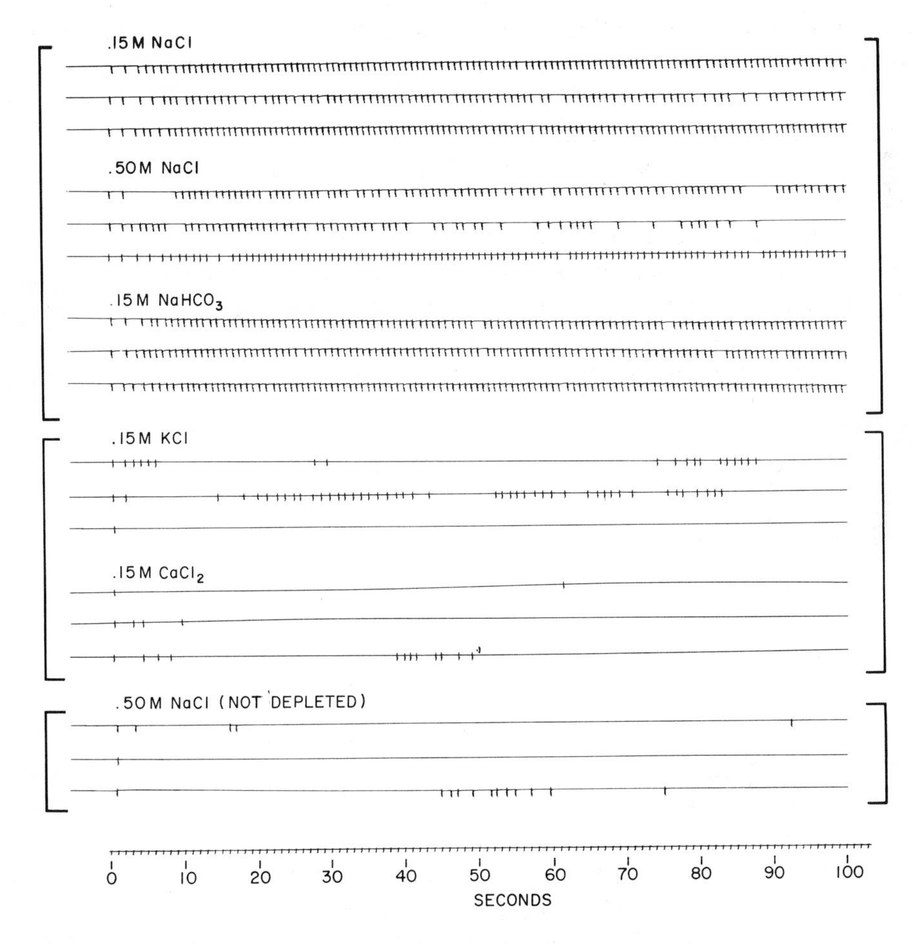

Figure 1 Drinkometer records of individual rats. Every fifth lick is represented by a vertical mark. Records within the upper brackets are from 9 sodium-depleted rats drinking solutions of sodium salts. Records within the middle brackets are from 6 sodium-depleted rats drinking solutions of nonsodium salts. Records within the lower brackets are from 3 nondepleted rats drinking a sodium salt solution. (From Handal[2])

ternal (body sodium deficiency) and external (sodium taste) stimuli. While we cannot be sure that the rats never experienced sodium deficiency prior to the experiment, we know that they never previously tasted sodium salts in isolation from other taste stimuli, because their only source of sodium was their laboratory chow and their excretions, both of which contain many other effective taste stimuli. Under these maintenance conditions, the only way the rat could specifically associate the sodium with the reduction of the need is by some innate mechanism which makes it especially sensitive to the taste of sodium during deficiency.

The next experiment, done by Quartermain, Miller, and Wolf,[7] indicates that the ingestion of sodium by inexperienced sodium-deficient rats is not merely a reflexive act but represents "voluntary motivated behavior" (see Teitelbaum[9] for definitions). Rats which had been pre-trained in bar pressing (by being given water as reinforcement when thirsty) were injected with various amounts of formalin to induce varying degrees of sodium depletion. The rats were prevented from tasting sodium prior to testing, but were given water and salt-free food ad libitum as in Handal's experiment. The next day they were put in the Skinner box where bar presses now yielded a saline solution.

The results of the experiment are shown in Figure 2. When bar pressing yielded saline, the rate of pressing increased with the degree of sodium depletion. This effect was statistically significant within 10 minutes after commencement of pressing, at which time only a small amount of sodium (about 0.3 mM) had been delivered, so that the differences between the groups could not have been due to repletion of body

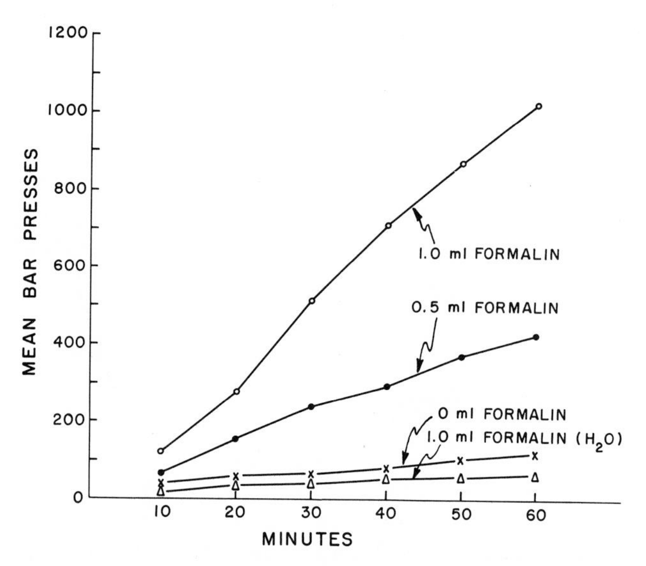

Figure 2 Mean cumulative bar presses for 4 groups of rats injected with 0 ml, 0.5 ml, or 1.0 ml of 1.5% formalin and given 0.33 M NaCl solution as reinforcement, and 1 group injected with 1.0 ml of 1.5% formalin and given water as reinforcement. There were 4 rats in each group. (From Quartermain, Miller, and Wolf[7])

sodium. A group of sodium-depleted rats given water instead of saline reinforcement did not press the bar. This experiment suggests that the tendency for the sodium-deficient rat to ingest sodium salts is not merely a reflexive act triggered by adequate need and taste stimuli, but that there is an innately determined relationship between the degree of deficiency and the degree of motivation to obtain sodium.

While the above experiment demonstrates a relationship between sodium need and motivated behavior, it does not elucidate the degree to which the increased propensity to press the bar (motivated behavior) is due to increased internal drive stimuli or to increased gratification from the taste of sodium. The final experiment[4] suggests that the initial sodium deficiency elicits a specific desire for sodium that precedes and hence is not dependent upon reinforcement from the taste of sodium. The experiment was designed to show that inexperienced sodium-deficient rats are motivated to obtain sodium prior to experiencing the effects of ingesting it while in the deficient state.

Six groups of rats which had presumably never experienced sodium deficiency were deprived of water and trained to press a bar to receive as reinforcement one of six aqueous solutions. Three of the solutions contained sodium salts, while the other three did not (sodium chloride, sodium phosphate, and sodium acetate vs. potassium chloride, calcium chloride, and pure water). After a few days of bar-press training, when rats in all groups were pressing at similar rates, the rats were injected with formalin to induce sodium depletion. Two additional groups of rats trained with sodium chloride solution or with water reinforcement were sham injected to serve as nondepleted controls. As in the previous experiments, all the rats were washed and given water ad libitum until the next day, when they were put back in the Skinner boxes under extinction conditions (i.e., bar presses did not yield any reinforcement).

Figure 3 shows that the groups which had previously learned that bar presses yield sodium salts and which were sodium-depleted during testing pressed the bar at a rate

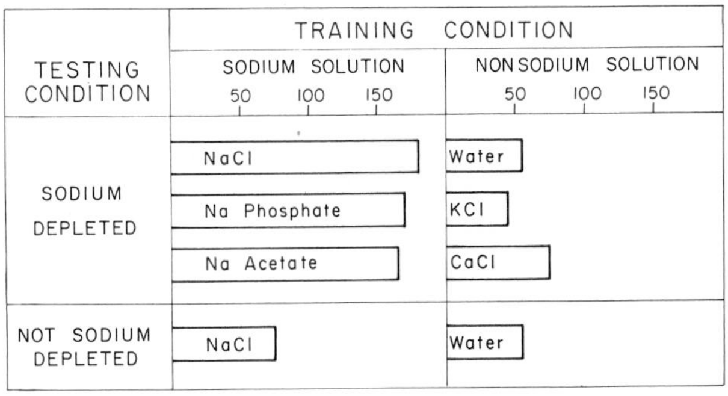

TESTING CONDITION	TRAINING CONDITION	
	SODIUM SOLUTION (50 100 150)	NONSODIUM SOLUTION (50 100 150)
SODIUM DEPLETED	NaCl	Water
	Na Phosphate	KCl
	Na Acetate	CaCl
NOT SODIUM DEPLETED	NaCl	Water

Figure 3 Horizontal bars indicate mean number of bar presses during 1-hour extinction session for various groups of 12 rats as a function of solution used as reinforcement during training (denoted inside individual bars) and of sodium balance during testing. The salt solutions are all 0.15 M and sodium and nonsodium solutions are balanced for palatability. The figure combines the results of experiments #2 and #3 of Krieckhaus and Wolf.[4]

two to three times faster than did groups in the other three conditions, even though they now received no reinforcement. Krieckhaus has recently succeeded in also demonstrating that rats can learn the location of sodium in a T-maze when thirsty and utilize the information when subsequently depleted of sodium (personal communication). (The initial study of Krieckhaus and Wolf[4] reports a failure to demonstrate this phenomenon. This was apparently attributable to insufficient training.) The above results are interpreted to show that rats have a specific desire for sodium that emerges spontaneously the very first time they are deficient in sodium and that precedes and directs behavior toward sodium acquisition. In addition, the results indicate that rats can remember how and where they obtained sodium even though they did not need it at the time of ingestion.

In summary, the three experiments investigated the innate mechanisms for the behavioral regulation of body sodium. According to the foregoing assumptions and interpretations of the results, the first experiment showed that rats experiencing sodium deficiency for the first time ingest sodium salts and reject nonsodium salts within seconds after the initial taste. This phenomenon precluded the possibility that such discriminative behavior is based upon learning that only sodium salts will relieve the symptoms of the deficiency. The phenomenon could, however, be explained on the basis of ingestive reflexes, which are released when stimuli of sodium need and sodium taste are simultaneously present. The second experiment showed that the alimentary behavior of the sodium-deficient rat is not simply reflexive but is motivated, because rats will perform a learned response to obtain sodium. The third experiment showed that the motivated behavior could not be attributed simply to an increase in the rewarding properties of sodium (e.g., increased palatability), but that sodium deficiency elicits an internal drive state which guides behavior towards acquisition of sodium.

The fact that the motivational response to sodium deficiency appears to be mediated in large part by innate mechanisms does not necessitate the conclusion that there exists a special neural system whose sole function is the regulation of sodium intake. While it seems likely that certain correlates of body sodium deficiency are monitored by a specialized receptor system,[8] all that is necessary to account for the major phenomena described in the present paper is some innately determined relationship between the coding of neural signals of body sodium deficiency and the coding of neural signals of the taste of sodium.

I would like to utilize the above concept of similarly coded drive and taste signals in a rather preliminary and incomplete hypothesis attempting to account for the phenomenon demonstrated in the third experiment—clearly the most inclusive and complex of the various phenomena described in this paper. Stimulation of the taste receptors by sodium elicits a certain consistent pattern of neural activity in the central nervous system. Each time the taste receptors are stimulated by sodium this pattern is recorded in memory in association with temporally related stimuli from other sensory modalities. Sodium deficiency, in addition to eliciting general tension and arousal, gives rise to a pattern of neural activity which has a configuration similar to the sensory pattern for sodium taste. Thus, via scanning and matching mechanisms, the motivational state is immediately related to the taste of sodium and to the complex of memories associated with previous experiences of sodium ingestion. Mechanisms for scanning and matching similar neural impulse configurations have been discussed by Pitts and McCulloch[6] and by John.[3] I sometimes think of the neural pattern of the sodium drive as a "negative" of the neural pattern of sodium taste or, more specifically, that the two patterns have a similar over-all configuration but have opposing facilitatory-inhibitory effects on common target neurons.

The above model provides a potential neural mechanism for an innately organized drive which has an experiential component associated with a specific taste sensation—for example, a craving to ingest a salty-tasting substance. The system can be summarized in behavioral terms as follows. When the sodium drive is activated, the taste of sodium is charged with reinforcement value and stimuli associated with previous sodium ingestion gain secondary reinforcement value in accordance with a spatial-temporal gradient of contiguity to the ingestive act. Sequential secondary reinforcements, increasing in intensity as the goal is approached, guide the sodium-deficient rat toward places or responses which previously yielded the primary reinforcer—sodium.

ACKNOWLEDGMENTS

Work on this paper was supported by USPHS grant MH 13189 and done during tenure of an Established Investigatorship of the American Heart Association.

REFERENCES

1. Bacchus, H. 1950. Cytochemical study of the adrenal cortex of the rat under salt stresses. *Amer. J. Physiol.* **163**, pp. 326–331.
2. Handal, P. J. 1965. Immediate acceptance of sodium salts by sodium-deficient rats. *Psychonomic Sci.* **3**, pp. 315–316.
3. John, E. R. 1962. Some speculations on the psychophysiology of the mind. *In* Theories of the Mind (J. M. Scher, editor). Free Press of Glencoe, New York, pp. 80–121.
4. Krieckhaus, E. E., and G. Wolf. 1968. Acquisition of sodium by rats: Interaction of innate mechanisms and latent learning. *J. Comp. Physiol. Psychol.* **65**, pp. 197–201.
5. Nachman, M. 1962. Taste preferences for sodium salts by adrenalectomized rats. *J. Comp. Physiol. Psychol.* **55**, pp. 1124–1129.
6. Pitts, W., and W. S. McCulloch. 1947. How we know universals. *Bull. Math. Biophys.* **9**, pp. 127–147.
7. Quartermain, D., N. E. Miller, and G. Wolf. 1967. Role of experience in relationship between sodium deficiency and rate of bar pressing for salt. *J. Comp. Physiol. Psychol.* **63**, pp. 417–420.
8. Stricker, E. M., and G. Wolf. 1969. Behavioral control of intravascular fluid volume: Thirst and sodium appetite. *Ann. N.Y. Acad. Sci.* **157**, pp. 553–568.
9. Teitelbaum, P. 1967. Physiological Psychology: Fundamental Principles. Prentice-Hall, Englewood Cliffs, N.J., pp. 56–59.
10. Wolf, G., and E. A. Steinbaum. 1965. Sodium appetite elicited by subcutaneous formalin: Mechanism of action. *J. Comp. Physiol. Psychol.* **59**, pp. 335–339.

PREFERENCE THRESHOLD AND APPETITE FOR NaCl SOLUTION AS AFFECTED BY PROPYLTHIOURACIL AND DESOXYCORTICOSTERONE ACETATE IN RATS

MELVIN J. FREGLY
Department of Physiology, University of Florida, College of Medicine, Gainesville, Florida

INTRODUCTION

Propylthiouracil (PTU), an antithyroid and mild diuretic agent, induced a spontaneous salt appetite in rats given a choice between water and 0.15 M NaCl solution to drink.[9] This compound also reduced the preference threshold for NaCl solution to a level one-sixth below that of controls (0.023 to 0.004 mEq/l) when tested by the two-bottle preference technique of Richter.[14] When graded doses of desoxycorticosterone acetate (DOCA) up to 100 μg/100 g body weight/day were administered to PTU-treated rats, spontaneous NaCl intake was reduced in a graded fashion.[9] However, doses of DOCA greater than 100 μg/100 g body weight/day *increased* NaCl intake toward the level of PTU-treated, control rats. Thus, a U-shaped dose-response curve exists between intake of NaCl solution and dose of DOCA administered, suggesting that blood concentration of mineralocorticoids may influence spontaneous NaCl intake.

Subsequent measurement of aldosterone secretion rate by the adrenal glands of PTU-treated rats showed that these animals secrete at approximately one-third the rate (mμg/min/adrenal) observed for control rats.[6] The shape of the dose-response curve mentioned above predicts that a reduction in aldosterone secretion rate would be accompanied by an increase in spontaneous NaCl intake, assuming that the physiological effects of aldosterone and DOCA are the same. Hence, further suggestive evidence was obtained linking changes in mineralocorticoid secretion rate (and presumably blood concentration) with NaCl intake. The possibility that changes in blood concentration of mineralocorticoids might also influence preference threshold for NaCl solution was tested here because a number of experimental conditions inducing a NaCl appetite also reduce preference threshold. Thus, bilateral adrenalectomy,[1,14] administration of PTU,[9] hydrochlorothiazide,[5] or DOCA[10] are reported to produce both effects in rats. The objective of this experiment was to determine whether graded doses of DOCA administered to PTU-treated rats affected their preference threshold for NaCl solution.

METHODS

Thirty male rats of the Carworth CFN strain that initially weighed from 250 to 300 g were used. They were administered PTU in their powdered Purina Chow diet at a concentration of 1 g/kg food for three weeks before the experiment was begun. During this time all rats were given tap water ad libitum. The animals were kept in a room maintained at 26 ± 1°C and illuminated from 8 a.m. to 6 p.m.

Ten days before the experiment began the rats were caged individually and given choice between two bottles of distilled water in containers consisting of cast aluminum fountains and infant nursing bottles.[12] The food containers were spillproof and have been described in detail.[4] The rats were divided randomly into six equal groups. Group 1 served as control while groups two through six respectively received daily subcutaneous injections of 40, 80, 160, 240, and 400 μg DOCA/100 g body weight/day beginning on the first day of the experiment. The DOCA was dissolved in peanut oil and the control group received an equal volume (0.2 ml) of peanut oil daily. Both body weight and intakes of water and NaCl solution were measured daily throughout the experiment. Positions of the bottles on each cage were changed daily to avoid habit formation in selection of drinking fluid. During the 10-day equilibration period, when both bottles contained distilled water, intakes were measured daily to ascertain that each rat drank roughly equal amounts of water from each bottle. During this time about 15 per cent of the original group were rejected as position drinkers and were replaced by other rats. At the end of the 10-day period, administration of DOCA was initiated and two-day test periods were begun. During the second test period, each rat was offered a choice between distilled water and 0.0006 M NaCl solution. All NaCl solutions used were made by serial dilution from a concentrated stock NaCl solution. Each dilution was checked for accuracy by determination of chloride concentration.[3] During subsequent two-day periods, each rat was given a choice between distilled water and the following molar NaCl solutions in chronological sequence: 0.001, 0.005, 0.010, 0.020, 0.050, 0.100, 0.150, and distilled water. In all, 20 days were required to complete the experiment. Every second day, when the solutions were changed, all bottles and drinking fountains were washed thoroughly. Daily intakes of water, NaCl solution and food are expressed as ml or g/100 g body weight/day to correct for differences in body weight.

Several criteria for determination of preference threshold concentration were used. The first was similar to that of Richter[14]; the concentration of NaCl solution at and above which simultaneous volumes taken from the test (NaCl) bottle exceeded that from the reference (water) bottle and differed significantly ($P < 0.05$) from it. Statistical comparison of the results obtained from the two groups was made by the "t" test.[16] A second criterion of preference threshold was cross-over concentration.[8] Simultaneous intakes of distilled water and NaCl solution were graphed for each individual rat. The concentration at which NaCl intake increased and remained elevated, while water intake decreased and remained low, was determined for each rat.

RESULTS

Figure 1 shows the effects of administration of graded doses of DOCA on spontaneous intakes of water and NaCl solution for the six groups. Preference thresholds, measured by the technique of Richter[14] for control, 40, 80, 160, 240 and 400 μg DOCA were: 5, 10, 20, 10, 10, and 5 mEq/liter respectively. These are shown by the × in each panel of Figure 1. Preference threshold, estimated by mean cross-over concentrations, and shown by the arrow in each panel in Figure 1, agrees closely with the preference threshold as estimated by the method of Richter. The greatest variation occurred in the group given 160 μg DOCA (Figure 1D) in which threshold measured by cross-over concentration was 8.6 mEq/liter (0.0086 M/liter) greater than the threshold measured by the method of Richter. For some unknown reason, the variability from rat to rat within this group was larger than in any of the other five groups.

Maximal intake of NaCl solution occurred when a concentration of 0.150 M/liter was offered to all six groups. Water intake was minimal at the same time.

When, at the end of the experiment, both bottles contained distilled water, the amount of water drunk from the test bottle was somewhat greater than during the first period of the experiment. The variabilities of the measurements were similar excepting for the group receiving 160 μg DOCA. This group had a much greater variability in intake at the end of the experiment than at the beginning. No explanation can be advanced for this.

When mean preference threshold for each group determined by the cross-over technique was graphed against the daily dose of DOCA administered to that group, an inverse U-shaped curve was observed (Figure 2). The reduced preference threshold of the control, PTU-treated group increased gradually with graded increases in the dose of DOCA administered and reached a maximum when approximately 120 μg DOCA/100 g body weight/day was administered. Doses greater than this gradually reduced preference threshold and reached the level of control, PTU-treated rats when 400 μg DOCA/100 g body weight/day were administered.

The percentage change from the control group in intake of 0.15 M NaCl solution is also graphed against dose of DOCA administered (Figure 3). The greatest reduction in NaCl intake occurred at approximately the same dose of DOCA which increased preference threshold concentration maximally. When percentage change in intake of 0.15 M NaCl solution is graphed against preference threshold (Figure 4), a linear relationship is observed. As preference threshold concentration increased, there was a greater and greater percentage reduction in NaCl intake from that of PTU-treated controls.

DISCUSSION

Earlier studies from this laboratory showed that administration of propylthiouracil to rats increased their spontaneous intake of 0.15 M NaCl solution and that graded doses of DOCA (up to 100 μg/100 g body weight/day), given simultaneously with PTU, reduced the elevated NaCl intake.[8] However, administration of DOCA at

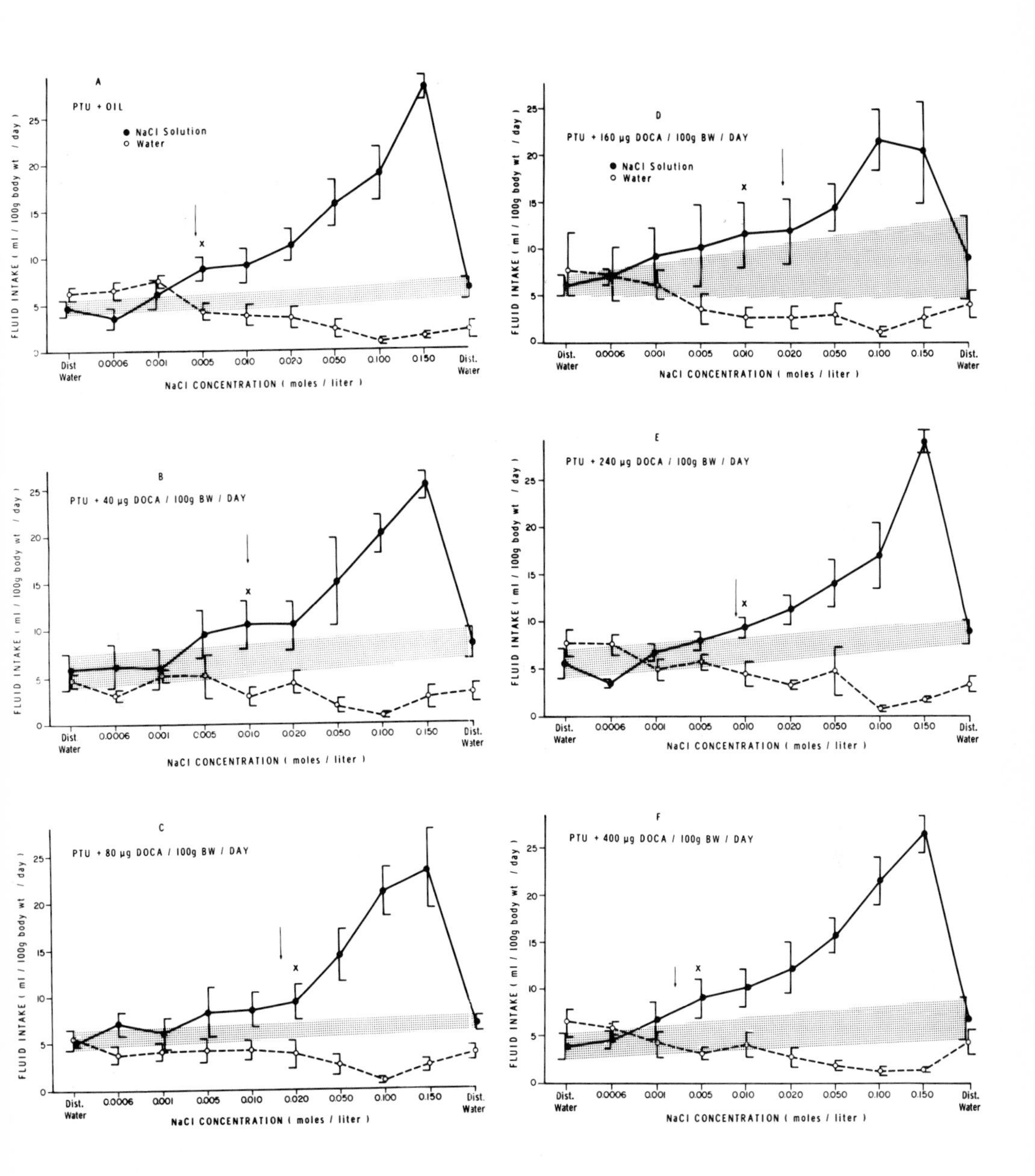

Figure 1 Spontaneous intakes of water and NaCl solution by the 6 groups of rats used in this experiment. Concentration of NaCl solution was changed every second day. One standard error is set off at each group mean. The X in each panel indicates the preference threshold for NaCl solution determined by the method of Richter,[14] while the arrow indicates the mean preference threshold determined by the cross-over concentration.[8] The shaded area connects the standard errors of the two periods during which both drinking bottles contained water. The dose of desoxycorticosterone administered to each rat is shown at the top of each panel. (A = the control group.)

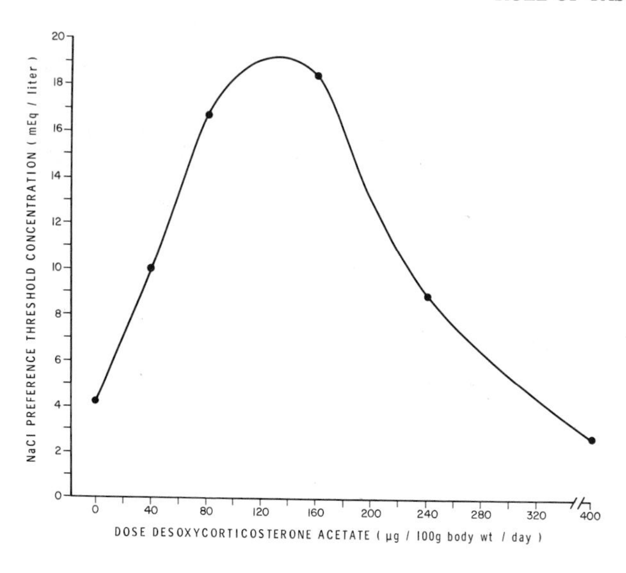

Figure 2 Mean preference threshold for NaCl solution of each group of rats determined by the cross-over concentration method is graphed against dose of desoxycorticosterone acetate administered.

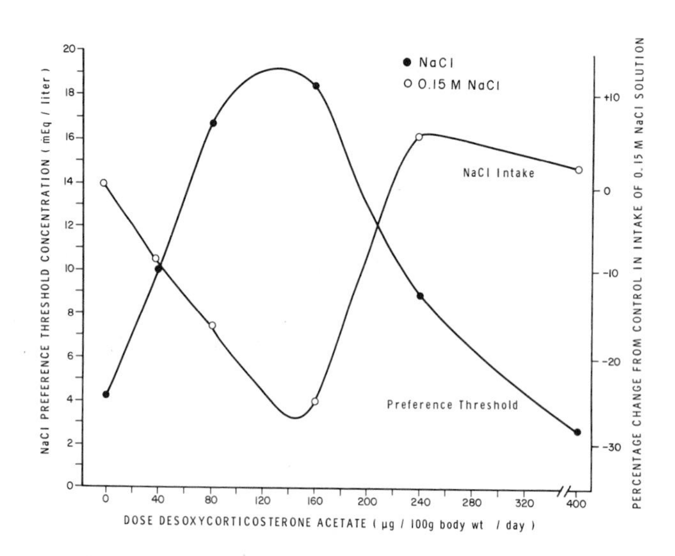

Figure 3 Mean preference threshold for NaCl solution and percentage change from control in intake of 0.15 M NaCl solution are graphed against dose of desoxycorticosterone acetate administered. Both maximal preference threshold and minimal intake of NaCl solution occur at approximately the same dose of DOCA.

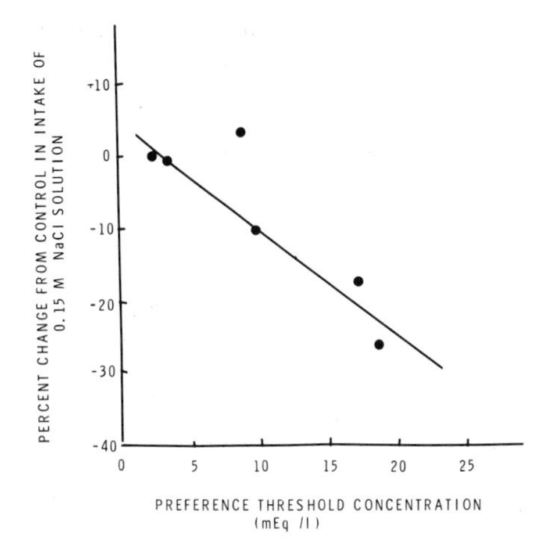

Figure 4 The percentage change from control in intake of 0.15 M NaCl solution is graphed against the corresponding preference threshold concentration for NaCl for each of the 6 groups of rats used here. A linear relationship appears to exist between the two.

doses greater than 100 μg/100 g body weight/day increased NaCl intake toward that of PTU-treated controls.[8] The relationship between NaCl intake and dose of DOCA administered was U-shaped. The possibility that both intake of 0.15 M NaCl solution and preference threshold might change together was rationalized from the published results of several investigators. When salt appetites occur, as they do following bilateral adrenalectomy,[14] administration of propylthiouracil,[8] hydrochlorothiazide,[1] or desoxycorticosterone acetate,[5] preference threshold is also reduced. It was particularly interesting to observe that a salt appetite and a reduced preference threshold for NaCl occurred both in adrenalectomized rats in the complete absence of mineralocorticoid hormone and in intact rats given large doses of mineralocorticoid hormone.[5,14] This suggested the possibility that graded doses of DOCA might affect preference threshold for NaCl solution inversely to their effect on NaCl intake. The results of our experiment add substance to this possibility.

The mechanism mediating the change in preference threshold is unknown. The schema shown in Figure 5 suggests some possibilities. A decrease in blood pressure,

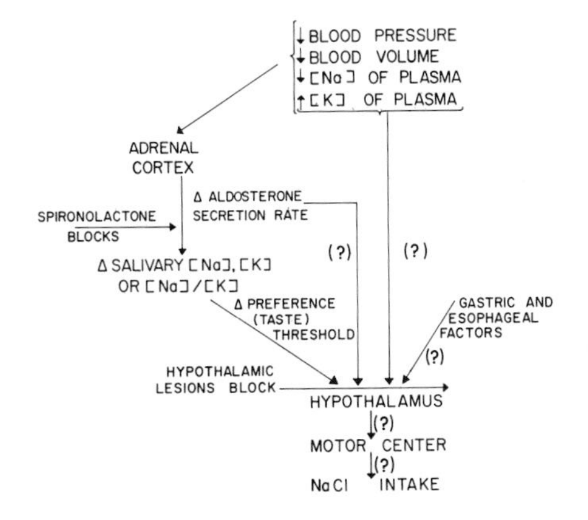

Figure 5 A schema is presented to suggest possible interactions that may be important in initiating the spontaneous NaCl intake in the rat.

or blood volume, a decrease in plasma sodium concentration, or an increase in plasma potassium concentration probably stimulates the adrenal cortex of the rat to increase its rate of secretion of aldosterone, perhaps by way of the renin mechanism.[6] The increase in sodium intake by either normal rats or sheep made sodium-deficient is well documented.[2,7] It is assumed that the increase in aldosterone secretion rate occurring under these conditions may affect the spontaneous NaCl intake by altering salivary concentrations of sodium, of potassium, or of the ratio of the two. The results of the study presented here suggest that preference (taste) threshold is also influenced by DOCA and presumably also by aldosterone. Whether the change in threshold is transmitted to the hypothalamus, in terms of a change in impulse traffic in the nerves mediating taste, is unknown. However, the suggestion must be tempered by the fact that abolition of an estimated 85 to 90 per cent of the total gustatory afferent input by peripheral nerve section alters very little the normal preference of rats for NaCl solutions.[13,15]

Aldosterone may also act directly on the hypothalamus to initiate NaCl intake. Hypothalamic lesions are reported to block the spontaneous NaCl appetite of adrenalectomized rats and to prevent the increased NaCl intake attendant upon either administration of large doses of DOCA or intraperitoneal dialysis against isotonic glucose solution.[11,17]

Although evidence is lacking, it is also possible that those factors initiating an increase in aldosterone secretion rate, e.g., change in sodium or potassium concentration of plasma, may affect the hypothalamus directly. The pathways from the hypothalamus to the motor center initiating the ingestion of NaCl solution are unknown.

As with thirst, a factor has been postulated to turn off sodium ingestion as well as turn it on. From earlier studies performed to determine the ability of rats to control their sodium intake,[7] it is suggested that this factor resides in the mouth, pharynx, and esophagus or, possibly, in the stomach. The schema presented here represents a particular view and should be considered only tentative.

SUMMARY

Dietary administration of the antithyroid drug, propylthiouracil (1.0 g/kg food), increased the spontaneous intake and reduced the preference threshold of rats for NaCl solution. Simultaneous administration of graded doses of desoxycorticosterone acetate to PTU-treated rats reduced NaCl intake and increased preference threshold up to doses of approximately 120 μg/100 g body weight/day. Doses greater than this gradually increased NaCl intake and reduced preference threshold for NaCl solution toward that of untreated controls. The results show a U-shaped dose-response relationship between percentage change in intake of 0.15 M NaCl solution and dose of DOCA administered. An inverse U-shaped dose-response curve was observed between preference threshold for NaCl solution and dose of DOCA administered. The results suggest an interrelationship between NaCl intake, blood level of mineralocorticoid and preference threshold for NaCl solution.

ACKNOWLEDGMENT

Supported by Grant AM-10772-02 from the National Institute of Arthritis and Metabolic Diseases.

REFERENCES

1. Bare, J. K. 1949. The specific hunger for sodium chloride in normal and adrenalectomized white rats. *J. Comp. Physiol. Psychol.* **42**, pp. 242–253.
2. Blair-West, J. R., J. P. Coghlan, D. A. Denton, J. R. Goding, M. Wintour, and R. D. Wright. 1963. The control of aldosterone secretion, *Recent Progr. Hormone Res.* **19**, pp. 311–383.
3. Cotlove, E., H. V. Trantham, and R. L. Bowman. 1958. An instrument and method for automatic, rapid, accurate, and sensitive titration of chloride in biologic samples. *J. Lab. Clin. Med.* **51**, pp. 461–468.
4. Fregly, M. J. 1960. A simple and accurate feeding device for rats, *J. Appl. Physiol.* **15**, p. 539.
5. Fregly, M. J. 1967. Effect of hydrochlorothiazide on preference threshold of rats for NaCl solutions. *Proc. Soc. Exp. Biol. Med.* **125**, pp. 1079–1084.
6. Fregly, M. J., J. R. Cade, I. W. Waters, J. A. Straw, and R. E. Taylor, Jr. 1965. Secretion of aldosterone by adrenal glands of propylthiouracil-treated rats. *Endocrinology* **77**, pp. 777–784.
7. Fregly, M. J., J. M. Harper, Jr., and E. P. Radford, Jr. 1965. Regulation of sodium chloride intake by rats. *Amer. J. Physiol.* **209**, pp. 287–292.
8. Fregly, M. J. and I. W. Waters. 1965. Effect of propylthiouracil on preference threshold of rats for NaCl solutions. *Proc. Soc. Exp. Biol. Med.* **120**, pp. 637–640.
9. Fregly, M. J. and I. W. Waters. 1966. Effect of desoxycorticosterone acetate on NaCl appetite of propylthiouracil-treated rats. *Physiol. Behav.* **1**, pp. 133–138.
10. Herxheimer, A. and D. M. Woodbury. 1960. The effect of desoxycorticosterone on salt and sucrose taste preference thresholds and drinking behavior in rats. *J. Physiol.* (*London*) **151**, pp. 253–260.
11. Kissileff, R. H. and A. N. Epstein. 1962. Loss of salt preference in rats with lateral hypothalamic damage. *Amer. Zool.* **2**, p. 533.
12. Lazarow, A. 1954. Methods for quantitative measurement of water intake, *Methods Med. Res.* **6**, pp. 225–229.
13. Pfaffman, C. 1952. Taste preference and aversion following lingual denervation. *J. Comp. Physiol. Psychol.* **45**, pp. 393–400.
14. Richter, C. P. 1939. Salt taste thresholds of normal and adrenalectomized rats. *Endocrinology* **24**, pp. 367–371.
15. Richter, C. P. 1939. Transmission of taste sensation in animals, *Trans. Amer. Neurol. Ass.* **65**, pp. 49–50.
16. Snedecor, G. W. 1962. Statistical Methods, 5th edition. Iowa State Univ. Press, Ames, p. 45.
17. Wolf, G. 1964. Effect of dorsolateral hypothalamic lesions on sodium appetite elicited by desoxycorticosterone and by acute hyponatremia, *J. Comp. Physiol. Psychol.* **58**, pp. 396–402.

SPONTANEOUS SODIUM APPETITE AND RED-CELL TYPE IN SHEEP

A. R. MICHELL *and* F. R. BELL

Department of Medicine, Royal Veterinary College, London, England (Dr. A. R. Michell is a Clement Stephenson Research Scholar)

INTRODUCTION

That animals may crave salt was recorded as early as 1600,[15] but the earliest attempt to explain the appetite was probably that of Von Bunge in 1901. Von Bunge's explanation of salt craving on the basis of the high potassium content of many diets has not been upheld[24] and although we may no longer concur with Cannon[10] that the nature of this hunger is quite unknown, it remains true that in the study of specific appetites in general and salt appetite in particular many essential problems remain unresolved.[23,25] We still find ourselves unable to refute Carlson's[11] suggestion that animals eat salt quite irrespective of need.

The sheep has recently become an important subject for the experimental study of sodium metabolism through the work of the Melbourne School, where techniques have been established for the controlled depletion of sodium by wastage of parotid saliva and the transplantation of the adrenal gland to an accessible subcutaneous site in the neck.[8] Such investigations have advanced the knowledge of salt appetite in the sodium-depleted animal.[9] Ostensibly normal sheep, however, also show salt appetite to various extents.[6,17] Little information is available about the factors involved in this spontaneous appetite, although salt preference has also been demonstrated in rat,[29] cat and rabbit,[12] goat,[3] calf,[5] pigeon,[18] and many other species.[16] Extensive studies of spontaneous salt appetite have been made only on the rat.

The lack of information on spontaneous salt appetite in sheep seems particularly important, because it could lead to erroneous experimental interpretations, especially because experiments with sheep usually involve small numbers of animals. Among the factors that have been studied with a view to demonstrating their possible effects[26] are the genetically determined characteristics of the blood, such as the erythrocyte potassium concentration and the hemoglobin type. These factors are already known to be associated with a number of physiological differences in sheep, although they do not necessarily cause them.[14,30] Although research has been restricted to a small number of mammals, Beidler[2] has suggested that interesting differences in the acuity of salt taste might exist between those that have potassium as the major cation of the red cell and those in which sodium predominates. The sheep presents an ideal subject for studies of this possibility because, unlike other species, it has individuals with either red-cell type.[19] Beidler's electrophysiological studies[1] on salt sensitivity in sheep did not show any difference between high potassium (HK) and low potassium (LK) individuals, but these investigations were confined to two subjects of the HK type and seven of the LK type. Since it was already known[20] that HK and LK sheep may differ in their water metabolism, it appeared that further studies in this area were overdue.

METHODS

The results presented were obtained from two series of experiments carried out from May to November, 1966 and 1967. In the first series, 64 female Cheviot sheep (20 HK, 44 LK) were examined for sodium preference, using concentrations of sodium bicarbonate from 15–240 mEq/l (Na^+). In the subsequent series, 32 of this flock were tested with sodium chloride concentrations from 30–480 mEq/l, and a new flock (9 HK, 39 LK) was tested with both salts; a total of 112 sheep were tested with bicarbonate and 64 with chloride.

The sheep were kept in outdoor pens with covered food hoppers and were fed a concentrate ration that provided about 20 mEq of Na^+ daily as well as chaffed hay (ad lib) containing about 50 mEq/kg of Na^+. The solutions were made up in tap water and offered in identical polythene buckets as the alternative choice to tap water. Any necessary corrections were made for evaporation or rain. In the first 24 hours tap water was offered in both buckets. Five two-day units with doubling concentrations of test solution were used for each batch of eight sheep. The test solution followed a left-right sequence in the first unit of the test, vice versa in the second, and so on throughout the five units.

RESULTS

The results of the 1966 experiment appeared to be clear-cut; the LK sheep showed a greater sodium preference, as indicated by a greater intake of sodium bicarbonate and a lower intake of water. Total intakes for the two groups were very similar (within 1.5 per cent). As a simple method of ranking the sheep for sodium preference, the mean preference for the five concentrations was used and is referred to as the "preference index." The preference index suggested that high extremes of sodium preference were uncommon among HK sheep, a point we will return to below. We were also interested by our observation that although the total intakes of the two groups were similar, the intake of water by LKs was only 79 per cent of the figure for HKs. Evans[20] had found a median value for water intake in LKs which was 76 per cent of that in HKs when water was the sole drinking fluid. The subsequent experiments did not substantiate the result, however—and, indeed, Evans' result has not been confirmed.[21]

With the inclusion of the 1967 results, preference data were available for 112 sheep, 29 of which were HK. Figure 1 shows histograms for the HK and LK sheep, indicating the percentage of each type with an over-all preference index in each range from 0–100 per cent. Clearly, it is improbable that the HK and LK sheep represent a com-

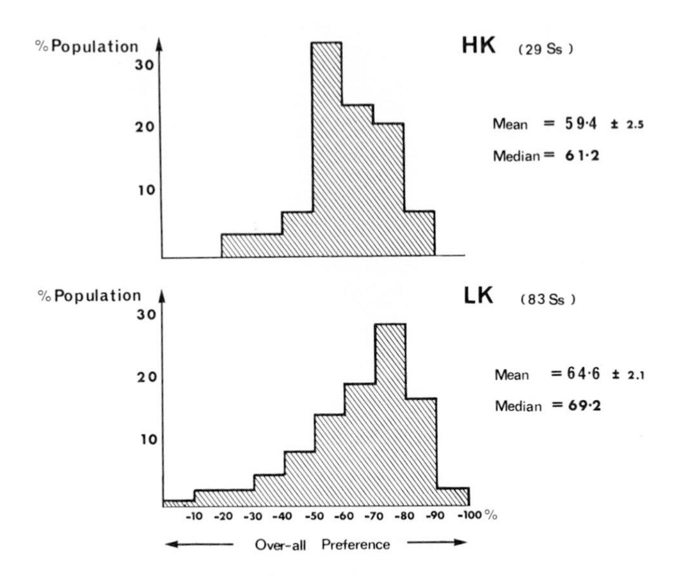

Figure 1 Over-all preference for sodium bicarbonate (15–240 mEq/l) shown by high potassium (HK) and low potassium (LK) sheep.

parable population with respect to their over-all sodium preference in this test. In particular, if the 112 sheep formed a flock from which an experimental group was to be drawn, the inclusion of HK sheep would most likely result in a preference index in the range of 50–60 per cent, whereas the figure for LKs would be 70–80 per cent. Moreover, a preference index above 70 per cent might occur in nearly 50 per cent of LKs but in less than 30 per cent of HKs.

The consideration of the rank occupied by the HK sheep within each batch supports the above conclusion; with 14 batches of sheep, 26 per cent of which are HK, the HK sheep should have 14 individuals in the top four positions of their batch. In fact, there are only 11, and no HK sheep showed the highest over-all preference in a batch. It was evident, therefore, that in this experiment high sodium preference was less common among HK sheep.

Figure 2 illustrates the relationship between sodium preference and sodium concentration among LK and HK sheep tested with chloride and bicarbonate. It includes the simplification that results for the two different flocks are combined. In LK sheep, which comprised the larger group, the over-all preference index was 2.9 per cent higher for the first flock than for the second with both chloride and bicarbonate. The first flock of HK sheep also showed a 4.3 per cent higher preference index when offered chloride, but with bicarbonate, results from the first and second flocks do not match well. Therefore, the amalgamation of results seemed a reasonable simplification, with the possible exception of the HK results for bicarbonate with the second flock. As these results tend to reduce the HK-LK differences that the experiment attempts to

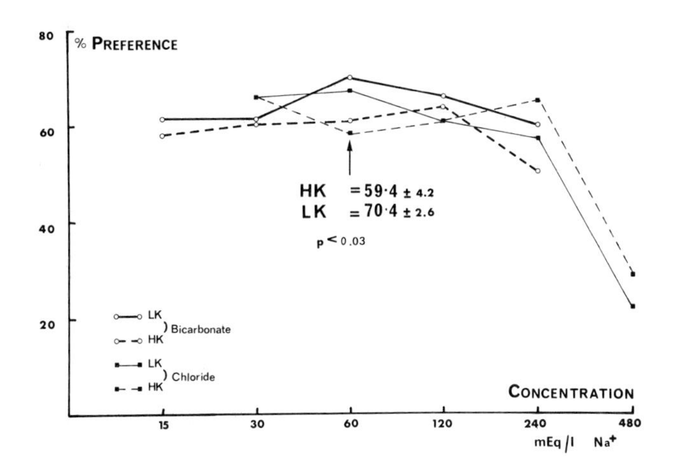

Figure 2 Preference for various concentrations of sodium chloride and bicarbonate shown by HK and LK sheep.

show, it was considered that they could also be included. The reason that values for preference, rather than for absolute intake, are used is that the absolute intakes of the two flocks differed considerably.

It is clear (Figure 2) that with LK sheep the maximum preference with both chloride and bicarbonate is at 60 mEq of Na^+/l, whereas with HK sheep the decline in preference begins only at a higher concentration (120 or 240 mEq/l). The juxtaposition of these effects results in a significant difference in sodium preference ($p < 0.03$) between HK and LK sheep at 60 mEq of Na^+/l, at which concentration the LK sheep show greater preference. The results also suggest, however, that at much higher concentrations the HK sheep might have the greater sodium preference. The fact that the preference line for LK sheep on both chloride and bicarbonate is similar but differs from that for HK sheep implies that influence on the level of preference depends less on the difference in sodium salt than on the difference in the type of sheep.

Unpublished data[26] suggest that the sodium concentration at which preference begins to decline may also be related to the hemoglobin type of the sheep; it is lowest in type A, highest in type B, and intermediate in type AB. This effect does not appear to depend on different numbers of HK individuals in the three groups, which were, in fact, very comparable.

It has often been pointed out that the sodium preference exhibited by the rat is maximal at concentrations of approximately 0.9 per cent salt,[29] this being close to the sodium concentration of plasma. Furthermore, this concentration is most rapidly released from the stomach to the duodenum.[28] In view of this, it is interesting that, except for HK sheep offered bicarbonate, the peak preference occurred at a concentration different from that in plasma (Figure 3). Moreover, during the preference test

			Concentration for PEAK PREFERENCE	PEAK SODIUM INTAKE mEq/l Total Fluid
Bicarbonate	HK	(29 Ss)	120 mEq/l	123
	LK	(83 Ss)	60	144
				Mean=141 mEq/l
Chloride	HK	(22 Ss)	(240)	156
	LK	(42 Ss)	60	138

Figure 3 Concentration of peak sodium preference and peak sodium intake, relative to water.

the maximum intake of sodium ions relative to water occurred at 240 mEq/l in HK and LK sheep with both salts. Nevertheless, this intake of sodium ion represented a concentration in the total fluid intake of about 140 mEq/1, which is very close to the plasma sodium concentration. It would also appear to be unlikely for sheep to show much spontaneous preference for sodium concentrations exceeding 280 mEq/l, and this fits the known findings for the rat, goat,[5] and, indeed, the sheep.[9] Therefore, the possibility of an innate salt appetite in sheep cannot be satisfactorily investigated with such high sodium concentrations.

DISCUSSION

The results obtained still represent a small number of individuals studied, as compared with other laboratory species; to obtain results with even this number has necessitated some concessions in experimental design, such as the use of outdoor pens. It could be argued, for example, that the introduction of the sheep to the drier laboratory diet could affect their sodium requirements, although one would expect this to affect all groups equally. Nevertheless, provided these results stand the test of independent repetition, a number of interesting possibilities suggest themselves.

1. Since breeds of sheep differ in their incidence of erythrocyte and hemoglobin types, different breed's susceptibilities to salt deficit or salt excess might be demonstrable and important. In this context, the possibility that HKs and LKs could differ in their renal physiology requires investigation. It is already known that the cation content of their renal cortexes are different.[27]

2. If the appearance of maximal sodium preference occurs with lower concentrations of sodium in LK sheep, it might be that such sheep are more sensitive to the taste of sodium, i.e., their preference curve lies further to the left than that of HKs. This would not fit Beidler's suggestion[2] that LK species are less sensitive to salt. On the other hand some LK species, such as cat and dog,[31] do possess water-responding fibers in the chorda tympani that may serve to extend the range of salt sensitivity.[13] In the ox (LK), Bernard[7] could find only a flow response, and Bell and Kitchell[4] could find no water fibers in calf, goat, or sheep. It would be interesting to reopen this

question with a view to the salt acuity and presence of water fiber in HK and LK sheep. Some HK species are already thought not to have water fibers, e.g., rat and man, although the pig (which is HK) does.

3. If any differences are demonstrable between salt acuity in HKs and LKs, it would be pertinent to extend the study to other sapid substances, including the thiurea drugs, in view of the work of Fischer[22] on Gaussian and non-Gaussian taste responses in man.

For the moment, however, the fact remains that the sheep may represent not only an interesting species in the study of salt appetite but also a number of discrete biochemical types worthy of investigation in themselves. This provides both a warning and an interest for the future.

ACKNOWLEDGMENT

This work is supported by the Agricultural Research Council.

REFERENCES

1. Beidler, L. M. 1962. Taste receptor stimulation. *Progr. Biophys. Biophys. Chem.* **12**, pp. 109–151.
2. Beidler, L. M., I. Y. Fishman, and C. W. Hardiman. 1955. Species differences in taste responses. *Amer. J. Physiol.* **181**, pp. 235–239.
3. Bell, F. R. 1959. Preference thresholds for taste discrimination in goats. *J. Agric. Sci.* 52, pp. 125–128.
4. Bell, F. R., and R. L. Kitchell. 1966. Taste reception in the goat, sheep and calf. *J. Physiol.* (*London*) **183**, pp. 444–451.
5. Bell, F. R., and H. L. Williams. 1959. Threshold values for taste in monozygotic twin calves. *Nature* **183**, pp. 345–346.
6. Beilharz, S., D. A. Denton, and J. R. Sabine. 1962. The effect of concurrent deficiency of water and sodium on the sodium appetite of sheep. *J. Physiol.* (*London*) **163**, pp. 378–390.
7. Bernard, R. A. 1964. An electrophysiological study of taste reception in peripheral nerves of the calf. *Amer. J. Physiol.* **206**, pp. 827–835.
8. Blair West, J. R., E. Bott, G. W. Boyd, J. P. Coghlan, D. A. Denton, J. R. Goding, S. Weller, M. Wintour, and R. D. Wright. 1955. General biological aspects of salivary secretion. *In* Physiology of Digestion of Ruminants (R. W. Dougherty, editor). Butterworths, Washington, D.C., pp. 198–221.
9. Bott, E., D. A. Denton, and S. Weller. 1967. The innate appetite for salt exhibited by sodium-deficient sheep. *In* Olfaction and Taste II (T. Hayashi, editor). Pergamon Press, Oxford, etc., pp. 415–429.
10. Cannon, W. B. 1932. The Wisdom of the Body. Kegan Paul, Trench, Trubner and Co. Ltd., London, pp. 91–97.
11. Carlson, A. J. 1916. The Control of Hunger in Health and Disease. Univ. of Chicago Press, Chicago, Illinois, p. 15.
12. Carpenter, J. A. 1956. Species differences in taste preferences. *J. Comp. Physiol. Psychol.* **49**, pp. 139–144.
13. Cohen, M. J., S. Hagiwara, and Y. Zotterman. 1955. The response spectrum of taste fibres in the cat: A single fibre analysis. *Acta Physiol. Scand.* **33**, pp. 316–332.
14. Dawson, T. J., and J. V. Evans. 1965. Effects of hemoglobin type of the cardiorespiratory system of sheep. *Amer. J. Physiol.* **209**, pp. 593–598.
15. Denton, D. A. 1965. Evolutionary aspects of the emergence of aldosterone secretion and salt appetite. *Physiol. Rev.* **45**, pp. 245–295.
16. Denton, D. A. 1967. Salt appetite. *In* Handbook of Physiology, Section VI, Alimentary Canal **1**, American Physiological Society, Washington, D.C., pp. 433–459.
17. Denton, D. A., and J. R. Sabine. 1961. The selective appetite for Na^+ shown by Na^+ -deficient sheep. *J. Physiol.* (*London*) **157**, pp. 97–116.

18. Duncan, C. J. 1962. Salt preference in birds and mammals. *Physiol. Zool.* **35**, pp. 120–132.
19. Evans, J. V. 1954. Electrolyte concentrations in red blood cells of British breeds of sheep. *Nature* **174**, pp. 931–932.
20. Evans, J. V. 1957. Water metabolism in sheep. *Nature* **180**, p. 756.
21. Evans, J. V. 1963. Inherited physiological individuality in ruminants. *In* Progress in Nutrition and Allied Sciences (D. P. Cuthbertson, editor). Oliver and Boyd, London, pp. 199–212.
22. Fischer, R., F. Griffin, and M. A. Rockey. 1966. Gustatory chemoreception in man: Multidisciplinary aspects and perspectives. *Perspectives Biol. Med.* **9**, pp. 549–577.
23. Fregly, M. J., J. H. Harper, and E. P. Radford. 1965. Regulation of sodium chloride intake by rats. *Amer. J. Physiol.* **209**, pp. 287–292.
24. Kaunitz, H. 1956. Causes and consequences of salt consumption. *Nature* **178**, pp. 1141–1144.
25. Lat, J. 1967. Self selection of dietary components. *In* Handbook of Physiology, Section VI, Alimentary Canal **1**, American Physiological Society, Washington, D.C., pp. 367–386.
26. Michell, A. R. 1969. A study of salt appetite in the sheep with special reference to the role of taste. Ph.D. Thesis. Univ. of London.
27. Mounib, M. S., and J. V. Evans. 1960. The potassium and sodium content of sheep tissues in relation to potassium content of the erythrocytes and the age of the animal. *Biochem. J.* **75**, pp. 77–82.
28. O'Kelly, L. I., J. L. Falk, and D. Flint. 1958. Water regulation in the rat. I. Gastrointestinal exchange rates of water and sodium chloride in thirsty animals. *J. Comp. Physiol. Psychol.* **51**, pp. 16–21.
29. Richter, C. P. 1956. Salt appetite of mammals: its dependence on instinct and metabolism. *In* L'Instinct dans le Comportement des Animaux et de l'Homme (M. Antouri, editor). Masson et Cie, Paris, pp. 577–632.
30. Watson, J. H., and A. G. H. Khattab. Effect of haemoglobin and potassium polymorphism on growth and wool production in Welsh mountain sheep. *J. Agric. Sci.* **63**, pp. 179–183.
31. Zotterman, Y. 1959. The nervous mechanism of taste. *Ann. N.Y. Acad. Sci.* **81**, pp. 358–366.

DISCRIMINATORY RESPONSES OF HYPOTHALAMIC OSMOSENSITIVE UNITS TO GUSTATORY STIMULATION IN CATS

S. NICOLAÏDIS

Laboratoire de Physiologie des Sensibilités chimiques et Régulations alimentaires, de l'École Pratique des Hautes Études, College de France, Paris, France

INTRODUCTION

In my previous work on rats,[10] I demonstrated the existence of diuretic and antidiuretic reflexes elicited by oral stimulation. In other words, after oral stimulation with water, rapid and dramatic diuresis occurred. On the other hand, the same stimulation with hypertonic saline was followed by an equally strong antidiuresis. A second reflex, which I found in humans,[11] was a rapid increase in sweating that was elicited by both oral and gastric stimulations. Since these persistent responses to brief peripheral stimulation act in the same direction as the postabsorptive effects of the stimulating liquids, they may be considered as anticipatory reflexes.

Neurons that modify their firing rate in response to intracarotid injection of hypertonic solutions have been found in the anterior hypothalamus, and have been extensively studied.[6,13] It is now generally accepted that these "osmosensitive" units regulate both release of the antidiuretic hormone (ADH) and urinary output under the physiological influence of blood osmotic pressure.[17] The participation of such osmosensitive units in the control system of water intake has also been suggested.

One might suspect that the afferent pathways active in the anticipatory reflexes could converge on the classic hypothalamic areas that are sensitive to systemic osmotic changes, which are also effective in producing diuretic adjustments.

To investigate this possibility of convergent intero- and exteroceptors, I first localized hypothalamic units responsive to intracarotid hypertonic NaCl injections and then investigated their responsiveness to oral stimulations with hypertonic saline and water.

MATERIALS AND METHODS

Twenty-four cats were studied under chloralose anesthesia. The stimulations were performed with normal NaCl solution or distilled water, applied randomly to the systemic compartment by intracarotid (IC) injections of 0.2 ml, after which 4 ml of the same solutions were used for oral stimulations.

The individual firing of a few units of various hypothalamic areas was picked up with a stainless steel concentric electrode with a tip diameter of 2 to 3 μ. After selecting the more specific units by using electronic gates, it was possible to integrate their activity in order to obtain a proportional deviation of the recording pens. Using the short time-constants of integration, one may obtain recordings of individual spikes. A simultaneous recording of the intra-arterial pressure, the integrated renal flow and the electrocorticogram was made. We used the Prussian blue technique to localize the placement of the recording electrode.

RESULTS

According to the particular area, two opposite and reciprocal types of response could be found:

(1) The IC injection of saline increased the previous frequency of spikes (Figure 1A). On the other hand, the injection of distilled water did not reproduce the same reaction, but instead elicited an opposite (inhibitory) effect. These responses might reach a plateau lasting several minutes before decreasing, or might last more than 30 sec and then decrease slowly. In this latter case a rebound of a high activity that lasted more than 1 min was usually found. When such responsive units were found, we investigated the effect of oral stimulations with the same liquids. At this level, also, the saline increased the previous activity and the distilled water decreased it. The patterns

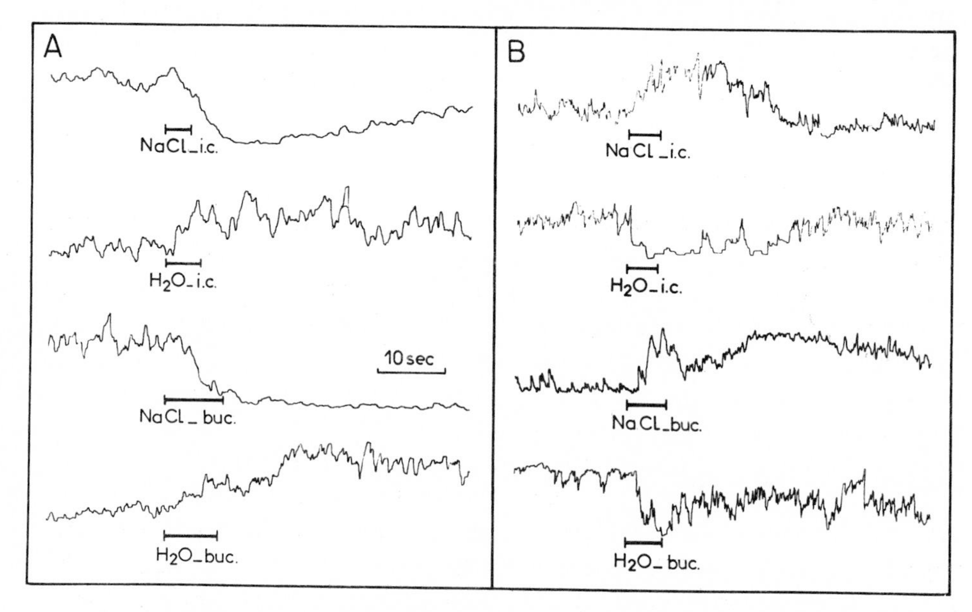

Figure 1 Integrated activities recorded from a limited number of units lying at the post-commissural level of the medial forebrain bundle: (A) in the medial third; (B) in the inferior third. The same activating or inhibitory effect was obtained after both intracarotid (i.c.) and buccal (buc.) stimulations. The type of response depends on the hypotonic (H_2O) or hypertonic (1 M NaCl) stimuli. Notice that, depending on the site, these convergent responses are antagonistic.

of these changes remained similar to those of the intracarotid stimulations. One characteristic of the oral responses is the slower acceleration of their activity, which reaches its maximum only after several seconds.

(2) The second type of response was almost the opposite of the first one (Figure 1B). Units showing this kind of response were inhibited by both the intracarotid and the oral hypertonic saline. Therefore, the responses were still discriminative. In other words, when the two stimulations were repeated with distilled water, the frequency of these same units decreased. The duration of the effect was as variable as that of the first type. The two opposite types of response could be found in two neighboring areas, separated by at least 50 μ in one case. The responses to the systemic stimulation were abolished by atropine, which only diminished those to the oral stimulations. The histological study shows that these two types of response have been obtained in the supraoptic nucleus (SON), in the medial forebrain bundle (MFB), and in the pallido-hypothalamic bundle (PHB) against the fornical column (FC).

Some remarks are in order: we have not been able to find either the first or the second type of response in many of our cats, in spite of exploring the same part of the supraoptic nucleus or the medial forebrain bundle, but the responsive areas showed a highly reproducible pattern. Some areas were responsive to the intracarotid stimulations only and others to both the intracarotid and oral ones, but lacked discriminatory patterns according to the stimuli. Nociceptive stimuli were not effective. Finally, like the potodiuretic and the potohidrotic (from the Greek for drink and sweat) reflexes, the oral stimulation could not be repeated too often without a gradual extinction of the hypothalamic response.

DISCUSSION

These results, completing preliminary data,[12] confirm that limited hypothalamic areas are responsive to both systemic and gustatory osmotic stimulations. These specific areas exhibit discriminatory and opposite responses to water and saline solutions. In general, close to these areas there are distinct populations of units in which the previous excitatory stimuli are inhibitory, and vice versa. Thus, it appears that two antagonistic and more or less intermingled systems of discriminative neurons are present. Their location in the SON, MFB, and PHB near the FC corresponds to the areas used by numerous investigators in regulation of water and NaCl intake, as well as of their renal and sudoral output. This responsiveness of some hypothalamic units to the direct humoral action of blood osmolarity and, via gustatory afferents, to hypertonic solutions and water present in the mouth, is of particular interest. Much evidence is accumulated, demonstrating that the hypothalamus is mainly involved in control systems insuring physiological regulations through the sensitivity of specific sites to variations of their humoral environment. Such internal sensors, responsive to blood temperature, to glucose supply, or to various metabolites and hormones, for example, permit homeostatic regulations through positive or negative feedback and through various effectors. But most of these control systems, to insure the effectiveness of such internal regulations, must integrate information coming from the external environment. These external sensory inputs conveyed to the control system permit more rapid and even anticipated commands of adjustments to variations in the external environment that will affect the internal milieu more slowly. Such a mechanism has recently been shown to be effective, for instance, in regulations of body temperature and of endocrine function under the influence of both external and humoral signals.[2-4,8,18]

Our present findings provide the first evidence of a parallel mechanism in the regulation of body fluid balance. The gustatory projections on hypothalamic osmosensitive neurons may explain anticipatory sudoral or renal responses (potodiuretic and potohidrotic reflexes). It may also explain anticipated and short-term control of these oral intakes, themselves insuring, through oropharyngeal metering, the adjustment of intake to present or future bodily deficits.[1,9,15,16]

The parallel of the temporal patterns of a unit's discharges and of functional responses should be pointed out. In the same way as long-lasting satiety or anticipatory diuretic and sweating responses are elicited by brief oral stimulations, long-lasting discharges of osmosensitive units are also brought about by the same brief stimulations.

When these neurons fire, the opposite effects of water and hypertonic solutions applied to the tongue are particularly instructive. At the level of regulatory centers, this emphasizes the significance and the functional importance of the gustatory discrimination of salt and water stimulations studied at the peripheral level.[5,7,14,19] Whether these osmosensitive neurons are the true osmoceptors and whether the same neurons are involved both in the regulation of outputs and in the control of thirst, are still open to discussion.

REFERENCES

1. Adolph, E. F., J. P. Barker, and P. Hoy. 1954. Multiple factors in thirst. *Amer. J. Physiol.* **178**, pp. 538–562.
2. Anand, B. K., G. S. Chhina, K. N. Sharma, S. Dua, and B. Singh. 1964. Activity of single neurons in the hypothalamic feeding centers: effect of glucose. *Amer. J. Physiol.* **207**, pp. 1146–1154.
3. Anand, B. K., and R. V. Pillai. 1967. Activity of single neurons in the hypothalamic feeding centers: effect of gastric distension. *J. Physiol. (London)* **192**, pp. 63–77.
4. Barraclough, C. A., and B. A. Cross. 1963. Unit activity in the hypothalamus of the cyclic female rat: effect of genital stimuli and progesterone. *J. Endocrinol.* **26**, pp. 339–359.
5. Bartoshuk, L. M., and C. Pfaffmann. 1965. Effects of pre-treatment on the water taste response in cat and rat. *Fed. Proc.* **24**, p. 207 (Abstract).
6. Cross, B. A., and J. D. Green. 1959. Activity of single neurones in the hypothalamus: effect of osmotic and other stimuli. *J. Physiol. (London)* **148**, pp. 554–569.
7. Halpern, B. P. 1959. Gustatory responses in the medulla oblongata of the rat. Unpublished Ph.D. thesis. Brown Univ., Providence, R.I.
8. Hammel, H. T., F. T. Caldwell, Jr., and R. M. Abrams. 1967. Regulation of body temperature in the blue-tongued lizard. *Science* **156**, pp. 1260–1262.
9. Le Magnen, J. 1955. Le rôle de la réceptivité gustative au chlorure de sodium dans le mécanisme de régulation de la prise d'eau chez le rat blanc. *J. Physiol. (Paris)* **47**, pp. 405–418.

10. Nicolaïdis, S. 1963. Effets sur la diurèse de la stimulation des afférences buccales et gastriques par l'eau et les solutions salines. *J. Physiol.* (*Paris*) **55**, pp. 309–310.
11. Nicolaïdis, S. 1964. Etude d'une réponse de sudation après ingestion d'eau chez le sujet déshydraté. *C.R. Acad. Sci.* **259**, pp. 4370–4372.
12. Nicolaïdis, S. 1969. Early systemic responses to oro-gastric stimulation in the regulation of food and water balance. Functional and electrophysiological data. Conference on Neural Regulation of Food and Water Intake, New York (February 1967). *Ann. N.Y. Acad. Sci.* (In press)
13. Novin, D., and R. Durham. 1969. D-C and unit potential studies in the supra-optic nucleus of the hypothalamus. Conference on Neural Regulation of Food and Water Intake, New York (February 1967). *Ann. N.Y. Acad. Sci.* (In press)
14. Pfaffmann, C. 1941. Gustatory afferent impulses. *J. Cell. Comp. Physiol.* **17**, pp. 243–258.
15. Stevenson, J. A. F. 1964. Current reassessment of the relative functions of various hypothalamic mechanisms in the regulation of water intake. *In* Thirst (M. J. Wayner, editor). Pergamon Press, Oxford. pp. 553–565.
16. Stevenson, J. A. F. 1965. Control of water exchange: regulation of content and concentration of water in the body. *In* Physiological Controls and Regulations (W. S. Yamamoto and J. R. Brobeck, editors). W. B. Saunders Co., Philadelphia, Pa. pp. 253–273.
17. Verney, E. B. 1947. The antidiuretic hormone and the factors which determine its release. *Proc. Roy. Soc. Ser. B* (*Biol. Sci.*) **135**, pp. 25–106.
18. Wit, A., and S. C. Wang. 1967. Effects of increasing ambient temperature on unit activity in the preoptic-anterior hypothalamus (PO/AH) region. *Fed. Proc.* **26**, p. 555.
19. Zotterman, Y. 1956. Species differences in the water taste. *Acta Physiol. Scand.* **37**, pp. 60–70.

THE METABOLIC REGULATION OF TASTE ACUITY

ROBERT I. HENKIN
National Heart Institute, Bethesda, Maryland

INTRODUCTION

One of the major problems in gustation has been to determine the manner by which the taste process occurs. Historically, this problem has been concerned with the changes in the neural events which accompany the application of sapid substances on the tongue of animals and man. During the past few years, however, a new body of knowledge has developed which suggests that the neurophysiological correlates of taste refer to only one aspect of gustation and that other aspects need to be included if the mechanism of gustation is to be fully understood. This new body of knowledge refers to the biochemical correlates of taste, and it is the purpose of this paper to outline the evidence for the existence of these correlates and to demonstrate how taste acuity is regulated through chemical reactions. For the purposes of this paper, the biochemical correlates of taste may be considered to affect two of the major systems of the taste process: the taste receptors themselves, those specialized cells which mediate the taste response; and the gustatory neurons, those nerves over which taste information is transmitted to higher integrative centers. For convenience, these effects will be artificially separated into the preneural events of taste, those primarily affecting the taste buds, and the neural events of taste, those primarily affecting the gustatory neurons.

BIOCHEMICAL CORRELATES OF THE NEURAL EVENTS OF TASTE

Many changes occur in the nervous system that affect taste acuity. Most of the pathological processes that affect nervous tissue in general can be observed to affect the gustatory neurons and thereby alter taste acuity. Infectious processes such as diphtheria,[15] demyelinating processes such as multiple sclerosis,[1] and granulomatous processes such as sarcoidosis[2] have been associated with decreases in taste acuity. The manner by which these and other pathological processes affect taste acuity is not well known. In addition, changes which occur in the multiplicity of biochemical constituents of nervous tissue can alter taste acuity. Thus, the biochemical precursors and metabolic products of catecholamines, acetylcholine, various phospholipids, etc. may all ultimately be shown to play a role in the biochemical control of taste acuity.

One group of chemical substances, the steroid hormones of the adrenal cortex, appear to influence taste acuity in a predictable fashion through their effect on neural function. This portion of the paper will describe that aspect of the biochemical control.

Adrenalectomized rats have been observed to exhibit a decrease in preference threshold for sodium chloride (NaCl).[18] Similarly, detection thresholds for NaCl in patients with untreated adrenal cortical insufficiency or panhypopituitarism, or after total adrenalectomy are significantly decreased below normal.[9] These observations were extended to other taste qualities to show that these patients also exhibited an increased ability to detect other salts such as potassium chloride and sodium bicarbonate, sweet substances such as sucrose, bitter substances such as urea, and sour substances such as hydrochloric acid (Figure 1). Thus, the increase in detection acuity demonstrated by patients with untreated adrenal cortical insufficiency, from any cause, is not limited to the detection of NaCl alone, but extends to the detection of each of the four qualities of taste.

This was emphasized by the results of further experiments in which patients with adrenal cortical insufficiency were treated with the salt-retaining hormone desoxycorticosterone acetate.[9] The results, shown for NaCl in Figure 2, demonstrated that taste acuity did not change after treatment produced a return of the hyponatremia, hyperkalemia, and lowered extracellular fluid volume toward or to normal levels. Thus, in spite of a return to a normal body electrolyte balance, there was no associated decrease in the increased taste acuity for any of the four taste qualities. How-

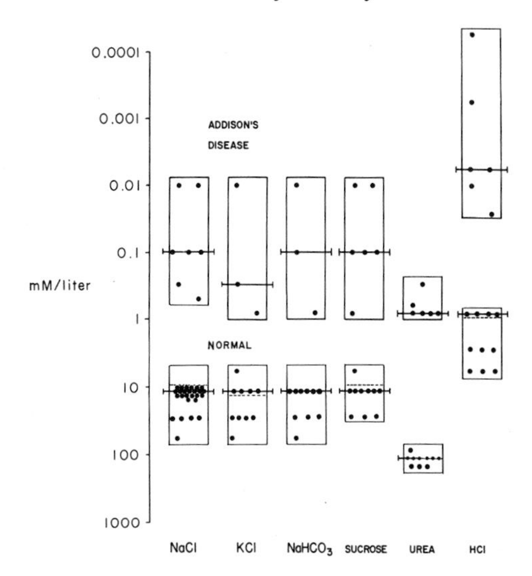

Figure 1 Detection thresholds for sodium chloride, potassium chloride, sodium bicarbonate, sucrose, urea, and hydrochloric acid in normal volunteers and in patients with untreated adrenal cortical insufficiency. Each dot represents individual detection thresholds, each box, the range of taste responsiveness, the line through each box, the median detection threshold. The dotted line within some boxes, defining the normal range, indicates detection thresholds determined by other investigators. Patients' median detection thresholds for each of the four taste qualities are approximately 1/100 those for the normal volunteers. There is no overlap between thresholds of patients and those of normal subjects.

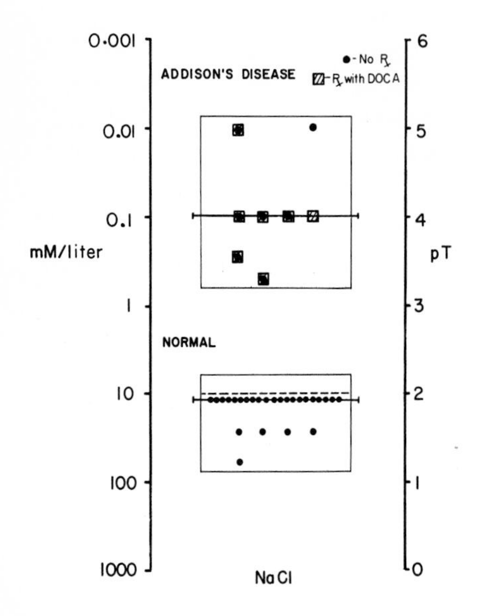

Figure 2 Detection thresholds for sodium chloride in normal volunteers and in patients with adrenal cortical insufficiency treated with desoxycorticosterone acetate (DOCA). The small hatched squares surrounding the individual detection thresholds of the patients, indicating treatment with DOCA, demonstrate that these thresholds are as low as those obtained in the untreated state. Similar differences were found for thresholds for each of the four taste qualities.

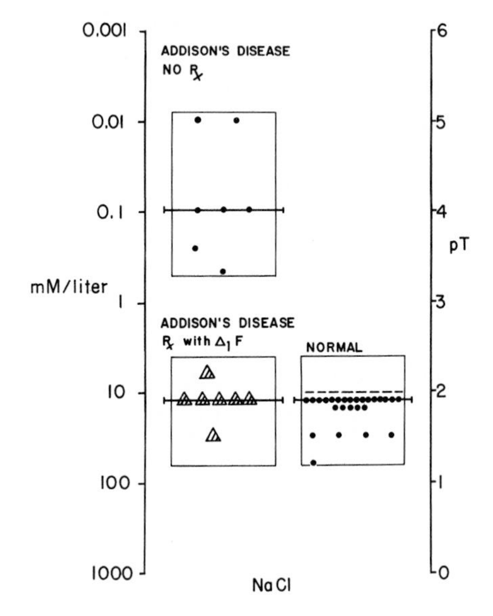

Figure 3 Detection thresholds for sodium chloride in normal volunteers and in patients with adrenal cortical insufficiency treated with the carbohydrate-active steroid prednisolone. The hatched triangles, indicating treatment with prednisolone, demonstrate that these thresholds are not different from those obtained in normal subjects. A similar return to normal taste acuity was found for thresholds for each of the four taste qualities.

ever, after carbohydrate-active steroid was administered by any route, a gradual return of taste-detection acuity to normal took place within 48 hr.[9] This occurred not only for NaCl thresholds, as shown in Figure 3, but also for each of the other three taste qualities.

Although these results demonstrate that removal of adrenal cortical hormone is associated with the production of increased detection acuity for each of the four taste qualities, and this is returned to normal only after treatment with carbohydrate-active steroids, they do not necessarily indicate that this effect is primarily through the action of these steroids on gustatory neurons.

Further experiments in man and other animals, however, have demonstrated that carbohydrate-active steroids probably affect taste-detection acuity through their effect on some of the homeostatic mechanisms in neural function that change after these steroids are removed. Thus, not only is taste-detection acuity increased in untreated adrenal cortical insufficiency; so, too, is detection acuity for the sensory modalities of olfaction,[5] audition,[13] and proprioception.[8] Still further experiments have demonstrated that carbohydrate-active steroids play a significant role in the conduction of neural impulses along axons and across synapses.[8,10,17] The absence of adrenal cortical steroids is associated with a decrease in axonal conduction velocity and an increase in synaptic delay, while replacement of these steroids returns conduction velocities to normal.[8,10,17] These hormones may be important for the normal maintenance of myelin as well as for the maintenance of a number of enzyme systems which are critical for normal energy metabolism in nervous tissue.[20] In this broad sense, these steroids play a significant role in the physiological and structural integrity of the nervous system.

In addition, adrenal cortical steroids cross the blood-brain–barrier[14] and are found in significant concentrations in tissues of the central and peripheral nervous systems.[7] Removal of these steroids after adrenalectomy causes a significant decrease in their concentration in neural tissue in a time pattern similar to that observed in the development of increased taste-detection acuity.[7]

Therefore, it seems likely that adrenal cortical steroids regulate taste acuity, through their effect on neural function, on the neural events of the taste process. This regulation is through inhibition, for the removal of carbohydrate-active steroids increases neural excitability[8] and thereby increases detection acuity. After the removal of these steroids, their normal inhibitory effect on neural function decreases and sensory stimuli not usually detected—those normally considered "subthreshold"—are detected. Replacement of carbohydrate-active steroids returns the tissue concentrations of this hormone to normal and allows for normal inhibition to occur. At this time both aspects of neural function regulated by carbohydrate-active steroids and taste-detection acuity return to normal.

BIOCHEMICAL CORRELATES OF THE PRENEURAL EVENTS OF TASTE

Many changes occur in the metabolic net that do not directly or indirectly affect neural function but do affect taste acuity. Pathologically these processes are associated with metabolic abnormalities, and information about the specific biochemical abnormalities in some are well known. Taste acuity is altered in some of these diseases and correction of the specific biochemical abnormality has been associated with the return of taste acuity to normal. Thus it has been possible to identify the anatomical

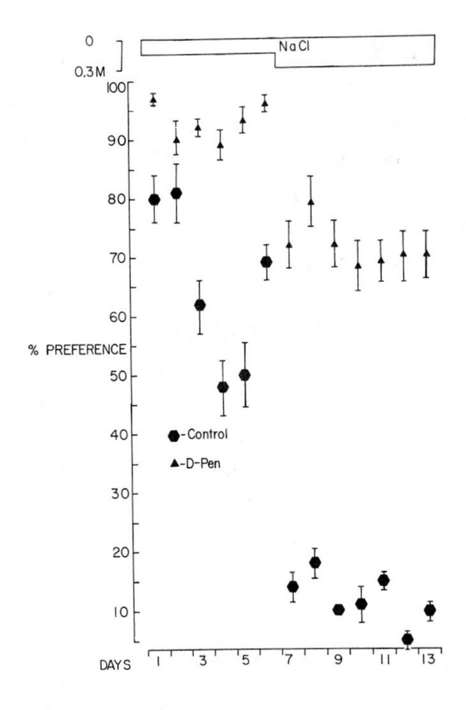

Figure 4 Preference for 0.15 and 0.30 M sodium chloride in control and D-penicillamine-treated rats. Each symbol represents the mean preference of four rats for each day; the line extending above and below the symbol, the standard error of the mean. The % preference for both 0.15 and 0.30 M sodium chloride among the D-penicillamine-treated rats is significantly higher than among the controls. There is no overlap in the preference of the two groups of animals. Similar differences were found for preference of 0.5% sucrose.

and physiological bases for the taste defect. In general, these biochemical effects have occurred at the taste bud, the structure that for convenience we have chosen to refer to as the chemical filter which regulates the preneural events of taste.

Treatment of patients with cystinuria, scleroderma (progressive systemic sclerosis), rheumatoid arthritis, and idiopathic pulmonary fibrosis with the thiol-containing drug D-penicillamine ($\beta\beta$-dimethyl cysteine) produces a predictable decrease in taste acuity (Table I) for each of the four qualities of taste.[11,16] Administration of this drug to rats is also associated with a decrease in taste acuity as measured by increased preference for sodium chloride (Figure 4) and taking food that is normally rejected.[12] D-penicillamine has a number of pharmacological actions in addition to increasing thiol concentration in various body tissues. One of these is its capacity to decrease the concentration of copper in serum and tissue. Serum copper concentrations in patients and animals in whom treatment with D-penicillamine had produced hypogeusia (a decrease in taste acuity) were significantly lower than normal and were associated with an increase in urinary excretion of the metal. Discontinuation of treatment resulted in a return of taste acuity to normal along with a decrease in thiol concentration and an increase in copper concentration.[11] Administration of copper, despite continued administration of D-penicillamine, also returned taste acuity to normal (Figure 5), provided the amount of copper given was sufficient to react with the concentration of D-penicillamine present.[6,11]

In another set of experiments hypogeusia was produced in both man and other animals after the administration of the thiol-containing drug 5-mercaptopyridoxal.[6]

TABLE I
DETECTION AND RECOGNITION THRESHOLDS FOR EACH OF THE FOUR QUALITIES OF TASTE IN 11 PATIENTS TREATED WITH D-PENICILLAMINE

The numerator of each fraction is the detection threshold.
The denominator of each fraction is the recognition threshold.
S denotes the ability to detect or recognize a saturated solution.

				Median threshold* (mM per l.) for:			
Patient	Age	Sex	Diagnosis	Salt	Sucrose	Hydrochloric acid	Urea
1	28	F	Scleroderma	150/150	60/150	30/30	300/300
2	55	F	Scleroderma	300/300	300/300	0.8/60	300/300
3	40	F	Scleroderma	300/300	300/300	0.8/6	90/150
4	27	F	Scleroderma	30/150	30/150	15/15	300/300
5	43	M	Scleroderma	30/150	60/150	30/30	500/500
6	18	M	Cystinuria	150/150	30/60	0.8/6	60/60
7	29	F	Cystinuria	60/60	150/S	6/15	60/60
8	38	M	Cystinuria	150/150	150/150	300/300	S/S
9	41	F	Idiopathic pulmonary fibrosis	60/60	60/60	6/15	500/500
10	57	M	Idiopathic pulmonary fibrosis	300/300	S/S	60/300	1000/S
11	25	M	Wilson's disease	150/150	150/150	30/30	300/S
Median (patients)				150/150	150/150	15/30	300/300
Median (normal)				12/30	12/30	0.8/3	90/120
Detection range (normal)				(0.5–60)	(0.5–60)	(0.5–6)	(60–150)
Recognition range (normal)				(3–60)	(3–60)	(0.5–6)	(60–150)

* Expressed in the form detection/recognition.

This hypogeusia was produced without any associated decrease in serum or urinary copper concentrations. Discontinuation of the 5-mercaptopyridoxal resulted in a return to normal taste acuity along with a decrease in thiol concentration.[6]

These results suggested that both thiols and copper were involved in the metabolic regulation of taste acuity.[6] The hypothesis was confirmed in another group of experiments. A patient with multiple myeloma of the rare G3 subclass spontaneously reported that she had lost her taste. This subjective report of hypogeusia was confirmed by quantitative taste tests; she was found to have elevated detection and recognition thresholds for each of the four qualities of taste. Since thiols and copper both seemed to be involved in the regulation of taste, acuity studies were undertaken to measure the concentrations of those substances in the patient. Results of the studies showed that serum and, presumably, tissue concentrations of copper were normal. However, due to the elevated serum concentration of abnormal gamma globulin with

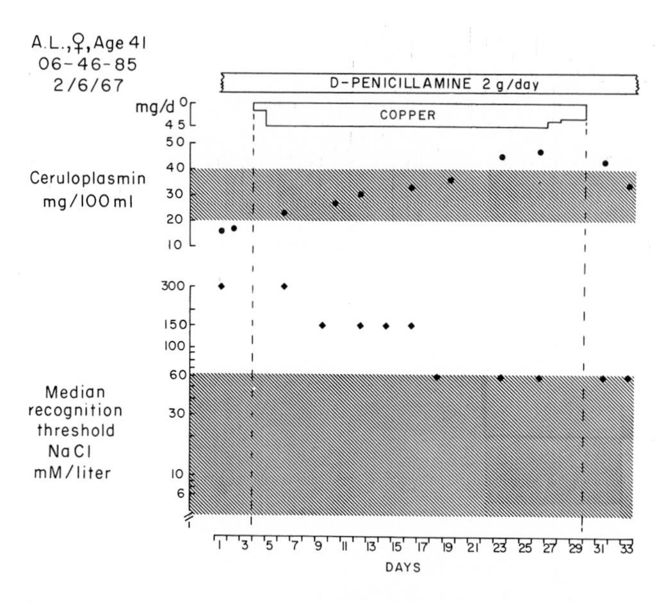

Figure 5 Effect of treatment with D-penicillamine and oral copper on median recognition thresholds for sodium chloride in patient A. L. This patient had scleroderma. After treatment with 60 mg of D-penicillamine and copper daily for 14 days (a diet high in copper content and oral copper sulfate), detection and recognition thresholds for sodium chloride returned to normal. The normal range is indicated by the lower hatched area. This return to normal also occurred for thresholds for each of the other taste qualities. Serum ceruloplasmin concentration (and hence also serum copper concentration) was below the normal range (upper hatched area) prior to copper therapy. Serum ceruloplasmin concentration returned to normal levels within three days after copper therapy was initiated.

its large number of exchangeable SH groups, it was hypothesized that thiol levels in the patient were elevated above normal. On the basis of this hypothesis she was treated with oral copper, with the expectation that the additional copper would tie up the excess thiols in blood and tissues. Within four days after therapy with liquid copper sulfate was initiated, her taste acuity for each of the four taste qualities had returned to normal and remained there throughout the administration of the drug (Figure 6). The return of normal taste acuity was easily detected subjectively by the patient and objectively by quantitative taste tests. Subjectively it was realized when the patient complained that the liquid copper sulfate had an obnoxious taste, whereas previously she took the drug without complaint. Hypogeusia recurred within 48 hr after copper therapy was discontinued.

Since copper sulfate returned taste acuity to normal, presumably by reacting with excess thiol, it was hypothesized that other metals that would chelate or complex with thiols would also reverse the patient's hypogeusia. To test this hypothesis, liquid zinc chloride was given orally; within four days her taste acuity for each of the four qualities of taste had subjectively and objectively returned to normal (Figure 6),

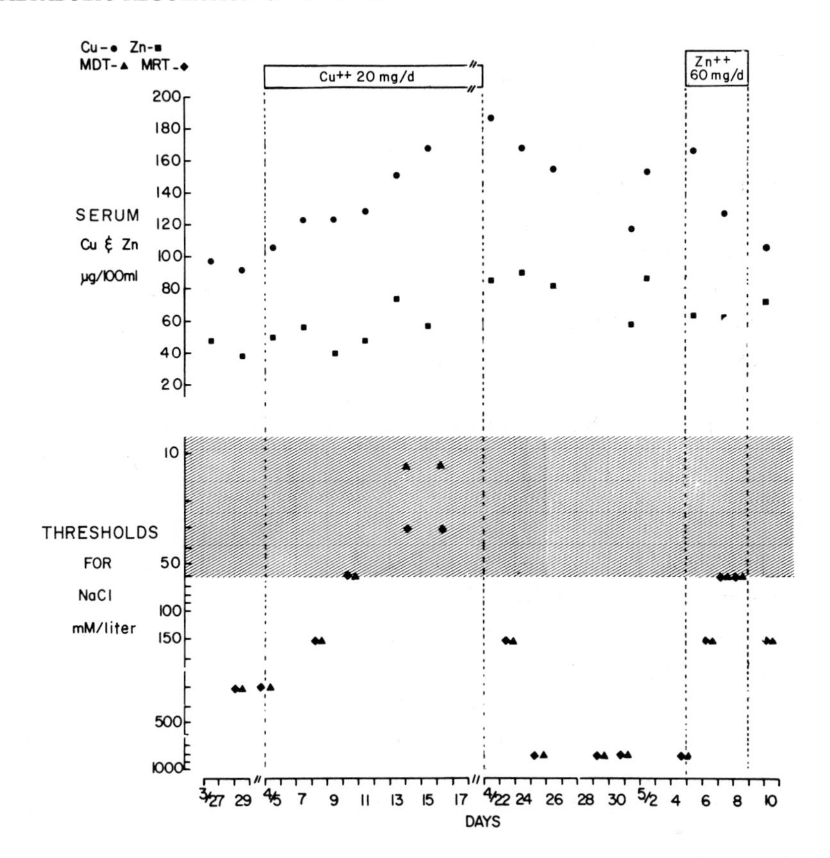

Figure 6 Effect of treatment with oral copper and zinc on detection and recognition thresholds for sodium chloride in patient G. B. This patient had multiple myeloma of the G3 subclass. After treatment with 20 mg of copper sulfate daily for four days, detection and recognition thresholds for sodium chloride returned to normal. The normal range is indicated by the hatched area. Serum copper concentration was within normal limits at the beginning of this therapy and increased during treatment. Similarly, treatment with 60 mg of zinc chloride daily for four days returned detection and recognition thresholds for sodium chloride to the normal range. A similar return of taste acuity to normal for each of the four taste qualities occurred after treatment with copper and zinc. MDT represents median detection threshold; MRT represents median recognition threshold.

and she rejected liquid zinc chloride as obnoxious. Within 24 hr after zinc chloride was discontinued her hypogeusia returned.

It is important to emphasize that this patient had hypogeusia without any associated abnormality of vision or proprioception. There were no associated abnormalities of neural conduction or synaptic delay. Therapy with copper or zinc also did not alter any sensory function except taste. Thus it seemed unlikely that hypogeusia was produced by a significant alteration in neural function per se. We therefore assumed that the patient had no significant changes in what we have termed the neural events of taste.

These results suggested that thiols and metals regulated taste acuity through their action on those components of the taste process which we termed the preneural events of taste, i.e., those events which affect the taste buds per se. This regulation occurs by means of the normal inhibitory action of thiols on the taste bud, similar in some ways to the inhibitory action of carbohydrate-active steroids on gustatory neuron function. If thiol concentration is increased, hypogeusia is produced. This is similar to the production of hypogeusia in patients with Cushing's syndrome secondary to their endogenous overproduction of carbohydrate-active steroid.[4]

Previous studies have indicated that taste buds act as "chemical filters" through which various tastants (substances which are detected or recognized by their taste quality) pass to reach the gustatory neurons in the taste bud or in the lingual papilla.[3,19] In this sense, taste threshold was redefined operationally as the net flux of tastant required to produce depolarization of the gustatory neurons and the subsequent detection and integration of this event by the higher neural centers. If the pores of the taste bud were smaller than normal, a higher concentration of tastant would be required to pass through the filter per unit time in order to provide that concentration of tastant that would produce depolarization of the neurons. This would correspond to the hypogeusia observed in patients with excess thiol or decreased metal concentrations. In some ways, this would be analogous chemically to the physical action of the iris in limiting the amount of light reaching the retina of the eye. If the pores of the taste bud were open, a smaller concentration of tastant would be required to produce the taste response. It is in this sense that I hypothesize that the taste bud membrane acts as a chemical filter, as a type of molecular sieve, because its open or closed state would either allow or restrict the influx of tastant to the gustatory membrane.

We have proposed that the opening and closing of the membrane is due to changes in the conformation of a protein which lines the pores of the taste-receptor membrane. We have called this protein the gate-keeper protein, for it limits the passage of tastant to control acuity for each of the four qualities of taste.[6]

SUMMARY

These studies indicate that taste acuity is regulated through a number of biochemical reactions. These reactions relate the physiological activity of the preneural and the neural events of taste. Normally, both carbohydrate-active steroids and thiols act in an inhibitory manner to control taste acuity (Figure 7). Thiols do this through their effects on the preneural events of taste; carbohydrate-active steroids, in part, through their effects on the neural events of taste. The neural events are regulated by many chemical reactions which determine neural function. One of the more explicit of these reactions involves the manner by which carbohydrate-active steroids inhibit neural excitability and conduction (Figure 8). The preneural events of taste are also regulated by many biochemical reactions, including those involving thiols and metals. One such reaction is shown in Figure 9. If thiol concentration is increased, as shown in Figure 10, taste acuity decreases. If metal concentration is decreased, as shown in Figure 11, taste

Figure 7 A diagrammatic representation of the manner by which carbohydrate-active steroids (CAS) and thiols (RSH) normally inhibit taste acuity. The size of the box represents normal taste acuity; the length and breadth of the arrows indicate normal inhibition.

Figure 8 A diagrammatic representation of the manner by which carbohydrate-active steroids (CAS) normally inhibit neural excitability and thereby regulate taste acuity through their effects on the neural events of taste. This figure demonstrates that if carbohydrate-active steroid concentration is decreased, the normal level of inhibition is decreased and taste detection acuity increases.

acuity also decreases. Reactions such as those shown in Figure 9 are postulated to determine the conformation of a protein, which we have chosen to call the gate-keeper protein. This protein is postulated to line the holes of the taste-bud membrane. These chemical reactions are hypothesized to cause the protein to unfold and decrease taste acuity if thiols are increased or metals are decreased; they may cause the protein to refold and return taste acuity to normal if thiols are decreased to normal or metals are increased. Through these concepts it is possible to relate taste acuity to a series of biochemical steps that may regulate the manner in which taste sensitivity occurs.

$$2\ RSH + Me^{++}\ (Cu^{++}, Zn^{++}) = (RS)_2\ Me + 2H^{+}$$

Figure 9 An example of one of the chemical interactions that occur between thiols and metal ions.

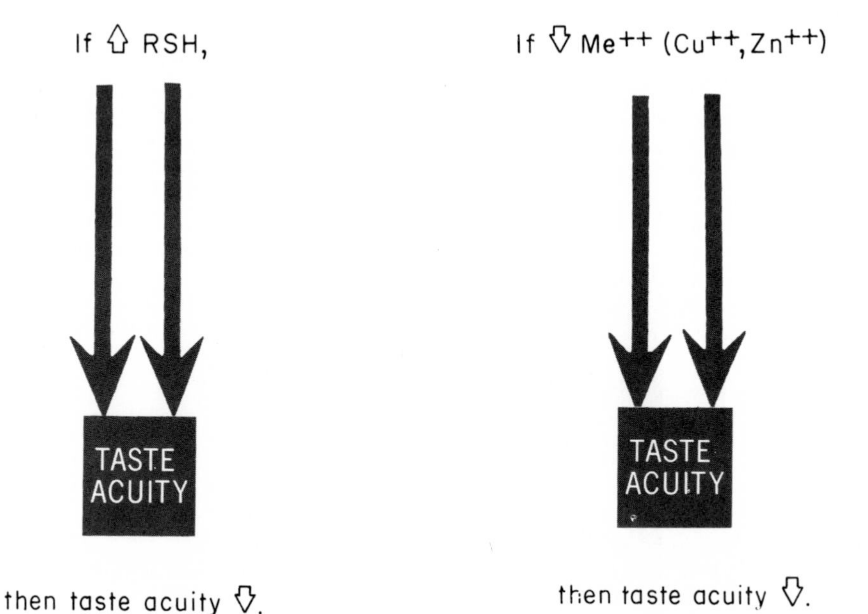

Figure 10 A diagrammatic representation of the manner by which thiols (RSH) affect taste acuity. Normally, thiols inhibit taste acuity and thereby directly or indirectly regulate acuity through their effects on the preneural events of taste. This figure demonstrates that if thiol concentration is increased the normal level of inhibition is increased and taste acuity decreased.

Figure 11 A diagrammatic representation of the manner by which metal ions (Me^{++}) affect taste acuity. If metal concentration decreases, either directly or indirectly, the normal level of inhibition is increased and taste acuity decreases.

REFERENCES

1. Cohen, L. 1964. Disturbance of taste as a symptom of multiple sclerosis. *Brit. J. Oral. Surg.* **2**, pp. 184–185.
2. Colover, J. 1948. Sarcoidosis with involvement of the nervous system. *Brain* **71**, pp. 451–475.
3. Henkin, R. I. 1969. On the role of unmyelinated nerve fibers in the taste process. *In* Oral Sensation and Perception II (J. F. Bosma, editor). Charles C Thomas, Springfield, Ill. (In press)
4. Henkin, R. I. 1969. The neuroendocrine basis of sensation. *In* Oral Sensation and Perception II (J. F. Bosma, editor). Charles C Thomas, Springfield, Ill. (In press)
5. Henkin, R. I., and F. C. Bartter. 1966. Studies on olfactory thresholds in normal man and in patients with adrenal cortical insufficiency: the role of adrenal cortical steroids and of serum sodium concentration. *J. Clin. Invest.* **45**. pp. 1631–1639.
6. Henkin, R. I., and D. F. Bradley. 1969. Regulation of taste acuity by thiols and metal ions. *Proc. Nat. Acad. Sci. U.S.* **62**, pp. 30–37.
7. Henkin, R. I., A. G. T. Casper, R. Brown, A. B. Harlan, and F. C. Bartter. 1968. Presence of corticosterone and cortisol in the central and peripheral nervous system of the cat. *Endocrinology* **82**, pp. 1058–1061.
8. Henkin, R. I., and R. L. Daly. 1968. Auditory detection and perception in normal man and in patients with adrenal cortical insufficiency: the role of adrenal cortical steroids. *J. Clin. Invest.* **47**, pp. 1269–1280.
9. Henkin, R. I., J. R. Gill, Jr., and F. C. Bartter. 1963. Studies on taste thresholds in normal man and in patients with adrenal cortical insufficiency: the role of adrenal cortical steroids and of serum sodium concentration. *J. Clin. Invest.* **42**, pp. 727–735.
10. Henkin, R. I., J. R. Gill, Jr., J. R. Warmolts, A. A. Carr, and F. C. Bartter. 1963. Steroid-dependent increase of nerve conduction velocity in adrenal insufficiency. *J. Clin. Invest.* **42**, p. 941.
11. Henkin, R. I., H. R. Keiser, I. R. Jaffe, I. Sternlieb, and I. H. Scheinberg. 1967. Decreased taste sensitivity after D-penicillamine reversed by copper administration. *Lancet II* pp. 1268–1271.
12. Henkin, R. I., H. R. Keiser, and M. R. Kare. 1968. The effects of D-penicillamine (D-pen) and of copper repletion on salt preference and fluid intake. *Fed. Proc.* **27**, p. 583.
13. Henkin, R. I., R. E. McGlone, R. Daly, and F. C. Bartter. 1967. Studies on auditory thresholds in normal man and in patients with adrenal cortical insufficiency: the role of adrenal cortical steroids. *J. Clin. Invest.* **46**, pp. 429–435.
14. Henkin, R. I., M. D. Walker, A. B. Harlan, and A. G. T. Casper. 1967. Dynamics of transport of cortisol from peripheral blood into cerebrospinal fluid (CSF), central and peripheral nervous system and other tissues of the cat. *Progr. Endocrinol. Soc.* p. 92.
15. Hertz, M., and P. Thygesen. 1948. Nervous complications in diphtheria. *Acta Med. Scand.* **206**, pp. 541–546.
16. Keiser, H. R., R. I. Henkin, F. C. Bartter, and A. Sjoerdsma. 1968. Loss of taste during therapy with penicillamine. *J. Amer. Med. Assoc.* **203**, pp. 381–383.
17. Ojemann, G. A., and R. I. Henkin. 1967. Steroid dependent changes in human visual evoked potentials. *Life Sci.* **6**, pp. 327–334.
18. Richter, C. P. 1936. Increased salt appetite in adrenalectomized rats. *Amer. J. Physiol.* **115**, pp. 155–161.
19. Robbins, N. 1969. Are taste bud nerve endings the site of gustatory transduction? *In* Oral Sensation and Perception II (J. F. Bosma, editor). Charles C Thomas, Springfield, Ill. (In press)
20. Vellis, J. de, and D. Inglish. 1968. Hormonal control of glycerolphosphate dehydrogenase in the rat brain. *J. Neurochem.* **15**, pp. 1061–1070.

DIGESTIVE FUNCTIONS OF TASTE STIMULI

MORLEY R. KARE
Monell Chemical Senses Center, University of Pennsylvania, Philadelphia, Pa.

Taste stimuli are usually studied in terms of the sensation and correlates of neural activity they evoke. However, there are other physiological actions from chemical stimuli that could relate to the taste response of the organism. To illustrate this point, two examples will be reported. The first is the influence of taste stimuli on a digestive secretion; the second, a possible extraneural pathway for gustatory information.

Pavlov[12] demonstrated that sham feeding in dogs resulted in a transient increase in the rate of pancreatic secretion. No cephalic phase of gastric secretion occurred when the dog chewed "neutral" tasting substances, whereas appealing foods resulted in a copious gastric secretion. Crittenden and Ivy[4] reported an increased rate of pancreatic secretion in enterectomized dogs when they were fed meat broth, but water or milk were without effect.

The role of taste stimuli on other parts of the digestive tract is not totally unexplored. For example, a strong taste stimulant will reduce gastric contractions but increase intestinal motility. Of all phases of digestion the relation of taste to salivary flow has received the most consideration. It is common knowledge that the nature of the taste stimulus can modify the volume and character of saliva.

A series of experiments were designed to consider how taste would affect gastrointestinal function. The procedure is reported in detail by Behrman and Kare.[3] In this experiment, interest was centered on the effect of palatable and aversive taste stimuli on pancreatic secretion.

Dogs were fitted with gastric and intestinal cannulae (Figure 1). The intestinal cannula permitted intubation of the major pancreatic duct (Figure 2) according to the method outlined by Thomas.[15] This preparation permitted collection of pancreatic juice from conscious animals uncontaminated with intestinal contents.

The animals were suspended in a canvas sling during the collection of pancreatic juice; they passively accepted this physical restriction after several preliminary trials. Basal secretion was measured for 15 min just prior to test. A second collection was then made for 15 min following the first oral stimulus. The food stimuli were administered as suspensions at 3-min intervals. A preliminary study, using a water-dye mixture, showed that this amount and frequency of presentation of stimuli allowed complete drainage from the gastric cannula with no overflow into the duodenum.

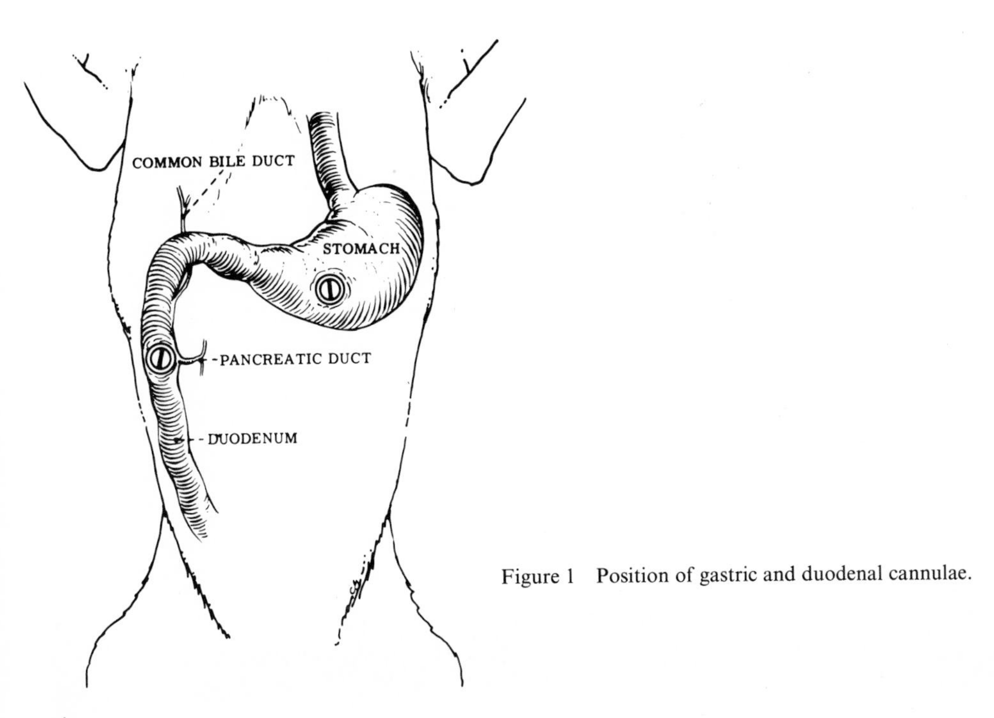

Figure 1 Position of gastric and duodenal cannulae.

Gastric juice will not induce a secretin response if it does not enter the duodenum. Therefore the pancreatic responses were attributed to oral stimulation.

The taste stimuli were 10 g of ground basal diet in 30 ml of distilled water, 10 per cent sucrose, 0.001 M quinine hydrochloride, or 0.1 M citric acid. Some preliminary experiments indicated that mixing a portion of basal diet with water would markedly stimulate pancreatic flow. It was felt that the stimuli could be better compared under the conditions of high flow occurring when some basal diet was included than under the probable low flow expected if the solutions were presented alone. Further, in-

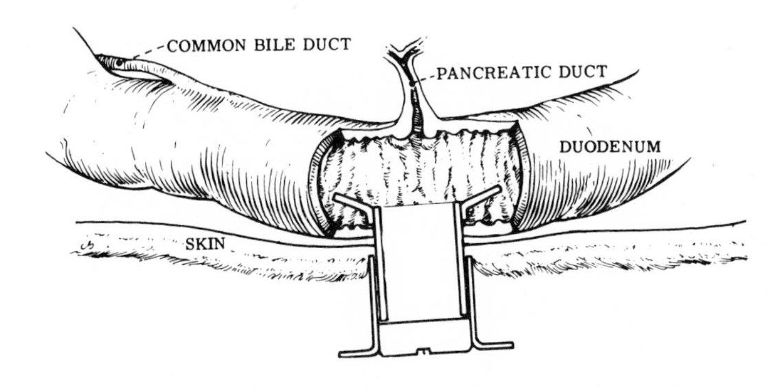

Figure 2 Position of intestinal cannulae in relation to pancreatic duct.

clusion of the basal diet would render the test conditions more "natural." Responses to treatment were computed as ratios, test volume/basal volume and test protein secretion/basal protein secretion.

Treatment effects on flow indicate that water and sucrose yielded similar responses that were significantly different from those for citric acid and quinine; the latter two in turn, were, similar. The average flow responses for water and sucrose were 12.1 and 10.9, respectively, whereas those for citric acid and quinine were 4.2 and 5.4. During the tests, the animals eagerly accepted the water and sucrose preparations but were reluctant to ingest the citric acid and quinine preparations.

The treatment effects on protein response were not parallel to those for flow. Sucrose and water tended to give a higher response than did citric acid and quinine, but the water response was considerably lower than for sucrose. Responses for water, citric acid, and quinine were 6.2, 2.6, and 5.5, respectively, whereas that for glucose was 10.4.

The initial event in pancreatic secretion excited by oral stimulation has been reported to be neural, operating via the vagal cholinergic fibers.[7] Grossman[9] speculated that the vagal effects may be due to acetylcholine release within the pancreatic tissue. Recent evidence indicates the existence of a gastric phase of pancreatic secretion that is mediated by gastrin.[8] Stimulation of gastrin secretion by vagal activity has been reported to occur in the dog during sham feeding,[13] and subsequent evidence indicates that a part of the pancreatic response to sham feeding arises as a result of gastrin release from the pyloric gland area of the stomach. In the present study, the food stimuli were drained by a gastric cannula and did not enter the duodenum. Preshaw et al.,[14] have reported that gastric juice will not induce a secretin response if it does not enter the duodenum. This is supported by the work of Crittenden and Ivy,[4] who reported that they were unable to stimulate pancreatic secretion by introducing food into the stomach. Thus, it appears safe to assume that the observed pancreatic responses can be attributed to oral stimulants.

This first example demonstrates that, in addition to evoking sensations and neural activity, specific taste stimuli can affect the volume of pancreatic flow and its enzyme (protein) content. It follows that taste stimuli may have additional influences on digestive functions. It is reasonable to consider that modified digestive function could, in turn, influence taste thresholds or preference behavior.

Our second experiment might be titled "a direct pathway to the brain." Usually a description of the function of the initial portion of the digestive tract is limited to that of a site for the preparation of nutrients for deglutition and eventual absorption in the intestine. The nature of food ingested has been implicated in the regulation of digestion. Several investigators have suggested some sort of metering in the oropharyngeal cavity.[1,5,6] Recently, it was reported that receptors for thiamine exist in the oral cavity.[2] The question can be raised as to whether this regulation is related solely to neural activity arising from the stimulation of taste receptors. Classically, "the oral area" has not been described as a zone for absorption of either nutrients or drugs. However, there are a number of indirect indicators of oral absorption with rapid movement to the brain, e.g., the toxic action of cyanide.

Experiments were designed to determine if nutrients can pass directly from the mouth to the brain, by-passing the gut. Rats were anesthetized and had their esophagus and trachea securely ligated medially to the submaxillary gland. The trachea was cannulated to permit breathing. A solution of uniformly labeled glucose-^{14}C or $^{24}NaCl$ was pipetted into the mouths of the rats, which were mounted in a normal upright position by a head holder. Control animals received similar quantities of isotope introduced into the stomach, ligated at the juncture of the duodenum, or into the initial portion of the duodenum. Cardiac punctures for blood samples were made just prior to the termination of the isotope placement period. The mouth was briefly rinsed with distilled water to remove the isotope solution. To assay activity, the animal was decapitated and the brain rapidly removed. A variety of isotopes and a series of times were studied. Table I illustrates typical results obtained. Details on the procedure are available in a report by Maller et al.[11]

Apparently, significant quantities of glucose and salt pass rapidly from the ligated oropharyngeal cavity to the brain. It was deemed desirable to use whole body radioautographic techniques to establish the pathway from the mouth to the brain.

For radioautograms, the whole animal was frozen by immersion in liquid nitrogen at the termination of the oral placement period and was then mounted on a sectioning block. Forty-micron sections were dried, then placed in contact with photographic paper for approximately six weeks. The entire procedure was carried out in a cold room at minus 9 $\pm$ 1 degrees C. Details on this experiment are available in a paper by Kare et al.[10]

Figure 3 is a radiogram of a control animal after duodenal administration of the

TABLE I

RADIOACTIVITY IN TISSUES AFTER ORAL AND DUODENAL PLACEMENT OF LABELED GLUCOSE-^{14}C dps/g DRY WEIGHT*

Placement	Brain	Liver	Blood
	(N)	(N)	(N)
Glucose			
Oral, 9–14 min	136 (3)	15 (3)	0 (3)
Oral, 4–5 min	32 (3)	11 (2)	0 (2)
Gut, 1 min	0 (2)	134 (2)	72 (2)
Control	0 (2)	0 (2)	0 (2)

* Normalized to correspond to a 400 g rat body weight and a 5μc per 0.25 ml application. Adjustments for background have been incorporated. Corrections were made for self-absorption.

isotope. A photograph of the same tissue section is above it. The distribution of activity throughout the body is apparent.

Figure 4 illustrates the massive activity in the head region after placing labeled glucose in the mouth. Above the radiograph is a photograph of the actual section from which it was made. It will be noted that there is no activity beyond the ligature, but it is present throughout the face region and clearly evident in the brain. The distribution

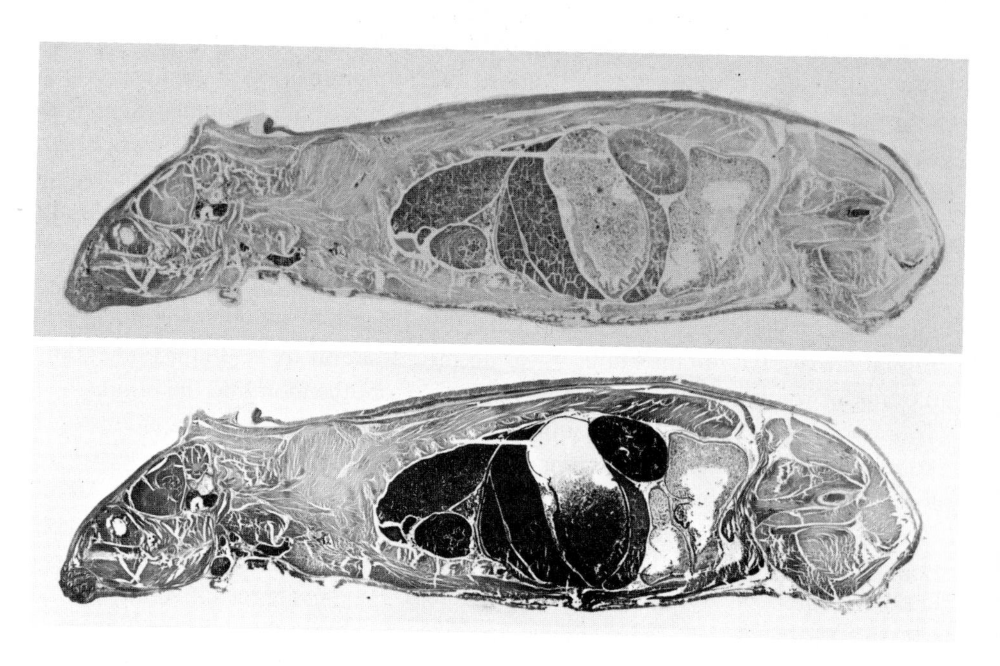

Figure 3 Glucose-^{14}C, duodenal administration, 4 min. Top: tissue section (unstained). Bottom: radioautogram of section.

of activity in the perfused animals was indistinguishable from the results with animals which were not perfused. The activity in the cardiac blood samples of the experimental animals was so low as to eliminate the possibility that the activity in the brain was primarily blood-borne.

The absence of detectable activity in the blood samples serves to confirm that circulation is not the critical avenue for movement. While the control suggests that the isotope absorbed through the intestinal mucosa traveled to the brain by way of the blood, the evidence supports a noncirculatory direct route from the oral mucosa to the brain in the experimental animals. The results in the perfused animals establish that the activity in the brain is within the cellular tissue rather than a reflection of activity in the blood.

Approximately half of the animals that received the orally administered labeled glucose provided radiograms similar to those in Figure 4. The results in other animals

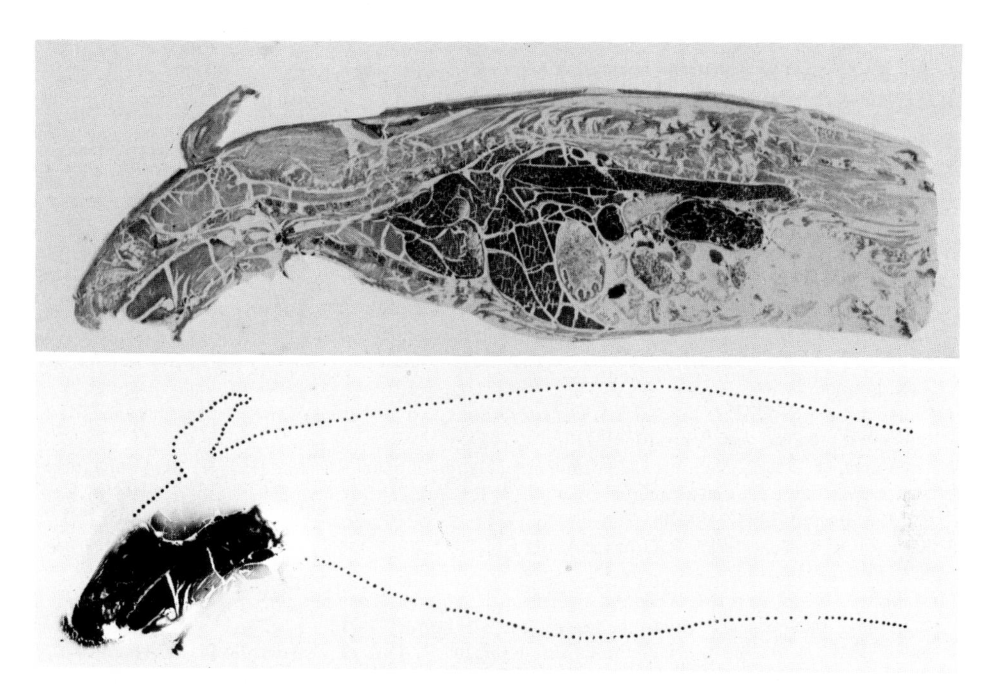

Figure 4 Glucose-^{14}C, oropharyngeal administration, 4 min, esophagus ligated. Top: tissue section (unstained). Bottom: radioautogram of section.

varied considerably, even to the extreme case in which the isotope did not appear to leave the oral cavity. The failure of the isotope to move could perhaps be related to technique, the character of the saliva, the animal, or other factors. It is tempting to attach physiological significance, such as that related to food-intake metering, to this individual variability.

The results suggest maximum concentration of isotope in the direct line of diffusion with a reduction in the area of the brain farthest from the wave of flow. The animal used for Figure 4 received the isotope for a 4 min interval. In preliminary trials where activity was measured in an entire brain, detectable activity was recorded in animals in which the isotope was applied for less than 30 sec.

At least some taste stimuli can move fairly rapidly from the oropharyngeal cavity directly to the brain. It is conceivable that these materials could contribute gustatory and other information to complement that obtained neurally from the taste receptors.

Taste stimuli apparently can have a specific and substantial influence on digestive secretions. Activity from orally placed labeled glucose was demonstrated to move rapidly directly to the brain. Physiological action of oral stimuli could be involved in the gustatory response to these stimuli. Further gustatory information obtained from extraneural pathways could be of consequence to the total sensation, function, and preference behavior.

REFERENCES

1. Adolph, E. F. 1950. Thirst and its inhibition in the stomach. *Amer. J. Physiol.* **161**, pp. 374–386.
2. Appledorf, H., and S. R. Tannenbaum. 1967. Specific hunger for thiamine in the rat: selection of low concentrations of thiamine in solution. *J. Nutr.* **92**, pp. 267–273.
3. Behrman, H. R., and M. R. Kare. 1968. Canine pancreatic secretion in response to acceptable and aversive taste stimuli. *Proc. Soc. Exp. Biol. Med.* **129**, pp. 343–346.
4. Crittenden, P. J., and A. C. Ivy. 1937. The nervous control of pancreatic secretion in the dog. *Amer. J. Physiol.* **119**, pp. 724–733.
5. Denton, D. A. 1965. Evolutionary aspects of the emergence of aldosterone secretion and salt appetite. *Physiol. Rev.* **45**, pp. 245–295.
6. Fregly, M. J., J. M. Harper, Jr., and E. P. Radford, Jr. 1965. Regulation of sodium chloride intake by rats. *Amer. J. Physiol.* **209**, pp. 287–292.
7. Gregory, R. A. 1962. Secretory Mechanisms of the Gastrointestinal Tract. Edward Arnold Publishers, Ltd., London.
8. Gregory, R. A., and H. J. Tracy. 1964. The constitution and properties of two gastrins extracted from hog antral mucosa. *Gut* **5**, pp. 103–117.
9. Grossman, M. I. 1961. Nervous and hormonal regulation of pancreatic secretion. *In* Ciba Foundation Symposium on the Exocrine Pancreas (A. V. S. De Reuck and M. P. Cameron, editors). Little, Brown and Co., Boston, Mass., pp. 208–224.
10. Kare, M. R., P. Schecter, S. P. Grossman, and L. J. Roth. 1969. Direct pathway to the brain. *Science* **163**, pp. 952–953.
11. Maller, O., M. R. Kare, M. Welt, and H. Behrman. 1967. Movement of glucose and sodium chloride from the oropharyngeal cavity to the brain. *Nature* **213**, pp. 713–714.
12. Pavlov, I. P. 1910. The Work of the Digestive Glands, 2nd ed. (Translated by W. H. Thompson). Charles Griffin and Co., Ltd., London.
13. Pethein, M., and B. Schofield. 1959. Release of gastin from the pyloric antrum following vagal stimulation by sham feeding in dogs. *J. Physiol.* (*London*) **148**, pp. 291–305.
14. Preshaw, R. M., A. R. Cooke, and M. I. Grossman. 1966. Sham feeding and pancreatic secretion in the dog. *Gastroenterology* **50**, pp. 171–178.
15. Thomas, J. E. 1950. The External Secretion of the Pancreas. Charles C Thomas, Springfield, Ill.

ANTIDIURESIS ASSOCIATED WITH THE INGESTION OF FOOD SUBSTANCES

JAN W. KAKOLEWSKI *and* ELLIOT S. VALENSTEIN
Fels Research Institute, Yellow Springs, Ohio

INTRODUCTION

A number of lines of investigation have pointed to the biological significance of sensory information accompanying ingestion. Sensory feedback originating in the oropharyngeal cavity may initiate neural, endocrine, and metabolic changes which make it possible for an organism to prepare for the forthcoming absorption. These stimuli also make it possible to evaluate the ingesta while it is still in the mouth and therefore they contribute to the detection and discrimination of potential food substances and the motivation to eat.[1] The adaptive significance of such mechanisms is easy to appreciate.

We have recently completed a series of experiments which emphasize the role of oropharyngeal sensations in the regulation of water balance. Animals kept in a state of constant water diuresis exhibited an antidiuresis that started almost immediately with the initiation of food ingestion.[4] The rapidity with which this "short-latency antidiuresis" occurred suggested that the process was initiated during the initial stages of food ingestion and therefore might be attributed to an oral or possibly a gastric signaling factor. This conclusion was supported by the work of Nicolaïdis,[5] who had reported that in acute preparations an antidiuresis could be evoked by application of 5 per cent NaCl to a rat's tongue. Stacy and Brook[6] reported an overall antidiuresis associated with feeding in sheep and concluded that the antidiuresis was due to secretion of antidiuretic hormone (ADH) in response to the postprandial effects of ingestion.

In the experiments reported here we investigated the antidiuresis associated with food ingestion in the rat. Our primary interests were (1) the development of a biological model for the study of antidiuresis in a chronic preparation; (2) to determine the relevance of oral factors in this process; (3) to provide preliminary data on the effectiveness of different food substances; and (4) to provide preliminary information on the contribution of gastric and other postingestional factors to the antidiuretic response. For these purposes, we emphasized the elapsed time (latency) from the initiation of food ingestion to the occurrence of antidiuresis, the persistence of the antidiuresis, and the effect of bypassing the oral cavity by gastric intubation. The effects of several different food substances were evaluated.

METHODS

In all experiments we utilized food-deprived rats that overhydrated themselves by consuming large quantities of a highly palatable mixture of 0.125 g of saccharin and 3.0 g glucose (S + G) dissolved in 100 ml of distilled water.[7] This technique avoids the complication of a possible antidiuretic hormone release associated with the stress imposed by the procedures generally used in acute preparations. During a 24 hr period, food-deprived rats ingest the S + G solution in amounts equal to 90 per cent of body weight.

Initially, animals were provided with Purina Lab Chow pellets ad libitum and the highly palatable solution of S + G. After four days of exposure to food and S + G solution, the rats were deprived of food for four days. Daily measurements of fluid consumption were obtained. After the fourth day of total food deprivation, the subjects were divided into two groups matched for consumption of the S + G solution during the preceding four days. The S + G solution was removed from the cages of all the animals and the experimental (food) group of 16 animals was provided with Purina Lab Chow pellets. The 13 control (no food) animals received no food.

Measurements of volume and concentration of total solids in each urine sample were then taken from both groups during the three hours after food was presented to the experimental animals. During this period no fluid was available to the animals. Each urine sample was collected on a cardboard covered with "Saran Wrap" placed under the cages, and the time of occurrence of each urine sample was recorded. The urine concentration was measured with a TS* Meter (American Optical Company) refractometer to the nearest 0.1 per cent total solids, and volume was determined with a 2 ml syringe to the nearest 0.05 ml. In addition to obtaining the total solid percentages with the refractometer, the osmolality of the urine was obtained from a number of specimens by means of an osmometer (Advanced Instruments, Inc.).

During the four days of food deprivation, the mean daily fluid consumption of the S + G solution for the experimental (food) and control (no food) animals was 245 ml. Consequently, prior to the beginning of urine measurements the animals in both groups consumed equally large volumes of fluid and were in a constant state of water diuresis, resulting in frequent voiding of a very dilute urine. Immediately after the presentation of Purina Lab Chow to the experimental animals, the volume of urine production decreased, as can be seen by comparison with the unfed control animals (Figure 1). The differences in volume persisted for about two hours and the total volume of urine excretion was significantly lower ($p < 0.01$, Wilcoxon test) for the fed group for the entire 3 hr observation period. Differences in urine concentration were observed beginning about 45 min after the food was presented, and persisted throughout the remaining period of observation. The occurrence of an over-all antidiuresis associated with food ingestion is obvious, but our primary interest was in the short latency of the antidiuretic response (as judged by the urine volume) observed in the experimental group. This effect, which appears at maximum strength

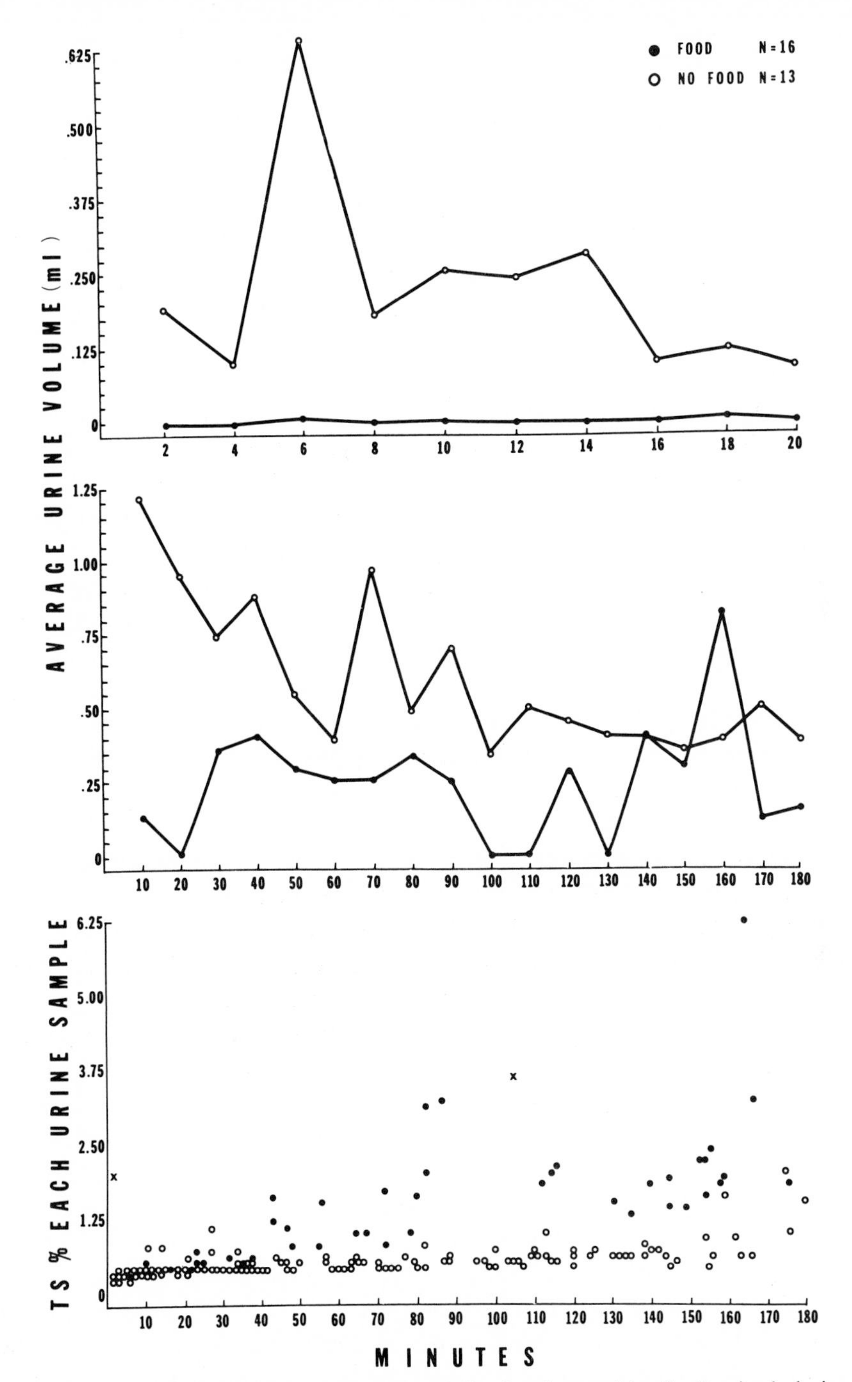

Figure 1 Average urine volume of experimental (food) and control (no food) animals during the initial 20 min (top) after presentation of food to the experimental animals and during the entire 180 min observation period (middle). The bottom graph depicts the percentage of total solids in each urine sample. The X on the bottom graph indicates the results from the one control animal that had consumed the least amount of the solution during the food deprivation period. ●, experimental animals (N = 16) that received food; ○, control animals (N = 13) with no food.

almost immediately after the introduction of food, cannot be adequately accounted for either by the postingestional absorption of food or the withdrawal of fluids into the stomach, as a gradual, rather than abrupt, onset of antidiuresis would have been expected. This latter argument would apply as well to the withdrawal of fluids in the mouth as a function of salivation.

In terms of evaluating this technique for studying antidiuresis there will be some interest in additional quantitative data on the urine excreted by the experimental and control animals. During the 3 hr observation, the average control animal excreted 10.7 ml of urine while the average experimental animal excreted 4.5 ml of urine. Table I summarizes the average urine volume and the per cent of total solids (TS) excreted per animal during the first 90 min of observation.

TABLE I

AVERAGE VOLUME AND TS% OF URINE OF EXPERIMENTAL (FOOD) AND CONTROL (NO FOOD) ANIMALS

	TIME IN MINUTES								
	0–10	10–20	20–30	30–40	40–50	50–60	60–70	70–80	80–90
Control:									
Volume	1.2	0.9	0.7	0.9	0.5	0.4	1.0	0.5	0.7
TS%	0.3	0.4	0.4	0.5	0.5	0.5	0.5	0.5	0.6
Experimental:									
Volume	0.1	0.1	0.4	0.4	0.3	0.3	0.3	0.3	0.3
TS%	0.4	0.4	0.5	0.6	1.2	1.2	1.0	1.3	2.8

In a separate experiment to evaluate the influence of smell we tested seven animals, which were treated identically to the present experimental animals except that the food was attached to the outside of the cage. No antidiuresis was observed. As animals in this group were very active in trying to reach the food, we have concluded that the "short-latency antidiuresis" observed in the present study is dependent upon actual ingestion of food, not smell, and does not result from any difference in activity between the groups. On the other hand, animals tested with the same food (Purina Lab Chow) but of different texture (pellets or powder) exhibited an identical short-latency antidiuresis.

In a subsequent study, using identically overhydrated animals, we examined the effect of a number of different diets. Table II summarizes the time course of the antidiuresis resulting from the ingestion of various substances. The substances used were limited by what the animal would be willing to ingest. Judging by the volume excreted, it can be seen that the addition of NaCl to the powdered Purina Lab Chow could not reveal any change in the time course of the antidiuresis, as the response to the Purina Lab Chow alone was almost immediate. The measures of urine concentration, however, did reveal a difference. Normally, Purina Lab Chow causes an

TABLE II
THE DEVELOPMENT OF ANTIDIURESIS FOLLOWING THE INGESTION OF VARIOUS FOOD SUBSTANCES

		TIME IN MINUTES											
Condition	N	0–10	10–20	20–30	30–40	40–50	50–60	60–70	70–80	80–90	90–100	100–110	110–120
Original Controls (no food)	13	—	—	—	—	—	—	—	—	—	—	—	—
Original Experimentals (Purina Pellets)	16	+	+	+	+	+	+	+	+	+	+	+	+
Purina Powder	10	+	+	+	+	+	+	+	+	+	+	+	+
Wesson Oil + Alphacel (1:2)	8	—	—	+	+	+	+	+	+	+	+	+	+
Purina + 5 or 10% NaCl*	3	+	+	+	+	+	+	+	+	+	+	+	+
Purina + 50% Glucose	4	—	+	+	+	+	+	+	+	+	+	+	+
Cane Sugar	4	—	—	—	+	+	+	+	+	+	+	+	+
Cane Sugar + 1% NaCl	2	—	—	+	+	+	+	+	+	+	+	+	+
Cane Sugar + 3% NaCl	2	—	—	+	+	+	+	+	+	+	+	+	+
Cane Sugar + 5% NaCl	2	—	—	+	+	+	+	+	+	+	+	+	+

— Water diuresis.
+ Antidiuresis (decrease in urine volume and/or increase in TS%).
* The time course of the antidiuresis in the case of animals receiving Purina Chow plus NaCl was identical to that of animals receiving only Purina Chow as indicated by the decrease in urine production, but the TS% and the osmolality increased at an earlier time (cf. text).

increase of TS% of the urine at about 45 min and an increase in osmolality at about 30 min from the initiation of eating. If NaCl is added to this food, the TS% starts to increase at about 25 min and the osmolality at about 10 to 15 min from the initiation of eating. These results were obtained although adding either 5 or 10 per cent of NaCl to the Purina Lab Chow decreases the palatability and, as a result, the animals eat less. The ingestion of all other diets shown in Table II produced an over-all antidiuresis; however, the latency of this effect was relatively long. Of special interest is the long latency observed following the ingestion of cane sugar, suggesting that the over-all antidiuresis might be due to postprandial factors only. It is possible that the failure of cane sugar to trigger an antidiuresis during the prandial phase of ingestion might be related to the fact that these animals were overhydrated with a sweet-tasting solution that may have interfered with the effectiveness of this stimulus. On the other hand, the sugar might actively interfere with the triggering of the short-latency antidiuresis at an oral level. Supporting this latter hypothesis is the suggestion of an increase in the latency of the antidiuresis obtained when glucose was added to the Purina powder. The authors feel that these results should be replicated with a large group of subjects. The data included in Table II also argue against the significance of osmolality as a trigger of antidiuresis at the oral level, because, of the substances tested, cane sugar has the highest osmotic potential but a long antidiuretic latency. Similarly, the significance of salivation during food ingestion appears to be minimal, as different dry diets (cane sugar vs. Purina Lab Chow) produce different latencies.

The last series of experiments was designed to investigate the development of antidiuresis in response to the introduction of different foods into the stomach by means of a gastric tube, thereby bypassing the oral cavity. To accomplish this, it was necessary to determine the effect of the various procedures related to gastric loading on antidiuresis. We employed a modification of our initial procedure in which animals were deprived of food for three days. Twenty-four hours after the removal of food, but not water, they were transferred from their home cages to a metabolic cage, where they were provided with a 20 min opportunity to drink the highly palatable saccharin-glucose solution employed in the initial experiment. The animals were then returned to their home cages and from then on deprived of food and water. On the third day they were transferred once again to the metabolic cages and offered the sweet solution to drink for 20 minutes. Once again the animals were returned to their home cages; on the fourth day (the test day) they were exposed to the saccharin-glucose solution for 20 min and urine volume and TS% were measured from the moment the animals were first placed in the metabolic cages. On such a regimen the animals drank an average of 17 ml during the 20 min on the test day. One half-hour after the initiation of drinking, a water diuresis developed that was characterized by a high volume and low concentration of urine, which persisted for the 90-min observation period. During these 90 min the average urine output of the 10 control animals was 9.7 ml, and the average TS% ranged from 0.1 to 0.9. (For purposes of comparison the TS% of the urine of animals maintained ad libitum was 18.5.) In successive 10 min periods the average volume excreted per control animal was 1.21, 1.29, 1.29, 1.39, 1.11, 0.9, 1.0, 0.9, 0.6. In addition to the 10 control animals, other groups were used to control for the procedure of intubation under ether anesthesia. Light ether anesthesia was employed because it was determined that it reduced the variability in the response elicited by the handling and insertion of the catheter during the intubation procedure.

It can be seen from Table III that the procedure of intubation under ether anesthesia produced an antidiuresis that lasted approximately 25 min after which the animals returned to the diuretic state. If the animals were permitted to ingest Purina Lab Chow when the diuresis was reinstated, an antidiuresis was produced almost immediately. Due to the antidiuretic effect produced by the intubation procedure, the results obtained with the stomach loads could not provide information about the role of the stomach in the short-latency antidiuresis. It was possible, however, to determine the influence of the stomach in triggering an over-all antidiuresis. The results summarized in Table III suggest that stimuli originating in the stomach are not very effective in producing an over-all antidiuresis. For purposes of comparison, data illustrating the effect of ingesting powdered Purina Lab Chow and the administration of 40 micro-units (μU) of Pitressin are also provided in the table. All food substances intubed, except those with a very high NaCl content, failed to produce an antidiuresis. Furthermore, the fact that the 30 per cent glucose solution failed to pro-

Table III
THE DEVELOPMENT OF ANTIDIURESIS FOLLOWING GASTRIC INTUBATION OF VARIOUS FOOD SUBSTANCES

	N	TIME IN MINUTES 0–10	10–20	20–30	30–40	40–50	50–60	60–70	70–80	80–90
					Control for Procedures					
Control (handling)	10	—	—	—	—	—	—	—	—	—
Stomach intubation under ether anesthesia	8	+	+	±	—	—	—	—	—	—
ADH 40 μU	2	+	+	+	±	—	—	—	—	—
Purina Chow powder (oral)	6	+	+	+	+	+	+	+	+	+
					Gastric Loads under Ether Anesthesia					
Inert bulk (5 ml): Alphacel + Mineral Oil 1.1:1	4	+	+	+	—	—	—	—	—	—
5% NaCl in inert bulk (5 ml)	2	+	+	+	+	+	+	+	+	OD
Wesson Oil + Alphacel 1:1 (5 ml)	4	+	+	—	—	—	—	—	—	—
30% Glucose (5 ml)	2	+	+	+	—	—	—	—	—	—
3% NaCl (2 ml)	2	+	+	—	—	—	—	—	—	—
5% NaCl (2 ml)	1	+	+	—	—	—	—	—	—	—
5% NaCl (4 ml)	2	+	+	—	—	—	+	+	+	OD

— Water diuresis.
+ Antidiuresis (decrease in volume, increase in total solids and osmolality).
OD Osmotic diuresis.

duce an antidiuresis while the 5 per cent NaCl did, indicates that the osmolality of the gastric content is not critical for triggering an over-all antidiuresis.

It was seen (Table II) that the ingestion of a Wesson Oil diet produced an over-all antidiuresis. In contrast, when Wesson Oil is introduced directly into the stomach the antidiuresis only lasts as long as that produced by the intubation procedure alone. We feel it is necessary to be cautious about the results because of the limited number of animals used in each condition; however, if these results are confirmed, it would appear that the over-all, as well as the short-latency antidiuresis, was dependent upon the initial passage of food through the oral cavity. Although Wesson Oil and the 30 per cent glucose solution would be expected to draw water into the stomach,[2,3] we did not observe an over-all antidiuresis within the 90 min observation period, when these substances were intubed directly into the stomach. In this experiment, the substances that caused an over-all antidiuresis when introduced directly into the stomach (inert bulk plus 5 per cent NaCl and 5 per cent NaCl solution), are generally aversive and, after an initial sampling, animals refuse to ingest them even when hungry or thirsty. Therefore, at this point it would seem that, under normal conditions, animals respond to acceptable foods with an antidiuresis initiated during the prandial phase of ingestion.

GENERAL CONCLUSIONS

The present series of experiments describe a model for studying the development of antidiuresis in a chronic preparation. The results with this technique permit the following conclusions:

(1) A short-latency antidiuresis can be triggered from oral stimulation by certain food substances. This antidiuresis does not appear to be related to salivation. The triggering of the short-latency response requires the direct stimulation of the oral cavity, as odors of effective foods do not constitute an adequate stimulus. The response does not appear to be dependent upon the texture or potential osmolality of the foodstuff. The ingestion of fats in the form of Wesson Oil and sugar (cane sugar) does not trigger the short-latency antidiuresis.

(2) In preliminary experiments, which bypassed the oral cavity by utilizing gastric loads, the results suggested that stimuli originating in the stomach may not play a significant role in the triggering of either short-latency or an over-all antidiuresis. Substances that were effective in eliciting an over-all antidiuresis when ingested were not effective when introduced directly into the stomach. Of the substances tested, only concentrations of salt above those the animals would ingest voluntarily were found to trigger a persistent antidiuresis when intubed directly into the stomach.

ACKNOWLEDGMENT

The work reported in this paper was supported by National Institute of Mental Health Research Grant M-4529, Research Scientist Award MH-4947, and National Aeronautic and Space Administration Research Grant NSG-437.

REFERENCES

1. Epstein, A. N. 1967. Feeding without oropharyngeal sensations. *In* The Chemical Senses and Nutrition (M. R. Kare and O. Maller, editors). The Johns Hopkins Press, Baltimore, Md., pp. 263–280.
2. Fenton, P. F. 1945. Response of the gastrointestinal tract to ingested glucose solutions. *Amer. J. Physiol.* **144**, pp. 609–619.
3. Harper, A. E., and H. E. Spivey. 1958. Relationship between food intake and osmotic effect of dietary carbohydrate. *Amer. J. Physiol.* **193**, pp. 483–487.
4. Kakolewski, J. W., V. C. Cox, and E. S. Valenstein. 1968. Short-latency antidiuresis following the initiation of food ingestion. *Science* **162**, pp. 458–460.
5. Nicolaïdis, S. 1963. Effets sur la diurèse de la stimulation des afférences buccales et gastriques par l'eau et les solutions salines. *J. Physiol.* (*Paris*) **55**, pp. 309–310.
6. Stacy, B. D., and A. H. Brook. 1965. Antidiuretic hormone activity in sheep after feeding. *Quart. J. Exp. Physiol. Cog. Med. Sci.* **50**, pp. 65–78.
7. Valenstein, E. S., V. C. Cox, and J. W. Kakolewski. 1967. Polydipsia elicited by the synergistic action of a saccharin and glucose solution. *Science* **157**, pp. 552–554.

DEPRESSION OF INTAKE BY SACCHARIN ADULTERATION: THE ROLE OF TASTE QUALITY IN THE CONTROL OF FOOD INTAKE

ROBERT L. GENTILE
Pioneering Research Laboratory, U.S. Army Natick Laboratories, and Department of Biology, Clark University, Worcester, Massachusetts

INTRODUCTION

Research on the role of taste input in the control of food intake has been concerned primarily with the role of hedonic value, or palatability. With the exception of research on specific hungers, investigators have tended to ignore the possible effects of specific taste qualities. Generally, therefore, stimuli have been treated as roughly interchangeable. The assumption has been that laws concerning the relative effects on intake of any two stimuli of known relative preference are generalizable to any other two stimuli of known relative preference. The only exception to this—which is not really an exception—is the recognition that positive and negative taste stimuli might require separate consideration.

The aim of the present investigation was to show the effects of taste quality—independent of preference (as measured in a two-choice test)—on food intake. More specifically, it was to examine a commonly made assumption that the more palatable or preferred the food stimulus the greater the intake—if taste has any effect at all. Classically, one has expected agreement between two-choice (preference) test results and one-choice test results. If an animal prefers one diet to another, its intakes in the single-choice test are expected to be in the same direction (i.e., greater intake of the preferred diet). When this agreement has not been found the animal has been said to be, in some sense, ignoring taste.[1,3,4] Such differences are usually attributable to the greater role of post-ingestional effects in one-choice test behavior. For example, the failure of animals to overeat preferred glucose has been interpreted as evidence for some kind of caloric regulation.[7,8]

The underlying assumption leading to this interpretation of conflicting one-choice and two-choice data seems to be that the hedonic response as measured by a two-choice preference test is *the* response to the taste stimuli being tested. The possibility that two-choice tests and one-choice tests might sample different kinds of taste-input effects has, therefore, been overlooked. It has even been suggested that the *true* taste effect is that reflected by brief exposure data.[10] Possible changes in taste effects following the initial reaction have, therefore, been all but ignored.

The fact that taste effects may change over time (e.g., between the beginning and the end of a meal) need not be of much importance if such changes are consistent with the initial effects. Suppose, for example, that eating rate is used as a measure of a facilitatory taste effect (or hedonic value).[11] One would expect that the rate of eating would decrease in the course of a meal, but as long as the *ratio* of eating rates of test-diets remains constant we may safely ignore absolute changes in eating rate caused by ingestion. One would not be justified in ignoring such changes, however, if they are not consistent with initial eating rates. Suppose one tests two metabolically equivalent diets—X and Y—in one-choice tests. If diet X is initially eaten more quickly than diet Y but, subsequently, less quickly to the extent that the total intake of Y is greater than that of X, it would appear that the taste effects of the two diets are differentially altered by ingestion. Diet Y, although less effective in initiating eating, is more effective in maintaining eating.

As an introduction to the present studies, let us compare an artificially sweetened diet to an unsweetened diet. If we look at preference results we find that the sweetened diet is preferred. We may then assume that the initial facilitatory effect on eating this diet is greater than that for the nonsweetened diet in a one-choice test.[10] However, this does not mean that the relationship will hold during the course of a meal. This is especially plausible in the case of sweetness. Sweet taste input initially facilitates eating, presumably because it is an ecologically valid indicator of caloric content. It is also likely that this ecological correspondence allows the animal to use sweet-taste input to monitor or limit its caloric intake. The artificially sweetened diet, even though isocaloric with the unsweetened diet, might be expected to satiate the animal more quickly because it is sweeter and, thus, tastes more calorically dense. Perhaps this satiation would merely be a faster decrease in the facilitatory effect on eating of the sweetened diet. Perhaps the animal would, in some sense, be less hungry after eating the sweet diet.

EXPERIMENTS

With the hypothesis of taste-induced satiety in mind, the effects of saccharin adulteration on food intake were investigated. Four measures were compared: (1) 1-hr, two-choice test; (2) six-day, ad lib., one-choice test; (3) six-day, 22-hr-deprived, one-choice test; and (4) eating behavior during the course of a meal in a two-hr, one-choice test.

If saccharin provides the kind of taste stimulus that is initially more highly facilitatory but subsequently less so (or satiates the animal more quickly) than does the control stimulus, the following predictions may be made. (1) The saccharin diet will be preferred to the control diet in the 1-hr, two-choice test. Also, the initial rate of eating in the one-choice meal will be at least as great for the saccharin diet as for the control diet. (2) The addition of preferred saccharin will tend to *depress* intake in a one-choice test. This effect should be primarily a decrease in saccharin diet meal size. To the extent that the ad lib. and 22-hr, one-choice tests reflect this decrease in meal size, total intake in the two situations should also be depressed. This should be corroborated by intrameal eating behavior: a faster decline in eating rate over time for the saccharin diet than for the control diet.

PROCEDURES AND RESULTS

Stimuli and Subjects. Subjects were Holtzman strain male albino rats weighing 300–350 g. The stimuli (diets) used were the same for all experiments:

(1) 0.10 per cent saccharin diet—499 g Wayne Mouse Breeder Meal + 500 ml tap water + 1.0 g saccharin (o-benzoic sulfimide, Eastman Organic Chemicals).

(2) H_2O diet—500 g Wayne meal + 500 ml tap water.

These diets were of equal caloric density and were, presumably, of equal physiological effect beyond any differences resulting from taste input.[5,9]

Preference Tests. Twenty-five subjects were given three trials of a 1-hr, two-choice preference test between the above diets. Of these, 21 were judged to prefer the 0.10 per cent saccharin diet. The criteria for preference were (1) that the animal show greater intake of the diet on at least two of three tests, and (2) that total relative intake of the diet for all three tests be greater than 1.0 per cent.

Only those animals preferring the 0.10 per cent saccharin diet were used in the following tests. We did not assess the one-choice test behavior of animals that did not prefer the saccharin diet.

One-choice tests. Having selected those animals for which saccharin was a preferred stimulus, it was necessary to show that for at least one of the one-choice tests intake would be depressed. Such a result would be in accordance with the hypothesis that hedonic value decreases—or satiety increases—more rapidly on the saccharin diet than on the H_2O diet. Failure to obtain the result would not, in itself, disprove the taste-satiety hypothesis; subjects could compensate for the predicted smaller meals by increasing meal frequency, especially in the ad lib. test.

The procedure was to present one of the diets to each subject for six days. Following this, the animals were returned to dry stock meal (Wayne) for three days and then each subject was given the other diet. Half of the animals were given the 0.10 per cent saccharin diet first. Thus, each subject was its own control, and order effects were controlled for. Preference for the 0.10 per cent saccharin diet was established *prior* to the one-choice tests, thus precluding the possibility of initial aversion to saccharin.

Fourteen subjects were run on the ad lib. condition. One month later 12 of these were tested in the 22-hr deprived condition. As a check for possible effects of prior experience on performance in the latter test, an additional seven animals were run (total N = 19). Since there were no qualitative differences, the two 22-hr groups were combined.

Figure 1 represents the results of the ad lib. and 22-hr deprivation experiments. Saccharin diet intake is plotted as a fraction of H_2O diet intake. Two features of these data indicate intake suppression. One is the relative intake of the 0.10 per cent saccharin and H_2O diets within days, and the other is the change in this ratio over days: saccharin diet intake increases relative to H_2O diet intake.

Table I gives the statistical results for the two experiments. Intake of the saccharin diet remains below that of the H_2O diet for 4 days (ad lib.) and 2 days (22-hr deprived). In neither case is the depression either very large or highly significant. The change in

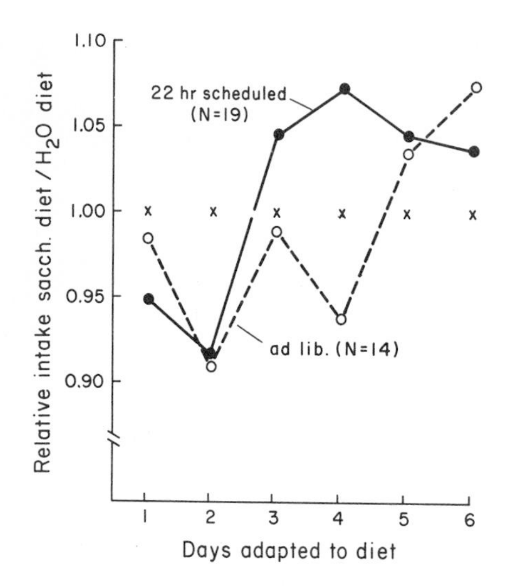

Figure 1 Ad lib. and 22-hr-deprived subjects' intake of 0.10% saccharin diet relative to intake of H_2O diet on 6 successive days of single-choice testing. Horizontal row of x's indicates baseline (1.00).

intake ratio over days is clearly the more impressive measure. Both measures, however, are consistent with the hypothesis that saccharin temporarily depresses food intake. Is one measure to be preferred over the other? The former measure is obviously dependent on the palatability of the control diet used and its effect on intake. The latter measure, however, may be interpreted as showing that subjects merely learn to "like" the 0.10 per cent saccharin diet. This interpretation is not tenable because of the preference for it established prior to the test.

Figure 2, from an earlier experiment of the same kind, illustrates both the effects of the palatability of the control diet on relative intake within days and the uniqueness of the relative rise in intake over days associated with a saccharin diet (in this case, 0.15 per cent saccharin Na). Since this rise in relative saccharin diet intake over days is independent of the control diet used (H_2O or 0.04 per cent QSO_4), it must be

TABLE I

COMPARISON OF 0.10% SACCHARIN (S) AND H_2O (W) DIET INTAKES WITHIN DAYS AND COMPARISON OF INTAKE RATIOS (S/W) BETWEEN DAYS

Numbers in columns represent two-tailed probabilities. Hypotheses being tested are: (1) S intake $<$ W intake and (2) S/W increases over days (n.s. = not significant).

	(1) S intake $<$ W intake					(2) S/W increases	
	DAY 1	DAY 2	DAY 3	DAY 4	DAYS 5–6	DAYS 1–2 vs 3–4	DAYS 1–4 vs 5–6
Ad lib.*	n.s.	.08	n.s.	.10	S $>$ W	n.s.	.002
22-hr.†	.06	.17	S $>$ W	S $>$ W	S $>$ W	.0001	n.s.

* Mann-Whitney U-Test (Siegel[6]).
† Binomial Test (Siegel[6]).

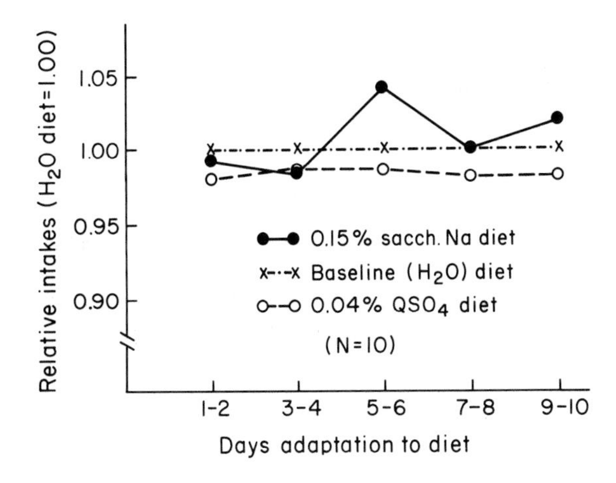

Figure 2 Ad lib. subjects' (N = 10) intakes of 0.15% saccharin Na and 0.04% quinine sulfate diets relative to H_2O diet over 10 consecutive days of single-choice testing. Results show both the increase in intake over days peculiar to a saccharin-adulterated diet and the effect of palatability of the control diet on the apparent ability of saccharin to depress intake. (After Gentile[2])

considered a better measure of saccharin adulteration effects. Also, since there is such a rise (which in all cases so far tested is absolute as well as relative), it is reasonable to consider that the initial intake of the saccharin diet is depressed, at least relative to its subsequent intake.

Intrameal eating behavior. Although the long-term, one-choice tests give results consistent with the present hypothesis, they are not conclusive. In order to say with some certainty that the specific taste quality associated with saccharin is influencing intake, one must know how meal size is affected. If saccharin-diet meals are larger but less frequent, it may be argued that changes in the relationship between meal size and frequency caused by increased meal size alone lead to decreased saccharin-diet intake. In addition, such a delayed saccharin effect is hard to account for in terms of taste input. On the other hand, if saccharin-diet meals are smaller, the hypothesis that the saccharin diet satiates the animal (or loses its facilitatory effect) more quickly that the control H_2O diet in the course of a meal becomes more reasonable.

Accordingly, seven 22-hr deprived animals were tested for meal-taking behavior during the first two days of adaptation to one of the two diets. Intake was measured by electronic scale (Farrell Instruments, Grand Island, Neb.) every 1.25 min for 45 min and again at the end of 2 hr. Each animal was given both diets so that the results include two trials on each diet for each subject. A meal was defined as a period of eating not interrupted by more than 10 min. In no case did a meal last more than 30 min. This period accounted for 81 per cent of the saccharin diet intake and 91 per cent of the H_2O diet intake in the 2-hr period.

Figure 3 shows intakes on the two diets after 15, 30, 45, and 120 min. As can be seen, intake differences after 30 min (first meal) are in the same direction as those after 120 min. In fact, the 30-min differences are greater. Saccharin appears to depress meal size. Such a depression was observed in six of the seven subjects.

In an effort to reconcile the two-choice data and the single-choice data, intrameal

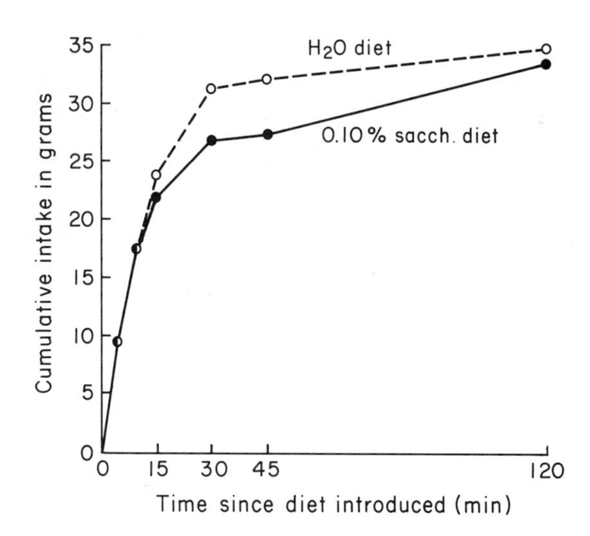

Figure 3 Cumulative intakes of 0.10% saccharin and H_2O diets as a function of time after diet was introduced. Subjects were deprived for 22 hr. Data include all meals for all 7 subjects tested, and represent eating patterns on the first 2 days' adaptation to the respective diets.

eating behavior was investigated. The following hypothesis was tentatively set forth: the two-choice data are accounted for by the initial, relatively facilitatory effect of the saccharin diet. Decreased saccharin-diet meal size in the one-choice test would be attributable to a relative decrease in this effect in the course of a meal or to satiety resulting from the saccharin taste input.

Therefore, it was predicted that initially the rate of saccharin diet intake would exceed that of the H_2O diet, even in meals at which total saccharin diet intake was less. In order to test this, only those meals were used in which (for each animal) saccharin diet intake was less than the H_2O diet intake. Figure 4 shows that this hypothesis is confirmed. At the very least, initial saccharin diet intake rate is not less than that of the H_2O diet. It should be re-emphasized that, because of small sample size, the data concerning intrameal behavior are primarily descriptive at this point and are not conclusive.

CONCLUSIONS

It has been the purpose of this investigation to examine the validity of the simple hedonic hypothesis concerning the role of taste in the control of food intake. This hypothesis assumes that the effect of taste is adequately measured by a two-choice preference test and that, if a substance is preferred, it will tend to be overeaten relative to the unpreferred substance in a one-choice test. If such a result is not found, the animal is said to be unresponsive to taste in the one-choice test situation.

Saccharin, however, results in decreased intake and, more importantly, decreased meal size, despite being preferred in a two-choice test and despite causing an initially high rate of intake in the one-choice test. These results are consistent only with a more complex hedonic hypothesis, which takes changes in hedonic value specific to

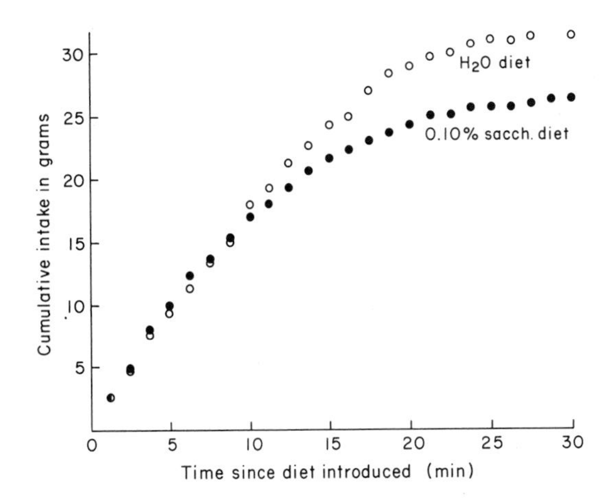

Figure 4 Amount eaten as a function of diet and time after start of meal. Meals included here were only those in which saccharin intake was less than H_2O intake for the corresponding meal. Results represent eating patterns on first 2 days' adaptation to the respective diets.

the saccharin stimulus into account, or with a hypothesis which allows for a sensory quality effect independent of hedonic value (e.g., the use of taste to monitor caloric intake).

In the present experiments these two theories are not discriminable. However, the insertion of preference tests at various times after the initiation of feeding both the H_2O and the saccharin diets would be decisive in determining whether, as the complex hedonic hypothesis predicts, the H_2O diet becomes preferred to the saccharin diet during the course of a meal or whether, as the sensory quality hypothesis predicts, prior saccharin ingestion inhibits any subsequent intake, regardless of type of diet.

REFERENCES

1. Adolph, E. F. 1947. Urges to eat and drink in rats. *Amer. J. Physiol.* **151**, pp. 110–125.
2. Gentile, R. L. The role of taste preference in the control of food intake and meal-taking in the albino rat. (In preparation)
3. Jacobs, H. L. 1962. Some physical, metabolic, and sensory components in the appetite for glucose. *Amer. J. Physiol.* **203**, 1043–1054.
4. Jacobs, H. L., and K. N. Sharma. 1969. Taste versus calories: sensory and metabolic signals in the control of food intake. *In* Neural Regulation of Food and Water Intake (P. J. Morgane, conference editor). *Ann. N.Y. Acad. Sci.* **157**, pp. 1089–1125.
5. Nicolaïdis, S. 1969. Early systemic responses to oro-gastric stimulation in the regulation of food and water balance. *In* Neural Regulation of Food and Water Intake (P. J. Morgane, conference editor). *Ann. N.Y. Acad. Sci.* **157**, pp. 1176–1203.
6. Siegel, S. 1956. Non-parametric Statistics for the Behavioral Sciences. McGraw-Hill Book Company, New York, Toronto, London.
7. Soulairac, A. 1967. Control of carbohydrate intake. *In* Handbook of Physiology (C. F. Code, editor) Vol. I, Section 6. American Physiological Society, Washington, D.C., pp. 387–398.
8. Teitelbaum, P. 1955. Sensory control of hypothalamic hyperphagia. *J. Comp. Physiol. Psychol.* **48**, 156–163.
9. Thompson, M. M., and J. Mayer. 1959. Hypoglycemic effects of saccharin in experimental animals. *Amer. J. Clin. Nutr.* **7**, 80–85.
10. Young, P. T. 1961. Motivation and Emotion. John Wiley and Sons, New York.
11. Young, P. T. 1967. Palatability: the hedonic response to foodstuffs. *In* Handbook of Physiology (C. F. Code, editor) Vol. I, Section 6. American Physiological Society, Washington, D.C., pp. 353–366.

VII ABSTRACTS SUBMITTED FOR DISCUSSION

VII · ABSTRACTS SUBMITTED FOR DISCUSSION

RESPONSES OF THE CHORDA TYMPANI TO ELECTROTONIC, ELECTROLYTIC, AND NONELECTROLYTIC STIMULATIONS ON THE TONGUE

AKIRA ADACHI

NAS-NRC Visiting Scientist Research Associate, U.S. Army Natick Laboratories (On leave from Osaka University, Japan)

Electrotonic stimulation of the cat's tongue surface has been shown to modulate chorda tympani activity.[1] The study reported here extended these observations to rats, comparing nonelectrolyte as well as electrolyte solutions to electrotonic stimulation on the tongue.

Salt fibers were found to be highly sensitive to anodal electrotonic stimulation. The response pattern was similar to that produced by a sodium chloride solution. Cathodal stimulation decreased spontaneous baseline activity in the resting fiber as well as attenuating the electrical response to salt solution.

These effects could not be demonstrated in sucrose fibers. The failure of sucrose fibers to respond to electrotonic stimulation on the tongue suggests that the current stimulates receptors sensitive to electrolytes and is not simply an artifact resulting from direct electrical stimulation of the nerve ending.

Further evidence for direct receptor effects was obtained from analysis of latency data.

REFERENCES

1. Pfaffmann, C. 1941. Gustatory afferent impulses. *J. Cell. Comp. Physiol.* **17**, pp. 243–258.

PSYCHOPHYSIOLOGICAL MEASURES OF TASTE PERCEPTION

G. L. FISHER

Department of Psychology, University of Maryland, College Park, Md.

The methods of psychophysiology are being applied to analyze processes implicated in acceptance/rejection behavior patterns of normal humans. P. T. Young[3] has suggested that affective arousal is an important factor determining the palatability of food substances. Lacey,[1] Sokolov,[2] and others have suggested that response profiles derived from measures of physiological activity of separate response systems may distinguish affective qualities.

Initially, magnitude estimates of subjective intensity and preference were obtained for concentration series of sucrose and quinine hydrochloride. Subjective intensity functions were significantly steeper than the relative preference curves, supporting a view that different perceptual systems may operate for the two judgmental dimensions.

In one investigation, dermal conductance change was recorded during adequate stimulation with water, 0.01 M QHCl, 0.001 M QHCl, 1 M sucrose and 0.1 M sucrose. The largest mean changes in conductance were obtained for the quinine solutions. The sugar solutions produced only modest reactions that were slightly larger than those to water.

Earlier work indicated that non-gustatory cues potentiated measured preference. Therefore, reactions were measured for 0.01 M QHCl and 1.0 M sucrose; each was paired with a tone using a classical (Pavlovian) differential-delay conditioning paradigm. Dermal conductance changes were much larger to onset of the low intensity tones than to the sapid chemicals they preceded. The requirement of a differential motor response to the two tones potentiated the conductance change to the tones but not to the subsequent chemical stimulation.

Measurements of other physiological reactions, e.g., pupillary changes and alpha-blocking, will also be discussed.

REFERENCES

1. Lacey, J. I. 1967. Somatic response patterning and stress: some revisions of Activation Theory. *In* Psychological Stress: Issues in Research (M. H. Appley and R. Trumbull, editors). Appleton-Century-Crofts, New York, pp. 14–43.
2. Sokolov, E. N. 1963. Perception and the Conditioned Reflex. Pergamon Press, Oxford, England.
3. Young, P. T. 1968. Evaluation and preference in behavioral development. *Psychol. Rev.* **75**, No. 3, pp. 222–241.

Centrifugal Modulation of Gustatory Responses

BRUCE P. HALPERN *and* ALAN D. BRUSH

Department of Psychology and Section on Neurobiology and Behavior, Cornell University, Ithaca, New York

Multiunit neural activity in the lingual branch of the glossopharyngeal nerve (IX) of anesthetized toads (*Bufo marinus*) was recorded with a platinum electrode[4] and led through an electronic digital summator.[5] Liquids (10 ml) at 24 ± 1 C were applied for ca 25 sec to the receptor side of the tongue, which was pinned in an open flow chamber. The stomach was stimulated by distending an intragastric balloon with 9 ml water at 24 ± 1 C. This volume is within the physiological range.[1] In the intact IX, differential modulation of gustatory responses occurred during gastric distention.[3] Modulation was seen primarily at the time of maximum response. Thus, during distention, the rapidly peaking responses to QHCl decreased (−) at 2 sec after stimulus onset, while the slowly rising responses to NaCl increased (+) at 5 sec after stimulus onset. Transection of contralateral IX did not prevent the effects of gastric distention. After bilateral transection or ligation, however, gastric distention produced no consistent change (0) in response to NaCl, QHCl, or glucose.

These observations suggest that consistent differential modulation acts primarily through ipsilateral IX.

	INTACT IX							BLOCKED IX							
Tongue Stimulus	Median Change %	Number of Animals			% Animals		P	Median Change %	Number of Animals			% Animals		P	P_{diff} I vs B
		+	−	0	+	−			+	−	0	+	−		
NaCl 0.5 M	+29	9	2	6	53	12	0.02*	−5	2	4	8	14	29	0.5	0.05*
QHCl 0.002 M	−29	1	8	8	6	47	0.01*	−9	2	3	8	15	23	0.6	0.02*
Glucose 0.5 M	+2	4	4	3	36	36	0.5	+12	6	4	3	46	21	0.36	—

P is Wilcoxon Sign-rank, two-tailed, except * = one-tailed. P_{diff} I vs B compares responses from blocked animals' intact median.

(Supported by NIH grant NB-06945)

REFERENCES

1. Brush, A. D. 1968. Control of gustatory responses in the toad. Honors Thesis, Cornell University, Ithaca, New York.
2. Chernetski, K. E. 1964. Sympathetic enhancement of peripheral sensory input in the frog. *J. Neurophysiol.* **27**, pp. 493–515.
3. Esakov, A. I., S. E. Margolis, and G. Y. Yur'eva. 1966. Nauchnyje doklady vysshei shkoly. *Biolog. Nauki* **4**, pp. 92–96.
4. Halpern, B. P. 1967. Some relationships between electrophysiology and behavior in taste. *In* The Chemical Senses and Nutrition (M. R. Kare and O. Maller, editors). The Johns Hopkins Univ. Press, Baltimore, Md., pp. 213–241.
5. Walsh, L. F., and B. P. Halpern. A digital summator. (In preparation)

Midbrain Chemoreceptor of Circadian Sleep-inducing Substance in Mammals and Birds

T. HAYASHI *and* S. SEKI

Dr. Hayashi is Professor Emeritus, Department of Physiology, School of Medicine, Keio University, Tokyo, Japan
Dr. Seki is Associate Professor, Department of Physiology, Kanagawa Dental College, Yokosuka, Japan

In the 1965 Olfaction and Taste conference, Hayashi, et al., advanced the theory that circadian sleep was the result of the intrinsic genesis of γ hydroxybutyrate ($OHCH_2CH_2CH_2COOH$) in the brain.[1]

In current studies, the authors show that, in such diurnal animals as fowls, the concentration of γ hydroxybutyrate in the brain is higher in the evening than in the morning. On the contrary, in nocturnal animals, such as bats, which were used in this work, concentration is higher in the morning than in the evening.

The authors also show that the substance diffuses out from the nerve cells of the brain into the cerebrospinal fluid, which, in turn, affects the midbrain chemoreceptor of the reticular formation.

REFERENCES

1. Hayashi, T., H. Hoshino, and T. Ootsuka. 1967. Chemoreceptors in brain to γ hydroxybutyrate through cerebrospinal fluid of dogs. *In* Olfaction and Taste II (T. Hayashi, editor). Pergamon Press, Oxford, etc., pp. 599–608.

Olfactory Perceptions of Human Novices

JAMES W. JOHNSTON, JR.

Department of Physiology and Biophysics, Georgetown University, Washington, D.C.

Good subjects for olfactory experimentation can be found by screening N volunteers among the naive residents of a given subpopulation. The research reported herein was performed at various times from 1965 to 1968 with young men and women between the ages of 17 and 25 years. The purpose was to obtain a corps of people with an adequate sense of smell for research on odor chemistry. An effort was made to relate the innate ability of about 130 persons to discriminate odor quality and odor strength within pairs of unknowns. The result was a successful two-stage screening procedure, in which the first stage was comprised of nine pairs of coded unknowns. They presented a gamut of subjective quality classes in identical and nonidentical pairs and were replicated three times. An eight-interval rating scale was employed to "measure" degrees of similarity and dissimilarity of quality. Persons who achieved a minimum weighted score of 12 points in 19, or better, were further tested by the next stage.

This second, or final, stage was comprised of ten pairs of odorants in which a putrid-smelling odorant was compared with a treated sample of the same chemical. A mixture of oxidants was added in order to determine its effect upon putridity. The subjects rated similarity-dissimilarity by means of a four-interval scale and strength by means of a novel three-interval scale. The individual performances were graded for interval consistency and centering tendency. Total weighted points were the basis of a successful prediction of the value of 16 novices for experimentation.

Aversion for Hypertonic Saline Solutions by Rats Ingesting Untasted Fluids

HARRY R. KISSILEFF
The Rockefeller University, New York, New York

Rats avoid hypertonic saline (e.g., sodium chloride) solutions in two-choice tests and drink less of them than water in single-stimulus tests. Postingestive events have been implicated in this behavior. Can saline aversion be exhibited without taste?

Normal rats were trained to ingest fluids without tasting them by pressing a bar for intragastric reinforcement (FR-1 to FR-5, 0.5 to 2.0 ml reinforcements). In single-stimulus tests, a day of continuous access to the bar, reinforced with intragastric hypertonic saline, was preceded and followed by three days of intragastric water reinforcement. Rats ingested twice as much total fluid as water with 1.5 per cent or 1.9 per cent saline reinforcements, but only as much total fluid as water with 2.5 or 3 per cent saline reinforcements. After 4 to 6 days, with two bars available ad libitum, rats pressed the bar reinforced with 3 per cent saline less often than they pressed the bar reinforced with water. Positions of bars and reinforcements were not changed daily. When positions of water and saline reinforcement were reversed, rats switched from the bar producing intragastric saline to the bar producing water. When two bars were available for only 1 hr per day, rats did not show preferential bar pressing, but either gradually stopped working over a 6-day period, or pressed indiscriminately.

Rats can learn to avoid ingestion of hypertonic saline solutions without tasting them, if given ample time, and will ingest less of 3 per cent saline than 1.5 per cent saline. Therefore, taste of a solution is not a necessary cue for learning its aversive postingestional consequences.

(Supported by U.S.P.H.S. Fellowship 1-F2-AM-34, 509-01; NSF grant GB-4198)

Odor Preferences in Men and Rats

JACK T. TAPP

Department of Psychology, Vanderbilt University, Nashville, Tennessee

The perceived pleasantness of odors contributes substantially to human judgments of the differences between odors,[3] and it has been demonstrated that odors have motivating properties for rats.[1] These observations suggested a series of experiments designed to: (a) assess the relative preferences for odors that differ widely in quality; (b) evaluate the similarities in the preferred aspects of odors of similar quality; (c) compare and evaluate different methods for assessing the pleasantness of odors; (d) compare the odor preferences of men and rats for the same odorous substances.

Two representative odors from each of Amoore's seven categories were selected for study in the human experiments. The pleasantness of odors was evaluated by the methods of paired comparisons, categorical judgments, and magnitude estimation. All three methods yielded similar results and the odors from each category were judged similarly pleasant.

One odor from each category was used in the rat experiments and the pleasantness of the odor was determined by assessing its rewarding properties in the apparatus described by Long and Tapp.[2] Procedures analogous to magnitude estimation and paired comparisons were employed. These methods did not yield similar results, suggesting that the "pleasantness" of the odors for rats are related to the context within which the judgments were made.

Comparisons between the odor preferences exhibited by rats and men indicated both similarities and differences in odor preferences. The results of these studies are compared with the results of other studies and discussed in terms of the role of experience as a determinant of odor preferences.

REFERENCES

1. Long, C. J., and J. T. Tapp. 1967. Reinforcing properties of odors for the albino rat. *Psychon. Sci.* **7**, pp. 17–18.
2. Long, C. J., and J. T. Tapp. 1968. An apparatus for the assessment of the reinforcing properties of odors in small animals. *J. Exp. Anal. Behav.* **11**, pp. 49–51.
3. Woskow, M. H. 1968. Multidimensional scaling of odors. *In* Theories of Odors and Odor Measurement (N. Tanyolac, editor). Robert College Research Center, Istanbul, Turkey, pp. 147–188.

PARTICIPANTS

A. ADACHI, Psychology Laboratories, Pioneering Research Laboratory, U.S. Army Natick Laboratories, Natick, Massachusetts 01760

LORD ADRIAN, Trinity College, Cambridge, CB2 ITQ, England

P. F. ALBRITTON, F.P.O. Box 2, Seattle, Washington 98109

J. E. AMOORE, U.S. Agricultural Research Service, 800 Buchanan Street, Albany, California 94710

P. K. ANOKHIN, Kutuzowsky av. 20, app. 38, Moscow, U.S.S.R.

K. AOKI, Department of Physiology, School of Medicine, Gunma University, 39–22, 3-Chome, Showa-Machi, Maebashi, Gunma, Japan

J. ATEMA, School of Natural Resources, The University of Michigan, Ann Arbor, Michigan 48104

J. E. BARDACH, School of Natural Resources and Department of Zoology, University of Michigan, Ann Arbor, Michigan 48104

L. M. BARTOSHUK, Pioneering Research Division, U.S. Army Natick Laboratories, Natick, Massachusetts 01760

L. M. BEIDLER, Department of Biological Science, Florida State University, Tallahassee, Florida 32306

F. R. BELL, The Royal Veterinary College, Department of Medicine, Hawkshead House, Hawkshead Lane, North Mimms, Hatfield, Hertfordshire, England

R. A. BERNARD, The Rockefeller University, New York, New York 10021

J. BOECKH, Zoologisches Institut der Universität, 6 Frankfurt/M., Siesmayerstrasse 70, W. Germany

J. N. BROUWER, Unilever Research Laboratorium, Vlaardingen, The Netherlands

A. D. BRUSH, School of Medicine, Harvard University, 25 Shattuck Street, Boston, Massachusetts 02115

R. L. BUTTRICK, Pioneering Research Laboratory, U.S. Army Natick Laboratories, Natick, Massachusetts 01760

W. S. CAIN, John B. Pierce Foundation Laboratory, New Haven, Connecticut 06510

C. CASELLA, Istituto di Fisiologia Generale, Università di Pavia, Pavia, Italy

V. C. COX, Fels Research Institute, Yellow Springs, Ohio 45387

F. R. DASTOLI, Monsanto Research Corporation, Everett, Massachusetts 02149

G. P. DATEO, Pioneering Research Laboratory, U.S. Army Natick Laboratories, Natick, Massachusetts 01760

D. DENTON, Howard Florey Laboratories of Experimental Physiology, University of Melbourne, Parkville 3052, Victoria, Australia

H. DIAMANT, Department of Otorhinolaryngology, Umeä University, Umeä 50, Sweden

G. S. DOETSCH, Comparative Psychology Division, 6571st Aeromedical Research Laboratory, Holloman A. F. B., New Mexico 88330

E. DZENDOLET, Department of Psychology, University of Massachusetts, Amherst, Massachusetts 01002

R. EMMERS, Department of Physiology, College of Physicians and Surgeons, 630 West 168th Street, New York, New York 10032

T. ENGEN, Psychology Department, Brown University, Providence, Rhode Island 02912

R. P. ERICKSON, Department of Psychology, Duke University, Durham, North Carolina 27706

G. L. FISHER, Psychology Department, Morrill Hall, University of Maryland, College Park, Maryland 20740

A. FRANCKE, Unilever Research Laboratorium, Vlaardingen, The Netherlands

M. FRANK, The Rockefeller University, New York, New York 10021

M. J. FREGLY, Department of Physiology, College of Medicine, University of Florida, Gainesville, Florida 32601

D. FRISCH, Department of Biostructure, Northwestern University, Chicago, Illinois 60611

J. J. GANCHROW, Department of Psychology, Duke University, Durham, North Carolina 27706

J. GARCIA, Psychology Department, S.U.N.Y. Stony Brook, Long Island, New York 11790

R. GENTILE, Behavioral Sciences Division, Pioneering Research Laboratory, U.S. Army Natick Laboratories, Natick, Massachusetts 01760

T. GETCHELL, Department of Biological Sciences, Northwestern University, Evanston, Illinois 60201

P. P. C. GRAZIADEI, Department of Biological Science, Florida State University, Tallahassee, Florida 32306

F. GREEN, California State College at Long Beach, Long Beach, California 90804

B. P. HALPERN, Department of Psychology and Division of Biology, Cornell University, Ithaca, New York 14850

K. HANSEN, Zoologisches Institut, 69 Heidelberg, Berliner Strasse 15, W. Germany

T. HAYASHI, Department of Physiology, School of Medicine, Keio University, Shinjuku-ku, Tokyo, Japan

G. HELLEKANT, Department of Physiology, Kungl, Veterinärhögskolan, Stockholm 50, Sweden

D. E. HENDRIX, Northwestern University, Chicago, Illinois 60611

R. HENKIN, Clinical Endocrinology Branch, National Heart Institute, Bethesda, Maryland 20014

G. J. HENNING, Unilever Research Laboratorium, Vlaardingen, The Netherlands

S. HIGASHINO, Department of Physiology, Columbia University, New York, New York 10032

R. M. HOGAN, Department of Physiology, Medical College of Virginia, Health Sciences Division, Virginia Commonwealth University, Richmond, Virginia 23219

J. HUGHES, Department of Neurology and Psychiatry, Northwestern University, Chicago, Illinois 60611

M. IINO, Department of Physiology, School of Medicine, Gunma University, 39-22, 3-Chome, Showa-Machi, Maebashi, Gunma, Japan

J. W. JOHNSTON, JR., Georgetown University Medical School, 3900 Reservoir Road, N.W., Washington, D.C. 20007

K.-E. KAISSLING, Max-Planck-Institut für Verhaltensphysiologie, Seewiesen und Erling-Andechs, W. Germany

J. W. KAKOLEWSKI, Fels Research Institute, Yellow Springs, Ohio 45387

M. R. KARE, University of Pennsylvania, Monell Chemical Senses Center, 529 Lippincott Building, Philadelphia, Pennsylvania 19104

J. G. KEPPLER, Unilever Research Laboratorium, Vlaardingen, The Netherlands

H. R. KISSILEFF, The Rockefeller University, New York, New York 10021

E. P. KÖSTER, Psychologisch Laboratorium der Rijksuniversiteit, Varkenmarkt 2, Utrecht, The Netherlands

K. KURIHARA, Department of Biological Science, Florida State University, Tallahassee, Florida 32306

Y. KURIHARA, Department of Biological Science, Florida State University, Tallahassee, Florida 32306

P. LAFFORT, Collège de France, 11 Place Marcelin Berthelot, Paris Ve, France

J. LEVETEAU, Laboratoire de Physiologie Générale, 9 Quai Saint-Bernard, Paris Ve, France

L. LONG, JR., Pioneering Research Laboratory, U.S. Army Natick Laboratories, Natick, Massachusetts 01760

D. MATTHEWS, The University of Calgary, Calgary, Alberta, Canada

D. H. MCBURNEY, Department of Psychology, University of Pittsburgh, Pittsburgh, Pennsylvania 15213

P. MACLEOD, Laboratoire de Neurophysiologie, Collège de France, 11 Place Marcelin Berthelot, 75, Paris Ve, France

K. MCGOWAN, Barrow Neurological Research Institute, Phoenix, Arizona 85026

A. R. MICHELL, The Royal Veterinary College, Department of Medicine, Hawkshead House, Hawkshead Lane, North Mimms, Hatfield, Hertfordshire, England

H. MORITA, Department of Biology, Faculty of Science, Kyushu University, Fukuoka, Japan

G. R. MORRISON, Department of Psychology, McMaster University, Hamilton, Ontario, Canada

D. G. MOULTON, Department of Biology, Clark University, Worcester, Massachusetts 01601

M. M. MOZELL, Department of Physiology, Upstate Medical Center, State University of New York, Syracuse, New York 13210

R. G. MURRAY, Department of Anatomy and Physiology, Indiana University, Bloomington, Indiana 47401

J. F. NELSON, Howard Florey Laboratories of Experimental Physiology, University of Melbourne, Parkville 3052, Victoria, Australia

L. M. NELSON, Department of Psychology, Duke University, Durham, North Carolina 27708

S. NICOLAÏDIS, Collège de France, 11 Place Marcelin Berthelot, Paris Ve, France

H. NOMURA, Department of Physiology, Tokyo Dental College, Misakicho, Chiyoda-ku, Tokyo, Japan

T. E. OBERJAT, UCLA Medical Center, L.A.C. Heart Laboratory, A3-381 BRI, Los Angeles, California 90024

H. OGAWA, Department of Physiology, Kumamoto University Medical School, 430 Honjo-Machi, Kumamoto City 860, Japan

E. ORCHARD, Howard Florey Laboratories of Experimental Physiology, University of Melbourne, Parkville 3052, Victoria, Australia

D. PFAFF, The Rockefeller University, New York, New York 10021

C. PFAFFMANN, The Rockefeller University, New York, New York 10021

S. PRICE, Department of Physiology, Medical College of Virginia, Richmond, Virginia 23219

E. PRIESNER, Max-Planck-Institut für Verhaltensphysiologie, Seewiesen, W. Germany

G. RAPUZZI, Istituto di Fisiologia Generale, Università di Pavia, Pavia, Italy

S. SAKADA, Department of Physiology, Tokyo Dental College, Misakicho, Chiyoda-ku, Tokyo, Japan

A. SALZMAN, Hospital Pirovano, Sala 9, Monroe 3555, Buenos Aires, Argentina

S. SEKI, Department of Physiology, Tokyo Dental College, Misakicho, Chiyoda-ku, Tokyo, Japan

D. SCHNEIDER, Max-Planck-Institut für Verhaltensphysiologie, Seewiesen über Starnberg (OBB), W. Germany

T. G. SCHULTZE-WESTRUM, Zoologisches Institut der Universität, 8 München 2, Luisenstrasse 4, W. Germany

T. SHIBUYA, Zoological Institute, Faculty of Science, Tokyo Kyoiku University, 3-29-1, Otsuka, Bunkyo-ku, Tokyo, Japan

J. C. SMITH, Department of Psychology and Biological Science, Florida State University, Tallahassee, Florida 32306

R. A. STEINBRECHT, Max-Planck-Institut für Verhaltensphysiologie, Seewiesen, W. Germany

H. STONE, Department of Biobehavioral Sciences, Stanford Research Institute, Menlo Park, California 94025

S. TAKAGI, Department of Physiology, School of Medicine, Gunma University, 39-22, 3-Chome, Showa-Machi, Maebashi, Gunma, Japan

H. TAKEUCHI, Department of Internal Medicine, Gunma University, 39-22, 3-Chome, Showa-Machi, Maebashi, Gunma, Japan

J. T. TAPP, Psychology Department, Vanderbilt University, Nashville, Tennessee 37203

J. H. TODD, Department of Zoology, San Diego State College, San Diego, California 92115

D. TUCKER, Department of Biological Science, Florida State University, Tallahassee, Florida 32306

E. S. VALENSTEIN, Department of Psychophysiology-Neurophysiology, Fels Research Institute, Yellow Springs, Ohio 45387

D. J. VANDENBELT, Pioneering Research Laboratory, U.S. Army Natick Laboratories, Natick, Massachusetts 01760

H. VAN DER WEL, Unilever Research Laboratorium, Vlaardingen, The Netherlands

A. VIOLANTE, Istituto di Fisiologia Generale, Università di Pavia, Pavia, Italy

S. WELLER, Howard Florey Laboratories of Experimental Physiology, University of Melbourne, Parkville 3052, Victoria, Australia

B. M. WENZEL, Department of Physiology, University of California, Los Angeles, California 90024

N. WETZEL, Passavant Memorial Hospital, 303 E. Superior Street, Chicago, Illinois 60611

W. K. WHITTEN, Jackson Laboratory, Bar Harbor, Maine 04609

G. WOLF, Department of Anatomy, Mount Sinai School of Medicine, 5th Avenue and 100th Street, New York, New York 10029

T. YAJIMA, Department of Physiology, School of Medicine, Gunma University, 39-22, 3-Chome, Showa-Machi, Maebashi, Gunma, Japan

S. YAMASHITA, Department of Physiology, Kumamoto University Medical School, 430 Honjo-Machi, Kumamoto City 860, Japan

Y. ZOTTERMAN, Department of Physiology, Kungl, Veterinärhögskolan, Stockholm 50, Sweden

ATTENDEES

N. AI, Department of Biology, Clark University, Worcester, Massachusetts 01610

K. AKERT, Institut für Hirnforschung der Universität Zürich, August Forel-Strasse 1, 8008 Zürich, Switzerland

H. R. BEHRMAN, Sensory Physiology Laboratory, North Carolina State University, Raleigh, North Carolina 27607

M. H. BENNETT, Department of Anatomy, School of Medicine, 1542 Tulane Avenue, Louisiana State University, New Orleans, Louisiana 70112

R. and C. BRADLEY, Department of Biological Science, Florida State University, Tallahassee, Florida 32306

D. F. BRADLEY, Department of Chemistry, Polytechnic Institute of Brooklyn, 333 Jay Street, Brooklyn, New York 11201

J. BRAUN, Psychology Department, Yale University, New Haven, Connecticut 06510

K. BROWN, Department of Health, Education, and Welfare, National Institute of Dental Research, Human Genetics Branch, Bethesda, Maryland 20014

W. J. CARR, Department of Psychology, Temple University, Philadelphia, Pennsylvania 19122

H. CHAUNCEY, Veterans Administration Building, Washington D.C. 20420

E. A. EDELSACK, Scientific Department, Office of Naval Research, Department of the Navy, 1076 Mission Street, San Francisco, California 94103

H. M. EDINGER, Department of Physiology, University of Pennsylvania, School of Medicine, Philadelphia, Pennsylvania 19104

E. FONBERG, Nencki Institute, 3 Pasteura, Warsaw 22, Poland

H. FERREYRA, Department of Physiology, State University of New York, Downstate Medical Center, 450 Clarkson Avenue, Brooklyn, New York 11203

D. FOSTER, Virginia Military Institute, Lexington, Virginia 24450

S. FREEMAN, International Flavors and Fragrances, Union Beach, New Jersey 07735

D. GANCHROW, Department of Psychology, Duke University, Durham, North Carolina 27706

F. GAULT, Department of Psychology, Yale University, New Haven, Connecticut 06510

H. L. GILLARY, Department of Biological Sciences, Stanford University, Stanford, California 94301

N. M. GILMORE, U.S.D.A. Agricultural Research Service, Human Metabolism Laboratory, Beltsville, Maryland 20705

J. GRACE, Department of Physiology, University of Western Ontario, London, Ontario, Canada

H. HARPER, The Rockefeller University, New York, New York 10021

L. HOFF, Behavioral Sciences Division, Pioneering Research Laboratory, U.S. Army Natick Laboratories, Natick, Massachusetts 01760

C. HOPKINS, The Rockefeller University, New York, New York 10021

H. JACOBS, Pioneering Research Division, U.S. Army Natick Laboratories, Natick, Massachusetts 01760

B. JAHAN-PAVAR, Mental Health Research Institute, University of Michigan, Ann Arbor, Michigan 48104

R. E. JOHNSTON, The Rockefeller University, New York, New York 10021

T. KIKUCHI, Department of Biological Science, Florida State University, Tallahassee, Florida 32306

H. L. KLOPPING, DuPont Experimental Station, Building 324, Wilmington, Delaware 19801

G. KRAUTHAMER, Department of Neurology, College of Physicians and Surgeons of Columbia University, 630 West 168 Street, New York, New York 10032

M. KUWABARA, Department of Biology, School of Physics, Kyushu University, Fukuoka, Japan

V. LACHER, Life Sciences Building 18, Stanford Research Institute, Menlo Park, California 94025

J. LEMAGNEN, Collège de France, 11 Place Marcelin Berthelot, Paris Ve, France

C. LEONARD, The Rockefeller University, New York, New York 10021

B. MCCUTCHEON, Department of Psychology, State University of New York, Albany, New York 12203

F. MACRIDES, Department of Psychology, Building E 10, Massachusetts Institute of Technology, Cambridge, Massachusetts 02139

O. MALLER, Veterans Administration Hospital, Coatesville, Pennsylvania 19320

A. MARCSTROM, Institute of Zoophysiology, University of Uppsala, Uppsala, Sweden

D. A. MARSHALL, Department of Biology, Clark University, Worcester, Massachusetts 01610

M. E. MASON, International Flavors and Fragrances, Union Beach, New Jersey 07735

H. MEISELMAN, Department of Psychology, Liddell Laboratory, 118 Freese Road, R.F.D.7, Cornell University, Ithaca, New York 14850

I. MILLER, Department of Biological Science, Florida State University, Tallahassee, Florida 32306

B. OAKLEY, Department of Zoology, University of Michigan, Ann Arbor, Michigan 48104

R. O'CONNELL, The Rockefeller University, New York, New York 10021

M. OGURA, Instituto Biologica, Universidad Central de Venezuela, Caracas, Venezuela

S. B. PACE, Radiopharmaceutical R & D, The Squibb Institute for Medical Research, New Brunswick, New Jersey 08903

S. Palfrey, The Rockefeller University, New York, New York 10021

R. M. Pangborn, Food Service and Technology, University of California, Davis, California 95616

P. B. Pearson, The Nutrition Foundation Inc., 99 Park Avenue, New York, New York 10016

N. Robbins, 3D47 Building 10, National Institute of Health, Bethesda, Maryland 20014

L. Schmitt, Givaudan Corporation, 125 Delawanna Avenue, Clifton, New Jersey 07090

J. Scott, The Rockefeller University, New York, New York 10021

M. H. Sieck, Department of Psychology, University of California, Riverside, California 92507

H. L. Sipple, The Nutrition Foundation Inc., 99 Park Avenue, New York, New York 10016

F. A. Steiner, Institut für Hirnforschung der Universität Zürich, August Forel-Strasse 1, 8008 Zürich, Switzerland

H. Tateda, Department of Biological Science, Florida State University, Tallahassee, Florida 32306

R. von Baumgarten, Mental Health Research Institute, University of Michigan, Ann Arbor, Michigan 48104

P. E. Voorhoeve, Department of Neurophysiology, Jan Swammerdan Institute, University of Amsterdam, Eerste Const. Huygensstraat 20, Amsterdam, The Netherlands

M. Wang, The Rockefeller University, New York, New York 10021

G. Wise, Department of Zoology, University of Massachusetts, Amherst, Massachusetts 01002

M. Wolbarsht, Department of Ophthalmology, Duke University Medical Center, Durham, North Carolina 27706

K. S. Yackzan, Box 263, MCV Station, Medical College of Virginia, College Hospitals, Richmond, Virginia 23219

T. Yokota, Department of Neurophysiology, Jan Swammerdan Institute, University of Amsterdam, Eerste Const. Huygensstraat 20, Amsterdam, The Netherlands

INDEX

Note: Numbers in italics refer to pages on which tables or figures appear

F

P

T